GREEN CHEMISTRY
2nd Edition

Fundamentals and Applications

GREEN CHEMISTRY
2nd Edition

Fundamentals and Applications

Edited by
Suresh C. Ameta, PhD
Rakshit Ameta, PhD

AAP | APPLE ACADEMIC PRESS

First edition published 2024

Apple Academic Press Inc.
1265 Goldenrod Circle, NE,
Palm Bay, FL 32905 USA

760 Laurentian Drive, Unit 19,
Burlington, ON L7N 0A4, CANADA

CRC Press
2385 NW Executive Center Drive,
Suite 320, Boca Raton FL 33431

4 Park Square, Milton Park,
Abingdon, Oxon, OX14 4RN UK

Library and Archives Canada Cataloguing in Publication

Title: Green chemistry : fundamentals and applications / edited by Suresh C. Ameta, PhD, Rakshit Ameta, PhD.
Other titles: Green chemistry (Palm Bay, Fla.)
Names: Ameta, Suresh C., editor. | Ameta, Rakshit, editor.
Description: 2nd edition. | Includes bibliographical references and index.
Identifiers: Canadiana (print) 20230497721 | Canadiana (ebook) 20230497799 | ISBN 9781774913901 (hardcover) | ISBN 9781774913918 (softcover) | ISBN 9781003431473 (ebook)
Subjects: LCSH: Environmental chemistry. | LCSH: Green chemistry.
Classification: LCC TP155.2.E58 G74 2024 | DDC 577/.14—dc23

Library of Congress Cataloging-in-Publication Data

CIP data on file with US Library of Congress

ISBN: 978-1-77491-390-1 (hbk)
ISBN: 978-1-77491-391-8 (pbk)
ISBN: 978-1-00343-147-3 (ebk)

Dedicated

to

Mrs. Anita Ameta

who always motivated us for completing this book on time.

About the Editors

Suresh C. Ameta, PhD

Suresh C. Ameta, PhD, has served as Founder Professor and Head, Department of Chemistry, North Gujarat University Patan (1994) and M. L. Sukhadia University, Udaipur (2002–2005) and Head, Department of Polymer Science (2005–2008). He also served as Dean, P.G. Studies for a period of 4 years (2004–2008) and Dean, Faculty of Science, PAHER University, Udaipur (2011–2019). Presently, he is working as a Professor of Eminence (2019 onwards) in the same university. Prof. Ameta has occupied the coveted position of President, Indian Chemical Society, Kolkata, and is now a lifelong advisor. He was awarded a number of prizes during his career like National Prize (twice) for writing chemistry books in Hindi, Prof. M. N. Desai Award, Prof. W. U. Malik Award, the National Teacher Award, Prof. G. V. Bakore Award, and Life Time Achievement Awards by the Indian Chemical Society, Kolkata, the Indian Council of Chemists and Association of Chemistry Teachers in 2011, 2015, and 2018, respectively. The Indian Chemical Society has instituted a national award in his honor from 2003 and it is awarded to a senior scientist every year. He has successfully guided more than 100 PhD students. Prof. Ameta has more than 400 research publications to his credit in journals of national and international repute. He is the author of around 40 UG and PG level books. He has published about 15 books with international publishers, such as Taylor & Francis, Apple Academic Press, Elsevier. He has completed five major research projects from different funding agencies. Prof. Ameta has delivered lectures and chaired sessions in various international and national conferences. He is also a reviewer of number of international journals. Prof. Ameta has around 51 years of experience in teaching and research.

Rakshit Ameta, PhD

Rakshit Ameta, PhD, has had a first-class career and was awarded a gold medal for achieving the first position at M. L. Sukhadia University, Udaipur, India. He was also awarded the Fateh Singh Award from the Maharana Mewar Foundation, Udaipur, for his meritorious performance. He has served at M. L. Sukhadia University, Udaipur; University of Kota, Kota; and Paher University, Udaipur. Presently, he is serving at J. R. N. Rajasthan Vidyapeeth

(Deemed to be University) University, Udaipur, as Director of the Faculty of Science. To date, fifteen PhD students have been awarded their degrees under his supervision on various aspects of green chemistry. He has around 150 research publications in journals of national and international repute. He also has a patent to his credit. Dr. Rakshit has organized 10 international and national conferences at the University of Kota, Paher University, and J. R. N. Rajasthan Vidyapeeth (Deemed to be University) University. Dr. Rakshit was elected as a council member of the Indian Chemical Society, Kolkata (2011–2013; 2019–2022) and Indian Council of Chemists, Agra (2012–2014). He has not only written 10 degree-level books but has contributed chapters in books published by Nova Publishers, Taylor & Francis, Springer, Trans-Tech Publications, Elsevier, and published about eight books with international publishers, such as Taylor & Francis, Apple Academic Press, Elsevier, etc.

Contents

Contributors .... *xi*

Abbreviations .... *xv*

Preface .... *xxi*

1. **Introduction** **1**
   Rakshit Ameta

2. **Benign Starting Materials** **9**
   Neetu Shorgar, Sanyogita Sharma, Neelam Kunwar, Sangeeta Kalal, and P. B. Punjabi

3. **Eco-Friendly Products and Reagents** **55**
   Jayesh P. Bhatt, Neelu Chouhan, Anil Kumar, Ajay Sharma, and Rameshwar Ameta

4. **Green Catalysts** **117**
   Monika Jangid, Shikha Panchal, Yuvraj Jhala, Anuradha Soni, and Suresh C. Ameta

5. **Ionic Liquids: Promising Solvents** **147**
   Avinash K. Rai, Arpit Pathak, Nirmala Jangid, and P. B. Punjabi

6. **Supercritical Solvents** **189**
   Ravi Changwal, Abhilasha Jain, Shikha Panchal, Shweta Sharma, and Rameshwar Ameta

7. **Other Green Solvents** **221**
   Neha Godha, Abhilasha Jain, Ritu Vyas, Aarti Ameta, and P. B. Punjabi

8. **Photocatalysis: An Emerging Technology** **279**
   Shubang Vyas, Indu Bhati, Paras Tak, H. S. Sharma, and Rakshit Ameta

9. **Photo-Fenton Reactions: A Green Chemical Route** **323**
   Meghavi Gupta, Noopur Ameta, Surbhi Benjamin, and P. B. Punjabi

10. **Sonochemistry: A Pollution-Free Pathway** **359**
    Neha Kapoor, Garima Ameta, Surbhi Benjamin, Vikas Sharma, and Shipra Bhardwaj

11. Microwave-Assisted Organic Synthesis: A Need of the Day 399
Seema Kothari, Chetna Ameta, K. L. Ameta, B. K. Sharma, Rajat Ameta, and Rakshit Ameta

12. Green Composites 453
Priyanka Jhalora, Narendra Pal Singh Chauhan, Yasmin, and Rohit Ameta

13. Green Manufacturing Processes 507
Shambhu Lal Agarwal, Jitendra Vardia, Dipti Soni, and Rakshit Ameta

14. Future Trends 523
Suresh C. Ameta

Index 527

Contributors

Shambhu Lal Agarwal
Department of Chemistry, PAHER University, Udaipur, Rajasthan, India

Aarti Ameta
Department of Chemistry, Guru Nanak Girls' P.G. College, Udaipur, Rajasthan, India

Chetna Ameta
Department of Chemistry, M. L. Sukhadia University, Udaipur, India

Garima Ameta
Department of Chemistry, M. L. Sukhadia University, Udaipur, India

K. L. Ameta
Department of Chemistry, Modi University, Lakshmangarh, India

Noopur Ameta
Department of Chemistry, M. L. Sukhadia University, Udaipur, India

Rajat Ameta
Zyfine Cadila, Ahmedabad, Gujarat, India

Rakshit Ameta
Department of Chemistry, J. R. N. Rajasthan Vidhyapeeth, (Deemed to be University), Udaipur, India

Rameshwar Ameta
Department of Chemistry, PAHER University, Udaipur, Rajasthan, India

Rohit Ameta
R&D Section, Apollo Tyres, Chennai, Tamil Nadu, India

Suresh C. Ameta
Department of Chemistry, PAHER University, Udaipur, Rajasthan, India

Surbhi Benjamin
Department of Chemistry, J. R. N. Rajasthan Vidhyapeeth, (Deemed to be University), Udaipur, Rajasthan, India

Shipra Bhardwaj
Department of Chemistry, Govt. Meera Girls' College, Udaipur, Rajasthan, India

Indu Bhati
Department of Chemistry, M. L. Sukhadia University, Udaipur, Rajasthan, India

Jayesh P. Bhatt
Department of Chemistry, PAHER University, Udaipur, Rajasthan, India

Ravi Changwal
Department of Chemistry, PAHER University, Udaipur, Rajasthan, India

Narendra Pal Singh Chauhan
Department of Chemistry, B. N. University, Udaipur, Rajasthan, India

Neelu Chouhan
Department of Pure and Applied, Chemistry, University of Kota, Kota, Rajasthan, India

Neha Godha
Department of Chemistry, PAHER University, Udaipur, Rajasthan, India

Meghavi Gupta
Department of Chemistry, PAHER University, Udaipur, Rajasthan, India

Abhilasha Jain
Department of Chemistry, St. Xavier's College, Mumbai, Maharashtra, India

Monika Jangid
Department of Chemistry, PAHER University, Udaipur, Rajasthan, India

Nirmala Jangid
Department of Chemistry, Banasthali Vidhyapith, Banasthali, Rajasthan, India

Yuvraj Jhala
Department of Chemistry, B. N. University, Udaipur, Rajasthan, India

Priyanka Jhalora
Department of Chemistry, PAHER University, Udaipur, Rajasthan, India

Sangeeta Kalal
Department of Chemistry, M. L., Sukhadia University, Udaipur, Rajasthan, India

Neha Kapoor
Department of Chemistry, PAHER University, Udaipur, Rajasthan, India

Seema Kothari
Department of Chemistry, PAHER University, Udaipur, Rajasthan, India

Anil Kumar
Department of Chemistry, M.P. Govt. P.G. College, Chittorgarh, Rajasthan, India

Neelam Kunwar
Department of Chemistry, PAHER University, Udaipur, Rajasthan, India

Shikha Panchal
Department of Chemistry, PAHER University, Udaipur, Rajasthan, India

Arpit Pathak
Department of Chemistry, G. G. Govt. P.G. College, Banswara, Rajasthan, India

P. B. Punjabi
Department of Chemistry, M. L. Sukhadia University, Udaipur, Rajasthan, India

Avinash K. Rai
Department of Chemistry, PAHER University, Udaipur, Rajasthan, India

Ajay Sharma
Department of Chemistry, Govt. P.G. College, Sirohi, Rajasthan, India

B. K. Sharma
Department of Chemistry, G. G. Govt. P.G. College, Banswara, Rajasthan, India

H. S. Sharma
Department of Chemistry, Govt. P.G. College, Kota, Rajasthan, India

Sanyogita Sharma
Department of Chemistry, PAHER University, Udaipur, Rajasthan, India

Shweta Sharma
Department of Chemistry, PAHER University, Udaipur, Rajasthan, India

Vikas Sharma
Jain Mandir Road, Kota Jn., Kota, Rajasthan, India

Neetu Shorgar
Department of Chemistry, PAHER University, Udaipur, Rajasthan, India

Anuradha Soni
Department of Chemistry, M. L. Sukhadia University, Udaipur, Rajasthan, India

Dipti Soni
Department of Chemistry, J. R. N. Rajasthan Vidhyapeeth, (Deemed to be University), Udaipur, Rajasthan, India

Paras Tak
Department of Chemistry, PAHER University Udaipur, Rajasthan, India

Jitendra Vardia
J. D. M. Scientific Research Organization, Vadodara, Gujarat, India

Ritu Vyas
Department of Chemistry, Pacific Institute of Technology, Udaipur, Rajasthan, India

Shubang Vyas
Department of Chemistry, PAHER University Udaipur, Rajasthan, India

Yasmin
Department of Chemistry, Geetanjali Institute of Technical Studies, Udaipur, Rajasthan, India

Abbreviations

AA	acetic acid
ABS	acrylonitrile–butadiene–styrene
ABs	animal bones
AC	activated carbon
AChE	acetylcholinesterase
ALA	aminolevulinic acid
AMP	aqua mesophase pitch
AOPs	advanced oxidation processes
API	active pharmaceutical ingredient
ASR	automobile shredder residue
AWS	aspen wood sawdust
BBHA	bovine bone-derived hydroxyapatite
BFD	blast furnace dust
BHC	bulk heterojunction
BPS	banana pseudostem
BVMOs	Baeyer–Villiger monooxygenases
BW	beeswax
CA	carbonic anhydrase
CAN	ceric ammonium nitrate
CCL	*Candida cylindracea*
CD	cyclodextrin
CDs	carbon dots
CG	coal gangue
CHP	combined heat and power
CIL	chiral ionic liquid
CLW	candelilla wax
CM	carbon modified
CM	compression molding
CNF	cellulose nanofibril
CNHs	carbon nanohorns
CNTs	carbon nanotubes
CP	coordination polymer
CPC	cetylpyridinium chloride
CPME	cyclopentyl methyl ether

CVD	chemical vapor deposition
CW	carnauba wax
DBS	dodecylbenzenesulfonate
DBSA	dodecylbenzenesulfonic acid
DBT	dibenzothiophene
DCP	dichlorophenol
DES	deep eutectic solvent
DMAD	dimethyl acetylenedicarboxylate
DMF	dimethylformamide
DMS	dimethyl sulfide
DPPH	diphenyl-1-picrylhydrazyl
EE	enantiomeric excess
EEI	environmental efficiency index
EG	ethylene glycol
EL	ethyl lactate
EPA	Environmental Protection Agency
FA	formic acid
FAME	fatty acid methyl esters
FAP	fluorapatite
FBS	fluorous biphasic system
FDCA	furandicarboxylic acid
FUR	furfural
GA	gum arabic
GE	gasification efficiency
GF	glass fiber
GF	graphene foam
GGBS	ground granulated blast furnace slag
GO	graphene oxide
GPC	geopolymer cement
GVL	gamma-valerolactone
HDS	hydrodesulfurization
HMF	hydroxymethylfurfural
HNTs	halloysite nanotubes
HS	*Hibiscus sabdariffa*
HSPI	hydrolyzed soybean protein isolate
HSPI/U/F	urea (U)/formaldehyde (F)
ILs	ionic liquids
IM	Irish moss
IPA	isopropanol

LAB	lactic acid bacteria
LABS	linear alkylbenzene sulfonates
LAS	linear alkyl sulfonate
LCA	life cycle assessment
LCC	liquid coordination complex
LDA	lithium diisopropylamide
LHs	lignocellulose hydrogels
LIBs	lithium–ion batteries
LOI	limiting oxygen index
LTMP	tetramethylpiperidide
LV	levofloxacin
MAL	*Mentha aquatica* leaf
MAOS	microwave-assisted organic synthesis
MB	methyl benzoate
MCL	medium chain length
MCP	monocrotophos pesticide
2-MeTHF	2-methyltetrahydrofuran
MFC	microfibrillated cellulose
MNZ	Mexican natural zeolite
MOFs	metal–organic frameworks
MOP	metal organic polymer
MRR	material removal rate
MSA	methanesulfonic acid
MW	microwave
NBS	N-bromosuccinimide
NC	nanocomposite
NCP	natural clay powder
NCW	near critical water
NFC	nanofibrillated cellulose
NMP	N-methyl-2- pyrrolidone
NPs	nanoparticles
NR	nanorod
NTA	nitrilotriacetate
NWs	nanowires
nZVI	nanoscale zero-valent iron
OA	oleyl alcohol
OPV	organic photovoltaic
PA	propionic acid
PAH	polycyclic aromatic hydrocarbons

PBM	Petasis borono–Mannich
PC	propylene carbonate
PCBM	phenyl-C61-butyric acid methyl ester
PCBs	polychlorinated biphenyls
PCBs	printed circuit boards
PCE	power conversion efficiency
PDI	perylene diimide
PE	polythene
PEG	polyethylene glycol
PET	polyethylene terephthalate
PHAs	polyhydroxyalkanoate
PHB	polyhydroxybutyrate
PHE	phenanthrene
PHH	polyhydroxyhexanoate
PHV	polyhydroxyvalerate
PICs	products of incomplete combustion
PLA	polylactic acid
PMA	phosphomolybdic acid
PMW	pulp mill wastewater
PPA	polyphosphoric acid
PPCP	pharmaceuticals and personal care products
PPD	p-phenylenediamine
PPG	procaine penicillin–G
PPL	*Porcine pancreas* lipase
PR	primary rolling
PS	polystyrene
PSF	polysulfone
PTC	phase-transfer catalyst
PTT	polytrimethylene terephthalate
PVC	polyvinylchloride
PVP	polyvinylpyrrolidone
QDs	quantum dots
RB	rhodamine-B
RBW	rice bran wax
RGO	reduced graphene oxide
RME	reaction mass efficiency
ROL	*Rhizopus oryzae* lipase
SA	sand aggregates
SA	sodium alginate

SA	sodium aluminate
SBR	sequencing batch reactor
SCB	sugarcane bagasse
SCFs	supercritical fluids
SCL	short-chain length
SCMs	supplementary cementitious materials
SCW	supercritical water
SCWG	supercritical water gasification
SCWO	supercritical water oxidation
SCWU	supercritical water upgrading
SDS	sodium dodecyl sulfate
SDZ	sulfadiazine
SF	straw fiber
SFC	supercritical fluid chromatography
SFE	supercritical fluids extraction
SFW	sunflower wax
SILAR	sequential ionic layer adsorption and reaction
SLS	sodium lauryl sulphate
SMX	sulfamethoxazole
SNPs	sulfur nanoparticles
SOFC	solid oxide fuel cells
SP	super plasticizer
SPB	sodium perborate
SPI	soy protein isolate
SSC	silica sodium carbonate
SSF	simultaneous saccharification and fermentation
TA	tartaric acid
TBA	tetrabutylammonium
TBH	tebuthuron
TBP	tetrabutylphosphonium
TBP	tributyl phosphate
TBPH	tetrabutyl phosphonium hydroxide
TBTO	tributyltin oxide
TC	tetracycline
TCC	triclocarban
TCE	trichloroethylene
TCS	triclosan
TDS	total dissolved solids
TEG	triethylene glycol

THMs	trihalomethanes
TKN	total Kjeldahl nitrogen
TMAA	tetramethyladipic acid
TMG	tetramethylguanidinium
TNT	titania nanotube
TNT	trinitrotoluene
TOA	trioctylamine
TOC	total organic carbon
TPA	terephthalic acid
TPC	total polyphenol content
TPS	thermoplastic starch
TPs	tomato peels
TSE	twin-screw extrusion
UAE	ultrasound assisted extraction
UASB	upflow anaerobic sludge blanket
UVO	UV–ozone
VOA	volatile organic acids
VOCs	volatile organic compounds
VTMS	vinyltrimethoxysilane
WGSR	water gas shift reaction
WVP	water vapor permeability
WWTP	waste water treatment plants

Preface

Chemical sciences have played an important role in the development of society, making significant contributions in various fields. This advancement, though, has come at the cost of our environment, with various adverse effects on humans, animals, plants, and aquatic organisms. Many toxic chemicals have passed into our food chain and have become a part of our ecosystem. This book is a step toward promoting sustainable development, with the goal of moving toward a "greener world."

Green chemistry is an emerging field that calls for developing less harmful processes in industrial manufacturing. It is a combination of 12 basic principles and, in a few words, can be defined as the design, invention, and application of either benign chemical processes or less harmful/hazardous products so as to reduce or eliminate the formation and distribution of hazardous substances in the environment. Efforts should be made to maximize the utilization of atoms of reactants into the final products.

Green chemistry is not at all a new branch of chemistry instead; it gives a newer dimension to our thinking so that one can provide/develop a newer chemical with desired properties to society without putting any adverse impact on our environment, keeping this beautiful planet Earth clean and green. Various green chemicals have been and are being developed today using newer pathways/routes, taking a step further toward a safer world. The crucial role of green chemistry in the sustenance of the environment has been presented here.

This book addresses various topics that fall in the arena of "Green Chemistry," such as, benign starting material, ecofriendly products, green catalysts, supercritical fluids, green solvents, ionic liquids, photocatalysis, photo-Fenton reaction, microwave and ultrasound-assisted reactions, green composites, green manufacturing processes, etc. The current and future impacts of green chemistry have also been discussed in this book.

The main aim of this book is to enlighten the scientific community to become aware of the green routes, which are both eco-friendly and useful on an industrial scale. There is a great need to put the majority of the 12 principles (as far as one can) of green chemistry into actual practice. The first edition of this book was published in 2014, and it was appreciated by the scientific community all over the globe. It has now been updated up to the end of 2022. The readers are requested to read, understand, and give suggestions for further improvement of the book.

—Suresh C. Ameta and Rakshit Ameta

CHAPTER 1

Introduction

RAKSHIT AMETA

Department of Chemistry, J. R. N. Rajasthan Vidhyapeeth (Deemed to be University), Udaipur, India

ABSTRACT

The world is facing an ever-increasing environmental pollution due to a large number of industries and heavy transportation. One cannot stop these activities as these are necessary for the overall development of society. Therefore, existing chemical processes creating pollution should be replaced by some other alternate processes that are harmless or less harmful to the environment. Hence, the importance of green chemistry. Traditional or gray chemistry is responsible for creating the pollution load while green chemistry takes care of the environment in advance. Thus, green chemical processes are replacing the chemical processes in use at present. Twelve basic principles of green chemistry have been discussed.

The environment is a surrounding for all of us, and it comprises a variety of physical and chemical components. One interacts with this environment and it is also a part of it. It is well established that science, particularly chemical sciences, are developing at a very fast pace. As a result, a huge amount of chemicals are used in manufacturing processes and they are increasingly showing up in the environment. These chemical components are increasing day by day in the environment, many of which are undegradable. This causes environmental pollution. Here, it may be concluded that the addition of these undegradable or recalcitrant substances or molecules cause disorder, harm, discomfort, or instability to our ecosystem and create nothing but pollution.

It can be very well said that green chemistry is expanding its wings step by step from small laboratories to pilot-scale plants and ultimately to large

Green Chemistry, 2nd Edition: Fundamentals and Applications. Suresh C. Ameta & Rakshit Ameta (Eds.)

manufacturing processes or units. It has been predicted that the turnover of a number of industries has increased many folds today and will further increase in coming decades by following the basic principles of green chemistry.

The entire world is in the cancerous grip of rapidly increasing environmental pollution in its various facets, such as water pollution, air pollution, soil pollution. This problem is further supported by global warming and the energy crisis. Environmental pollution makes our life miserable on this beautiful planet, "The Earth." This is all because of the use of gray chemistry to fulfill different demands of materials with varied applications, such as metallurgy, synthesis of pharmaceuticals and other chemicals, use of volatile organic solvents, polymers, dry cleaning, agriculture, use of detergents, which creates different kinds of pollutions. The men cannot survive without using many of these toxic chemicals to make their life more comfortable even at the cost of their health. Therefore, there is a pressing demand all over the globe to either reduce the use of more toxic materials or to replace them with less toxic or less harmful alternates.

This can be achieved by transforming from gray chemistry to green chemistry. The term green chemistry was used first by Paul T. Anastas in the beginning of the last decade of the 20th century. The field of green chemistry has been excellently presented by several workers (Tundo and Anastas, 1999, 2000; Anastas et al., 2000; Ameta, 2002; Lancaster, 2002; Ameta et al., 2004; Matlack, 2010; Ameta et al., 2012; Ameta and Ameta, 2013). The green chemistry is in no way different from gray chemistry except that the approach toward a chemical process, may be manufacture design or applications. Green chemistry is totally different from environmental chemistry. In environmental chemistry, one takes care of the kind of pollution, the extent of pollution and methods to combat against this pollution, whereas green chemistry takes care of all these factors in advance. It is something like diagnosis of any disease and its treatment is environmental chemistry while prevention from that disease is like green chemistry. There is a well-known proverb that "Precaution is better than Cure." Prevention is a green chemical pathway while cure is the environmental pathway.

Many useful materials have been invented by chemists to provide comforts to the society on its own demand, but it not only yields the desired products in larger amounts but also other toxic (harmful) and undesirable materials as by-products or wastes. This is really a great challenge to chemists to maintain a sustainable environment. It requires much more efficient technologies to get rid off from these waste materials. Green chemistry can provide a big platform to us for finding more efficient processes to minimize the toxic and harmful waste along with maximizing the yield of the desired products.

Apart from the existing forms of pollution, we are going to face and even we are facing these emerging faces of pollution today also, that is, polymer and detergent pollution. Almost all materials, such as metal, wood, textile are slowly being replaced by one or other kind of polymer, which has resulted into the accumulation of this polymeric material in the dumping yards. The disposal of this dumped material or its recycling is a burning problem of the day.

Soap is biodegradable but as it does not work in hard water, and therefore, it is rapidly being substituted by detergent. We have almost forgotten the use of washing soap leaving aside bathing soap. These detergents are not biodegradable, and therefore, remain for years together in nearby water resources; thus, making this water unfit for its use as portable water. It is utmost necessary to find out either some substitute for these detergents or develop methods to degrade the accumulated detergents in water.

Pharmaceutical industries are facing a problem, which is like a double-headed arrow. The drugs should be toxic to bacteria or fungi, but they should not be harmful to human beings, animals, plants, etc. If an effort is being made to increase the efficacy of a drug, insecticides, weedicides, etc., it may also increase its toxicity, which is undesirable. To keep the toxicity low and increasing the efficiency is a challenging task for a chemist because one is working to achieve two totally opposing objectives. However, green chemistry may provide some feasible solutions to this problem.

Green chemistry utilizes a set of 12 principles that either reduces/eliminates the use/generation of any hazardous substances in designing, manufacture, and application of chemicals (Anastas and Warner, 1998). This is an approach, which is based on reducing the amount of waste generated at the source rather than treating this waste after it has been formed. We as the chemists are normally blamed for creating pollution, but the green chemical approach not only solves the problem of pollution but it will also provide the methods to synthesize/utilize substances in an eco-friendly manner. Green chemical approach is holistic in nature and encompasses almost all the major branches of chemical science, such as organic/inorganic synthesis, catalysis, drug discovery, material sciences, polymer, nanochemistry, supramolecular chemistry, treatment of wastewater.

Green chemistry is also known synonymously as:

- Clean chemistry
- Atom economy
- Benign by design chemistry
- Eco-friendly chemistry

- Environmentally benign chemistry
- Sustainable chemistry
- E-chemistry

Achieving green chemical pathways in the laboratory as well as industrial level still exists as a challenge for chemists. Collaborative efforts are urgently needed from Government, industries, academic, and NGOs to face this challenge.

In a presidential address to Indian Chemical Society, Ameta (2002) has very rightly given the slogan:

- Green chemistry: green earth
- Clean chemistry: clean earth

There may be confusion that the green chemical pathway is almost benign, but it is not a perfectly true statement because there cannot be any chemical, which is perfectly benign, and therefore, green chemistry diverts the use of chemicals from malign to benign manner. Common salt is necessary for life, but it may develop hypertension if taken in excess. The same is the case with carbohydrates (sugar), which is required for providing energy for daily routine life but if given in excess, it may be harmful to humans. Therefore, shifting from less benign (more malign) to more benign (less malign) process may be considered a green chemical approach. Someone has well said that "A matter may act as poison if given in a large amount, and a poison, if given in very small amount may act as a nector." This is the basic concept of homeopathy, which deals with very small concentrations of toxic chemicals to surprisingly cure many dreadful diseases and also the efficiency of these homeopathic medicines increases on dilutions.

The green chemical approach is governed by 12 principles given by Anastas and Warner (1998). These principles are important in combating environmental pollution and for the betterment of human health. These principles are:

i) **Prevention:** It is better to prevent wastes rather than treating or cleaning up wastes after they are produced.
ii) **Atom economy:** Syntheses should be so designed, wherever possible, to maximize the incorporation of all materials used in the process into their final products.
iii) **Less hazardous chemical syntheses:** Wherever practicable, synthetic methods should be designed to use and generate substances possessing little or no toxicity to human health and the environment.

iv) **Designing safer chemicals:** Chemical products should be designed to exhibit their desired function while minimizing their toxicity.
v) **Safer solvents and auxiliaries:** The use of auxiliary substances (e.g., solvents, separating agents, etc.) should be made unnecessary wherever possible, and innocuous when used.
vi) **Design for energy efficiency:** Energy requirements of chemical processes should be recognized for their environmental and economic impacts and should be minimized. If possible, synthetic methods should be conducted at ambient temperature and pressure.
vii) **Use of renewable feedstocks:** A raw material or feedstock should be renewable rather than depleting whenever technically and economically practicable or feasible.
viii) **Unnecessary derivatization:** Blocking group, protection/deprotection, and temporary modification of physical/chemical processes should be avoided whenever possible.
ix) **Catalysis:** Catalytic reagents (as selective as possible) are superior to stoichiometric reagents.
x) **Design for degradation:** Chemical products should be designed so that at the end of their function, they break down into innocuous or harmless degradation products and do not persist in the environment.
xi) **Real-time analysis for pollution prevention:** Analytical methodologies need to be further developed to allow for real-time, in-process monitoring, and control prior to the formation of hazardous substances.
xii) **Inherently safer chemistry for accident prevention:** Substances and the form of a substance used in a chemical process should be chosen to minimize the potential for chemical accidents, including releases of chemicals, explosions, and fires.

Atom economy is an important concept in the philosophy of green chemistry (Trost, 1995; Sheldon, 2000). It is important to utilize the maximum number of atoms of the reactant to minimize the generation of waste products. It is defined as an atom economy in green chemistry meaning that one has to be economic in using atoms. The atom economy is defined as–

$$\%\text{ Atom Economy} = \frac{\text{Molecular weight of desired product}}{\text{Molecular weight of all reactants}} \times 100 \quad (1.1)$$

Addition reactions and rearrangements normally follow the atom economy but chemistry is not complete with only these reactions but some substitution and elimination reactions are also required. Thus, the generation of wastes is bound to be there but the efforts of the chemists should be to produce minimum by-product.

"Gray" process can be made "green" by making a judicious selection of green substrate, green solvents, green reagents, green catalysts, green conditions, etc. to synthesize a green product.

Principles of green chemistry are beautifully condensed by Tang et al. (2008) as **PRODUCTIVELY:**

Principles of Green Chemistry

P—Prevent wastes
R—Renewable materials
O—Omit derivatization steps
D—Degradable chemical products
C—Catalytic reagents
T—Temperature and pressure ambient
I—In-process monitoring
V—Very few auxiliary substances
E—E-factor, maximize feed in product
L—Low toxicity of chemical products
Y—Yes it's safe

Efforts are being made to fulfill all the 12 conditions to make a chemical process perfectly green, but it is not practicable to satisfy all the requirements of 12 principles of green chemistry. Therefore, a chemical process is better defined as greener than the other chemical processes, which fulfills more conditions and further research may make it still greener and this process will go on.

KEYWORDS

- **environment**
- **eco-friendly chemistry**
- **green chemistry**
- **pollution**
- **atom economy**

REFERENCES

Ameta, R.; Ameta, C.; Tak, P.; Benjamin, S.; Ameta, R.; Ameta, S. C. Green Chemistry: A Step Towards Clean and Sustainable Development. *J. Indian Chem. Soc.* **2012,** *89*, 992–1018.

Ameta, S. C. Green Chemistry: The Chemistry of New Millenium. *J. Indian Chem. Soc.* **2002,** *79*, 305–307.

Ameta, S. C.; Ameta, R., Eds. *Green Chemistry: Fundamentals and Applications*; Apple Academic Press: New Jersey, 2013.

Ameta, S. C.; Mehta, S.; Sancheti, A.; Vardia, J. Green Chemical Pathways: A Need of the Day. *J. Indian Chem. Soc.* **2004,** *81*, 1127–1140.

Anastas, P. T. Green Chemistry and the Role of Analytical Methodology Development. *Crit. Rev. Anal. Chem.* **1999,** *29*, 167–175.

Anastas, P. T.; Warner, J. C., Eds. *Green Chemistry: Theory and Practice*; Oxford University Press: New York, 1998.

Anastas, P. T.; Heine, L. G.; Williamson, T. C. *Green Chemical Synthesis and Processes*; American Chemical Society: Washington, 2000.

Lancaster, M. *Green Chemistry: An Introductory Text*; Royal Society of Chemistry: London, 2002.

Matlack, A. *Introduction to Green Chemistry*, 2nd ed.; CRC Press: London, 2010.

Sheldon, R. A. Atom Efficiency and Catalysis in Organic Synthesis. *Pure Appl. Chem.;* **2000,** *72* (7), 1233–1246.

Tang, S. Y.; Bourne, R. A.; Smith, R. L.; Poliakoff, M. The 24 Principles of Green Engineering and Green Chemistry: "Improvements Productively". *Green Chem.* **2008,** *10* (3), 268–269.

Trost, B. M. Atom Economy—A Challenge for Organic Synthesis: Homogeneous Catalysis Leads the Way. *Angew. Chem. Int. Ed. Engl.* **1995,** *34*, 259–281.

Tundo, P.; Anastas, P. T. *Green Chemistry: Challenging Perspectives*; Oxford University Press: New York, 2000.

CHAPTER 2

Benign Starting Materials

NEETU SHORGAR[1], SANYOGITA SHARMA[1], NEELAM KUNWAR[1], SANGEETA KALAL[2], and P. B. PUNJABI[2]

[1]*Department of Chemistry, PAHER University, Udaipur, India*

[2]*Department of Chemistry, M. L., Sukhadia University, Udaipur, India*

ABSTRACT

Society needs a number of useful products and these are prepared by either petroleum based or some other harmful starting materials. It is necessary to find some substitute for these starting materials, which are green chemicals in nature or bio-based natural products. Selection of such material should be such that these are renewable, pose no hazard, can be prepared in a few number of steps, and that their 100% atom economy with high yields. These can be derived from biomass such as cellulose, glucose, lactic acid, etc. Hydrogen can be generated from photosplitting of water. Carbon dioxide can be used to synthesize energy rich-products. Biodiesel can be used as substitute for conventional petroleum based fuels. Various bio-transformations have been described in this chapter.

2.1 INTRODUCTION

Our environment is composed of the atmosphere, earth, water, and space. Under normal circumstances, it remains clean and, therefore, enjoyable. However, with the increasing world population and with limited natural resources, the composition and complex nature of our environment has changed. Our world is beautiful, but the increasing use and improper disposal of the effluents from various industries are creating pollution of the environment.

Green Chemistry, 2nd Edition: Fundamentals and Applications. Suresh C. Ameta & Rakshit Ameta (Eds.)

The time is approaching for natural gas and petroleum production to peak, plateau, and then decline. Prices are also increased substantially in the last few decades contributing to the almost uncertainty. These trends and the uncertain future inevitably influence industrial nations. As our fossil raw materials are irreversibly decreasing supported by pollution pressure on our environment, the progressive changeover of the chemical industry to renewable feedstocks for their raw materials has become an inevitable necessity (Okkerse and Bekkum, 1999; NRC, 2000).

We have to proceed increasingly to the removable raw material basis before natural gas oil and all other sources are completely exhausted. The over-reliance of the chemical industry on fossil raw materials has its limitations; as these are depleting at a rapid pace and are not renewable. Now there is a question, when will fossil fuels be exhausted? or when will fossil raw materials become so expensive that bio-feedstocks will become an economically competitive alternative? Experts opine the end of cheap oil by 2040 at the latest (Umbach, 1996; Klass, 1998). This is a development that one can witness by the fact that chemical industries are now combating the increasing costs of natural oil and gas (Campbell and Laherrere 1998).

The future is quite promising. Scientists and technologists following the trends on sustainability and natural resources are persuading industries either to use alternative resources or to develop some new approaches toward more efficient chemical processes. Thus, there is a pressing demand for the transition to a more bio-based production system, but this is hampered by different obstacles. Fossil-based raw materials are relatively more economic at present and the process technology for their conversion into organic chemicals is well-developed. It is basically a different form that is required for transforming carbohydrates into products with industrial applications.

Green chemistry is a philosophy and study of the design of products or substances that will not involve materials harmful to the environment. It is a modern science of chemistry that deals with the application of environmentally friendly chemical compounds in the various areas of our life such as industrial uses and many others. This area of chemistry has been developed by the need to avoid chemical hazards that organic and inorganic compounds have on the body of humans and animals. Chemistry plays a pivotal role in determining the quality of modern life. The chemical and other related industries supply us with a huge variety of essential products, from plastics to pharmaceuticals. However, these industries also have the potential to seriously damage our environment. Therefore, green chemistry serves to promote the design and efficient use of environmentally benign chemicals and chemical processes.

2.2 SUSTAINABILITY

The most commonly used definition for sustainable development comes from a report by the World Commission on Environment and Development, that is "To meet the needs of the present without compromising the ability of future generations to meet their own needs, green chemistry and green engineering are striving hard to develop new methodologies for sustainable development. Their proposals focus on:

2.2.1 *RENEWABLE FEEDSTOCKS AND RAW MATERIALS*

Green chemistry needs to change the starting materials into renewable feedstocks. The most desired property of basic starting material is its lower toxicity and low environmental impact. Health and safety protection of workers involved and the environment is on top priority. Green chemistry is just proposing a change of direction from fossil-based feedstocks into biological raw materials. There are many problems in using these materials, but in the last few years, there are some encouraging new results for large-scale production and the use of alternative renewable materials. The terrestrial biomass is quite complex containing high molecular weight products, such as sugars, hydroxy and amino acids, lipids, biopolymers (cellulose, hemicellulose, chitin, starch, and lignin) proteins. The most important class of biomolecules produced is carbohydrates (~ 75% of the annually renewable biomass approx. 200 billion tons). A minor fraction (4%) of this is used by man, while the rest decays or recycles by natural pathways. The bulk of these annually renewable feedstocks (carbohydrate biomass) is polysaccharides, and their nonfood utilization is limited to textile, paper, and coating industries (Dewulf and Lagenhove, 2006; Benaglia, 2009).

2.2.2 *OLEOCHEMISTRY*

Oleochemicals, such as fats and oils (from plants and animals) are becoming a new source of chemical feedstocks as raw materials (Hill, 2000; Gutsche, et al., 2008). A series of new raw materials exist in the market with a variety of applications, such as cosmetics, polymers, lubricating oils.

2.2.3 PHOTOCHEMISTRY

Green chemistry also puts a lot of emphasis on photochemical reactions in chemical processes (Albini and Fagnoni, 2004; Ravelli et al., 2009). Light (ultraviolet and visible region) can catalyze many reactions. Photochemistry has a great potential, and quite a few interesting research findings were introduced in the last few decades including some applications. Ultraviolet as well as visible light from the Sun is considered as renewable energy source and thus, photochemistry can contribute to some of the green synthetic chemistry applications.

2.2.4 BIOCATALYSIS AND BIOTRANSFORMATIONS

Biocatalysis is particularly a green technology with many applications, which are considered benign for the environment and energy efficient (Ran et al., 2008; Whittall and Sutton, 2009; Cheng and Gross, 2011; Tao and Kazlauskas, 2011). Enzymes have been used for many synthetic chemical routes with great advantages in the food and pharmaceutical industries. Biocatalysis is at the interface of fermentation techniques (food and alcoholic drink industries) with some other industrial processes, where enzymes are used for higher yields and low energy consumption. Biotransformations can be achieved through biocatalysis and these are considered good green techniques for a series of chemical industries and a variety of chemical products. An eco-friendly and economic biocatalytic route with lipase extracted from *Aspergillus niger* as an efficient biocatalyst has been suggested for glycerol carbonate synthesis (Tudorache et al., 2012).

2.2.5 CAPTURE OR SEQUESTRATION OF CARBON DIOXIDE

Green chemistry is involved in carbon dioxide reduction in chemical industries. Climate change and the phenomenon of greenhouse effects due to CO_2 emission is considered very important environmental problem by green chemists. Any effort to reduce CO_2 emissions during the industrial process is an important goal from the green chemistry point of view. Also, any design in chemical processes, which sequesters or captures or can use CO_2 is worthy of the aims of green chemistry (Holtz, 2003; Hester and Harrison, 2009; Allen and Brent, 2010; Leimkuhler, 2010).

2.2.6 WASTE BIOMASS AS CHEMICAL FEEDSTOCK, BIOMATERIALS AND BIOFUEL

The advances in the last decade were to use biomass for the production of various materials and these were quite impressive. It was known for decades that biomass from agricultural processes goes almost to waste. Biomass is considered a solution to a very important problem of sustainability with increasing fossil fuel prices. In recent years, many new technologies showed the use of biomass as biofuel, raw material for the production of biomaterials, polymers, and various other applications (Ravindranath and Hall, 1995; Ragauskas et al., 2006; Soetaert and Vendamme, 2009). Selective hydrogenation of alternative oils is a useful tool for the production of biofuels. Highly selective hydrogenation of nonfood oils like flax over nontoxic heterogeneous catalysts can be used to make them suitable for biodiesel formulation (Zaccheria et al., 2009). Renewable gasoline was prepared directly from aqueous phase hydrodeoxygenation of aqueous sugar solution in a two-bed reactor (Li et al., 2011). Catalytic upgrading of bio-oil using 1-octene and 1-butanol over sulfonic acid catalysts was done by Zhang et al. (2011). It is an atom economic route for upgrading bio-oil to oxygenated fuels by simultaneous acid-catalyzed reactions with olefins and alcohols. A route to liquid hydrocarbon fuels has been suggested by Case et al. (2012) by pyrolyzing mixtures of levulinic acid and formic acid salts. This one-step process is operated at atmospheric pressure without catalyst or hydrogen addition.

2.2.7 BIODEGRADATION OF BIOMASS TO BIOGAS AND BIODIESEL

Biomass is well known for its use in biofuel, especially from organic wastes in landfills. Biomass can be used for the production of biodiesel through some chemical and physical processes. Akbar et al. (2009) prepared Na-doped SiO_2 solid catalyst by sol–gel method for the production of biodiesel from Jatropha oil. The biodiesel was produced by transesterification of Jatropha oil using a solid catalyst, Na/SiO_2 to form fatty acid methyl ester with a very high yield under mild conditions of operation. Biodiesel was also produced by palladium catalyzed decarboxylation of higher aliphatic esters (Han et al., 2010). It is an effective and highly selective decarboxylation approach to convert higher aliphatic esters into diesel-like paraffins. The methodology of this process provides a new protocol to utilization of biomass-based resources, especially to the second-generation biodiesel production. Biomass offered an opportunity for the production of 19% of energy on a global scale. Now, it is estimated that 4% of all fuel products in cars are produced from biomass. Dicyclohexylguanidine group covalently attached to silica gel is an efficient basic heterogeneous catalyst for the production of biodiesel in a continuous flow reactor (Balbino et al., 2011). Crude glycerol obtained from biodiesel waste was found suitable for the production of intracellular non-reducing sugar trehalose and relatively pure propionic acid simultaneously (Ruhal et al., 2011).

2.3 SUSTAINABLE MATERIALS

Materials produced and used in modern society are quite diverse and evolving. Approximately 75,000 chemicals are used commercially. Because no one formulation has a unique sustainability, it is useful to provide an operational definition. A sustainable material is a material that fits within the constraints of a sustainable material system. In order to be sustainable, a material must be appropriate for the system and vice versa.

Two strategies have been identified to support a sustainable materials economy: (i) Dematerialization, which involves developing ways to use less material to provide the same service in order to satisfy human needs and (ii) Detoxification of materials used in products and industrial processes.

Chemistry plays a pivotal role in the production of food, materials supply for clothing and shelter, preventing disease, and providing health care products. Organic chemicals are some of the important starting materials for a large number of major chemical industries. The production of organic chemicals as raw materials or reagents for other applications is a major sector of manufacturing polymers, pharmaceuticals, pesticides, paints, artificial fibers, food additives, etc. Organic synthesis on a large scale as compared with the laboratory scale involves the use of energy and basic chemical ingredients from the petrochemical sector like catalysts, and after the end of the reaction, it also involves separation, purification, storage, packaging, distribution, etc. During these processes, there are many problems of health and safety for workers in addition to the environmental problems caused by their use and disposal as waste.

For a long time, the most important goal of a chemist was to prepare a compound in suitable amounts and desirable high purity from available starting materials. In a world with a continuously increasing population and limited resources, the idea of sustainable development is of major importance for the future in the 21st century.

In the last three decades, much more attention has been paid to the effect of chemical products on the environment. It is clear that it is much better, less difficult, and less expensive to develop processes and compounds that are sustainable from scratch than to change an existing chemical process or to remove a toxic chemical from the environment to reduce its potential hazards and pollution created by it. In order to do so, chemists, biochemists, engineers, and pharmacists working together in drug development or constructing new materials must think always about sustainability, when they transform their ideas into any products and processes.

They must learn to judge the suitability of a chemical transformation or the use of a chemical compound within a limit of different parameters. It is not only the yield of the reaction, which counts but also which starting materials are required or used? Whether one can make these from renewable resources? Do these generate toxic by-products and how these by-products can be avoided? How much waste material is generated by this process and whether it is an energy-efficient process? Asking such questions at the beginning of any chemical compound, process, and technology development will lead to a proper, more efficient, and sustainable use of chemistry.

Chemistry is the science of material and its transformation, which plays a key role in the process and acts as the bridge between physics, material sciences, and life sciences. Only those chemical processes, which have reached maximum efficiency (after careful optimization), will lead to more sustainable compounds and processes. The awareness, creativity, and progressive attitude of a scientist are necessary to bring these reactions and chemical processes to maximum efficiency. The term "Green Chemistry" has been coined for all such efforts to achieve this goal. Therefore, attempts have been made by them to design synthesis and manufacturing processes in such a way that the waste products are minimized so that they have no or negligible effect on the environment and their disposal is also convenient.

Therefore, it is necessary that the starting materials, solvent, and catalyst should be carefully chosen for carrying out reactions, for example, the use of benzene as solvent must be avoided at any cost sas it is carcinogenic. If possible, it is better to carry out reactions in the aqueous phase.

2.4 CHOICE OF STARTING MATERIALS

It is very important to make a proper selection of the appropriate starting materials. Till now, most of the organic synthesis processes make use of petrochemicals and other hazardous or toxic chemicals, which affect the workers handling these starting materials. Petrochemicals are nonrenewable and these also require considerable amount of energy, therefore, it is important to reduce the use of such petrochemicals by using alternative starting materials of agricultural or biological origin.

Feedstock selection largely dictates the reactions and conditions that will be employed in chemical synthesis, and it should come from renewable sources rather than from depletable resources as far as possible. The ideal feedstock must be

- Renewable
- Poses no hazards
- Converted to the desired product using a few steps
- 100% Yield
- 100% Atom economy

2.4.1 CHEMICAL SUBSTITUTES AND REPLACEMENTS

Methylene chloride, benzene, and xylenes are among the top 20 starting materials produced in 1990 (Relsch, 1991). These are still used because economic losses are associated with phasing out these chemicals, but the chemical industries are introducing alternates at a rapid pace. N-Methyl-2-pyrrolidone has been commercialized as a promising replacement for methylene chloride.

Many times, simple substitutes cannot replace these chemicals, which are unique precursors for the synthesis of secondary derivatives and materials. Formaldehyde and vinyl chloride have unique properties and tremendous industrial value, but these are also hazardous to humans. Such chemicals do not bioaccumulate, but have well-known harmful threshold levels for humans. These are controlled by some regulations for their storage, transport, handling, and workplace exposure. Therefore, some of the major initiatives have focused to replace them by utilizing some other renewable and environmentally benign starting materials, which are obtained from agricultural, animal, and microbial resources.

2.4.2 RENEWABLE FEEDSTOCK FROM AGRICULTURE (BIOMASS)

Some major benefits of using biomass include

- It provides renewable feedstock.
- It does not contribute to net CO_2 to the atmosphere.

- It conserves fossil fuel leading to a secure domestic supply.
- It provides a platform for making use of chemical products, which are otherwise considered waste.

(i) Chemicals from fermentation processes (Glucose fermentation)

- **Glucose**

Glucose can be obtained from various carbohydrates, such as starch, cellulose, sucrose, and lactose. On a large scale, glucose is produced from starch by enzymatic hydrolysis. Corn is the main source of glucose. Another important source of producing glucose is woody biomass. Improvement in processes for harvesting and processing wood cellulose could result in an alternate source of glucose, which is relatively much less expensive than corn.

Mesoporous silica nanoparticles (MSNs) with different pore sizes were synthesized and used as hosts to physically adsorb or chemically link cellulose for its conversion to glucose. The results show that chemically linked cellulose onto large pore MSNs exhibits a glucose yield of more than 80% with excellent stability (Chang et al., 2011).

- **Lactic acid**

Lactic acid (2-hydroxypropionic acid) can be produced either by chemical synthesis or by fermentation of different carbohydrates, such as glucose (obtained from starch), maltose (produced by specific enzymatic starch conversion), sucrose (obtained from syrups, juices, and molasses), lactose (produced from whey) (Ravindranath and Hall, 1995). Lactic acid is produced on industrial scale today mainly through the fermentation of glucose. An important step in lactic acid production is the recovery from fermentation broth. The traditional process for the recovery of lactic acid is still far from ideal. Lactic acid exists in two optically active isomeric forms, L (+) and D (–). It is used in the food, chemical, pharmaceutical, and cosmetic industries. It is a bifunctional compound bearing a hydroxyl group and an acid group, and is utilized for a number of chemical conversions to useful products.

Nowadays, there is an increasing demand for biodegradable polymers that can replace conventional plastic materials and can also be used as new materials like controlled drug delivery devices or artificial prostheses. Thus, polylactic acid polymers could be an environment-friendly substitute for such plastics derived from petrochemical materials.

It esterifies with itself to give two primary esterification products: (i) linear lactic acid lactate (2-lactyloxypropanoic acid) and (ii) cyclic lactide (3,6-dimethyl-1,4-dioxane-2,5-dione). This lactide is an important compound, because it is a monomer for the production of poly (lactic acid) or polylactide, and other copolymers. Effective conversion of D-glucose into lactic acid has been described by Epane et al. (2010) using microwave irradiation in solventless conditions with alumina potassium hydroxide.

Esters of lactic acid and different alcohols (particularly methanol, ethanol, and butanol) are nontoxic and biodegradable. These are high boiling liquids and have excellent solvent properties, and therefore, replace toxic and halogenated solvents for a large range of industrial uses. Lactate esters are also used as plasticizers in cellulose and vinyl resins and they enhance the detergent properties of ionic surfactants. Direct hydrogenation of lactic acid or lactates produces propylene glycol and it can be an alternative green route to the petroleum-based process. Propylene glycol (1,2-propanediol) is a commodity chemical, which can be used as a solvent for the production of unsaturated polyester resins, drugs, cosmetics, and foods (Corma et al., 2007).

Dehydration of lactic acid gives acrylic acid. This acid, its amide, and its ester derivatives are the primary building blocks in the manufacture of acrylate polymers, which find numerous applications in surface coatings, textiles, adhesives, paper treatment, leather, fibers, detergents, etc.

Advances in fermentation and especially in separation technology have reduced the potential production cost of lactic acid. The production of lactic acid from waste sugarcane bagasse-derived cellulose was reported by Adsul et al. (2007). It deals with the simultaneous saccharification and fermentation (SSF) of sugarcane bagasse cellulose to lactic acid using *Penicillium janthinellum* mutant EUI and cellobiose utilizing *Lactobacillus delbrueckii* mutant Uc3. Salt-assisted organic acid-catalyzed depolymerization of cellulose was carried out by Stein et al. (2010). Dicarboxylic acids combined with inorganic salts (NaCl or $CaCl_2$) afforded the depolymerization of crystalline cellulose under mild conditions in water. The mechanical force and layered catalysts can efficiently depolymerize cellulosic materials (up to 84% conversion) (Hick et al., 2010). The most effective mechanocatalysts were aluminosilicates based on the kaolinite structure.

Catalytic upgrading of lactic acid to fuels and chemicals by dehydration and C-C coupling reactions was observed by Serrano-Ruiz and Dumesic (2009). They described a single reactor catalytic process to convert an aqueous solution of lactic acid into a spontaneously separating organic phase that can serve as a source of valuable chemicals (propanoic acid and C_4-C_7

ketones) and can be used to produce high energy density fuels. Simultaneous saccharification and fermentation (SSF) of cellulosic substrate to D-lactic acid using EUI cellulases and *Lactobacillus* lactic mutant RM2-24 was also reported (Singhvi et al., 2010). The SSF was carried out in screw-cap flasks at 42°C with shaking at 150 rpm.

- **Succinic acid**

Succinic acid is also produced in this way. Succinic acid reacts with alcohols in the presence of acid catalysts to form dialkyl succinates (Fumagalli, 1997). The esters of succinic acid with low molecular weight alcohols (methyl and ethyl succinates) find applications as solvents and synthetic intermediates for very many important compounds.

Direct hydrogenation of succinic acid, succinic anhydride, and succinates leads to the formation of products, such as 1,4- butanediol (BDO), tetrahydrofuran (THF), and γ-butyrolactone (GBL). 1,4-Butanediol is a compound of quite common interest as a starting material for the production of some important polymers, such as polyesters, polyurethanes, and polyethers. (Weissermel and Arpe, 2008). γ-Butyrolactone was synthesized more selectively from biomass-derived 1,4-butanediol by vapor-phase dehydrocyclization over novel copper–silica nanocomposite catalyst (Hwang et al., 2011). A wide variety of aldehydes and ketones can undergo the Stobbe condensation with succinic ester giving a variety of compounds. These products find applications in the technical and medical fields and biological activities (Moussa et al., 1982; Baghos et al., 1993).

- **3-Hydroxypropionic acid**

1,3-Propanediol is a starting material for the production of polyesters. It is used together with terephthalic acid to produce polytrimethylene terephthalate (PTT), which is in turn used in the manufacture of some fibers and resins. 3-Hydroxypropionic acid on dehydration gives acrylic acid. Acrylic acid and its derivatives (esters, salts, or amides) are important compounds and are used as monomers in the manufacture of some polymers and copolymers. These polymers have numerous applications, such as surface coatings, absorbents, textiles, paper making, sealants, adhesives.

3-Hydroxypropanoic acid is oxidized to give malonic acid (propanedioic acid). This acid and its esters are utilized to yield a large number of condensation products. They are important intermediates in the synthesis of vitamins B_1 and B_6, barbiturates, nonsteroidal anti-inflammatory agents, other numerous pharmaceuticals, agrochemicals, flavor, and fragrance compounds.

- **Glutamic acid**

Glutamic acid is a nonessential amino acid and it is used in food, drugs, dietary supplements, cosmetics, personal care products, fertilizers, etc.

(ii) Chemical transformations of monosaccharides

Thermal dehydration of pentoses and hexoses in acid media gives three important basic nonpetroleum chemicals (i) Furfural (2-furancarboxaldehyde) from dehydration of pentoses, (ii) 5-(Hydroxymethyl)furfural (HMF) from dehydration of hexoses, and (iii) Levulinic acid from hydration of HMF.

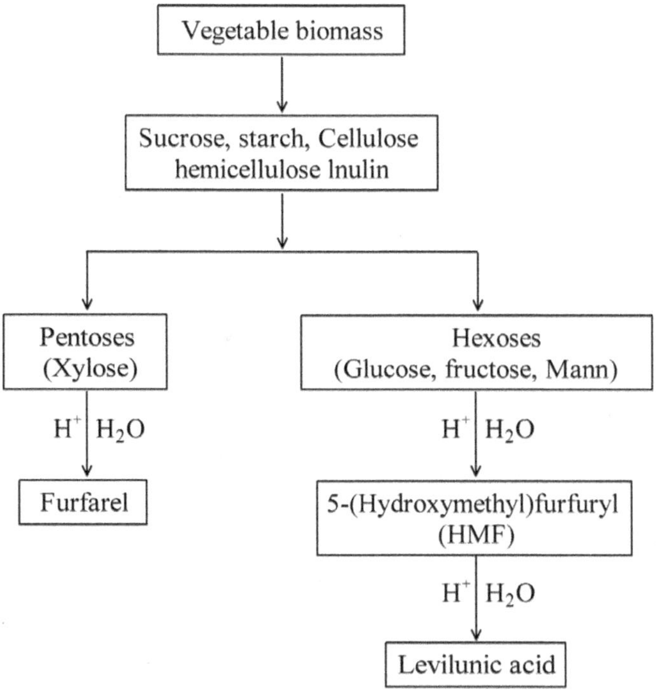

FIGURE 2.1 Chemical transformation of monosaccharides.

5-(Hydroxymethyl)furfural possesses a high potential industrial demand, and therefore, it has been judiciously called a sleeping giant (Bicker et al., 2005) and it is one of the new petrochemicals readily accessible from renewable resources (Lichtenthaler et al., 1991). Some surfactants were synthesized by Gassama et al. (2010) using furfural derived 2[5H]-furanone and fatty amines. Furfural obtained from biomass has been transformed into surfactants belonging to a new betain family with two hydrophobic moieties.

5-Hydroxymethylfurfural and furfural were produced by dehydration of biomass-derived mono and polysaccharides by Chheda et al. (2007). They present a biphasic system for acid-catalyzed dehydration of various biomass-derived carbohydrates to form furan derivatives, which have the potential to be sustainable substitutes for petroleum-based building blocks used in the production of fuels, polymers, and drugs. A simple procedure for the conversion of HMF (5-alkyl-and 5-arylaminomethyl-furan-2-yl) to methanol has been developed by Cukalovic and Stevens (2010). In this process, reactions were conducted without the use of a catalyst and under very mild conditions. Marcotullio and Jong (2010) used chloride ions to enhance furfural formation from D-xylose in dilute aqueous acidic solution. The reaction mechanism leads from xylose to furfural in acidic solutions. The simple addition of NaCl to an aqueous acidic solution significantly improved the yield and selectivity of furfural. Xylitol hydrogenolysis occurs efficiently on Ru/C with $Ca(OH)_2$ involving kinetically relevant dehydrogenation of xylitol to xylose and its subsequent retro-aldol reaction (Sun and Liu, 2011). Wang et al. (2012a) observed the direct conversion of glucose-based carbohydrates into 5-hydroxymethylfurfural (HMF) via an easily prepared Sn-mont catalyst with high efficiency and good stability. The conversion of carbohydrates and lignocellulosic biomass into 5-hydroxymethylfurfural was observed by Yang et al. (2012) by using $AlCl_3.6H_2O$ catalyst in a biphasic solvent system. Low-cost and nontoxic $AlCl_3.6H_2O$ in a biphasic medium of water/THF with NaCl additive converts raw biomass directly to high yields of furfural and modest yields of HMF.

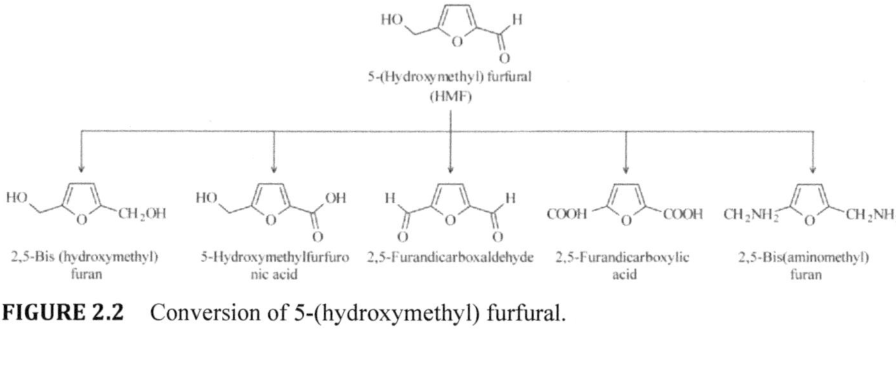

FIGURE 2.2 Conversion of 5-(hydroxymethyl) furfural.

These intermediates produced from HMF could replace some petrochemical-based monomers. The 2,5-furandicarboxylic acid may replace terephthalic, isophthalic, and adipic acids in the manufacture of polyamides, polyesters, and polyurethanes (Gandini and Belgacem, 1997; Moreau et al., 2004). The 2,5-furancarboxaldehyde is a starting material for the preparation of Schiff bases, and 2,5-bis(aminomethyl)furan can replace hexamethylene-diamine in the preparation of polyamides while 2,5-bis(hydroxymethyl) furan is used in the manufacture of polyurethane foams (Pentz, 1970). The fully saturated 2,5-bis- (hydroxymethyl)tetrahydrofuran can be used like alkanediol in the preparation of polyesters.

Lopes et al. (2017) synthesized modified tin oxide catalysts and followed by H_2SO_4 treatment (SO_4^{2-}/SnO_2) and alumina doping (SO_4^{2-}/Al_2O_3-SnO_2) at different calcination temperatures. It was observed that the chemoselectivity of the conversion of glucose to 5-hydroxymethylfurfural was enhanced due to the incorporation of alumina in the sulfated tin oxide catalyst when low calcination temperatures were used, increasing the acidity of catalyst and surface area.

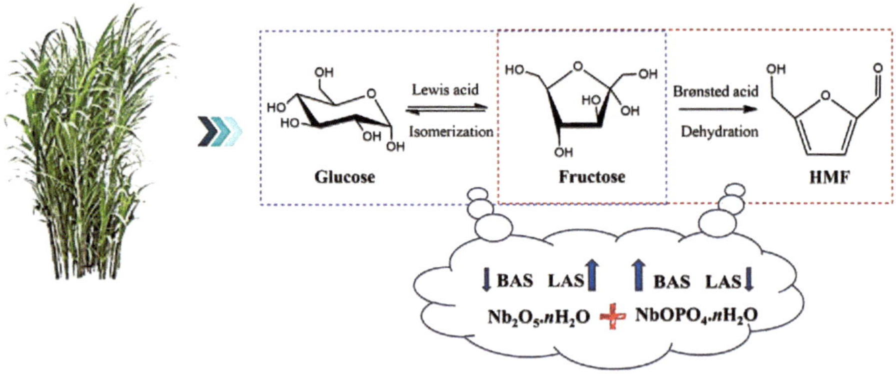

FIGURE 2.3 Conversion of glucose to 5-hydroxymethylfural.

Source: Reprinted with permission from Catrinck et al., 2017. © 2017 Elsevier.

The 5-hydroxymethyl-2-furfural is used as a bio-based platform for producing biofuels and renewable monomers. The production of HMF from fructose and inulin was reported by Antonetti et al. (2017) through a green chemical approach, which includes an aqueous medium, appreciable low loading of substrate concentration (10 wt%) of zirconium phosphate or niobium (heterogeneous acid catalyst), and microwave heating. It was reported that both the catalysts were quite active and promising. Based on statistical modeling, highest yield HMF (about 40 mol.%) was observed at 190°C and that too in a short reaction time (8 min). It was also revealed that catalysts can be reused and their original activity can be maintained.

The acid-catalyzed conversion of glucose into 5-hydroxymethylfurfural in an aqueous medium was reported by Catrinck et al. (2017). They used niobic acid, niobium phosphate, and a mixture of both the solid acid catalysts. It was revealed that there was excellent glucose conversion (55%) and HMF selectivity (56%), on keeping weight ratio 1:1 of NbO and NbP.

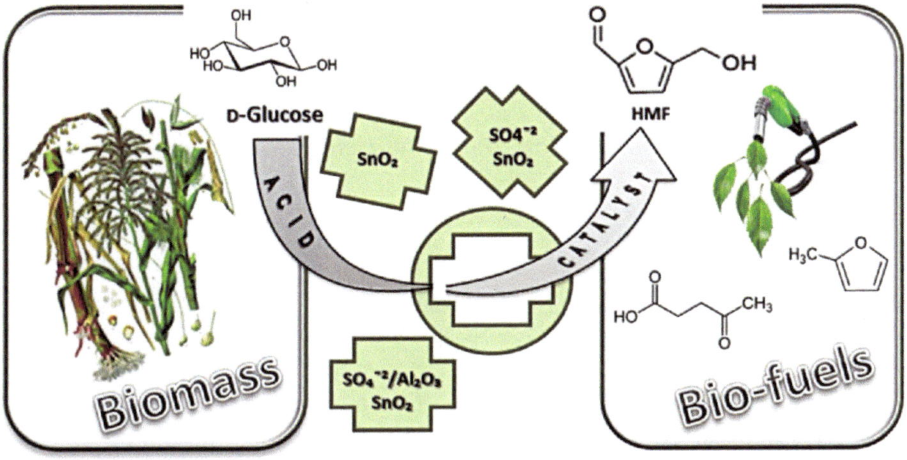

FIGURE 2.4 Acid-catalyzed conversion of glucose into 5-hydroxymethylfurfural.
Source: Reprinted from Lopes et al. 2017. © 2017 Elsevier.

Sun et al. (2018) synthesized a series of metal-containing silicoaluminophosphate (MeSAPOs) molecular sieves via hydrothermal method. They used bauxite as the source for Al, Si, and Ti or Fe. The catalytic performance of as-prepared MeSAPOs was evaluated for the dehydration of fructose to 5-hydroxymethylfurfural. It was indicated that this catalytic performance increased in the following order:

$$\text{MeSAPO-44} < \text{MeSAPO-34} < \text{MeSAPO-5} < \text{MeSAPO-11}$$

It was found to depend on the order of the total number of acid sites. It was found that out of all MeSAPOs molecular sieves, MeSAPO-11 gave the highest yield of HMF (65.1%) at 170°C after 2.5 h, and this may be due to more acid sites.

As value-added biochemicals from cellulosic renewable biomass are possible from the isomerization of glucose to fructose, Zhang et al. (2019) developed a strategy for isomerization (aqueous) of glucose, which was catalyzed with basic ionic liquids (ILs). They could achieve fructose yield and selectivity of 36.8 and 73.8%, respectively and in 30 min in the presence of tetrabutyl ammonium proline ([N4,4,4,4]Pro) at 80°C. This IL catalyst exhibited good reusability by retaining a satisfactory performance even after four consecutive runs.

A green processing approach for the production of furfural from xylose and xylan was reported by Zhang et al. (2014) under very mild conditions with the addition of metal chlorides in ChCl–oxalic acid as a deep eutectic solvent (DES). It was observed that ChCl–oxalic acid can act as both. Brønsted acid catalyst and reaction medium.

(iii) Chemical transformation of disaccharides

- **Sucrose**

Sucrose is the main carbohydrate feedstock of low molecular weight chemicals. It is present in honey, sugar, fruits, berries, and vegetables. Sucrose can be functionalized and converted into different interesting additives. The hydrolysis of sucrose allows its conversion into inverted hexoses, that is, glucose and fructose. These are widely used in the food industry. Sucrose can also find applications in some polymers such as polyurethanes. The use of sucrose for the preparation of phenolic or alkyd resins as well as polyesters, polycarbonates, and polyurethanes has been reviewed by Kollonitsch (1970).

(iv) Conventional route

- **Formation of formic acid**

When methanol and carbon monoxide are treated in the presence of a strong base, methyl formate is obtained.

$$CH_3OH + CO \rightarrow HCO_2CH_3$$

This reaction is performed in the liquid phase at elevated pressure in industries. Typical reaction conditions are 80°C and 40 atm and the base is sodium methoxide. Hydrolysis of the methyl formate produces formic acid.

$$HCO_2CH_3 + H_2O \rightarrow HCO_2H + CH_3OH$$

Efficient hydrolysis of methyl formate requires a large excess of water. Some routes proceed indirectly by first treating the methyl formate with ammonia to give formamide, which is then hydrolyzed with sulfuric acid.

$$HCO_2CH_3 + NH_3 \rightarrow HC(=O)NH_2 + CH_3OH$$

$$2\,HC(=O)NH_2 + 2\,H_2O + H_2SO_4 \rightarrow 2\,HCO_2H + (NH_4)_2SO_4$$

This approach suffers from the need to dispose off ammonium sulfate, which is formed as a by-product. This problem has led some manufacturers to develop some other energy-efficient means to separate formic acid as the large excess amount of water is used in direct hydrolysis. In one of these processes, the formic acid is removed from the water via liquid–liquid extraction with an organic base. This method is now being replaced by a green chemical route.

- **Green formation of formic acid**

A new method to transform carbohydrate-based biomass to formic acid has been reported by Wölfel et al. (2011). This process involves oxidation with molecular oxygen in an aqueous solution using a Keggin-type $H_5PV_2Mo_{10}O_{40}$ polyoxometalate as a catalyst. Several water-soluble carbohydrates were fully and selectively converted to formic acid and CO_2 under very mild conditions. The complex biomass mixtures, such as wood sawdust was transformed to formic acid, giving 19 wt% yield (11% based on the carbon atoms in the feedstock) under nonoptimized conditions.

Vegetable oils and animal fats can give two types of reactions. These are i) Reactions of carboxyl group and (ii) Reactions of fatty chains.

Fats and oils are obtained from vegetable and animal fats, which are mainly formed by mixed triglycerides having fatty acid moieties. They are chemically not very different from some petroleum fractions in the sense that they contain a large paraffinic or olefinic chain. A large proportion of vegetable oils, such as coconut, palm, and palm kernel oils, come from countries with tropical climates. soybean, rapeseed, and sunflower oils come from moderate climates. Animal fat is obtained from the meat industry, with beef tallow being the most abundant fat, and fish oil comes from the fishing industry. Biodiesel, lubricants, surfactants, surface coatings, polymers, pharmaceuticals, cosmetics etc., can be produced from these animal fats and vegetable oils.

Biodiesel is a diesel fuel made from renewable resources like oils derived from farm crops such as soybeans or even from recycled vegetable oils like that was left over from making fries at fast food restaurants. It is synthesized by removing glycerine from soybean or other vegetable oil. The by-product glycerine is also useful for making soap.

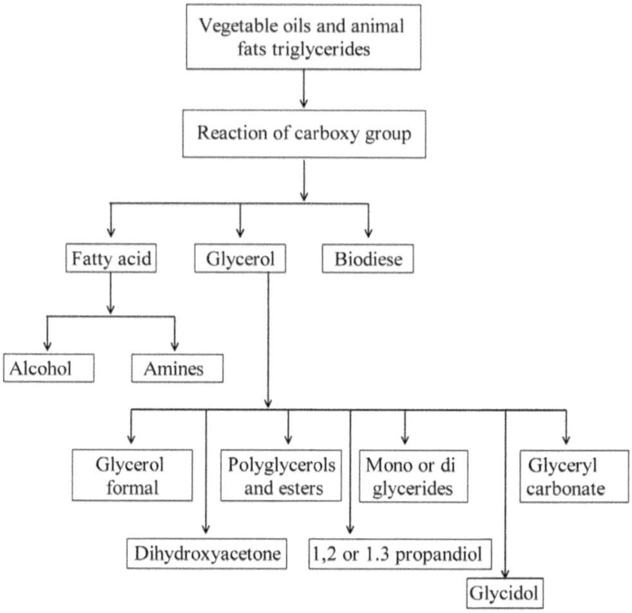

FIGURE 2.5 Reactions of carboxyl group.

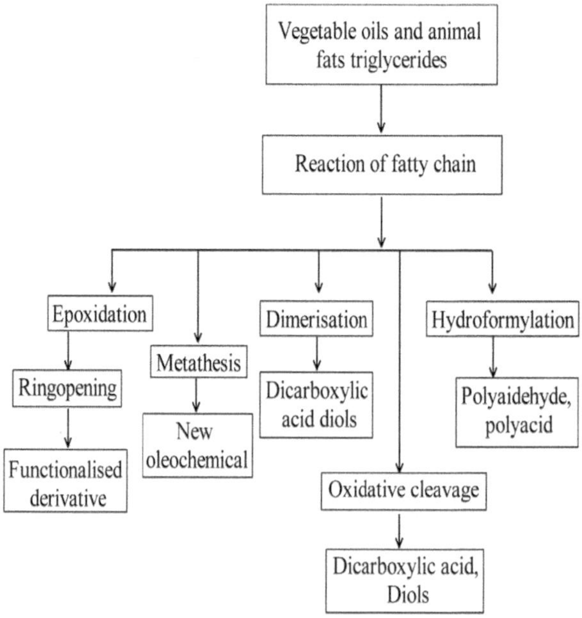

FIGURE 2.6 Reactions of fatty chains.

$$\text{Vegetable oil} + \text{Methanol} \xrightarrow{\text{KOH}} \text{Biodiesel} + \text{Glycerine}$$

(v) Advantages

- It is a renewable source of energy, unlike fossil fuel-derived petroleum diesel.
- Biodiesel on burning neither emits sulfur, nor it increases the overall amount of CO_2 in the atmosphere (the CO_2 released from biodiesel is balanced by the CO_2 taken up by plants.

Propylene glycol is synthesized by the utilization of waste glycerol from biodiesel production. This is a conversion involving dehydration of glycerol to acetol, followed by hydrogenation to yield propylene glycol.

$$\text{Glycerol} \xrightarrow[-H_2O]{} \text{Acetol} \xrightarrow[H_2]{} \text{Propylene glycol}$$

(vi) Cellulose, hemicellulose, and lignin

- **Preparation of cellulose from plant sources**

  i) Separation from the matrix of lignocellulose (hemicellulose and lignin) is required.
  ii) Harsh chemical processing is required.
  iii) Cellulose products may require bleaching.

- **Chemical modification of cellulose**

  i) Cellulose may be modified, as it contains a number of -OH groups, where other groups can be bonded to impart a variety of properties.
  ii) Rayons are made by treating cellulose with base and carbon disulfide and then extruding the product through fine holes to make the thread.
  iii) A similar process is followed to make cellophane extruding it through a long narrow slot.

- **Cellulose acetate**

An ester is obtained, when most of the -OH groups present in cellulose are replaced by acetate groups by treating cellulose with acetic anhydride.

- **Cellulose nitrate**

Cellulose nitrate is obtained, when most of the -OH groups on cellulose are replaced by $^{-}ONO_2$ groups by treating cellulose with a mixture of nitric acid (HNO_3) and sulfuric acid (H_2SO_4)

i) It is used as an explosive.
ii) In the early days of moving pictures, transparent film was used for movie films, which resulted in some disastrous fires, giving highly toxic fumes of NO_2 gas.

- **Synthesis of propargylcellulose**

A fast and simple reaction on exposure to microwave irradiation permits the synthesis of propargylcellulose in an aqueous alkaline medium and it is considered green. The influence of various reaction parameters, such as amount of propargyl bromide, reaction time, or microwave activation on the reaction efficiency has been reported (Faugeras et al., 2012).

- **Conversion of cellulose-derived 5-(chloromethyl)furfural into δ-aminolevulinic acid**

The 5-(chloromethyl)furfural (CMF) was prepared by the conversion of cellulose, and it was further converted into δ-aminolevulinic acid in three simple chemical steps involving conversion to (i) 5-(Azidomethyl)furfural, (ii) photooxidation, and (iii) catalytic hydrogenation in reasonably good yield 68%. The δ-aminolevulinic acid (ALA) is a natural product with important agrochemical and pharmaceutical applications (Mascal and Dutta, 2011).

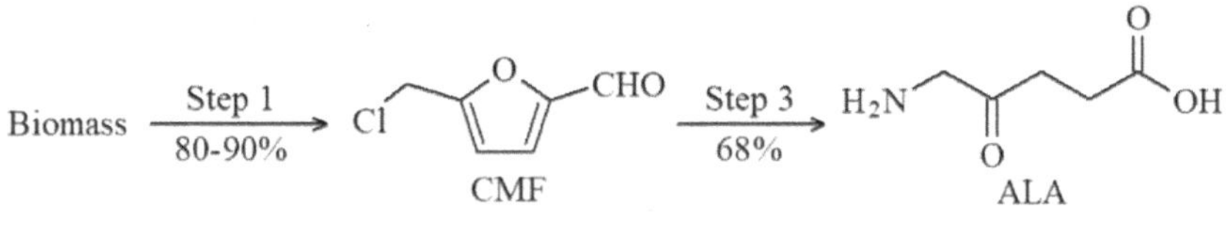

- **Various other feedstocks from cellulose wastes**

Rumen bacteria act on cellulose wastes and then a treatment followed with lime in large fomenters (in absence of oxygen) produced calcium acetate, calcium propionate, and calcium butyrate. These esters can be acidified to produce corresponding acids, such as acetic, propionic, and butyric acids. These carboxylic acids can be hydrogenated to give corresponding alcohols, such as ethanol, propanol, and butanol. Organic acids on treatment at 450°C produce ketones, such as acetone, methylethyl ketone, and diethyl ketone.

Sugar alcohol can be synthesized by hydrolytic hydrogenation of cellulose over supported metal catalysts. Pt/carbon black BP2000 is an effective and durable catalyst for the conversion of cellulose to sugar alcohols, which promotes both the hydrolysis of glycosidic bonds and the hydrogenation of glucose (Kobayashi et al., 2011).

Lignin is a chemically complex biopolymer that is associated with cellulose in plants. It is difficult to use lignin because of its inconsistent, and widely variable molecular structure. It was felt that catalytic transformation of readily available, widely distributed, and renewable nonfood lignocelluloses to some value-added chemicals is necessary, but it still remains a great challenge as the bulk of these resources and/or the energy available from such biomass are in the form of lignocellulose, which is tight, covalent, hydrogen bond-linked matrix of carbohydrate polymers (cellulose and hemicellulose) and phenolic polymers (lignin). This results in its insolubility in common solvents.

It was first found that the ionic liquid 1-butyl-3-methylimidazolium chloride (bmimCl) was able to dissolve cellulose with 10 wt% of solubility at 100°C. Li and Zhao (2007) and Li et al. (2008) reported that cellulose and original biomass can be efficiently hydrolyzed to reducing sugars in bmimCl with the help of a mineral acid. The catalytic conversion of lignocellulose and/or its constituents to chemicals such as 5-hydroxymethylfurfural (HMF) using ionic liquid as solvent has also been studied. It has been reported that cellulose and lignocelluloses such as corn stove could be also converted to 5-hydroxymethylfurfural in 1-ethyl-3-methylimidazolium chloride (emimCl), while N,N-dimethylacetamide and LiCl were applied as co-solvents in the presence of Cr (II) salt and a mineral acid catalyst (Binder and Raines, 2009; Zakrzewska et al., 2011). Production of HMF and furfural from other lignocellulosic biomass, such as corn stalk, rice straw, and pine wood in bmimCl and 1-butyl-3-methylimidazolium bromide (bmimBr) solvents under microwave irradiation has also been studied (Zhang and Zhao, 2010).

Therefore, transformation of these lignocelluloses into useful chemicals has been recognized as an effective approach for improving upon the problem of increasing energy crisis and climatic change. An efficient catalytic transformation process for agricultural residual lignocelluloses in cooperative ionic liquid pairs was also achieved by Long et al. (2012). The promotion of the dissolution equilibrium, combined with fast, *in situ* acid-catalyzed degradation of cellulose and hemicellulose, resulted in significantly greater conversion of the biomass to some important biochemicals and selective delignification. Acid treatment of carbohydrates was reported by Hu and

Li (2011) in methanol-rich medium, which stabilizes reactive intermediates and suppresses polymer formation and greatly promotes methyl levulinate production. A catalytic process was presented by Braden et al. (2011) for the conversion of lignocellulosic biomass to liquid alkenes. Its technoeconomic analyses show that the economics is comparable to that of cellulosic ethanol.

An eco-friendly route has been developed by Zhang et al. (2012) for converting various cellulose-based biomasses to glycolic acid directly in a water medium and oxygen atmosphere. Here, heteromolybdic acid was used as a multifunctional catalyst to catalyze the hydrolysis of cellulose, fragmentation of monosaccharides, and selective oxidation of fragmented products. It was observed that the yield of glycolic acid was 49.3% when commercial α-cellulose powder was used as the substrate. In this catalytic system, some raw cellulosic biomass (bagasse or hay) can also be used as the starting materials, giving a good yield of glycolic acid (~30%).

(vii) Synthesis of some heterocycles

Two completely benzoxazine monomers, 3-octadecyl-8-methoxy-3,4-dihydro-2*H*-1,3-benzoxazine (Bzs) and 3-furfuryl-8-methoxy-3,4-dihydro-2*H*-1,3-benzoxazine (Bzf), successfully prepared by Wang et al. (2012b) *via* a solventless method. The results revealed that these monomers undergo homogeneous copolymerization when the Bzf–Bzs molar ratio was more than 1:2. It was observed that furan moiety of Bzf was beneficial for this copolymerization, which includes improvement of their cross-linking density and enhanced thermal properties of copolymerized resins. These effects may be due to the electrophilic substitution (aromatic) of the furan ring. These findings have significant importance in designing some polybenzoxazines (fully bio-based) with tunable properties.

Liao et al. (2012) reported one-pot synthesis of 2-arylbenzofurans at ambient temperature. This reaction involves a Wittig reaction of salicylaldehyde (substituted), which was followed by oxidative cyclization (in situ). The use of nonhazardous materials and benign reaction conditions (green solvent and room temperature) made it environment-friendly.

Microwave-assisted one-pot three-component reaction between phenylacetylene, aromatic amines, and aromatic aldehydes was carried out in the presence of potassium dodecatungstocobaltate trihydrate by Anvar et al. (2012). It afforded the corresponding quinolines as well as bis-quinolines with high yields. It was also reported that this catalyst could be reused again without any major loss in the catalytic activity.

Sahu et al. (2012) synthesized octahydroquinazolinones 4H-pyrimido[2,1-b][1,3]benzothiazole, 1,2,4-triazoloquinazolines, and fused thiazolo[2,3-b] quinazolinone by reacting dicarbonyl, aldehydes and 2-amino benzothiazole/3-amino-1,2,4-triazole/urea/thiourea. They used Mg–Al–CO_3 and Ca–Al–CO_3 hydrotalcite as heterogeneous catalysts.

Some 2-alkynylindoles were prepared by Mothe et al. (2012) using a tandem reaction alkynylation/heterocyclization of 1-(2-tosylamino)aryl)but-2-yne-1,4-diols, which was catalyzed by AgOTf reaction can be carried out under mild conditions. The alkyne side chain of the N-heterocycle and indole ring; both can be formed from readily available, low-cost, and eco-friendly starting materials.

The synthesis of 3,4-dihydropyrimidin-2(1H)-ones was reported by Pouramiri et al. (2014) using three-component, one-pot condensation reaction of urea or N-methylurea, aromatic aldehydes, and β-keto ester using $LaCl_3/ClCH_2COOH$ (catalyst) under solvent-free conditions. This proposed method provides good to excellent yields (80–99%) in a shorter reaction time as compared with classical Biginelli reaction conditions.

A three-component domino reaction of thiophene-2-carbaldehyde, 2-aminobenzothiazoles and carbonyl compounds has been reported by Khandelwal et al. (2015). They used p-toluenesulfonic acid (p-TSA) as a reusable eco-friendly catalyst in aqueous–alcoholic medium (1:5). The diverse heterocycles were obtained in very high yields (92–97%) in 10–48 min.

A three-component synthesis of triazolo[1,2-a]indazole-triones in a single pot was observed by Shekouhy et al. (2015) under catalyst-free conditions. They used glycerol as a nontoxic, benign, and biodegradable medium. A wide variety of substrates was condensed with carbonyl compounds with a reactive-methylene group and urazole derivatives.

Natarajan et al. (2016) could get 6-arylphenanthridines (moderate with good yields), when aryl diazonium salts in situ $[Ru(bpy)_3]Cl_2$ was used as a photoredox catalyst and blue LED as source of irradiation, which is formed in the reaction of tert-butyl nitrite (tBuONO) and N-(2-aminoaryl) benzoimines. The present work provides a newer method to get 6-arylphenanthridines with low cost and environmentally benign starting materials.

Zolfigol et al. (2016) employed bifunctional nanostructured molten salt [4,4-bipyridine]-1,1-diium tricyanomethanide as an efficient catalyst for preparing 2-amino-4H-chromenes. A number of aromatic aldehydes were condensed with 1-naphthol or 2-naphthol, malononitrile, and resorcinol under mild and solvent-free conditions affording high to excellent yields.

An eco-friendly protocol has been developed by Xie et al. (2017) for one-pot synthesis of various functionalized 2-sulfonylquinolines in water under open-air conditions. It was reported that the use of ultrasound improved the reaction efficiency and rate along with minimizing the side reactions. They evaluated the E-factor and eco-scale scores as 1.17 and 71, respectively.

A green one-pot synthesis of 2-phenylimidazo[1,2-a]quinoline and 2-phenylimidazo[1,2-a]pyridines from aromatic carbonyl compound, succinamide, 2-aminopyridine, and in situ generated -iodo acetophenone has been reported by Jadhav et al. (2017). They used green water (2:1) and solvent PEG-400 under microwave exposure. This protocol avoids the use of lachrymatric -chloro and -bromocarbonyl compounds, toxic organic, and hazardous solvents.

Saeidian and Moradnia (2017) developed a synthesis of N-aryl-3, 10-dihydroacridin-1(2H)-ones based on Knoevenagel condensation of dimedone to different 2-(N-arylamino)benzaldehydes, which was followed by intramolecular enamination. The 20 mol% of nanocrystalline ZnO was used as a catalyst. It gave relatively high yields using low-cost starting materials.

Two facile and environmentally friendly routes were reported by Fatahpour et al. (2017) for the synthesis of active compounds. They used aspirin as a novel and green catalyst for this purpose. They prepared target dihydropyrano[2,3-c]pyrazoles and spiropyranopyrazoles by one-pot, four-component reaction.

An efficient and green protocol has been reported by Eskandari et al. (2018) for the synthesis of acyl-substituted bis(pyrazolyl)methane derivatives. The reaction was carried out in one-pot, three-component (2 + 1) condensation reaction by using arylglyoxal derivatives, 3-methyl-1-phenyl-2-pyrazolin-5-one, and silica sodium carbonate (SSC) as a recyclable green catalyst.

Snieckus and Jiao (2018) observed a direct one-step synthesis of 2-substituted indazolones. They used readily available and benign starting materials, such as α-nitrobenzyl alcohols and alkyl amines. It involves the use of hydrazine and afforded a high yield (89%).

Porous TiO_2 nanosheets-supported Pt nanoparticles (Pt/P-TiO_2) were developed by Xie et al. (2019) as the heterogeneous catalyst. As-prepared Pt/P-TiO_2 was found to be highly efficient for reductive amination of LA to various N-substituted pyrrolidones at ambient temperature and H_2 pressure. The Pt/P-TiO_2 showed good applicability for the reductive amination of a number of compounds (levulinic esters, 4-acetylbutyric acid, 2-acetylbenzoic acid, and 2-carboxybenzaldehyde).

(viii) Biofuel waste streams (Amino acids)

The use of biomass as a feedstock provides an alternative with a closed carbon cycle. Amino acids are quite abundant in biomass and these amino acids are highly functionalized. These properties make some efficient routes possible, toward bulk chemicals saving energy as well as reagents. The amino acid contents of several sources of biomass were recently reviewed by Lammens (2011). European guidelines were that 10% of all transportation fuels should be originated from biomass by 2020. Combining with the estimated worldwide biofuel production (IEA, 2012), this will lead to the production of about 100 million tons of protein per year, on average to 5% (5 million tons) of each amino acid, including phenylalanine. The 13, 000 tons of phenylalanine were produced for food applications (Demain, 2000). Substantial amounts of phenylalanine could be available as a feedstock for the production of bulk chemicals without competing with the food market in years to come.

Cinnamic acid could also be obtained from plant residue rest streams, so that a strategy can be made for a potential route to biostyrene and bioacrylate compounds, which are important monomers for the plastic industry (Spekreijse et al., 2012).

(ix) Production of γ-valerolactone by levulinic acid

Hydrogenation of levulinic acid to γ-valerolactone has been reported by Galletti et al. (2012). It can be easily obtained in high yields with very mild reaction conditions, by hydrogenation of an aqueous solution of levulinic acid, where a commercial ruthenium supported catalyst in combination with a heterogeneous acid co-catalyst, such as the ion exchange resins Amberlyst A70 or A15, niobium phosphate, or oxide is used. All the hydrogenations were carried out at 70–50°C and at low hydrogen pressure (3–0.5 MPa). The combined effect of acid and hydrogenating heterogeneous components was also verified for the hydrogenation of aliphatic ketones to the corresponding alcohols. Acid treatment of carbohydrates was reported by Hu and Li (2011) in methanol-rich medium, which stabilizes reactive intermediates and suppresses polymer formation and greatly promotes methyl levulinate production.

The efficient catalytic hydrogenation of levulinic acid is a key step in biomass conversion. It was shown that the levulinic acid could be reduced efficiently to γ-valerolactone in the presence of a catalyst in situ generated from $Ru(acac)_3$ and different sulfonated phosphine $R_nP(C_6H_4\text{-m-}SO_3Na)_{3-n}$ (n = 1,2; R = Me, Pr. iPr, nBu, Cp) ligands in a solvent and promoter-free reaction (Tukacs et al., 2021). A highly efficient and recyclable heterogeneous $Cu–ZrO_2$ nanocomposite catalyst was developed for selective hydrogenation of levulinic acid to γ-valerolactone with complete conversion and 90–100% selectivity (Hengne and Rode, 2012).

(x) Biodiesel from vegetable oils vs. algae

Biodiesel is one of the most prominent renewable alternative fuels, which can be derived from a variety of sources, including vegetable oils, animal fats and used cooking oils, as well as alternative sources such as algae. Issues, such as land use change, food *vs.* fuel, feedstock availability, and production potential have greatly influenced the search for the best feedstocks, but an issue that will ultimately determine or decide the usability of any biodiesel fuel properties as cold flow and oxidative stability have been problematic issues for biodiesel.

The fatty acid profile of any biodiesel fuel largely depends on the feedstock, which significantly influences these properties. The comparison has also been done between biodiesel derived from vegetable oils and biodiesel obtained from algae particularly with respect to fuel properties. The fatty acid profiles of many algal oils possess high amounts of saturated and polyunsaturated fatty acids. Thus, biodiesel fuels derived from algae is likely to possess poor fuel properties, in many cases, *that is*, both poor cold flow and low oxidative stability. This observation showed that potential of production of biodiesel only is not sufficient to decide suitability of any feedstock. It is also important to know, how the fuel properties of a biodiesel can be improved through modification of its fatty ester content? Algal oils are probably best produced under tightly controlled conditions since the fatty acid profile of algal oils is quite susceptible to changes under these conditions. Algal oils yielding biodiesel with the least problematic properties have been determined by reported fatty acid profiles (Knoth, 2011).

(xi) Esterification of glycerol to methyl glycerate

The oxidation of glycerol in methanol has been carried out by gold nanoparticles on different oxide supports (TiO_2, Al_2O_3, and ZnO) using molecular oxygen as the oxidant in a batch set-up. The main oxidation products of glycerol are methyl glycerate and dimethyl mesoxalate, which indicates that C–C bond scission occurs to a limited extent as compared with glycerol oxidations in water. Highest selectivity to methyl glycerate was observed in the case of Au/TiO_2 as the catalyst. The use of a base is not essential for the glycerol oxidation reaction to occur, although for TiO_2 and Al_2O_3, higher initial activities were found in the presence of sodium methoxide.

Au/ZnO gives comparable activity and selectivity both; in the presence and absence of a base. Oxidation with reaction intermediates indicates that oxidation of methyl glycerate to higher oxygenate does not occur to a significant extent in methanol. An alternative pathway for the formation of dimethyl mesoxalate involving dihydroxyacetone has been proposed (Purushothaman et al., 2012).

(xii) Production of hydrogen

Standards on the quality of hydrocarbon fuels do not compromise particularly on sulfur and aromatic contents. This is one of the major force behind increasing hydrogen demands by petroleum refineries. The fuel standards are often based on keeping a control on environmental pollution. However, most of the commercial hydrogen production processes are based on nonrenewable resources, which are associated with high carbon footprints. With increasing demands of hydrogen as the fuel, the carbon footprint associated with hydrogen production will increase accordingly. Incentives for hydrogen production by green technologies will be encouraging for smooth functions of industrial processes from high to low carbon footprint. It will facilitate the entry of green reforming technologies into the hydrogen market. James et al. (2011) reviewed the potential of some emerging reforming technologies for hydrogen production from renewable resources.

Hydrogen production by water splitting still remains a golden dream of a scientist, because it is a route to non-fossil fuel as well as a potential source of clean energy. It is quite essential to use a nonpolluting energy source like sunlight to drive the reaction. The production of hydrogen by the photocatalytic cleavage of water can overcome all the hurdles. This area is briefly reviewed by Bowker (2011) focusing mainly on two routes: (i) Use of sacrificial agents, such as alcohols to act as oxygen scavengers and liberate hydrogen only and (ii) Direct water splitting to produce both hydrogen and oxygen. The factors, which are important in determining the characteristics of effective photocatalytic water splitting systems have been discussed. Chouhan et al. (2012) have generated hydrogen from water splitting.

- **Triacetic acid lactone**

Triacetic acid lactone has been established to be a biorenewable molecule with much potential as a platform chemical for the production of commercially valuable bifunctional chemical intermediates and end products, such as sorbic acid (Chia et al., 2012).

- **Hydrogenolysis of γ-valerolactone into 1,4-pentanediol or 2-methyltetrahydrofuran**

The γ-valerolactone is one of the most significant cellulose-derived compounds, which is directly converted into 1,4-pentanediol (Du et al., 2012). This conversion was carried out by chemoselective hydrogenolysis catalyzed by a simple but quite versatile copper–zirconia catalyst. Depending on the reaction conditions, 2-methyltetrahydrofuran could also be obtained in excellent yields.

(xiii) Valorization of corn cob residues to porous carbonaceous materials

A strategy to utilize corn cobs to generate porous carbonaceous materials has been reported, which are used to generate biodiesel from waste oils. The focus has been increased on reducing organic wastes in industry and for providing and utilizing renewable chemicals and fuels. Valorization of wastes is attracting considerable attention now a days, providing an alternative to the disposal of a range of waste materials in landfill sites. Particularly, the valorization of food wastes is considered to be quite promising.

Arancon et al. (2011) utilized corn cobs, a common food waste to generate microporous carbonaceous material. This material was then subsequently sulfonated to give a solid acid catalyst, which exhibited excellent activity in the simultaneous esterification/transesterification of waste oils.

- **Trapping of radical in neat conditions**

Shapiro et al. (2010) reported that radicals generated during oxidation of aldehyde to carboxylic acids can be efficiently trapped under environmentally friendly conditions, either in neat conditions or in water.

(xiv) Use of carbon dioxide

A large amount of carbon dioxide is available in atmosphere and oceans, but its utilization as feedstock for the chemical industry is often prevented by its thermodynamic stability. Only a limited processes based on carbon dioxide as a raw material have been realized on a technical scale so far, for example, production of urea, methanol or salicylic acid.

A few catalytic reactions of a lactone platform chemical have been discussed, where CO_2 has been used as a feedstock. Numerous other reactions can be also carried out starting from this molecule leading to versatile interesting products, such as acids, alcohols or diols, aldehydes, amino acids, amines, etc. in high yields. Furthermore, esters, silanes or even some polymers are obtained using the δ-lactone as a building block. Thus, in the presence of efficient catalytic systems leading to high selectivities, a new approach for the utilization of CO_2 as a reasonable feedstock for chemical reactions has been described by Behr and Henze (2011). Copolymerization of carbon dioxide was also reported by Darensbourg and Wilson (2012). Carbon dioxide and epoxides (oxiranes) produce polycarbonates to afford high selectivity for copolymer *versus* cyclic carbonate formation.

(xv) Synthesis of polyoxygenated compounds

Air, light, water, and spirulina are utilized in a green method to transform readily accessible furan substrates into a diverse range of synthetically useful polyoxygenated motifs commonly found in natural products (Noutsias et al., 2012).

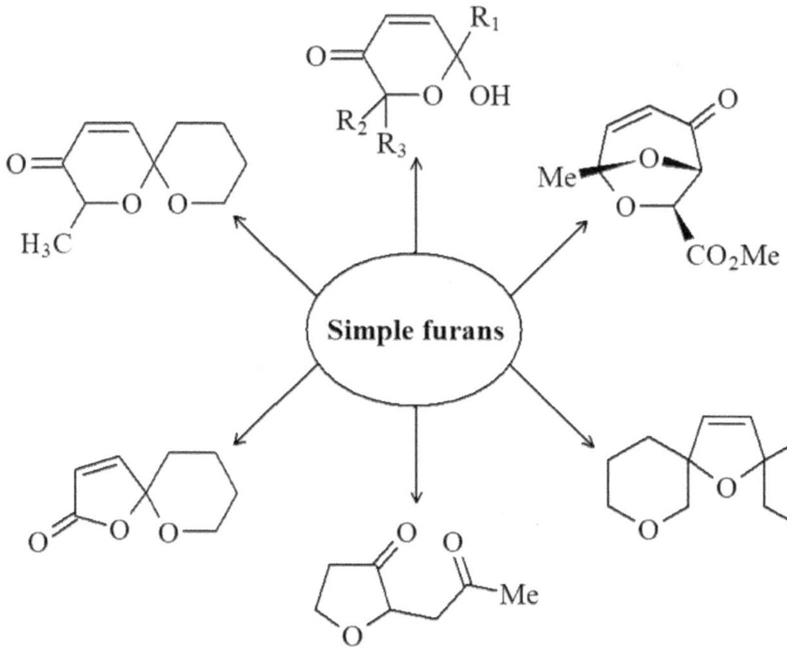

FIGURE 2.7 Synthesis of polyoxygenated compounds.

(xvi) Waste materials as resources for catalytic applications

The use of high volume waste materials in catalysis or for the synthesis of a catalyst has been studied by Balakrishnan et al. (2011). Waste materials derived from both industrial and biological sources have attracted attention of chemists. The materials include red mud, aluminum dross, fly ash, blast furnace slag, rice husk, and various kinds of shell.

(xvii) Transforming collagen wastes into doped nanocarbons

Leather industry produces huge quantities of biowaste that can be utilized as raw material for the bulk synthesis of carbonaceous materials. The synthesis of multifunctional carbon nanostructures from pristine collagen wastes by a simple high temperature treatment was reported Ashokkumar et al. (2012). It was observed that the nanocarbons derived from this biowaste have a partially graphitized structure with onion-like morphology. These are naturally doped with nitrogen and oxygen, resulting in some multifunctional properties. This route from biowaste raw material provides a cost-effective alternative to existing chemical vapor deposition (CVD) methods for the synthesis of functional nanocarbon materials. It also presents a sustainable approach to tailor nanocarbons for various applications.

(xviii) Transforming animal fats into biodiesel using charcoal and CO_2

A simple methodology for producing biodiesel has been reported, confirming that the main driving force of biodiesel conversion through the noncatalytic transesterification reaction is temperature-dependent. Noncatalytic biodiesel conversion can be achieved in the presence of a porous material *via* a thermochemical process and a real continuous flow system. In addition, this noncatalytic conversion of biodiesel can be enhanced by the presence of carbon dioxide. The transformation of animal fat (beef tallow and lard) into biodiesel was achieved by Kwon et al. (2012) using charcoal and CO_2 under ambient pressure. This methodology for producing biodiesel combines esterification of free fatty acids and transesterification of triglycerides into a single process and leads to a 98.5 (± 0.5%) conversion efficiency of biodiesel within 1 min at 350–500°C. This new process has very high potential to achieve a breakthrough in minimizing the cost of biodiesel production owing to its simplicity and technical advantages.

(xix) Alternative electronic chemicals

Due to the acute toxicity of arsine (even more toxic than hydrogen cyanide), a search is on for alternatives to arsine in cylinders. One of the possible solutions to the hazards associated with the sudden release of arsine stored in compressed cylinders is contaminanation of the gas at atmospheric pressure. Another solution is the replacement of arsine by completely nontoxic reagents. Lethally toxic arsine is made less toxic by alkylation and the fully substituted reagent trimethyl arsine, $As(CH_3)_3$ is practically nontoxic according to contemporary chemical standards.

(xx) Biomass

Owens et al. (2015) synthesized a pentamethine cyanine dyes with increased yields (89–98%) as compared with conventional heating method (18–64%).

Yang et al. (2018) fabricated heterogeneous Se/C catalyst using carbohydrates as biomass carbon source and used it for the regiospecific epoxidation of β-ionone. It has very high catalyst turnover numbers (3.9×10^5). They were able to produce (E)-4-(2,2,6-trimethyl-7-oxabicyclo[4.1.0]heptan-1-yl) but-3-en-2-one in ethyl acetate.

An efficient catalytic strategy (metal-free) for the synthesis of β-thioketones was developed by Zhou et al. (2019) from enone derivatives and thiols using thia-Michael addition, which was enabled by heterogeneous prolinamide. Up to 98% yield of β-thioketones could be obtained over solid UiO-66-NH-proline catalyst at room temperature. It was reported that a promotional role is played by a cooperative effect of proline (amino group) and UiO-66-NH2 (amide species) via thia-Michael addition. It resulted in a high TOF value (1124.3 h^{-1}). It was interesting to note that UiO-66-NH-proline could be separated from the reaction mixture easily and it can reuse five times with no significant loss of activity.

Di(isononyl)1,2-cyclohexanedicarboxylate (DINCH) is widely used as safe plasticizers in the polyvinylchloride (PVC) industry. They are industrially manufactured from petroleum-derived o-xylene. This was achieved by Hu et al. (2020) using a protocol to 1,2-cyclohexanedicarboxylates with bio-based fumarates and 1,3-dienes just to reduce Diels–Alder (D–A) cycloaddition followed by Pd/C-catalyzed hydrogenation in one-pot under solvent-free conditions with good yield (> 98%).

Jeong and Kim (2021) reported the conversion of *Chlorella* sp. to 5-hydroxymethylfurfural (5-HMF) via thermochemical reaction with levulinic acid (LA) and aluminum sulfate. They could obtain 5-HMF (23.46%) at 165°C with 0.75 g-catalyst/g-biomass and 5% biomass amount in 30 min while 32.89% LA yield was obtained at 205°C under 1.0 g-catalyst/g-biomass and 2.5% biomass in 60 min.

Kanakikodi et al. (2021) proposed protocol for the condensation of furfural (FUR) with acetone (Ac) over γ-alumina supported sodium aluminate (SA) (catalyst) SA/γ-Al_2O_3. It was observed that 25 wt.% SA/-Al_2O_3 had the conversion of FUR (99.5%) with good selectivity to 4-(2-furyl)-3-buten-2-one (FAc) (90%) and that too under mild reaction conditions. It was also revealed that this catalyst is stable and can be reused with consistent catalytic activity.

One of wastes generated in banana plantation is banana pseudostem (BPS). The BPS was used by Taib et al. (2021) as raw feedstock in biomass conversion. It was pyrolyzed in a fluidized bed reactor with a nominal capacity of 300 g h1 with residence time in the range (0.91.2 s) and temperature in the range (470–540°C). The optimum residence time and pyrolysis temperature were found to be 1.02 s and 500°C, respectively. It was reported that high heating value of BPS liquid was 5.35 MJ kg1.

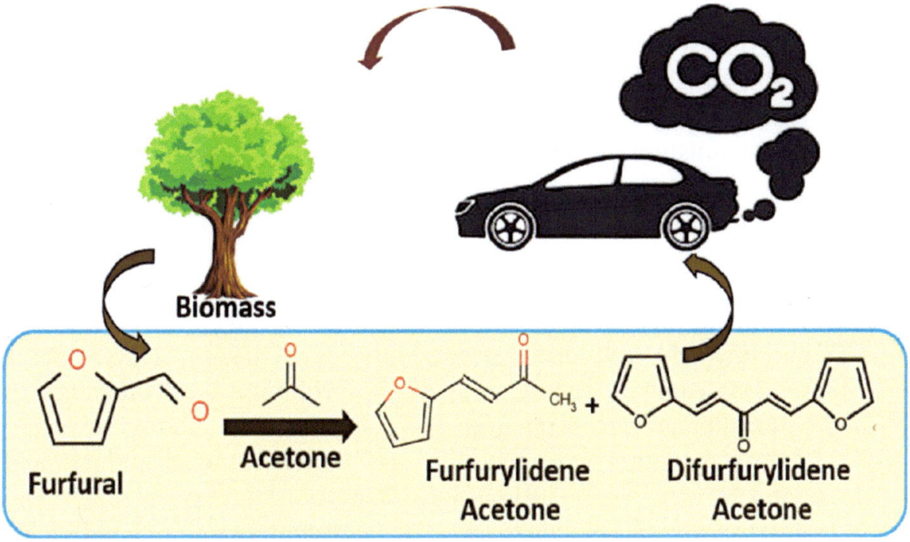

FIGURE 2.8 Catalyzed condensation of furfural with acetone.

Source: Reprinted from Kanakikodi et al., 2021. © 2021 Elsevier.

The influence of Fe, Ni, and Se, important cofactors for hydrogenase was evaluated by Braga et al. (2021) in the production of H_2, alcohols, and volatile organic acids (VOA) using *Clostridium butyricum* as inoculum and cellulose, glucose or sugarcane bagasse (SCB) 5 gL^{-1} in batch reactors at 37°C. It was observed that H_2 production was better, that is, 16 and 6 mmol L^{-1} for glucose and cellulose, respectively.

(xxi) Carbonization

Ma et al. (2021) obtained citric acid-modified hydrochar from pomelo peel (PP) using hydrothermal carbonization in the temperature range (180–240°C). It was observed that pomelo peel hydrochar prepared at 220 and 240°C (PPH240-CA/PPH220-CA) have better energy yield and HHV contents, but low residual ash contents of 1.48 and 1.09%, respectively. It was reported that CO_2 emissions was significantly reduced and greenhouse effect was alleviated, when PPH220-CA was added into fuel blends of coal.

Murillo et al. (2021) reported co-hydrothermal carbonization of pine sawdust (PS) and rapeseed meal (RM) to biofuels. These were optimized on the basis of energy yield-ash content (EY-AC) and mass yield-higher heating value-ash content (MY-HHV-AC). The optimal conditions were obtained as 209.3°C for MY-HHV-AC and 34.5% of RM set, whereas it was 193.4°C and 30.0% for RM of EY-AC set. Then the resulting hydrochars were pelletized before burning. These EY-AC hydrochar pellets decreased the emissions of incomplete combustion (CO and particulate matter).

(xxii) Others

Biobased N-methylpyrrolidone was prepared by the cyclization of γ-aminobutyric acid (obtained from glutamic acid) to 2-pyrrolidone and subsequent catalytic methylation of 2-pyrrolidone with methanol to N-methylpyrrolidone was done in a one-pot procedure (Lammens et al., 2010). Bio-based synthesis of secondary arylamines from (-)-shikimic acid was reported by Wu et al. (2012). 3-Arylamino-4-hydroxybenzoates and 3,4-dihydroxy-5-alkylaminobenzoates have been synthesized in good yields starting from the biomass-based feedstock (-)-shikimic acid via the tandem cross-coupling and aromatization reaction with primary amines.

Molybdenum trioxide has been used as catalyst by Alagiri et al. (2012) via oxidative cross-dehydrogenative coupling for CH functionalization of N-aryl tetrahydroisoquinolines. Cui et al. (2013) developed a route for the synthesis of N-alkylamines via N-alkylation of ammonia or amines using alcohols. They used NiCuFeOx catalyst for this purpose.

Environment-friendly surfactants were explored by Hurkes et al. (2014) based on new silanols, which can be used as as substitutes for isoelectronic phosphonates. It was observed that surface tensions were dramatically reduced in aqueous solutions. They also investigated their potential as coating agents for glass substrates as hydrophilic oxide surfaces. Liu et al. (2016) used hydrogen peroxide as an economic and environmentally friendly oxidant, avoiding metal as a catalyst. They carried out a two-step synthesis of -ketoamides and -amino ketones using secondary benzylic alcohols and amines via direct oxidative coupling.

Dong et al. (2017) synthesized a series of O-aryl(alkyl) N,N-dimethylthio-carbamates in good yields (7077%). They treated alkyl alcohols or substituted phenols with low cost, stable, and eco-friendly tetramethylthiuram disulfide

(TMTD) using NaH. This method provides a green and facile preparation and avoids use of the toxic and corrosive N,N-dialkylthiocarbamoyl chloride. Chen et al. (2018) carried out the oxidation of 5-hydroxymethylfurfural to 2,5-furandicarboxylic acid (2,5-FDCA) under base-free conditions. An improvement of catalytic activity of Pt/ZrO2 in the oxidation of 5-HMF was noticed.

Dai et al. (2019) investigated the feasibility of glucosaminic acid (GlcNA) production from chitosan through tandem hydrolysis and oxidation reactions. The Au/MgO and Amberlyst-15 were found as suitable catalysts for the tandem (hydrolysis and oxidation) step, respectively. It was observed that Au catalyzed oxidation reaction enhanced the GlcNA yield from 17% for untreated hydrolysate to 63% producing GlcNA from chitosan with 36% yield. Hu et al. (2019) developed a route to synthesize cyclohexane-based plasticizers using fumarates and diacetone alcohol as the green starting materials. They used mild Raney Ni-catalyzed hydrogenation of diacetone alcohol, one-step dehydration/DielsAlder reaction of as-produced 2-methyl-2,4-pentandiol and fumarates in a choline chloride (ChCl) followed by Pd/C-catalyzed hydrogenation. It was indicated that the as-prepared di(2-ethylhexyl) 3,5-dimethylcyclohexane-1,2-dicarboxylate can serve as a safe plasticizer for PVC materials.

A synergistic nanostructured MnO_x/TiO_2 catalyst for the synthesis of valuable aromatic imines was developed by Sudarsanam et al. (2021). The highest activity was exhibited with 99.9% imine selectivity and 95.6% benzylamine conversion. This catalyst can be recycled four times. Chestnut shells extracts were prepared by Pinto et al. (2021) using three different green techniques: Supercritical fluids extraction (SFE), ultrasound-assisted extraction (UAE), and subcritical water extraction (SWE). It was reported that there was antimicrobial activity in extracts (particularly against *Staphylococcus aureus*). These extracts also inhibited activities of elastase and hyaluronidase. It was also revealed that UAE and SWE extracts gave best results.

Pirmoradi et al. (2020) used three activated carbon (AC) supported catalysts ($PdTiO_2$, PdCu and PdFe) for hydrogenation of furfural. It was reported that weak acid sites, which were developed by the Pd–TiO_2 catalyst, opened the furan ring with 39% selectivity to 5-hydroxy-2-pentanone (5H2P) in a limited residence time of 7.6 min at 180°C. The PdCu catalyst was found selective toward furfuryl alcohol with 42% selectivity. Yan et al. (2021) used barium lanthanum ferrite, $La_{1-x}Ba_xFeO_3$, using perovskite structure, which acts as oxygen carrier for algae chemical looping catalytic steam gasification method. It was observed that activity of $LaFeO_3$ was dramatically improved on substituting Ba, may be due to the formation of oxygen vacancies and Fe^{4+}/

Fe^{3+}. Among all, the $La_{0.3}Ba_{0.7}FeO_3$ gave highest syngas yield (1.00 m^3 kg^{-1} algae), gasification efficiency (98.86%), and carbon conversion (96.71%).

Starting material should be benign in nature. This is possible by taking biomass and biomass-derived derivatives as the starting materials because these are provided by nature itself. This is not always possible as far as variety of chemical reactions are concerned, and therefore, such feed stocks are to be search down, which are either harmless or less toxic as compared with conventional reactants.

2.5 RECENT DEVELOPMENTS

Zhong et al. (2022) developed solar-driven catalysis system for converting cellulose into lactic acid in presence of Cu modified natural palygorskite (Pal) catalyst. Remarkable photocatalytic lactic acid selectivity under visible light was observed, when 10 wt% Cu_2O QDs/Cu–Pal nanocomposite was used. The reaction pathways and catalytic activity for the production of lactic acid from glycerol was studied by Chen et al. (2022) on using iridium–heterocyclic carbene (iridium–NHC) complex as catalyst. It was reported that the energy barrier for this conversion sharply decreased from 75.2 to 16.8 kcal mol^{-1}, when iridium–NHC complex was introduced. Here, the reaction involved dehydrogenation of glycerol to 2,3-dihydroxypropanal, which is finally isomerization to lactic acid.

Li et al. (2022) synthesized MnO_x catalysts synthesized via hydrothermal method and used for conversion of glucose into formic acid. They could achieve 81.1% yield of formic acid with MnO_x catalyst, which was almost as obtained with vanadium-based heterogeneous catalysts. Other (glucose, sucrose, maltose, xylose, cellobiose) catalysts produce the yield of formic acid greater than 50%; thus, MnO_x is a promising catalyst for converting biomass into formic acid. Meinita et al. (2022) produced levulinic acid (LA) and formic acid (FA) (high-value chemicals used in industries) using biomass *Kappaphycus alvarezii* (*K. alvarezii*). The LA and FA were produced through conversion of macroalgae *K. alvarezii* via a thermochemical reaction using methanesulfonic acid (MSA), which is an environmental-friendly and strong acidic catalyst. It was reported that highest LA and FA yield of 14.69 and 5.35%, respectively could be attained at 180°C with 0.6 M MSA catalyst concentration in 30 min using 2.5% biomass load.

Maleic acid is an important intermediate in chemical industries. Ayoub et al. (2022) reported a catalyst-free process for synthesis of maleic acid from furfural under high frequency ultrasound (HFUS) irradiations. About

70% selectivity of maleic acid with 92% of furfural conversion could be achieved without using any catalyst and that too under mild conditions using hydrogen peroxide as an oxidant. Their approach enables the use of biomass an alternative to petroleum for synthesis of maleic acid from furfural in an eco-friendly and energy-efficient process.

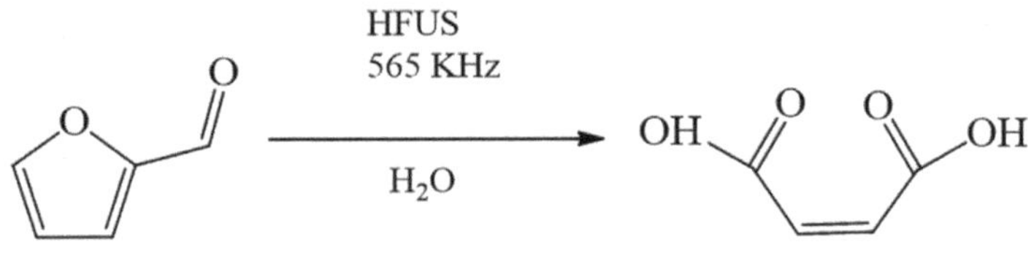

Furfural 20 wt% Catalyst-free Maleic Acid

KEYWORDS

- **biodegradation**
- **biofuels**
- **biomass**
- **sustainable process**
- **valorisation**
- **chemical transformation**

REFERENCES

Adsul, M. G.; Varma, A. J.; Gokhale, D. V. Lactic Acid Production from Waste Sugarcane Bagasse Derived Cellulose. *Green Chem.* **2007,** *9* (1), 58–62.

Akbar, E.; Binitha, N.; Yaakob, Z.; Kamarudin, S. K.; Salimon, J.; Preparation of Na Doped SiO_2 Solid Catalysts By the Sol-Gel Method for the Production of Biodiesel from Jatropha Oil. *Green Chem.* **2009,** *11*, 1862–1866.

Alagiri, K.; Devadig, P.; Prabhu, K. R. Molybdenum Trioxide Catalyzed Oxidative Cross-Dehydrogenative Coupling of Benzylic sp^3 C–H Bonds: Synthesis of α-Aminophosphonates Under Aerobic Conditions. *Tetrahedron Lett.* **2012,** *53* (12), 1456–1459.

Albini, A.; Fagnoni, M. Green Chemistry and Photochemistry Were Born at the Same Time. *Green Chem.* **2004,** *6* (1). DOI: 10.1039/B309592D.

Allen, D. J.; Brent, G. F. Sequestering CO_2 by Mineral Carbonation: Stability Against Acid Rain Exposure. *Environ. Sci. Technol.* **2010,** *44*, 2735–2739.

Antonetti, C.; Melloni, M.; Licursi, D.; Fulignati, S.; Ribechini, E.; Rivas, S. et al. Microwave-Assisted Dehydration of Fructose and Inulin to HMF Catalyzed by Niobium and Zirconium Phosphate Catalysts. *Appl. Catal. B: Environ.* **2017,** *206*, 364–377.

Anvar, S.; Mohammadpoor-Baltork, I.; Tangestaninejad, S.; Moghadam, M.; Mirkhani, V.; Khosropour, A. R. et al. Efficient and Environmentally-Benign Three-Component Synthesis of Quinolines and Bis-Quinolines Catalyzed By Recyclable Potassium Dodecatungstocobaltate Trihydrate Under Microwave Irradiation. *RSC Adv.* **2012,** *2* (23), 8713–8720.

Arancon, R. A.; Barros Jr, H. R.; Balu, A. M.; Vargas, C.; Luque, R.; Valorisation of Corncob Residues to Functionalised Porous Carbonaceous Materials for the Simultaneous Esterification/Transesterification of Waste Oils. *Green Chem.* **2011,** *13*, 3162–3167.

Ashokkumar, M.; Narayanan, N. T.; Reddy, A. L. M.; Gupta, B. K.; Chandrasekaran, B.; Talapatra, S. et al. Transforming Collagen Wastes Into Doped Nanocarbons for Sustainable Energy Applications. *Green Chem.* **2012,** *14* (6), 1689–1695.

Ayoub, N.; Toufaily, J.; Guénin, E.; Enderlin, G. Catalyst-Free Process for Oxidation of Furfural to Maleic Acid By High Frequency Ultrasonic Activation. *Green Chem.* **2022,** *24* (10), 4164–4173.

Baghos, V. B.; Doss, S. H.; Eskander, E. F.; Synthesis of Potentially Biologically Active Compounds via the Stobbe Condensation. *Org. Prep. Proced. Int.* **1993,** *25* (3), 301–307.

Balakrishnan, M.; Batra, V. S.; Hargreaves, J. S. J.; Pulford, I. D. Waste Materials–Catalytic Opportunities: An Overview of the Application of Large Scale Waste Materials as Resources for Catalytic Applications. *Green Chem.* **2011,** *13* (1), 16–24.

Balbino, J. M.; de Menezes, E. W.; Benvenutti, E. V.; Cataluña, R.; Ebeling, G.; Dupont, J. Silica-Supported Guanidine Catalyst for Continuous Flow Biodiesel Production. *Green Chem.* **2011,** *13* (11), 3111–3116.

Behr, A.; Henze, G. Use of Carbon Dioxide in Chemical Syntheses via a Lactone Intermediate. *Green Chem.* **2011,** *13* (1), 25–39.

Benaglia, M. *Renewable and Recyclable Catalysts*; Advancing Green Chemistry Series; Wiley-VCH, West Sussex, 2009.

Bicker, M.; Kaiser, D.; Ott, L.; Vogel, H. Dehydration of D-Fructose to Hydroxymethylfurfural in Sub-and Supercritical Fluids. *J. Supercrit. Fluids*, **2005,** *36* (2), 118–26.

Binder, J. B.; Raines, R. A. Simple Chemical Transformation of Lignocellulosic Biomass Into Furans for Fuels and Chemicals. *Chem. Soc.* **2009,** *131*, 1979–1985.

Bowker, M. Sustainable Hydrogen Production by the Application of Ambient Temperature Photocatalysis. *Green Chem.* **2011,** *13* (9), 2235–2246.

Braden, D. J.; Henao, C. A.; Heltzel, J.; Maravelias, C. C. Dumesic, J. A. Production of Liquid Hydrocarbon Fuels by Catalytic Conversion of Biomass-Derived Levulinic Acid. *Green Chem.* **2011,** *13* (7), 1755–1765.

Braga, J. K.; Stancari, R. A.; Motteran, F.; Malavazi, I.; Varesche, M. B. A. Metals Addition for Enhanced Hydrogen, Acetic and Butyric Acids Production from Cellulosic Substrates by Clostridium Butyricum. *Biomass Bioenergy* **2021,** 150. DOI: 10.1016/j.biombioe.2020.105679.

Campbell, C. J.; Laherrère, J. H. The End of Cheap Oil. *Sci.Am.* **1998,** *278*, 60–65.

Case, P. A.; van Heiningen, A. R. Wheeler, M. C. Liquid Hydrocarbon Fuels from Cellulosic Feedstocks via Thermal Deoxygenation of Levulinic Acid and Formic Acid Salt Mixtures. *Green Chem.* **2012,** *14* (1), 85–89.

Catrinck, M. N.; Ribeiro, E. S.; Monteiro, R. S.; Ribas, R. M.; Barbosa, M. H.; Teófilo, R. F. Direct Conversion of Glucose to 5-Hydroxymethylfurfural Using a Mixture of Niobic Acid and Niobium Phosphate as a Solid Acid Catalyst. *Fuel* **2017,** *210*, 67–74.

Chang, R. H. Y.; Jang, J.; Wu, K. C. W. Cellulase Immobilized Mesoporous Silica Nanocatalysts for Efficient Cellulose-to-Glucose Conversion. *Green Chem.* **2011,** *13* (10), 2844–2850.

Chanshetti, U. Green Chemistry: Environmentally Benign Chemistry. *Int. J. Adv. Res. Chem. Sci.* **2014,** *1* (1), 110–115.

Chen, H.; Shen, J.; Chen, K.; Qin, Y.; Lu, X.; Ouyang, P.; Fu, J. Atomic Layer Deposition of Pt Nanoparticles on Low Surface Area Zirconium Oxide for the Efficient Base-Free Oxidation of 5-Hydroxymethylfurfural to 2, 5-Furandicarboxylic Acid. *Appl. Catal. A: Gen.* **2018,** *555*, 98–107.

Chen, S.; Xu, S.; Ge, C.; Hu, C. Mechanistic Investigations of the Synthesis of Lactic Acid from Glycerol Catalyzed by an Iridium–NHC Complex. *Processes* **2022,** *10* (4). DOI: 10.3390/pr10040626

Cheng, H.; Gross, R., Eds. *Green Polymer Chemical Biocatalysis Biomaterials*; Oxford University Press: Oxford, 2011.

Chheda, J. N.; Román-Leshkov, Y. Dumesic, J. A.; Production of 5-Hydroxymethylfurfural and Furfural by Dehydration of Biomass-Derived Mono-and Poly-Saccharides. *Green Chem.* **2007,** *9* (4), 342–350.

Chia, M.; Schwartz, T. J.; Shanks, B. H.; Dumesic, J. A. Triacetic Acid Lactone as a Potential Biorenewable Platform Chemical. *Green Chem.* **2012,** *14* (7), 1850–1853.

Chouhan, N.; Chen, C. K.; Chang, W. S.; Chang, K. W.; Liu, R. S. Photoelectrochemical Cells for Hydrogen Generation in Electrochemical Technologies for Energy Storage and Conversion; Liu, R. S., Zhang, L., Sun, X., Liu, H., Zhang, J., Eds.; Wiley-VCH: Weinheim, 2012.

Corma, A.; Iborra, S.; Velty, A. Chemical Routes for the Transformation of Biomass into Chemicals. *Chem. Rev.* **2007,** *107* (6), 2411–2502.

Cui, X.; Dai, X.; Deng, Y.; Shi, F. Development of a General Non-Noble Metal Catalyst for the Benign Amination of Alcohols with Amines and Ammonia. *Chem. Eur. J.* **2013,** *19* (11), 3665–3675.

Cukalovic, A.; Stevens, C. V. Production of Biobased HMF Derivatives by Reductive Amination. *Green Chem.* **2010,** *12* (7), 1201–1206.

Dai, J.; Gözaydın, G.; Hu, C.; Yan, N. Catalytic Conversion of Chitosan to Glucosaminic Acid by Tandem Hydrolysis and Oxidation. *ACS Sustain. Chem. Eng.* **2019,** *7* (14), 12399–12407.

Darensbourg, D. J. Wilson, S. J. What's New with CO_2? Recent Advances in Its Copolymerization with Oxiranes. *Green Chem.* **2012,** *14* (10), 2665–2671.

Demain, A. L.; Small Bugs, Big Business: The Economic Power of the Microbe. *Biotechnol. Adva.* **2000,** *18* (6), 499–514.

Dewulf, J.; Van Langenhove, H., Eds. *Renewables-Based Technology: Sustainability Assessment*; John Wiley & Sons: New York, 2006.

Dong, Z. B.; Wang, M.; Zhu, H.; Liu, X.; Chang, C. Z. Base-Promoted Synthesis of O-Aryl/Alkyl N, N-Dimethylthiocarbamates Starting from Inexpensive and Environmentally Benign Disulfide. *Synthesis* **2017,** *49* (23), 5258–5262.

Du, X. L.; Bi, Q. Y.; Liu, Y. M.; Cao, Y.; He, H. Y.; Fan, K. N. Tunable Copper-Catalyzed Chemoselective Hydrogenolysis of Biomass-Derived γ-valerolactone into 1, 4-pentanediol or 2-Methyltetrahydrofuran. *Green Chem.* **2012,** *14* (4), 935–939.

Epane, G.; Laguerre, J. C.; Wadouachi, A.; Marek, D. Microwave-Assisted Conversion of D-Glucose Into Lactic Acid Under Solvent-Free Conditions. *Green Chem.* **2010,** *12* (3), 502–506.

Eskandari, K.; Karami, B.; Pourshojaei, Y.; Asadipour, A. An Eco-Compatible, Three-Component Synthesis of Acyl-Substituted Bis (pyrazolyl) Methanes by Employing Recyclable Silica

Sodium Carbonate as an Efficient and Environmentally Benign Catalyst in Aqueous Medium. *Monatsh Chem.* **2018,** *149* (6), 1075–1081.

Fatahpour, M.; Sadeh, F. N.; Hazeri, N.; Maghsoodlou, M. T.; Lashkari, M. Aspirin: An Efficient Catalyst for Synthesis of bis (pyrazol-5-ols), Dihydropyrano[2, 3-c]pyrazoles and Spiropyranopyrazoles in an Environmentally Benign Manner. *J. Iran. Chem. Soc.* **2017,** *14* (9), 1945–1956.

Faugeras, P. A.; Elchinger, P. H.; Brouillette, F.; Montplaisir, D. Zerrouki, R. Advances in Cellulose Chemistry-Microwave-Assisted Synthesis of Propargylcellulose in Aqueous Medium. *Green Chem.* **2012,** *14* (3), 598–600.

Fumagalli, C. in *Kirk-Othmer Encyclopedia of Chemical Technology*, 4th ed.; Kroschwitz, J.; HomeGrant, M.; Eds.; John Wiley and Sons: New York, 1997.

Galletti, A. M. R.; Antonetti, C.; De Luise, V.; Martinelli, M. A Sustainable Process for the Production of γ-Valerolactone by Hydrogenation of Biomass-Derived Levulinic Acid. *Green Chem.* **2012,** *14* (3), 688–694.

Gandini, A.; Belgacem, M. N. Furans in Polymer Chemistry. *Prog. Polym. Sci.* **1997,** *22* (6), 1203–1379.

Gassama, A.; Ernenwein, C.; Hoffmann, N.; Synthesis of Surfactants from Furfural Derived 2 [5H]-Furanone and Fatty Amines. *Green Chem.* **2010,** *12* (5), 859–865.

Gutsche, B.; Rößler, H.; Würkert, S. Heterogeneous Catalysis in Oleochemistry. *Handbook of Heterogeneous Catalysis: Online*, **2008,** 3329–3359.

Han, J.; Sun, H.; Ding, Y.; Lou, H.; Zheng, X. Palladium-Catalyzed Decarboxylation of Higher Aliphatic Esters: Towards a New Protocol to the Second Generation Biodiesel Production. *Green Chem.* **2010,** *12* (3), 463–467.

Hengne, A. M.; Rode, C. V. Cu–ZrO_2 Nanocomposite Catalyst for Selective Hydrogenation of Levulinic Acid and Its Ester to γ-Valerolactone. *Green Chem.* **2012,** *14* (4), 1064–1072.

Hester, R. E.; Harrison, R. M., Eds. *Air quality in Urban Environments*; Royal Society of Chemistry: London, 2009.

Hick, S. M.; Griebel, C.; Restrepo, D. T.; Truitt, J. H.; Buker, E. J.; Bylda, C. et al. Mechanocatalysis for Biomass-Derived Chemicals and Fuels. *Green Chem.* **2010,** *12* (3), 468–474.

Hill, K. Special Topic Issue on Green Chemistry. *Pure Appl. Chem*, **2000,** *72* (7), 1255–1264.

Holtz, M. H. *Second Annual Conference on Carbon Sequestration: Developing the Technology Base to Reduce Carbon Intensity*; Alexandria: Virginia, May, 2003.

Hu, X.; Li, C. Z. Levulinic Esters from the Acid-Catalysed Reactions of Sugars and Alcohols as Part of a Bio-Refinery. *Green Chem.* **2011,** *13* (7), 1676–1679.

Hu, Y.; Liu, C.; Wang, P.; Li, G.; Wang, A.; Cong, Y. et al. Sustainable Production of Safe Plasticizers with Bio-Based Fumarates and 1, 3-Dienes. *Ind. Eng. Chem. Res.* **2020,** *59*, 7367–7374.

Hu, Y.; Yuan, L.; Zhang, X.; Zhou, H.; Wang, P.; Li, G. et al. Production of 1, 2-Cyclohexanedicarboxylates from Diacetone Alcohol and Fumarates. *ACS Sustain. Chem. Eng.* **2019,** *7* (3), 2980–2988.

Hurkes, N.; Ehmann, H. M.; List, M.; Spirk, S.; Bussiek, M. et al. Silanol-Based Surfactants: Synthetic Access and Properties of an Innovative Class of Environmentally Benign Detergents. *Chem. Eur. J.* **2014,** *20* (30), 9330–9335.

Hwang, D. W.; Kashinathan, P.; Lee, J. M.; Lee, J. H.; Lee, U. H.; Hwang, J. S. Production of γ-Butyrolactone from Biomass-Derived 1, 4-Butanediol Over Novel Copper-Silica Nanocomposite. *Green Chem.* **2011,** *13* (7), 1672–1675.

International Energy Agency. 2012. http://www.iea.org/press/pressdetail.asp Technology Roadmap, Biofuels for Transport, http://www.iea.org/papers/2011/biofuels_roadmap.pdf.

Jadhav, S. A.; Shioorkar, M. G.; Chavan, O. S.; Sarkate, A. P.; Shinde, D. B. Rapid and Efficient One-Pot Microwave-Assisted Synthesis of 2-Phenylimidazo [1, 2-a] Pyridines and 2-Phenylimidazo [1, 2-a] Quinoline in Water–PEG-400. *Synth. Commun.* **2017,** *47* (4), 285–290.

James, O. O.; Maity, S.; Mesubi, M. A.; Ogunniran, K. O.; Siyanbola, T. O.; Sahu, S.; et al. Towards Reforming Technologies for Production of Hydrogen Exclusively from Renewable Resources. *Green Chem.* **2011,** *13*, 2272–2284.

Jeong, G. T.; Kim, S. K. Hydrothermal Conversion of Microalgae Chlorella sp. into 5-Hydroxymethylfurfural and Levulinic Acid by Metal Sulfate Catalyst. *Biomass Bioenergy*, **2021,** *148*, DOI: 10.1016/j.biombioe.2021.106053.

Kanakikodi, K. S.; Churipard, S. R.; Bai, R.; Maradur, S. P. Upgrading of Lignocellulosic Biomass-Derived Furfural: An Efficient Approach for the Synthesis of Bio-Fuel Intermediates Over γ-Alumina Supported Sodium Aluminate. *Mol. Catal.* **2021,** 510. DOI: 10.1016/j.mcat.2021.111716.

Khandelwal, S.; Rajawat, A.; Kumar Tailor, Y.; Kumar, M. Diversity Oriented p-TSA Catalyzed Efficient and Environmentally Benign Synthetic Protocol for the Synthesis of Structurally Diverse Heteroannulated Benzothiazolopyrimidines. *Curr. Organocatal.* **2015,** *2* (1), 37–43.

Klass D. H. *Biomass for Renewable Energy, Fuels and Chemicals. Fossil Fuel Reserves and Depletion*; Academic Press: SanDiego, 1998; pp. 10–19.

Knothe, G. A Technical Evaluation of Biodiesel from Vegetable Oils vs. Algae: Will Algae-Derived Biodiesel Perform? *Green Chem.* **2011,** *13* (11), 3048–3065.

Kobayashi, H.; Ito, Y.; Komanoya, T.; Hosaka, Y.; Dhepe, P. L.; Kasai, K.; Hara, K.; Fukuoka, A. Synthesis of Sugar Alcohols by Hydrolytic Hydrogenation of Cellulose Over Supported Metal Catalysts. *Green Chem.* **2011,** *13* (2), 326–333.

Kollonitsch, V. *Sucrose Chemicals: A Critical Review of a Quarter Century of Research by the Sugar Research Foundation*; International Sugar Research Foundation: Washington, DC, 1970.

Kwon, E. E.; Seo, J.; Yi, H. Transforming Animal Fats Into Biodiesel Using Charcoal and CO_2. *Green Chem.* **2012,** *14* (6), 1799–1804.

Lammens, T. M. Wageningen University and Research Centre; PhD Thesis, 2011.

Lammens, T. M.; Franssen, M. C.; Scott, E. L.; Sanders, J. P.; Synthesis of Biobased N-Methylpyrrolidone by One-Pot Cyclization and Methylation of γ-Aminobutyric Acid. *Green Chem.* **2010,** *12* (8), 1430–1436.

Leimkuhler, H. J., Ed. *Managing CO_2 Emissions in the Chemical Industry*; Wiley-VCH: West Sussex, 2010.

Li, C.; Wang, Q.; Zhao, Z. K. Acid in Ionic Liquid: An Efficient System for Hydrolysis of Lignocellulose. *Green Chem.* **2008,** *10* (2), 177–182.

Li, C.; Zhao, Z. K.; Efficient Acid-Catalyzed Hydrolysis of Cellulose in Ionic Liquid. *Adv. Synth. Catal.* **2007,** 349, 1847–1850.

Li, J.; Smith, R. L.; Xu, S.; Yang, J.; Zhang, K.; Shen, F. Manganese Oxide as an Alternative to Vanadium-Based Catalysts for Effective Conversion of Glucose to Formic Acid in Water. *Green Chem.* **2022,** *24* (1), 315–324.

Li, N.; Tompsett, G. A.; Zhang, T.; Shi, J.; Wyman, C. E.; Huber, G. W. Renewable Gasoline from Aqueous Phase Hydrodeoxygenation of Aqueous Sugar Solutions Prepared by Hydrolysis of Maple Wood. *Green Chem.* **2011,** *13*, 91–101.

Liao, L. Y.; Shen, G.; Zhang, X.; Duan, X. F. A Practicable Environmentally Benign One-Pot Synthesis of 2-Arylbenzofurans at Room Temperature. *Green Chem.* **2012,** *14* (3), 695–701.

Lichtenthaler, F. W.; Cuny, E.; Martin, D.; Ronninger, S. In *Carbohydrates as Organic Raw Materials*; Lichtenthaler, F. W., Ed.; VCH: Weinheim, New York, 1991.

Liu, C.; Fang, Z.; Yang, Z.; Li, Q.; Guo, S.; Guo, K. A Two-Step Continuous Synthesis of α-Ketoamides and α-Amino Ketones from 2° Benzylic Alcohols Using Hydrogen Peroxide as an Economic and Benign Oxidant. *RSC Adv.* **2016,** *6* (30), 25167–25172.

Long, J.; Li, X.; Guo, B.; Wang, F.; Yu, Y.; Wang, L. Simultaneous Delignification and Selective Catalytic Transformation of Agricultural Lignocellulose in Cooperative Ionic Liquid Pairs. *Green Chem*, **2012,** *14* (7), 1935–1941.

Lopes, M.; Dussan, K.; Leahy, J. J.; Da Silva, V. T. Conversion of d-Glucose to 5-Hydroxymethylfurfural Using Al_2O_3-Promoted Sulphated Tin Oxide as Catalyst. *Catal. Today*, **2017,** *279*, 233–243.

Ma, R.; Fakudze, S.; Shang, Q.; Wei, Y.; Chen, J.; Liu, C. et al. Catalytic Hydrothermal Carbonization of Pomelo Peel for Enhanced Combustibility of Coal/Hydrochar Blends and Reduced CO_2 Emission. *Fuel* **2021,** 304. DOI: 10.1016/j.fuel.2021.121422.

Marcotullio, G.; De Jong, W. Chloride Ions Enhance Furfural Formation from D-xylose in Dilute Aqueous Acidic Solutions. *Green Chem.* **2010,** *12* (10), 1739–1746.

Mascal, M. and Dutta, S. Synthesis of the Natural Herbicide δ-Aminolevulinic Acid from Cellulose-Derived 5- (Chloromethyl) Furfural. *Green Chem.* **2011,** *13* (1), 40–41.

Meinita, M. D. N.; Amron, A.; Harwanto, D.; Trianto, A.; Jeong, G. T. The Production of Levulinic Acid and Formic Acid from Red Macroalga Kappaphycus Alvarezii Using Methanesulfonic Acid. *Bioresour. Technol. Rep.* **2022,** *17*. DOI: 10.1016/j.biteb.2022.100954.

Moreau, C.; Belgacem, M. N.; Gandini, A.; Recent Catalytic Advances in the Chemistry of Substituted Furans from Carbohydrates and in the Ensuing Polymers. *Top. Catal.* **2004,** *27* (1), 11–30.

Mothe, S. R.; Kothandaraman, P.; Lauw, S. J. L.; Chin, S. M. W.; Chan, P. W. H. Silver Triflate Catalyzed Tandem Heterocyclization/Alkynylation of 1- ((2-Tosylamino) aryl) But-2-yne-1, 4-Diols to 2-Alkynyl Indoles. *Chem. Eur. J.* **2012,** *18* (20), 6133–6137.

Moussa, H. H.; Meguid, S. A.; Atalla, M. M. Studies in Stobbe Condensation. Part 7: Synthesis and Biological Activity of Some New Cyclopentadienones and Cyclopentadienes. *Pharmazie*, **1982,** *37* (5), 352–354.

Murillo, H. A.; Díaz-Robles, L. A.; Santander, R. E.; Cubillos, F. A. Conversion of Residual Biomass Into Valuable Biofuels by Co-Hydrothermal Carbonization for Utilization in Household Pellet Stoves. *Biomass Bioenergy* **2021,** 151. DOI: 10.1016/j.biombioe.2021.106153.

Natarajan, P.; Kumar, N.; Sharma, M. Visible Light-Mediated Intramolecular C–H Arylation of Diazonium Salts of N- (2-Aminoaryl) Benzoimines: A Facile Synthesis of 6-Arylphenanthridines. *Org. Chem. Front.* **2016,** *3* (10), 1265–1270.

National Research Council. USA, Priorities for Research and Commercialization of Biobased Industrial Products. *Natl Acad. Sci.* **2000,** *97*, 11140.

Noutsias, D.; Alexopoulou, I.; Montagnon, T.; Vassilikogiannakis, G. Using Water, Light, Air and Spirulina to Access a Wide Variety of Polyoxygenated Compounds. *Green Chem.* **2012,** *14* (3), 601–604.

Okkerse, C.; Van Bekkum, H. From Fossil to Green. *Green Chem.* **1999,** *1* (2), 107–114.

Owens, E. A.; Bruschi, N.; Tawney, J. G.; Henary, M. A Microwave-Assisted and Environmentally Benign Approach to the Synthesis of Near-Infrared Fluorescent Pentamethine Cyanine Dyes. *Dyes Pigments* **2015,** *113*, 27–37.

Pentz, W. J. *GB Patent* 2131014, 198425, 1970.

Pinto, D.; de la Luz Cádiz-Gurrea, M.; Garcia, J.; Saavedra, M. J.; Freitas, V.; Costa, P. et al. From Soil to Cosmetic Industry: Validation of a New Cosmetic Ingredient Extracted from Chestnut Shells. *Sustain. Mater. Technol.* **2021,** 29. DOI: 10.1016/j.susmat.2021.e00309.

Pirmoradi, M.; Gulotty, R. J.; Kastner, J. R. Continuous Hydroxyketone Production from Furfural Using Pd–TiO_2 Supported on Activated Carbon. *Catal. Sci. Technol.* **2020,** *10* (20), 7002–7015.

Pouramiri, B.; Kermany, E. T.; Khajesamani, H.; Khabazzadeh, H. An Efficient, Three-Component Synthesis of 3, 4-Dihydropyrimidin-2 (1H)-Ones Using $LaCl_3$/$ClCH_2COOH$ as Environmentally Benign and Green Catalytic System. *J. Sci. Islam. Repub.* **2014,** *25* (4), 323–327.

Purushothaman, R. K. P.; van Haveren, J.; van Es, D. S.; Melian-Cabrera, I.; Heeres, H. J. The Oxidative Esterification of Glycerol to Methyl Glycerate in Methanol Using Gold on Oxidic Supports: An Insight in Product Selectivity. *Green Chem.* **2012,** *14*, 2031–2037.

Ragauskas, A. J.; Williams, C. K.; Davison, B. H.; Britovsek, G.; Cairney, J.; Eckert, C. A. et al. The Path Forward for Biofuels and Biomaterials. *Science* **2006,** *311*, 484–489.

Ran, N.; Zhao, L.; Chen, Z.; Tao, J. Recent Applications of Biocatalysis in Developing Green Chemistry for Chemical Synthesis at the Industrial Scale. *Green Chem.* **2008,** *10* (4), 361–372.

Ravelli, D.; Dondi, D.; Fagnoni, M.; Albini, A.; Photocatalysis. A Multi-Faceted Concept for Green Chemistry. *Chem. Soc. Rev.* **2009,** *38*, 1999–2011.

Ravindranath, N. H.; Hall, D. O. *Biomass, Energy and Environment: A Developing Country Perspective from India*; Oxford University Press, 1995.

Relsch, M. S. Air Products Baker Airs Views on Trade, Industry Image, Education. *Chem. Eng. News* **1991,** *69*, 13–16.

Ruhal, R.; Aggarwal, S.; Choudhury, B. Suitability of Crude Glycerol Obtained from Biodiesel Waste for the Production of Trehalose and Propionic Acid. *Green Chem.* **2011,** *13* (12), 3492–3498.

Saeidian, H.; Moradnia, F. Benign Synthesis of N-Aryl-3, 10-Dihydroacridin-1 (2H)-One Derivatives via ZnO Nanoparticle-Catalyzed Knoevenagel Condensation/Intramolecular Enamination Reaction. *J. Iran. Chem. Commun.* **2017,** *5*, 252–261.

Sahu, P. K.; Sahu, P. K.; Jain, R.; Yadav, R.; Agarwal, D. D. Hydrotalcite: Recyclable, Novel Heterogeneous Catalyst for Facile, Environmentally Benign and High Yielding Multi-Component Synthesis and Mechanistic Study Under Solvent Free Conditions. *Catal. Sci. Technol.* **2012,** *2* (12), 2465–2475.

Serrano-Ruiz, J. C.; Dumesic, J. A. Catalytic Upgrading of Lactic Acid to Fuels and Chemicals by Dehydration/Hydrogenation and C–C Coupling Reactions. *Green Chem.* **2009,** *11* (8), 1101–1104.

Shapiro, N.; Kramer, M.; Goldberg, I.; Vigalok, A. Straightforward Radical Organic Chemistry in Neat Conditions and "On Water". *Green Chem.* **2010,** *12*, 582–584.

Shekouhy, M.; Sarvestani, A. M.; Khajeh, S.; Khalafi-Nezhad, A. Glycerol: A More Benign and Biodegradable Promoting Medium for Catalyst-Free One-Pot Multi-Component Synthesis of Triazolo [1, 2-a] Indazole-Triones. *RSC Adv.* **2015,** *5* (78), 63705–63710.

Singhvi, M.; Joshi, D.; Adsul, M.; Varma, A.; Gokhale, D. D- (−)-Lactic Acid Production from Cellobiose and Cellulose by Lactobacillus Lactis Mutant RM2–2 4. *Green Chem.* **2010,** *12* (6), 1106–1109.

Snieckus, V.; Jiao, W. One-Step Synthesis of 2-Substituted Indazolones. *Synfacts* **2018,** *14* (11), DOI: 10.1055/s-0037–1611260.

Soetaert, V. E.; *Biofuels*; Wiley-VCH: Weinheim, 2009.

Spekreijse, J.; Le Nôtre, J.; van Haveren, J.; Scott, E. L.; Sanders, J. P. Simultaneous Production of Biobased Styrene and Acrylates Using Ethenolysis. *Green Chem.* **2012,** *14* (10), 2747–2751.

Stein, T. V.; Grande, P.; Sibilla, F.; Commandeur, U.; Fischer, R.; Leitner, W.; et al. Salt-Assisted Organic-Acid-Catalyzed Depolymerization of Cellulose. *Green Chem.* **2010,** *12* (10), 1844–1849.

Sudarsanam, P.; Köckritz, A.; Atia, H.; Amin, M. H.; Brückner, A. Synergistic Nanostructured MnOx/TiO_2 Catalyst for Highly Selective Synthesis of Aromatic Imines. *Chem. Cat. Chem.* **2021,** *13*, 1990–1997.

Sun, J.; Liu, H. Selective Hydrogenolysis of Biomass-Derived Xylitol to Ethylene Glycol and Propylene Glycol on Supported Ru Catalysts. *Green Chem.* **2011,** *13* (1), 135–142.

Sun, X.; Wang, J.; Chen, J.; Zheng, J.; Shao, H.; Huang, C. Dehydration of Fructose to 5-Hydroxymethylfurfural Over MeSAPOs Synthesized from Bauxite. *Micropor Mesopor Mater*. **2018,** *259*, 238–243.

Taib, R. M.; Abdullah, N. Aziz, N. S. M.; Bio-Oil Derived from Banana Pseudo-Stem via Fast Pyrolysis Process. *Biomass Bioenergy* **2021,** 148. DOI: 10.1016/j.biombioe.2021.106034.

Tao, J.; Kazlauskas, R. J., Eds. *Biocatalysis for Green Chemistry and Chemical Process Development*; John Wiley and Sons Inc: West Sussex, 2011.

Tudorache, M.; Protesescu, L.; Coman, S.; Parvulescu, V. I. Efficient Bio-Conversion of Glycerol to Glycerol Carbonate Catalyzed by Lipase Extracted from *Aspergillus niger*. *Green Chem.* **2012,** *14* (2), 478–482.

Tukacs, J. M.; Király, D.; Strádi, A.; Novodarszki, G.; Eke, Z.; Dibó, G.; Kégl, T.; Mika, L. T. Efficient Catalytic Hydrogenation of Levulinic Acid: A Key Step in Biomass Conversion. *Green Chem.* **2021,** *14* (7), 2057–2065.

Umbach, W. In Eierdanz, H., Ed., *Perspektiven Nachwach- sender Rohstoffe in der Chemie*; VCH Publ.: Weinheim, 1996; p XXIX.

Union, E. Directive 2009/28/EC of the European Parliament and of the Council of 23 April 2009 on the Promotion of the Use of Energy from Renewable Sources and Amending and Subsequently Repealing Directives 2001/77/EC and 2003/30/EC. *Off. J. Eur. Union*, 2009, 5.

Wang, C.; Sun, J.; Liu, X.; Sudo, A.; Endo, T.; Synthesis and Copolymerization of Fully Bio-Based Benzoxazines from Guaiacol, Furfurylamine and Stearylamine. *Green Chem.* **2012a,** *14* (10), 2799–2806.

Wang, J.; Ren, J.; Liu, X.; Xi, J.; Xia, Q.; Zu, Y. et al. Direct Conversion of Carbohydrates to 5-Hydroxymethylfurfural Using Sn-Mont Catalyst. *Green Chem.* **2012b,** *14* (9), 2506–2512.

Weissermel, K.; Arpe, H. J., Eds. *Industrial Organic Chemistry*; John Wiley & Sons: New Jersey, 2008.

Whittall, J.; Sutton, P. W., Eds. *Practical Methods for Biocatalysis and Biotransformations*; John Wiley & Sons: New Jersey, 2009.

Wölfel, R.; Taccardi, N.; Bösmann, A.; Wasserscheid, P. Selective Catalytic Conversion of Biobased Carbohydrates to Formic Acid Using Molecular Oxygen. *Green Chem.* **2011,** *13* (10), 2759–2763.

Wu, W.; Zou, Y.; Chen, Y.; Li, J.; Lv, Z.; Wei, W. Bio-Based Synthesis of Secondary Arylamines from (−)-Shikimic Acid. *Green Chem.* **2012,** *14* (2), 363–370.

Xie, C.; Song, J.; Wu, H.; Hu, Y.; Liu, H.; Zhang, Z. et al. Ambient Reductive Amination of Levulinic Acid to Pyrrolidones Over Pt Nanocatalysts on Porous TiO_2 Nanosheets. *J. Am. Chem. Soc.* **2019,** *141* (9), 4002–4009.

Xie, L. Y.; Li, Y. J.; Qu, J.; Duan, Y.; Hu, J.; Liu, K. J.; et al. A Base-Free, Ultrasound Accelerated One-Pot Synthesis of 2-Sulfonylquinolines in Water. *Green Chem.* **2017,** *19* (23), 5642–5646.

Yan, J.; Jiang, S.; Song, T.; Shen, L. Chemical Looping Catalytic Steam Gasification (CLCSG) of Algae Over $La_{1-x}Ba_xFeO_3$ Perovskites for Syngas Production. *Biomass Bioenerg.* **2021,** 151. DOI; 10.1016/j.biombioe.2021.106154.

Yang, Y.; Fan, X.; Cao, H.; Chu, S.; Zhang, X.; Xu, Q.; Yu, L. Fabrication of Se/C Using Carbohydrates as Biomass Starting Materials: An Efficient Catalyst for Regiospecific Epoxidation of β-Ionone with Ultrahigh Turnover Numbers. *Catal. Sci. Technol.* **2018,** *8* (19), 5017–5023.

Yang, Y.; Hu, C. W.; Abu-Omar, M. M. Conversion of Carbohydrates and Lignocellulosic Biomass into 5-Hydroxymethylfurfural Using $AlCl_3 \cdot 6H_2O$ Catalyst in a Biphasic Solvent System. *Green Chem.* **2012,** *14* (2), 509–513.

Zaccheria, F.; Psaro, R.; Ravasio, N. Selective Hydrogenation of Alternative Oils: A Useful Tool for the Production of Biofuels. *Green Chem.* **2009,** *11* (4), 462–465.

Zakrzewska, M. E.; Bogel-Łukasik, E.; Bogel-Łukasik, R. Ionic Liquid-Mediated Formation of 5-Hydroxymethylfurfural: A Promising Biomass-Derived Building Block. *Chem. Rev.* **2011,** *111* (2), 397–417.

Zhang, J.; Liu, X.; Sun, M.; Ma, X.; Han, Y. Direct Conversion of Cellulose to Glycolic Acid with a Phosphomolybdic Acid Catalyst in a Water Medium. *ACS Catal.* **2012,** *2* (8), 1698–1702.

Zhang, L. X.; Yu, H.; Yu, H. B.; Chen, Z.; Yang, L. Conversion of Xylose and Xylan Into Furfural in Biorenewable Choline Chloride–Oxalic Acid Deep Eutectic Solvent with the Addition of Metal Chloride. *Chin. Chem. Lett.* **2014,** *25* (8), 1132–1136.

Zhang, X.; Li, H.; Li, X.; Liu, Y.; Li, X.; Guan, J. Glucose Aqueous Isomerization Catalyzed by Basic Ionic Liquids. *ACS Sustain. Chem. Eng.* **2019,** *7* (15), 13247–13256.

Zhang, Z.; Wang, Q.; Tripathi, P.; Pittman Jr, C. U. Catalytic Upgrading of Bio-Oil Using 1-Octene and 1-Butanol Over Sulfonic Acid Resin Catalysts. *Green Chem.* **2011,** *13* (4), 940–949.

Zhang, Z.; Zhao, Z. K. Microwave-Assisted Conversion of Lignocellulosic Biomass Into Furans in Ionic Liquid. *Bioresour. Technol.* **2010,** *101* (3), 1111–1114.

Zhong, M.; Li, X.; Chu, X.; Zuo, S.; Yao, C.; Li, Z. et al. Solar Driven Catalytic Conversion of Cellulose Biomass Into Lactic Acid Over Copper Reconstructed Natural Mineral. *App. Catal. B: Environ.* **2022,** 317. DOI; 10.1016/j.apcatb.2022.121718.

Zhou, X.; Xu, F.; Wu, Z.; Li, H.; Yang, S. Heterogeneous Prolinamide-Catalyzed Atom-Economical Synthesis of β-Thioketones from Bio-Based Enones. *ACS Omega*, **2019,** *4* (5), 8588–8597.

Zolfigol, M. A.; Yarie, M.; Baghery, S. Application of $\{[4,4'\text{-BPyH}][C(CN)_3]_2\}$ as a Bifunctional Nanostructured Molten Salt Catalyst for the Preparation of 2-Amino-4H-Chromene Derivatives Under Solvent-Free and Benign Conditions. *Synlett.* **2016,** *27* (09), 1418–1422.

CHAPTER 3

Eco-Friendly Products and Reagents

JAYESH P. BHATT[1], NEELU CHOUHAN[2], ANIL KUMAR[3], AJAY SHARMA[4], and RAMESHWAR AMETA[1]

[1]*Department of Chemistry, PAHER University, Udaipur, India*

[2]*Department of Pure & Applied, Chemistry, University of Kota, Kota, India*

[3]*Department of Chemistry, M.P. Govt. P.G. College, Chittorgarh, India*

[4]*Department of Chemistry, Govt. P.G. College, Sirohi, India*

ABSTRACT

There is an increasing demand of the society for more effective and comfortable products for use such as cosmetics, drugs, textiles, fuels, etc. Presently, it is fulfilled by some toxic chemicals also. These products are harmful to human health or may have some side effects when used for a longer period of time. Therefore, these should be replaced by less toxic or almost harmless compounds. Bioplastic, green fuel, green pesticides, green drugs, green detergents/surfactants, green dyes, green construction materials, etc. are good substitutes for presently available products. Biodiesel can be a good alternate of conventional fuels. Reduction of carbon dioxide to energy-rich products can solve the problems of energy crisis and global warming both.

3.1 INTRODUCTION

Our atmosphere is in a constant state of chaos/disorder and is never being static. Internal and external transformations in our planet Earth that is either made by nature or man bring drastic changes in weather pattern, availability of natural resources, and living conditions. Scientific evidences highlighted

Green Chemistry, 2nd Edition: Fundamentals and Applications. Suresh C. Ameta & Rakshit Ameta (Eds.)

the role of man in environmental degradation as a result of industrialization exploiting mother Earth insanely. Industrial revolution put momentum to improve the quality of human life at many levels by improving global health, decreasing death rate, scientific gadgets that made our life more comfortable, mode of convenience that saves our time by keeping our comfort zones intact, mode of entertainments to relax us, carries our voices over telephone lines or through cell phones to keep in touch with dear ones or to expand business activities, quality of potable water, high production of varieties of agriculture products, that is, corn, wheat, and rice, paired with fertilizers and pesticides, dramatically increased the food stuff, the lights to enlighten our lives, advancement in satellite research telling us about when and where natural disasters (hurricane, earthquakes, etc.) might strike and with how much impact, and many more, all have been brought to us from courtesy of science.

One can say, science and technology have become an essential and integral part of daily lives of masses, as it serves us 24 × 7 in a variety of avatars and eventually or gradually transforms our life style. Hence, the importance of nations in the 21st century shall be judged not by their economic strength alone, but also by their power to conceptualize inventions and bring their benefits to people taking care of the environment. Unfortunately, this development generates a remarkable gap between human race and nature. In the past few decades, the size of this gap is gaining heights day by day. Until recently, almost negligible or no effort has been made to bridge this gap. All the hi-tech progresses of the human race done during the past few decades have provided us an easy lifestyle that is full of modern facilities at the very high cost of resources consumption and environmental degradation. A great turmoil in the delicate balance of our ecological system originated the global warming and its associated climate changes, increasing ocean temperature, changes in terrestrial geography, scarcity of drinking water, rainfall ratio, temperature, type of soil, etc. Tremendous critical issues such as the extinction of rare species of flora and fauna from Earth, various incurable or semi-curable diseases, acid rain, ozone layer depletion, excessive pollution, nuclear winter, and photochemical smog, especially in and around the urban areas, arise due to these changes that have proved to be dangerous to global life.

Out of these, fuels are the growing concern of masses, and Governments are the most eminent need of the time, that is, a persistent action for environmental protection. In this reference, the great Chinese Philosopher Mencius (20,000 years ago) observed that "Refraining from overfishing will ensure fishing last forever" and "Cutting woods according to the season ensures the health of the forest" are the means to achieve good harmony between man and nature. In this context, the words of Elsa Reichmanis, the former President of the American Chemical Society, were appropriate, "The days are passed, when we can trade environmental contamination for economical prosperity that is only a temporary bargain and now the cost of pollution on both ecosystem and human health is too high" (Ritter, 2003). Even then, we still do not wish to quit the comfortable lifestyle, but simultaneously, we cannot afford to continue along this path. Therefore, it is high time to rethink and set our accountability and shake hands with nature to satisfy our livelihood's demands in an eco-friendly manner. Although, we cannot single-handedly cleanup our environment, but we can make choices in our everyday life that will benefit the health of the planet and our community. The role of the policymakers also became very crucial in this reference. They have to pay serious attention to the various burning issues that arise due to the modernization, it includes developing research agendas driven by social challenges, engaging citizens through building constituencies, and cultivating scientists with a clear sense of civic responsibility. Therefore, Governments, industrialists, and researchers have to put their heads together on this leading issue with their careful concerns about the challenges of modernization and renewed the interest in the development of eco-friendly alternatives using breakthrough concepts and accelerated application of cutting-edge scientific, engineering, and analytical tools.

Simple changes such as finding green alternatives for the everyday products we consume can have a beneficial impact on water and soil quality, reduce energy uses and the amount of pollution/waste by using breakthrough concepts (catalysis by design, biodegradable consumer products) and accelerated application of cutting-edge green technology and products. With the customers getting gradually concerned about the environment-friendly products, the world market is now drifting more and more toward the recyclable/decomposable home appliances, hardware equipment, and daily life products—(i) Biodegradable bicycle: Marco Facciola has created a bike that is entirely made of glue and wood, (ii) Sony presented the world's first eco-friendly camera Odo. This camera is made of biodegradables and does not need any batteries. The power is produced by kinetic means, and (iii) Chinese mobile company Je-Hyun Kim has designed an eco-friendly mobile phone that is half recyclable and half biodegradable. The keypad and the screen for the mobile phone are made to be recycled after usage, while the rest of the mobile phone is made of grass-like carbonic components, making it fully decomposable.

Efforts of governments, industrialists, and researchers have been focused primarily to reducing the impact of toxic chemicals and exposure-based initiatives. To meet the goals of pollution prevention, few important legislative decisions come in the form of Water Act 1974, Air Act 1981, Environmental Prevention Act 1986, Hazardous Waste Rules 1989, Pollution Prevention Act 1990, Coastal Regulation Zones 1991, Biomedical Waste Rules 1998, Rules for Recycled Plastics 1999, Fly Ash Notification 2000, Municipal Solid Waste Rules 2000, Battery Mgt and Handling Rules 2000, Environment Protection (Amendment) Rules 2012, etc. One of its important initiatives is "Designing Safer Chemicals." It means designing a chemical that will not affect adversely the normal biochemical and physiological process of any organism. However, considering the complex, diverse, and dynamic nature of living organisms, in practice, this becomes a formidable challenge. The design of safer chemicals will require the ready availability of data and information on the relationship between chemical structure and industrial/commercial function. Chemical toxicity may be reduced by isosteric replacement of a carbon atom and toxic substances (e.g., DDT, etc.) can be redesigned such that they will retain their commercial efficacy but will decompose rapidly under physiological conditions to innocuous and readily excretable products.

Chemical substances that are toxic because of their ability to persist in the environment can be redesigned such that they biodegrade readily. Either technical, economical, and commercial feasibility of designing safer biocides, paint constituents or carcinogenic properties of many commercial aromatic amines (e.g., benzidines, anilines) can be greatly reduced by molecular modifications that facilitate excretion in the urine or prevent bioactivation. These initiatives represent a new approach to designing chemicals that emphasizes safety to human health and the environment as well as the efficacy of use. Implementing the concept will require major changes in the current practices of all social, academia, and industry regimes. To accomplish these changes, the concept must be understood, accepted, and practiced by all those associated with the development, manufacture, and use of industrial chemicals.

3.2 BIOPLASTICS

Plastics are the most demanded materials as they played an important role in food packaging and biomedical applications. Across the nations, there has been a public outcry for action to ban the plastic polybags to get rid of the almost nonbiodegradable plastic waste. In sunlight, a small portion of plastic waste degrades into toxic parts that contaminate soil and water. Accidentally, it can be ingested by animals and thereby enter into the food chain. The death of terrestrial animals, such as cow, buffalo was reported due to the consumption of such polythene carry bags. To the innocent marine and terrestrial life, polythene waste is recognized as a major threat, as it could be fatal for fishes, birds, and mammals. As per a report, nearly, 267 species are being affected in the marine environment due to this plastic pollution, which includes all mammals, sea turtles (86%), and seabirds (44%) (Derraik, 2002).

Bioplastics are a form of plastics derived from renewable biomass sources, such as vegetable fats and oils, corn starch, cellulose, biopolymers, or microbiota (Hong et al., 1999–2003). They have a high market potential because of their additional advantage of biodegradability in 10–15 years. Although currently plastics derived from petrochemicals constitute a sustainable alternative to conventional oil-based plastics, which degrade in 100–150 years. Common plastics rely more on scarce fossil fuels and produce more greenhouse gases. Bioplastics, which are designed to biodegrade can break down in either anaerobic or aerobic environments, depending on how they were manufactured? There are a variety of bioplastics being made. Aromatic polyesters are almost totally resistant to microbial attack, but most of the aliphatic polyesters are biodegradable due to their potentially hydrolyzable ester bonds. Naturally produced bioplastics are polyhydroxyalkanoates (PHAs) poly-3-hydroxybutyrate (PHB), polyhydroxyvalerate (PHV), and polyhydroxyhexanoate (PHH), renewable resource, polylactic acid (PLA), etc. Some common applications of bioplastics are packaging materials, dining utensils, food packaging, and insulation (Chen and Patel, 2012). Their production is expected to be more than 1.5 million tons per year in 2020. A novel and cost-effective polymerization technology has been developed to produce high-quality bioplastics with improved thermal stability up to 200°C.

Bilo et al. (2018) produced a new bioplastic from rice straw, an agricultural waste. This exhibited good mechanical properties. It is completely decomposed and embedded in soil for 105 days. Depending on the environmental humidity, the as-prepared material exhibited a dual mechanical behavior, which could be exploited to obtain shrink sheet and films or to drive shape memory effect. Ramakrishnan et al. (2018) synthesized a bioplastic. They used keratin from chicken feathers for this purpose. It was reported that this bioplastic (with 2% of glycerol) exhibited good thermal and mechanical properties. Apart from this, all these bioplastics are biodegradable.

Bioplastic fibers have been fabricated by Vinod et al. (2019) via electrospinning method. They used gum arabic (GA) for preparing these fibers.

Untreated and γ-ray-irradiated GA bioplastics show hydrophilic nature. They evaluated various properties, such as biodegradability water resistance, antibacterial properties, antioxidant potency, gas barrier attributes, and food contact migration through as-prepared all the GA bioplastic fibers (untreated, γ-ray-irradiated, and plasma-treated). It was reported that they have a potential as a viable option in food packaging (greener), environmental, and medically related products.

The bioplastics consisting of soy protein were prepared by Yamada et al. (2020). These bioplastics were found stable in water. It was reported that the bending strength of as-prepared bioplastics was similar to polyethylene. There was approximately 30% weight loss in this bioplastic on incubation for 6 days. Biodegradable and renewable materials obtained from biomass have the potential to replace nonbiodegradable petrochemical plastics. Lignocellulosic resources (wood) have been used by Xia et al. (2021) for the synthesis of a high-performance bioplastic. It was found that the as-prepared lignocellulosic bioplastic exhibited excellent water stability, ultraviolet light resistance, high mechanical strength, and improved thermal stability. It has a lower environmental impact because it can be easily recycled or biodegraded in nature.

3.3 GREEN FUEL (HYDROGEN)

Light fuel hydrogen has attracted the great attention of environmentalists, scientists, and industrialists as a benign fuel of the future because of its capability to produce pollution-free energy (no carbon emission and useful by-product of hydrogen fuel combustion that is only water) with highest energy density, that is, values per mass of 140 MJ kg^{-1}, is the beauty associated with hydrogen fuel. Hydrogen fuel is quite contemporary and relevant in the present energy scenario, that is "We are at the peak of the oil age but the beginning of the hydrogen age. Anything else is only an interim solution. The transition will be very messy and will take many technological paths, but the future will be hydrogen-fuel cells" (Chouhan et al., 2012). The majority of hydrogen used in industries is derived from fossil fuels or by the cleavage of water. Currently, the majority of industrial hydrogen need is satisfied from conventional sources (coal, oil, and natural gas), which contains about 10% CO_2 with hydrogen gas, and only 4% of H_2 comes from electrolysis, which is the low-cost method to generate hydrogen (US$ 3.51 per kg). The success of hydrogen technology will depend on the efficient generation of hydrogen from water cleavage powered by renewable sources (such as solar/wind). These are the important steps necessary for overall water cleavage (eqs. 3.1–3.4).

i) Absorption of light near the surface of the semiconductor creates electron–hole pairs (Chouhan and Liu, 2012).

$$\text{Photocatalyst} + h\nu \rightarrow \text{Photocatalyst } (e^-) + \text{Photocatalyst } (h^+) \quad \ldots(3.1)$$

ii) Holes (minority carriers) drift to the surface of the semiconductor (the photoanode), where they react with water to produce oxygen.

$$2\,h^+ + H_2O \rightarrow \tfrac{1}{2}\,O_2\,(g) + 2\,H^+ \ (1.23\text{ V vs. NHE at pH} = 0,\ +0.82\text{ V vs NHE pH} = 7) \quad \ldots(3.2)$$

iii) Electrons (majority carriers) are conducted toward a counter metal electrode (typically Pt), where they combine with H^+ ions in the electrolyte solution to generate H_2 :

$$2\,e^- + 2\,H^+ \rightarrow H_2\,(g)\ (0.00\text{ V, vs. NHE at pH} = 0, -0.41\text{ V vs. NHE pH} = 7) \quad \ldots(3.3)$$

iv) Transport of H^+ from the anode to the cathode through the electrolyte completes the electrochemical circuit. The overall reaction is

$$2\,h\nu + H_2O \rightarrow H_2\,(g) + \tfrac{1}{2}\,O_2\,(g)\ (1.23\text{ eV}, \Delta G^0 = 237\text{ kJ mol}^{-1}) \ldots(3.4)$$

The use of hydrogen as an energy carrier has enough potential to reduce energy dependence on gasoline and also reduces pollution and greenhouse gas emissions. Therefore, advancement in hydrogen technologies is under progress. The hydrogen economy can provide renewable energy solutions to the energy crisis, such as emergency backup power, heating, and electricity for commercial and residential purposes and hybrid electric vehicles (no matter whether storage of hydrogen still remains a "critical path" barrier, and it is one of the primarily focused areas). Market transformation activities aim to promote their adoption in stationary, portable, and specialty vehicle applications, such as forklifts, municipal vehicles, lawn mower (Yvon and Lorenzoni, 2006), municipal supply of clean fuel hydrogen to run electrical and other utility appliances (heater, air conditioner, fan, etc.). Hydrogen burner is one more interesting example of portable hydrogen appliance that can be used indoors safely without risk and the reaction product is water, which is actually beneficial for the room climate.

A panel of evaluators who confirmed hydrogen-roasted meat was undistinguishable in taste from that of propane-roasted meat. Likewise eco-friendly portable hydrogen house is another use of PEC hydrogen, where hydrogen is used as an energy source for most of the appliances, for example, gas engine-driven generators for the seasonal compensation (Wingens et al., 2008). Another most promising application of hydrogen is that produced by PEC cells in micro-combined heat and power (CHP) supply of residential houses (Hollmuller et al., 2000) and exploiting both gas-engine generated electric power and waste heat (Matics and Krost, 2007). Hydrogen-driven city buses are some commercial means of transportation, which utilize PEC hydrogen. Highly efficient clean fuel hydrogen is the most famous fuel for spacecraft. In 1990, the world's first solar-powered hydrogen production plant (research and testing facility) became operational at Solar-Wasserstoff-Bayern in southern Germany. In 1994, Daimler Benz demonstrated its first NECA I (New Electric CAR) fuel cell vehicle at a press conference in Ulm, Germany. In 1999, Europe's first hydrogen fueling stations were opened in the German cities of Hamburg and Munich.

Agrafiotis et al. (2005) synthesized iron oxide-based redox materials. These are capable of operating under a complete redox cycle, that is, they take oxygen from water producing pure hydrogen at 800°C and regenerate at 1300°C. A series of Cd1-xZnxS (x = 0–0.92) photocatalysts were prepared by Xing et al. (2006) via coprecipitation followed by calcination. The band gaps of such photocatalysts were found to be in the range of 2.20–3.12 eV. These can produce hydrogen in the presence of Cd1-xZnxS by photocatalytic splitting of water under ultraviolet and visible light irradiation. It was revealed that the photocatalyst (Cd0.62Zn0.16S) exhibited the highest rate of hydrogen evolution with a quantum efficiency of 2.17 and 0.60% under ultraviolet and visible light irradiation, respectively.

Abanades and Flamant (2006) used a thermochemical cycle for the production of H_2, which is based on CeO_2/Ce_2O_3 oxides. It proceeds in two chemical steps: (i) reduction, and (ii) hydrolysis. In this method, only pure hydrogen is produced and there is no impurity of carbon products, such as CO_2 and CO_2. A process analysis of Fe_3O_4/FeO, ZnO/Zn, and Fe_2O_2/Fe_3O_4 thermochemical cycles on large-scale potential with high efficiency, and eco-friendly routes have been suggested by Charvin et al. (2008) to produce hydrogen by concentrated solar energy. The thermochemical process was coupled with a solar tower plant and an economic assessment was made for the cost of H_2 production.

TiO_2–ZnO-mixed oxides photoconductors (1.0, 3.0, 5.0, and 10.0 wt.% Zn) were prepared by Pérez-Larios et al. (2012). They used these oxides for the production of H_2 from water splitting. High specific surface areas (85–159 m^2g^{-1}) could be obtained in all the samples as compared with bare TiO_2 (64 m^2g^{-1}). Their band gaps were found to be in the range of 3.05–3.12

eV. These solids can generate H_2 (1300 μmol h^{-1}) by photocatalytic water splitting as compared with TiO_2 (190 μmol h^{-1}), which is more than six times.

Zhu et al. (2015) reduced black TiO_2 by CaH_2 with 81% solar absorption at low temperatures (300, 400°C). The as-prepared TiO_2 (black) exhibited higher photocatalytic properties and an excellent performance in photoelectrochemical water-splitting, which was 1.7 and 4.5-fold more than that in the presence of pristine TiO_2, respectively. The water-splitting reaction was investigated for H_2 production by Tentu and Basu (2017) using a thermochemical cycle and ferrites. Among all the prepared samples of ferrites, Ni-ferrite was found to be the most suitable photocatalyst for H_2 production. It was reported that $NiFe_2O_4$ produced an average of 0.442 cm^3g^{-1} cycle of H_2. It was indicated that this is an excellent material in terms of structural durability and stability.

Landman et al. (2020) fabricated a PEC system (separate cell) with decoupled oxygen and hydrogen cells for the production of hydrogen. They used 100 cm^2 hematite (α-Fe_2O_3) photoanodes and nickel hydroxide/nickeloxyhydroxide electrodes as redox mediators. It was demonstrated that such a PEC water splitting system (decoupled) can have the potential for the generation of hydrogen with separate oxygen and hydrogen cells.

A photocatalytic fuel cell system with a green supercapacitor with a chemical bias has been fabricated by Zhang et al. (2021). It was suggested that this system could achieve a regular production of hydrogen under illumination (32 mol L^{-1}) and in dark (13 mol L^{-1}) with improved efficiency of the photoelectrocatalytic conversion. This system can generate electricity with an appreciable degradation capacity of ethylene glycol in the presence of light.

3.4 GREEN PESTICIDES

Worldwide one-third of harvest losses are from weeds, diseases, and insects, which were gaining heights day by day. Every US$ 1 spent by farmers on pesticides, saved US$ 3–5 from crop loss. But, the role of modern pesticides (organochlorines such as DDT, chlorobenzene, etc.) ChE inhibitors, organophosphates, carbamates, phenoxyherbicides, pyrethroids, bromine-based, phenol derivatives, dipyridyl derivatives, etc. are hazardous to our human health and ecosystem. As most of them are nonselective, endocrine disrupter, reproductive toxins, neurotoxins (lindane), CNS toxic, increases hepatocellular tumors, kill more than just the target, persistent and move around in the environment. They can cause cancer (lymphoma, leukemia, brain/lung/

testicle/breast), sterility, higher rates of miscarriage, greater risk in children with birth defects/stunted limbs, immune system suppression, a potential link to Parkinson's, etc. Green pesticides are derived from organic sources, which are considered environmentally friendly and causing less harm to human and animal health, and to habitats and the ecosystem (Lai et al., 2006). Pesticides or biocides include germicides, antibiotics, antibacterials, antivirals, antifungals, antiprotozoals, and antiparasites, which come in the form of sprays and dusts. Biopesticides are generally safer than synthetic pesticides, but they are not always more safer or environmentally friendly than synthetic pesticides. Likewise, natural pesticides, including sulfur, rotenone, mixture of copper, lime and water, nicotine sulfate, strychnine, arsenic-containing pesticides, and pyrethrums are used as pesticides from ancient times, but most of them are banned in organic farming.

Phytoalexin elicitor glucohexatose has been called a green pesticide (Ning et al., 2003) along with a new class of insecticides called spinosad, which shows a remarkable selectivity in destroying harmful pests and leaving beneficial insects alive. Few of the insecticides are Wormwood extract, Chive extract, Summer tansy dust, Stinging nettle extract, Daffodil extract, Garlic extract, Rhubarb extract, Onion extract, Sambucus extract, Tobacco extract, Stale beer, Sulfur (organic fungicide, pesticide, and acaricide), Bio-S (sulfur mixture), Pilzvorsoge, Spruzit, Carbolineum, basalt dusting powder, gene silencing pesticide, and steam (thermal pest control). For the past few decades, there has been a considerable research interest in the area of natural product delivery using particulate system for controlling plant diseases. The secondary metabolites in plants have been used in the formulation of nanoparticles, increasing the effectiveness of therapeutic compounds used to reduce the spreading of plant diseases, while minimizing the side effects for being a rich source of bioactive chemicals, biodegradable in nature, and non-polluting (eco-friendly). There are myriad of nanomaterials, including polymeric nanoparticles, iron oxide nanoparticles, gold nanoparticles, and silver nanoparticles, which can be easily synthesized and exploited as pesticide. Promising results of ZnO nanoparticles' antibacterial activity suggest its usage in food systems as a preservative agent (Tayel et al., 2011). Inhibition effects of 100 ppm silver nanoparticles against powdery mildew disease was reported on Cucumber and Pumpkin (Lamsa et al., 2011).

The efficiency of sulfur nanoparticles (SNPs) was evaluated by Rao and Paria (2013) against two phytopathogens, *Venturia inaequalis* and *Fusarium solani*. It was observed that small-sized particles (35 nm) were quite effective in preventing fungal growth. Feng and Zhang (2017) reported that methyl benzoate (MB) may have toxicity against various stages of a number of insect pests, such as *Halyomorpha halys*, brown marmorated stinkbug, diamondback moth, *Manduca sexta, Plutella xylostella, Drosophila suzukii* and tobacco hornworm. This is at least five to twenty times more toxic than the other conventional pesticide. As it is considered environment-friendly, it has a great potential for the management of insect/pest in crop production.

Patzke and Schieber (2018) used five phenolic compounds (ferulic acid, phlorizin, 5-n-alkylresorcinols, resveratrol, and quercetin) as active ingredients for preparing a bioactive emulsion. This was then screened for its ability to inhibit the growth of four phytopathogenic fungi *Fusarium culmorum, Penicillium expansum, Botrytis cinerea*, and *Aspergillus niger*. Out of these, ferulic acid was found to be highly effective against the growth of *Botrytis cinerea.* The toxicity of essential oil lemongrass (*Cymbopogon citratus* DC. Stapf) was evaluated by Moustafa et al. (2021) against the black cutworm (*Agrotis ipsilon*) and it was revealed that lemongrass essential oil exhibited insecticidal activity against the black cutworm (*A. ipsilon*).

3.5 GREEN DRUGS

Pharmaceutical industries became the most dynamic sector of the chemical industries of 21st century as the sales of medicines and other pharmaceutical products have increased many folds from 1985. It has been estimated that the number of potential drug targets may be between 5000 and 10,000 out of the estimated 10^{60} possible compounds (Drews and Ryser, 1997). Their production process generates huge amount of waste, which typically included volatile organic solvents and other hard-to cleanup agents, as the volumes were daunting. According to the systematic data gathered by the Environmental Protection Agency (EPA), only in the United States, around 278 million tonnes of hazardous waste were generated in 1991 at more than 24,000 sites.

More effective, efficient, and elegant solution to this problem in the words of Anastas et al., 2001 is "Simply better chemistry" or "Green chemistry." Hence, pharmaceutical sector has welcomed green chemistry very warm-heartedly, perhaps because no company can afford to ignore green chemistry's potential savings in terms of money and eco-health. Typical pharmaceutical plants generate 25–100 kilograms of waste per kilogram of product, a ratio known as the environmental factor, or "E-factor," in comparison to their other industrial counterparts like oil refineries (E-factor < 0.1), bulk chemicals

(E-factor < 1–5), and fine chemicals (E-factor ~ 5–50). Consequently, there is plenty of room to increase the efficiency and cut costs. Re-drug designing in such a way that drugs must have a degradable chemical structure with acceptable safety profile, stability in synthesis, formulation, storage, and use is welcome.

The technical advancement such as the introduction of apt catalyst (Nagendrappa, 2002), microwave heating (Liu and Zhang, 2011), ultrasonic energy (Mason, 1997; Cintas and Luche, 1999; Cravotto and Cintas, 2006), combinatorial chemistry (Weller et al., 2006), recycling of by-products or solvents, use of solvent-free or mild solvents, biocatalyst, use of ionic liquids in place of conventional solvents (Martínez-Palou, 2007), offers simple, clean, fast, efficient, and economic synthesis of a large number of pharmaceutical products, which provide enough momentum for many chemists to switch from traditional to green chemical pathway. To catalyze the implementation of green chemistry and engineering in the pharmaceutical industry globally, there is an immense need for modification in current (conventional) manufacturing practices.

Ibuprofen was discovered by Adams and Nicholson (1960) and it was commercially synthesized by Boots in 1960 as an eminent case study, which belongs to the category of analgesic, nonsteroidal anti-inflammatory drug with very high sales. The conventional synthesis of ibuprofen was performed in six steps and it has many disadvantages, under the "green" principles such as low production of secondary by-products and waste, very poor atom economy, and 40% final yield. Afterward, a new synthetic route with only three steps and increased efficiency was discovered for BHC. In both the synthetic routes, the starting chemical is 2-methylpropylbenzene, a product of petrochemical industry. But in this new innovated synthetic approach, 77% yield was obtained and the catalyst used is Raney nickel (Ni/CO/Pt), which can be recycled; thus, decreasing the substantially of the steps (Cann and Connelly, 2000).

Similarly, Simvastatin, the second best-selling drug for treating high cholesterol, was manufactured from a natural product (lovastatin). The traditional multistep synthesis was wasteful and used large amounts of hazardous reagents (Alberts et al., 1980). Xie and Tang (2007) conceived a single-step synthesis using an engineered enzyme and a practical low-cost feedstock. Furthermore, Codexis (LovD) optimized both; the enzyme and the chemical process. The resulting process greatly reduces E-factor and generates low hazardous waste, cost-ffective and meets the needs of customers.

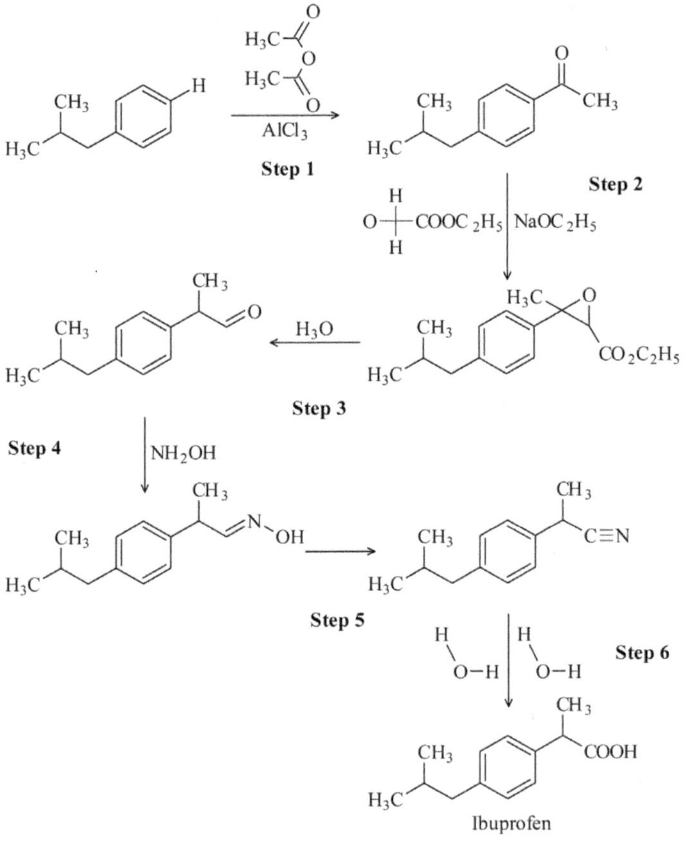

Conventional route

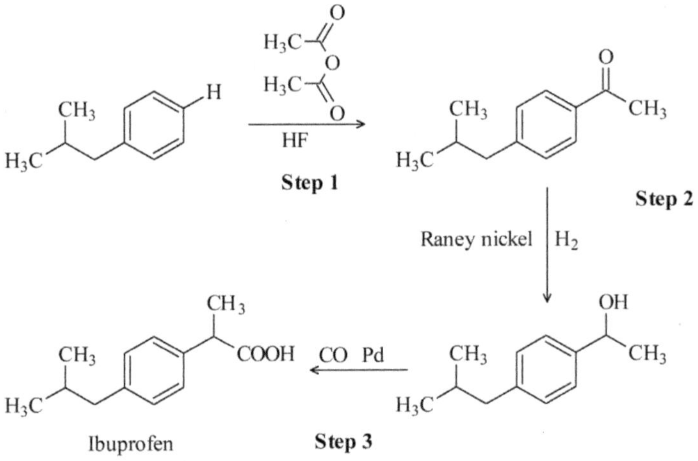

Geen chemical route

In this context, drug maker Pfizer, rigorously re-examining every step of the synthesis of anti-impotence drug sildenafil citrate (Viagra) had an E-factor of 105. They have replaced all the chlorinated solvents with less toxic alternatives, and then introduced measures to recover and reuse these solvents. The need to use hydrogen peroxide was also eliminated, which can cause burns. They also eliminated any requirement for oxalyl chloride, a reagent that produces carbon monoxide in reactions, a major safety concern. Eventually, later, Pfizer's team cut Viagra's E-factor to 8 (Dunn et al., 2004). After that success, Pfizer has also reduced the E-factor of the anticonvulsant pregabalin (Lyrica) from 86 to 9 by biocatalytic mechanism with low (~0.5%) protein loading, conducting all four reactions in water, resolution at first step (wrong enantiomer can be recycled). They have also made similar improvements for the antidepressant sertraline and the nonsteroidal anti-inflammatory celecoxib. These three drugs altogether eliminated more than half a million metric tons of chemical waste (Sanderson, 2011).

Microwave heating has proved its efficiency in dramatic reduction of reaction time, and high-yield product formation almost solventless and these are potentially important factors in drug discovery. Various advanced techniques have been introduced to microwave synthesis, such as click chemistry (Simone et al., 2011), green chemistry (Kümmerer, 2010), multicomponent reactions (Murray et al., 2005), combinatorial synthesis (Lin et al., 2011), parallel synthesis (Zhou et al., 2010), and automated library production (Hsiao et al., 2010) to increase the output of pharmaceutically active chemical entities. All these techniques and strategies hold their distinguished advantages as well as shortcomings.

On the other way, microwave heating has proved its efficiency in dramatic reduction of reaction time, which is potentially important in drug discovery. Therefore, one of the ideal options to accelerate the synthetic processes is to combine microwave chemistry with these techniques in drug discovery. That is why many pharmaceutical companies are incorporating microwave chemistry into their drug discovery efforts (Kappe and Dallinger, 2006). Tryptamines were synthesized by the reduction of glyoxalylamide precursors with lithium aluminum deuteride via microwave-enhanced single-mode system under elevated pressure, anhydrous tetrahydrofuran as solvent at 150°C for 5 min (Brandt et al., 2008). Few microwave-assisted green drugs, such as the psychoactive drug N,N-dialkylated tryptamines, dihydropyrimidinones, and dihydropyrimidinethiones as antibacterial, antiviral, antihypertensive, and anticancer agents (used in AIDS therapies) can be produced with improved atom economy under microwave heating and fluorous solid-phase extractions (F-SPE) (Piqani and Zhang, 2011).

Microwave-enhanced efficient and convenient, combinatorial method was developed for the preparation of 2,4-(1*H*,3*H*)-quinazolinediones and 2-thioxoquinazolinones by substituting methyl anthranilate with various iso(thio)cyanates in DMSO/H_2O without any catalyst or base in 20 min (Li et al., 2008).

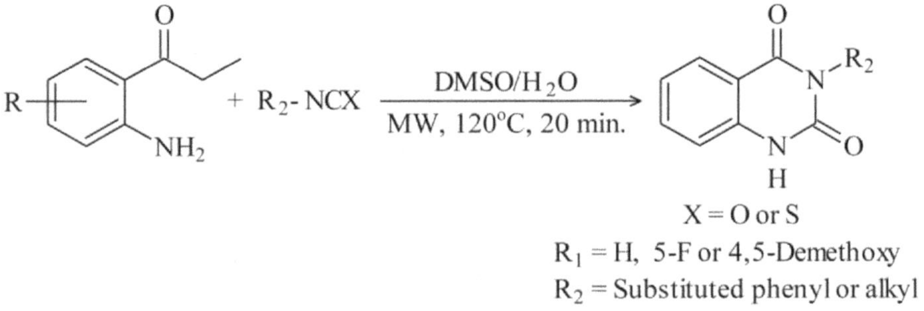

Green microwave synthetic approaches to diverse fused tricyclic xanthines have been reported as anticonvulsants to treat chemically induced seizures. It has good atom economy and high functional group tolerance (Ye et al., 2009a).

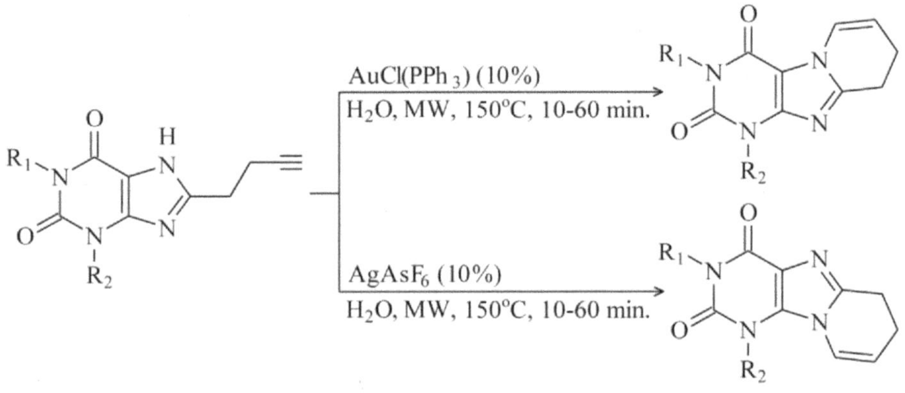

Biologically interesting compounds indole-1-carboxamides were prepared in moderate to high yields in 5 min by Au (I) catalyzed 5-endo-dig cyclization in water under microwave irradiation (Ye et al., 2009b). Microwave technology can also address the challenges of the rapid labeling of radiopharmaceuticals (Jones and Lu, 2006). New HIV-1 protease inhibitors with more potency (56 times) and excellent antiviral activities were obtained by introducing microwave irradiation to accelerate Stille or Suzuki cross-couplings at 120°C in 30–50 min.

The manufactured drugs have ultimately come to the waterway. The occurrence of a large number of pharmaceuticals and personal care products (PPCP) in the environment is a serious multifaceted issue as many of which are highly bioactive and perpetually present in water locales that badly affects the surface water and groundwater quality. So far, about 100 different pharmaceuticals have been detected in the aquatic resources, usually in STW effluents (Vanderford and Snyder, 2006; Batt et al., 2008; Kasprzyk-Hordern, et al., 2008). Those pharmaceuticals cover many different therapeutic classes, including analgesics, beta-blockers, selective serotonin reuptake inhibitors (SSRIs), fibrates, antiepileptics, and steroids. Indirect continuous exposure of multiple PPCP to humans and the environment (no matter how much is low dose?), needs implementing of a wide range of proactive actions in near and long term to minimize the introduction of these PPCP into our ecosystem.

If we still do not realize the risk paradigm that arises from the magnitude of this issue and the size of the problem, then the future generations will not forgive us. Although, there are very few evidences of these pharmaceuticals in the environment, which are visualized to result in acute effects such as organ damage. There are two cases, where drugs have had drastic effects (i) where the anti-inflammatory drug diclofenac has virtually wiped out the 97% of the vulture population of Asia in 2004 (Rattner et al., 2008) and (ii) steroid estrogens, both natural (e.g., estradiol) and man-made (ethinyl estradiol; EE2), which has caused feminization of male fish worldwide in surface water receiving domestic wastewater treatment plant effluent (Sumpter and Johnson, 2008). Similarly, the presence of the most popular analgesic drug ibuprofen show antibacterial activity in water; furazolidone-medicated fish feed causes acute toxicity on crustaceans/copepods; antibiotic drugs streptomycin and chlortetracycline/oxytetracycline prevent the growth of blue-green algae and pinto beans, respectively.

Tinidazole prescribed for protozoa infections results in bacterial mutagens in the urine of patients in treatment. The volume of pharmaceuticals that are used is expected to continue increasing worldwide as population density increases, per capita incomes rise, and new disease target groups as well as more potent compounds are identified (Jjemba, 2008). Therefore, the unintended consequences of pharmaceuticals in the environment cannot continue to be overlooked, thus minimizing their potential impact is an enormous task that is not going to be accomplished easily. It could also improve medical healthcare outcomes for customers and reduce healthcare cost.

Sincere research on this major thrust area is needed to improve the risk assessment from pharmaceuticals at different trophic levels. Drinking water is a cocktail of chemical compounds (more than 95 organic wastewater contaminants, including pharmaceuticals, personal care products, and other extensively used chemicals, such as detergent metabolites, insecticides). However, studies have only focused on the effects of individual substances. Therefore, the mixtures of substances in water could be studied with respect to their chronic exposure to low doses. Presently, there is no regulation for pharmaceuticals. Thus, there is an urgent need to set a national/area-wise primary drinking water regulation/policy for pharmaceuticals because it is highly important to make healthcare providers and patients made aware of the medical and environmental consequences of our medication practices, including overprescribing. It also aims at minimizing pharmaceutical use by creating awareness about the linkages between human health and ecological health.

Hou et al. (2021) reported the preparation of paracetamol via a biosynthetic pathway in *Escherichia coli*. It was also established that 4.2 g L^{-1} (27.7 mM) paracetamol can be obtained in 9 h (95% conversion rate) while using p-aminobenzoate as the substrate.

Borsoi et al. (2021) reported a new synthetic approach providing efficient access to a series of 4-alkoxy-6-methoxy-2-methylquinolines in 15 min using ultrasound energy. It afforded the target products at good yields (45–84%) and with high purity (≥95%).

3.6 GREEN DETERGENTS/SURFACTANTS

Synthetic detergents are surfactants or a mixture of surfactants, which lower surface tension in water and break down fatty materials. In other words, they decrease the fabric's hold on the dirt and they also dissolve the dirt particles. Detergents with cleaning properties in dilute solutions were manufactured in the United States from the early 1930s. The core components of the detergents were usually alkylbenzenesulfonates, sodium tripolyphosphate, etc. (Hampel and Blasco, 2002; León et al., 2006) in particular, which are obtained from petroleum. A benzene ring can be added to the tetramer of propylene dodec-1-ene (petroleum product) via Fridal-Crafts reaction in the presence of hydrogen chloride/aluminum chloride as catalyst. The dodecylbenzene is sulfonated by refluxing it with concentrated sulfuric acid to produce 4-dodecylbenzene sulfonic acid, which generates detergent after neutralization with sodium hydroxide.

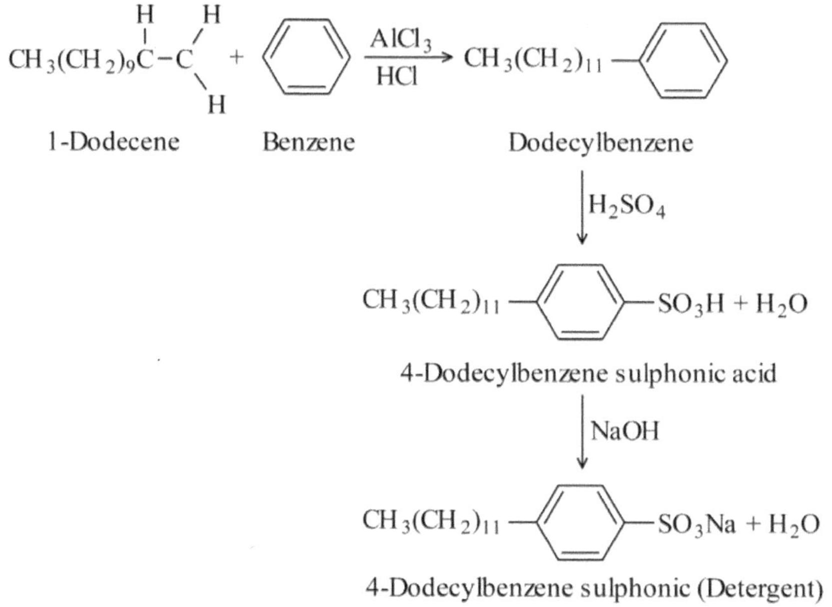

Water hardness is a significant factor of modern detergents. The harder is the water, the more detergent is required. The ions in hard water, particularly calcium and magnesium, bind to the surfactant components. So, manufacturers add builders, which bind to and essentially remove these ions and soften the water. Phosphates have been most commonly used as the builder in detergents that create freely available phosphates. However, excess phosphates can cause problems in our waterways, such as excessive toxic algal growth (blue-green algae) or decomposer organisms requiring oxygen may increase, which can deplete the amount of oxygen dissolved in the water. Therefore, some detergent manufacturers have developed phosphate-free detergents. As an alternative to phosphates, manufacturers can use a builder, or combination of builders, including zeolites (aluminosilicates), sodium citrate, and nitrilotriacetate (NTA). Detergent waste waters containing alternative builders also have environmental impacts and must be treated by sewage treatment process. Some of them (alkyl phenols) are estrogen mimics that can have serious detrimental effects on populations of aquatic animals, such as decreasing their ability to reproduce. Even after treatment, the environmental impacts of some alternative builders remain.

Most of the detergents, no matter, if they are with phosphate builder or not, found in the market are chemical-based and they are resistant to the action of biological agents (bacteria, the decomposers). It is very difficult to eliminate them from the municipal wastewaters. They can pose a serious problem to aquatic life. These problems are overcome by biodegradable detergents (Jan, 2007). The Australian standard for biodegradability of surfactants (AS1792) requires 80% of the mixture to be degraded within 21 days if the product is to carry the label "biodegradable." There is a growing market for green products now and these detergents are preferred by several environmental lovers. As they are effective, low cost, can remove all types of stains without damaging paint fastness, do not harm hands and fabric, nontoxic, and safe for the environment. Eco-friendly detergents are plant-based and use absolutely zero petrochemicals, ammonia bleach, animal ingredients (or testing), paraben, phosphates, dyes, artificial fragrances, etc. Instead, most of the ingredients would include water, natural salts, essential oils, and plant-based cleaners.

Vegetable oil-derived detergents contain straight-chain linear alkyl sulfonate (LAS), making them suitable for detergent production. Therefore, because of the availability and the environmental friendliness of these oils, detergents produced from neem seed oil (Ameh et al., 2010) (low surface tension as 0.00523 N/m, Foamability 3 cm per 10 min.) and castor oil (Isah, 2006) are preferred. There are reports on the possible anaerobic degradation of LAS under methanogenic conditions (Mogensen et al., 2003; Angelidaki et al., 2004; Løbner et al., 2005). The experiments were performed mainly with UASB reactors operating either in mesophilic (37°C) or in thermophilic (55°C) conditions. Removal of 40–80% of the initial LAS concentration was reported. Although, doubts still remain about the actual abatement of LAS through biological reactions under anaerobic conditions because of the LAS-induced inhibition of autotrophic nitrification in soil and water, which can be monitored by the account of ammonia-oxidizing bacteria (Nielsen et al., 2004). Moreover, synthetic surfactants or oil in marine aerosol magnify the toxic effects of pollutants (ozone, polyaromatic hydrocarbons, phthalates, n-alkanes, etc.) on plants, especially in the winter season, when degradation of LAS in seawater is considerably inhibited at lower temperatures (León et al., 2004). Some recent data indicate that an improvement of the coastal vegetation (in Spain) is observed in areas, where wastewater is adequately treated by wastewater treatment plants (WWTP).

Chu and Feng (2013) reported a green route for the preparation of vegetable-derived long-chain surfactants. They used bioresource-derived erucic acid (leftovers of rapeseed oil) as starting material without using any solvent and no chemical waste was produced. It was revealed that high-yield products could be obtained with lesser reaction time. These erucic acid-derived surfactants were found to be more eco-friendly as compared with surfactants via a saturated hydrophobic tail shorter than C18. It is all due to their lower dosages in practical applications and also due to the presence of the degradable unsaturated bond and amido group.

Brazilian coal fly ash was used by Cardoso et al. (2015) for the synthesis of zeolites 4A and Na–P1 by two different routes and that too with moderate operating conditions (temperature and pressure). It was revealed that synthetic zeolite 4A used builder in detergents exhibited conformity parameters equal to or more than commercial zeolite.

Alpha methyl ester sulfonate (α-MES) is an anionic surfactant, which is derived from palm oil-based methyl ester. It has lower manufacturing cost, higher tolerance to hard water, excellent biodegradability, good detergency with less dosage, and lower ecotoxicity as compared with linear alkylbenzene sulfonates (LABS). This palm-based α-MES was used as the sole surfactant in powder detergent by Low et al. (2021); however, there are still unsettled issues related to viscosity and phase stability in using it.

The use of green surfactants is welcome for the treatment of contaminated soils because it is biocompatible and efficient as compared with the synthetic surfactants (as additives), which have problems like toxicity and pile up of by-products. Three green surfactants and microbial surfactants were produced from the *Starmerella bombicola* ATCC 22214 (da Silva et al. 2021). These were used for the remediation of soil. The results were compared with a synthetic surfactant (Tween 80). It was reported that commercial biosurfactant formulation was quite effective in removing motor oil, particularly from beach sand (65.0 ± 0.14%) and contaminated sandy soil (80.0 ± 0.46%).

Umemura et al. (2021) reported the preparation of CNT suspensions with green detergents, which were prepared from bamboo and coconuts. Single-walled CNTs (SWCNTs) with a few carboxylic acid groups (35%) and pristine multiwalled CNTs (MWCNTs) were first mixed in each detergent solution and then sonicated with a bath-type sonicator. It was found that the stability of suspensions prepared with coconut detergents was superior as compared with that prepared with SDS. It was revealed that the use of these green detergents is associated with the advantage of dispersion of carbon nanotubes as well as SDS.

Yang et al. (2021) explored halloysite clay nanotube (HNT) with unique hollow structure as green, multifunctional, nontoxic, and low-cost detergent. HNT has a small size large aspect ratio and high adsorption ability, has a probability to remove various stains. HNT detergent was also found to be effective in removing stains of ink, and tea from textiles with a cleaning efficiency of more than 88%. It was reported that HNT can effectively eliminate chili oil from stainless steel, ceramic, plastic plate, and glass in the kitchen and can be used as kitchen detergent.

3.7 GREEN DYES

Until the discovery of the first synthetic dye mauveine, the plant and animal kingdom provided all the materials for dyeing, textiles, paints and cosmetics, and colored food stuffs for making food more attractive, visually. The use of synthetic dyes involves the use and release of large amount of hazardous chemicals in the environment during their production, subsequent use, and creating worker safety concerns. Petroleum is the starting material for all synthetic dyes. Thus, the price of dyes is sensitive to the petroleum price. Synthetic dyes are clearly superior with better fastness properties, good fixation, and shade reproducibility. Preparation of synthetic dyes has substantial drawbacks, such as relatively strenuous reaction conditions, namely, refluxing reactants for several hours in organic solvents. A lot of organic solvents are not friendly to the environment and the complexity of isolation of the products and their nonbiodegradability is also a problem. Recently, the awareness of the environment and increasing disputes about the risks of synthetic dyes (more than 4400 organic dyes are available) resulted in the growing interest in natural coloring agents for textiles, leather, plastic newspaper, magazines, decorative items, films, every day utility products, and to satisfy other human needs.

Presently, the dye industry has started paying great attention to newer products, which has a nice blend of fashion trends as well as environmental specifications. A recent ban on the use of azo dyes, due to their carcinogenic properties, has led all major dyestuff manufacturers to search for benign alternatives to toxic dyes. The safest way is to use the eco-friendly natural dyes. From the sustainable point of view, environmentally benign or natural dyeing stuff is becoming a top priority in the recent years because the medium in which these plant cells or fungi or bacteria grow contains no expensive or toxic chemicals. They require low processing temperature (around 30°C), neutral pH of synthetic process; they produce high yield with high purity and show their biodegradability. Along this line, dyeing using the extract of green tea is particularly of interest due to its antibacterial and UV-protection nature. Black tea (*Camellia sinensis* variety *assamica*) is used as a source of colorant to dye cotton and nylon fabrics.

Chitosan and commercially available cationic fixing agents to control the fastness and dye uptaking capacity were used as the potential substitute for the non-eco-friendly heavy metal salts. Chitosan and the cationic fixing agents are utilized as the linkers between tea catechin and cellulose of cotton and polyamide of nylon via the formation of hydrogen bonding and ionic bonding (Nuramdhani et al., 2011). Tannin extracted from the tamarind seed coat is employed as a natural mordant with 0.5–1% metal mordant mainly copper sulfate for dyeing cotton, wool, and silk fabrics using natural dyes, such as turmeric and pomegranate rind. This resulted in good antibacterial activity up to 20 washes (Prabhu et al., 2011). The cranberry fruits contain anthocyanin and flavonol pigments (Boulanger and Singh, 1998), which play an important protective function against the damaging effects of UV radiation on widely used polyamide fabrics, Nylon-66 and wool (Takahashi et al., 1991). Canadian golden rod plant (Bechtold et al., 2007a), barberry, madder, hollyhock, privet, walnut, sticky alder tree (Bechtold et al., 2003), ash tree (Bechtold et al., 2007b), *Hibiscus mutabilis* (Shanker and Vankar, 2007), *Terminalia arjuna, Punica granatum, Rheum emodi* (Vankar et al., 2007), *Coffea arabica L.* (Lee, 2007), *Garcinia mangostana L.* (Chairat et al., 2007), *Rhizoma coptidis* (Ke et al., 2006), curcumin (Fithriyah, 2013), *Rubia cordifolia* (Vankar et al., 2008), etc. were used as natural dyes for silk, wool, and cotton fabrics with different mordant. Natural dyes are extracted from some abundantly occurring plant materials of forest origin. These dyes may be used for imparting different shades on silk, wool, and cotton using common mordants, such as alum, salts of iron, tin, and chrome. Environmental-friendly syntheses, such as microwave irradiation, sonochemical/ultrasonic techniques are the powerful tools toward the organic synthesis of dyes. Solvent-free microwave irradiation is well known as an environmentally benign method, which offers several advantages including shorter reaction times, cleaner reaction profiles, and simple experimental/product isolation procedures. For example, styryl dyes are widely used as sensitizers, additives in the photographic industry, a biologically active compound in the pharmaceutical industry, and novel successful fluorescent probes in bioanalytical methods.

However, conventional synthesis of such dyes is often carried out by the reaction of 2- or 4-methyl quaternary salts and aromatic aldehydes at high temperatures for a relatively longer reaction time of 8 h. which usually leads to the formation of undesired side products, low yields, and considerable power consumption. Therefore, the environmentally benign procedure under solvent-free conditions and microwave irradiation in the presence of different basic or acidic reagents was adopted to synthesize the series of styryl dyes (Vasilev et al., 2008).

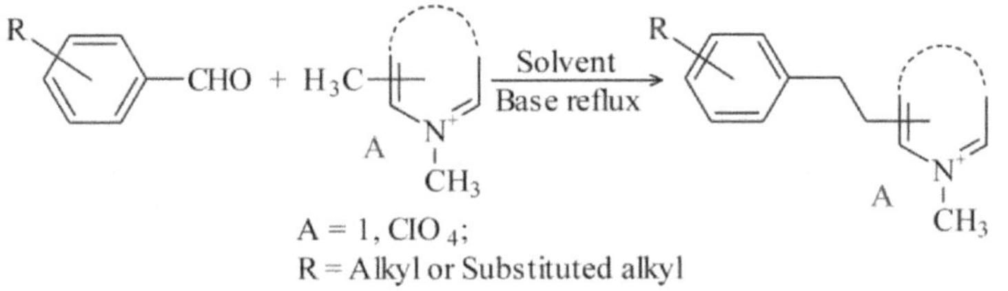

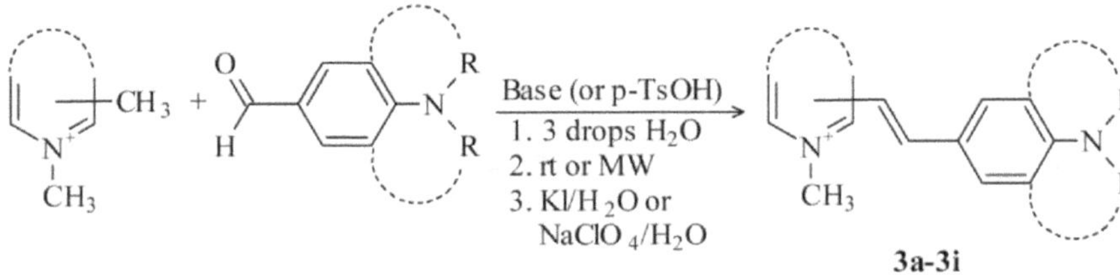

3a-3i

R = Alkyl, Y = 1; ClO_4

In this revolutionary way, environmentally benign yellow rare earth inorganic pigments, that is, $NaCe_{0.5}(MoO_4)$ (Sreeram et al., 2007), $Ce(MoO_4)_2$, and $Ce_{1-(x+y)}Zr_xTa_yO_2^{+\delta}$ (x ranges from 0.15 to 0.2 and y ranges from 0 to 0.05) (Vishnu et al., 2009) displaying colors ranging from white to yellow have been synthesized by designed yellow pigments consisting of nontoxic elements as alternatives to lead, cadmium, and chromium pigments.

Numerous efforts are made to avoid or reduce risks due to acute toxicity of dyes. Dye manufacturers and tanneries are confronted with effluents, wastes, and contaminated containers or packaging material that require carefully thought-out disposal. The contamination of soils and waters by dye-containing effluents is a serious environmental concern. Due to the increasing awareness and concern of the global community over the discharge of synthetic dyes into the environment and their persistence there, much attention has been focused on the remediation of these pollutants. Among the current pollution control technologies, the biodegradation of synthetic dyes by different microbes is emerging as an effective and promising approach. The biodegradation of synthetic dyes is an economic, effective, bio-friendly, and environmentally benign process (Ali, 2010). Synthetic dyes, such as bisphenol A, bromophenol blue, remazol brilliant blue R, methyl orange, relative black 5, Congo red, and acridine orange were decolorized by the effective enzyme *Trametes* laccase obtained from *Trametes polyzona* and the percentage of decolorization increased, when 2 mM HBT was added in the reaction mixture (Chairin et al., 2013).

Although Government, industry, and advocacy groups have taken significant actions to solve the related problems, including restricting the use of certain substances, the response remains inadequate. The EEC has promulgated at "EC Control of Substance Hazardous to Health Act, 1989" and published a red list enumerating a number of chemicals, the presence of which in any kind of fabric has been banned. An ordinance in Germany stipulates that no garment or any other article that comes into contact with the skin shall contain any of the 20 aromatic amines. Similar restrictions have been imposed in many other countries. In India, the ban on the use of azo dyes has been imposed by the Union Ministry of Environment and Forest under section 6(2) (d) of the Environment (Protection) Act, 1986 read with the Rule 13 of the Environment (Protection) Rules, 1986.

Bamoniri et al. (2014) synthesized a series of azo dyes by mixing aromatic amines and sodium nitrate in the presence of nano-silica supported boron trifluoride (nano $BF_3 \cdot SiO_2$). It was followed by diazo coupling with 1-naphthol under solvent-free conditions at room temperature. It was revealed that aryl diazonium salts supported on nano-$BF_3 \cdot SiO_2$ were found quite stable at room temperature for several months.

An attempt was made by Rather et al. (2016) for dyeing of wool using *Adhatoda vasica* extract as a natural dye. They could obtain a beautiful color palette of shades of varied tone and hue using different mordants. Shabbir et al. (2016) used *Terminalia chebula* (Myrobalan/Harda) natural dye extract for eco-friendly shades on woolen yarn. It was observed that different metal salts, such as ferrous sulfate, alum, and stannous chloride enhanced the fastness properties (wash, light, dry, and wet rubs) of dyed woolen yarn. It was revealed that when woolen yarn samples were pretreated with metal salts, then they exhibited enhanced better fastness and color strength properties.

Polyphenolic dyes were extracted from henna leaves, pomegranate rind, and *Pterocarya fraxinifolia* leaves (Ebrahimi and Gashti, 2016). The dyeing ability of these extracts on nylon 6 fabric was investigated. They used three compounds, such as aluminum sulfate, tin chloride, and tannic acid as mordants before dyeing. It was reported that mordants improved the fastness properties and increased the color strength of the fibers.

Hussaan et al. (2016) have carried out microwave-mediated extraction of natural colorants from the leaves of milkweed (*Calotropis procera* L.). The extraction from *C. procera* leaves (in alkali and aqueous medium) and cotton fabrics were irradiated with microwaves for 2–10 min. It was revealed that better extraction was observed when microwave radiations were applied for 4 min in alkali as compared with aqueous solution. Among the chemical mordants, iron was found to be effective for better color strength, while The bio-extract of *Acacia nilotica* bark improved the fastness properties and color strength as pre-mordant and *Curcuma longa* tuber as post-mordant.

The dyeing ability of anthraquinone colorants was observed by Yusuf et al. (2017) with *R. cordifolia* Lin. roots extracts. They used $CaCl_2$ and $AlCl_3$ as metallic mordants. The optimum conditions for extraction and dyeing were pH = 2 and 4, time = 45–90 min at temperature 90°C, respectively. It was reported that radiant red shades can be obtained by madder root extract for dyeing of wool (with or without mordants) with fastness properties. Adeel et al. (2018a) evaluated the coloring potential of harmala (*Peganum harmala*) seeds. They were able to improve the color (strength) of dye on microwave exposure followed by a mordanting process. The coloring efficacy of Arjun bark (*T. arjuna*) has also been evaluated by Adeel et al. (2018b) for silk fabric in the presence of ultrasonic radiation. Extracts and silk fabrics were irradiated ultrasonically for 15–60 min at 60°C. They also observed that 9% of pomegranate and turmeric extracts (biomordants) used as pre- and post-biomordants showed excellent coloring properties.

After textile dyeing, hair coloring is another popular application of colors. Some experts estimate that between 65 and 75% of women dye their hairs, and it is believed that the practice goes all the way back to ancient Egyptians. Kingsley says that vegetable dyes do not adhere to the hair and cause more damage because they must be used more frequently, while more effective formula simply uses less irritating and without odor substitute involving different chemicals. Synthetic hair dyes contain ammonia, which binds the color (blue/green); p-phenylenediamine (PPD), which increases the risk of cancer and cell mutation that is linked to problems with the immune and nervous systems with increased risk of diseases such as non-Hodgkin's lymphoma parabens; hydrogen peroxide, which lightens the existing color; plastics, sulfites, sodium lauryl suflates, and other baddies, as common ingredients that causes allergic reactions, scalp sensitivity, and inhaled fumes can lead to respiratory issues. Healthier hair color is certainly on the minds of industry professionals, as even the best organic hair colors are going to have some chemicals that we may not like.

After all these notorious salons outcomes, a promising news is coming from research laboratories that a new hair bleach may be on the way, and it will be gentler for both; people and Earth. Japanese scientists are on their way to creating an eco-friendly hair bleach, whose roots are in fungus. An enzyme from a strain of forest fungus naturally breaks down melanin, the substance that gives hair its color. Compared with bleaching with hydrogen peroxide, this enzyme would probably be easier on hair and skin. But researchers are still pretty far from turning theories into a product. More studies are required to make this dream into a reality.

3.8 ECO-WAXES

Waxes are commercially produced at large scale for use in cosmetics, polishes, surface coatings, and many other applications. Waxes are mainly derived from three major sources. These are Jojoba, carnauba, and beeswax. The composition of the waxes varies greatly according to the plant species and the site of wax deposition (leaf, flower, fruit, etc.). Plant waxes usually provide a hydrophobic coating that reduces water loss and protects the surface. A greener approach to obtain waxes is to utilize the low cost and abundant by-products, such as cereal straw, carnauba, and sheep wool as a raw material and to use benign extraction techniques such as supercritical CO_2 for their selective removal from the plant matrix with lipids and pigments. Selection of the temperature and pressure of the supercritical CO_2 used in the extraction can vary their physical properties. Importantly, the waxes from straws have a microcrystalline structure and this property is important for many cosmetic uses. These waxes can be used as replacements for a wide range of existing products.

Tambe et al. (2016) observed that moisture resistance of a Kraft paper can be improved on coating with silylated soybean oil (cured via silanol condensation). This type of moisture barrier coating was developed using LowSat® Soybean oil grafted with vinyltrimethoxysilane (VTMS). Dibutyltin dilaurate (DBTDL) was used as condensation catalyst. It was revealed that the coated paper had a fine coating with better adhesion of the coating material to the paper oxidant.

Eggplant has 3 days of shelf life only. Singh et al. (2016) evaluated the functional quality of carnauba wax (CW) with various additives for increasing the shelf life in eggplant packed in 35 μ polypropylene pouches. It was revealed that the packaged eggplant treated with polyethylene glycol and sodium alginate in CW emulsion remained quite acceptable upto 12 days.

Morrissette et al. (2018) fabricated fluorine- and silane-free, water-based coating formulations with plant-based filler materials (e.g., cellulose, lycopodium). These can be used for making superhydrophobic surfaces (water contact angles > 150°). All these formulations contain a plant-based filler material imparting surface roughness, and an aqueous polymeric dispersion, which provides low surface energy coating component. It was reported that a coating formulation comprising of lycopodium and a natural wax (e.g., carnauba wax, beeswax) exhibited highest performance (hydrophobicity and water mobility).

The process involved in producing hand-drawn batik was simplified by Ariani and Pandanwangi (2021). They used natural cold wax, which was made from tamarind seed powder (*Tamarindus indica* L). It was reported that polysaccharide content in tamarind seed powder exhibited good coagulant properties, and therefore, it can be used as a barrier in the process of coloring batik in an environmentally friendly manner.

3.9 GREEN BUILDING CONSTRUCTION MATERIALS

Green cements/concrete are resource-saving structures with reduced environmental impact in terms of energy saving, CO_2 emissions, and wastewater. It was first invented in Denmark in the year 1998. Otherwise, the entire cement making is environmentally destructive process as it includes the extraction and mining of limestone, transportation of materials, and energy intensive. It was reported in Chronicle that each ton of cement emits about 1,763 pounds of CO_2 during manufacture. Combustion in kilns creates air pollution (worse when burning tires, 37% in U.S. or hazardous waste), resulting in toxic ash (cement kiln dust, i.e., 9 tons per 100 tons of clinker). Cement industries pollution generates carbon dioxide (global warming gas), acid gases (H_2SO_4, HF, HCl), nitrogen oxides, sulfur dioxide, particulate matter (dioxins and furans), 19 heavy metals that include lead, mercury, cadmium, and chromium as the products of incomplete combustion (PICs), including dioxins, furans, and polycyclic aromatic hydrocarbons (PAHs). Over 7% of all greenhouse gas emissions worldwide are caused from the manufacturing of Portland cement.

To develop new green cements and binding materials, the use of alternative raw materials and energy saving strategies are in practice (Phair, 2006) that includes fly ash (Yazici and Hasan, 2012), waste glass (Nassar and Soroushian, 2011), waste fiberglass (Wang et al., 2000), blast furnace slag (Videla and Gaedicke, 2004), calcined shale, municipal solid waste incinerated product (Horiguchi and Saeki, 2004), and alternative fuels (sewage sludge, etc.) (Zabaniotou and Theofilou, 2008) to develop/improve cement with low-energy consumption. Residual products from the concrete industry, that is, stone dust (from crushing of aggregate) and concrete slurry (from washing of mixers and other equipments) and ceramic wastes are used as green aggregates. These new types of cement with reduced environmental impact are more cost-effective, and it provides tremendous environmental benefits. Few more ecological benign cementitious materials are

(i) Geopolymer concrete

Geopolymers are amorphous alumino-silicate binding materials, unlike natural crystalline zeolitic materials. These can be synthesized by polycondensation reaction of geopolymeric precursor, and alkali polysilicates. Basic ingredients of geopolymers include fly ash C (FA), sand aggregates (SA), alkaline liquid sodium silicate and sodium hydroxide solution (AL), water and super plasticizer (SP). They contain the molar ratio of Si-to-Al about 1–3. Since no limestone is used in geopolymer cement (GPC), they possess excellent properties to survive within the tough environments (both acid and salt). Compared with Portland cement, the geopolymer has a relative higher strength, excellent volume stability with better durability and low cost. Geopolymer concrete based on pozzolana is a new material that does not need the presence of Portland cement as a binder. The polymerization process involves substantially a fast chemical reaction under alkaline condition on Si-Al minerals that results in a three-dimensional polymeric chain and ring structure consisting of Si-O-Al-O bonds. A geopolymer can take one of these three basic forms.

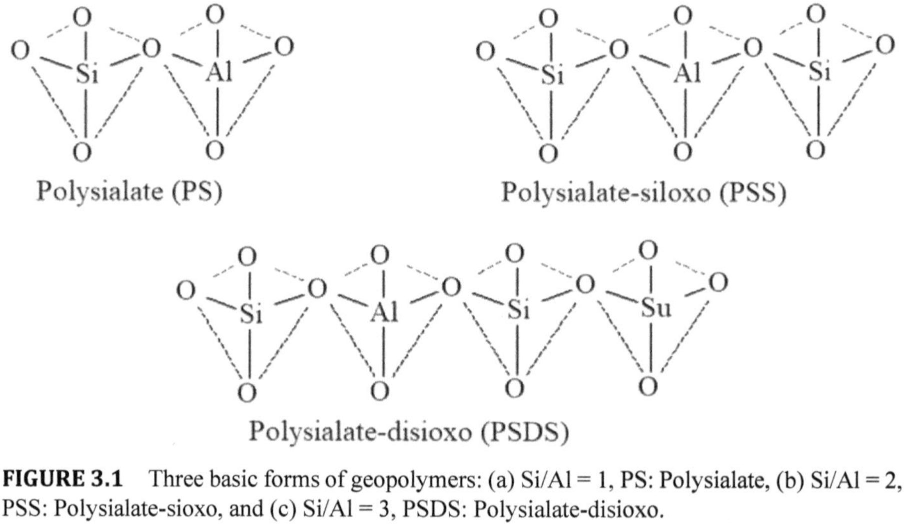

FIGURE 3.1 Three basic forms of geopolymers: (a) Si/Al = 1, PS: Polysialate, (b) Si/Al = 2, PSS: Polysialate-sioxo, and (c) Si/Al = 3, PSDS: Polysialate-disioxo.

This technology makes concrete more corrosion resistant and durable than Portland-based concrete. Sea water can be used for the blending of the GPC, which was impossible with Portland cement.

(ii) Novacem's cement

Novacem is a carbon negative cement. It is a composite of hydrated MgO and silicate. Its strength developed through the formation of magnesium silicate hydrates (M-S-H) rather than carbonation with atmospheric CO_2. Novacem technology is strongly protected by more than four patents. Non-carbonate feedstock (uses magnesium silicates) is used in Novacem, so that no CO_2 is there from the raw material. No absorption of CO_2 in cement production and lower process temperature (700°C) can make us able to utilize biomass as fuel. A 0–150 kg CO_2 is created per ton of cement, depending on the fuel mix used, while conventional cement produced almost 800 kg CO_2 per ton of cement (Global weighted average figures from International Energy Agency, 2007). Cement composition includes a carbonate created during production process by absorbing CO_2.

(iii) Mineral admixtures or supplementary cementitious materials (SCMs)

SCMs are alumino-siliceous materials that possess pozzolanic (materials containing reactive silica and/or alumina, which have little or no binding property of their own), reactivity and/or latent hydraulic reactivity. CaO, silicates and H_2O, are the main constituents of SCMs.

$$2\,C_3S + 6\,H \rightarrow C_3S_2H_3 + 3\,CH$$

$$2\,C_2S + 4\,H \rightarrow C_3S_2H_3 + CH$$

where C = CaO; S = SiO_2; H = H_2O and C-S-H; molar ratios can vary with strength-giving phase. CH has no cementitious properties (does not contribute to strength), it is easily leached and prone to chemical attack. SCMs can improve concrete properties by many of the beneficial effects. SCM has the pore structure effect such as microfiller effect that increased the packing of cementitious particles, as the porous CH is replaced with C-S-H; wall effect that densifies the ITZ (interfacial transition zone) at the cement aggregate interface causes pore blocking, which occurs because of a combination of these factors. These effects refine the pore structure and reduce the permeability of concrete, thereby making it more resistant to the penetration of deleterious agents. SCMs are of two types: (i) natural (ASTM C 618 Class N); it is produced from natural mineral deposits (e.g., volcanic ash or pumicite, diatomaceous earth, opaline cherts and shales).

It also requires heat treatment (e.g., metakaolin or calcined clay, calcined shale, rice hull ash, calcined shale) and (ii) processed/manufactured, which contains silica fume (ASTM C 1240), fly ash (ASTM C 215), slag (ASTM C 989), etc. Sustainable construction practices also include blended cements, a combination of Portland cement and SCMs, which has improved strength, workability, and durability. It is less expensive than the pure cement. These are currently more common in Europe and South America than in the US. Partial replacement for cement reduces energy consumption and CO_2 emissions, when used as productive use of industrial waste, which may be land filled. Contractors can derive some financial benefits to green construction.

- **Recent Advances**

High density polyethylene waste was mixed with Portland cement by Jassim et al. (2017). Waste of polyethylene packages including food crates and bottle in the range (10–80% by volume) was used. It was reported that plastic cement can be produced from polyethylene waste and Portland cement using 60 and 40%, respectively.

El-Gamal and Selim (2017) used ground granulated blast furnace slag (GGBS) to produce the GPC. The influence of replacement of slag by 5 and 10% of clay-bricks or fly ash wastes (Homra) has evolved the properties of the produced geopolymer.

The use of different demolished bricks wastes, rice husk ash, and fly ash was investigated by Hossain et al. (2019) as raw materials for fabricating sustainable, and eco-friendly fired building bricks. It was reported that compressive strength and water absorption were comparable as per devised standards. The 70 wt% of other waste and 10 wt% of glass cullet exhibited better properties at 800°C.

Coal gangue (CG) has low reactivity, which restricts its large-scale applications as supplementary cementitious materials (SCMs), and therefore, Zhao et al. (2021) conducted wet-grinding treatment to increase the reactivity of CG. It was revealed that compressive strength of CG/OPC composites was increased by around 71% using CG12 as SCMs.

3.10 BIO-BASED MATERIAL STARBONS

Starbons are a novel family of mesoporous materials derived from polysaccharides, which retained their organized structure during pyrolysis with tunable surface properties.

(i) $\text{Starch} \xrightarrow{\text{Expansion}} \underset{\text{gel block}}{\text{Porous}} \xrightarrow{\text{Drying}} \underset{\text{starch}}{\text{Mesoporous}} \xrightarrow{\text{Pyrolysis}} \text{Starbon}$

Starbon products deliver a step change in performance over existing systems, for example, acid resins in catalysis, porous graphitic carbons in separation, and activated carbon in adsorption. They are more selective, efficient, and effective because of their unmatched high mesoporosity (0.4—0.7 $cm^3\ g^{-1}$), high surface areas (150–500 $m^2\ g^{-1}$), readily functionalizable (range of heteroatoms, acid/base, functionality, metal complexation), adjustable surface properties (different energies, hydrophilic to hydrophobic), excellent solvent stability, good chemical and heat resistance, controllable electrical conductivity, availability of starbons in different forms, from homogeneous micronized powders to beads to monoliths, etc. Starbons are used for various applications such as catalysis of biorefinery downstream processes including esterification reactions in aqueous systems.

a) Starbon > DArco@SO_3H > β-25 > ZrO_2 > KSF
b) Starbon > Darco > Norit > Blank

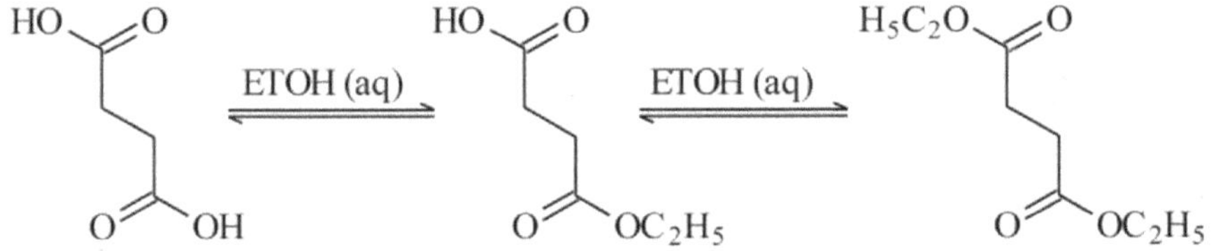

Biomass fermentation produces a wide range of organic acids, which can be utilized as platform molecules in various applications, such as polymers and higher value intermediates. Esterification is one of the key upgrading steps for these acids. The fermentation process is carried out in aqueous media. Therefore, it requires intensive separation steps before the acids can be upgraded. Here, Starbon-supported sulfonic acid catalysts developed by the Green chemistry center overcome this problem and are able to perform esterification reactions with very high conversion yields and rates.

Other reactions, where these catalysts have excelled, include aromatic amidiations and acylations. Starbon/nanometals perform well in other aqueous phase reactions including reductions with H_2 and oxidations with H_2O_2.

- Starbon C Series—Catalysis/catalyst support (solid acid/base, immobilized metal or active in aqueous media).

- Starbon P Series—Gas trapping and water purification, in particular, removal of harmful organics and heavy metals to purify water and cleanup waste streams.
- Starbon S Series—Separation of complex mixtures for production and analysis with Starbon as the stationary phase in chromatographic systems.
- Starbon R Series—Recovery of precious metals through reductive adsorption.

These have also been used as biomedical devices, separation media, absorbency, remediation, effluent treatment, fuel cells, etc. (Clark et al., 2012).

García et al. (2015) reported the preparation and application of starch-derived carbonaceous mesoporous materials (Starbon®) for the recovery of some noble metals (Au^{3+}, Pt^{2+} and Pd^{2+}). It was observed that Starbon® separated these metals from a mixture containing other metal ions (Ni^{2+}, Cu^{2+} and Zn^{2+}).

Kim et al. (2018) investigated alginic acid-derived mesoporous carbonaceous materials (Starbon® A800 series). They used these as negative electrodes for lithium ion batteries. The highest electrochemical performance was obtained with the material A800, which was having high pore volume (A800HPV). The highest pore volume was 0.9 cm^3g^{-1} and the highest electronic conductivity was 84 S m^{-1}. The A800HPV exhibited better long-term stability, and markedly improved rate capability as compared with commercial mesoporous carbon.

The Starbon composite with molybdenum oxide nanoparticles was reported by Kaur et al. (2020). As-prepared Mo-containing composite was found to be an efficient catalyst for reducing 4-nitrophenol to 4-aminophenol with sodium borohydride.

3.11 BIODIESEL

Green chemistry tries to utilize benign and renewable feedstocks as raw materials, wherever it is possible. Therefore, combustion of fuels obtained from renewable feedstocks would be more preferable than combustion of the fossil fuels from depleting finite sources. Worldwide, there are many vehicles fuelled with diesel oil, and the production of biodiesel oil is a promising green option. Fat embedded plant's oil can be converted into the biodiesel via a transesterification reaction using methanol and caustic or acid catalysts. During these reactions, the triglycerides are converted into the methyl ester and glycerol, which is a valuable raw material for soap production. Same can be achieved by utilizing supercritical methanol without a catalyst (Bunyakiat et al., 2006). This has the advantage of allowing a greater range of feedstocks (in particular, used cooking oil), the product does not need to be washed to remove catalyst, and it is easier to design as a continuous process.

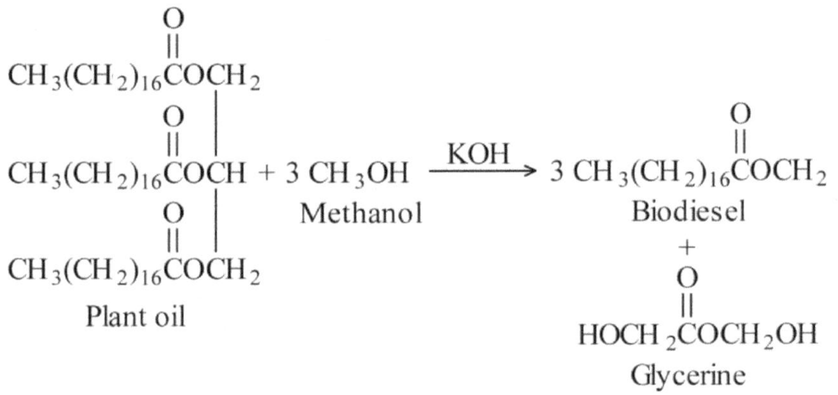

Use of calcium oxide and potassium phosphate for biodiesel production was reported by Choedkiatsakul et al. (2013) in an ultrasound-assisted reactor (US). It was observed that high methyl ester yields could be achieved in the US reactor, that is, 90% with CaO and 80% with K_3PO_4. It was revealed that the US reactor not only provides a high yield of methyl ester in lesser reaction time, but there was hindrance in dissolution as compared with the mechanical stirred reactor.

Calcined waste starfish was used by Jo et al. (2013) as a base catalyst for producing biodiesel from soybean. The content of fatty acid methyl esters (FAME) was used to determine the biodiesel yield. Highest activity for the transesterification reaction was obtained with starfish-derived catalyst calcined at 750°C or more.

Bala et al. (2015) explored the use of silicotungstic acid, which was anchored to mesoporous siliceous support KIT-6 (Korean Institute of Technology-6) as a catalyst to produce biodiesel. The four feedstock oils (coffee oil, algae oil, used cooking oil as well as palmitic acid) were used. It was reported that the resulting materials exhibited stable as well as more efficient (catalytic performance) in the production of biodiesel with 99% conversion than that of solid acid catalysts. In addition, the catalyst can be reused for four cycles.

Lv et al. (2019) prepared biodiesel using a novel free liquid lipase A from *Candida antarctica* (CALA) as a catalyst in the presence of excess of water.

The conditions of transesterification were optimized as: CALA load 5%, substrate molar ratio (soybean oil/methanol) 1:7, water load 14%, reaction time 26 h, and temperature 38°C. They could achieve maximum biodiesel yield (92.4 ± 0.8%) under optimal conditions.

The efficiency of seaweeds was evaluated by El-Sheekh et al. (2021) as a feedstock for biodiesel. Fifteen macroalgal species were used and it was recommended that *U. compressa* and *U. fasciata* are the promising biodiesel feedstocks, which gave high net energy output, that is, it reached 1.24 and 1.30 GJ ton^{-1}, respectively.

3.12 ENVIRONMENTAL BENIGN SUPERCRITICAL FLUIDS

The use of supercritical fluids (SCFs) in chemical processes became more and more rampant (Jessop and Leitner, 1999). Planets of our solar system are good examples of systems that possess SCFs. Out of these Sun is the best, which is mainly composed of the hydrogen and helium at temperatures well above their critical points. The gaseous outer atmospheres of Jupiter and Saturn transit smoothly into the interior SCFs. The term supercritical fluids comprises the liquids and gases at temperatures and pressures higher than their critical temperatures and pressures. It makes easy to adjust density and solution ability by a small change in temperature or pressure. Due to this, the supercritical fluids are able to dissolve many compounds with different polarity and molecular mass. Therefore, SCFs are suitable as a substitute for organic solvents in a wide range of industrial and laboratory processes. Among them, supercritical carbon dioxide ($scCO_2$) and water (scH_2O) are quite popular. Easily available (from natural sources) and low cost, supercritical carbon dioxide is a new surfactant with high surface activity and it has opened a way to new processes in textile and metal industries and for dry cleaning of clothes (Wardencki et al., 2005). Supercritical carbon dioxide can be used instead of PERC (perchloroethylene) or other undesirable solvents for dry cleaning. Moreover, CO_2 at high pressures has antimicrobial properties also (Fraser, 1951).

Room temperature ionic liquids are considered to be environmentally benign reaction media because they are low viscosity liquids with no measurable vapor pressure. However, due to the lack of sustainable techniques for the removal of products from the room temperature, ionic liquids have limited their application. Fortunately, $scCO_2$ dissolves in the ionic liquid at appreciable extent to facilitate their extractions. Moreover, no measurable cross-contamination of the $scCO_2$ by the ionic liquids is seen, so the product can be recovered in pure form. Hence, it can be used on a large scale for the decaffeination of green coffee beans, the extraction of hops for beer production, and the production of essential oils and pharmaceutical products from plants.

Supercritical water can be used to decompose biomass via supercritical water gasification of biomass. This type of biomass gasification can be used to produce hydrocarbon and hydrogen fuels. Supercritical fluid chromatography (SFC) can be used on an analytical scale, in a few cases, such as chiral separations and analysis of high molecular weight hydrocarbons. Impregnation is, in essence, the converse of extraction. A substance is dissolved in the supercritical fluid, the solution flows past a solid substrate, and is deposited on or dissolves in the substrate. Dyeing of polyester fiber can also be performed using SCFs. Formation of the small particles (nano systems, quantum dots, etc.) of a substance is an important process in the pharmaceutical and many other industries. Supercritical fluids provide a number of ways for their production by rapidly exceeding saturation point of a solute by dilution, depressurization, or a combination of these. This promotes nucleation or spinodal decomposition over crystal growth and harvests very small and regular-sized particles.

Supercritical fluids have shown the capability to reduce particles up to a range of 5–2000 nm (Yeob and Kirana, 2005). Furthermore, supercritical fluids can be used to deposit functional nanostructured films and nanometer size particles of metals onto surfaces (Bart, 2005). Supercritical drying of the archeological and biological samples is a method of removing solvent from main products without surface tension effects. As a liquid dries, the surface tension drags on small structures within a solid, causing distortion and shrinkage. Under supercritical conditions, there is no surface tension, and the supercritical fluid can be removed without distortion.

Supercritical water oxidation uses supercritical water along with molecular oxygen as an oxidizing agent that gives up oxygen atoms, which oxidizes the hazardous waste and eliminate the production of the toxic combustion products that the burning can produce (Oakes et al., 1999). The efficiency of a heat engine (via raise in operating temperature of the power station) for subcritical operation is raised from 39 to 45%, when scH_2O is used as the working fluid. The efficiency of the power stations will be improved by raising the operating temperature. Supercritical water or carbon dioxide reactors (SCW/CRs) are the promising advanced nuclear systems that offer the thermal efficiency gains. Supercritical water/$scCO_2$ reactors are promising advanced nuclear systems that offer thermal efficiency gains to power stations (Dostal et al., 2004). Supercritical carbon dioxide is being used in domestic heat pumps as an emerging new refrigerant material for low-carbon solutions by small inputs of electric power by moving heat into the system from their surroundings.

3.13 NANOPARTICLES

Routine techniques for nanoparticle production, such as photochemical reduction (Eustis et al., 2005), laser ablation (Mafune et al., 2002), electrochemistry (Rodríguez-Sánchez et al., 2000), lithography or high energy irradiation (Zhang and Wang, 2008; Treguer et al., 1998), either remain expensive or employ hazardous substances, such as organic solvents, and toxic reducing agents, such as sodium borohydride and N,N-dimethylformamide. The biosynthetic procedures involve living organisms, such as bacteria (Joerger et al., 2000), fungi (Bhainsa and D'Souza, 2006), and plants (Gardea-Torresdey et al., 2002), or plants extracts (Vilchis-Nestor et al., 2008), biocatalyst (enzymes/engineered enzyme) and a practical low-cost natural feedstock, and these are used to perform many chemical reactions of great importance. Resulting process greatly reduces hazard and waste, cost-effective and meets the needs of customers. Biological synthetic processes have emerged as a simple and viable alternative to more complex physicochemical approaches to obtain nanomaterials with adequate control of size and shape (Shankar et al., 2004).

Ahmed et al. (2016) synthesized silver nanoparticles. They used *Azadirachta indica* aqueous leaf extract for this purpose. The extract of plant was used as capping agent as well as reducing agent both. These silver nanoparticles exhibited antibacterial activities against *Staphylococcus aureus* and *E. coli*. It is a simple, rapid, eco-friendly, one step, and nontoxic method, which requires only 15 min for preparing silver nanoparticles from silver ions at room temperature.

Li et al. (2016) prepared graphitic carbon nitride quantum dots (GCNQDs) which were doped with sulfur and oxygen bright green luminescent via microwave treatment of thiourea and citric acid. It was revealed that these GCNQDs with low cytotoxicity can be used as a fluorescent probe effectively for HeLa cell imaging with great potential in bioanalysis. Blosi et al. (2016) developed a green and versatile synthesis of stable mono- and bimetallic colloids using microwave irradiation and eco-friendly reagents, such as water (solvent), glucose (mild and nontoxic reducer), and polyvinylpirrolidone (chelating agent). It was reported that all these materials were found to serve as effective catalysts in the reduction of p-nitrophenol.

Synthesis of fluorescent CQDs has been reported by Yu et al. (2018) using triethylenediamine hexahydrate and phthalic acid as precursors via microwave-assisted method. The reaction was complete in 60 s. The CQDs with the feed ratio (1:0.5) exhibited a strong green-yellow fluorescence, which may used in fabricating optoelectronic devices. As-prepared CQDs showed high pH sensitivity so that these have great potential as a fluorescence nanosensor for sensing of pH.

Zinc oxide nanoparticles (ZnONPs) were synthesized by Ogunyemi et al. (2019) using plant extracts of red tomato fruit (*Lycopersicon esculentum* M.),chamomile flower (*Matricaria chamomilla* L.), and olive leave (*Olea europaea*). It was reported that ZnONPs synthesized by *O. europaea* was having the size range of 40.5–124.0 nm.

A facile and efficient method was reported by Alvand et al. (2019) for the green synthesis of cadmium telluride quantum dots (CdTe QDs). They used an aqueous extract of *Ficus johannis* plant for this purpose. Two extraction methods were used, such as microwave-assisted extraction (90 and 270w; 15 min) and ultrasonic-assisted extraction (15 min; 45°C). The average particle size of as-obtained QDs was found to be 1.2 nm. The antimicrobial properties, antioxidant, genotoxicity, toxicity, and antifungal activities of as-prepared CdTe QDs were also investigated.

Nouri et al. (2020) synthesized ultra-small Ag nanoparticles (AgNPs) by an eco-friendly technique using *Mentha aquatica* leaf (MAL) extract as capping and reducing agent. It was reported that the smaller AgNPs (8 nm) could be obtained on application of ultrasound. These particles have MIC values against *Bacillus cereus*, *Pseudomonas aeruginosa*, and *S. aureus*, *E. coli* as 20, 2.2, 198 and 58 μg mL^{-1}, respectively.

3.14 ANTIFOULANTS

Rohm and Haas Company received a Presidential Green Chemistry Challenge Award for designing the environmentally safe marine antifoulant called Sea-Nine™. The unwanted growth of plants and animals on a ship's surface causes fouling and it costs the shipping industry approximately US$ 3 billion a year. The main compounds used worldwide to control this fouling are organotin antifoulants, such as tributyltin oxide (TBTO). This agent has widespread environmental problems due to its persistence in the environment and the side effects. They cause acute toxicity, bioaccumulation, decreased reproductive viability, and increased shell thickness in shellfish.

This company selected 4,5-dichloro-2-*n*-octyl-4- isothiazolin-3-one as an alternative. TBTO bioaccumulates as much as 10,000 times, while Sea-Nine™ bioaccumulation is essentially zero. Both TBTO and Sea-Nine™ were acutely toxic to marine organisms, but TBTO had widespread chronic toxicity, but Sea-Nine™ antifoulant showed no chronic toxicity.

Pérez et al. (2016) synthesized 7-hydroxy-4-methylcoumarin as environmental-friendly antifoulants to replace metallic biocides. Its antifouling properties was evaluated on the bivalve Mytilus edulis platensis. It was claimed that as-synthesized green coumarin is a promising candidate as a antifoulant, particularly for marine protective coatings. Feng et al. (2018) evaluated 18 alkaloids derived from terrestrial plants for antifouling activity. It was revealed that 4 out of these 18 alkaloids were found effective to inhibit larval settlement of *B. neritina* while fifteen can inhibit the larval settlement of *B. albicostatus*.

Sathicq et al. (2019) synthesized a series of furylchalcones, which was obtained by biomass building block (furfural). The field trials of these antifouling paints revealed that paints containing furylchalcones exhibited strong antifouling effect and were most active against calcareous tubeworms and algae. The antifouling activity of peracetylated cholic acid, a bile acid derivative, was evaluated by Pérez et al. (2019). They reported that it has a good antifouling activity and low toxicity.

3.15 REDUCTION OF CARBON DIOXIDE

Carbon dioxide has been considered the largest contributor among greenhouse gases or the main culprit of global warming. Its amount is regularly increasing in the atmosphere causing natural catastrophes. It can be reduced to alternate synthetic fuels, such as HCOOH, HCHO, CH_3OH, and CH_4. These fuels can be used as a source of energy and replace conventional fuels, and their use will not increase CO_2 in the atmosphere.

The photoelectrocatalytic reduction of carbon dioxide was studied by Inoue et al. (1979) for methane, methyl alcohol, formaldehyde, and formic acid in the presence of photocatalyst suspension in water. Aliwi and Al-Jubori (1989) reported photoelectrocatalytic reduction of carbon dioxide to formic acid and formaldehyde in aqueous solution in the presence of n-CdS and n-Bi_2S_3 semiconductor powders. It was observed that the presence of hydrogen sulfide increase rate of this photoreduction. It was revealed that highest concentrations of HCHO and HCOOH were produced at pH 6 over

Bi_2S_3 semiconductor in 1 h. They also evaluated the effects of pH, light intensity and reaction temperature.

Pure TiO_2 anatase particles (diameters = 4.5–29 nm) were prepared by Kočí et al. (2009) via precipitation and sol–gel method. These were used for photocatalytic reduction of CO_2, where methane and methanol were found to be the main reduction products. Zhang et al. (2009) carried out a gas–solid heterogeneous system for solar–chemical energy conversion of selective catalytic reduction of CO_2 with H_2O on using nano-TiO_2 and Pt-metal supported photocatalysts. The yield of CH_4 was quite remarkable (4.8 μmol h−1 gTi−1).

Photoelectrochemical reduction of CO_2 to HCOO– (formate) was observed by Sato et al. (2011) in the presence of over p-type InP/Ru complex polymer hybrid photocatalyst. Here, water was used as both; a proton source and an electron donor. It was revealed that selectivity for HCOO− production was more than 70%, and the conversion efficiency was in the range of 0.03–0.04% (Figure 3.2).

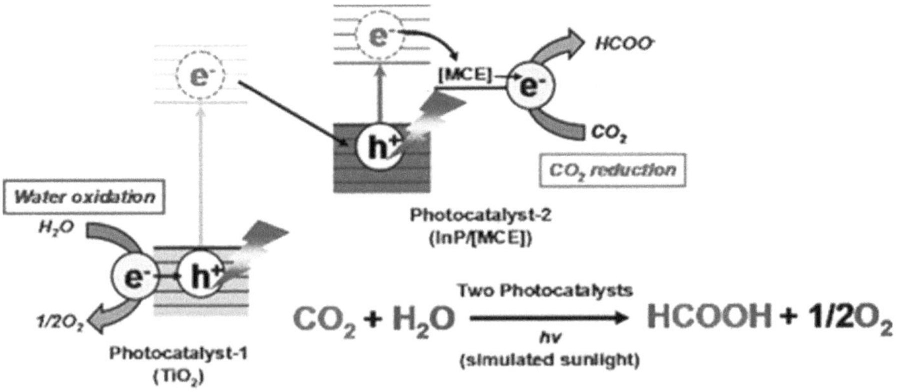

FIGURE 3.2 Photoelectrochemical reduction of CO_2 to HCOO–.

Source: Reprinted with permission from Sato et al., 2011. © 2011 American Chemical Society.

Le et al. (2011) observed direct reduction of carbon dioxide to methanol in the presence of air-oxidized and anodized Cu electrodes. The yield of CH_3OH was found to be 43 mol cm^2 h^{-1} at cuprous oxide electrodes, which was significantly higher than anodized or air-oxidized Cu electrodes, which suggests that Cu(I) species may play an important role in selectivity to methanol.

Carbon-supported, nitrogen-based organometallic silver was prepared by Tornow et al. (2012) and used as catalyst for the electrochemical reduction of CO_2. It was observed that Faradaic efficiencies of this organometallic catalyst was found to be higher than 90%. It was also found that on adding an amine ligand to Ag/C, there was a significant increase in the partial current density for CO, which suggested a possible co-catalyst mechanism.

Zhu et al. (2013) reported electrocatalytic reduction (selective) of carbon dioxide to carbon monoxide in the presence of gold nanoparticles in 0.5 M $KHCO_3$. The maximum Faradic efficiency (up to 90% at −0.67 V vs. reversible hydrogen electrode, RHE) was obtained with 8 nm Au NPs. It was observed by Habisreutinger et al. (2013) that copper(I) boryl complex [(IPr) Cu(Bpin)] (where pin = pinacolate: 2,3-dimethyl-2,3-butanediolate and IPr = 1,3-bis(2,6-diisopropylphenyl)imidazol-2-ylidene) can deoxygenate CO_2 rapidly affording CO quantitatively. It was also reported that the use of a stoichiometric diboron reagent and copper(I) alkoxide precatalyst resulted in catalytic CO_2 reduction depending on the reaction conditions and supporting ligand.

Morikawa et al. (2013) that developed a hybrid photocatalyst (InP/Ru complex) for CO_2 reduction. The CO_2 was reduced to formate using water as an electron donor as well as proton source with conversion efficiency as 0.04%. The electrochemical reduction of CO_2 was reported by Sen et al. (2014) at copper foams. They revealed that the distribution of products in this reaction and their Faradaic efficiencies were different as obtained at copper electrodes (smooth electropolished). These differences were attributed to hierarchical porosity, high surface roughness, and confinement of reactive species.

Highly efficient metal-free catalytic reduction of CO_2 to methanol was observed by Gomes et al. (2014) using hydroboranes as the reductant and catalysts (guanidines and amidines). They used hydroboranes like catechol-borane (catBH) and 9-borabicyclo[3.3.1]nonane (9-BBN), 1,5,7-triaz-abicyclo[4.4.0]dec-5-ene (TBD), 1,8-diazabicycloundec-7-ene (DBU), and 7-methyl-1,5,7-triazabicyclo[4.4.0]dec-5-ene (Me-TBD), as catalysts. It was reported that Me-TBD can catalyze the reduction of carbon dioxide to methoxyborane with TONs (648 h–1) and TOFs up to 33 h−1 at room temperature.

Gao et al. (2015) reported a prominent size-dependent activity/selectivity in the electrocatalytic reduction of CO_2 in the presence of Pd NPs (2.4–10.3 nm). The Faradaic efficiency for CO production from the reduction of CO_2 varies from 5.8% at −0.89 V (vs. reversible hydrogen electrode) and 10.3 nm NPs to 91.2% as compared with over 3.7 nm NPs, which is about 18.4 times increase in current density.

Photoelectrochemical reduction of CO_2 over a hybrid photocathode was investigated by Sahara et al. (2016). They used Ru(II)–Re(I) supramolecular metal complex (photocatalyst), which was immobilized on a NiO electrode (NiO–RuRe). The NiO–RuRe photocathode generated carbon monoxide with high selectivity, and turnover number 32. The $CuGaO_2$ (p-type) semiconductor electrode was used by Kumagai et al. (2017) for the construction of a new hybrid photocathode with this Ru(II)–Re(I) supramolecular photocatalyst (RuRe/$CuGaO_2$). It was observed that photoelectrochemical cell using this as photocathode and a CoOx/TaON photoanode does not require any external bias for visible-light-driven catalytic reduction of CO_2 to CO using water.

Ru(II)Re(I) supramolecular photocatalyst was used by Kamata et al. (2018) with a difference that they used Ru(II) redox photosensitizer deposited NiO electrode with photocatalyst. A modified poly-RuRe/NiO could produce almost 2.5 times more CO. The role of surface roughening was investigated on the reduction of carbon dioxide by Jiang et al. (2020) over Cu. It was revealed that undercoordinated Cu sites with a higher fraction are available on the roughened surface, which can preferentially bind CO.

3.16 OTHER GREEN CHEMICALS

Term green chemistry was first coined to design chemicals and chemical processes that will be less harmful to human health and environment by implementing sustainable development in chemistry and chemical technologies. But the irony is that the analytical methods used to assess the state of environmental pollution may in fact be the great source of pollutants emission, which influences the environment adversely. Therefore, the basic principles of green chemistry were proposed to protect the environment from pollution. These principles help to increase the efficiency of synthetic methods, use of less toxic and renewable solvents or starting materials, reduce the stages of the synthetic routes, lower the energy used, minimizing waste as far as practically possible and more biodegradable by-products (Tundo and Anastas et al., 2000; Sheldon and Arends, 2006; Poliakoff and Licence, 2007). In this way, organic synthesis will be part of our effort for sustainable development. Green chemistry along with the green engineering provides us the potential tools, alternative materials, processes, and systems. Figure 3.3 shows not only the sustainability of the chemicals/materials production, but also their environmental credentials by reducing toxicity and increasing recyclability.

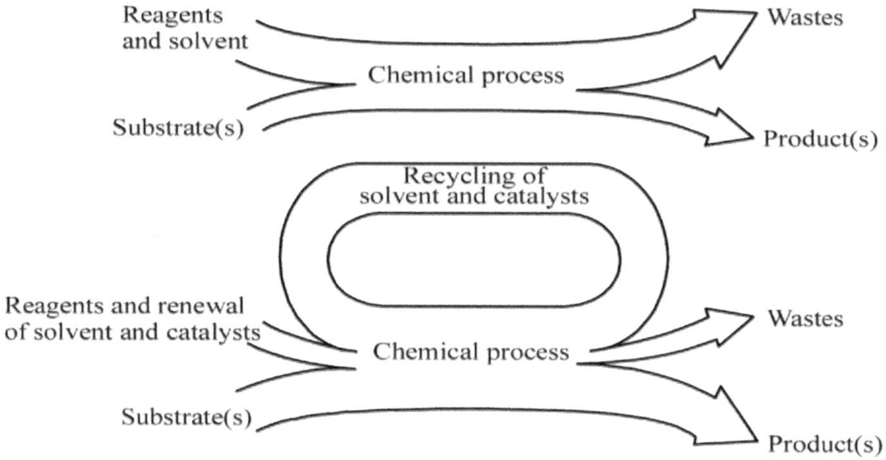

FIGURE 3.3 Chemical processes: (i) conventional way, (ii) green chemical method, and (iii) green process of material synthesis.

One interesting example is of adipic acid formation. Adipic acid is a very popular starting material for Nylon-6,6 and catechol synthesis (widely used in the pharmaceutical and pesticide industries). Traditionally, the adipic acid was produced by the oxidation of cyclohexanone/cyclohexanol with nitric acid in solvent benzene and catalyst copper/vanadium (Cu 0.1–05% and V 0.02–0.1%).

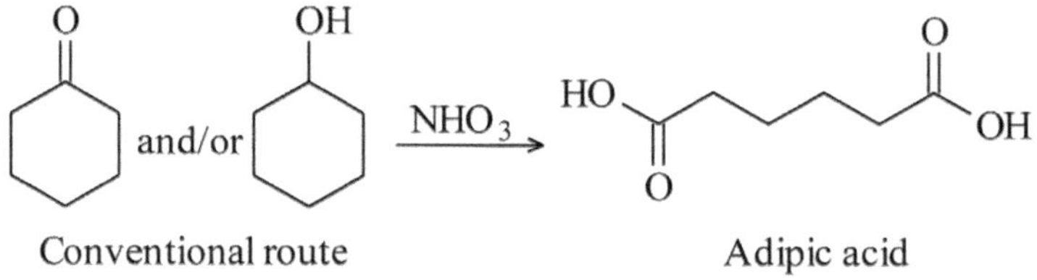

Benzene is mainly obtained from petrochemical industry and is also known for its carcinogenic properties. Oxidant nitric acid also generate toxic fumes of nitric oxides (NO_x), which is one of the contributors to the greenhouse effect and the destruction of the ozone layer in the stratosphere. The yield of this reaction and its reaction mass efficiency (RME) was found to be 93 and 55.7%, respectively. Finally, a greener chemical route with green oxidizing agent H_2O_2, without solvent, using a new generation of catalysts, phase transfer catalyst (PTC) was presented (Anastas et al., 2001). The starting chemical is cyclohexene and its oxidation was performed by

30% hydrogen peroxide. The tungsten salts (Na_2WO_4 /$KHSO_4$/Aliquat 336 (Stark's catalyst) was dissolved in a special organic solvent (Aliquat 336). The yield was 45–86% with cyclohexene or 1,2-cyclohexanediol (Sato et al., 1998; Usui and Sato, 2003).

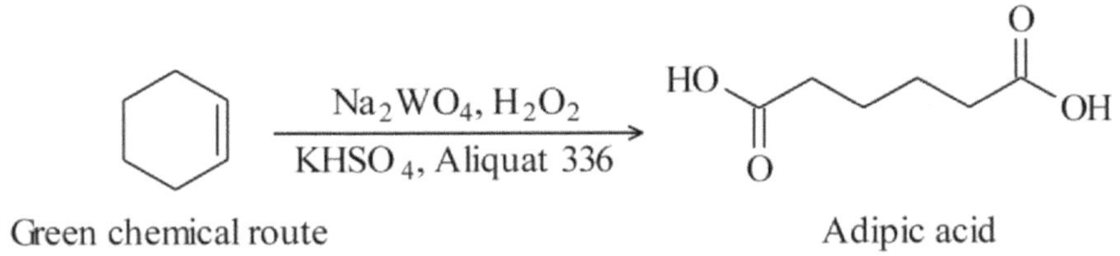

Solvent Aliquat 336 is a mixture of octyl C8 and decyl C10 chains (with C8 predominating). It is a quaternary ammonium salt used as PTC and metal extraction reagent with ability to dissolve metal complexes. Here, one can also use W as oxpoeroxo tungsten complexes with molybdenum.

This eco-friendly method did not produce toxic waste. Moreover, its yield and RME is 90 and 67%, respectively. RME of this green chemical method is 11% higher than the traditional one. Recently, biocatalytic method for the synthesis of adipic acid from D-glucose has also been promoted, where genetically transgenic bacteria *Klebsiella Pneumoniae*, a nontoxic strain of *E. coli or Enterobacteriaceae* was used as biocatalyst (Draths and Frost, 1994).

Maleic anhydride (MA or cis-butenediol acid) is used as a starting material for the production of polyimides, polyester resins, surface coatings, lubricant additives, phthalic-type alkyd, plasticizers and copolymers. It is also used as an important intermediate in the synthesis of 1,4-butanediol (in the industry of polyurethane and butyrolactone). Traditionally, maleic anhydride was produced using benzene, butene, or butane as a starting material and air as an oxidizing gas in the presence of catalyst, which was composed of oxides of vanadium and molybdenum, V_2O_5 and MoO_3 (fixed-bed reactor) under 3–5 bar pressure and 350–450°C temperature.

Reaction yield and RME was found to be 95 and 44.4%, respectively. In the 1990s, two very big industrial enterprises UCB chemicals (Belgium) and BASF (Germany) started producing this as a by-product of the oxidation of naphthalene into phthalic acid and phthalic anhydride. Another method was proposed with starting material, n-butane, catalyst $(VO)_2P_2O_5$, or new catalysts (special complexes of vanadium–phosphorous; fixed-bed reactor, which is now converted into circulating fluidized-bed reactor) in air, at temperatures 0–200°C (RME = 57.6% and yield = 60%) without using any solvent (Contractor et al., 1994). Therefore, these methods are much greener as compared with the original, which uses benzene and the atom economy of the reaction was better without much waste. The new greener method did not produce toxic waste and the yield was 90% with RME 67%.

There are few more examples of eco-benign chemical reactions, such as mild N-formylation in the presence of indium metal as a catalyst under solvent-free conditions. It is a chemoselective reaction of amines and α-amino acid esters without epimerization (Kim and Jang, 2010).

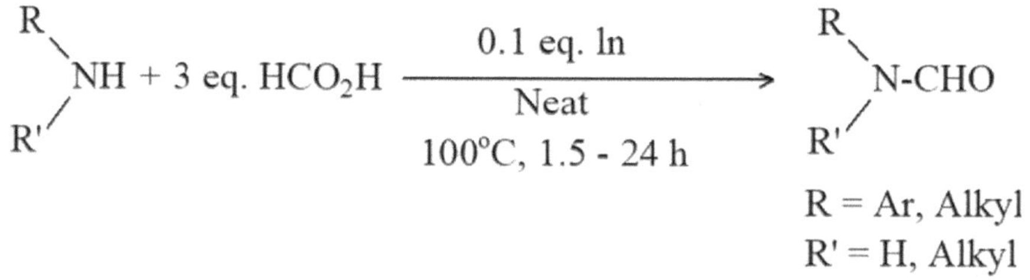

Similarly, the use of phosgene and methylene chloride in the synthesis of polycarbonates has been replaced by diphenylcarbonate. Borono-Mannich reactions can be performed in solvent-free conditions under microwave irradiation with shorter reaction time. Full conversion of the starting materials toward the expected product was achieved, starting from stoichiometric quantities of reactants, and avoiding column chromatography. No purification step other than an aqueous washing was required (Nun et al., 2010).

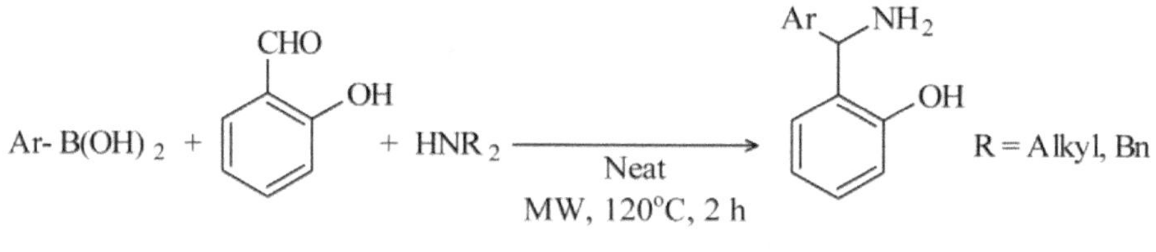

Intermolecular addition of perfluoroalkyl radicals on electron-rich alkenes and alkenes with electron withdrawing groups in water mediated by silyl radicals gives perfluoroalkyl-substituted compounds in good yields. The radical triggering events employed consist of thermal decomposition of 1,1′-azobis(cyclohexanecarbonitrile) (ACCN) or dioxygen initiation (Barata-Vallejo and Postigo, 2010).

Oxidation is the most polluting reaction in industries. Implementation of green chemistry has led to alternative less polluting reagents viz., molecular O_2 as the primary oxidant with extremely high oxidation state transition metal complexes. A convenient green synthesis of acetaldehyde is by Wacker oxidation of ethylene with O_2 in the presence of a catalyst, instead of its synthesis by oxidation of ethanol or hydration of acetylene with H_2SO_4. Conventional methylation reactions employ toxic alkyl halides or methyl sulfate that leads to environmental hazards. These are replaced by dimethyl carbonate with no deposit of inorganic salts. Microwave-assisted eco-friendly syntheses have also become a novel approach in green synthesis of chemicals. Organic synthesis under the microwave irradiation has many advantages as compared with the conventional reactions, which need very high temperatures and hazardous solvents. Microwave-assisted reactions are cleaner, last only for very few minutes, no solvent was used, gave high yield and produced minimum waste (Lidstrom et al., 2001).

Dinitrosalicylic acid-functionalized chitosan, CHN-DNSA, was prepared by Shoueir (2020). It showed enhanced adsorption property against rhodamine B (RhB) and chromium Cr(VI). It exhibited significant adsorption potency of Cr(VI) at pH 3.0 (98.4%) within 1 h and the adsorption performance was found to be 91.1% for RhB.

Ultrasound-assisted organic syntheses were also found as another green synthetic routes with great advantages for high efficiency, low waste, and low-energy requirements. Ultrasonic region of 20 kHz to 1 MHz has many applications due to its high energy and the ability to disperse reagent in small particles and accelerate reactions. Acoustic cavitation occasionally occurs (growth, and implosive collapse of bubbles). These cavitations can create extreme physical and chemical conditions (bubbles have temperatures around 5000 K and/or pressures of roughly 1000 atm) in otherwise cold liquids (Mason, 1997; Cintas and Luche, 1999; Cravotto and Cintas, 2006). Sonochemical engineering is a newly emerging field involving the application of sonic and ultrasonic waves to chemical processing. Sonochemistry enhances or promotes chemical reactions and mass transfer. It offers the potential for shorter reaction cycles, low-cost reagents, and less extreme physical conditions. The traditional multistep synthesis was wasteful and used large amounts of hazardous reagents.

Soltani et al. (2018) prepared amino-modified MCM-41/poly(vinyl alcohol) nanocomposite through ultrasonic-assisted. M-MCM-41/PVOH NC. They found 46.73 mg g^{-1} as maximum adsorption capacity of this composite for Cd(II) at 298 K. Chang et al. (2018) prepared a nanoparticle of cobalt(II) coordination polymer (CP), [Co(L)(npht)]n (H2npht = 4-Nitrophthalic acid, L = 1,3-Bis(5,6-dimethylbenzimidazol-1-ylmethyl)benzene) and its nanocomposite (Ag/CP) via sonochemical method: These nanoparticles

(Ag/CP) exhibited excellent photocatalytic activity for the degradation of methylene blue under UV and visible light.

Elhamifar et al. (2018) prepared magnetic iron oxide supported phenylsulfonic acid. They used it as an efficient catalyst for synthesizing tetrahydrobenzo[b]pyrans via a green chemical route using water as green solvent on ultrasonic exposure. The products of this reaction were found to give excellent yields in lesser time. The recoverability, durability, and reusability of this nanocatalyst were also tested.

Chemistry has provided us comfort from all angles, but it has also added to environmental pollution. One cannot change his life style without using these chemicals in some or the other ways. However, these products can be replaced by some other chemicals or products, which are eco-friendly. In this context, biodegradable materials are welcome. Nature has provided numerous examples, such as corn husk, wrapper, and beautiful network of pipelines for water supply in leaves and so on, which are biodegradable. It is presumed that most of the chemical products will be replaced by some or the other biodegradable or less toxic materials in years to come.

3.17 RECENT DEVELOPMENTS

Ahmad et al. (2022) carried out biodiesel synthesis from nonedible seed oil of *Monotheca buxifolia.* They used green nanoparticles of CaO synthesized with aqueous leaves extract of *Boerhavia procumbens* for this purpose. They reported that high yield of biodiesel (95%) could be achieved at optimum reaction conditions, that is, methanol to oil molar ratio (9:1), catalyst loading = 0.83 (wt.%), reaction time = 180 min and temperature = 85°C. GC/MS data of biodiesel revealed four distinct peaks of methyl esters. The formation of methyl esters in this biodiesel sample was confirmed. Fuel properties of as-prepared biodiesel were found to be almost equivalent to international standards biodiesel, such as viscosity (5.35 $mm^2 s^{-1}$), density (0.821 kg m^{-3}), pour point (–9°C), flash point (95°C), and cloud point (–8°C).

Almadiy and Nenaah (2022) hydrodistilled and analyzed the essential oil of *Origanum ulgare L.*. They identified monoterpenes, such as p-cymene (5.7%), thymol (8.3%), carvacrol (61.2%), and γ-terpinene (9.6%). It was reported that the use of nanoemulsion (40 μg mL^{-1}) resulted in 100% adult and larval mortality after 12 h. It was also observed that LC50's ranged between 13.2 and 75.7 μg mL^{-1} against larvae, and 10.1 and 36.0 μg mL^{-1} against adults after 24 h exposure. Test products were safe toward the nontarget aquatic species. *Gambusia affinis* and *Diplonychus indicus*.

An eco-friendly synthesis of α-hydroxyphosphonates was reported by Gundluru et al. (2022) by reacting aldehydes and diethyl phosphite in solvent-free conditions using nano–NiO as catalyst (Figure 3.4). It was observed that these as-prepared compounds exhibited good antimicrobial and antioxidant activities.

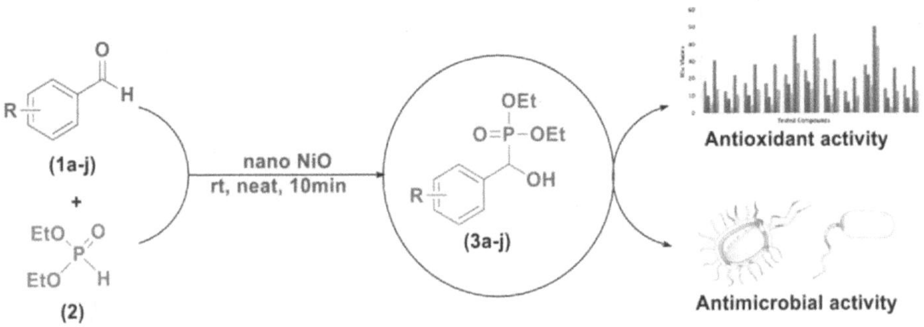

FIGURE 3.4 Green synthesis of α-hydroxyphosphonates using nano-NiO.
Source: Reprinted with permission from Gundluru et al. © 2022 Elsevier.

Habib et al. (2022) extracted a natural reddish-brown colorant from peepal (*Ficus religiosa*) and used for silk dyeing in the presence of microwave radiations. This colorant was extracted in aqueous and acidic media, heated with microwave for 1–5 min, and was given to fabric and changes in the intensity of color was observed. It was reported that the aqueous extract irradiation with microwaves for 3 min gave high color intensity onto silk fabric. They reported that the use of 3% Al, 4% Fe, and 2% tannic acid as pre-chemical mordant gave good color characteristics. The same was true for 4% Al, 4% Fe, and 3% tannic acid as post-chemical mordant. Apart from this, 4% acacia and 3% turmeric and pomegranate, while 3% acacia and turmeric and 4% pomegranate extracts also have resulted in excellent color characteristics when used as post-biomordant.

The microwave-assisted isolation of alkannin dye from *Alkanna tinctoria* was carried out by Adeel et al. (2022) and used for dyeing of biomordanted silk. They treated acid solubilized, water solubilized, and acid–methanol solubilized extracts and silk fabrics under microwave irradiation for 2, 4, and 6 min, respectively. It was also reported that acid solubilized extract gave excellent higher color strength when applied at 55°C for 55 min with 7 g 100 mL^{-1} NaCl salt as the exhausting agent.

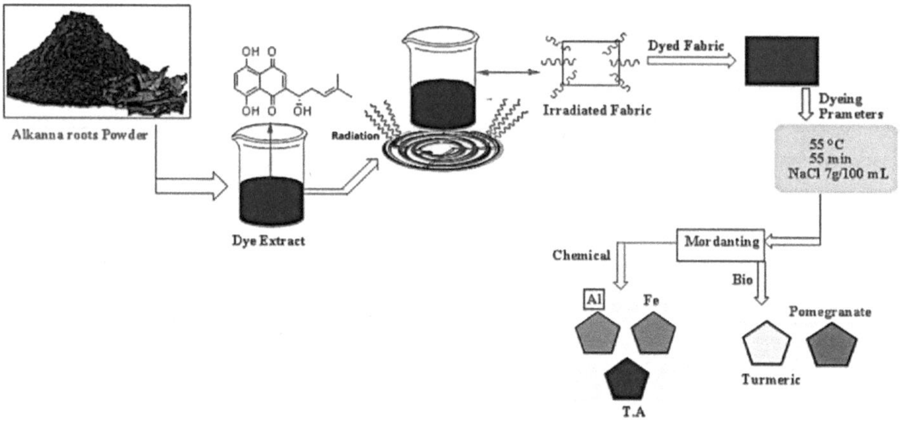

FIGURE 3.5 Microwave-assisted isolation of alkannin dye from *Alkanna tinctoria*.
Source: Reprinted with permission from Fazal-ur-Rehman et al., 2022. © 2022 Elsevier.

Romdhani et al. (2022) extracted natural dyestuff from *Citrus Sinensis L.* and used it as a dye on wool fibers. It was observed that optimum extraction conditions were Temperature = 90°C, extraction time = 90 min and pH = 5, while optimal conditions for dyeing of wool fibers were Temperature = 90°C, dyeing time = 90 min, and pH3. It was reported that iron sulfate is the most active mordant providing deeper color when used with concentration of 5% (w/w).

Winkler-Moser et al. (2022) stabilized natural peanut butter with 1.0–2.0% (w/w) beeswax (BW), sunflower wax (SFW), rice bran wax (RBW), or candelilla wax (CLW). It was reported that the use of waxes and their blend ratio significantly influenced the mouthfeel, appearance, firmness, spreadability, and flavor attributes. It was revealed that samples with 1.5–2.0% CLW, or 1.0–1.5% RBW had little differences in texture and appearance from the reference and commercial samples. It was observed that samples, which were stabilized with 1.0–1.5% RBW were observed close to commercial and reference samples.

Ghazani et al. (2022) characterized olive oil oleogels structured binary blends of sunflower wax (SFW), rice bran wax (RBW), candelilla wax (CLW), and beeswax (BW) (< 4%). It was reported that samples structured with 3% (w/w) of binary mixtures of SFW and RBW, and CDW and BW, exhibited a high oil binding capacity as compared with other mixtures. It was observed that the melting point of binary mixtures of waxes in olive oil ranged between 43.2°C and 67.4°C. It was revealed that hardness and plasticity (brittleness) of 2% and 3% binary wax mixtures in olive oil were found to be almost similar to commercial soft margarine.

KEYWORDS

- **bioplastics**
- **eco-waxes**
- **geopolymer concrete**
- **green chemicals nanoparticles**
- **green dyes**

REFERENCES

Abanades, S.; Flamant, G. Thermochemical Hydrogen Production from a Two-Step Solar-Driven Water-Splitting Cycle Based on Cerium Oxides. *Solar Energy* **2006,** *80* (12), 1611–1623.

Adams, S.; Nicholson, J. US Patent No. 3,228,831 (1960).

Adeel, S.; Liaqat, S.; Hussaan, M.; Mia, R.; Ahmed, B.; Wafa, H. Environmental Friendly Bio-Dyeing of Silk Using Alkanna Tinctoria Based Alkannin Natural Dye. *Ind. Crops Prod.* **2022,** *186*. DOI: 10.1016/j.indcrop.2022.115301

Adeel, S.; Rehman, F.; Iqbal, M.; Habib, N.; Kiran, S.; Zuber, M. et al. Ultrasonic Assisted Sustainable Dyeing of Mordanted Silk Fabric Using Arjun (Terminalia Arjuna) Bark Extracts. *Environ. Prog. Sustain. Energy* **2018b,** *38* (s1), S331–S339.

Adeel, S.; Zuber, M.; Fazal-ur-Rehman, Zia, K. Microwave-Assisted Extraction and Dyeing of Chemical and Bio-Mordanted Cotton Fabric Using Harmal Seeds as a Source of Natural Dye. *Environ. Sci. Poll. Res.* **2018a,** *25* (11), 11100–11110.

Agrafiotis, C.; Roeb, M.; Konstandopoulos, A. G.; Nalbandian, L.; Zaspalis, V. T.; Sattler, C. et al. Solar Water Splitting for Hydrogen Production with Monolithic Reactors. *Solar Energy* **2005,** *79* (4), 409–421.

Ahmad, M.; Elnaggar, A. Y.; Teong, L. K.; Sultana, S.; Zafar, M.; Munir, M. et al. Sustainable and eco-friendly synthesis of biodiesel from novel and non-edible seed oil of Monotheca buxifolia using green nano-catalyst of calcium oxide. *Energy Convers. Manag.* **2022,** *13*. DOI: 10.1016/j.ecmx.2021.100142.

Ahmed, S.; Saifullah Ahmad, M.; Swami, B. L.; Ikram, S. Green synthesis of silver nanoparticles using Azadirachta indica aqueous leaf extract. *J. Radiation Res. Appl. Sci.* **2016,** *9* (1). DOI: 10.1016/j.jrras.2015.06.006.

Alberts, A. W.; Chen, J.; Kuron, G.; Hunt, V.; Huff, J.; Hoffman, C. et al. Mevinolin: a highly potent competitive inhibitor of hydroxymethylglutaryl-coenzyme A reductase and a cholesterol-lowering agent. *Proc. Nat. Acad. Sci.* **1980,** *77* (7), 3957–3961.

Ali, H. Biodegradation of synthetic dyes—a review. *Water, Air, Soil Pollut.* **2010,** *213* (1), 251–273.

Aliwi, S. M.; Al-Jubori, K. Photoreduction of CO_2 by metal sulphide semiconductors in presence of H_2S. *Solar Energy Mater.* **1989,** *18* (3–4), 223–229.

Almadiy, A. A.; Nenaah, G. E. Essential oil of Origanum vulgare, its nanoemulsion and bioactive monoterpenes as eco-friendly novel green pesticides for controlling Aedes aegypti, the common vector of Dengue virus. *J. Essent. Oil Res.* **2022,** 1–15. DOI: 10.1080/10412905.2022.2091673.

Alvand, Z. M.; Rajabi, H. R.; Mirzaei, A.; Masoumiasl, A.; Sadatfaraji, H. Rapid and green synthesis of cadmium telluride quantum dots with low toxicity based on a plant-mediated approach after microwave and ultrasonic assisted extraction: Synthesis, characterization, biological potentials and comparison study. *Mater. Sci. Eng.: C* **2019,** *98*, 535–544.

Ameh, A. O.; Isa, M. T.; Udoka, E. K. Biodegradable detergents from Azadirachta Indica (neem) seed oil. *Leonardo Electronic Journal of Practices and Technologies* **2010,** *16*, 69–74.

Anastas, P. T.; Kirchhoff, M. M.; Williamson, T. C. Catalysis as a foundational pillar of green chemistry. *Appl. Catal. A. Gen.* **2001,** *221* (1–2), 3–13.

Angelidaki, I.; Toräng, L.; Waul, C. M.; Schmidt, J. E. Anaerobic bioprocessing of sewage sludge, focusing on degradation of linear alkylbenzene sulfonates (LAS). *Water Sci. Technol.* **2004,** *49* (10), 115–122.

Ariani, A.; Pandanwangi, A. Eco-friendly batik painting wax made from tamarind seed powder (Tamarindus indica L). *IOP Conf. Series: Earth Environ. Sci.* **2021,** *737* (1), doi:10.1088/1755–1315/737/1/012069.

Bala, D. D.; de Souza, K.; Misra, M.; Chidambaram, D. Conversion of a variety of high free fatty acid containing feedstock to biodiesel using solid acid supported catalyst. *J. Clean. Prod.* **2015,** *104*, 273–281.

Bamoniri, A.; Mirjalili, B. B. F.; Moshtael-Arani, N. Environmentally green approach to synthesize azo dyes based on 1-naphthol using nano $BF_3 \cdot SiO_2$ under solvent-free conditions. *Green Chem. Lett. Rev.* **2014,** *7* (4), 393–403.

Banik, B.; Sahoo, B.; Ravi Kumar, B.; Panda, J.; Kumar, A. Green chemistry and synthetic approaches in the development of antidepressant and antipsychotic agents. *Green Approaches In Medicinal Chemistry For Sustainable Drug Design* **2020,** 713–736.

Barata-Vallejo, S.; Postigo, A. (Me3Si) 3SiH-mediated intermolecular radical perfluoroalkylation reactions of olefins in water. *J. Org. Chem.* 2010, *75* (18), 6141–6148.

Bart, C. J. (Ed.) *Additives in Polymers: Industrial Analysis and Applications;* John Wiley: Chichester, U. K.; **2005,** 171–281.

Batt, A. L.; Kostich, M. S.; Lazorchak, J. M. Analysis of ecologically relevant pharmaceuticals in wastewater and surface water using selective solid-phase extraction and UPLC− MS/MS. *Anal. Chem.* **2008,** *80* (13), 5021–5030.

Bechtold, T.; Mahmud-Ali, A.; Mussak, R. A. Reuse of ash-tree (Fraxinus excelsior L.) bark as natural dyes for textile dyeing: Process conditions and process stability. *Color. Technol.* **2007a,** *123* (4), 271–279.

Bechtold, T.; Mahmud-Ali, A.; Mussak, R. Natural dyes for textile dyeing: A comparison of methods to assess the quality of Canadian golden rod plant material. *Dyes Pigments* **2007b,** *75* (2), 287–293.

Bechtold, T.; Turcanu, A.; Ganglberger, E.; Geissler, S. Natural dyes in modern textile dyehouses—how to combine experiences of two centuries to meet the demands of the future?. *J. Cleaner Prod.* **2003,** *11* (5), 499–509.

Bhainsa, K. C.; D'souza, S. F. Extracellular biosynthesis of silver nanoparticles using the fungus Aspergillus fumigatus. *Colloids Surf. B Biointerf.* 2006, *47* (2), 160–164.

Bilo, F.; Pandini, S.; Sartore, L.; Depero, L.; Gargiulo, G.; Bonassi, A. et al. A sustainable bioplastic obtained from rice straw. *J. Clean. Prod.* **2018,** *200*, 357–368.

Blosi, M.; Ortelli, S.; Costa, A. L.; Dondi, M.; Lolli, A.; Andreoli, S. et al. Bimetallic nanoparticles as efficient catalysts: Facile and green microwave synthesis. *Materials* **2016,** *9* (7). DOI: org/10.3390/ma9070550.

Borsoi, A.; Paz, J.; Pissinate, K.; Rambo, R.; Pestana, V.; Bizarro, C. et al. Ultrasound-Assisted Synthesis of 4-Alkoxy-2-Methylquinolines: An Efficient Method Toward Antitubercular Drug Candidates. *Molecules* **2021,** *26* (5). DOI: 10.3390/molecules26051215.

Boulanger, R. R.; Singh, B. R. Light Regulation of Anthocyanin and Flavonol Biosynthesis in Cranberry Plants. In *Abstracts of Papers of the American Chemical Society* (Vol. 215, pp. U194-U194). 1155 16TH ST, NW, Washington, DC 20036: American Chemical Society, **1998,** April.

Brandt, S. D.; Tirunarayanapuram, S. S.; Freeman, S.; Dempster, N.; Barker, S. A.; Daley, P. F. et al. Microwave-Accelerated Synthesis of Psychoactive Deuterated N, N-dialkylated-[α, α, β, β-d4]-Tryptamines. *J. Labelled Compd. Radiopharma.* **2008,** *51* (14), 423–429.

Bunyakiat, K.; Makmee, S.; Sawangkeaw, R.; Ngamprasertsith, S. Continuous Production of Biodiesel via Transesterification from Vegetable Oils in Supercritical Methanol. *Energy Fuels* **2006,** *20* (2), 812–817.

Cann, M. C.; Connelly, M. E. *Real World Cases in Green Chemistry.* American Chemical Society: Washington, DC, 2000.

Cardoso, A.; Horn, M.; Ferret, L.; Azevedo, C.; Pires, M. Integrated Synthesis of Zeolites 4A and Na–P1 Using Coal Fly Ash for Application in the Formulation of Detergents and Swine Wastewater Treatment. *J. Hazard. Mater.* **2015,** *287*, 69–77.

Chairat, M.; Bremner, J. B.; Chantrapromma, K. Dyeing of Cotton and Silk Yarn with the Extracted Dye from the Fruit Hulls of Mangosteen, Garcinia Mangostana Linn. *Fibers Polym.* **2007,** *8* (6), 613–619.

Chairin, T.; Nitheranont, T.; Watanabe, A.; Asada, Y.; Khanongnuch, C.; Lumyong, S. Biodegradation of Bisphenol a and Decolorization of Synthetic Dyes by Laccase from White-Rot Fungus, Trametes polyzon A. *Appl. Biochem. Biotechnol.* **2013,** *169* (2), 539–545.

Chang, H. N.; Hou, S. X.; Hao, Z. C.; Cui, G. H. Ultrasonic Green Synthesis of an Ag/CP Nanocomposite for Enhanced Photodegradation Effectiveness. *Ultrason. Sonochem.* **2018,** *40*, 1039–1048.

Charvin, P.; Stéphane, A.; Florent, L.; Gilles, F. Analysis of Solar Chemical Processes for Hydrogen Production from Water Splitting Thermochemical Cycles. *Energy Conv. Manag.* **2008,** *49* (6), 1547–1556.

Chen, G. Q.; Patel, M. K. Plastics Derived from Biological Sources: Present and Future: A Technical and Environmental Review. *Chem. Rev.* **2012,** *112* (4), 2082–2099.

Chen, L.; Duan, Y.; Cui, M.; Huang, R.; Su, R.; Qi, W. et al. Biomimetic surface coatings for marine antifouling: Natural antifoulants, synthetic polymers and surface microtopography. *Sci. Total Environ.* **2021,** *766*. DOI: 10.1016/j.scitotenv.2020.144469.

Choedkiatsakul, I.; Ngaosuwan, K.; Assabumrungrat, S. Application of Heterogeneous Catalysts for Transesterification of Refined Palm Oil in Ultrasound-Assisted Reactor. *Fuel Proce. Technol.* **2013,** *111*, 22–28.

Chouhan, N.; Chen, C. K.; Chang, W. S.; Chang, K.W.; Liu, R. S. Photoelectrochemical Cells for Hydrogen Generation in Electrochemical Technologies for Energy Storage and Conversion, *2*, Liu, R. S. Zhang, L. Sun, X. Liu, H. Zhang, J. Eds.; Wiley-VCH: Weinheim 2012.

Chu, Z.; Feng, Y. Vegetable-Derived Long-Chain Surfactants Synthesized via a "Green" Route. *ACS Sustain. Chem. Eng.* **2013,** *1* (1), 75–79.

Chua, H.; Peter, H. F.; Ma, C. K. Accumulation of Biopolymers in Activated Sludge Biomass. *Appl. Biochem. Biotechnol.* **1999,** *78* (1), 389–399.

Cintas, P.; Luche, J. L. Green Chemistry: The Sonochemical Approach. *Green Chem.* **1999,** *1* (3), 115–125.

Clark, J. H.; Budarin, V. L.; Macquarrie, D. J.; Breeden, S. W. *Nanotechnology 2012: Bio Sensors, Instruments, Medical, Environment and Energy*, Vol. 3, 2012; Chapter 7.

Contractor, R. M.; Garnett, D. I.; Horowitz, H. S.; Bergna, H. E.; Patience, G. S.; Schwartz, J. T. et al. A New Commercial Scale Process for N-Butane Oxidation to Maleic Anhydride Using a Circulating Fluidized Bed Reactor. *Stud. Surface Sci. Catal.* **1994,** *82*, 233–242.

Cravotto, G.; Cintas, P. Power Ultrasound in Organic Synthesis: Moving Cavitational Chemistry from Academia to Innovative and Large-Scale Applications. *Chem. Soc. Rev.* **2006,** *35* (2), 180–196.

da Silva, I. G. S.; de Almeida, F. C. G.; da Rocha e Silva, N. M. P.; de Oliveira, J. T. R.; Converti, A.; Sarubbo, L. A. Application of Green Surfactants in the Remediation of Soils Contaminated by Hydrocarbons. *Processes* **2021,** *9* (9). DOI: org/10.3390/pr9091666.

Derraik, J. G. The pollution of the Marine Environment by Plastic Debris: A Review. *Mar. Pollut. Bull.* **2002,** *44* (9), 842–852.

Dostal, V.; Driscoll, M. J.; Hejzlar, P. A Supercritical Carbon Dioxide Cycle for Next Generation Nuclear Reactors, MIT-ANP-TR-100, MIT-ANP-Series, 2004.

Draths, K. M.; Frost, J. W. Environmentally Compatible Synthesis of Adipic Acid from D-Glucose. *J. Am. Chem. Soc.* **1994,** *116* (1), 399–400.

Drews, J.; Ryser, S. A Symposium On: The Genetic Basis of Human Disease 1996, Human disease: From Genetic Causes to Biochemical Effects: Proceedings of the Symposium "The Genetic Basis of Human Disease" Held to Celebrate the 100th Anniversary of the Founding of F. Hoffmann-La Roche Ltd; Blackwell Science: Basel, October 2–3, 1997.

Dunn, P. J.; Galvin, S.; Hettenbach, K. The Development of an Environmentally Benign Synthesis of Sildenafil Citrate (Viagra™) and Its Assessment by Green Chemistry Metrics. *Green Chem.* **2004,** *6* (1), 43–48.

Ebrahimi, I.; Gashti, M. P. Extraction of Polyphenolic Dyes from Henna, Pomegranate Rind, and Pterocarya fraxinifolia for Nylon 6 Dyeing. *Color. Technol.* **2016,** *132* (2), 162–176.

El-Gamal, S.; Selim, F. Utilization of Some Industrial Wastes for Eco-Friendly Cement Production. *Sustain. Mater. Technol.* **2017,** *12*, 9–17.

Elhamifar, D.; Ramazani, Z.; Norouzi, M.; Mirbagheri, R. Magnetic Iron Oxide/Phenylsulfonic Acid: A Novel, Efficient and Recoverable Nanocatalyst for Green Synthesis of Tetrahydrobenzo [b] Pyrans Under Ultrasonic Conditions. *J. Colloid Interf. Sci.* **2018,** *511*, 392–401.

El-Sheekh, M. M.; Bases, E. A.; El-Shenody, R. A.; El Shafay, S. M. Lipid Extraction from Some Seaweeds and Evaluation of Its Biodiesel Production. *Biocatal. Agricult. Biotechnol.* **2021,** *35*. DOI: org/10.1016/j.bcab.2021.102087.

Eustis, S.; Hsu, H. Y.; El-Sayed, M. A. Gold Nanoparticle Formation from Photochemical Reduction of Au^{3+} by Continuous Excitation in Colloidal Solutions. A Proposed Molecular Mechanism. *J. Phys. Chem. B.* **2005,** *109* (11), 4811–4815.

Feng, D. Q.; He, J.; Chen, S. Y.; Su, P.; Ke, C. H.; Wang, W. The Plant Alkaloid Camptothecin as a Novel Antifouling Compound for Marine Paints: Laboratory Bioassays and Field Trials. *Marine Biotechnol.* **2018,** *20* (5), 623–638.

Feng, Y.; Zhang, A. A Floral Fragrance, Methyl Benzoate, Is an Efficient Green Pesticide. *Sci. Rep.* **2017,** *7* (1). DOI: 10.1038/srep42168.

Fithriyah, N. H. The Application of Chitosan for Environmentally Benign Process of Curcumin Dyeing of Silk Fabrics. *J. Basic Appl. Scient. Res.* **2013,** *3* (1), 5–14.

Fraser, D. Bursting Bacteria by Release of Gas Pressure. *Nature,* **1951,** *167* (4236), 33–34.

Gao, D.; Zhou, H.; Wang, J.; Miao, S.; Yang, F.; Wang, G. et al. Size-Dependent Electrocatalytic Reduction of CO_2 Over Pd Nanoparticles. *J. Am. Chem. Soc.* **2015,** *137* (13), 4288–4291.

García, A. M.; Hunt, A.; Budarin, V.; Parker, H.; Shuttleworth, P.; Ellis, G. et al. Starch-Derived Carbonaceous Mesoporous Materials (Starbon®) for the Selective Adsorption and Recovery of Critical Metals. *Green Chem.* **2015,** *17* (4), 2146–2149.

Gardea-Torresdey, J. L.; Parsons, J. G.; Gomez, E.; Peralta-Videa, J.; Troiani, H. E.; Santiago, P. et al. Formation and Growth of Au Nanoparticles Inside Live alfalfa Plants. *Nano Lett.* **2002,** *2* (4), 397–401.

Ghazani, S. M.; Dobson, S.; Marangoni, A. G. Hardness, Plasticity, and Oil Binding Capacity of Binary Mixtures of Natural Waxes in Olive Oil. *Curr. Res. Food Sci.* **2022,** *5*, 998–1008.

Gomes, C. N.; Blondiaux, E.; Thuéry, P.; Cantat, T. Metal-Free Reduction of CO_2 with Hydroboranes: Two Efficient Pathways at Play for the Reduction of CO_2 to Methanol. *Chem. Eur. J.* **2014,** *20* (23), 7098–7106.

Gundluru, M.; Mallu, K. K. R.; Sarva, S.; Cirandur, S. R. Green and Eco-Friendly Synthesis of α-Hydroxyphosphonates as Antioxidant and Antimicrobial Agents. *J. Mol. Struct.* **2022,** 1256. DOI: 10.1016/j.molstruc.2022.132554.

Habib, N.; Akram, W.; Adeel, S.; Amin, N.; Hosseinnezhad, M. Environmental-Friendly Extraction of Peepal (*Ficus Religiosa*) Bark-Based Reddish Brown Tannin Natural Dye for Silk Coloration. *Environmen. Sci. Pollut. Res.* **2022,** *29* (23), 35048–35060.

Habisreutinger, S. N.; Schmidt-Mende, L.; Stolarczyk, J. K. Photocatalytic Reduction of CO_2 on TiO_2 and Other Semiconductors. *Angew. Chemie Int. Ed.* **2013,** *52* (29), 7372–7408.

Hampel, M.; Blasco, J. Toxicity of Linear Alkylbenzene Sulfonate and One Long-Chain Degradation Intermediate, Sulfophenyl Carboxylic Acid on Early Life-Stages of Seabream (Sparus aurata). *Ecotoxicol. Environ. Saf.*, **2002,** *51* (1), 53–59.

Hollmuller, P.; Joubert, J.-M.; Lachal, B.; Yvon, K. Evaluation of a 5 kWp Photovoltaic Hydrogen Production and Storage Installation for a Residential Home in Switzerland. *Inter. J. Hydrogen Energy*, **2000,** *25*, 97–109.

Horiguchi, I.; Saeki, N. *Compressive Strength and Leachate Characteristics of New Green CLSM with Eco-Cement and Melted Slag from Municipal Solid Waste* Special Publication, **2004,** *221*, 529–558.

Hossain, S. S.; Mathur, L.; Majhi, M. R.; Roy, P. K. Manufacturing of Green Building Brick: Recycling of Waste for Construction Purpose. *J. Mater. Cycles Waste Manag.* **2019,** *21* (2), 281–292.

Hou, F.; Xian, M.; Huang, W. De Novo Biosynthesis and Whole-Cell Catalytic Production of Paracetamol on a Gram Scale in *Escherichia coli. Green Chem.* **2021,** *23*. DOI: 10.1039/d1gc02591k.

Hsiao, Y. S.; Yellol, G. S.; Chen, L. H.; Sun, C. M. Multidisciplinary Synthetic Approach for Rapid Combinatorial Library Synthesis of Triaza-Fluorenes. *J. Comb. Chem.* **2010,** *12* (5), 723–732.

Hussaan, M.; Iqbal, N.; Adeel, S.; Azeem, M.; Tariq Javed, M.; Raza, A. Microwave-Assisted Enhancement of Milkweed (*Calotropis procera* L.) Leaves as an Eco-Friendly Source of Natural Colorants for Textile. *Environ. Sci. Poll. Res.* **2016,** *24* (5), 5089–5094.

Inoue, T.; Fujishima, A.; Konishi, S.; Honda, K. Photoelectrocatalytic Reduction of Carbon Dioxide in Aqueous Suspensions of Semiconductor Powders. *Nature*, **1979,** *277* (5698), 637–638.

Isah, A. G. Production of Detergent from Castor Oil. *Leonardo El. J. Pract. Technol.* **2006,** *9*, 153–160.

Jan, W.G. *Encylopedic Dictionary of Polymers*; Springer Science: New York, 2007; pp 108–109.

Jassim, A. Recycling of Polyethylene Waste to Produce Plastic Cement. *Proc. Manuf.* **2017,** *8*, 635–642.

Jessop, P. G.; Leitner, W., Eds. *Chemical Synthesis Using Supercritical Fluids*; Wiley-VCH: Weinheim, 1999.

Jiang, K.; Huang, Y.; Zeng, G.; Toma, F. M.; Goddard III, W. A.; Bell, A. T. Effects of Surface Roughness on the Electrochemical Reduction of CO2 Over Cu. *ACS Energy Lett.* **2020,** *5* (4), 1206–1214.

Jjemba, P. K. *Pharma-Ecology: The Occurrence and Fate of Pharmaceuticals and Personal Care Products in the Environment*; Hoboken, 2008.

Jo, Y. B.; Park, S. H.; Jeon, J. K.; Ko, C. H.; Ryu, C.; Park, Y. K. Biodiesel Production via the Transesterification of Soybean Oil Using Waste Starfish (Asterina pectinifera). *Appl. Biochem. Biotechnol.* **2013,** *170* (6), 1426–1436.

Joerger, R.; Klaus, T.; Granqvist, C. G. Biologically Produced Silver–Carbon Composite Materials for Optically Functional Thin-Film Coatings. *Adv. Mater.* **2000,** *12* (6), 407–409.

Jones, J. R.; Lu, S., Eds. *Microwaves in Organic Synthesis*, 2nd ed.; Wiley-VCH: Weinheim, Germany, 2006; Chapter 18, pp 820–859.

Kamata, R.; Kumagai, H.; Yamazaki, Y.; Sahara, G.; Ishitani, O. Photoelectrochemical CO_2 Reduction Using a Ru (II)–Re (I) Supramolecular Photocatalyst Connected to A Vinyl Polymer on a NiO Electrode. *ACS Appl. Mater. Interf.* **2018,** *11* (6), 5632–5641.

Kappe, C. O.; Dallinger, D. The Impact of Microwave Synthesis on Drug Discovery. *Nat. Rev. Drug Discov.* **2006,** *5* (1), 51–63.

Kasprzyk-Hordern, B.; Dinsdale, R. M.; Guwy, A. J. The Occurrence of Pharmaceuticals, Personal Care Products, Endocrine Disruptors and Illicit Drugs in Surface Water in South Wales, UK. *Water Res.* **2008,** *42* (13), 3498–3518.

Kaur, J.; Kaur, K.; Mehta, S.; Matharu, A. A NOVEL MOLYBDENUM OXIDE–STARBON CATALYST for Wastewater Remediation. *J. Mater. Chem. A*, **2020,** *8* (29), 14519–14527.

Ke, G.; Yu, W.; Xu, W. Color Evaluation of Wool Fabric Dyed with Rhizoma Coptidis Extract. *J. Appl. Polym. Sci.* **2006,** *101* (5), 3376–3380.

Kim, J. G.; Jang, D. O. Indium-Catalyzed N-Formylation of Amines Under Solvent-Free Conditions. *Synlett*, 2010, *2010* (08), 1231–1234.

Kim, S.; De bruyn, M.; Alauzun, J.; Louvain, N.; Brun, N.; Macquarrie, D. et al. Alginic Acid-Derived Mesoporous Carbonaceous Materials (Starbon®) as Negative Electrodes for Lithium Ion Batteries: Importance of Porosity and Electronic Conductivity. *J. Power Sources*, **2018,** *406*, 18–25.

Kočí, K.; Obalová, L.; Matějová, L.; Plachá, D.; Lacný, Z.; Jirkovský, J. et al. Effect of TiO_2 Particle Size on the Photocatalytic Reduction of CO_2. *Appl. Catal. B: Environ.* **2009,** *89* (3–4), 494–502.

Kumagai, H.; Sahara, G.; Maeda, K.; Higashi, M.; Abe, R.; Ishitani, O. Hybrid Photocathode Consisting of a $CuGaO_2$ p-Type Semiconductor and a Ru (II)–Re (I) Supramolecular Photocatalyst: Non-Biased Visible-Light-Driven CO_2 Reduction with Water Oxidation. *Chem. Sci.* **2017,** *8* (6), 4242–4249.

Kümmerer, K. Why Green and Sustainable Pharmacy? In *Green and Sustainable Pharmacy*. Springer: Berlin, Heidelberg, 2010; pp 3–10. DOI: 10.1007/978-3-642-05199-9_1

Lai, F.; Wissing, S. A.; Müller, R. H.; Fadda, A. M. Artemisia Arborescens L Essential Oil-Loaded Solid Lipid Nanoparticles for Potential Agricultural Application: Preparation and Characterization. *Aaps Pharm. Sci. Tech.* **2006,** *7* (1), E10–E18.

Lamsa, K.; Kim, S. W.; Jung, J. H.; Kim, Y. S.; Kim, K. S.; Lee, Y. S. Inhibition Effects of Silver Nanoparticles Against Powdery Mildews on Cucumber and Pumpkin. *Mycobiology* **2011,** *39* (1), 26–32.

Landman, A.; Halabi, R.; Dias, P.; Dotan, H.; Mehlmann, A.; Shter, G. E. et al. Decoupled Photoelectrochemical Water Splitting System for Centralized Hydrogen Production. *Joule* **2020,** *4* (2), 448–471.

Le, M.; Ren, M.; Zhang, Z.;Sprunger, P. T.; Kurtz, R. L.; Flake, J. C. Electrochemical Reduction of CO_2 to CH_3OH at Copper Oxide Surfaces. *J. Electrochem. Soc.* **2011,** *158* (5). DOI: 10.1149/1.3561636.

Lee, Y. H. Dyeing, Fastness, and Deodorizing Properties of Cotton, Silk, and Wool Fabrics Dyed with Coffee Sludge (Coffea arabica L.) extract. *J. Appl. Polym. Sci.* **2007,** *103* (1), 251–257.

León, V. M.; Gómez-Parra, A.; González-Mazo, E. Biodegradation of Linear Alkylbenzene Sulfonates and Their Degradation Intermediates in Seawater. *Environ. Sci. Technol.* **2004,** *38* (8), 2359–2367.

León, V. M.; López, C.; Lara-Martín, P. A.; Prats, D.; Varó, P.; González-Mazo, E. Removal of Linear Alkylbenzene Sulfonates and Their Degradation Intermediates at Low Temperatures During Activated Sludge Treatment. *Chemosphere* **2006,** *64* (7), 1157–1166.

Li, H.; Shao, F. Q.; Huang, H.; Feng, J. J.; Wang, A. J. Eco-Friendly and Rapid Microwave Synthesis of Green Fluorescent Graphitic Carbon Nitride Quantum Dots for Vitro Bioimaging. *Sens. Actuat. B: Chem.* **2016,** *226*, 506–511.

Li, Z.; Huang, H.; Sun, H.; Jiang, H.; Liu, H. Microwave-Assisted Efficient and Convenient Synthesis of 2, *4* (1 H, 3 H)-Quinazolinediones and 2-Thioxoquinazolines. *J. Comb. Chem.* **2008,** *10* (3), 484–486.

Lidström, P.; Tierney, J.; Watheyb, B.; Westmana, J. Microwave Assisted Organic SynthesisÐa Review. *Tetrahedron* **2001,** *57*, 9225–9283.

Lin, P. T.; Salunke, D. B.; Chen, L. H.; Sun, C. M. Soluble Polymer Supported Divergent Synthesis of Tetracyclic Benzene-Fused Pyrazino/Diazepino Indoles: An Advanced Synthetic Approach to Bioactive Scaffolds. *Org. Biomol. Chem.* **2011,** *9* (8), 2925–2937.

Liu, H.; Zhang, L.; Eds.; *Microwave Heating*; InTech: New York, USA, 2011; Chapter 16.

Liu, R. S.; Zhang, L.; Sun, X.; Liu, H.; Zhang, J., Eds. *Electrochemical Technologies for Energy Storage and Conversion*, Vol. 2; Wiley-VCH: Weinheim, Germany, 2012; Chapter 1.

Løbner, T.; Toräng, L.; Batstone, D. J.; Schmidt, J. E.; Angelidaki, I. Effects of Process Stability on Anaerobic Biodegradation of LAS in UASB Reactors. *Biotechnol. Bioeng.* **2005,** *89* (7), 759–765.

Low, S. Y.; Tan, J. Y.; Ban, Z. H.; Siwayanan, P. Performance of Green Surfactants in the Formulation of Heavy-Duty Laundry Liquid Detergents (HDLD) with Special Emphasis on Palm Based Alpha Methyl Ester Sulfonates (α-MES). *J. Oleo Sci.* **2021,** *70* (8), 1027–1037.

Lv, Y.; Sun, S.; Liu, J. Biodiesel Production Catalyzed by a Methanol-Tolerant Lipase a from Candida Antarctica in the Presence of Excess Water. *ACS Omega* **2019,** *4* (22), 20064–20071.

Mafuné, F.; Kohno, J. Y.; Takeda, Y.; Kondow, T. Full Physical Preparation of Size-Selected Gold Nanoparticles in Solution: Laser Ablation and Laser-Induced Size Control. *J. Phys. Chem. B*, **2002,** *106* (31), 7575–7577.

Martínez-Palou, R. Ionic Liquid and Microwave-Assisted Organic Synthesis: A "Green" and Synergic Couple. *J. Mexican Chem. Soc.* **2007,** *51* (4), 252–264.

Mason, T. J. Ultrasound in Synthetic Organic Chemistry. *Chem. Soc. Rev.* **1997,** *26* (6), 443–451.

Matics, J.; Krost, G. Prospective and Adaptive Management of Small Combined Heat and Power Systems in Buildings. Proceeding of the 9th REHVA World Congress CLIMA, Helsinki Finland, 2007.

Mogensen, A. S.; Haagensen, F.; Ahring, B. K. Anaerobic Degradation of Linear Alkylbenzene Sulfonate. *Environ. Toxicole. Chem.* **2003,** *22* (4), 706–711.

Morikawa, T.; Sato, S.; Arai, T.; Uemura, K.; Yamanaka, K. I.; Suzuki, T. M. et al. Selective CO_2 Reduction Conjugated with H_2O Oxidation Utilizing Semiconductor/Metal-Complex Hybrid Photocatalysts. *AIP Conf. Proc.* **2013,** *1568* (1), 11–15.

Morrissette, J.; Carroll, P.; Bayer, I.; Qin, J.; Waldroup, D.; Megaridis, C. A Methodology to Produce Eco-Friendly Superhydrophobic Coatings Produced from All-Water-Processed Plant-Based Filler Materials. *Green Chem.* **2018,** *20* (22), 5169–5178.

Moustafa, M.; Awad, M.; Amer, A.; Hassan, N.; Ibrahim, E.; Ali, H. et al. Insecticidal Activity of Lemongrass Essential Oil as an Eco-Friendly Agent Against the Black Cutworm Agrotis ipsilon (Lepidoptera: Noctuidae). *Insects*, **2021,** *12* (8). DOI: 10.3390/insects12080737.

Murray, J. K.; Farooqi, B.; Sadowsky, J. D.; Scalf, M.; Freund, W. A.; Smith, L. M. et al. Efficient Synthesis of a β-Peptide Combinatorial Library with Microwave Irradiation. *J. Am. Chem. Soc.* **2005,** *127* (38), 13271–13280.

Nagendrappa, G. Organic Synthesis Using Clay Catalysts. *Resonance* **2002,** *7* (1), 64–77.

Nassar, R.-Ud-D.; Soroushian P. Field Investigation of Concrete Incorporating Milled Waste Glass. *J. Solid Waste Technol. Manag.* **2011,** *37* (4), 307–319.

Nielsen, K. B.; Brandt, K. K.; Jacobsen, A. M.; Mortensen, G. K.; Sørensen, J. Influence of Soil Moisture on Linear Alkylbenzene Sulfonate-Induced Toxicity in Ammonia-Oxidizing Bacteria. *Environ. Toxicol. Chem. Int. J.* **2004,** *23* (2), 363–370.

Ning, J.; Kong, F.; Lin, B.; Lei, H. Large-Scale Preparation of the Phytoalexin Elicitor Glucohexatose and Its Application as a Green Pesticide. *J. Agricul. Food Chem.* **2003,** *51* (4), 987–991.

Nouri, A.; Yaraki, M. T.; Lajevardi, A.; Rezaei, Z.; Ghorbanpour, M.; Tanzifi, M. Ultrasonic-Assisted Green Synthesis of Silver Nanoparticles Using Mentha Aquatica Leaf Extract for Enhanced Antibacterial Properties and Catalytic Activity. *Colloid Interf. Sci. Commun.* **2020,** *35*. DOI: org/10.1016/j.colcom.2020.100252.

Nun, P.; Martinez, J.; Lamaty, F. Microwave-Assisted Neat Procedure for the Petasis Reaction. *Synthesis* **2010,** *2010* (12), 2063–2068.

Nuramdhani, I.; Widodo, M.; Harnirat, H.; Juhana, J. The Application of Chitosan and Cationic Fixing Agent for Environmentally Benign Process of Black Tea Dyeing of Cotton and Polyamide Fabrics. *Proceedings of the 2nd International Seminar on Chemistry*; Jatinangor, 2011; pp 347–352.

Oakes, R. S.; Clifford, A.; Bartle, K.; Pett, M.; Rayner, C. Sulfur Oxidation in Supercritical Carbon Dioxide: Dramatic Pressure Dependant Enhancement of Diastereoselectivity for Sulfoxidation of Cysteine Derivatives. *Chem. Commun.* **1999,** *3*, 247–248.

Ogunyemi, S. O.; Abdallah, Y.; Zhang, M.; Fouad, H.; Hong, X.; Ibrahim, E. et al. Green Synthesis of Zinc Oxide Nanoparticles Using Different Plant Extracts and Their Antibacterial

Activity Against Xanthomonas oryzaepv. oryzae. *Artif. Cells Nanomed. Biotechnol.* **2019,** *47* (1), 341–352.

Padil, V. V. T.; Senan, C.; Wacławek, S.; Černík, M.; Agarwal, S.; Varma, R. S. Bioplastic Fibers from Gum Arabic for Greener Food Wrapping Applications. *ACS Sustain. Chem. Eng.* **2019,** *7* (6), 5900–5911.

Patzke, H.; Schieber, A. Growth-Inhibitory Activity of Phenolic Compounds Applied in an Emulsifiable Concentrate—Ferulic Acid as a Natural Pesticide Against Botrytis Cinerea. *Food Res. Int.* **2018,** *113*, 18–23.

Pérez, M.; García, M.; Ruiz, D.; Autino, J. C.; Romanelli, G.; Blustein, G. Antifouling Activity of Green-Synthesized 7-Hydroxy-4-Methylcoumarin. *Mar. Environ. Res.* **2016,** *113*, 134–140.

Pérez, M.; Sánchez, M.; García, M.; Patiño, C. L.; Blustein, G.; Palermo, J. Antifouling Activity of Peracetylated Cholic Acid A Natural Bile Acid Derivative. *Steroids*, **2019,** *149*. DOI: 10.1016/j.steroids.2019.05.006.

Pérez-Larios, A.; Lopez, R.; Hernandez-Gordillo, A.; Tzompantzi, F.; Gómez, R.; Torres-Guerra, L. M. Improved Hydrogen Production from Water Splitting Using TiO_2–ZnO Mixed Oxides Photocatalysts. *Fuel* **2012,** *100*, 139–143.

Phair, J. W. Green Chemistry for Sustainable Cement Production and Use. *Green Chem.* **2006,** *8* (9), 763–780.

Piqani, B.; Zhang, W. Synthesis of Diverse Dihydropyrimidine-Related Scaffolds by Fluorous Benzaldehyde-Based Biginelli Reaction and Post-Condensation Modifications. *Beilstein J. Org. Chem.* **2011,** *7* (1), 1294–1298.

Poliakoff, M.; Licence, P. Green chemistry. *Nature*, **2007,** *450* (7171), 810–812.

Prabhu, K. H.; Teli, M. D.; Waghmare, N. G. Eco-Friendly Dyeing Using Natural Mordant Extracted from *Emblica officinalis* G. Fruit on Cotton and Silk Fabrics with Antibacterial Activity. *Fibers Polym.* **2011,** *12* (6), 753.

Ramakrishnan, N.; Sharma, S.; Gupta, A.; Alashwal, B. Keratin Based Bioplastic Film from Chicken Feathers and Its Characterization. *Int. J. Biol.Macromol.* **2018,** *111*, 352–358.

Rao, K.; Paria, S. Use of Sulfur Nanoparticles as a Green Pesticide on *Fusarium solani* and *Venturia inaequalis* Phytopathogens. *RSC Adv.* **2013,** *3* (26). DOI: 10.1039/c3ra40500a.

Rather, L. J.; Shahid-ul-Islam; Shabbir, M.; Bukhari, M. N.; Shahid, M.; Khan, M. A.; Mohammad, F. Ecological Dyeing of Woolen Yarn with *Adhatoda vasica* Natural Dye in the Presence of Biomordants as an Alternative Copartner to Metal Mordants. *J. Environ. Chem. Eng.* **2016,** *4* (3), 3041–3049.

Rattner, B. A.; Whitehead, M. A.; Gasper, G.; Meteyer, C. U.; Link, W. A.; Taggart, M. A. et al. Apparent Tolerance of Turkey Vultures (Cathartes aura) to the Non-Steroidal Anti-Inflammatory Drug Diclofenac. *Environ. Toxicol. Chem. Int. J.* **2008,** *27* (11), 2341–2345.

Ritter, S. K. As Quoted in "Green Solution to Global Problems". *Chem. Eng. News* **2003,** *81*, 31–33.

Rodriguez-Sanchez, L.; Blanco, M. C.; López-Quintela, M. A. Electrochemical Synthesis of Silver Nanoparticles. *J. Phys. Chem. B.* **2000,** *104* (41), 9683–9688.

Romdhani, Z.; Sakji, N.; Hamdaoui, M. Eco-Friendly Dyeing of Wool Fabrics with Natural Dye Extracted from Citrus Sinensis L Peels. *Fibers Polym.* **2022,** *23*, 1621–1630.

Sahara, G.; Kumagai, H.; Maeda, K.; Kaeffer, N.; Artero, V.; Higashi, M. et al. Photoelectrochemical Reduction of CO_2 Coupled to Water Oxidation Using a Photocathode with a Ru (II)–Re (I) Complex Photocatalyst and a CoO_x/TaON Photoanode. *J. Am. Chem. Soc.* **2016,** *138* (42), 14152–14158.

Sanderson, K. Chemistry: It's Not Easy Being Green. *Nat. News* **2011,** *469* (7328), 18–20.

Sathicq, Á.; Paola, A.; Pérez, M.; Dallesandro, O.; García, M.; Roldán, J. et al. Furylchalcones as New Potential Marine Antifoulants. *Int. Biodeter. Biodegrad.* **2019,** *143*. DOI: 10.1016/j.ibiod.2019.104730

Sato, K.; Aoki, M.; Noyori, R. A "Green" Route to Adipic Acid: Direct Oxidation of Cyclohexenes with 30 Percent Hydrogen Peroxide. *Science* **1998,** *281* (5383), 1646–1647.

Sato, S.; Arai, T.; Morikawa, T.; Uemura, K.; Suzuki, T. M.; Tanaka, H. et al. Selective CO_2 Conversion to Formate Conjugated with H_2O Oxidation Utilizing Semiconductor/Complex Hybrid Photocatalysts. *J. Am. Chem. Soc.* **2011,** *133* (39), 15240–15243.

Schowanek, D.; David, H.; Francaviglia, R.; Hall, J.; Kirchmann, H.; Krogh, P. H. et al. Probabilistic Risk Assessment for Linear Alkylbenzene Sulfonate (LAS) in Sewage Sludge Used on Agricultural Soil. *Regulatory Toxicol. Pharmacol.* **2007,** *49* (3), 245–259.

Sen, S.; Liu, D.; Palmore, G. T. R. Electrochemical Reduction of CO_2 at Copper Nanofoams. *ACS Catal.* **2014,** *4* (9), 3091–3095.

Shabbir, M.; Islam, S.; Bukhari, M.; Rather, L.; Khan, M.; Mohammad, F. Application of Terminalia Chebula Natural Dye on Wool Fiber—Evaluation of Color and Fastness Properties. *Textiles Cloth. Sustain.* **2016,** *2* (1). DOI: 10.1186/s40689–016–0011–8

Shankar, S. S.; Rai, A.; Ankamwar, B.; Singh, A.; Ahmad, A.; Sastry, M. Biological Synthesis of Triangular Gold Nanoprisms. *Nat. Mater.* **2004,** *3* (7), 482–488.

Shanker, R.; Vankar, P. S. Dyeing Cotton, Wool and Silk with Hibiscus Mutabilis (Gulzuba). *Dyes Pigm.* **2007,** *74* (2), 464–469.

Sheldon, R. A.; Arends, I. *Green Chemistry and Catalysis*; Wiley–VCH: Indianapolis, 2006.

Shoueir, K. R. Green Microwave Synthesis of Functionalized Chitosan with Robust Adsorption Capacities for Cr (VI) and/or RhB in Complex Aqueous Solutions. *Environ. Sci. Pollut. Res.* **2020,** *27* (26), 33020–33031.

Simone, R. D.; Chini, M. G.; Bruno, I.; Riccio, R.; Mueller, D.; Werz, O. et al. Structure-Based Discovery of Inhibitors of Microsomal Prostaglandin E2 Synthase– 1, 5-Lipoxygenase and 5-Lipoxygenase-Activating Protein: Promising Hits for the Development of New Anti-Inflammatory Agents. *J. Med. Chem.* **2011,** *54* (6), 1565–1575.

Simone, R. D.; Chini, M. G.; Bruno, I.; Riccio, R.; Mueller, D.; Werz, O.; Bifulco, G. *J. Med. Chem.* **2011,** *54*, 1565–1575.

Singh, S.; Khemariya, P.; Rai, A.; Rai, A.; Koley, T.; Singh, B. Carnauba Wax-Based Edible Coating Enhances Shelf-Life and Retain Quality of Eggplant (Solanum melongena) Fruits. LWT, **2016,** *74*, 420–426.

Soltani, R.; Dinari, M.; Mohammadnezhad, G. Ultrasonic-assisted synthesis of novel nanocomposite of poly (vinyl alcohol) and amino-modified MCM-41: A green adsorbent for Cd (II) removal. *Ultrason. Sonochem.* **2018,** *40*, 533–542.

Sreeram, K. J.; Srinivasan, R.; Devi, J. M.; Nair, B. U.; Ramasami, T. Cerium Molybdenum Oxides for Environmentally Benign Pigments. *Dyes Pigm.* **2007,** *75* (3), 687–692.

Sumpter, J. P.; Johnson, A. C. 10th Anniversary Perspective: Reflections on Endocrine Disruption in the Aquatic Environment: From Known Knowns to Unknown Unknowns (and Many Things in Between). *J. Environ. Monitor.* **2008,** *10* (12), 1476–1485.

Takahashi, A.; Takeda, K.; Ohnishi, T. Light-Induced Anthocyanin Reduces the Extent of Damage to DNA in UV-Irradiated Centaurea Cyanus Cells in Culture. *Plant Cell Physiol.* **1991,** *32* (4), 541–547.

Tambe, C.; Graiver, D.; Narayan, R. Moisture Resistance Coating of Packaging Paper from Biobased Silylated Soybean Oil. *Prog. Org. Coatings* **2016,** *101*, 270–278.

Tayel, A. A.; EL-TRAS, W. F.; Moussa, S.; EL-BAZ, A. F.; Mahrous, H.; Salem, M. F.; Brimer, L. Antibacterial Action of Zinc Oxide Nanoparticles Against Foodborne Pathogens. *J. Food Safety*, **2011,** *31* (2), 211–218.

Tentu, R. D.; Basu, S. Photocatalytic Water Splitting for Hydrogen Production. *Curr. Opin. Electrochem.* **2017,** *5* (1), 56–62.

Tornow, C. E.; Thorson, M. R.; Ma, S.; Gewirth, A. A.; Kenis, P. J. Nitrogen-Based Catalysts for the Electrochemical Reduction of CO_2 to CO. *J. Am. Chem. Soc.* **2012,** *134* (48), 19520–19523.

Treguer, M.; de Cointet, C.; Remita, H.; Khatouri, J.; Mostafavi, M.; Amblard, J. et al. Dose Rate Effects on Radiolytic Synthesis of Gold– Silver Bimetallic Clusters in Solution. *J. Phys. Chem. B* **1998,** *102* (22), 4310–4321.

Tundo, P.; Anastas, P. T., Eds. *Green Chemistry: Challenging Perspectives*; Oxford University Press: Oxford, 2000.

Umemura, K.; Hamano, R.; Komatsu, H.; Ikuno, T.; Siswoyo, E. Dispersion of Carbon Nanotubes with "Green" Detergents. *Molecules* **2021,** *26* (10). DOI: 10.3390/molecules 26102908.

Usui, Y.; Sato, K. A Green Method of Adipic Acid Synthesis: Organic Solvent-and Halide-Free Oxidation of Cycloalkanones with 30% Hydrogen Peroxide. *Green Chem.* **2003,** *5* (4), 373–375.

Vanderford, B. J.; Snyder, S. A. Analysis of pharmaceuticals in Water by Isotope Dilution Liquid Chromatography/Tandem Mass Spectrometry. *Environ. Sci. Technol.* **2006,** *40* (23), 7312–7320.

Vankar, P. S.; Shanker, R.; Mahanta, D.; Tiwari, S. C. Ecofriendly Sonicator Dyeing of Cotton with *Rubia cordifolia* Linn. Using Biomordant. *Dyes Pigm.* **2008,** *76* (1), 207–212.

Vankar, P. S.; Shanker, R.; Verma, A. Enzymatic Natural Dyeing of Cotton and Silk Fabrics Without Metal Mordants. *J. Clean. Prod.* **2007,** *15* (15), 1441–1450.

Vasilev, A.; Deligeorgiev, T.; Gadjev, N.; Kaloyanova, S.; Vaquero, J. J.; Alvarez-Builla, J.; Baeza, A. G. Novel Environmentally Benign Procedures for the Synthesis of Styryl Dyes. *Dyes Pigm.* **2008,** *77* (3), 550–555.

Videla, C.; Gaedicke, C. Modeling Portland Blast-Furnace Slag Cement High-Performance Concrete. *Mater. J.* **2004,** *101* (5), 365–375.

Vilchis-Nestor, A. R.; Sánchez-Mendieta, V.; Camacho-López, M. A.; Gómez-Espinosa, R. M.; Camacho-López, M. A.; Arenas-Alatorre, J. A. Solventless Synthesis and Optical Properties of Au and Ag Nanoparticles Using Camellia Sinensis Extract. *Mater. Lett.* **2008,** *62* (17–18), 3103–3105.

Vishnu, V. S.; George, G.; Divya, V.; Reddy, M. L. P. Synthesis and Characterization of New Environmentally Benign Tantalum-Doped $Ce_{0.8}Zr_{0.2}O_2$ Yellow Pigments: Applications in Coloring of Plastics. *Dyes Pigm.* **2009,** *82* (1), 53–57.

Wang, Y.; Wu, H. C.; Li, V. C. Concrete Reinforcement with Recycled Fibers. *J. Mater. Civil Eng.* **2000,** *12* (4), 314–319.

Wardencki, W.; Cury, J.; Namieoenik, J. Green Chemistry—Current and Future Issues. *Polish J. Environ. Studies* **2005,** *14* (4), 389–395.

Weller, H. N.; Nirschl, D. S.; Petrillo, E. W.; Poss, M. A.; Andres, C. J.; Cavallaro, C. L.; Echols, M. M.; Grant-Young, K. A.; Houston, J. G.; Miller, A. V.; Swann, R. T. J. *Comb. Chem.* **2006,** *8*, 664–669.

Wingens, J.; Krost, G.; Ostermann, D.; Damm, U.; Hess, J. Hydrogen Production for Autonomous Solar Based Electricity Supply, DRPT2008, Nanjing, China, 2008.

Winkler-Moser, J. K.; Anderson, J. A.; Hwang, H. S. Texture and Flavor Evaluation of Peanut Butter Stabilized with Natural Waxes. *J. Food Scie.* **2022,** *87* (4), 1851–1864.

Xia, Q.; Chen, C.; Yao, Y.; Li, J.; He, S.; Zhou, Y. et al. A Strong, biodegradable and Recyclable Lignocellulosic Bioplastic. *Nat. Sustain.* **2021,** *4* (7), 627–635.

Xie, X.; Tang, Y. Efficient Synthesis of Simvastatin by Use of Whole-Cell Biocatalysis. *Appl. Environ. Microbio.* **2007,** *73* (7), 2054–2060.

Xing, C.; Zhang, Y.; Yan, W.; Guo, L. Band Structure-Controlled Solid Solution of $Cd_{1-x}Zn_xS$ Photocatalyst for Hydrogen Production by Water Splitting. *Int. J. Hydrogen Energy*, **2006,** *31* (14), 2018–2024.

Yamada, M.; Morimitsu, S.; Hosono, E.; Yamada, T. Preparation of Bioplastic Using Soy Protein. *Int. J. Biol. Macromol.* **2020,** *149*, 1077–1083.

Yang, X.; Cai, J.; Chen, L.; Cao, X.; Liu, H.; Liu, M. Green Detergent Made of Halloysite Nanotubes. *Chem. Eng. J.* **2021,** *425*. DOI: 10.1016/j.cej.2021.130623.

Yazici, S.; Hasan, S. A. Effects of Fly Ash Fineness on the Mechanical Properties of Concrete, Sadhana. *Indian Acad. Sci.* **2012,** *37* (3), 389–403.

Ye, D.; Wang, J.; Zhang, X.; Zhou, Y.; Ding, X.; Feng, E. et al. Gold-Catalyzed Intramolecular Hydroamination of Terminal Alkynes in Aqueous Media: Efficient and Regioselective Synthesis of Indole-1-Carboxamides. *Green Chem.* **2009a,** *11* (8), 1201–1208.

Ye, D.; Zhang, X.; Zhou, Y.; Zhang, D.; Zhang, L.; Wang, H. et al. Gold- and Silver-Catalyzed Intramolecular Hydroamination of Terminal Alkynes: Water-Triggered Chemo- and Regioselective Synthesis of Fused Tricyclic Xanthines. *Adv. Synth. Catal.* **2009b,** *351* (17), 2770–2778.

Yeob, S.-D.; Kirana, E. Formation of Polymer Particles with Supercritical Fluids: A Review. *J. Supercrit. Fluids* **2005,** *34* (3), 287–308.

Yu, T.; Wang, H.; Guo, C.; Zhai, Y.; Yang, J.; Yuan, J. A Rapid Microwave Synthesis of Green-Emissive Carbon Dots with Solid-State Fluorescence and pH-Sensitive Properties. *Royal Soc. Open Sci.* **2018,** *5* (7). DOI: 10.1098/rsos.180245.

Yusuf, M.; Mohammad, F.; Shabbir, M. Eco-Friendly and Effective Dyeing of Wool with Anthraquinone Colorants Extracted from Rubia Cordifolia Roots: Optimization, Colorimetric and Fastness Assay. *J. King Saud Uni.–Sci.* **2017,** *29* (2), 137–144.

Yvon, K.; Lorenzoni, J. L. Hydrogen-Powered Lawn Mower: 14 Years of Operation. *Int. J. Hydro. Energ.* **2006,** *31* (12), 1763–1767.

Zabaniotou, A.; Theofilou, C. Green Energy at Cement Kiln in Cyprus—Use of Sewage Sludge as A Conventional Fuel Substitute. *Renew. Sustain. Energy Rev.* **2008,** *12* (2), 531–541.

Zhang, G.; Wang, D. Fabrication of Heterogeneous Binary Arrays of Nanoparticles via Colloidal Lithography. *J. Am. Chem. Soc.* **2008,** *130* (17), 5616–5617.

Zhang, J.; Zheng, J.; Yang, W. Green Supercapacitor Assisted Photocatalytic Fuel Cell System for Sustainable Hydrogen Production. *Chem. Eng. J.* **2021,** *403*. DOI: org/10.1016/j.cej.2020.126368.

Zhang, Q. H.; Han, W. D.; Hong, Y. J.; Yu, J. G. Photocatalytic Reduction of CO_2 with H_2O on Pt-loaded TiO_2 Catalyst. *Catal. Today* **2009,** *148* (3–4), 335–340.

Zhao, Y.; Qiu, J.; Ma, Z.; Sun, X. Eco-Friendly Treatment of Coal Gangue for Its Utilization as Supplementary Cementitious Materials. *J. Clean. Prod.* **2021,** *285*. DOI: 10.1016/j.jclepro.2020.124834.

Zhou, Y.; Zhai, Y.; Li, J.; Ye, D.; Jiang, H.; Liu, H. Metal-Free Tandem Reaction in Water: An Efficient and Regioselective Synthesis of 3-Hydroxyisoindolin-1-Ones. *Green Chem.* **2010,** *12* (8), 1397–1404.

Zhu, G.; Yin, H.; Yang, C.; Cui, H.; Wang, Z.; Xu, J. et al. Black Titania for Superior Photocatalytic Hydrogen Production and Photoelectrochemical Water Splitting. *ChemCatChem* **2015,** *7* (17), 2614–2619.

Zhu, W.; Michalsky, R.; Metin, O.; Lv, H.; Guo, S.; Wright, C. J. et al. Monodisperse Au Nanoparticles for Selective Electrocatalytic Reduction of CO_2 to CO. *J. Am. Chem. Soc.* **2013,** *135* (45), 16833–16836.

CHAPTER 4

Green Catalysts

MONIKA JANGID[1], SHIKHA PANCHAL[1], YUVRAJ JHALA[2], ANURADHA SONI[3], and SURESH C. AMETA[1]

[1]*Department of Chemistry, PAHER University, Udaipur, India*

[2]*Department of Chemistry, B. N. University, Udaipur, India*

[3]*Department of Chemistry, M. L. Sukhadia University, Udaipur, India*

ABSTRACT

Many chemical reactions are relatively slow and it is important to increase their rate by the use of some catalysts. This will reduce the time taken for completion of that reaction. This is also a green chemical approach. This can be achieved by using some catalysts. Oragnocatalysts and metallocatalysts have been used successfully in a number of reactions, but the use of biocatalysts is a preferred one for this purpose. Biocatalysts such as Baker's yeast, Rhizopus oryzae lipase (ROL), lysozyme, lyase, Lactobacillus kefiri P2, phytase, etc. have been quite commonly used. Apart from these, the role of organocatalysts and metallocatalysts cannot be ignored. The use of different catalysts in a variety of chemical reactions has been discussed.

4.1 INTRODUCTION

Many organic reactions of synthetic importance are very slow, and it is very important to enhance their reaction rates. The rate of the reaction can be enhanced by using a catalyst. This catalyst may be toxic in nature and it is important to find out some alternative catalyst, which is harmless or less toxic. Just to avoid the environmental pollution, such a job can be done by any enzyme also. These enzymes are called biocatalyst or in general

Green Chemistry, 2nd Edition: Fundamentals and Applications. Suresh C. Ameta & Rakshit Ameta (Eds.)

green catalyst. Enzymes are used in the chemical industries when extremely specific catalysts are required. However, the use of enzymes is limited because of their stability in organic solvents and higher temperatures. Searching or creating new enzymes with novel properties, either through rational designing or in vitro evolution (Hult and Berglund, 2003; Renugopalakrishnan et al., 2005) is a challenging task for chemists. A few enzymes have now been designed from scratch to catalyzed reactions that do not occur in nature (Jiang et al., 2008).

4.2 BIOCATALYSTS

Lipase catalyzed double-enantioselective transesterifications of racemic carboxylic esters and cyclic *meso*-diols to give the hydroxy esters have been investigated by Theil et al. (1994). The penicillin acylase catalyzed synthesis of ampicillin, *via* the acylation of 6-aminopenicillanic acid with D-phenylglycine amide, is accompanied by the formation of the hydrolysis product D-phenylglycine (Langen et al., 2001). The immobilized resting-cell of *Geotrichum candidum* was used as a catalyst for the reduction of a ketone in a semicontinuous flow process using supercritical carbon dioxide (Matsuda et al., 2003) while Shao et al. (2002) carried out biocatalytic synthesis of uridine 5′-diphosphate *N*-acetylglucosamine by multiple enzymes co-immobilized on agarose beads.

Using Bakers' yeast as a biocatalyst, the chemoselective reduction of aromatic nitro compounds bearing electron-withdrawing groups gave the corresponding hydroxylamines with good to excellent conversion under mild conditions (Li et al., 2004). Bohn et al. (2007) investigated that caffeine affects the stereoselectivity of microbial high cell density reductions with commercial grade *Saccharomyces cerevisiae* (Baker's yeast), while a novel Baker's yeast (*Saccharomyces cerevisiae*) catalyzed protocol for Knoevenagel condensation of aldehydes and active methylene compounds including 2,4-thiazolidinedione in an organic solvent at ambient temperature has been developed (Pratap et al., 2011).

New bacterial alcohol dehydrogenases with high and complementary enantioselectivity for the reduction of ethyl 3-keto-4,4,4-trifluorobutyrate and methyl 3-keto-3-(3′-pyridyl)-propionate have been investigated by Zhang et al. (2004). Stereoselective reductase-catalyzed asymmetric deoxygenation of racemic alkylaryl, dialkyl, and phenolic sulfoxides was observed by Boyd et al. (2004). Cytochrome $P450_{BM3}$, from *Bacillus megaterium*, catalyzes the epoxidation of linolenic acid yielding 15,16-epoxyoctadeca-9,12-dienoic acid with complete regio- and moderate enantioselectivity (60% ee) (Çelik et al., 2005). Wu et al. (2005) have reported a new enzymatic process, where penicillin G acylase from *Escherichia coli* displays a promiscuous activity in catalyzing the Markovnikov addition of allopurinol to vinyl ester.

A mediatorless microbial fuel cell based on the direct biocatalysis of *E. coli* shows significantly enhanced performance by using bacteria electrochemically evolved in fuel cell environments through a natural selection process and a carbon/PTFE composite anode with an optimized PTFE content (Zhang et al., 2006). The enantioselective synthesis of (2*S*)-2-phenylpropanol and (2*S*)-2-(4-*iso*-butylphenyl)propanol ((*S*)-Ibuprofenol) has been achieved by horse liver alcohol dehydrogenase through dynamic kinetic resolution (Giacomini et al., 2007) Ríos et al. (2007) investigated green method for Baeyer–Villiger oxidation of substituted cyclohexanones via lipase-mediated perhydrolysis utilizing urea–hydrogen peroxide in ethyl acetate.

DNAzyme cascades activated by Pb^{2+}- or L-histidine-dependent DNAzymes yield the horseradish peroxidase-mimicking catalytic nucleic acids that enable the colorimetric or chemiluminescence detection of Pb^{2+} or L-histidine (Elbaz et al., 2008). A green procedure for the kinetic resolution of chiral amines via enzymatic acylation and deacylation has been demonstrated by Ismail et al. (2008). Asymmetric dihydroxylation of aryl olefins has been carried out by sequential enantioselective epoxidation and regioselective hydrolysis with tandem biocatalysts (Xu et al., 2009). Dupont et al. (2009) discussed multiphase conditions using classical acid or base catalysts as well as biocatalysts with some recent catalytic approaches and achievements, such as alcoholysis of triglycerides. The flavoprotein catalyzed reduction of aliphatic nitro compounds represents a biocatalytic equivalent to the Nef reaction (Durchschein et al., 2010).

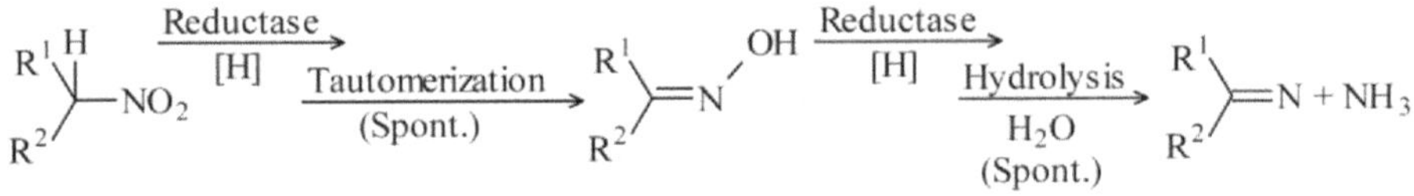

Three enzymes, *Rhizopus oryzae* lipase (ROL), lysozyme and phytase are reported to catalyze the condensation of the model compound, trimethylsilanol, formed in situ from trimethylethoxysilane and produced hexamethyldisiloxane in aqueous media at 25°C and pH 7 (Abbate et al., 2010). A straightforward, high-yielding, and chemoenzymatic total synthesis of enantiopure (*S*)-rivastigmine was developed by Fuchs et al. (2010) using various ω-transaminases for the asymmetric amination of appropriate acetophenone precursors. The sponge-restricted enzyme silicatein-α catalyzes *in vivo* silica formation from monomeric silicon compounds from sea water (*i.e.*, silicic acid) and plays the pivotal role during synthesis of the siliceous sponge spicules (Wolf et al., 2010). Znabet et al. (2010) synthesized a very important drug candidate telaprevir, featuring a biocatalytic desymmetrization and two multicomponent reactions as the key steps.

Mitochondria, considered as the "powerhouse" of the living cell, can do bioelectrocatalysis of pyruvate, fatty acids, and amino acids at electrode surfaces for biofuel cell applications (Bhatnagar et al., 2011). Li et al. (2011a) discovered the unnatural ability of nuclease p1 from *Penicillium citrinum* to catalyze asymmetric aldol reactions between aromatic aldehydes and cyclic ketones under solvent-free conditions.

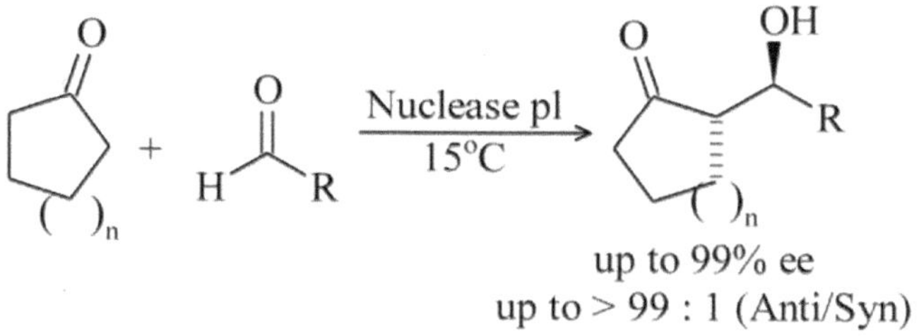

An enzymatic oxidation of methanol combined with lyase for the hydroxymethylation of aldehydes, coupled with a further enzymatic reduction afforded enantiopure diols in one pot (Shanmuganathan et al., 2012).

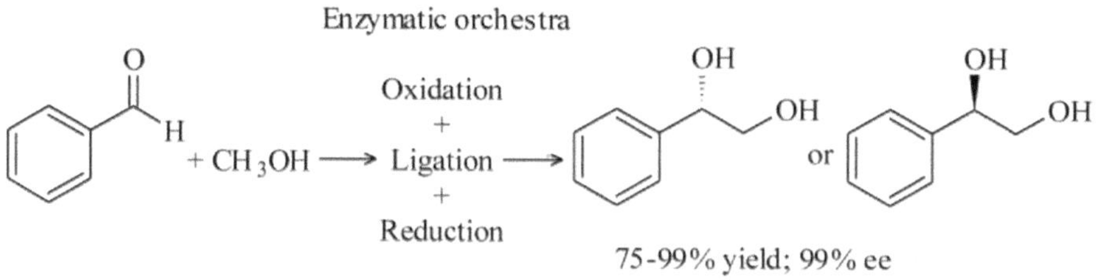

Stereoselective benzylic hydroxylation of alkylbenzenes and epoxidation of styrene derivatives catalyzed by the peroxygenase of *Agrocybe aegerita* has been reported (Kluge et al., 2012). Balke et al. (2012) reported that Baeyer–Villiger monooxygenases (BVMOs) are useful enzymes for organic synthesis as they enable the direct and highly regioselective and stereoselective oxidation of ketones to esters or lactones simply with molecular oxygen. A hybrid biocatalyst, where a synthetic rhodium complex is covalently inserted into a β-barrel protein scaffold, undergoes a stereoregular polymerization of poly(phenylacetylene) containing trans-stereostructure (Onoda et al., 2012). Yara-Varón et al. (2012) synthesized poly(ethyl acrylate-*co*-allyl acrylates) from acrylate mixtures prepared by a continuous solvent-free enzymatic process.

Enzymatic synthesis of amoxicillin by penicillin G acylase in the presence of ionic liquids has been reported (Pereira et al., 2012). Ni et al. (2012) reported biohydrogenation (under mild reaction conditions) of carboxylic acids to the corresponding alcohols or aldehydes using *Pyrococcus furiosus*. Lozano et al. (2012) have developed a clean biocatalytic approach for producing flavor esters in switchable ionic liquid/solid phases by using an iterative cooling/centrifugation protocol. Highly selective semisynthetic lipases have been prepared by site-specific incorporation of tailor-made peptides on the lipase-lid site (Romero et al., 2012). Strategies have been reported for reducing the side reactions of chemoenzymatic DKR tuning the organometallic catalyst and entrapping the biocatalyst (Pollock et al., 2012). Dioxygenase catalyzed stereoselective dihydroxylation of benzo[*b*] thiophenes and benzo[*b*]furans yielded cis- and trans-diols having synthetic potential (Boyd et al., 2012).

Deshmukh et al. (2012) used lemon juice under solvent-free conditions for the Knoevenagel condensation reaction as an efficient homogeneous acid catalyst. It was observed that when different aldehydes and malononitrile were mixed with lemon juice and stirred at room temperature for 5–120 min., then condensation product was obtained in good yields.

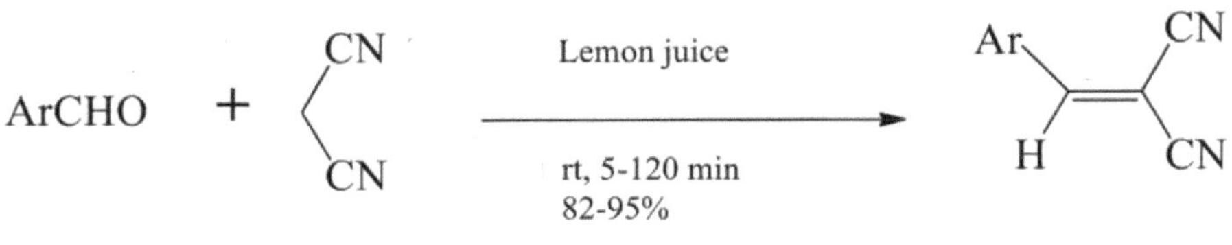

A carbonic anhydrase (CA)-based biocatalyst was developed by Jo et al. (2014), which was encapsulated in a biosilica matrix and used for CO_2 sequestration. It was observed that encapsulated CA exhibited high thermostability and could retain 80% of its activity even after 5 days at 50°C.

El-Gendy et al. (2016) prepared fluorapatite (FAP) via a simple calcination process of waste animal bones (ABs) and mixed with ZnO nanoparticles to get ZnO/FAP binary oxide nanobiocomposite. They exhibited major photocatalytic degradation of 3-chlorophenol and 2,3-dichlorophenol under UV-irradiation and also for remediating petroleum refineries wastewater.

The *Chondrus crispus* is a family of sulfated polysaccharide. The Fe_3O_4 nanoparticles were synthesized by Hemmati et al. (2016) in the presence of natural Irish moss (IM) giving Fe_3O_4@IM. The as-prepared Fe_3O_4@IM was used as a biocatalyst in the synthesis of imidazopyrimidine derivatives. It was claimed that Fe_3O_4@IM is an efficient, recyclable, green in nature and magnetically separable.

Diphenyldiselenide was stabilized on magnetic chitosan by Rangraz et al. (2018) and was used as a green biocatalyst. This biocatalyst was used for the chemoselective oxidation of sulfides to sulfoxides with H_2O_2 (green oxidant) at ambient temperature affording good to high yields. Riyadh et al. (2018) prepared a chitosan–MgO hybrid nanocomposite. It was reported that this nanocomposite can act as a powerful ecofriendly catalyst in the synthesis of 2-hydrazono[1,3,4]thiadiazoles and 5-arylazo-2-hydrazonothiazoles under microwave irradiation incorporating a sulfonamide group.

Laccase and 4-amino-2,2,6,6-tetramethylpiperidine-1-oxyl were separately immobilized by Mogharabi-Manzari et al. (2018) on amine functionalized iron (II, III) oxide nanoparticles. They used glutaraldehyde as a coupling reagent. Twelve benzoxazole and benzimidazole derivatives were prepared using this catalyst. It was observed that this catalyst could retain more than 85% of its activity even after 10 runs and it can be separated with the help of an external magnet.

A simple protocol was used by Salehi and Mirjalili (2018) to prepare nano-ovalbumin from egg white. It is a retrievable and metal-free biocatalyst, which exhibited a high activity for efficient and green synthesis of tetrahydrodipyrazolo pyridines through multicomponent reaction of hydrazine, ethyl acetoacetate, ammonium acetate, and aldehyde in water.

Baydaş et al. (2020) evaluated three different lactic acid bacteria (LAB) strains for bioreduction of acetophenone. Out of these strains, *Lactobacillus kefiri* P2 strain was found to be the best asymmetric reduction biocatalyst. It was reported that secondary chiral alcohols were obtained as the product of bioreduction with results up to 99% enantiomeric excess.

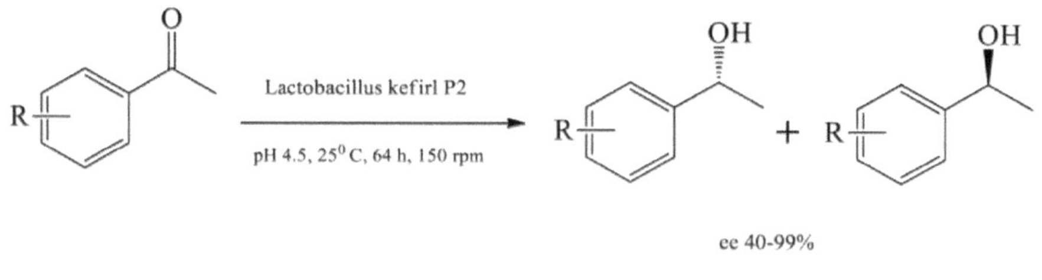

Bavandi et al. (2020) used *Porcine pancreas* lipase (PPL) as an efficient green biocatalyst for the preparation of bis-4-hydroxycoumarin compounds. It was reported that bis-4-hydroxy coumarin compounds could be successfully synthesized under mild reaction conditions and with high yields (81–88%).

Marandi et al. (2021) prepared Indium(III) immobilized on Fe_3O_4@apple seed starch (Fe_3O_4@apple seed starch-In(III)) using apple seed starch. This catalyst was then used in the synthesis of isochromeno[4,3-c]pyrazole-5(1H)-one derivatives under solvent-free conditions.

4.3 ORGANOCATALYSTS

Ranu et al. (1999) developed environment-friendly procedure for acylation of ferrocene with direct use of carboxylic acid in the presence of trifluoroacetic anhydride on the solid phase of alumina. Tanaka et al. (2000) reported that solvent-free condensation of cyclohexanone and diethyl succinate in the presence of t-BuOK at room temperature gives cyclohexylidenesuccinic acid, while heating mixture and t-BuOK at 80°C gives only cyclohexenylsuccinic acid. Jiang et al. (2000) carried out Wacker reaction in supercritical carbon dioxide or ROH/supercritical carbon dioxide and they observed that both $scCO_2$ and co-solvent can remarkably affect the selectivity toward methyl ketone and the presence of ROH accelerates the reaction.

Tetramethyladipic acid (TMAA), a starting monomer for several technically important polymers (polyester resins and polyamide fibers) was synthesized by direct carbon–carbon bond formation between the saturated primary carbon atoms of pivalic acid using a sonoelectrochemical Fenton process (Bremner et al., 2001). Solvent-free intermolecular and intramolecular Thorpe reactions proceeded efficiently to give acyclic and cyclic enamines (Yoshizawa et al., 2002). A simple and convenient system for benzylic bromination of toluene has been developed by Mestres and Palenzuela (2002) using a two-phase mixture (sodium bromide, aqueous hydrogen peroxide/carbon tetrachloride or chloroform) under visible light. Substitution of the chlorinated solvents by other more environmentally benign organic solvents has been attempted and good results were obtained for methyl pivalate.

Perchlorinated aryl compounds were efficiently dechlorinated and de-aromatized in hydrogen during 0.5–2 h at 50–90 °C (Yuan et al., 2003).

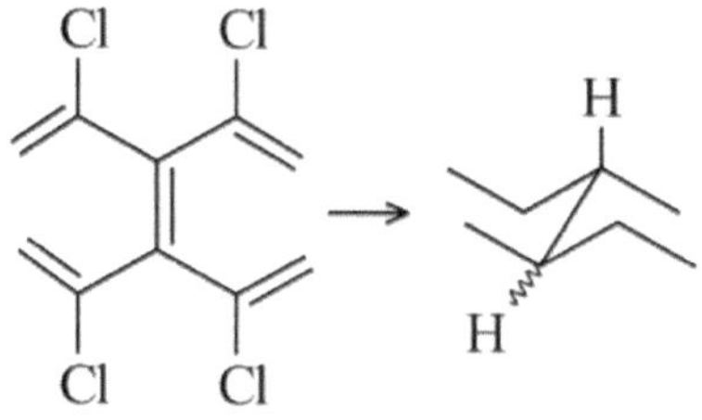

Zhu et al. (2003) reported that esterification of carboxylic acids occurs with alcohols in the ionic liquid $[Hmim]^{+}BF_4^{-}$ without any organic solvents and the ionic liquid could be reused over eight times.

$$RCOOH + R'OH \xrightarrow[110^{\circ}C]{[Hmim]^{+}BF_4^{-}} RCOOR'$$

Laitinen et al. (2004) carried out ene reaction of allylbenzene and N-methylmaleimide in subcritical water and ethanol. A novel emulsifier-free copolymerization can be achieved by the ultrasonic radiation (Yan et al., 2004). Comisar and Savage (2005) carried out base-catalyzed benzil rearrangement at conventional conditions, which proceeds in high temperature water without added base. Acid catalysis also occurred under these conditions.

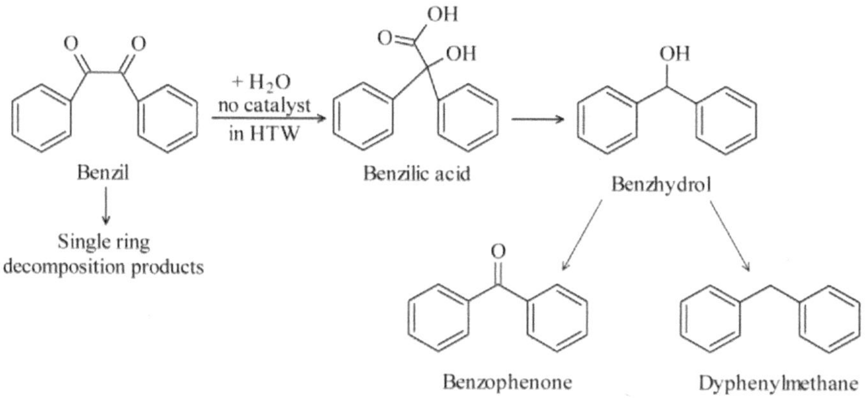

Zhang et al. (2005) reported that biotin methyl ester can be synthesized using catalytic carbonylation to generate urea, avoiding the traditional phosgene and phosgene derivative methodology.

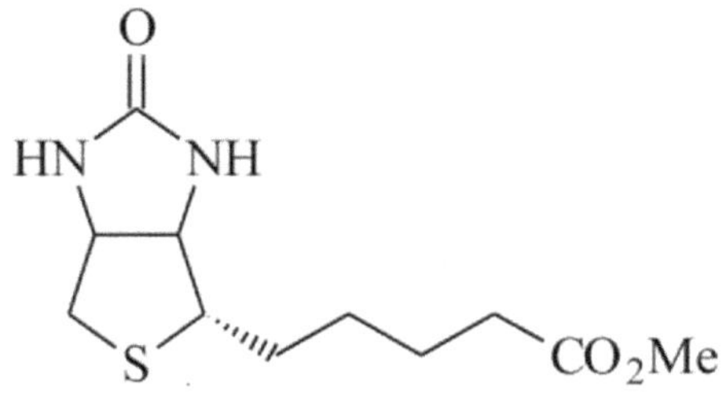

84% Yield

Bonnet et al. (2006) investigated that a single-step air oxidation of cyclohexane, based on a new lipophilic catalytic system, leads to the production of adipic acid with excellent results.

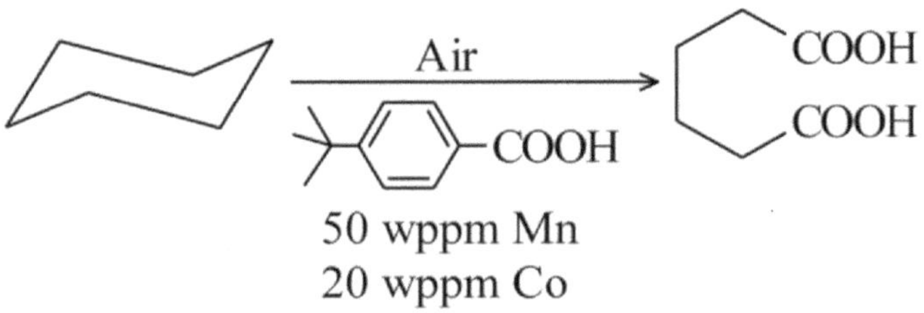

Guo et al. (2006) observed the Beckmann rearrangement of cyclohexanone oxime to afford caprolactam in a novel caprolactam-based Brønsted acidic ionic liquid as catalyst, and this reaction proceeded with high conversion and selectivity at 100°C. Baylis–Hillman products were produced in 98% yield in as little as 30 min by solvent-free mechanochemistry. This represents one of the fastest methods of Baylis–Hillman reactions under neat conditions in the presence of 20% DABCO (Mack and Shumba, 2007).

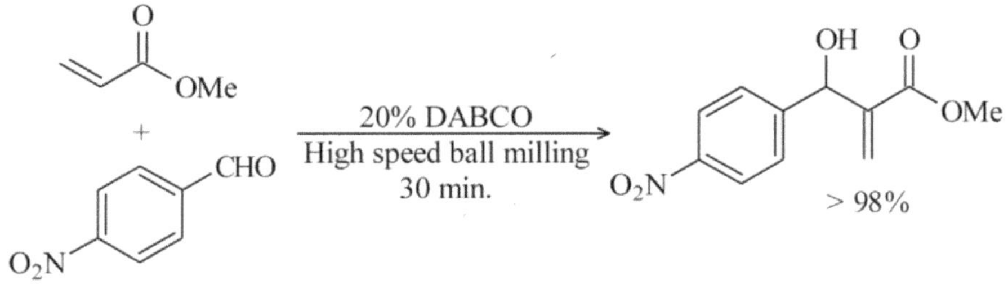

Chahbane et al. (2007) observed that orange II oxidation by peroxides catalyzed by Fe^{III}–TAMLs at pH 9–11 leads to CO_2, CO, phthalic, and smaller aliphatic acids as nontoxic major mineralization products, while Jin et al. (2008) reported the hydrothermal conversion of carbohydrate biomass into formic acid in an excellent yield at mild temperatures.

$$\text{Glucose} \xrightarrow[\text{250°C, 60 s}]{O_2 + OH^-} \underset{75\%}{\text{HCOOH}}$$

KI and β-cyclodextrin (β-CD) show excellent synergetic effect in promoting cycloaddition of CO_2 with epoxides to produce cyclic carbonates (Song et al., 2008).

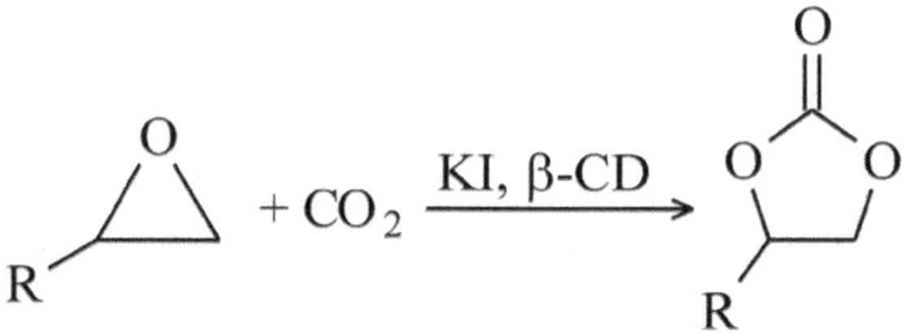

Tee et al. (2008) investigated the activity of effects of 10 ionic liquids on cytochrome P450 BM-3 by evaluating the influence of hydrophobicity and ion pairs on P450 BM-3. Stevens et al. (2009) have explored the aldol reactions of propionaldehyde and butyraldehyde in supercritical carbon dioxide over a variety of heterogeneous acidic and basic catalysts.

Kong et al. (2009) developed a green way to synthesize allyl phenols. Quantitative yield of 2-allyl-4-methoxyphenol was obtained via a fast Claisen rearrangement in a microreactor system without solvent and work-up. Polshettiwar and Varma (2009) reported an economical and sustainable transfer hydrogenation for aldehydes and ketones. The general protocol is mild, chemoselective and important. It uses neither precious or non-precious metals nor even ligands.

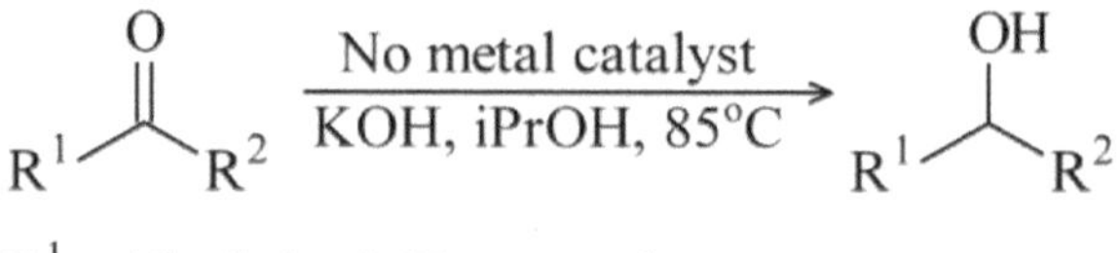

R^1 - Alkyl, Aryl, Heteroaryl
$R^2 = H, R^1$

Baj et al. (2009) reported the Baeyer–Villiger oxidation of ketones with bis(trimethylsilyl) peroxide in the presence of ionic liquids as the solvent and catalyst.

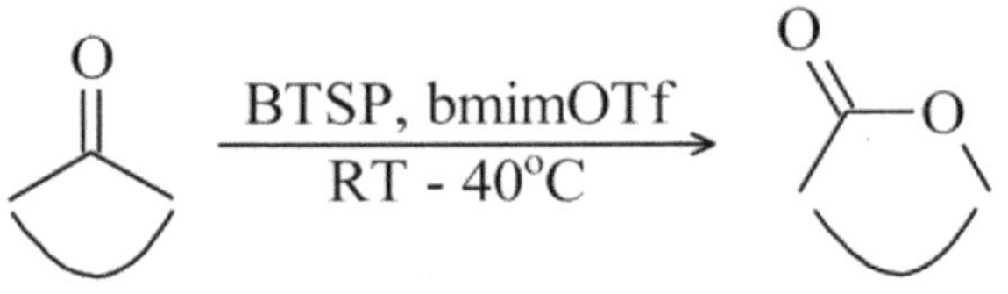

Huertas et al. (2009) used dicyclopentadiene as a source of in situ generated cyclopentadiene for Diels–Alder reactions under solvent-free conditions.

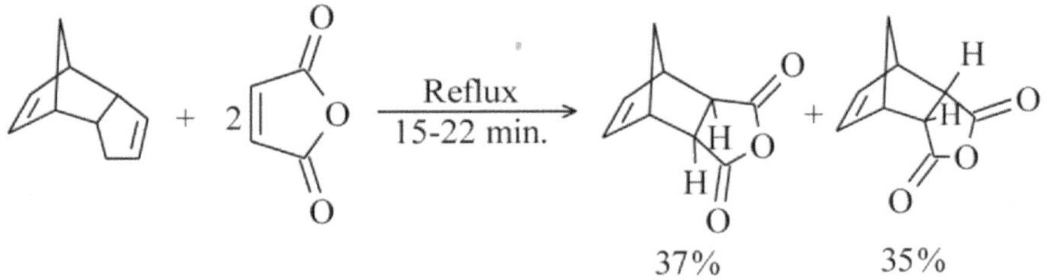

Phadtare and Shankarling (2010) have described that the biodegradable solvents provide an effective green method for the bromination of 1-aminoanthra-9,10-quinone under mild conditions and these are recyclable also.

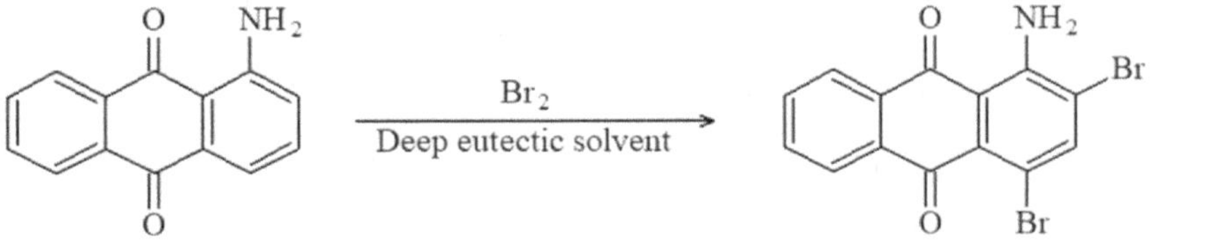

Immobilizing a mixture of palladium–guanidine complex and guanidine in the nanocages of SBA-16 provides an efficient solid catalyst for Suzuki coupling, The aerobic oxidation of alcohols was performed by Yang et al. (2010). Highly efficient, one-pot and three-component reactions of amines and carbon disulfide with alkyl vinyl ethers via Markovnikov addition reaction were carried out by Halimehjani et al. (2010) in water under a mild and green procedure with excellent yields and complete regiospecificity. Duval and Lever (2010) reported that mildly acidic aqueous conditions are suitable for Fischer syntheses of naltriben, naltrindole, and naltrindole analogs. Methyltetrahydrofuran is a useful bio-based (co)solvent for benzaldehyde-catalyzed reactions, affording a straightforward work-up with high yields and enantioselectivities (Shanmuganathan et al., 2010).

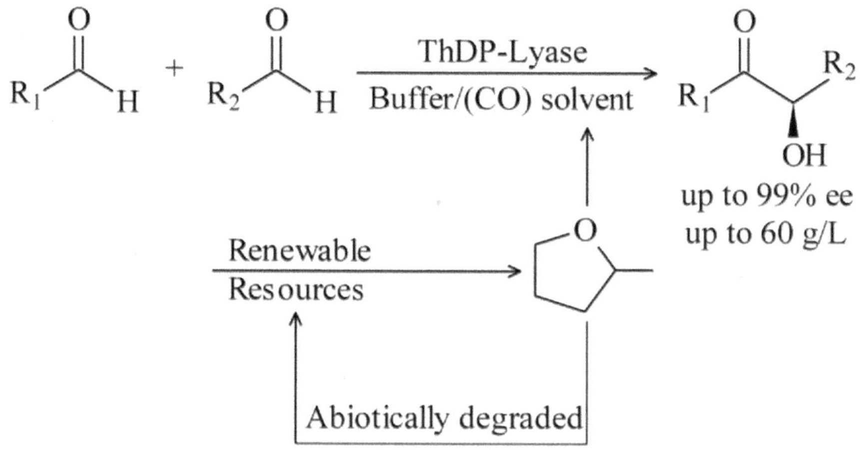

A rational design of phosphonium ionic liquid for ionic liquid-coated lipase (IL1-PS)-catalyzed reaction has been investigated by Abe et al. (2010). A very rapid transesterification of secondary alcohols occurs when IL1-PS was used as a catalyst in 2-methoxyethoxymethyl(tri-n-butyl) phosphonium bis(trifluoromethanesulfonyl)amide ([P_{444MEM}][NTf_2]) as a solvent while perfect enantioselectivity was maintained. Free hemoglobin (Hb) in water at pH 5 was able to oxidize 11 polycyclic aromatic hydrocarbons (PAH) (300 nM each) in the presence of H_2O_2 amounting to 75% PAH removal. PAH are carcinogenic, mutagenic and xenobiotic pollutants found in wastewaters of oil refineries (Laveille et al., 2010). Balbino et al. (2011) reported that dicyclohexylguanidine group covalently attached on silica gel is an efficient basic heterogeneous catalyst for the production of biodiesel in a continuous flow reactor. Malonic acid half esters were used as the equivalent of ester carbanions for the practical one-step synthesis of β-hydroxy and β-amino esters from aryl aldehydes and arylimines via decarboxylative aldol and Mannich-type reactions. Two mechanisms were unveiled depending on the substitution of the malonyl substrate (Baudoux et al., 2010).

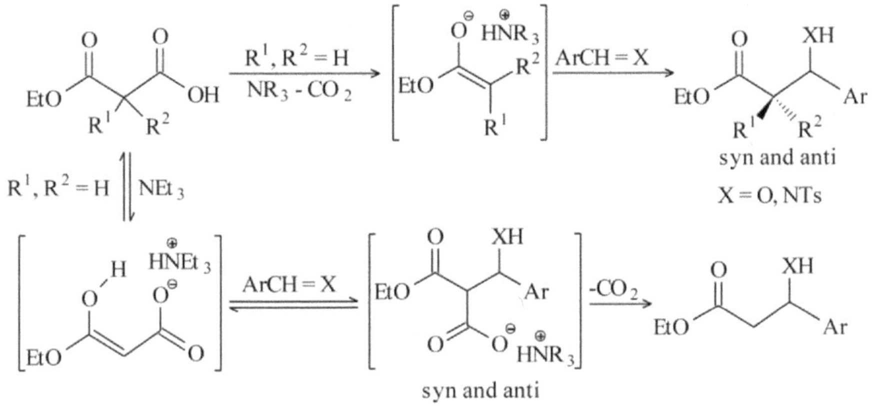

Bromination of industrially important aromatics using an aq. $CaBr_2$–Br_2 system as an instant and renewable brominating agent was carried out by Kumar et al. (2011).

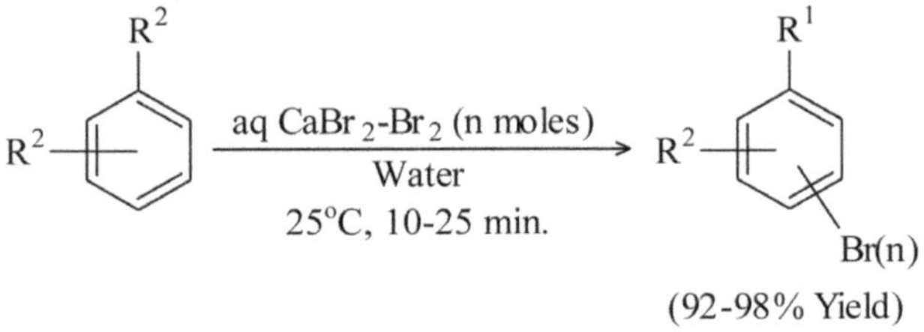

R^1 = OH, NH_2, NHCOMe, NHCOPh, CHO, COOH
R^2 = H, OH, NO_2, SO_2NH_2

Kotlewska et al. (2011) used hydrogen peroxide and a lipase dissolved in ionic liquids for epoxidation and Baeyer–Villiger oxidation.

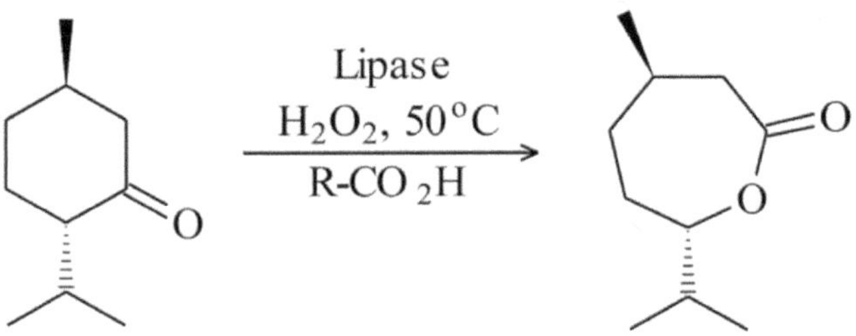

Kamimura et al. (2011) investigated the treatment of waste nylon-6 with supercritical MeOH resulting in smooth depolymerization, giving methyl 6-capronate and methyl 5-hexenoate in good yields.

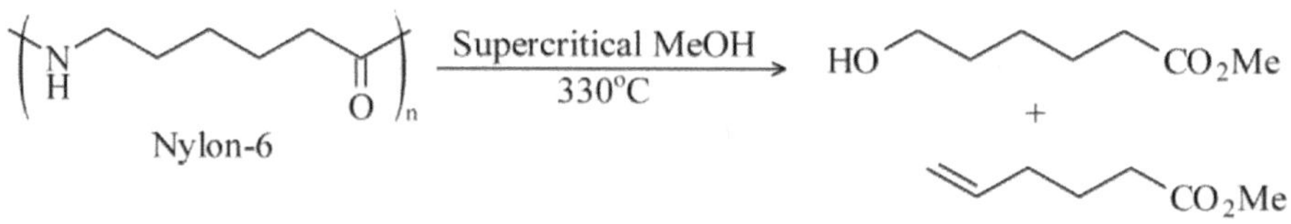

A highly efficient Knoevenagel condensation was catalyzed by a tertiary amine functionalized polyacrylonitrile fiber with excellent recyclability and reusability (Li et al., 2011b).

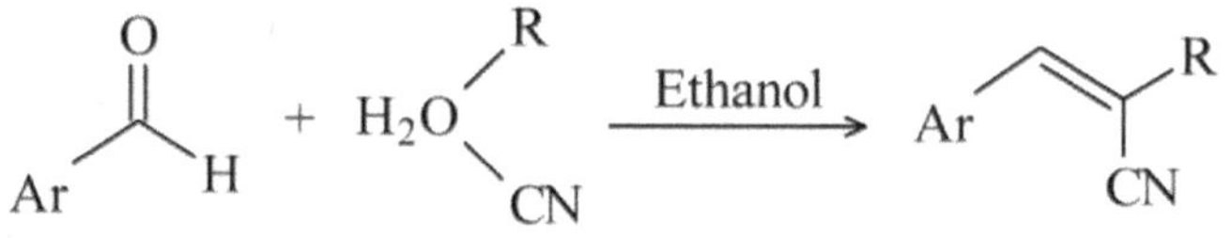

A simple and scalable organocatalytic aldol reaction of acetol and aromatic aldehydes has been developed by Czarnecki et al. (2011). Ando and Yamada (2011) have investigated solvent-free Horner–Wadsworth–Emmons reaction catalyzed by DBU in the presence of K_2CO_3 or Cs_2CO_3 affording olefins with high *E*-selectivity.

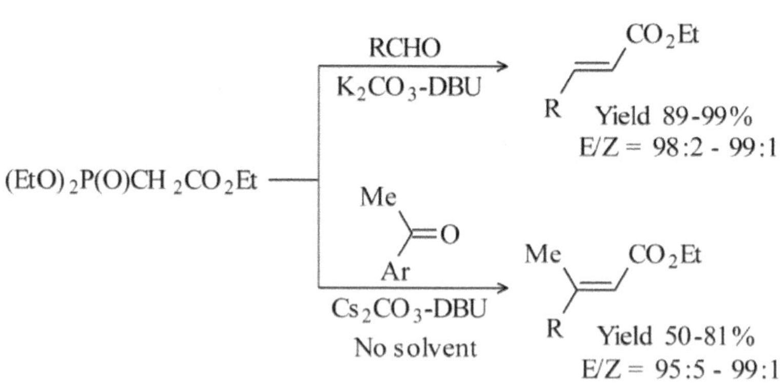

Hajimohammadi et al. (2011) used air and sunlight in the presence of porphyrins for a highly selective, green and economical conversion of alcohols into aldehydes and ketones. Radical benzyl bromination in diethyl carbonate under microwave-assisted reaction condition was performed by Pingali et al. (2011). Both the solvent and the reagent (NBS) are recyclable.

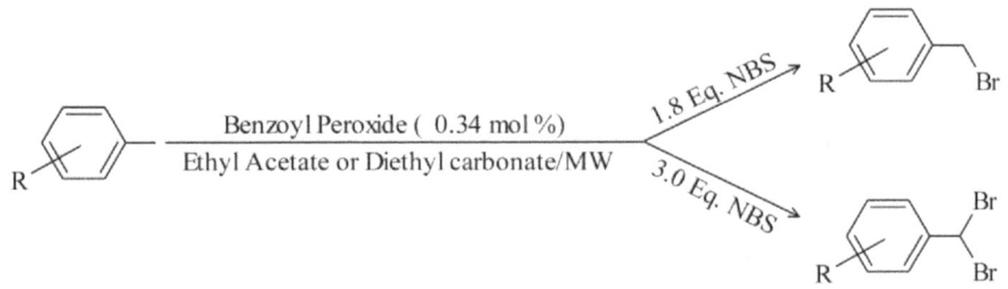

2,4-Disubstituted-1,2-dihydroquinazolines and quinazolines can be readily obtained from 2-aminobenzophenone, aldehyde, and urea under microwave irradiation in the absence of any solvent or catalyst. The reaction is simple, clean, and excellent yields are obtained within minutes (Sarma and Prajapati, 2011).

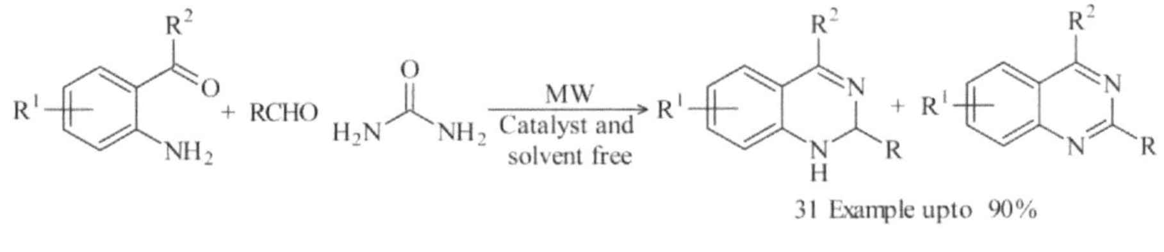

Ntainjua et al. (2012) observed that Au–Pd ion exchanged heteropoly-acid catalysts are considerably more effective in achieving high H_2O_2 yields in the absence of promoters than previously reported catalysts. Organocatalyzed direct aldol reactions were efficiently performed by Bellomo et al. (2012) in aqueous solutions of facial amphiphilic carbohydrates with high diastereoselectivity and yields. Mitsudome et al. (2012) reported titanium-exchanged montmorillonite (Ti^{4+}-mont) as an efficient heterogeneous catalyst for the etherification of various alcohols under mild reaction conditions. Solvent-free brominations of 1,3-dicarbonyl compounds, phenols and alkenes were achieved by employing sodium bromide and oxone under mechanical milling conditions (Wang et al., 2012a).

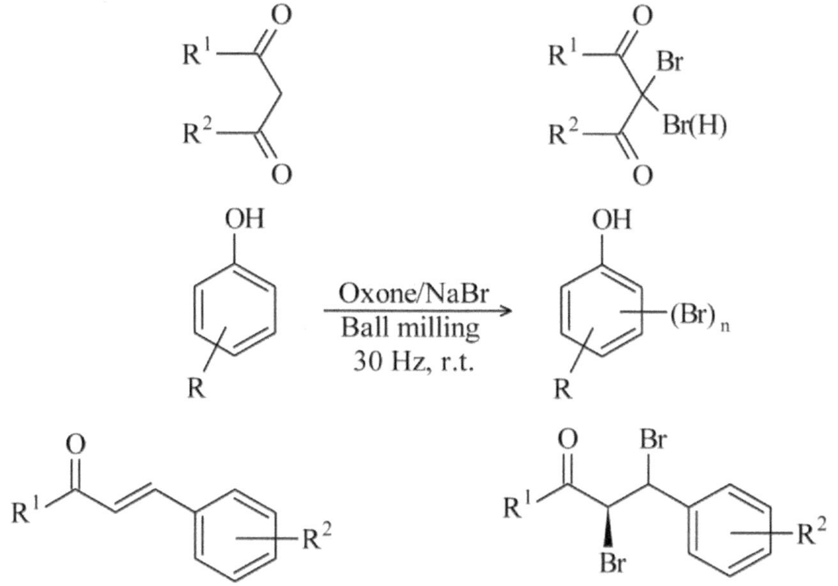

Wang et al. (2012b) also reported water as a green additive, which enhances the ring opening and contraction reactions of aromatics. A key pharmaceutical intermediate for the production of edivoxetine·HCl was prepared in > 99% ee via a continuous Barbier reaction, which improved the greenness of process relative to a traditional Grignard batch process (Kopach et al., 2012).

Shen et al. (2012) used ionic liquid supported imidazolidinone catalyst (an efficient and recyclable organocatalyst) for mediating highly enantioselective Diels–Alder reactions involving α,β-unsaturated aldehydes and cyclopentadiene.

Efficient and continuous flow methylation was performed with dimethylcarbonate using a basic ionic liquid as a catalyst (Glasnov et al., 2012). Synthesis of α-acyloxy amides has been developed by Cui et al. (2012) by ultrasound-promoted sterically congested Passerini reactions under solvent-free conditions.

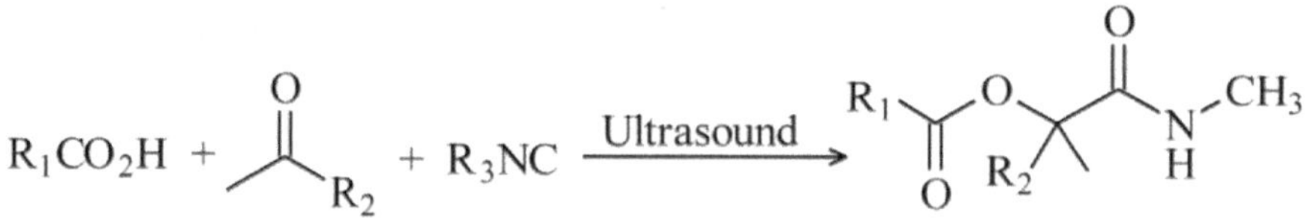

A novel magnetic nanoparticle supported acidic catalyst was prepared and used as a highly efficient and magnetically recoverable catalyst for the one-pot synthesis of benzoxanthenes (Zhang et al., 2012).

Imidazol-1-yl-acetic acid has been introduced as green bifunctional organocatalyst by Nazari et al. (2014) to synthesize 1,8-dioxooctahydroxanthenes under solvent-free conditions. This organocatalyst can be recycled up to 8 consecutive runs without any significantly loss of its efficiency.

Kaplaneris et al. (2015) synthesized *trans*- and *cis*-diastereomers of pyrrolidinine-thioxotetrahydropyrimidinone (having fluorine or hydroxyl group). It was assumed that catalysts could efficiently enhance the reactions in brine, without using any organic solvent, and that too with stoichiometric amount of reagents. Products were isolated by extraction with diastereoselectivities and enantioselectivities.

Cho et al. (2016) introduced tertiary amines as benign organocatalysts to activate carbon dioxide and insert into epoxides to afford cyclic carbonates. These amines were not having any halide and metal additives. It was reported that N,N,N,N-tetraethylethylenediamine exhibited a good efficiency and this organocatalyst does not contain toxic metals or halides and only 0.1 mol% loading was required.

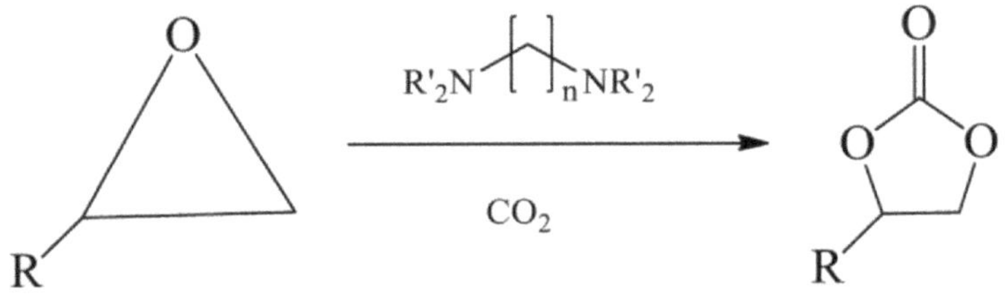

Franconetti et al. (2016) prepared a series of methylenemalononitriles and ethyl cyanoacrylates both via Knoevenagel condensation catalyzed by pure and modified chitosan-based heterogeneous catalysts. An efficient but metal- and solvent-free catalytic system (2-pyridinemethanollnBu4Ni) was developed by Wang et al. (2016) for converting epoxides and carbon dioxide to cyclic carbonates with high-to-excellent yields. This catalyst could be easily recycled.

Ishikawa et al. (2017) developed a bioinspired two-component redox organocatalyst system using 1,10-bridged flavinium and NH_4I. It was further used for the aerobic oxidative ring formation of 1,2,3-thiadiazoles from sulfur and *N*-tosylhydrazones in an environment-friendly manner. A green synthesis of isoxazolines from allyloximes was reported by Triandafillidi and Kokotos (2017). They utilized 2,2,2-trifluoroacetophenone as an organocatalyst for the oxidation of allyloximes with H_2O_2. Liu et al. (2018) prepared pincer-type compounds and used these as organocatalysts for the cycloaddition reaction of epoxides with CO_2.

An efficient tandem reaction has been observed by Olyaei et al. (2018) to obtain benzo[f]chromenes. They used 10 mol% guanidine hydrochloride (catalyst) (solvent-free conditions) using 2,3-dihydroxynaphthalene, malononitrile, and aldehydes. They also used this for synthesizing novel 12H-benzo[5,6] chromeno[2,3-b]pyridines. An environmentally benign protocol has been reported by Valiey et al. (2019) for the synthesis of a variety of bioactive benzylpyrazolyl coumarin (functionalized) and dihydropyrano[2,3-c]pyrazole derivatives using aryl aldehydes, malononitrile, ethyl acetoacetate, 4-hydroxycoumarin, and hydrazine derivatives in the presence of melamine-modified chitosan materials as bifunctional organocatalyst, which can be reused again.

Bifunctional organocatalysts were used by Zhang et al. (2021) for the synthesis of cyclic carbonates by some cycloaddition reactions of epoxides and carbon dioxide. It was reported that diamine can activate CO_2 via carbamate formation and enhance its transformation to cyclic carbonates. It was revealed that high yields (9299%) and selectivities (99%) were there in this reaction.

4.4 METALLOCATALYSTS

Palladium(II)-catalyzed Heck arylation of both electron-poor and electron-rich olefins with arylboronic acids as arylpalladium precursors were conducted under air (Enquist et al., 2006). A novel Pd/Ph–Al-MCM-41 catalyst was designed by Li et al. (2007), which exhibited excellent activity, selectivity, and hydrothermal stability in an aqueous medium Ullmann reaction. This may be due to the promoting effects of the ordered mesoporous structure, and the Ph– and Al–modifications. The Pd catalyst supported on 1,1,3,3-tetramethylguanidinium (TMG)-modified molecular sieve SBA-15 is a very active and stable catalyst for the Heck coupling reaction in solvent-free conditions (Ma et al., 2008).

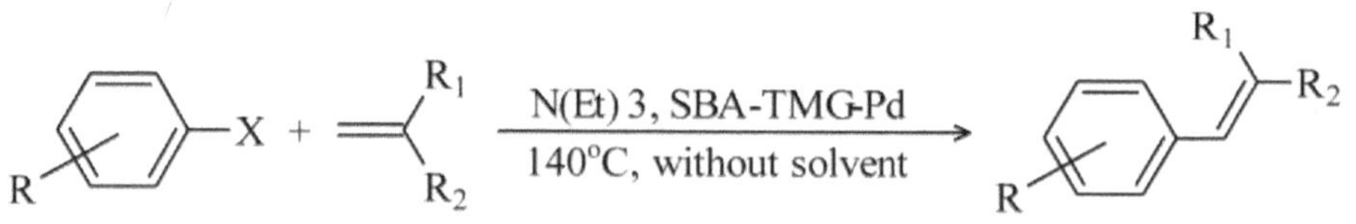

Wang et al. (2009) used ethylene carbonate as a unique solvent for the Wacker oxidation of higher alkenes and aryl alkenes. They have successfully used molecular oxygen as the sole oxidant, when colloidal Pd nanoparticles stabilized in ethylene carbonate facilitate its reoxidation under cocatalyst-free conditions.

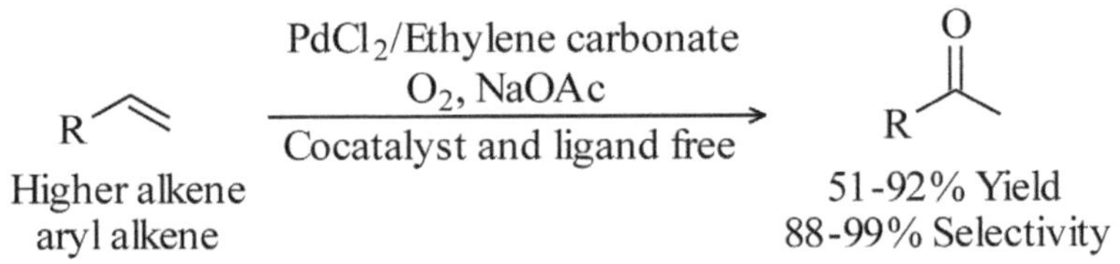

Waddell and Mack (2009) reported Tishchenko reaction of aryl aldehydes using high speed ball milling and a sodium hydride catalyst, in high yields in 0.5 h.

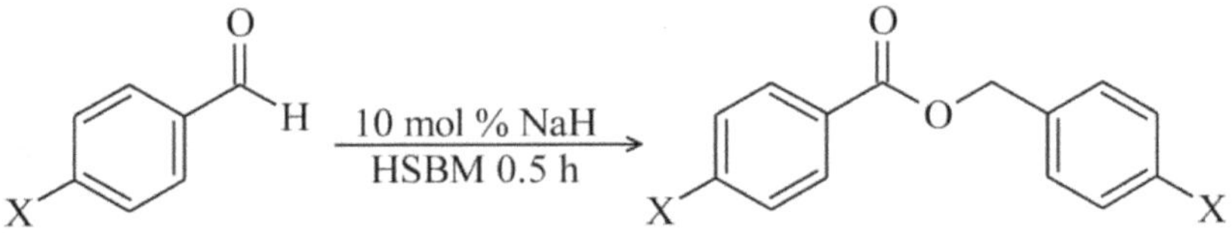

Cu–Mn spinel oxide catalyst has been investigated by Yousuf et al. (2010) for the ligand-free Huisgen [3+2] cycloaddition. Zhang et al. (2011) described the reductive cleavage of the C–O bond of aromatic and aliphatic acetals to ethers, catalyzed by $Cu(OTf)_2$ or $Bi(OTf)_3$ in excellent yields and selectivity.

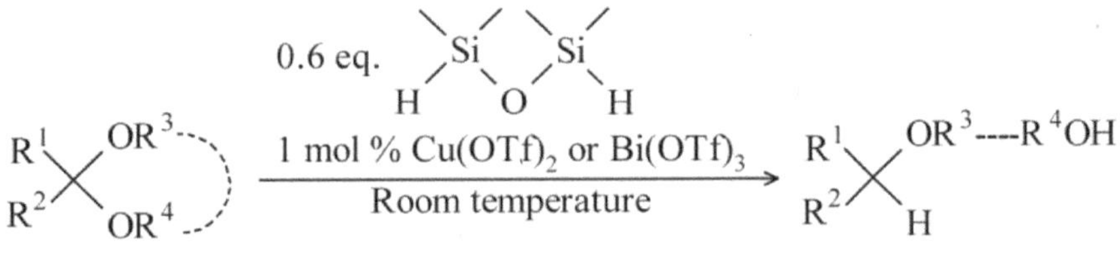

A green process using nickel-loaded $La_xNa_{1-x}TaO_3$ prepared by hydrogen peroxide–water-based solvent for hydrogen production has been reported by Husin et al. (2011). Direct transformation of cellulose and a lignocellulosic biomass raw material into 5-hydroxymethyl-2-furfural (HMF) using a specific combination of Cr(II) and Ru(III) metals in [emim]Cl has been reported (Kim et al., 2011). The Ru/ZnO–$ZrO_x(OH)_y$ catalyst is very efficient for the selective hydrogenation of benzene to cyclohexene, and the yield of cyclohexene could reach 56% without using any additive (Liu et al., 2011).

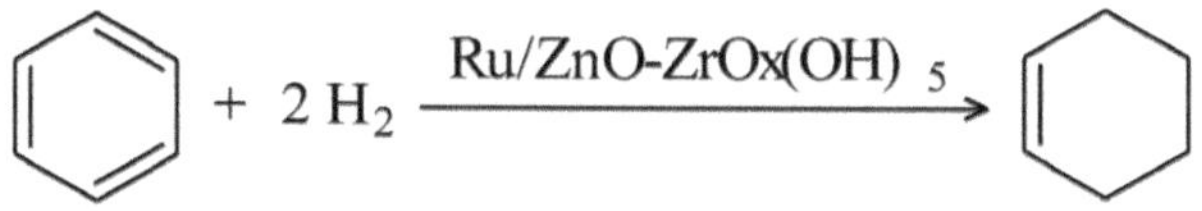

Füldner et al. (2011) reported that colored metal oxides ($PbBiO_2X$) are efficient visible light photocatalysts for nitrobenzene reduction. A fluorous oxime-based palladacycle, which promotes carbon–carbon bond formation reactions (Suzuki–Miyaura, Sonogashira and Stille) in aqueous media under microwave irradiation was developed by Susanto et al. (2012).

Kisukuri et al. (2016) prepared bimetallic nanoshells (AgAu, AgPd, and AgPt), which exhibited catalytic activities. These bimetallic nanoshells were used for the oxidation of silane to silanols and production of hydrogen.

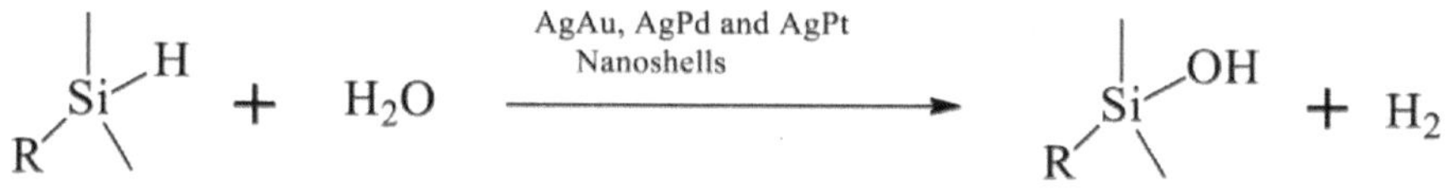

Most of the chemical reactions require some or the other catalyst to enhance their rates, but some of these are toxic in nature as these contain some metal, metal derivatives or harmful compounds; off course, leaving aside the biochemical reaction involving enzymes as catalyst. Although catalysts have been used by some workers, but search is still on for environmentally friendly catalyst, which can serve the purpose of a catalyst taking care of environment.

4.5 RECENT DEVELOPMENTS

Shaikh et al. (2022) reported the synthesis of bis-coumarin derivatives only by grinding 4-hydroxy coumarin (20 mM) with various aromatic aldehydes (10 mM) in the presence of different biocatalysts. These biocatalysts include kiwi (*Actinidia deliciosa*) juice fruit dragon fruit (*Hylocereus undatus*, the white-fleshed pitahaya) juice and buttermilk in the aqueous medium within 10–15 min under ambient conditions. The product (crude) was obtained by filtration and recrystallized by ethanol. It was revealed that acidic nature of an aqueous medium plays a dominating role in increasing the rate of the reaction. The proposed protocol is a greener approach for such organic transformations, which efficiently synthesize bioactive compounds in an eco-friendly way under mild reaction conditions, shorter reaction time, nonhazardous nature, and with excellent yield of product.

Azzallou et al. (2022) loaded waste bovine bone-derived hydroxyapatite (BBHA) aqueous solution of zinc chloride (0.01 M). They used it as a catalyst in the synthesis of 1-amidoalkyl-2-naphthols derivatives in solvent-free manner using 2-naphthol, aldehydes, and urea/acetamides at 80°C. They first extracted natural hydroxyapatite from waste bovine bone by decomposing it at 800°C for 2 h. Then, hydroxyapatite was loaded with $ZnCl_2$. The reaction conditions were also optimized to achieve high yields (86–96%) in 25–40 min only and that too with a small amount of catalyst (50 mg).

Moazzam et al. (2022) developed a mild method for the catalytic arylation of imidazo[1,2-*a*]pyridine with diazonium salt derivatives. They used chlorophyll as a biocatalyst in the presence of visible light. Natural pigments chlorophyll was used as a green photosensitizer and an eco-friendly catalyst. This procedure is easy and it provides a transition-metal-free alternative for the synthesis of 2,3-diaryl imidazo[1,2-*a*]pyridine derivatives in good to excellent yields at ambient temperature.

Different wood ash catalysts were prepared by Rostamian et al. (2022) from different natural resources, which were comprised of calcium- and potassium-rich carbonates. They observed their catalytic efficiency in the synthesis of benzochromene derivatives. It was reported that the catalyst prepared at 850°C could efficiently synthesize benzochromene derivatives in water at 80°C and that too with high yields. Some oxides, such as MgO, Fe_2O_3, SiO_2, CaO, and Al_2O_3 are commonly utilized as both catalyst as well as catalyst support. Wood ash catalysts with these oxides could effectively enhance the electrophilic activity of carbonyl groups for nucleophilic attack.

An efficient green route for the synthesis of biologically active pyrazolopyranopyramidine derivatives has been developed by Dharmendra et al. (2022) by four-component reaction of aromatic aldehydes, ethyl acetoacetate, hydrazine hydrate, and barbituric acid in the presence of ultrasound. The reaction was catalyzed by starch@Fe_3O_4 using water as a solvent at room temperature. The proposed method has many advantages, such as mild conditions, easily spreadable catalyst, reusability.

Zanda et al. (2022) used polystyrene-supported 1,5,7-triazabicyclodec-5-ene (TBD) as a catalyst for converting epoxy amines into various 2-oxazolidone scaffolds. This approach utilized CO_2 as low cost and abundant source of carbon (C_1), which is halide-free for the synthesis of a pharmaceutically important 2-substituted oxazolidinones, including drug toloxatone. It was reported that the immobilized catalyst is stable and recyclable and it can be used for more than two weeks without any significant loss in its catalytic activity.

Yilmaz et al. (2022) synthesized some novel proline and 1,2,3,4-tetrahydroisoquinoline-3-carboxylic acid (THIQA)-based diamides by amidation reactions. The catalytic activities of these as-prepared catalyst in asymmetric aldol reaction were evaluated by the reactions of aliphatic ketones with different aromatic aldehydes. It was observed that particularly (*S*)-*N*-((*S*)-1-(morpholinoamino)-1-oxo-3-phenylpropan-2-yl)pyrrolidine-2-carboxamide exhibited good catalytic activities in the presence of benzoic acid cocatalyst in water at room temperature. It was reported that the yields could be achieved with good enantioselectivities (89.6%), diastereomeric ratios (92/8) and yield could be achieved up to 97.3%. As this catalyst has high catalytic activities in water, and therefore, this is an important contribution to green chemistry requirements also.

KEYWORDS

- **biocatalysts**
- **oraganocatalysts**
- **metallocatalysts**
- **solvent-free conditions**

REFERENCES

Abbate, V.; Bassindale, A. R.; Brandstadt, K. F.; Lawson, R.; Taylor, P. G. Enzyme Mediated Silicon–Oxygen Bond Formation: The Use of Rhizopus Oryzae Lipase, Lysozyme and Phytase Under Mild Conditions. *Dalton. Trans.* **2010,** *39*, 9361–9368.

Abe, Y.; Yoshiyama, K.; Yagi, Y.; Hayase, S.; Kawatsura M.; Itoh, T. A Rational Design of Phosphonium Salt Type Ionic Liquids for Ionic Liquid Coated-Lipase Catalyzed Reaction. *Green Chem.* **2010,** *12*, 1976–1980.

Ando, K.; Yamada, K. Highly E-Selective Solvent-Free Horner–Wadsworth–Emmons Reaction Catalyzed by DBU. *Green Chem.* **2011,** *13*, 1143–1146.

Azzallou, R.; Ouerghi, O.; Geesi, M. H.; Riadi, Y.; Taleb, M. A.; Mamouni, R. et al. Bovine Bone-Derived Natural Hydroxyapatite-Supported $ZnCl_2$ as a Sustainable High Efficiency Heterogeneous Biocatalyst for Synthesizing Amidoalkyl Naphthols. *J. Phys. Chem. Solids,* **2022,** *163*, doi.10.1016/j.jpcs.2021.110533.

Baj, S.; Chrobok, A.; Słupska, R. The Baeyer–Villiger Oxidation of Ketones with Bis (Trimethylsilyl) Peroxide in the Presence of Ionic Liquids as the Solvent and Catalyst. *Green Chem.* **2009,** *11*, 279–282.

Balbino, J. M.; Menezes, E. W. D.; Benvenutti, E. V.; Cataluña, R.; Ebeling, G.; Dupont, J. Silica-Supported Guanidine Catalyst for Continuous Flow Biodiesel Production. *Green Chem.* **2011,** *13* (11), 3111–3116.

Balke, K.; Kadow, M.; Mallin, H.; Saß, S.; Bornscheuer, U. T. Discovery, Application and Protein Engineering of Baeyer–Villiger Monooxygenases for Organic Synthesis. *Org. Biomol. Chem.* **2012,** *10* (31), 6249–6265.

Baudoux, J.; Lefebvre, P.; Legay, R.; Lasne, M.; Rouden, J. Environmentally Benign Metal-Free Decarboxylative Aldol and Mannich Reactions. *Green Chem.* **2010,** *12* (2), 252–259.

Bavandi, H.; Habibi, Z.; Yousefi, M. Porcine Pancreas Lipase as a Green Catalyst for Synthesis of Bis-4-Hydroxy Coumarins. *Bioorg. Chem.* **2020,** *103.* doi.org/10.1016/j.bioorg.2020. 104139.

Baydaş, Y.; Dertli, E.; Şahin, E. Green Synthesis of Chiral Aromatic Alcohols with *Lactobacillus kefiri* P2 as a Novel Biocatalyst. *Synth. Commun.* **2020,** *50* (7), 1035–1045.

Bellomo, A.; Daniellou, R.; Plusquellec, D. Aqueous Solutions of Facial Amphiphilic Carbohydrates as Sustainable Media for Organocatalyzed Direct Aldol Reactions. *Green Chem.* **2012,** *14* (2), 281–284.

Bhatnagar, D.; Xu, S.; Fischer, C.; Arechederra, R. L.; Minteer, S. D. Mitochondrial Biofuel Cells: Expanding Fuel Diversity to Amino Acids. *Phys. Chem.* **2011,** *13* (1), 86–92.

Bohn, M.; Leppchen, K.; Katzberg, M.; Lang, A.; Steingroewer, J.; Weber, J. et al. Effects of Caffeine on Stereoselectivities of High Cell Density Biotransformations of Cyclic β-Keto Esters with Saccharomyces Cerevisiae. *Org. Biomol. Chem.* **2007,** *5* (21), 3456–3463.

Bonnet, D.; Ireland, T.; Fache, E.; Simonato, J. Innovative Direct Synthesis of Adipic Acid by Air Oxidation of Cyclohexane. *Green Chem.* **2006,** *8* (6), 556–559.

Boyd, D. R.; Sharma, N. D.; Brannigan, I. N.; Evans, T. A.; Haughey, S. A.; McMurray, B. T. et al. Toluene Dioxygenase-Catalyzed Cis-Dihydroxylation of Benzo [b] Thiophenes and Benzo [b] Furans: Synthesis of Benzo [b] Thiophene 2, 3-oxide. *Org. Biomol. Chem.* **2012,** *10* (36), 7292–7304.

Boyd, D. R.; Sharma, N. D.; King, A. W.; Shepherd, S. D.; Allen, C. C.; Holt, R. A. et al. Stereoselective Reductase-Catalysed Deoxygenation of Sulfoxides in Aerobic and Anaerobic Bacteria. *Org. Biomol. Chem.* **2004,** *2* (4), 554–561.

Bremner, D. H.; Burgess, A. E.; Li, F. A Sonoelectrochemical FENTON PROCESS for Efficient Synthesis of Tetramethyladipic Acid from Pivalic Acid. *Green Chem.* **2001,** *3* (3), 126–130.

Çelik, A.; Sperandio, D.; Speight, R. E.; Turner, N. J. Enantioselective Epoxidation of Linolenic Acid Catalysed by Cytochrome P450BM3 from Bacillus Megaterium. *Org. Biomol. Chem.* **2005,** *3* (15), 2688–2690.

Chahbane, N.; Popescu, D.; Mitchell, D. A.; Chanda, A.; Lenoir, D.; Ryabov, A. D.; Schramm, K.; Collins, T. J. Fe III–TAML-Catalyzed Green Oxidative Degradation of the Azo Dye Orange II by H_2O_2 and Organic Peroxides: Products, Toxicity, Kinetics, and Mechanisms. *Green Chem.* **2007,** *9* (1), 49–57.

Cho, W.; Shin, M. S.; Hwang, S.; Kim, H.; Kim, M.; Kim, J. G. et al. Tertiary Amines: A New Class of Highly Efficient Organocatalysts for CO_2 Fixations. *J. Indust. Eng. Chem.* **2016,** *44*, 210–215.

Comisar, C. M.; Savage, P. E. The Benzil–Benzilic Acid Rearrangement in High-Temperature Water. *Green Chem.* **2005,** *7* (11), 800–806.

Cui, C.; Zhu, C.; Du, X.; Wang, Z.; Li, Z.; Zhao, W. Ultrasound-Promoted Sterically Congested Passerini Reactions Under Solvent-Free Conditions. *Green Chem.* **2012,** *14* (11), 3157–3163.

Czarnecki, P.; Plutecka, A.; Gawroński, J.; Kacprzak, K. Simple and Practical Direct Asymmetric Aldol Reaction of Hydroxyacetone Catalyzed by 9-Amino Cinchona Alkaloid Tartrates. *Green Chem.* **2011,** *13* (5), 1280–1287.

Deshmukh, M. B.; Patil, S. S.; Jadhav, S. D.; Pawar, P. B. Green Approach for Knoevenagel Condensation of Aromatic Aldehydes with Active Methylene Group. *Synth. Commun.* **2012,** *42* (8), 1177–1183.

Dharmendra, D.; Chundawat, P.; Vyas, Y.; Ameta, C. Ultrasound-Assisted Efficient Synthesis and Antimicrobial Evaluation of Pyrazolopyranopyrimidine Derivatives Using Starch Functionalized Magnetite Nanoparticles as a Green Biocatalyst in Water. *J. Chem. Sci.* **2022,** *134* (2), 1–22.

Dupont, J.; Suarez, P. A. Z.; Meneghetti, M. R.; Meneghetti, S. M. P. Catalytic Production of Biodiesel and Diesel-Like Hydrocarbons from Triglycerides. *Energy Environ. Sci.* **2009,** *2* (12), 1258–1265.

Durchschein, K.; Silva, B. F.; Wallner, S.; Macheroux, P.; Kroutil, W.; Glueck, S. M.; Faber, K. The Flavoprotein-Catalyzed Reduction of Aliphatic Nitro-Compounds Represents a Biocatalytic Equivalent to the Nef-Reaction. *Green Chem.* **2010,** *12* (4), 616–619.

Duval, R. A.; Lever, J. R. Fischer Indole Synthesis in Water: Simple, Efficient Preparation of Naltrindole, Naltriben and Analogs. *Green Chem.* **2010,** *12* (2), 304–309.

Elbaz, J.; Shlyahovsky, B.; Willner, I. A DNAzyme Cascade for the Amplified Detection of Pb^{2+} Ions or L-Histidine. *Chem. Commun.* **2008,** *2008* (13), 1569–1571.

El-Gendy, N. S.; El-Salamony, R. A.; Younis, S. A. Green Synthesis of Fluorapatite from Waste Animal Bones and the Photo-Catalytic Degradation Activity of a New ZnO/Green Biocatalyst Nano-Composite for Removal of Chlorophenols. *J. Water Proc. Eng.* **2016,** *12*, 8–19.

Enquist, P.; Lindh, J.; Nilsson, P.; Larhed, M. Open-Air Oxidative Heck Reactions at Room Temperature. *Green Chem.* **2006,** *8* (4), 338–343.

Franconetti, A.; Domínguez-Rodríguez, P.; Lara-García, D.; Prado-Gotor, R.; Cabrera-Escribano, F. Native and Modified Chitosan-Based Hydrogels as Green Heterogeneous Organocatalysts for Imine-Mediated Knoevenagel Condensation. *App. Catal. A: Gen.* **2016,** *517*, 176–186.

Fuchs, M.; Koszelewski, D.; Tauber, K.; Kroutil, W.; Faber, K. Chemoenzymatic Asymmetric Total Synthesis of (S)-Rivastigmine Using ω-Transaminases. *Chem. Commun.* **2010,** *46* (30), 5500–5502.

Füldner, S.; Pohla, P.; Bartling, H.; Dankesreiter, S.; Stadler, R.; Gruber, M. et al. Selective Photocatalytic Reductions of Nitrobenzene Derivatives Using $PbBiO_2X$ and Blue Light. *Green Chem.* **2011,** *13* (3), 640–643.

Giacomini, D.; Galletti, P.; Quintavalla, A.; Gucciardo, G.; Paradisi, F. Highly Efficient Asymmetric Reduction of Arylpropionic Aldehydes by Horse Liver Alcohol Dehydrogenase Through Dynamic Kinetic Resolution. *Chem. Commun.* **2007,** *39*, 4038–4040.

Glasnov, T. N.; Holbrey, J. D.; Kappe, C. O.; Seddon, K. R.; Yan, T. Methylation Using Dimethylcarbonate Catalysed by Ionic Liquids Under Continuous Flow Conditions. *Green Chem.* **2012,** *14* (11), 3071–3076.

Guo, S.; Du, Z.; Zhang, S.; Li, D.; Li, Z.; Deng, Y. Clean Beckmann Rearrangement of Cyclohexanone Oxime in Caprolactam-Based Brønsted Acidic Ionic Liquids. *Green Chem.* **2006,** *8* (3), 296–300.

Hajimohammadi, M.; Safari, N.; Mofakham, H.; Deyhimi, F. Highly Selective, Economical and Efficient Oxidation of Alcohols to Aldehydes and Ketones by Air and Sunlight or Visible Light in the Presence of Porphyrins Sensitizers. *Green Chem.* **2011,** *13* (4), 991–997.

Halimehjani, Z. A.; Marjani, K.; Ashouri, A. Synthesis of Dithiocarbamate by Markovnikov Addition Reaction in Aqueous Medium. *Green Chem.* **2010,** *12* (7), 1306–1310.

Hemmati, B.; Javanshir, S.; Dolatkhah, Z. Hybrid Magnetic Irish Moss/Fe_3O_4 as a Nano-Biocatalyst for Synthesis of Imidazopyrimidine Derivatives. *RSC Adv.* **2016,** *6* (56), 50431–50436.

Huertas, D.; Florscher, M.; Dragojlovic, V. Solvent-Free Diels–Alder Reactions of In Situ Generated Cyclopentadiene. *Green Chem.* **2009,** *11* (1), 91–95.

Hult, K.; Berglund, P. Engineered Enzymes for Improved Organic Synthesis. *Curr. Opin. Biotechnol.* **2003,** *14* (4), 395–400.

Husin, H.; Su, W.; Chen, H.; Pan, C.; Chang, S.; Rick, J. et al. Photocatalytic Hydrogen Production on Nickel-Loaded $La_xNa_{1-x}TaO_3$ Prepared by Hydrogen Peroxide-Water Based Process. *Green Chem.* **2011,** *13* (7), 1745–1754.

Ishikawa, T.; Kimura, M.; Kumoi, T.; Iida, H. Coupled Flavin-Iodine Redox Organocatalysts: Aerobic Oxidative Transformation from N-Tosylhydrazones to 1,2,3-Thiadiazoles. *ACS Catal.* **2017,** *7* (8), 4986–4989.

Ismail, H.; Lau, R. M.; Langen, L. M. V.; Rantwijk, F. V.; Švedas, V. K.; Sheldon, R. A. A Green, Fully Enzymatic Procedure for Amine Resolution, Using a Lipase and a penicillin G Acylase. *Green Chem.* **2008,** *10* (4), 415–418.

Jiang H.; Jia, L.; Li, J. Wacker reaction in supercritical carbon dioxide. ***Green Chem.* 2000,** *2* (4), 161–164.

Jiang, L.; Althoff, E. A.; Clemente, F. R.; Doyle, L.; Röthlisberger, D.; Zanghellini, A.et al. De novo Computational Design of Retro-Aldol Enzymes. *Science* **2008,** *319* (5868), 1387–1391.

Jin, F.; Yun, J.; Li, G.; Kishita, A.; Tohji, K.; Enomoto, H. Hydrothermal Conversion of Carbohydrate Biomass Into Formic Acid at Mild Temperatures. *Green Chem.* **2008,** *10* (6), 612–615.

Jo, B. H.; Seo, J. H.; Yang, Y. J.; Baek, K.; Choi, Y. S.; Pack, S. P. et al. Bioinspired Silica Nanocomposite with Auto Encapsulated Carbonic Anhydrase as a Robust Biocatalyst for CO_2 Sequestration. *ACS Catal.* **2014,** *4* (12), 4332–4340.

Kamimura, A.; Kaiso, K.; Suzuki, S.; Oishi, Y.; Ohara, Y.; Sugimoto, T.; Kashiwagi, K.; Yoshimoto, M. Direct Conversion of Polyamides to ω-Hydroxyalkanoic Acid Derivatives by Using Supercritical MeOH. *Green Chem.* **2011,** *13* (8), 2055–2061.

Kaplaneris, N.; Koutoulogenis, G.; Raftopoulou, M.; Kokotos, C. G. 4-Fluoro and 4-Hydroxy Pyrrolidine-Thioxotetrahydropyrimidinones: Organocatalysts for Green Asymmetric Transformations in Brine. *J. Org. Chem.* **2015,** 80(11), 5464–5473.

Kim, B.; Jeong, J.; Lee, D.; Kim, S.; Yoon, H.; Lee, Y.; Cho, J. K. Direct Transformation of Cellulose Into 5-Hydroxymethyl-2-Furfural Using a Combination of Metal Chlorides in Imidazolium Ionic Liquid. *Green Chem.* **2011,** *13* (6), 1503–1506.

Kisukuri, C. M.; Palmeira, D. J.; Rodrigues, T. S.; Camargo, P. H.; Andrade, L. H. Bimetallic Nanoshells as Platforms for Metallo-and Biometallo-Catalytic Applications. *ChemCatChem*, **2016,** *8* (1), 171–179.

Kluge, M.; Ullrich, R.; Scheibner, K.; Hofrichter, M. Stereoselective Benzylic Hydroxylation of Alkylbenzenes and Epoxidation of Styrene Derivatives Catalyzed by the Peroxygenase of Agrocybe Aegerita. *Green Chem.* **2012,** *14* (2), 440–446.

Kong, L.; Lin, Q.; Lv, X.; Yang, Y.; Jia, Y.; Zhou, Y. Efficient Claisen Rearrangement of Allyl Para-Substituted Phenyl Ethers Using Microreactors. *Green Chem.* **2009,** *11* (8), 1108–1111.

Kopach, M. E.; Roberts, D. J.; Johnson, M. D.; Groh, J. M.; Adler, J. J.; Schafer, J. P. et al. The Continuous Flow Barbier Reaction: An Improved Environmental Alternative to the Grignard Reaction? *Green Chem.* **2012,** *14* (5), 1524–1536.

Kotlewska, A. J.; Rantwijk, F. V.; Sheldon, R. A.; Arends, I. W. C. E. Epoxidation and Baeyer–Villiger Oxidation Using Hydrogen Peroxide and a Lipase Dissolved in Ionic Liquids. *Green Chem.* **2011,** *13* (8), 2154–2160.

Kumar, L.; Mahajan, T.; Agarwal, D. D. An Instant and Facile Bromination of Industrially-Important Aromatic Compounds in Water Using Recyclable $CaBr_2$–Br_2 system. *Green Chem.* **2011,** *13* (8), 2187–2196.

Laitinen, A.; Takebayashi, Y.; Kylänlahti, I.; Yli-Kauhaluoma, J.; Sugeta, T.; Otake, K. Ene Reaction of Allylbenzene and N-Methylmaleimide in Subcritical Water and Ethanol. *Green Chem.* **2004,** *6* (1), 49–52.

Langen, L. M. V.; Vroom, E. D.; Rantwijk, F. V.; Sheldon, R. A. Enzymatic Coupling Using a Mixture of Side Chain Donors Affords a Greener Process for Ampicillin. *Green Chem.* **2001,** *3* (6), 316–319.

Laveille, P.; Falcimaigne, A.; Chamouleau, F.; Renard, G.; Drone, J.; Fajula, F. et al. Hemoglobin Immobilized on Mesoporous Silica as Effective Material for the Removal of Polycyclic Aromatic Hydrocarbons Pollutants from Water. *New J. Chem.* **2010,** *34* (10), 2153–2165.

Li, F.; Cui, J.; Qian, X.; Żhang, R. A Novel Strategy for the Preparation of Arylhydroxylamines: Chemoselective Reduction of Aromatic Nitro Compounds Using Bakers' Yeast. *Chem. Commun.* **2004,** (20), 2338–2339.

Li, H.; He, Y.; Yuan, Y.; Guan, Z. Nuclease p1: A New Biocatalyst for Direct Asymmetric Aldol Reaction Under Solvent-Free Conditions. *Green Chem.* **2011a,** *13*, 185–189.

Li, G.; Xiao, J.; Zhang, W. Knoevenagel Condensation Catalyzed by a Tertiary-Amine Functionalized Polyacrylonitrile Fiber. *Green Chem.* **2011b,** *13* (7), 1828–1836.

Li, H.; Chen, J.; Wan, Y.; Chai, W.; Zhang, F.; Lu, Y. Aqueous Medium Ullmann Reaction Over a Novel Pd/Ph–Al-MCM-41 as a New Route of Clean Organic Synthesis. *Green Chem.* **2007,** *9* (3), 273–280.

Liu, H.; Jiang, T.; Han, B.; Liang, S.; Wang, W.; Wu, T.; Yang, G. Highly Selective Benzene Hydrogenation to Cyclohexene Over Supported Ru Catalyst Without Additives. *Green Chem.* **2011,** *13* (5), 1106–1109.

Liu, N.; Xie, Y. F.; Wang, C.; Li, S. J.; Wei, D.; Li, M.; Dai, B. Cooperative Multifunctional Organocatalysts for Ambient Conversion of Carbon Dioxide Into Cyclic Carbonates. *ACS Catal.* **2018,** *8* (11), 9945–9957.

Lozano, P.; Bernal, J. M.; Navarro, A. A Clean Enzymatic Process for Producing Flavour Esters by Direct Esterification in Switchable Ionic Liquid/Solid Phases. *Green Chem.* **2012,** *14* (11), 3026–3033.

Ma, X.; Zhou, Y.; Zhang, J.; Zhu, A.; Jiang, T.; Han, B. Solvent-Free Heck Reaction Catalyzed by a Recyclable Pd Catalyst Supported on SBA-15 via an Ionic Liquid. *Green Chem.* **2008,** *10* (1), 59–66.

Mack, J.; Shumba, M. Rate Enhancement of the Morita–Baylis–Hillman Reaction Through Mechanochemistry. *Green Chem.* **2007,** *9* (4), 328–330.

Marandi, A.; Nasiri, E.; Koukabi, N.; Seidi, F. The Fe_3O_4@ apple Seed Starch Core-Shell Structure Decorated In (III): A Green Biocatalyst for the One-Pot Multicomponent Synthesis of Pyrazole-Fused Isocoumarins Derivatives Under Solvent-Free Conditions. *Int. J. Biol. Macromole.* **2021,** *190*, 61–71.

Matsuda, T.; Watanabe, K.; Kamitanaka, T.; Harada, T.; Nakamura, K. Biocatalytic Reduction of Ketones by a Semi-Continuous Flow Process Using Supercritical Carbon Dioxide. *Chem. Commun.* **2003,** *10*, 1198–1199.

Mestres, R.; Palenzuela, J. High Atomic Yield Bromine-Less Benzylic Bromination. *Green Chem.* **2002,** *4* (4), 314–316.

Mitsudome, T.; Matsuno, T.; Sueoka, S.; Mizugaki, T.; Jitsukawa, K.; Kaneda, K. Direct Synthesis of Unsymmetrical Ethers from Alcohols Catalyzed by Titanium Cation-Exchanged Montmorillonite. *Green Chem.* **2012,** *14* (3), 610–613.

Moazzam, A.; Farid, S. M.; Alizadeh, N.; Mahdavi, M. Photochemical Regioselective CH Arylation of Imidazo [1, 2-a] Pyridine Derivatives Using Chlorophyll as Biocatalyst and Diazonium Salts. *J. Chem.* **2022,** *46* (22), 10814–10819.

Mogharabi-Manzari, M.; Kiani, M.; Aryanejad, S.; Imanparast, S.; Amini, M.; Faramarzi, M. A. Magnetic Heterogeneous Biocatalyst Composed of Immobilized Laccase and 2, *2*, 6, 6-Tetramethylpiperidine-1-Oxyl (TEMPO) for Green One-Pot Cascade Synthesis of 2-Substituted Benzimidazole and Benzoxazole Derivatives under Mild Reaction Conditions. *Adv. Synth. Catal.* **2018,** *360* (18), 3563–3571.

Nazari, S.; Keshavarz, M.; Karami, B.; Iravani, N.; Vafaee-Nezhad, M. Imidazol-1-yl-Acetic Acid as a Novel Green Bifunctional Organocatalyst for the Synthesis of 1, 8-Dioxooctahydroxanthenes Under Solvent-Free Conditions. Chinese Chem. Lett. 2014, 25 (2), 317–320.

Ni, Y.; Hagedoorn, P.; Xu, J.; Arends, I. W. C. E.; Hollmann, F. A Biocatalytic Hydrogenation of Carboxylic Acids. *Chem. Commun.* **2012,** *48* (99), 12056–12058.

Ntainjua, E. N.; Piccinini, M.; Freakley, S. J.; Pritchard, J. C.; Edwards, J. K.; Carley, A. F. et al. Direct Synthesis of Hydrogen Peroxide Using Au–Pd-Exchanged and Supported Heteropolyacid Catalysts at Ambient Temperature Using Water as Solvent. *Green Chem.* **2012,** *14* (1), 170–181.

Olyaei, A.; Shahsavari, M. S.; Sadeghpour, M. Organocatalytic Approach Toward the Green One-Pot Synthesis of Novel Benzo [f] Chromenes and 12 H-benzo [5, 6] Chromeno [2, 3-b] Pyridines. *Res. Chem. Intermed.* **2018,** *44* (2), 943–956.

Onoda, A.; Fukumoto, K.; Arlt, M.; Bocola, M.; Schwaneberg, U.; Hayashi, T. A Rhodium Complex-Linked β-Barrel Protein as a Hybrid Biocatalyst for Phenylacetylene Polymerization. *Chem. Commun.* **2012,** *48* (78), 9756–9758.

Pereira, S. C.; Bussamara, R.; Marin, G.; Giordano, R. L. C.; Dupont, J.; Giordano, R. D. C. Enzymatic Synthesis of Amoxicillin by Penicillin G Acylase in the Presence of Ionic Liquids. *Green Chem.* **2012,** *14* (11), 3146–3156.

Phadtare, S. B.; Shankarling, G. S. Halogenation Reactions in Biodegradable Solvent: Efficient Bromination of Substituted 1-Aminoanthra-9, 10-Quinone in Deep Eutectic Solvent (Choline Chloride: Urea). *Green Chem.* **2010,** *12* (3), 458–462.

Pingali, S. R. K.; Upadhyay, S. K.; Jursic, B. S. Microwave-Assisted Benzyl Mono-and Dibromination in Diethyl Carbonate as Environmentally Friendly Alternative to Radical Bromination in Carbon Tetrachloride. *Green Chem.* **2011,** *13* (4), 928–933.

Pollock, C. L.; Fox, K. J.; Lacroix, S. D.; McDonagh, J.; Marr, P. C.; Nethercott, A. M. et al. Minimizing Side Reactions in Chemoenzymatic Dynamic Kinetic Resolution: Organometallic and Material Strategies. *Dalton Trans.* **2012,** *41* (43), 13423–13428.

Polshettiwar, V.; Varma, R. S. Revisiting the Meerwein–Ponndorf–Verley Reduction: A Sustainable Protocol for Transfer Hydrogenation of Aldehydes and Ketones. *Green Chem.* **2009,** *11* (9), 1313–1316.

Pratap, U. R.; Jawale, D. V.; Waghmare, R. A.; Lingampalle, D. L.; Mane, R. A. Synthesis of 5-Arylidene-2, 4-Thiazolidinediones by Knoevenagel Condensation Catalyzed by Baker's Yeast. *New J. Chem.* **2011,** *35* (1), 49–51.

Rangraz, Y.; Nemati, F.; Elhampour, A. Magnetic Chitosan Composite as a Green Support for Anchoring Diphenyl Diselenide as a Biocatalyst for the Oxidation of Sulfides. *Int. J. Biol. Macromol.* **2018,** *117*, 820–830.

Ranu, B. C.; Jana, U.; Majee, A. Selective Monoacylation of Ferrocene: An Eco-Friendly Procedure on the Solid Phase of Alumina. *Green Chem.* **1999,** *1* (1), 33–34.

Renugopalakrishnan, V; Garduno-Juarez, R; Narasimhan, G; Verma, C. S.; Wei X.; Li P. Rational Design of Thermally Stable Proteins: Relevance to Bionanotechnology. *J. Nanosci. Nanotechnol.* **2005,** *5* (11), 1759–1767.

Ríos, M. Y.; Salazar, E.; Olivo, H. F. Baeyer–Villiger Oxidation of Substituted Cyclohexanones via Lipase-Mediated Perhydrolysis Utilizing Urea–Hydrogen Peroxide in Ethyl acetate. *Green Chem.* **2007,** *9* (5), 459–462.

Riyadh, S. M.; Khalil, K. D.; Aljuhani, A. Chitosan-MgO Nanocomposite: One Pot Preparation and Its Utility as an Ecofriendly Biocatalyst in the Synthesis of Thiazoles and [1, *3*, 4] Thiadiazoles. *Nanomaterials* **2018,** 8(11), 928.

Romero, O.; Filice, M.; Rivas, B. D. L.; Carrasco-Lopez, C.; Klett, J.; Morreale, A. et al. Semisynthetic Peptide–Lipase Conjugates for Improved Biotransformations. *Chem. Commun.* **2012,** *48* (72), 9053–9055.

Rostamian, R.; Khalilzadeh, M. A.; Zareyee, D. Wood Ash Biocatalyst as a Novel Green Catalyst and Its Application for the Synthesis of Benzochromene Derivatives. *Sci. Rep.* **2022,** *12* (1), DOI: 10.1038/s41598–022–05133-x.

Salehi, N.; Mirjalili, B. B. F. Nano-Ovalbumin: A Green Biocatalyst for Biomimetic Synthesis of Tetrahydrodipyrazolo Pyridines in Water. *Res. Chem. Intermed.* **2018,** *44* (11), 7065–7077.

Sarma, R.; Prajapati, D. Microwave-Promoted Efficient Synthesis of Dihydroquinazolines. *Green Chem.* **2011,** *13* (3), 718–722.

Shaikh, Y. I.; Shaikh, V. S.; Nazeruddin, G. M.; Shekh, Z.; Gugale, G. S.; Prasad, N. R. A Green Chemistry Approach Towards Synthesis of Biscoumarins Catalyzed by Different Biocatalysts Such as Dragon Fruit Juice, Kiwi Fruit Juice, and Buttermilk Separately by Grind Stone Technique: A Comparative Study. *ES Food Agroforestry.* **2022,** *7*, 25–29.

Shanmuganathan, S.; Natalia, D.; Greiner, L.; María, P. D. D. Oxidation-Hydroxymethylation-Reduction: A One-Pot Three-Step Biocatalytic Synthesis of Optically Active α-aryl Vicinal Diols. *Green Chem.* **2012,** *14* (1), 94–97.

Shanmuganathan, S.; Natalia, D.; Wittenboer, A. V. D.; Kohlmann, C.; Greiner, L.; de María, P. D. Enzyme-Catalyzed C–C Bond Formation Using 2-Methyltetrahydrofuran (2-MTHF) as (co) Solvent: Efficient and Bio-Based Alternative to DMSO and MTBE. *Green Chem.* **2010,** *12* (12), 2240–2245.

Shao, J.; Zhang, J.; Nahálka, J.; Wang, P. G. Biocatalytic Synthesis of Uridine 5′-Diphosphate N-Acetylglucosamine by Multiple Enzymes Co-Immobilized on Agarose Beads. *Chem. Commun.* **2002,** *21*, 2586–2587.

Shen, Z. L.; Cheong, H. L.; Lai, Y. C.; Loo, W. Y.; Loh, T. P. Application of Recyclable Ionic Liquid-Supported Imidazolidinone Catalyst in Enantioselective Diels–Alder Reactions. *Green Chem.* **2012,** *14* (9), 2626–2630.

Song, J.; Zhang, Z.; Han, B.; Hu, S.; Li, W.; Xie, Y. Synthesis of Cyclic Carbonates from Epoxides and CO_2 Catalyzed by Potassium Halide in the Presence of β-cyclodextrin. *Green Chem.* **2008,** *10* (12), 1337–1341.

Stevens, J. G.; Bourne, R. A.; Poliakoff, M. The Continuous Self Aldol Condensation of Propionaldehyde in Supercritical Carbon dioxide: A Highly Selective Catalytic Route to 2-Methylpentenal. *Green Chem.* **2009,** *11* (3), 409–416.

Susanto, W.; Chu, C. Y.; Ang, W. J.; Chou, T. C.; Lo, L. C.; Lam, Y. Development of a Fluorous, Oxime-Based Palladacycle for Microwave-Promoted Carbon–Carbon Coupling Reactions in Aqueous Media. *Green Chem.* **2012,** *14* (1), 77–80.

Tanaka, K.; Sugino, T.; Toda, F. Selective Stobbe Condensation Under Solvent-Free Conditions. *Green Chem.* **2000,** *2* (6), 303–304.

Tee, K. L.; Roccatano, D.; Stolte, S.; Arning, J.; Jastorff, B.; Schwaneberg, U. Ionic Liquid Effects on the Activity of Monooxygenase P450 BM-3. *Green Chem.* **2008,** *10* (1), 117–123.

Theil, F.; Kunath, A.; Ramm, M.; Reiher, T.; Schick, H. Double Enantioselective Trans-esterification of Racemic Carboxylic Esters and Cyclic Meso-Diols by Lipase Catalysis. *J. Chem. Soc. Perkin Trans.* **1994,** *1*, 1509–1516.

Triandafillidi, I.; Kokotos, C. G. Green Organocatalytic Synthesis of Isoxazolines via a One-Pot Oxidation of Allyloximes. *Org.Lett.* **2017,** 19(1), 106–109.

Valiey, E.; Dekamin, M. G.; Alirezvani, Z. Melamine-Modified Chitosan Materials: An Efficient and Recyclable Bifunctional Organocatalyst for Green Synthesis of Densely Functionalized Bioactive Dihydropyrano [2, 3-c] Pyrazole and Benzylpyrazolyl Coumarin Derivatives. *Int. J. Biol. Macromol.* **2019,** *129*, 407–421.

Waddell, D. C.; Mack, J. An Environmentally Benign Solvent-Free Tishchenko Reaction. *Green Chem.* **2009,** *11* (1), 79–82.

Wang, G.; Gao, J. Solvent-Free Bromination Reactions with Sodium Bromide and Oxone Promoted by Mechanical Milling. *Green Chem.* **2012a,** *14* (4), 1125–1131.

Wang, Q.; Fan, H.; Wu, S.; Zhang, Z.; Zhang, P.; Han, B. Water as an Additive to Enhance the Ring Opening of Naphthalene. *Green Chem.* **2012b,** *14* (4), 1152–1158.

Wang, J. L.; He, L. N.; Miao, C. X.; Li, Y. N. Ethylene Carbonate as a Unique Solvent for Palladium-Catalyzed Wacker Oxidation Using Oxygen as the Sole Oxidant. *Green Chem.* **2009,** *11* (9), 1317–1320.

Wang, L.; Zhang, G.; Kodama, K.; Hirose, T. An Efficient Metal-and Solvent-Free Organo-catalytic System for Chemical Fixation of CO_2 Into Cyclic Carbonates Under Mild Conditions. *Green Chem.* **2016,** *18* (5), 1229–1233.

Wolf, S. E.; Schlossmacher, U.; Pietuch, A.; Mathiasch, B.; Schröder, H. C.; Müller, W. E. et al. Formation of Silicones Mediated by the Sponge Enzyme Silicatein-α. *Dalton Trans.* **2010,** *39* (39), 9245–9249.

Wu, W. B.; Wang, N.; Xu, J. M.; Wu, Q.; Lin, X. F. Penicillin G Acylase Catalyzed Markovnikov Addition of Allopurinol to Vinyl Ester. *Chem. Commun.* **2005,** *18*, 2348–2350.

Xu, Y.; Jia, X.; Panke, S.; Li, Z. Asymmetric Dihydroxylation of Aryl Olefins by Sequential Enantioselective Epoxidation and Regioselective Hydrolysis with Tandem Biocatalysts. *Chem. Commun.* **2009,** *12*, 1481–1483.

Yan, L.; Wu, H.; Zhu, Q. Emulsifier-Free Ultrasonic Emulsion Copolymerization of Styrene with Acrylic Acid in Water. *Green Chem.* **2004,** *6* (2), 99–103.

Yang, H.; Han, X.; Ma, Z.; Wang, R.; Liu, J.; Ji, X. Palladium-Guanidine Complex Immobilized on SBA-16: A Highly Active and Recyclable Catalyst for Suzuki Coupling and Alcohol Oxidation. *Green Chem.* **2010,** *12* (3), 441–451.

Yara-Varón, E.; Joli, J. E.; Balcells, M.; Torres, M.; Canela-Garayoa, R. Synthesis of Poly (Ethyl Acrylate-Co-Allyl Acrylates) from Acrylate Mixtures Prepared by a Continuous Solvent-Free Enzymatic Process. *RSC Adv.* **2012,** *2* (24), 9230–9236.

Yilmaz, D. G.; Aydogan, F.; Yolacan, C. An Investigation of Chiral Diamides as Organocatalysts in Asymmetric Aldol Reaction. *J. Heterocycl. Chem.* **2022,** *59*(7), 1169–1179.

Yoshizawa, K.; Toyota, S.; Toda, F. Efficient Solvent-Free Thorpe Reactions. *Green Chem.* **2002,** *4* (1), 68–70.

Yousuf, S. K.; Mukherjee, D.; Singh, B.; Maity, S.; Taneja, S. C. Cu–Mn Bimetallic Catalyst for Huisgen [3+ 2]-Cycloaddition. *Green Chem.* **2010,** *12* (9), 1568–1572.

Yuan, T.; Majid, A.; Marshall, W. D. Detoxification of Aryl-Organochlorine Compounds by Catalytic Reduction in Supercritical Carbon Dioxide. *Green Chem.* **2003,** *5* (1), 25–29.

Zanda, N.; Zhou, L.; Alza, E.; Kleij, A. W.; Pericàs, M. À. Continuous Organocatalytic Flow Synthesis of 2-Substituted Oxazolidinones Using Carbon Dioxide. *Green Chem.* **2022,** *24* (11), 4628–4633.

Zhang, F.; Bulut, S.; Shen, X.; Dong, M.; Wang, Y.; Cheng, X. et al. Halogen-Free Fixation of Carbon Dioxide Into Cyclic Carbonates via Bifunctional Organocatalysts. *Green Chem.* **2021,** *23* (3), 1147–1153.

Zhang, J.; Duetz, W. A.; Witholt, B.; Li, Z. Rapid Identification of New Bacterial Alcohol Dehydrogenases for (R)-and (S)-Enantioselective Reduction of ß-Ketoesters. *Chem. Commun.* **2004,** *18*, 2120–2121.

Zhang, Q.; Su, H.; Luo, J.; Wei, Y. A Magnetic Nanoparticle Supported Dual Acidic Ionic Liquid: A "Quasi-Homogeneous" Catalyst for the One-Pot Synthesis of Benzoxanthenes. *Green Chem.* **2012,** *14* (1), 201–208.

Zhang, T.; Cui, C.; Chen, S.; Ai, X.; Yang, H.; Shen, P.; Peng, Z. A Novel Mediatorless Microbial Fuel Cell Based on Direct Biocatalysis of *Escherichia coli. Chem. Commun.* **2006,** *21*, 2257–2259.

Zhang, Y.; Dayoub, W.; Chen, G.; Lemaire, M. Environmentally Benign Metal Triflate-Catalyzed Reductive Cleavage of the C–O Bond of Acetals to Ethers. *Green Chem.* **2011,** *13* (10), 2737–2742.

Zhang, Y.; Forinash, K.; Phillips, C. R.; McElwee-White, L. Preparation of Biotin Derivatives by Catalytic Oxidative Carbonylation of Diamines. *Green Chem.* **2005,** *7* (6), 451–455.

Zhu, H.; Yang, F.; Tang, J.; He, M. Brønsted Acidic Ionic Liquid 1-Methylimidazolium Tetrafluoroborate: A Green Catalyst and Recyclable Medium for Esterification. *Green Chem.* **2003,** *5* (1), 38–39.

Znabet, A.; Polak, M. M.; Janssen, E.; Kanter, F. J. J.; Turner, N. J.; Orru, R. V. A.; Ruijter, E. *Chem. Commun.* **2010,** *46*, 7918–7920.

CHAPTER 5

Ionic Liquids: Promising Solvents

AVINASH K. RAI[1], ARPIT PATHAK[2], NIRMALA JANGID[3], and P. B. PUNJABI[4]

[1]*Department of Chemistry, PAHER University, Udaipur, India*

[2]*Department of Chemistry, G. G. Govt. P.G. College, Banswara, India*

[3]*Department of Chemistry, Banasthali Vidhyapith, Banasthali, India*

[4]*Department of Chemistry, M. L. Sukhadia University, Udaipur, India*

ABSTRACT

Ionic liquids have many important properties, which make them an ideal solvent. These properties include low vapor pressure, high viscosity, low combustibility, good solvating nature, poor conductor of electricity, good thermal stability, etc. These ionic liquids have been used in many chemical reactions. These solvents are made up of cations such as tetraalkylammonium, N-alkylpyridinium, N-alkylquinolinium, dialkylpyrrolidinium, tetraalkylphosphonium, trialkylsulfonium, etc. and anions like chloride, bromide, nitrates, hydrogen sulfate, hexafluoro phosphate, tetrafluoroborate, tosylate, chloroaluminate, etc. A number of chemical reactions have been carried out using different combinations of these cations and anions such as Friedel-Crafts reaction, Mannich reaction, Diels-Alder reaction, Beckmann rearrangement, and so on. The use of some ionic liquids in a number of chemical reactions has been summarized.

5.1 INTRODUCTION

The first ionic liquid was reported by Gabriel and Weiner (1888), but it was debated for a long. It was ethanol ammonium nitrate (m.p. = 52–55°C).

Green Chemistry, 2nd Edition: Fundamentals and Applications. Suresh C. Ameta & Rakshit Ameta (Eds.)

Room temperature ionic liquid ethyl ammonium nitrate ($C_2H_5NH_3^+NO_3^-$) was synthesized by Walden (1914), which melts at 12°C. In the 1970s and 1980s, some more alkyl substituted imidazolium and pyridinium cations with halide or tetrahaloaluminate anions were developed and were used in batteries as electrolytes (Chum et al., 1975: Wilkes et al., 1982).

Solvents are required for the synthesis of organic compounds but their vapor creates air pollution. Therefore, efforts are being made to use solvents with high boiling points or to avoid solvent (solvent-free reaction). A new class of solvents has emerged, which are fluid at room temperature. The thermodynamics and kinetics of reaction was carried out, which is however different from the traditional molecular solvents. These solvents are high boiling, which means vapor pressure of that solvent is low and hence, no volatile organic compounds (VOCs) are escaped from these liquids at lower temperatures.

It makes these solvents interesting to chemists. An important quality of these green solvents is their physical properties, such as melting point, viscosity, low combustibility, good solvating property, which could be adjusted by varying the cations and anions. As these solvents can be designed to accommodate majority of conditions and therefore these are named as "Designer Solvents" by Freemantle (1998). But there remain two major disadvantages, and these are sensitivity toward moisture and acidity/basicity. Only in the early 1990s, Wilkes and Zaworotko (1992) prepared anionic liquid, which can be used in a variety of applications. They used hexafluorophosphate (PF_6^-) and tetrafloroborate (BF_4^-) as anions. Less toxic cations can be used for better ionic liquids. Search is still on for various cations and anions, which can improve the desired properties of ionic liquids. The field of ionic liquids has been excellently reviewed by several authors from time to time (Seddon, 1997: Welton, 1999: Holbrey and Seddon, 1999: Earle and Seddon, 2000).

These ionic liquids are synonymously known as liquid salts, liquid electrolytes, ionic fluids, ionic melts, fuse salts, or ionic glass. The term ionic liquid is most commonly used since 1942 (Barrer, 1943). Thus, ionic liquids contain large ions (cation and/or anions) and the cations with a low degree of symmetry. As a result, the lattice energy of the crystalline from of ionic salts is decreased resulting into low melting points. Most of the ionic liquids are important because of the following reasons:

i) They have negligible vapor pressure than conventional solvents.
ii) They have higher thermal stability than conventional organic molecular solvents.

iii) They have wide range of solubilities and miscibility. Some ionic liquids are hydrophobic while others are hydrophilic in nature.
iv) They have quite wider liquid range than molecular solvents.
v) They can be easily recycled.
vi) They are nonflammable as compared with other organic solvents.
vii) They are useful in reaction media and catalyst for many types of chemical reactions.
viii) They have wide electrochemical window.
ix) They are also used for the separation and extraction of a chemical from aqueous solutions as well as organic solvents.
x) They also contribute toward the development of green chemistry technology because they are replacing flammable, volatile, and toxic conventional solvent and also reduce chemical wastage.

Ionic liquids are excellent solvent to volatile and hazardous organic compounds because these are having low vapor pressure, thermal and chemical stability, noncorrosive, act as catalyst, and are nonflammable. The ionic liquids are ionic in nature, that is, they are salt-like materials. There is one organic cation and one anion. These are liquids below 100°C (Deetfefo and Seddon, 2003). Ionic liquids are of two types. First type is simple salts, which are made by the combination of simple cations and anions and the second type is binary ionic liquid salts, where an equilibrium is involved, for example, [$EtNH_3$] [NO_3] is a simple ionic liquid, whereas 1, 3-dialkyl imidazolium chloride is a binary ionic liquid.

5.2 IONIC LIQUID AS GREEN SOLVENT

In the present time, many studies have described the use of ionic liquid as a green solvent to develop more eco-friendly and efficient chemical synthesis in both; academic and industrial area. Ionic liquids are excellent alternative solvents to volatile organic solvents.

5.2.1 DESIGNER SOLVENTS FOR A CLEANER WORLD

Ionic liquids have been described as designer solvents (Freemantle, 1998), as these are made up of two components: (i) cations and (ii) anions, which vary with different types of groups. Property of ionic liquids depends on this group. Hence, the term "designer solvent" has been justified for ionic liquids. The nature of the cations and anions has a large influence on the properties of these ionic liquids. The most employed ionic liquid anions are polyatomic inorganic species, halogens, organic, and perfluorinated anions, such as $[BF_4]^-$, $[PF_6]^-$, $[SbF_6]^-$, $[NO_3]^-$, $[AcO]^-$, $Cl]^-$, $Br]^-$.

5.2.2 NOTATIONS

Name of these ionic liquids includes various cations and anions, but particularly containing alkyl chains attached to nitrogen, or phosphorous, and therefore, in scientific literature, various short notations are used for such cations, 1-ethyl-3-methyl imidazolium cation in ionic liquids is represented by $[emim]^+$, $[EtMeim]^+$ or $[C_2C_1im]^+$ with or without bracket. Ethyl-pyridinium cation is represented by $[C_2Py]^+$ or $[EtPy]^+$ cation. Anions are normally denoted in square bracket, if these are polyatomic like $[PF_6]^-$, $[BF_4]^-$ or $[NO_3]^-$ while monoatomic anions are denoted in simple manner, some of the examples are given as follows:

5.2.3 ANIONS

- Halides: bromide $[Br^-]$, chloride $[Cl^-]$
- Nitrate $[NO_3^-]$
- Hexafluoro phosphate $[PF_6]^-$
- Tetrafluoroborate $[BF_4]^-$
- Chloroaluminate $[AlCl_4]^-$
- Hydrogen sulfate $[HSO_4]^-$
- Alkyl sulfate $[RSO_4]^-$
- Acetate $[CH_3COO]^-$
- Tosylate $[OTs]^-$
- Tetrafluoroantimonate $[SbF_6]^-$

5.2.4 CATIONS

Tetraalkylammonium

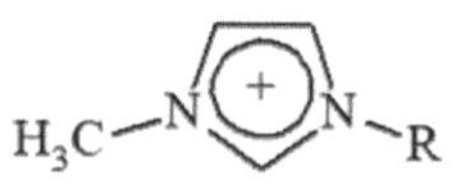

1-Alkyl-3-methylimidazolium

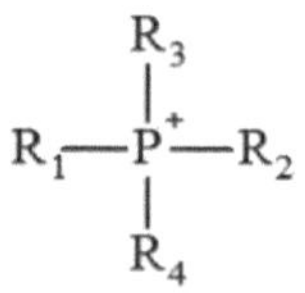

Tetraalkylphosphonium

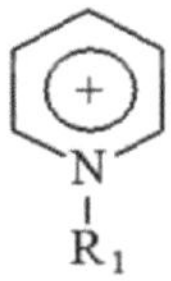

N-Alkylpyridinium

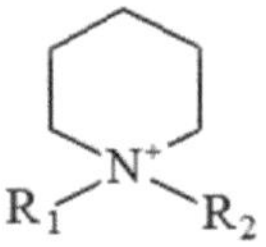

N, N-Dialkylpiperidinium

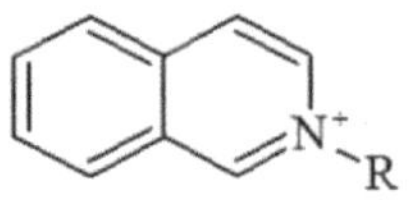

N-Alkylquinolinium

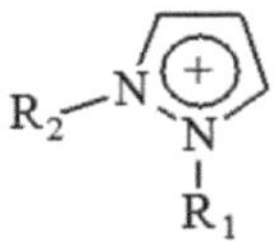

1, 2-Dialkylpyrazolium

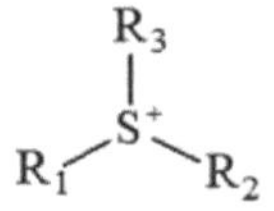

Trialkylsulfonium

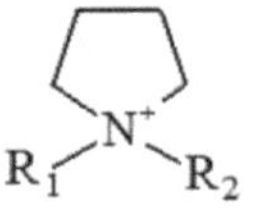

Dialkylpyrrolidinium

5.2.5 CHARACTERISTICS

Ionic liquids have following characteristics:

- Ionic liquids are often moderate to poor conductors of electricity
- Low vapor pressure
- High viscosity
- Non-ionizing
- Low combustibility
- Good thermal stability
- Wide liquid regions
- Good solvating properties

Due to favorably solvating properties by the selection of a proper cation and an anion, ionic liquids can dissolve a wide range of polar and nonpolar compounds, and due to higher viscosity, the ionic liquids are not emitted as vapor like other volatile organic solvents. Due to these different properties, ionic liquids are attracting the attention of synthetic organic chemists as well as electrochemists, chemical engineers, etc. These ionic liquids find their use in biotechnology (Wasserschied and Keim, 2000), chemical engineering (Chum et al., 1975), fuel cells (Wilkes et al., 1982), etc.

5.2.6 IONIC LIQUID GENERATION

Ionic liquids are classified into three generations. The first-generation ionic liquids are haloaluminate room temperature ionic liquids. Haloaluminate ionic liquids have been studied as solvent and catalyst in Friedel-Craft and other organic reactions.

The second-generation ionic liquids are non-haloaluminate ionic liquids, which are used as the best solvent in organic synthesis. The first example of second-generation ionic liquids was described by Wilkes and Zaworotko (1992). Other examples of this generation include dialkylimidazolium cation coordinated with hexafluorophosphate $[PF_6]^-$ and tetrafluoroborate $[BF_4]^-$.

The third-generation of ionic liquids are known as task-specific ionic liquid (TSIL) and chiral ionic liquid. It is also known as functionalized ionic liquid. These ionic liquids are designed with functionalized cation or anion with special property (Davis, 2004). Examples of this type of ionic liquid are imidazolium cation functionalized with sulfonic acid group ($-SO_3H$). Many types of ionic liquids are synthesized with chiral cation or chiral anions (Baudequin et al., 2005). Chiral ionic liquids are the most useful solvent or catalyst in the asymmetric synthesis. Other use of these ionic liquids is that they are used for the resolution of racemate.

5.3 SYNTHESIS OF IONIC LIQUIDS

Synthesis of ionic liquid consists of two steps: (i) synthesis of the desired cation by the simple alkylation or quarternization of nitrogen or phosphorus-containing compounds and (ii) anion resulting from the alkylation can be exchanged by metathesis reaction or by direct combination with Lewis acid or ion exchange resin or acid–base procedure. A general preparation of imidazolium ionic liquid is:

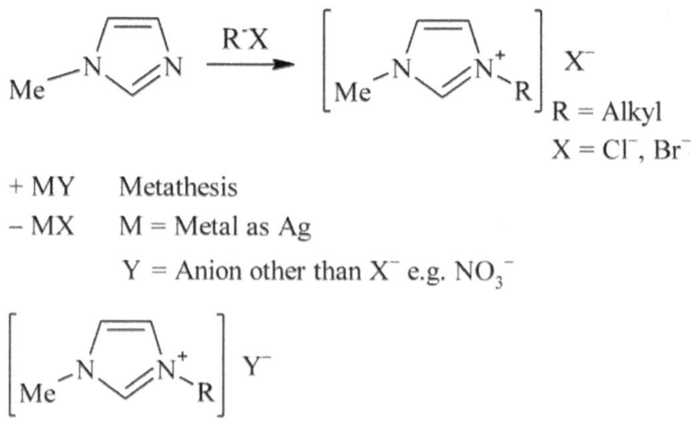

In the first step, preparation of ionic liquid has two disadvantages: (i) it is time-consuming and (ii) an excess of haloalkane is required to gain high yields. Since high-boiling alkyl halides are difficult to remove and the quarternization reaction is dirty, especially when higher chain derivatives are prepared. So, to reduce this problem, ionic liquids are prepared by using microwave radiations.

The second step in the synthesis of ionic liquid involves the preparation by metathesis process or acid–base reaction. In the metathesis reaction, metal salt is used, such as silver nitrate, silver thiocyanate or acid–base neutralization reaction is carried out producing by-product MX.

Now many alkylammonium halides are commercially available or they are made by the reaction of alkyl halide and amine. The use of salt with longer chain substituents such as [bmim]Cl has become popular, which can be prepared by conventional method under reflux (Dyson et al., 1997).

Men et al. (2015) made use of 1-ethyl-3-methylimidazolium acetate ([EMIM]Ac) as solvent (ionic liquid) for synthesizing a co-polymer of polystyrene grafted on starch (starch-g-PS). Potassium persulfate as an initiator. Dialkylcarbonates are considered low-cost and low-toxicity reagents. Homogeneous alkoxycarbonylation of cellulose was carried out by Labafzadeh et al. (2015) using dialkycarbonates (dimethyl and diethyl carbonate) in the ionic liquid electrolyte 1-ethyl-3-methylimidazolium acetate ([emim][OAc]) or trioctylphosphonium acetate ([P8881][OAc])/DMSO.

A one-step method was reported by Pang et al. (2016) to obtain water-soluble cellulose acetate (WSCA) having degrees of polymerization values higher. The 2,6-di-O-acetyl cellulose and 6-mono-O-acetyl moiety 3,6-di-O-acetylcellulose were synthesized by this method. The reactivity of hydroxyl groups in anhydroglucose units was found to be in the order: C-6 > C-3 > C-2. Twenty N-alkyl pyrimidinium ionic liquids were prepared by a two-step synthetic route by Goel et al. (2016), which can be used as reaction media as well as antibacterial agents. It was revealed that the [C9Pyr]BF_4 and [C10Pyr]BF_4 containing longer alkyl chains exhibited excellent bioactivity against both Gram-negative *E. coli*, and gram-positive *S. aureus* bacteria.

The extractive desulfurization of liquid fuel was carried out by Dharaskar et al. (2016) using 1-butyl-3-methylimidazolium tetrafluoroborate [BMIM]BF_4. The extractive desulfurization process could achieve 73.02% removal of dibenzothiophene in n-dodecane for mass ratio (1:1) in half an hour under mild reaction conditions. It was revealed that ionic liquids could be reused four times without almost no decrease in the activity. This work provides significant insights of imidazole-based ILs for desulfurization (extractive) of liquid fuels.

Liu et al. (2018) synthesized 1-ethyl-3-methylimidazolium fluoride ([EMIm]F) ionic liquid via an anion metathesis process. [EMIm]F had glass transition point and decomposition temperature at −71.3°C and 135°C, respectively. Liu et al. (2019) synthesized 1-ethyl-3-methylimidazolium fluoride ([EMIm]F) by a facile route via an anion metathesis process. It was observed that imidazole ring structure was not damaged during the synthesis process.

The pyrimidine-containing ionic liquids were synthesized by Goel et al. (2019) coupled with tetrafluoroborate and bis(trifluoromethane)sulfonamide as counter anions. Some of these ionic liquids showed antibacterial activities against gram +ve bacteria, such as *S. aureus* and *B. pumilis*. Cellulose nanocrystals were prepared by Babicka et al. (2020) using ionic liquids (ILs), 1-propyl-3-methylimidazolium chloride [PMIM][Cl] and 1-ethyl-3-methylimidazolium chloride [EMIM][Cl] from microcrystalline cellulose. It was reported that as-obtained nanocellulose existed in rod-like structure as is evident from SEM images.

5.3.1 USE OF MICROWAVE IRRADIATION

The use of microwave oven as a tool in synthetic chemistry is one of the fastest growing areas of research (Gloe et al., 1982; Blasius et al., 1985). Microwave irradiation for the synthesis of ionic liquids offers some advantages as compared with conventional heating method. In conventional method, synthesis of ionic liquid requires several hours of reaction (10–80 h) at high temperature and large amounts of reactants are required while use of microwaves in the synthesis of ionic liquids reduced dramatically the time required to obtain these compounds.

The first synthesis of ionic liquids under microwave was reported by Varma and Namboodiri (2001) for the synthesis of 1-alkyl-3-methylimidazolium halide (Cl^- and Br^-). An ionic liquid was obtained in 2 min with yields higher than 70%. Because of an increasing interest in ionic liquid (Fu and Liu, 2006; Deetlefs and Seddon, 2003), scientists have started focusing on chiral ionic liquids to use their potential in chiral synthesis (Thang et al., 2004).

Many amino acids-based ionic liquids were synthesized by using microwaves in recent year (Ohno and Fukumoto, 2007). Some new members of amino acid-based ionic liquids were recently synthesized under microwave irradiation by neutralization of L-glutamic acid with nitric sulfuric and hydrochloric acid. Recently, 1, 3-dialkylimidazolium tetrachloroindates were prepared by Kim and Varma (2005a, 2005b) and these ionic liquids were used in the protection of alcohol, diols, and in the synthesis of cyclic carbonates.

Ionic liquids have abilities to dissolve organic as well as inorganic substances. Due to these properties, ionic liquids offer numerous opportunities for the modification of existing and also for the development of new extraction processes. A series of 17 functionalized picolinium-based ionic liquids were prepared by Messali (2016).

The 1-butyl-3-methylimidazolium hydroxide ([bmim][OH]) ionic liquid (IL) was synthesized by Aliabadi and Mahmoodi (2016) and is used for the synthesis of pyranopyrazoles as a catalyst (heterogeneous). This organosolid catalyst was used in a three-component reaction of pyrazolone malononitrile and different benzaldehydes for synthesis of colorful pyrano[2,3-c]-pyrazoles microwave radiations. They also evaluated antioxidant activity of the products via a ABTS (2,2-azinobis(3-ethylbenzothiazoline-sulfonate) assay. More antioxidant activity was observed in this case.

A green and efficient preparation of 20 1-propylimidazolium-based ionic liquids has been reported by Alqurashy (2020) using microwave irradiation. The *in vitro* antitumor activities of all these ILs were investigated, out of which 3-(3-phenylpropyl)-1-propyl-1H-imidazol-3-ium tetrafluoroborate exhibited better activity toward human hepatocellular carcinoma (HEPG-2) as well as colon carcinoma (CACO2), and 3-(3-phenoxypropyl)-1-propyl-1H-imidazol-3-ium tetrafluoroborate against human breast adenocarcinoma (MCF7). The results were compared with that of standard drug (5-fluorouracil 5-FU).

Two imidazolium-based ionic liquids, 3-hexadecyl-1-methyl-1H-imidazol-3-ium bromide [$C_{16}M_1$Im] [Br] and 3-hexadecyl-1,2-dimethyl-1H-imidazol-3-ium bromide [$C_{16}M_2$Im] [Br] were prepared by Subasree and Selvi (2020). They revealed that inhibition efficiency was enhanced due to a rise in the concentration of inhibitor and it was better with [$C_{16}M_2$Im] [Br] than [$C_{16}M_1$Im] [Br], may be due to increased alkyl substituents.

5.4 USE OF IONIC LIQUIDS IN ORGANIC REACTIONS

Ionic liquids are attracting a great deal of attention as an alternate solvent replacing conventional molecular solvents for many organic reactions.

5.4.1 OXIDATION

Although ionic liquids are highly stable and have been evaluated as media for oxidation reactions (Bonhôte et al., 1996), even then little attention has been focused on carrying out catalytic oxidations in ionic liquids. Asymmetric epoxidations were performed with NaOCl in [bmim][PF_6] catalyzed by a chiral Mn complex (Jacobsen's catalyst) (Song and Roh, 2000). An improvement in the catalytic activity was observed by adding the ionic liquid to the dichloromethane solvent. The ionic liquid containing the catalyst was reused in four consecutive runs without any significant loss in yield; however, after the 5th run, the conversion dropped from 83 to 53%. This drop in conversion was believed to be due to degradation of [Mn (III) (salen)] complex.

A more exciting study by Gaillon and Bedioui (2001) was carried out utilizing a chiral Mn(salen) complex in [bmim][PF_6] for the electro-assisted biomimetic activation of molecular oxygen. It was observed that a highly reactive oxomanganese (V) intermediate could transfer its oxygen to an olefin, which hints at a promising future for clean oxidation with molecular oxygen in ionic liquid media.

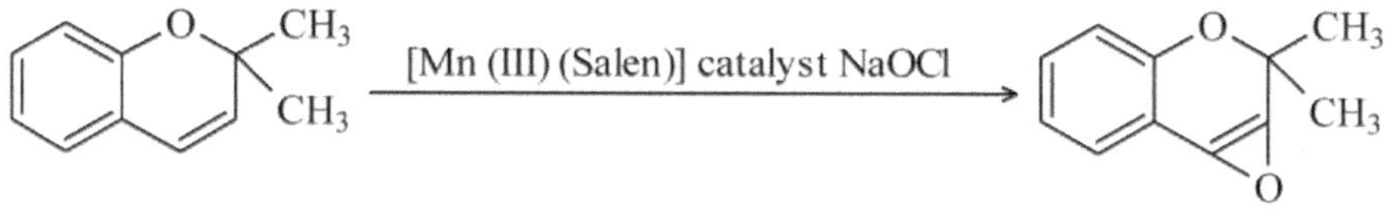

An efficient method for lactone synthesis was reported by Baj et al. (2009) that utilizes bis(trimethylsilyl) peroxide as an oxidant and ionic liquid 1-butyl-3-methylimidazolium trifluoromethanesulfonate as both, that is, the solvent and catalyst.

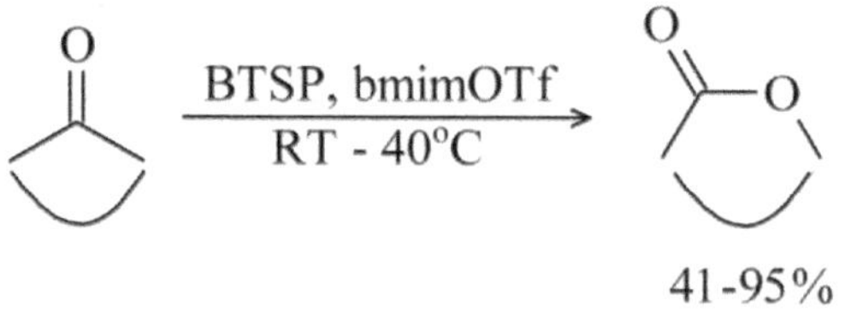

A spectroscopic investigation of the complexes involved in the cobalt catalyzed oxidation of lignin model compounds in ionic liquids was conducted using in situ ATR-IR, Raman, and UV-Vis spectroscopy (Zakzeski et al., 2011). Hydrophilic N,N-dimethylpyrrolidinium- and N,N-dimethylpiperidinium-based ionic liquids provide a highly effective and selective environment for styrene oxidation (Chiappe et al., 2011).

Lü et al. (2014) carried out extraction and catalytic oxidatase desulfurization (ECODS) model diesel in the presence of 30% H_2O_2 and 1-butyl-3-methylimidazolium hexafluorophosphate ([bmim]PF6), under mild conditions in the presence of an Anderson-type catalyst [(C4H9)4N]6Mo7O24. Four sulfur-containing compounds were selected: benzothiophene, dibenzothiophene, 4-methyldibenzothiophene, and 4,6-dimethyldibenzothiophene for extraction into ionic liquid from the model oil and oxidized. It was claimed that the sulfur removal of dibenzothiophene could reach 99.0%. The reactivity of these sulfur-containing compounds in presence of ECODS followed the order:

Dibenzothiophene > 4-Methyldibenzothiophene >
4,6-Dimethyldibenzothiophene > Benzothiophene

N-Hydroxyphthalimide was immobilized by Dobras et al. (2020a) via ester bond on silica gel (SiOCONHPI) and then it was coated with various ionic liquids containing dissolved $CoCl_2$ (SiOCONHPI@CoCl2@IL). It was used for ethylbenzene oxidation with oxygen under solvent-free conditions. A two-fold increase was reported in the conversion of ethylbenzene (4.7–8.6%).

Song et al. (2020) prepared a series of molybdate-based ionic liquids (Mo-ILs), which were thermally regulated. They carried out syntheses of three diverse one-pot oxidative cascade processes of various imines, flavones, and benzyl benzoates with moderate to excellent yields using the Mo-IL $[Bmim]_2[MoO_4]$ as a catalyst in the presence of air.

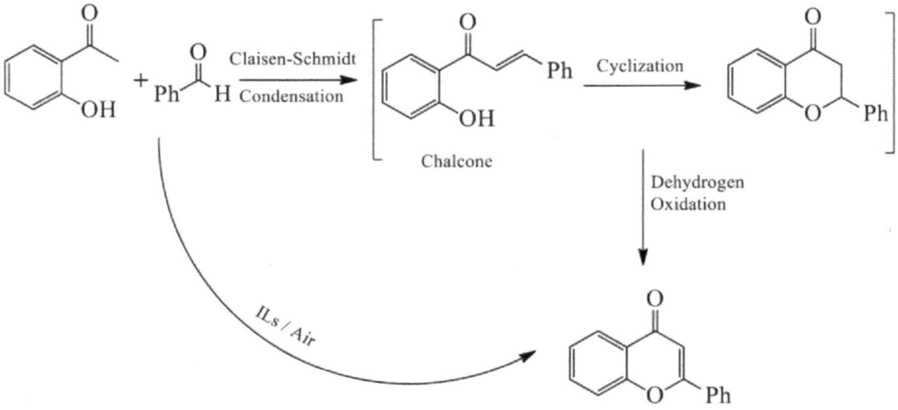

Some catalysts were synthesized by Wu et al. (2020) for oxidation of benzyl alcohol, which were based on polymeric ionic liquids. First, polymeric ionic liquid microspheres (PILM) were prepared, where cations are exchanged in the imidazole ring. These metal anions were then reduced to Pd nanoparticles supported on the surface of PILM. It was observed that the introduction of CeO_2 prevented the aggregation of these Pd nanoparticles and finally formed a core-shell structure of PILM/Pd/CeO_2 catalysts. The 48 and 98% conversion and selectivity could be achieved under the optimal reaction conditions, respectively. As-prepared catalysts were found effective and reusable even after five cycles.

Akopyan et al. (2020) synthesized hybrid materials consisting of polyoxometalate-based ionic liquid (having pyridinium cation)-containing Brönsted acid sites. Later, they used these as catalysts for the oxidation of real and model diesel fuels using Keggin-type polyoxometalates ([PW12O40]3, [PMo12O40]3, [PV2Mo10O40]4, and [PVMo11O40]4) as anions. It was revealed that best results were obtained when catalyst contained phosphomolybdate as anion and nicotinic acid derivative as cation. It was reported that dibenzothiophene was completely oxidized and 90% desulfurization degree of real diesel fuel could be achieved under optimal conditions. It was reported that as-synthesized catalysts could be used 5 times with only a very small decrease in its activity.

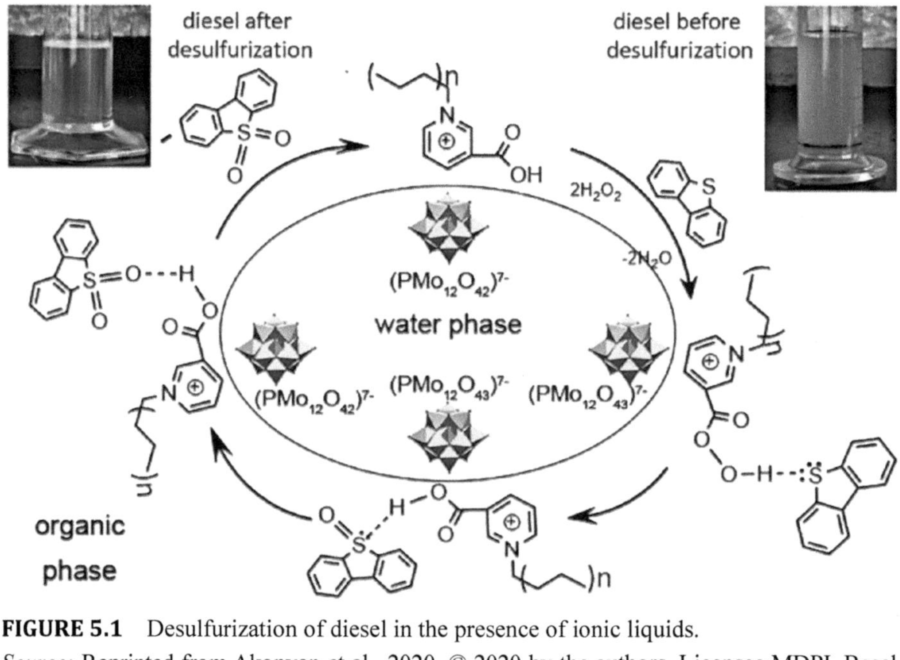

FIGURE 5.1 Desulfurization of diesel in the presence of ionic liquids.

Source: Reprinted from Akopyan et al., 2020. © 2020 by the authors. Licensee MDPI, Basel, Switzerland. http://creativecommons.org/licenses/by/4.0/

A synergistic action between Co(II)/N-hydroxyphthalimide (NHPI) and imidazolium-based ionic liquid (IL) [bmim][OcOSO3] systems was observed by Dobras et al. (2020b) in the catalytic oxidation of ethylbenzene (aerobic) in the absence of solvent. They could achieve above 35% conversion of acetophenone with 83% selectivity.

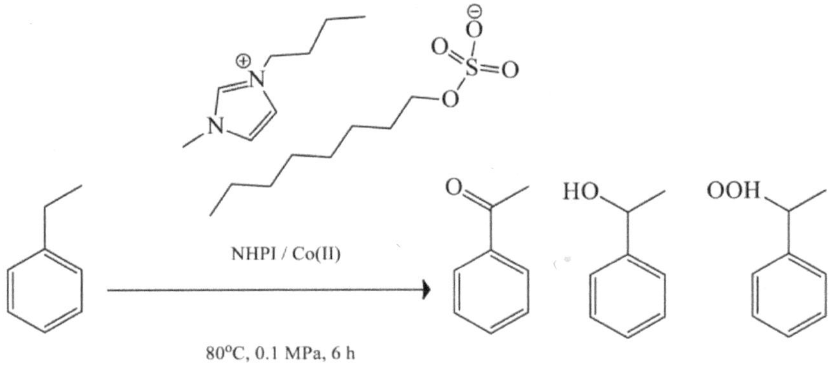

5.4.2 *OXIDATIVE CARBONYLATION OF ANILINE*

Isocyanates and 4,4'-diphenylmethyldiisocyanate are currently manufactured by phosphogenation of the corresponding amines with toxic phosgene, which may cause serious environmental pollution and also corrosion of equipment. Therefore, these are produced either by oxidative carbonylation of amines or by reductive carbonylation of nitro compounds in the presence of an alcohol for the synthesis of isocyanates with non-phosgene routes. These have been extensively studied for the last three decades involving carbamates as intermediates. The Pd, Ru, Rh, Au, and other transitional metal complexes were employed as the catalysts, but the corresponding catalytic turnover frequency is still not high enough for industrial applications.

Palladium complexes coordinated with N-containing compounds is normally insoluble in most of the conventional organic solvents, which played an important role in the homogeneous catalysis. Moreover, organic solvents, which are available to establish a suitable homogenous catalyst system for the carbonylations of N-containing compounds, are very limited. The diversity of ionic liquids may form an optimal homogeneous catalyst system with a specific organometallic complex toward a specific reaction, for example, carbonylation of amine (Shi et al., 2004).

5.4.3 REDUCTION

Howarth et al. (2001) reported the use of ionic liquid in the reduction of aldehyde and ketone. They used $NaBH_4$ as reducing agent in [bmim][PF_6]. This ionic liquid can be recycled.

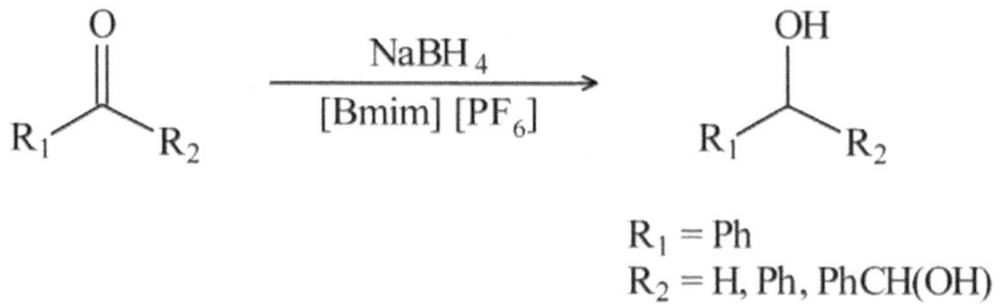

Kabalka and Malladi (2000) also reported the reduction of carbonyl group into alcohol by using trialkylborane catalyst in the presence of [bmim][BF_4], [emim][BF_4] and [emim][PF_6] ionic liquids as solvent.

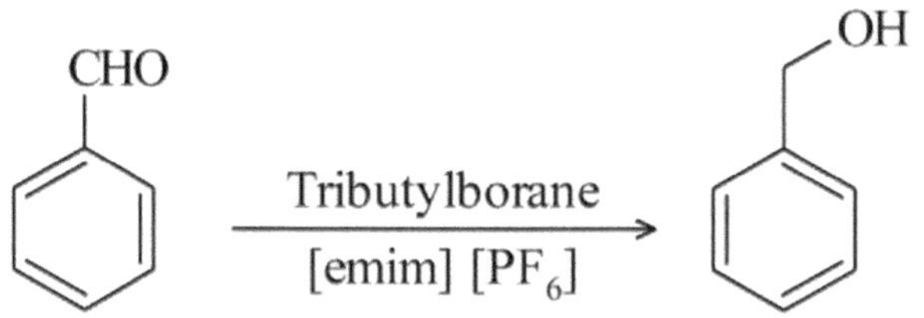

Recently, synthesis of (S)-naproxen has been reported by Moneiro et al. (1997) in the presence of [bmim][BF_4] ionic liquid involving asymmetric hydrogenation. Trihexyl(tetradecyl) phosphonium ionic liquids were found to support the formation of Pd(0) nanoparticles without the addition of reducing agents, such as $NaBH_4$. The resulting particles are highly crystalline and the particle shape is highly dependent on the anion of the ionic liquid (Kalviri and Kerton, 2011).

5.4.4 BECKMANN REARRANGEMENT

The rearrangement of a ketoxime to the corresponding amide is known as Beckmann rearrangement. It is utilized in the manufacture of caprolactam in chemical industries in the presence of excess or stoichiometric amounts of concentrated sulfuric acid or hydrogen chloride in a mixture of acetic acid and acetic anhydride. The application of inorganic acids usually causes a large amount of by-products and serious corrosion resulting in the environment problems. Development of clean and highly efficient catalytic process for Beckmann rearrangement is still necessary. It was reported by Peng and Deng (2001) that ionic liquid could be a catalytic reaction media for Beckmann rearrangement with high conversion and selectivity.

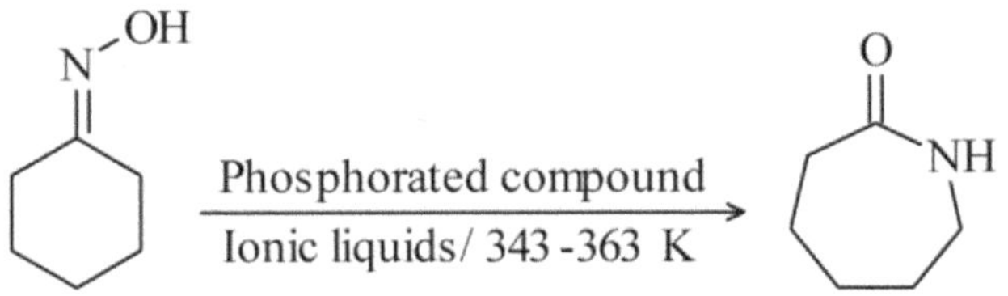

The use of a novel task-specific ionic liquids in the Beckmann rearrangement was reported by Gui et al. (2004). It was reported by Mahajan et al. (2015) that $FeCl_3 \cdot 6H_2O$ can catalyze the conversion of ketones to amides via Beckmann rearrangement with good to excellent yields in the presence of hydroxylamine hydrochloride. Here, no organic solvent is required. The $FeCl_3 \cdot 6H_2O$ is stable, inexpensive, eco-friendly, and easy to handle.

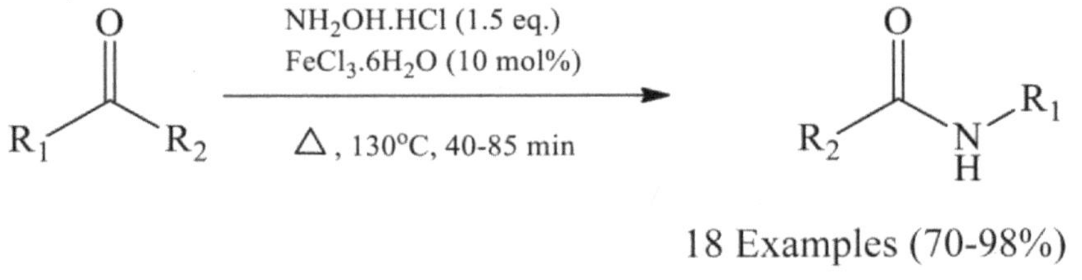

Xie et al. (2016) carried out a microwave-assisted N-fluorobenzenesulfonimide (NFSI)/Lewis acid-catalyzed Beckmann rearrangement. A remarkable promotion to the electrophilicity of NFSI was observed on using Lewis acids. A design of stable and ionic liquid hybrids (IL-MO) was reported by Annath et al. (2018). It facilitates enhanced mass transport and afforded higher performance of catalyst in the Beckmann rearrangement of cyclic oximes. The hybrids played an important role in modifying the reaction pathway, while synergistic enhancements in their catalytic properties was due to increased hydrophobicity of ionic liquid.

Hu et al. (2018) carried out Beckmann rearrangement of ketoxime, which was catalyzed by acidic ionic liquid-N-methylimidazolium hydrosulfate. It was reported that rearrangement of benzophenone oxime gave the product with 45% yield, but the yield could be increased to 91%, when co-catalyst P_2O_5 was added. It was also revealed that the catalyst could be reused for three cycles retaining its efficiency.

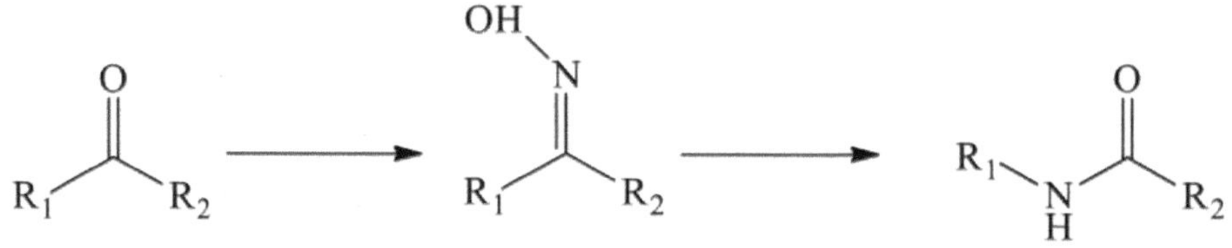

Yu et al. (2019) used a catalytic system consisting of metal chloride ionic liquids for converting ketones to amides via Beckmann rearrangement under solvent-free condition. The effect of type of ILs and metal chlorides, dosage of ILs, length of the chain on the cation of ILs and reaction temperature was investigated. It was reported that different combinations of metal chlorides and ILs were quite efficient for this reaction.

5.4.5 DIELS–ALDER REACTION

The Diels-Alder reaction of cyclopentadiene and methyl acrylate ester has been reported. Diels–Alder reaction is one of the most important tools for carbon–carbon bond formation. In this reaction, 1-ethyl-3-methylimidazolium or chloroaluminate was used as ionic liquid and it was found that the ratio of exo/endo products depends on the ratio of emimCl/$(AlCl_3)_x$ (Lee, 1999). Diels–Alder reactions in neutral ionic liquids (such as 1-butyl-3-methylimidazolium trifluoromethanesulfonate, 1-butyl-3-methylimidazolium hexafluorophosphate, 1-butyl-3-methylimidazolium tetrafluoroborate, and 1-butyl-3-methylimidazolium lactate) were reported by Earle et al. (1999).

The Diels-Alder reaction in the presence of pyridinium-based ionic liquid has also been reported (Xiao and Malhotra, 2004). Two ionic liquids, for example, 1-ethyl-pyridinium tetrafluoroborate ([EtPy][BF_4]) and 1-ethyl-pyridinium trifluoroacetate ([EtPy][CF_3COO]) were used in the Diels-Alder reaction between isoprene and acrylonitrile, acrylic acid, and methacrylic acid. Ionic liquid supported imidazolidinone catalyst was found to be an efficient and recyclable organocatalyst for mediating highly enantioselective Diels–Alder reactions involving α,β-unsaturated aldehydes and cyclopentadiene (Shen et al., 2012).

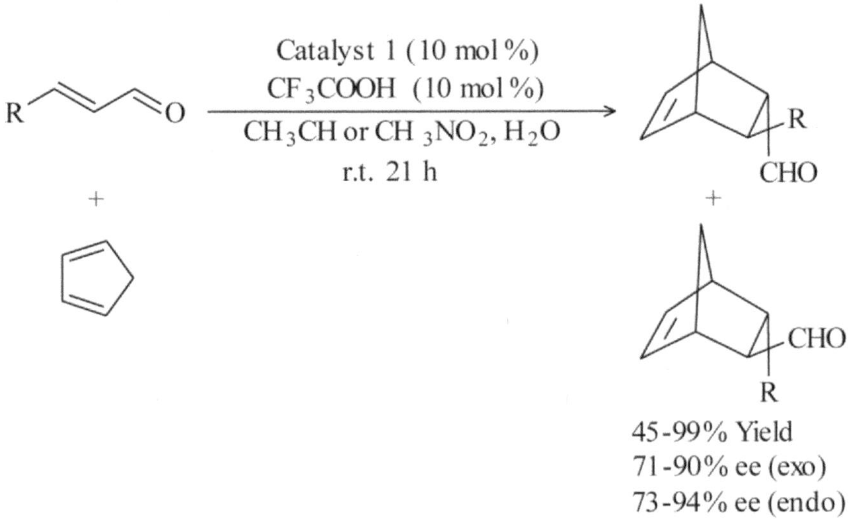

Matuszek et al. (2016) evaluated different metal chlorides (Lewis acids) as homogeneous catalysts in a Diels-Alder reaction. It was observed that $AlCl_3$ and $GaCl_3$ were the best for conversion and endo:exo selectivity. These supported chlorometallate(III) ionic liquids were then used under solventless conditions, where high yields (99%) and high endo-selectivities (95%) could be achieved only in a 5 min and that too at near-ambient temperature (25°C).

Ionic liquid/*n*-hexane were used by Beniwal et al. (2016) as reaction media for a Diels–Alder reaction. Their use gives large rate enhancements of the order varying from 10^6–10^8 times with high stereoselectivity. An enhancement in rate was attributed to polarities and H-bonding abilities of ionic liquids, while their hydrophobicity was considered responsible for controlling stereoselectivity.

Liquid-crystalline ionic liquids (LCILs) are ordered materials and can be used as a reaction medium. Duncan et al. (2016) evaluated the potential of LCILs to affect the stereochemical outcome of the Diels–Alder reaction between cyclopentadiene and methyl acrylate. It was suggested that the proportion of exo-product was found to increase as the reaction media was changed from an isotropic IL to a LCIL.

A facile and simple protocol has been proposed by Kumar and Rajendran (2016) for the Diels–Alder reactions, which were catalyzed by ionic liquid in conjunction with Lewis acid. They obtained Diels–Alder adduct of 1,4-benzoquinone with isoprene and 2,3-dimethyl-1,3-butadiene with good to excellent yields and that too in short time span of 4–5 min under microwave irradiation. The 3-methyl-1-octyl-imidazolium tetrachloroaluminate mixed with Lewis acids can not only be recycled, but it can be reused for some consecutive cycles also.

Goodrich et al. (2017) complexed bis-oxazoline ligand with Zn(II) and Cu(II) trifluoromethanesulfonate and used a range of chiral ionic liquid (CIL) additives (natural products based) as a co-catalyst for Diels-Alder reaction. They compared the catalytic performance of these systems with and without the presence of CIL additive. Significant enhancements were found in endo enantioselectivity up to 50% on addition of CIL in the case of Zn-based catalyst as compared with Cu-based catalyst.

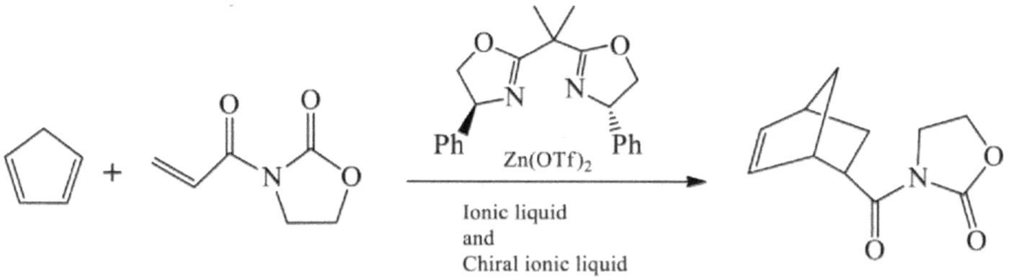

Matuszek et al. (2017) used ionic liquids, which were based on tricoordinate borenium cation as Lewis acid catalysts for Diels-Alder reaction. These borenium ionic liquids exceeded the performance of other catalysts. It was revealed that the conversion of the dienophile can be correlated with ionic liquids (Gutmann acceptor number values).

5.4.6 MANNICH REACTION

Mannich reaction is one of the most important C-C bond-forming reactions in the organic synthesis for the preparation of secondary and tertiary amine derivatives (Arend et al., 1998). Mannich bases are versatile synthetic intermediates, which are traditionally synthesized via acid (base) catalyzed reaction. Brønsted acidic ionic liquid can be used as a catalyst and solvent for this type of reaction. In this case, high yield was achieved and the ionic liquid can be easily reused (Zhao et al., 2004).

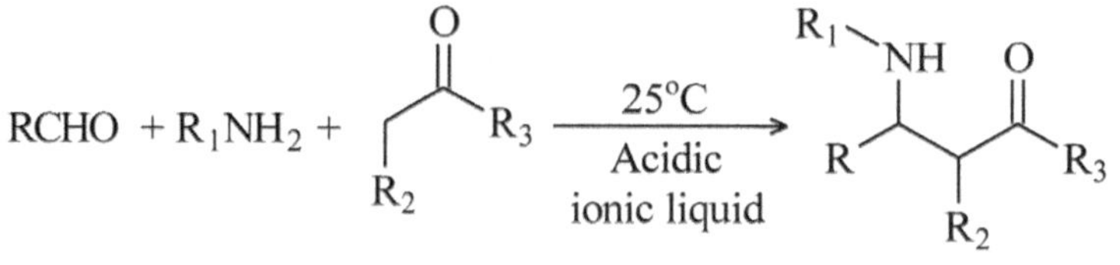

The Mannich reaction of the aldehyde, ketone, and amine in the presence of Bronsted acidic ionic liquid has been reported by Sahoo et al. (2006).

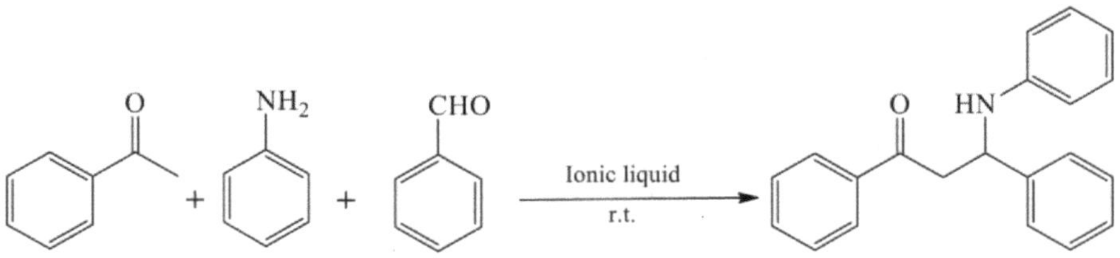

Qian et al. (2016) reported facile and efficient process for the synthesis of β-amino carbonyl compounds via Mannich reactions. They used caprolactam-based Brønsted acidic ionic liquids as catalyst under ultrasonic irradiation. The [Capl][BF_4] was found to be the most effective catalyst for affording good yield within 2–6 h. It was also observed that ultrasound can effectively reduce the reaction time along with increasing yield at mild conditions.

Huan et al. (2016) catalyzed Mannich-type reaction by a choline-based acidic ionic liquid. It was reported that this catalyst is a Lewis-Brønsted dual acid catalyst and it is also water-tolerant. It offered *β*-amino carbonyl compounds at room temperature in good yields (63–98%).

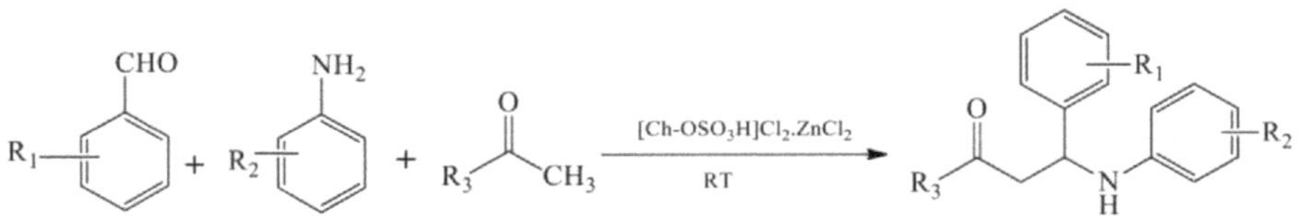

Sardar et al. (2017) observed the reaction between three Mannich components (ketones and aromatic aldehydes, and amines), which was catalyzed by four Bronsted acidic ionic liquids at room temperature. Here, ionic liquids were used as catalysts as well as solvents to produce a series of Mannich bases with high yield (75%) and in shorter reaction time (20 min.). This catalyst can be recycled up to four times without any significant loss of activity.

Ghomi and Zahedi (2017) prepared some ionic liquids with choline chloride and l-alanine and grafted these on Fe_3O_4 nanoparticles. They examined the catalytic activities of this catalyst in the Mannich reaction to synthesize β-aminocarbonyl compounds in the presence of ultrasonic irradiation. It was also indicated that these catalysts can be recycled six times without any loss of activity.

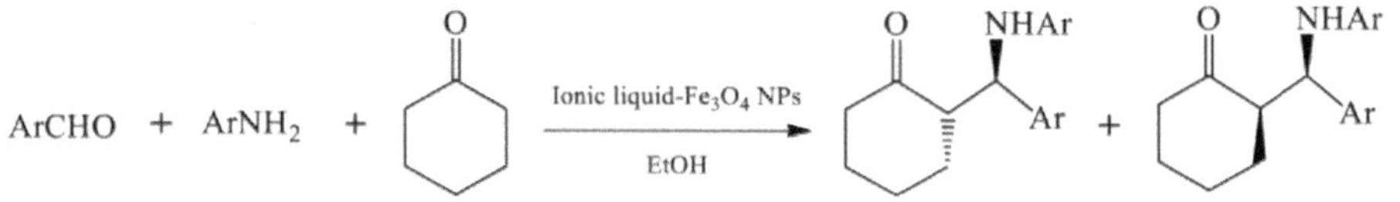

Novel mesoporous silica supported ILs were prepared by Wang et al. (2017) and were used in the Mannich reactions as a heterogeneous catalyst.

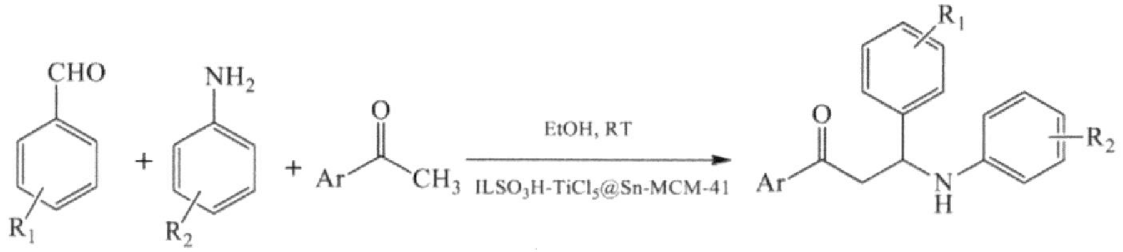

Khanapure et al. (2019) synthesized cellulose-supported ionic liquid phase (SILP) catalyst. This contains a pendant ferrocenyl group and camphor sulfonate anion. It was also reported that SILP catalyst exhibited excellent catalytic activity in the synthesis of β-amino carbonyl compounds by Mannich reaction.

An efficient, simple, and clean method for one-pot Mannich reaction was developed by Prabhakara and Maiti (2020) under solvent-free conditions using ionic liquid-immobilized proline(s) organocatalyst. Mannich reaction contains three components, such as acetophenones, substituted aromatic amines, and aromatic aldehydes in the presence of 7 mol% of ionic liquid-immobilized proline(s) organocatalyst. Mannich reaction provided β-amino carbonyl compounds within 2–3 h at room temperature and that too with excellent yields. It was revealed that this organocatalyst can be reused five times without any significant loss of its catalytic activity.

5.4.7 HECK REACTION

The first use of ionic liquids as reaction media for the palladium-catalyzed Heck coupling was reported by Kaufmann et al. (1996). In the Heck reaction, palladium catalyst, polar solvent, and an aryl iodide were used. 1-Butyl-3-methylimidazolium bromide ionic liquid ([bmim]Br) was used as a solvent, where aryl bromide reacts with styrene to produce stilbenes in high yields without adding phosphine ligand (Xu et al., 2000).

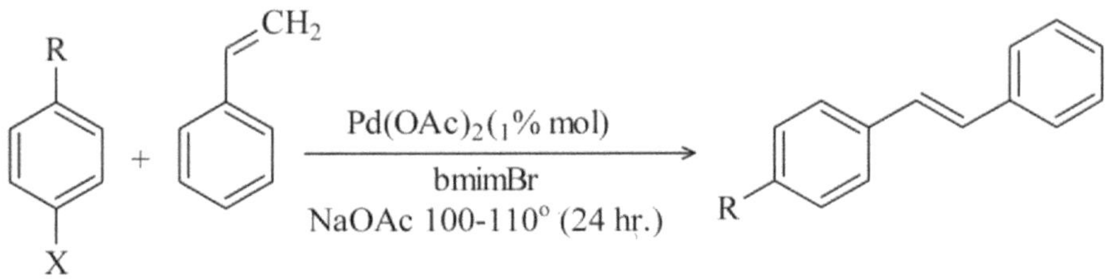

Elhamifar et al. (2013) used palladium-containing mesoporous organosilica (periodic) with ionic liquid (Pd@PMO-IL-I) as a catalyst in the Heck cross-coupling reaction. The as-prepared catalyst exhibited excellent catalytic activity with different aryl halides, and with good to excellent yields along with excellent E-selectivities.

An organic biphasic system and thermoregulated ionic liquid [$CH_3(OCH_2CH_2)_{16}N+Et_3$][$CH_3SO_3$] (ILPEG750) with ILPEG750-stabilized Pd nanoparticles and organic solvents was used for Heck reaction by Zeng et al. (2013). The As-synthesized catalyst exhibited high efficiency and it could be easily separated from products using phase separation. It was also revealed that it could be recycled for six times without any significant loss in its activity.

The use of amine-based ionic liquids (R_3N + PPh_2) was reported by Nowrouzi et al. (2014) as a reusable medium, reducing agent, and suitable Pd(II) ligand. They used it for Heck coupling reaction of aryl halides (bromides and iodides) with n-butyl acrylate and styrene. A green cobalt nanoparticle catalyst was prepared by Hajipour et al. (2015), which was supported on ionic liquid-functionalized MWCNTs. They evaluated its performance as a heterogeneous catalyst for Mizoroki Heck reaction.

Tetrabutylammonium (TBA) and tetrabutylphosphonium (TBP) ionic liquids were synthesized by Hayouni et al. (2017) via an acido-basic method. These were used as solvent for Heck reaction between halogenoaromatics or iodoarene with *tert*-butylacrylate in the presence of $PdCl_2$ as a catalyst and a base. The synthesis of surface active ionic liquids was reported by Taskin et al. (2017). Then, they used it for application in palladium-catalyzed cross-coupling reactions. Various ionic liquids were used in the Heck reaction of ethyl acrylate and iodobenzene producing high yields (> 90%) in water.

5.4.8 FRIEDEL–CRAFTS REACTION

The Friedel–Crafts acylation of benzene has been conducted in acidic chloroaluminate (III) ionic liquid (Boon et al., 1986). Other example is the benzoylation of anisole in the presence of copper triflate in $bmimBF_4$ and methoxybenzophenone was obtained (Ross and Xiao, 2002).

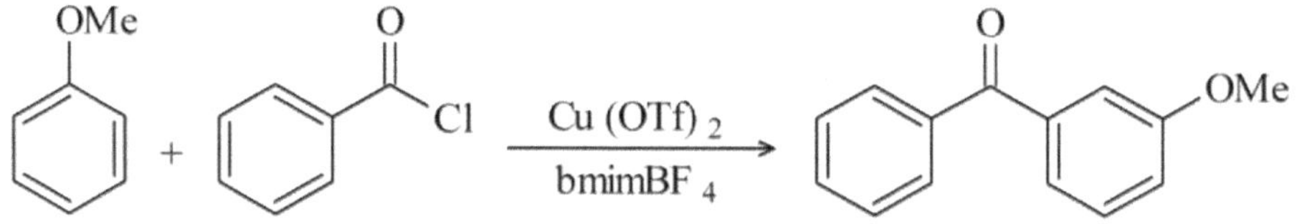

Naphthalene has been acetylated in ionic liquid with highest known selectivity for 1-position (89%) as compared to 2-position (2%) (Adams et al., 1998). Another interesting development is the use of [bmim][chloroaluminate] as Lewis acid catalyst for Friedel–Crafts sulfonylation of benzene and substituted benzenes with TsCl (Nara et al., 2001).

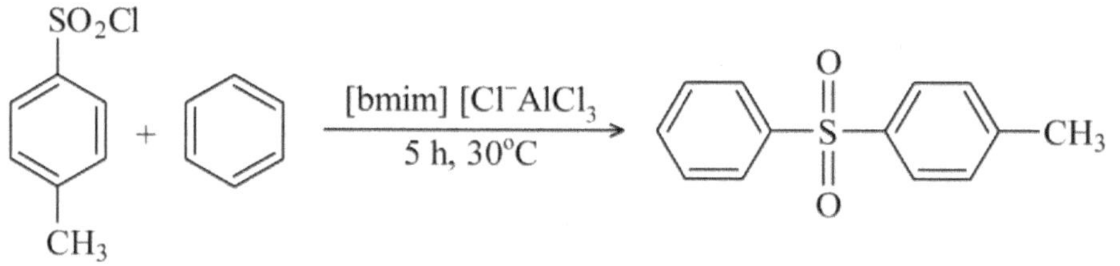

Gordon and Ritchie (2002) investigated the application of ionic liquids as solvents for indium- and tin-mediated allylation of carbonyl compounds. Solvent-controlled Friedel–Crafts reactions of indoles with various isatins were catalyzed by Tong et al. (2017) using 1,4-diazobicyclo[2.2.2]octane (DABCO)-base ionic liquids. The 3-indolyl-3-hydroxy oxindoles and 3, 3-diindolyl oxindoles were obtained using this catalyst. It was reported that monoindolylation products of 3-indolyl-3-hydroxy oxindoles were obtained in good to excellent yields (70–96%) when THF was used as a solvent while 3,3-diindolyl oxindoles were obtained in excellent yields (85–98%) in water.

Yang et al. (2019) made use of aryl imidazolium magnetic ionic liquids, which served two purposes: catalyst–solvent and dual Brønsted-Lewis acidity (B-L MILs) for Friedel-Crafts alkylation. They observed the catalytic properties of B-L MILs in the Friedel-Crafts alkylation of p-xylene using benzyl chloride. It was reported that B-L MIL exhibited the selectivity of product and conversion (> 99%) in 30 min.

A cross-linked dual Brønsted acidic ternary mesoporous poly(ionic liquids) (MPILs) was synthesized by Sha et al. (2019). They used divinylbenzene 1-vinyl-3-butyl imidazole bromide as cross-linker. A sponge-like mesoporous tunnel structure was obtained, which was used for the reaction of alkylation of o-xylene with styrene to 1-diphenylethane (PXE). They could achieve 100% conversion of styrene, and 93.7% PXE yield was obtained in 3 h.

Lin et al. (2020) catalyzed the Friedel–Crafts acylation and thioesterification using tunable aryl imidazolium ionic liquids. It was revealed that these reactions can form C−C and C−S bonds with high atom economy giving good to excellent yields and retaining their catalytic activities.

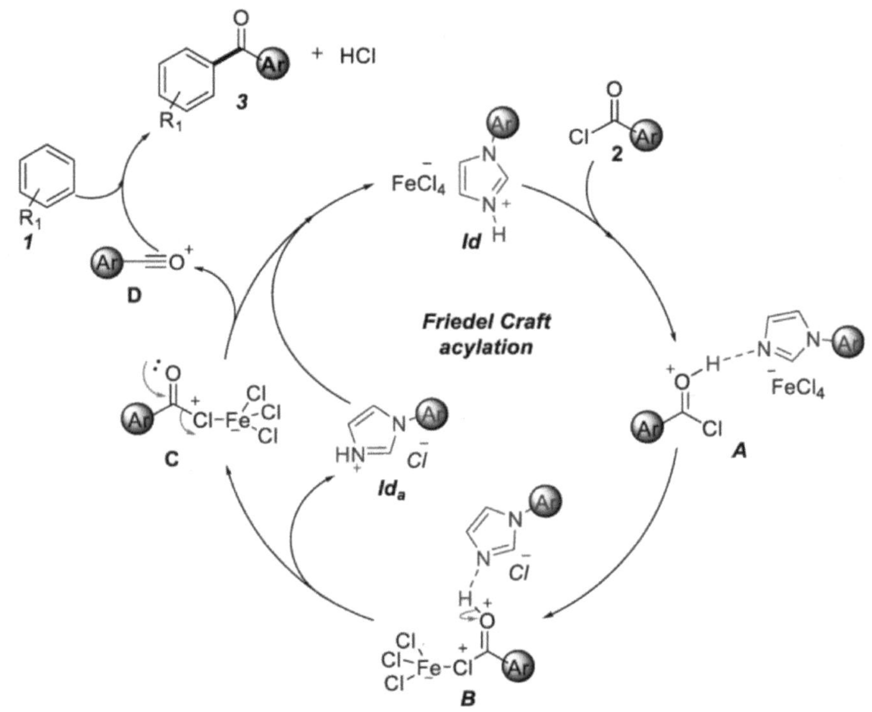

FIGURE 5.2 Friedel–Crafts acylation and thioesterification using tunable aryl imidazolium ionic liquids.

Source: Reprinted from Lin et al., 2020), Molecules, 2020. © 2020 by the authors. Licensee MDPI, Basel, Switzerland. http://creativecommons.org/licenses/by/4.0/

Two ionic liquids $[HN_{222}][Al_2Cl_7]$,where $[HN_{222}]$ = Triethylammonium and the liquid coordination complex (LCC) $AlCl_3$/O-NMPχ_{AlCl3}0.6 (O-NMP = *N*-Methyl-2-pyrrolidone) were used by Kore et al. (2020) for the acylation of isobutylbenzene to synthesize ibuprofen intermediate (major), 4-isobutylacetophenone in the presence of HF and $AlCl_3$; however, both catalysts have some demerits such as acute safety issues, which are associated with the requirement for additional processing for deacidification of $AlCl_3$ and use of volatile/toxic HF. It was observed that LCC, with a Lewis acidic cation ($[AlCl_2(O\text{-}NMP)_2]^+$) and anion ($[Al_2Cl_7]^-$) exhibited the highest catalytic performance (99%) conversion with 96% selectivity.

Gao et al. (2021) synthesized bismuth SBA-16 catalyst and four kinds of *p*-toluenesulfonic acid, which were functionalized with imidazole (ionic liquids). Then Bi(10)-SBA-16 silicon mesoporous material was functionalized with the ionic liquids to composite. The as-prepared composite catalyst was used in Friedel–Crafts acylation of anisole with acetic anhydride. It was reported that 1.2ILc@Bi(10)-SBA-16 can convert 85.41% anisole to yield aromatic ketone (69.19%) at 100°C.

5.4.9 WITTIG REACTION

Wittig reaction is a useful reaction for C=C formation. In the Wittig reaction, separation of product from the triphenylphosphine oxide is problematic. But when the reaction was carried out in the ionic liquid as solvent, the product is easily separated from the triphenylphosphine oxide by ether extraction (Boulaire and Gree, 2000).

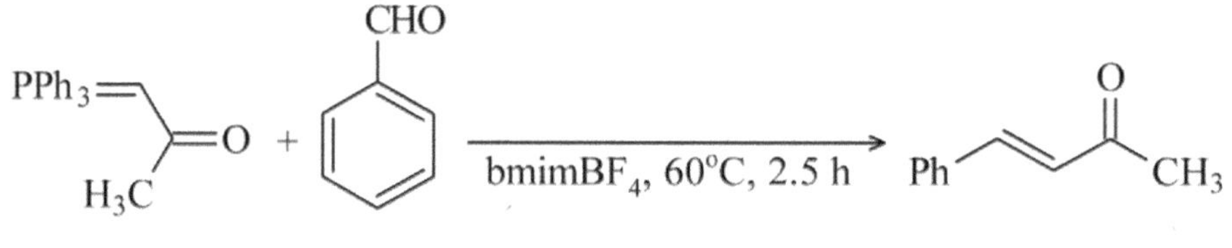

5.4.10 MICHAEL ADDITION REACTION

Zare et al. (2007) described the Michael addition reaction of sulfonamide to α, β-unsaturated esters using ZnO in [bmim] Br as a recyclable solvent.

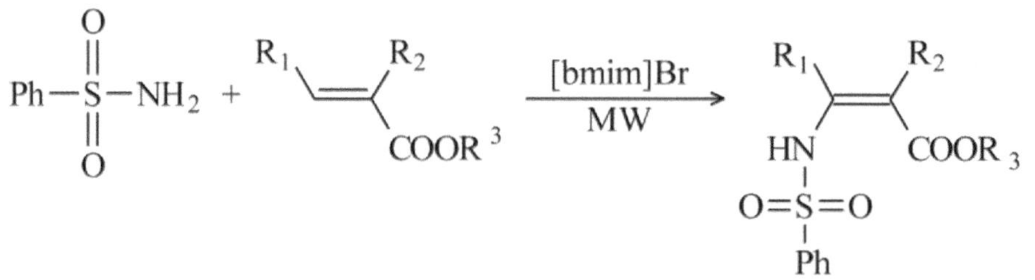

5.4.11 FISCHER INDOLE SYNTHESIS

Fischer indole synthesis in ionic liquid was reported using chloroaluminate ionic liquid as a catalyst and solvent (Rebeiro and Khadilkar, 2001).

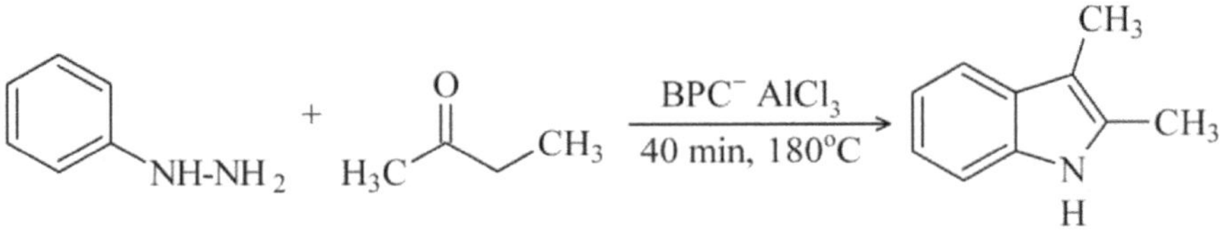

n-BPC = 1-(N-Butyl)pyridinium chloride

5.4.12 CONDENSATION REACTION

Benzoin condensation is one of the oldest C-C bond-forming reaction in organic chemistry, and in this reaction, cyanide ions were used as a catalyst. Benzoin condensation in imidazolium-based room temperature ionic liquids has been reported by Fang et al. (2005).

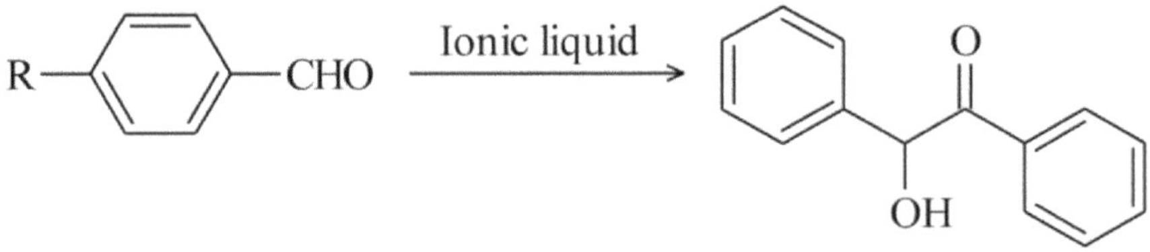

Xu et al. (2010) reported a simple, efficient and green procedure for Knoevenagel condensation catalyzed by $[C_4dabco][BF_4]$ ionic liquid in water.

5.4.13 COUPLING REACTION

Chiappe et al. (2004) investigated the Stille cross-coupling reaction in 10 different ionic liquids. The physicochemical properties of ionic liquids affect the transfer of vinyl and alkyl groups. Palladium acetate was immobilized in ionic liquid layers on the mesopore wall of hierarchical MFI zeolite, and tested as a catalyst for Suzuki coupling reaction in water (Jin et al., 2009).

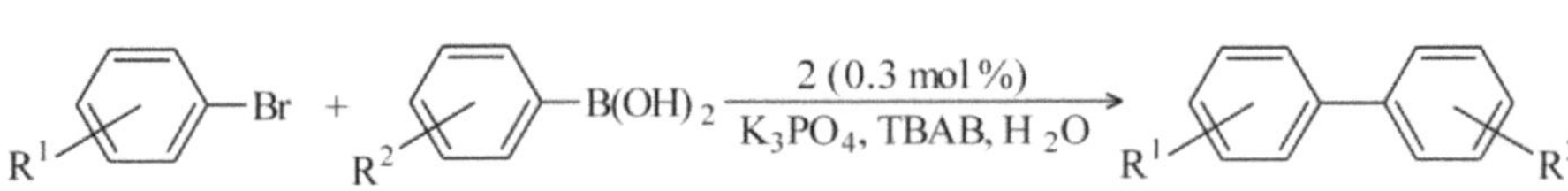

5.4.14 ESTERIFICATION REACTION

The application of ionic liquids as solvents for the enantioselective esterification of (R,S)-2-chloropropanoic acid with butan-1-ol using *Candida rugosa* lipase was reported by Gubicza et al. (2003). Liu et al. (2005) observed that the use of carbohydrates as renewable feedstocks was greatly hampered by their low solubility in any solvent but water. Ionic liquids that contain the dicyanamide ion (dca) can dissolve approx. 200 g L^{-1} of glucose, sucrose, and cyclodextrin. *Candida antarctica* lipase B (CaLB)-mediated the esterification of sucrose with dodecanoic acid in [bmIm][dca].

Candida antarctica lipase B maintained the transesterification activity upon dissolution in the ionic liquid [Et_3MeN][$MeSO_4$], but not in other ionic liquids that dissolved it, such as [bmIm][[dca]. In contrast, cross-linked enzyme aggregates of CaLB remained active in this latter ionic liquid (Rantwijk et al., 2006). Esterification reaction of alcohol and acetic acid in the presence of ionic liquid has been reported using 1-butylpyridinium chloride-aluminum chloride ionic liquid (Deng et al., 2001). Liu et al. (2011) used double-SO_3H-functionalized ionic liquids for the esterification of glycerol with acetic acid.

5.4.15 CYCLIZATION

The utility of ionic liquids as a safe recyclable reaction media at 200°C was reported by Zulfiqar and Kitazume (2000) in the presence of anhydrous scandium trifluoromethanesulfonate for a sequential reaction involving a Claisen rearrangement and cyclization. A Rh(I) catalyzed cycloisomerization of dienes with alkenes using ionic liquids as reaction media was investigated by Oonishi et al. (2009). It was found that the structure of ionic liquids strongly affected the recyclability of the catalyst. In this cycloisomerization, $[BDMI]^+$-based ionic liquid was found to be more effective than a $[BMI]^+$-based one.

5.4.16 BIOCHEMICAL REACTION

Docherty and Kulpa (2005) examined the toxicity and antimicrobial activity of imidazolium and pyridinium ionic liquids. It provides information for future cross-disciplinary studies to test the toxicity of ionic liquids, develop models, and create green solvents. Chiappe et al. (2007) investigated the effect of ionic liquids on epoxide hydrolase-catalyzed synthesis of chiral 1,2-diols while Cho et al. (2008) reported that the toxicity of fluoride-containing ionic liquid

[BMIM][BF_4] to algal growth can increase due to hydrolysis of anions. The effects of ionic liquids on two freshwater algae were reported by Kulacki and Lamberti (2008). They examined *Scenedesmus quadricauda* and *Chlamydomonas reinhardtii*, under high and low nutrient test conditions. Li et al. (2008) developed an efficient system for hydrolysis of lignocellulosic materials in ionic liquids with improved sugar yields at 100°C under atmospheric pressure.

5.4.17 BIODEGRADATION

With the goal of carbon-free production of hydrogen from fossil fuels, a supported ionic liquid membrane for separating CO_2 from H_2 has been suggested for use in a separation enhanced reactor (Raeissi and Peters, 2009). Docherty et al. (2010) reported that pyridinium-based ionic liquids are biodegraded via different pathways depending upon the length of the substituted alkyl chain. Biodegradation products are nontoxic to aquatic test organisms.

5.4.18 BIGINELLI REACTION

The manganese-containing mesoporous organosilica (periodic) with ionic liquid framework (Mn@PMO-IL) was used by Elhamifar and Shabani (2014) in the Biginelli reaction as a catalyst. The Biginelli condensation of different aldehydes (in one pot) with alkyl acetoacetates and urea was developed in the presence of Mn@PMO-IL affording dihydropyrimidone products in high to excellent yields and selectivities in shorter reaction times under solvent-free conditions.

Nagarajan et al. (2015) synthesized three Bronsted acid-based ionic liquids (1-Ethyl-1,2,4-triazolium triflate, 1-propyl-1,2,4-triazolium triflate and 1-butyl-1,2,4-triazolium triflate). These ionic liquids were then used as catalysts for one-pot synthesis of 3,4-dihydropyrimidin-2(1H)-thiones and 3,4-dihydropyrimidin-2(1H)-ones (Biginelli reaction products).

5.4.19 MISCELLANEOUS

The complete synthesis of the pharmaceutical drug, pravadoline, in ionic liquid was reported by Earle et al. (1998). The best results were obtained [bmim][PF_6] at 150°C. The studies of carbonylation reactions of nitrogen-containing compounds using CO_2 are relatively fewer in comparison than using CO, most probably due to its chemical inert nature. Good yield (60–90%) of desired products were achieved in large amounts of dehydrating agents, such as PCl_5, $POCl_3$, dicyclohexylcarbodiimide, etc., but the worse thing was that the large amounts of by-products were formed, such as Et_3NHCl, $Et_3NHPOCl$, and HCl etc. This indicates that using CO_2 as a carbonyl source to substitute phosgene in the synthesis of isocyanate and its derivatives was not only economically less favorable but it is also less eco-friendly.

The use of ionic liquid afforded new opportunity to solve this problem because of the special solubility of carbon dioxide in ionic liquids. A series of aliphatic amines and even aromatic amines could react with CO_2 to afford corresponding urea with moderate to high yields in the presence of CsOH / ionic liquid catalyst system (Shi et al., 2003). Both cations and anions of the ionic liquids had a strong impact on the formation of urea when cyclohexylamine was used as a substrate. When the conversion was high enough, the desired product would precipitate when water was added into the resulting liquid mixture after reaction because urea is insoluble in water, while the CsOH/ionic liquid remained still dissolved into the ionic liquid. Therefore, the desired solid product with good yield could be recovered. An isolated yield of 98% was achieved after filtration and dryness when bmimCl ionic liquid containing CsOH was employed.

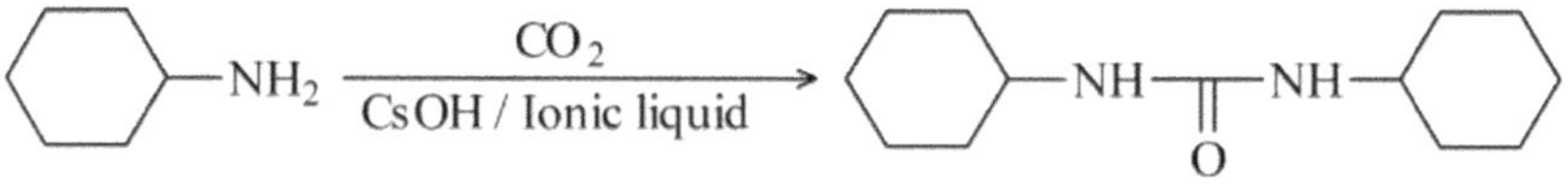

The syntheses and reactions of alkynyl zinc reagents and Reformatsky-type reactions in ionic liquids (as a safe recyclable reaction medium) were reported by Kitazume and Kasai (2001). A simple method for the mono-N-alkylation of primary amines in ionic liquids was developed by Chiappe and Pieraccini (2003).

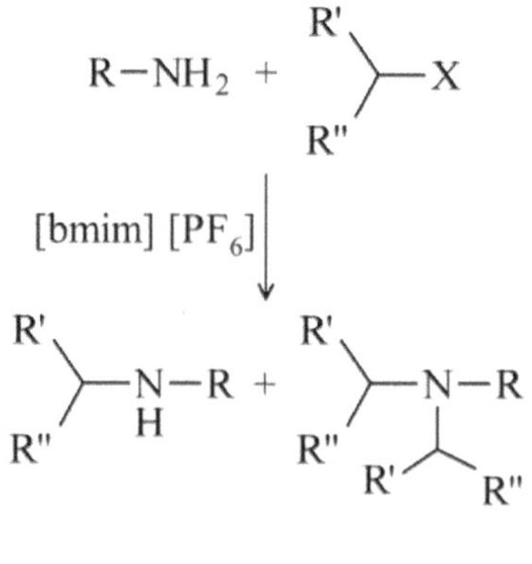

Sarca and Laali (2004) carried out transacylation and deacylation of sterically crowded acetophenones in various imidazolium ionic liquids using triflic acid as a catalyst. BF_4^- based ionic liquids have potential as extractants in the recovery of some amino acids from fermentation broth. Hydrophobicity of the amino acid, pH of the aqueous phase, and water solubility in the ionic liquid phase are the key factors affecting this extraction (Wang et al., 2005). Weng et al. (2006) used novel quaternary ammonium ionic liquids as acidic catalysts for the synthesis of cinnamic acid.

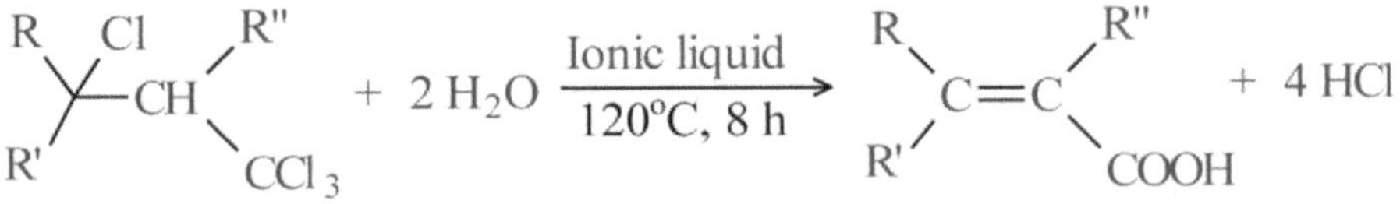

A successful ionic liquid-based aqueous biphasic system was proposed for lipase extraction with recovery values around 80%. This will allow the use of ionic liquids as withdrawal solvents and media for catalytic applications (Deive et al., 2011). Diego et al. (2011) reported the efficient production of biodiesel in hydrophobic ionic liquids using immobilized lipase with the straightforward extraction of the products using the properties of appropriate ionic liquids.

Ressmann et al. (2012) extracted pharmaceutically active triterpene betulin from biomass utilizing ionic liquids with significantly improved extraction yield and purity. A new functionalized ionic liquid [eimCH$_2$CONHBu]NTf$_2$ has been designed and synthesized by Huaxi et al. (2012) with high extraction efficiency and special selectivity for tryptophan. Duarte et al. (2012) investigated ionic liquids as foaming agents of semicrystalline natural-based polymers. Tetrahalogenoaurate anions are removed from water by precipitation with water-soluble ionic liquids or by extraction using hydrophobic ionic liquids (Papaiconomou et al., 2012).

Arce et al. (2007a) presented rigorous thermodynamic analyses to evaluate the ability of the ionic liquid 1-ethyl-3-methylimidazolium ethylsulfate as an extracting solvent or a distillation entrainer for the separation of ethanol and ethyl *tert*-butyl ether (ETBE). They also reported that ionic liquid 1-ethyl-3-methylimidazoliumbis{(trifluoromethyl)sulfonyl}amide can selectively remove benzene from its mixtures with hexane (Arce et al., 2007b). The product, methyl-(Z)-α-acetamido cinnamate can be crystallized from the ionic liquid 1-butyl-3-methylimidazolium tetrafluoroborate either by a shift to lower temperatures or by a shift to higher carbon dioxide concentrations (Kroon et al., 2008). Shen et al. (2008) developed an efficient method for the synthesis of optically active O-acetyl cyanohydrins via the one-pot lipase catalyzed kinetic resolution of in situ generated cyanohydrins and O-acetyl cyanohydrins in ionic liquid.

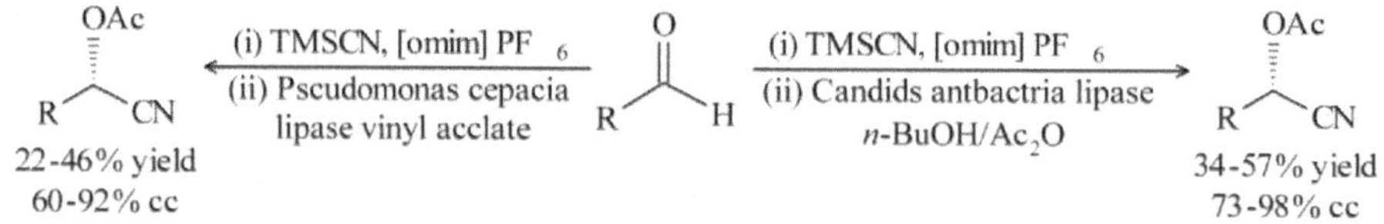

Chiappe et al. (2010) achieved high concentrations of metal ions by dissolving metal salts in ionic liquids with common anions. Ståhlberg et al. (2010) investigated the conversion of glucose to 5-(hydroxymethyl)furfural (HMF) in alkylimidazolium-based ionic liquids together with lanthanide catalysts, which speeded up by the higher hydrophobicity of the imidazolium cation. Gas-liquid acetylene hydrochlorination proceeds efficiently in the presence of catalysts of non-mercuric metal chlorides using ionic liquids as reaction media (Qin et al., 2011). The direct nucleophilic substitution reactions of alcohols can be promoted and modulated by a clean and recyclable reaction medium (zinc-based ionic liquid) $[CHCl][ZnCl_2]_2$ (Zhu et al., 2011).

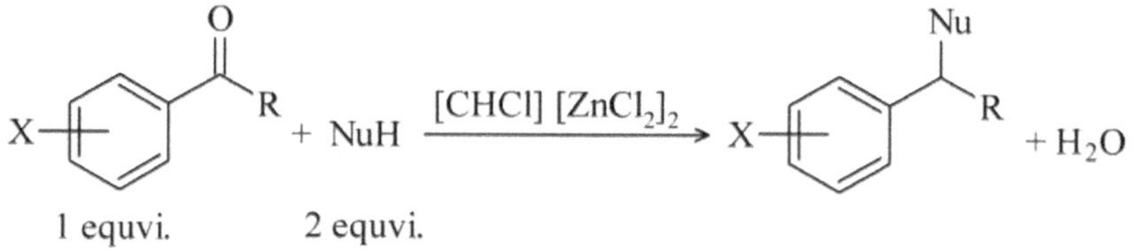

NuH = Amine, sulfonamide, amide, 1,3-dicarbonyl compound

Berger et al. (2011) used a simple ionic liquid-based system for the selective decomposition of formic acid to hydrogen and carbon dioxide. Gao et al. (2011) prepared Fe (III)-derived Lewis acid ionic liquid as an efficient and recyclable catalyst for the selective benzylation of arenes.

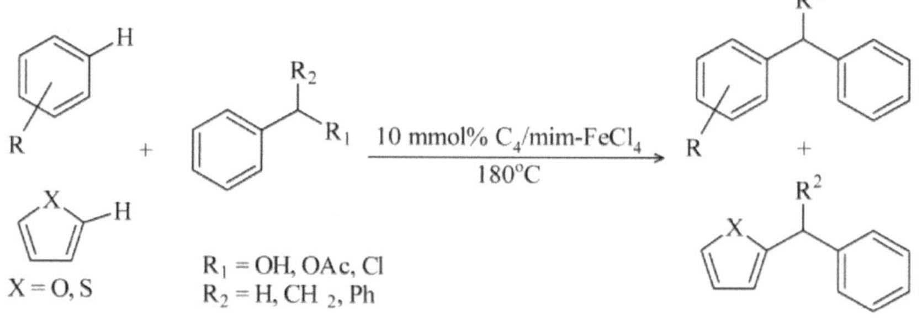

Natrajan and Wen (2011) reported ionic liquids as unique reaction media for the efficient introduction of N-sulfopropyl groups in the synthesis of hydrophilic, chemiluminescent acridinium esters using reduced quantities of the carcinogenic reagent 1,3-propane sultone.

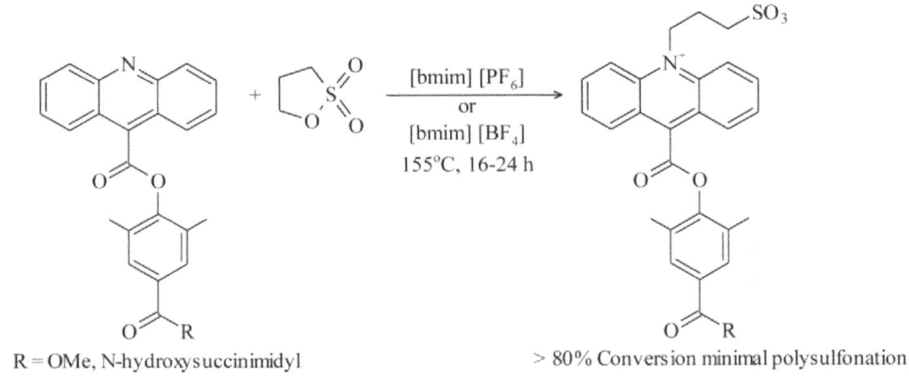

A novel one-pot synthesis of 6-aminouracils via in situ generated ureas and cyanoacetylureas in the presence of ionic liquid 1,1,3,3-tetramethylguanidine acetate as a recyclable catalyst has been reported by Chavan and Degani (2012).

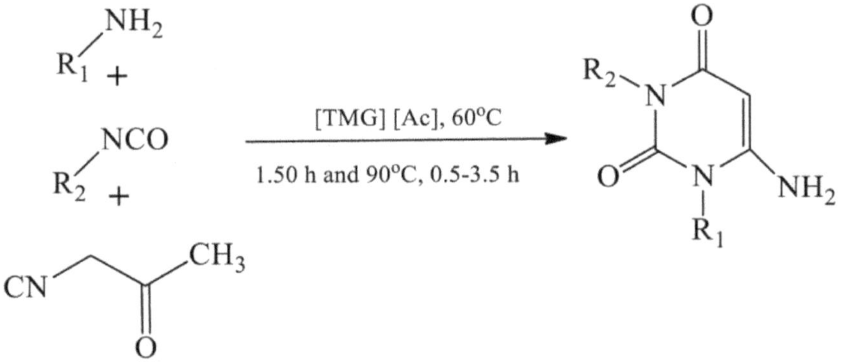

where R_1 = Me, n-Pr, n-Bu, Ph, p-OMe-Ph, p-MePh, o-MePh, m-MePh, bn, p-OMe Bn, $PhCH_2CH_2$. p-nMe_2Ph, R_3-K
where R_1 = Me, R_3 = Me and R_1 = n-Pr, R_2 = n-Pr
where R_1 = Pb, Bn, R_2 = Allyl

PEG-functionalized basic ionic liquids were proved to be efficient catalysts for dimethyl carbonate synthesis under atmospheric CO_2 through a "one-pot, two-stage" process (Yang et al., 2012). Sun et al. (2012) synthesized cyclic carbonates via cycloaddition of CO_2 to epoxides with bifunctional chitosan supported ionic liquid catalysts. Acidic ionic liquids functionalized, ordered and stable mesoporous polymers have been synthesized at high temperature with excellent catalytic activities (Liu et al., 2012).

Wellens et al. (2012) reported that cobalt can efficiently be separated from nickel by solvent extraction with the ionic liquid tri(hexyl)tetradecylphosphonium chloride. Li et al. (2012) reported hydrogen bonding-promoted dehydration of fructose to 5-hydroxymethylfurfural (HMF) in pure ionic liquid without any other additive or catalyst. Fehér et al. (2012) used Brønsted acidic ionic liquids supported on silica gel in the oligomerization of isobutene. The supported catalysts could be used several times without loss of activity or change in selectivity.

Most of the chemical reactions require a solvent and vapors of these solvents create environmental pollution. It is therefore necessary to find out a solvent with high boiling point; whereas most of the organic solvents are having a low boiling point. Ionic liquids enter the scene here. These are defined as designer solvents. Their properties can be so designed by selecting a proper cation and anion that they can serve our purpose.

5.5 RECENT DEVELOPMENTS

Biobutanol has high energy content, and hence, it is a promising fuel. A fermentative technique is used for generation of butanol by using solventogenic *Clostridium*, has serious limitations as it represses microbial movement (normally $\geq 10\ g L^{-1}$), which influences its production. Motghare et al. (2022) utilized ionic liquids for the separation of butanol from aqueous media using 1-hexyl-3-methylimidazolium hexafluorophosphate (HMIM[PF_6]) and 1-hexyl-3-methylimidazolium bis(trifluoromethanesulfon imide) (HMIM[NTF_2]), along with oleyl alcohol (OA), tri-*n*-butyl phosphate (TBP), and their blends. The order of extraction efficiency as compared with TBP was observed in the range:

$$\text{TBP} > 85\text{–}87\% \text{ for OA} > 55\text{–}85 \text{ for HMIM[NTF}_2] > 26\text{–}66\% \text{ for HMIM[PF}_6],$$

Rajadurai and Anguraj (2022) studied extraction behavior of halogen-free ionic liquid, 1-butyl-3-methylimidazolium octyl sulfate for separating lead from the aqueous solution. It was reported that the maximum extraction efficiency of lead (97%) could be achieved with the feed concentration of

10 ppm in 60 min at pH 5 using 2 g of the ionic liquid. It was also reported that the recovered ionic liquid can be effectively reused five times.

Shirzaei and Shaterian (2022) used ionic liquid, $[(EtO)_3Si(CH_2)_3NH_3^+]$ $[CH_3COO^-]$ as a catalyst under solvent-free conditions in an efficient procedure for preparing 2-(phenylsulfonyl)-1*H*-benzo[*a*]pyrano[*2,3-c*]phenazin-3-amine derivatives. They used four-component condensation reaction consisting of aromatic aldehydes, *o*-phenylenediamine, 2-hydroxynaphthalene-1,4-dione, and (phenylsulfonyl)acetonitrile for this purpose. Simple procedure, high yields, shorter reaction times and an environmentally benign method are advantages of this protocol. It was also observed that this ionic liquid can be recovered again and used number of times without any significant loss of its activity.

Ionic liquids ($[EMIM][TF_2N]$) was used as green solvents by Gui et al. (2022) to capture 1,2-dimethoxyethane (DMET) from exhaust gases. It was observed that ionic liquid exhibited good separation performance for capturing DMET with absorption ratio as 87.32%. This was concluded that use of ionic liquid has a potential for capturing volatile organic compounds (VOCs), with good energy-saving and improved economic benefits as compared with triethylene glycol (TEG) as a conventional solvent.

Generally, carbon dots (CDs) are synthesized via hydrothermal or solvent-thermal reaction at high pressure and temperature. Cao et al. (2022) used an ionic liquid for the rapid synthesis of CDs with high photoluminescent quantum yield (98.5%) and that too at low temperature ($\leq$ 100°C) and atmospheric pressure. Tunable multicolor emissive CDs were successfully obtained and used in preparing high-performance white light-emitting diodes.

Shan et al. (2022) synthesized two ionic liquid-type UiO-66 (Mod-UiO-66@(IL)) using hydrothermal method. They used these as an adsorbent for carrying out adsorption desulfurization performance of dibenzothiophene (DBT) in model oil. The conditions for this adsorption process were optimized as: adsorbent amount (1 g), adsorption time (10 min), and 100 mL of model oil (2000 mgL^{-1}). It was reported that the maximum adsorption capacity of Mod-UiO-66@(IL1) for dibenzothiophene was 43.52 mg g^{-1}. It was also observed that Mod-UiO-66@(IL1) can be regenerated by methanol five times and there is no significant loss in its adsorption capacity. The removal order of Mod-UiO-66@(IL1) of other sulfur compounds followed the order:

Dibenzothiophene > Benzothiophene > Thiophene.

KEYWORDS

- **Beckmann rearrangement**
- **Diels–Alder reaction**
- **Fischer indole synthesis**
- **Michael addition reaction**
- **Wittig reaction**
- **Mannich reaction**
- **Heck reaction**
- **Biginelli reaction**

REFERENCES

Adams, C. J.; Earle, M. J.; Roberts, G.; Seddon, K. R. Friedel–Crafts Reactions in Room Temperature Ionic Liquids. *Chem. Commun.* **1998,** *19*, 2097–2098.

Akopyan, A.; Eseva, E.; Polikarpova, P.; Kedalo, A.; Vutolkina, A.; Glotov, A. Deep Oxidative Desulfurization of Fuels in the Presence of Brönsted Acidic Polyoxometalate-Based Ionic Liquids. *Molecules* **2020,** *25* (3), 536.

Aliabadi, R. S.; Mahmoodi, N. O. Green and Efficient Synthesis of Pyranopyrazoles Using [bmim][OH−] as an Ionic Liquid Catalyst in Water Under Microwave Irradiation and Investigation of Their Antioxidant Activity. *RSC Adv.* **2016,** *6* (89), 85877–85884.

Alqurashy, B. A. Ecofriendly Microwave-Assisted Preparation, Characterization and Antitumor Activity of Some Propylimidazolium-Based Ionic Liquids Derivatives. *J. Taibah Univ. Sci.* **2020,** *14* (1), 1457–1462.

Annath, H.; Chapman, S.; Donnelly, G. F.; Marr, P. C.; Marr, A. C.; Raja, R. Heterogenized Ionic-Liquid Metal-Oxide Hybrids: Enhanced Catalytic Activity in the Liquid-Phase Beckmann Rearrangement. *ACS Sustain. Chem. Eng.* **2018,** *6* (12), 16797–16805.

Arce, A.; Earle, M. J.; Rodríguez, H.; Seddon, K. R. Separation of Aromatic Hydrocarbons from Alkanes Using the Ionic Liquid 1-Ethyl-3-Methylimidazolium Bis { (Trifluoromethyl) Sulfonyl} Amide. *Green Chem.* **2007a,** *9* (1), 70–74.

Arce, A.; Rodríguez, H.; Soto, A.; Arce, A.; Rodríguez, H.; Soto, A. Use of a Green and Cheap Ionic Liquid to Purify Gasoline Octane Boosters. *Green Chem.* **2007b,** *9* (3), 247–253.

Arend, M.; Westermann, B.; Risch, N. Modern Variants of the Mannich Reaction. *Angew. Chem. Int. Ed.* **1998,** *37* (8), 1044–1070.

Babicka, M.; Woźniak, M.; Dwiecki, K.; Borysiak, S.; Ratajczak, I. Preparation of Nanocellulose Using Ionic Liquids: 1-Propyl-3-Methylimidazolium Chloride and 1-Ethyl-3-Methylimidazolium Chloride. *Molecules* **2020,** *25*. DOI: 10.3390/molecules25071544

Baj, S.; Chrobok, A.; Słupska, R. The Baeyer–Villiger Oxidation of Ketones with Bis (Trimethylsilyl) Peroxide in the Presence of Ionic Liquids as the Solvent and Catalyst. *Green Chem.* **2009,** *11* (2), 279–282.

Baudequin, C.; Bregon, D.; Levillain, J.; Guillen, F.; Plaquevent, J. C.; Gaumont, A. C. Chiral Ionic Liquids, a Renewal for the Chemistry of Chiral Solvents? Design, Synthesis and Applications for Chiral Recognition and Asymmetric Synthesis. *Tetrahedron: Asymmetry* **2005,** *16* (24), 3921–3945.

Beniwal, V.; Manna, A.; Kumar, A. Spectacular Rate Enhancement of the Diels–Alder Reaction at the Ionic Liquid/n–Hexane Interface. *ChemPhysChem*, **2016,** *17* (13), 1969–1972.

Berger, M. E. M.; Assenbaum, D.; Taccardi, N.; Spiecker, E.; Wasserscheid, P. Simple and Recyclable Ionic Liquid Based System for the Selective Decomposition of Formic Acid to Hydrogen and Carbon Dioxide. *Green Chem.* **2011,** *13* (6), 1411–1415.

Blasius E.; Klein, W.; Schon, U. Separation of Strontium from Nuclear Waste Solutions by Solvent Extraction with Crown Ethers. *J. Radioanal Nucl. Chem.* **1985,** *89* (2), 389–398.

Bonhôte, P.; Dias, A.-P.; Papageorgiou, N.; Kalyanasundaram, K.; Grätzel M. Hydrophobic, Highly Conductive Ambient-Temperature Molten Salts. *Inorg. Chem.* **1996,** *35*, 1168–1178.

Boon, J. A.; Levisky, J. A.; Pflug, J. L.; Wilkes, J. S. Friedel-Crafts Reactions in Ambient-Temperature Molten Salts. *J. Org. Chem.* **1986,** *51*, 480–483.

Boulaire, V. L.; Gree, R. Communications-Wittig Reactions in the Ionic Solvent (bmim) (BF4). *Chem. Commun. Letchworth* **2000,** *22*, 2195–2196.

Cao, M.; Zhao, X.; Gong, X. Ionic Liquid-Assisted Fast Synthesis of Carbon Dots with Strong Fluorescence and Their Tunable Multicolor Emission. *Small* **2022,** *18* (11). DOI: 10.1002/smll.202106683.

Chavan, S. S.; Degani, M. S. Ionic Liquid Mediated One-Pot Synthesis of 6-Aminouracils. *Green Chem.* **2012,** *14* (2), 296–299.

Chiappe, C.; Imperato, G.; Napolitano, E.; Pieraccini, D. Ligandless Stille Cross-Coupling in Ionic Liquids. *Green Chem.* **2004,** *6* (1), 33–36.

Chiappe, C.; Leandri, E.; Hammock, B. D.; Morisseau, C. Effect of Ionic Liquids on Epoxide Hydrolase-Catalyzed Synthesis of Chiral 1, 2-Diols. *Green Chem.* **2007,** *9* (2), 162–168.

Chiappe, C.; Malvaldi, M.; Melai, B.; Fantini, S.; Bardi, U.; Caporali, S. An Unusual Common Ion Effect Promotes Dissolution of Metal Salts in Room-Temperature Ionic Liquids: A Strategy to Obtain Ionic Liquids Having Organic–Inorganic Mixed Cations. *Green Chem.* **2010,** *12* (1), 77–80.

Chiappe, C.; Pieraccini, D. Direct Mono-N-Alkylation of Amines in Ionic Liquids: Chemoselectivity and Reactivity. *Green Chem.* **2003,** *5* (2), 193–197.

Chiappe, C.; Sanzone, A.; Dyson, P. J. Styrene Oxidation by Hydrogen Peroxide in Ionic Liquids: The Role of the Solvent on the Competition Between Two Pd-Catalyzed Processes, Oxidation and Dimerization. *Green Chem.* **2011,** *13* (6), 1437–1441.

Cho, C. W.; Pham, T. P. T.; Jeon, Y. C.; Yun, Y. S. Influence of Anions on the Toxic Effects of Ionic Liquids to a Phytoplankton Selenastrum Capricornutum. *Green Chem.* **2008,** *10* (1), 67–72.

Chum, H. L.; Koch, V. R.; Miller, L. L.; Osteryoung, R. A. Electrochemical Scrutiny of Organometallic Iron Complexes and Hexamethylbenzene in a Room Temperature Molten Salt. *J. Am. Chem. Soc.* **1975,** *97* (11), 3264–3265.

Davis Jr.; J. H. Task-Specific Ionic Liquids. *Chem. Lett.* **2004,** *33* (9), 1072–1077.

Deetlefs, M.; Seddon, K. R. Improved Preparations of Ionic Liquids Using Microwave Irradiation. *Green Chem.* **2003,** *5* (2), 181–186.

Deive, F. J.; Rodríguez, A.; Pereiro, A. B.; Araújo, J. M. M.; Longo, M. A.; Coelho, M. A. Z. et al. Ionic Liquid-Based Aqueous Biphasic System for Lipase Extraction. *Green Chem.* **2011,** *13* (2), 390–396.

Deng, Y.; Shi, F.; Beng, J.; Qiao, K. Ionic liquid as a green catalytic reaction medium for esterifications. *J. Mol. Catal. A: Chem.* **2001,** *165* (1–2), 33–41.

Dharaskar, S. A.; Wasewar, K. L.; Varma, M. N.; Shende, D. Z.; Yoo, C. Synthesis, Characterization and Application of 1-Butyl-3-Methylimidazolium Tetrafluoroborate for Extractive Desulfurization of Liquid Fuel. *Arab. J. Chem.;* **2016,** *9* (4), 578–587.

Diego, T. D.; Manjón, A.; Lozano, P.; Vaultier, M.; Iborra, J. L. An Efficient Activity Ionic Liquid-Enzyme System for Biodiesel Production. *Green Chem.* **2011,** *13* (2), 444–451.

Dobras, G.; Kasperczyk, K.; Jurczyk, S.; Orlińska, B. N-hydroxyphthalimide Supported on Silica Coated with Ionic Liquids Containing $CoCl_2$ (SCILLs) as New Catalytic System for Solvent-Free Ethylbenzene Oxidation. *Catalysts* **2020a,** *10* (2), 252. DOI: 10.3390/catal10020252.

Dobras, G.; Sitko, M.; Petroselli, M.; Caruso, M.; Cametti, M.; Punta, C.; Orlińska, B. Solvent-Free Aerobic Oxidation of Ethylbenzene Promoted by NHPI/Co (II) Catalytic System: The Key Role of Ionic Liquids. *ChemCatChem* **2020b,** *12* (1), 259–266.

Docherty, K. M.; Joyce, M. V.; Kulacki, K. J.; Kulpa, C. F. Microbial Biodegradation and Metabolite Toxicity of Three Pyridinium-Based Cation Ionic Liquids. *Green Chem.* **2010,** *12* (4), 701–712.

Docherty, K. M.; Kulpa, C. F. Toxicity and Antimicrobial Activity of Imidazolium and Pyridinium Ionic Liquids. *Green Chem.* **2005,** *7* (4), 185–189.

Duarte, A. R. C.; Silva, S. S.; Mano, J. F.; Reis, R. L. Ionic Liquids as Foaming Agents of Semi-Crystalline Natural-Based Polymers. *Green Chem.* **2012,** *14* (7), 1949–1955.

Duncan W. B.; Yanan, G.; Lopes, C.; Nuno, J.; Karina, S.; John, M. S. Liquid-Crystalline Ionic Liquids as Ordered Reaction Media for the Diels-Alder Reaction. *Chem. Eur. J.* **2016**. DOI: 10.1002/chem.201602965.

Dyson, P. J.; Grossel, M. C.; Srinivasan N.; Vine, T.; Welton T.; William A. J. P.; et al. Organometallic Synthesis in Ambient Temperature Chloroaluminate (III) Ionic Liquids: Ligand Exchange Reactions of Ferrocene. J. *Chem. Soc. Dalton Trans* **1997,** (19), 3465–3469.

Earle, M. J.; McCormac, P. B.; Seddon, K. R. Diels–Alder Reactions in Ionic Liquids. A Safe Recyclable Alternative to Lithium Perchlorate–Diethyl Ether Mixtures. *Green Chem.* **1999,** *1* (1), 23–25.

Earle, M. J.; Seddon, K. R. Ionic Liquids: Green Solvents for the Future. *Pure Appl. Chem.* **2000,** *72* (7), 1391–1398.

Earle, M. L.; McCormac, P. B.; Seddon, K. R. Regioselective Alkylation in Ionic Liquids. *Green Chem.* **1998,** 2245–2246

Elhamifar, D.; Karimi, B.; Rastegar, J.; Banakar, M. H. Palladium-Containing Ionic Liquid-Based Ordered Mesoporous Organosilica: An Efficient and Reusable Catalyst for the Heck Reaction. *ChemCatChem*, **2013,** *5* (8), 2418–2424.

Elhamifar, D.; Shabani, A. Manganese-Containing Periodic Mesoporous Organosilica with Ionic–Liquid Framework (Mn@ PMO-IL): A Powerful, Durable, and Reusable Nanocatalyst for the Biginelli Reaction. *Chem. Eur. J.* **2014,** *20* (11), 3212–3217.

Fang, S. J.; Hua Y.; Ge, G.; Ru, G. X. *Chines. Chem. Lett.* **2005,** *16*, 321–324.

Fehér, C.; Kriván, E.; Hancsók, J.; Skoda-Földes, R. Oligomerisation of Isobutene with Silica Supported Ionic Liquid Catalysts. *Green Chem.* **2012,** *14*, 403–409.

Freemantle, M. Designer solvents. *Chem. Eng. News* **1998,** *76*, 32–37.

Fu, S. k.; Liu, S. T. Preparation of Functionalized Imidazolium Salts Under Microwave Irradiation. *Synth. Commun.* **2006,** *36* (14), 2059–2067.

Gabriel, S.; Weiner, J. Ueber einige Abkömmlinge des Propylamins. *Chemische Berichte.* **1888,** *21* (2), 2669–2679.

Gaillon, L.; Bedioui, F. First Example of Electroassisted Biomimetic Activation of Molecular Oxygen by a (salen) Mn Epoxidation Catalyst in a Room-Temperature Ionic Liquid. *Chem. Commun.* **2001,** *16,* 1458–1459.

Gao, G.; Zhao, Q.; Yang, C.; Jiang, T. p-Toluenesulfonic Acid Functionalized Imidazole Ionic Liquids Encapsulated into Bismuth SBA-16 as High-Efficiency Catalysts for Friedel–Crafts Acylation Reaction. *Dalton Trans.* **2021,** *50* (17), 5871–5882.

Gao, J.; Wang, J.; Song, Q.; He, L. Iron (III)-Based Ionic Liquid-Catalyzed Regioselective Benzylation of Arenes and Heteroarenes. *Green Chem.* **2011,** *13* (5), 1182–1186.

Ghomi, J. S.; Zahedi, S. Novel Ionic Liquid Supported on Fe_3O_4 Nanoparticles and Its Application as a Catalyst in Mannich Reaction Under Ultrasonic Irradiation. *Ultrasonics Sonochem.* **2017,** *34,* 916–923.

Gloe, K.; Muchi, P.; Kholkin, A; I. J. Meerbote, M.; Beger J. Extraction of Metal Salts by Means of Polycyclic Crown Ethers. *Isotopenpraxis* **1982,** *18* (5), 170–175.

Goel, K.; Bera, S.; Singh, M.; Mondal, D. Synthesis of Dual Functional Pyrimidinium Ionic Liquids as Reaction Media and Antimicrobial Agents. *RSC Adv.* **2016,** *6* (108), 106806–106820.

Goel, K.; Bera, S.; Singh, M.; Mondal, D. Synthesis, Photophysical Properties, and Biological Importance of Pyrimidinium Ionic Liquids. *ChemistrySelect* **2019,** *4* (23), 6888–6895.

Goodrich, P.; Gunaratne, H. N.; Hall, L.; Wang, Y.; Jin, L.; Muldoon, M. J.; et al. Using Chiral Ionic Liquid Additives to Enhance Asymmetric Induction in a Diels–Alder Reaction. *Dalton Trans.* **2017,** *46* (5), 1704–1713.

Gordon, C. M.; Ritchie, C. Indium and Tin-Mediated Allylation in Ionic Liquids. *Green Chem.* **2002,** *4* (2), 124–128.

Gubicza, L.; Nemestóthy, N.; Fráter, T.; Bélafi-Bakó, K. Enzymatic Esterification in Ionic Liquids Integrated with Pervaporation for Water Removal. *Green Chem.* **2003,** *5* (2), 236–239.

Gui, C.; Li, G.; Zhu, R.; Liu, Q.; Lei, Z. Ionic Liquids for Capturing 1, 2-Dimethoxyethane (DMET) in VOCs: Experiment and Mechanism Exploration. *Ind. Eng. Chem. Res.* **2022,** *61* (5), 2257–2267.

Gui, J.; Deng, Y.; Hu, Z.; Sun, Z. A Novel Task-Specific Ionic Liquid for Beckmann Rearrangement: A Simple and Effective Way for Product Separation. *Tetrahedron Lett.* **2004,** *45* (12), 2681–2683.

Hajipour, A. R.; Khorsandi, Z.; Karimi, H. Cobalt Nanoparticles Supported on Ionic Liquid-Functionalized Multiwall Carbon Nanotubes as an Efficient and Recyclable Catalyst for Heck Reaction. *App. Org. Chem.* **2015,** *29* (12), 805–808.

Hayouni, S.; Ferlin, N.; Bouquillon, S. High Catalytic and Recyclable Systems for Heck Reactions in Biosourced Ionic Liquids. *Mol. Catal.* **2017,** 437, 121–129.

Holbrey, J.; Seddon, K. R. Ionic Liquids. *Clean Prod. Proc.* **1999,** *1* (4), 223–236.

Howarth, J.; James, P.; Ryan, R. Sodium Borohydride Reduction of Aldehydes and Ketones in the Recyclable Ionic Liquid [BMIM] PF6. *Synth. Commun.* **2001,** *31* (19), 2935–2938.

Hu, H.; Cai, X.; Xu, Z.; Yan, X.; Zhao, S. Beckmann Rearrangement of Ketoxime Catalyzed by N-Methyl-Imidazolium Hydrosulfate. *Molecules* **2018,** *23* (7). DOI: 10.3390/molecules 23071764.

Huan, P.; Yulin, H.; Rong, X.; Dong, F. Choline-Based Biodegradable Ionic Liquid Catalyst for Mannich-Type Reaction. *J. Chem. Sci.* **2016,** *128* (12), 1855–1860.

Huaxi, L.; Zhuo, L.; Jingmei, Y.; Changping, L.; Yansheng, C.; Qingshan, L. et al. Liquid–Liquid Extraction Process of Amino Acids by a New Amide-Based Functionalized Ionic Liquid. *Green Chem.* **2012,** *14* (6), 1721–1727.

Jin, M. J.; Taher, A.; Kang, H. J.; Choi, M.; Ryoo, R. Palladium Acetate Immobilized in a Hierarchical MFI Zeolite-Supported Ionic Liquid: A Highly Active and Recyclable Catalyst for Suzuki Reaction in Water. *Green Chem.* **2009,** *11* (3), 309–313.

Kabalka, G. W.; Malladi, R. R. Reduction of Aldehydes Using Trialkylboranes in Ionic Liquids. *Chem. Commun.* **2000,** *22*. DOI: 10.1039/B007190K

Kalviri, H. A.; Kerton, F. M. Synthesis of Pd Nanocrystals in Phosphonium Ionic Liquids Without Any External Reducing Agents. *Green Chem.* **2011,** *13* (3), 681–686.

Kaufmann, D. E.; Nouroozian, M.; Henze, H. Molten Salts as an Efficient Medium for Palladium Catalyzed CC Coupling Reactions. *Synlett* **1996,** *1996* (11), 1091–1092.

Khanapure, S.; Jagadale, M.; Kale, D.;Gajare, S.; Rashinkar, G. Cellulose-Supported Ionic Liquid Phase Catalyst-Mediated Mannich Reaction. *Aust. J. Chem.* **2019,** *72* (7), 513–523.

Kim, Y. J.; Varma, R. S. Microwave-Assisted Preparation of Imidazolium-Based Tetrachloro-indate (III) and Their Application in the Tetrahydropyranylation of Alcohols. *Tetrahedron Lett.* **2005a,** *46* (9), 1467–1469.

Kim, Y. J.; Varma, R. S. Tetrahaloindate (III)-Based Ionic Liquids in the Coupling Reaction of Carbon Dioxide and Epoxides to Generate Cyclic Carbonates: H-Bonding and Mechanistic Studies. *J. Org. Chem.* **2005b,** *70* (20), 7882–7891.

Kitazume, T.; Kasai, K. The Synthesis and Reaction of Zinc Reagents in Ionic Liquids. *Green Chem.* **2001,** *3* (1), 30–32.

Kore, R.; Uppara, P. V.; Rogers, R. D. Replacing HF or $AlCl_3$ in the Acylation of Isobutylbenzene with Chloroaluminate Ionic Liquids. *ACS Sustain. Chem. Eng.* **2020,** *8* (28), 10330–10334.

Kroon, M. C.; Toussaint, V. A.; Shariati, A.; Florusse, L. J.; Spronsen, J. V.; Witkamp, G.; Peters, C. J. *Green Chem.* **2008,** *10*, 333–336.

Kulacki, K. J.; Lamberti, G. A. Toxicity of Imidazolium Ionic Liquids to Freshwater Algae. *Green Chem.* **2008,** *10* (1), 104–110.

Kumar, G. V.; Rajendran, A. Sustainable Diels–Alder Syntheses in Imidazolium Ionic Liquids. *Synth. Commun.* **2016,** *46* (6), 483–496.

Labafzadeh, S. R.; Helminen, K. J.; Kilpeläinen, I.; King, A. W. Synthesis of Cellulose Methyl-carbonate in Ionic Liquids Using Dimethylcarbonate. *ChemSusChem* **2015,** *8* (1), 77–81.

Lee, C. W. Diels-Alder Reactions in Chloroaluminate Ionic Liquids: Acceleration and Selectivity Enhancement. *Tetrahedron Lett.* **1999,** *40* (13), 2461–2464.

Li, C.; Wang, Q.; Zhao, Z. K. Acid in Ionic Liquid: An Efficient System for Hydrolysis of Lignocellulose. *Green Chem.* **2008,** *10* (2), 177–182.

Li, Y. N.; Wang, J. Q.; He, L. N.; Yang, Z. Z.; Liu, A. H.; Yu, B.; et al. Experimental and Theoretical Studies on Imidazolium Ionic Liquid-Promoted Conversion of Fructose to 5-Hydroxymethylfurfural. *Green Chem.* **2012,** *14* (10), 2752–2758.

Lin, Y. J.; Wu, Y. P.;Thul, M.; Hung, M. W.; Chou, S. H.; Chen, W. T.; et al. Tunable Aryl Imidazolium Recyclable Ionic Liquid with Dual Brønsted–Lewis Acid as Green Catalyst for Friedel–Crafts Acylation and Thioesterification. *Molecules* **2020,** *25* (2), 352.

Liu, F.; Zhong, X.; Xu, J.; Kamali, A.; Shi, Z. Temperature Dependence on Density, Viscosity, and Electrical Conductivity of Ionic Liquid 1-Ethyl-3-Methylimidazolium Fluoride. *Appl. Sci.* **2018,** *8*, 356. DOI: org/10.3390/app8030356.

Liu, F.; Zhong, X.; Xu, J.; Wang, Z.; Shi, Z. Facile Synthesis and Characterization of 1-Ethyl-3-Methylimidazolium Fluoride Ionic Liquid. In *J. Phy.: Conf. Ser.* **2019,** *1347* (1), 012109.

Liu, F.; Zuo, S.; Kong, W.; Qi, C. High-temperature synthesis of Strong Acidic Ionic Liquids Functionalized, Ordered and Stable Mesoporous Polymers with Excellent Catalytic Activities. *Green Chem.* **2012,** *14* (5), 1342–1349.

Liu, Q.; Janssen, M. H. A.; Rantwijk, F. V.; Sheldon, R. A. Room-Temperature Ionic Liquids That Dissolve Carbohydrates in High Concentrations. *Green Chem.* **2005,** *7* (1), 39–42.

Liu, X.; Ma, H.; Wu, Y.; Wang, C.; Yang, M.; Yan, P. et al. Esterification of Glycerol with Acetic Acid Using Double SO_3 H-Functionalized Ionic Liquids as Recoverable Catalysts. *Green Chem.* **2011,** *13* (3), 697–701.

Lü, H.; Deng, C.; Ren, W.; Yang, X. Oxidative Desulfurization of Model Diesel Using $[(C_4H_9)_4N]_6Mo_7O_{24}$ as a Catalyst in Ionic Liquids. *Fuel Proces. Technol.* **2014,** 119, 87–91.

Mahajan, S.; Sharma, B.; Kapoor, K. K. A Solvent-Free One Step Conversion of Ketones to Amides via Beckmann Rearrangement Catalysed by $FeCl_3{\cdot}6H_2O$ in Presence of Hydroxylamine Hydrochloride. *Tetrahedron Lett.* **2015,** *56* (14), 1915–1918.

Matuszek, K.; Chrobok, A.; Latos, P.; Markiton, M.; Szymańska, K.; Jarzębski, A. et al. Silica-Supported Chlorometallate (III) Ionic Liquids as Recyclable Catalysts for Diels–Alder Reaction Under Solventless Conditions. *Catal. Sci. Technol.* **2016,** *6* (22), 8129–8137.

Matuszek, K.; Coffie, S.; Chrobok, A.; Swadźba-Kwaśny, M. Borenium Ionic Liquids as Catalysts for Diels–Alder Reaction: Tuneable Lewis Superacids for Catalytic Applications. *Catal. Sci. Technol.* **2017,** *7* (5), 1045–1049.

Men, Y.; Du, X.; Shen, J.; Wang, L.; Liu, Z. Preparation of Corn Starch-G-Polystyrene Copolymer in Ionic Liquid: 1-Ethyl-3-Methylimidazolium Acetate. *Carbohydr. Polym.* **2015,** *121*, 348–354.

Messali, M. A Facile and Green Microwave-Assisted Synthesis of New Functionalized Picolinium-Based Ionic Liquids. *Arab. J. Chem.* **2016,** *9*, S564–S569.

Moneiro, A. L.; Zinn, F. K.; de Souza, R. F.; Dupont, J. *Tetrahedron Asymmetry* **1997,** *8*,177–179.

Motghare, K. A.; Shende, D. Z.; Wasewar, K. L. Butanol Recovery Using Ionic Liquids as Green Solvents. *J. Chem. Tech. Biotech.* **2022,** *97* (4), 873–884.

Nagarajan, S.; Shaikh, T. M.; Kandasamy, E. Synthesis of 1-Alkyl Triazolium Triflate Room Temperature Ionic Liquids and Their Catalytic Studies in Multi-Component Biginelli Reaction. *J. Chem. Sci.* **2015,** *127* (9), 1539–1545.

Nara, S. J.; Harjani, J. R.; Salunkhe, M. M. Friedel−Crafts Sulfonylation in 1-Butyl-3-Methylimidazolium Chloroaluminate Ionic Liquids. *J. Org. Chem.* **2001,** *66* (25), 8616–8620.

Natrajan, A.; Wen, D. Facile N-Alkylation of Acridine Esters with 1, 3-Propane Sultone In Ionic Liquids. *Green Chem.* **2011,** *13* (4), 913–921.

Nowrouzi, N.; Tarokh, D.; Motevalli, S. Amine-Based Ionic Liquids ($R_3N^+ PPh_2$) as a Reusable Reaction Medium and Pd (II) Ligand in Heck Reactions of Aryl Halides with Styrene and N-Butyl Acrylate. *J. Mol. Catal. A: Chem.* **2014,** 385, 13–17.

Ohno, H.; Fukumoto, K. Amino Acid Ionic Liquids. *Acc. Chem. Res.* **2007,** *40* (11), 1122–1129.

Oonishi, Y.; Saito, A.; Sato, Y. Study on Rh (I)-Catalyzed Cycloisomerization of Dienes with Alkenes in Ionic Liquids: Effect of the Structure of ILs on Catalyst Recyclability. *Green Chem.* **2009,** *11* (3), 330–333.

Pang, J.; Liu, X.; Yang, J.; Lu, F.; Wang, B.; Xu, F.; et al. Synthesis of Highly Polymerized Water-Soluble Cellulose Acetate by the Side Reaction in Carboxylate Ionic Liquid 1-Ethyl-3-Methylimidazolium Acetate. *Sci. Rep.* **2016,** *6* (1). DOI: 10.1038/srep33725.

Papaiconomou, N.; Vite, G.; Goujon, N.; Lévêque, J.; Billard, I. Efficient Removal of Gold Complexes from Water by Precipitation or Liquid–Liquid Extraction Using Ionic Liquids. *Green Chem.* **2012,** *14* (7), 2050–2056.

Peng, J.; Deng, Y. Catalytic Beckmann Rearrangement of Ketoximes in Ionic Liquids. *Tetrahedron Lett.* **2001,** *42* (3), 403–405.

Prabhakara, M. D.; Maiti, B. Ionic Liquid-Immobilized Proline (s) Organocatalyst-Catalyzed One-Pot Multi-Component Mannich Reaction Under Solvent-Free Condition. *Res. Chem. Intermed.* **2020,** *46* (4), 2381–2401.

Qian, H.; Kai, W.; Zhen, J. Ultrasound-Promoted Synthesis of β-Amino Carbonyl Compounds via a Mannich Reaction Catalyzed by Ionic Liquids. *Lett. Org. Chem.* **2016,** *13* (2), 143–147.

Qin, G.; Song, Y.; Jin, R.; Shi, J.; Yu, Z.; Cao, S. Gas-Liquid Acetylene Hydrochlorination Under Nonmercuric Catalysis Using Ionic Liquids as Reaction Media. *Green Chem.* **2011,** *13*, 1495–1498.

Raeissi, S.; Peters, C. J. A Potential Ionic Liquid for CO_2-Separating Gas Membranes: Selection and Gas Solubility Studies. *Green Chem.* **2009,** *11* (2), 185–192.

Rajadurai, V.; Anguraj, B. L. A Systematic Approach of Using Green Solvent for the Extraction of Pb (II) from Aqueous Solution. *Int. J. Environ. Sci. Tech.* **2022,** *19* (3), 1369–1382.

Rantwijk, F. V.; Secundo, F.; Sheldon, R. A. Structure and Activity of Candida Antarctica Lipase B in Ionic Liquids. *Green Chem.* **2006,** *8* (3), 282–286.

Rebeiro, G. L.; Khadilkar, B. M. Chloroaluminate Ionic Liquid for Fischer Indole Synthesis. *Synthesis* **2001,** *2001* (103), 0370–0372.

Ressmann, A. K.; Strassl, K.; Gaertner, P.; Zhao, B.; Greiner, L.; Bica, K. New Aspects for Biomass Processing with Ionic Liquids: Towards the Isolation of Pharmaceutically Active Betulin. *Green Chem.* **2012,** *14* (4), 940–944.

Ross, J.; Xiao, J. Friedel–Crafts Acylation Reactions Using Metal Triflates in Ionic Liquid. *Green Chem.* **2002,** *4* (2), 129–133.

Sahoo, S.; Joseph, T.; Halligudi, S. B. Mannich Reaction in Brönsted Acidic Ionic Liquid: A Facile Synthesis of β-Amino Carbonyl Compounds. *J. Mol. Catal. Chem.* **2006,** *244* (1–2), 179–182.

Sarca, V. D.; Laali, K. K. Triflic Acid-Promoted Transacylation and Deacylation Reactions in Ionic Liquid Solvents. *Green Chem.* **2004,** *6* (5), 245–248.

Sardar, S.; Wilfred, C. D.; Leveque, J. M. One-Pot Mannich Base Synthesis Using Task Specific Protic Ionic Liquids. *Malay. J. Anal. Sci.* **2017,** *21* (5), 1203–1209.

Seddon, K. R. Ionic Liquids for Clean Technology. *J. Chem. Technol. Biotechnol.: Int. Res. Proc. Environ. Clean Technol.* **1997,** *68* (4), 351–356.

Sha, X.; Sheng, X.; Zhou, Y.; Wang, B.; Liu, Y.; Bao, J. High Catalytic Performance of Mesoporous Dual Brønsted Acidic Ternary Poly (Ionic Liquids) for Friedel-Crafts Alkylation. *Appl. Organometal. Chem.* **2019,** *33* (11). DOI: 10.1002/aoc.5180.

Shan, Q.; Zhang, J.; Wang, Y.; Liu, W. Preparation of Ionic Liquid-Type UiO-66 and Its Adsorption Desulfurization Performance. Fuel, **2022,** 312. DOI: 10.1016/j.fuel.2021.122945.

Shen, Z. L.; Zhou, W. J.; Liu, Y. T.; Ji, S. J.; Loh, T. P. One-Pot Chemoenzymatic Syntheses of Enantiomerically-Enriched O-Acetyl Cyanohydrins from Aldehydes in Ionic Liquid. *Green Chem.* **2008,** *10* (3), 283–286.

Shen, Z.; Cheong, H.; Lai, Y.; Loo, W.; Loh, T. Application of Recyclable Ionic Liquid-Supported Imidazolidinone Catalyst in Enantioselective Diels–Alder Reactions. *Green Chem.* **2012,** *14* (9), 2626–2630.

Shi, F.; Deng, Y.; SiMa, T.; Peng, J.; Gu, Y.; Qiao, B. Alternatives to Phosgene and Carbon Monoxide: Synthesis of Symmetric Urea Derivatives with Carbon Dioxide in Ionic Liquids. *Angew. Chem.* **2003,** *115* (28), 3379–3382.

Shi, F.; Gu, Y.; Zhang, Q.; Deng, Y. Development of Ionic Liquids as Green Reaction Media and Catalysts. *Catal. Surv. Asia* **2004,** *8* (3), 179–186.

Shirzaei, F.; Shaterian, H. R. [$(EtO)_3Si\ (CH_2)3NH_3^+][CH_3COO^-]$ as a Novel Basic Ionic Liquid Catalyzed Green Synthesis of New 2- (Phenylsulfonyl)-1H-benzo [a] Pyrano [2, 3-c] Phenazin-3-Amine Derivatives. *J. Mol. Struct.* **2022,** 1256. DOI: 10.1016/j.molstruc. 2022.132558.

Song, C. E.; Roh, E. J. Practical Method to Recycle a Chiral (Salen) Mn Epoxidation Catalyst by Using an Ionic Liquid. *Chem. Commun.* **2000,** *10*, 837–838.

Song, Z.; Huang, W.; Zhou, Y.; Tian, Z. Q.; Li, Z. M.; Tao, D. J. Thermally Regulated Molybdate-Based Ionic Liquids Toward Molecular Oxygen Activation for One-Pot Oxidative Cascade Catalysis. *Green Chem.* **2020,** *22* (1), 103–109.

Ståhlberg, T.; Sørensen, M. G.; Riisager, A. Direct Conversion of Glucose to 5- (Hydroxymethyl) Furfural in Ionic Liquids with Lanthanide Catalysts. *Green Chem.* **2010,** *12* (2), 321–325.

Subasree, N.; Selvi, J. A. Imidazolium Based Ionic Liquid Derivatives; Synthesis and Evaluation of Inhibitory Effect on Mild Steel Corrosion in Hydrochloric Acid Solution. *Heliyon* **2020,** *6* (2). DOI: 10.1016/j.heliyon.2020.e03498.

Sun, J.; Wang, J.; Cheng, W.; Zhang, J.; Li, X.; Zhang, S.; She, Y. Chitosan Functionalized Ionic Liquid as a Recyclable Biopolymer-Supported Catalyst for Cycloaddition of CO_2. *Green Chem.* **2012,** *14* (3), 654–660.

Taskin, M.; Cognigni, A.; Zirbs, R.; Reimhult, E.; Bica, K. Surface-Active Ionic Liquids for Palladium-Catalysed Cross Coupling in Water: Effect of Ionic Liquid Concentration on the Catalytically Active Species. *RSC Adv.* **2017,** *7* (65), 41144–41151.

Thang, G. V.; Pegot, B.; Loupy, A. Solvent-Free Microwave-Assisted Preparation of Chiral Ionic Liquids from (−)-N-Methylephedrine. *Eur. J. Org. Chem.* **2004,** *2004* (5), 1112–1116.

Tong, J.; Li, Y. W.; Xu, D. Z. Solvent-Controlled Friedel–Crafts Reaction for the Synthesis of 3-Indolyl-3-Hydroxy Oxindoles and 3, 3-Diindolyl Oxindoles Catalyzed by DABCO-Base Ionic Liquids. *ChemistrySelect* **2017,** *2* (13), 3799–3803.

Varma, R. S.; Namboodiri, V. V. An Expeditious Solvent-Free Route to Ionic Liquids Using Microwaves. *Chem. Commun.* **2001,** *7*, 643–644.

Walden P.; Molecular Weights and Electrical Conductivity of Several Fused Salts. *Bull. Acad. Sci. St. Petersburg* **1914,** *1800*, 405–422.

Wang, H. B.; Yao, N.; Wang, L.; Hu, Y. L. Brønsted–Lewis Dual Acidic Ionic Liquid Immobilized on Mesoporous Silica Materials as an Efficient Cooperative Catalyst for Mannich Reactions. *New J. Chem.* **2017,** *41* (19), 10528–10531.

Wang, J.; Pei, Y.; Zhao, Y.; Hu, Z. Recovery of Amino Acids by Imidazolium Based Ionic Liquids from Aqueous Media. *Green Chem.* **2005,** *7* (4), 196–202.

Wasserschied, P.; Keim, W. Ionic Liquids—New "Solutions" for Transition Metal Catalysis. *Angew. Chem. Int. Ed. Engl.* **2000,** *39* (21), 3773–3789.

Wellens, S.; Thijs, B.; Binnemans, K. An Environmentally Friendlier Approach to Hydrometallurgy: Highly Selective Separation of Cobalt from Nickel by Solvent Extraction with Undiluted Phosphonium Ionic Liquids. *Green Chem.* **2012,** *14* (6), 1657–1665.

Welton, T. Room-Temperature Ionic Liquids. Solvents for Synthesis and Catalysis. *Chem. Rev.* **1999,** *99* (8), 2071–2083.

Weng, J.; Wang, C.; Li, H.; Wang, Y. Novel Quaternary Ammonium Ionic Liquids and Their Use as Dual Solvent-Catalysts in the Hydrolytic Reaction. *Green Chem.* **2006,** *8* (1), 96–99.

Wilkes, J. S.; Levisky, J. A.; Wilson, R. A.; Hussey, C. L.; *Inorg. Chem.* **1982,** *21*, 1236–1264.

Wilkes, J. S.; Zaworotko, M. J. Air and Water Stable 1-Ethyl-3-Methylimidazolium Based Ionic Liquids. *J. Chem Soc. Chem. Commun.* **1992,** *13*, 965–967.

Wu, Y.; Zhang, Y.; Lv, X.; Mao, C.; Zhou, Y.; Wu, W.; et al. Synthesis of Polymeric Ionic Liquids Mircrospheres/Pd Nanoparticles/CeO_2 Core-Shell Structure Catalyst for Catalytic Oxidation of Benzyl Alcohol. *J. Taiwan Inst. Chem. Eng.* **2020,** *107*, 161–170.

Xiao, Y.; Malhotra, S. V. Diels–Alder Reactions in Pyridinium Based Ionic Liquids. *Tetrahedron Lett.* **2004,** *45* (45), 8339–8342.

Xie, F.; Du, C.; Pang, Y.; Lian, X.; Xue, C.; Chen, Y.; et al. Lewis Acid-Assisted N-Fluorobenzenesulfonimide-Based Electrophilic Fluorine Catalysis in Beckmann Rearrangement. *Tetrahedron Lett.* **2016,** *57* (51), 5820–5824.

Xu, D.; Liu, Y.; Shi, S.; Wang, Y. A simple, Efficient and Green Procedure for Knoevenagel Condensation Catalyzed by [C_4dabco][BF_4] Ionic Liquid in Water. *Green Chem.* **2010,** *12* (3), 514–517.

Xu, L.; Chen, W.; Xiao, J. Heck Reaction in Ionic Liquids and the In Situ Identification of N-Heterocyclic Carbene Complexes of Palladium. *Organometallics* **2000,** *19* (6), 1123–1127.

Yang, C. H.; Chang, J. C.; Wu, T. Y.; Sun, I. W.; Wu, J. H.; Ho, W. Y. Novel Aryl-Imidazolium Ionic Liquids with Dual Brønsted/Lewis Acidity as Both Solvents and Catalysts for Friedel–Crafts Alkylation. *Appl. Sci.* **2019,** *9* (22), 4743.

Yang, Z.; Zhao, Y.; He, L.; Gao, J.; Yin, Z. Highly Efficient Conversion of Carbon Dioxide Catalyzed by Polyethylene Glycol-Functionalized Basic Ionic Liquids. *Green Chem.* **2012,** *14* (2), 519–527.

Yu, L.; Guo, L.; Hu, W.; Zhang, Z.; Bai, Y.; Ye, J.; et al. One-Pot Conversion of Ketones to Amides via Beckmann Rearrangement Catalyzed by Metal Chloride-Ionic Liquids Under Solvent-Free Condition. *Catal. Commun.* **2019,** 123, 119–123.

Zakzeski, J.; Bruijnincx, P. C. A.; Weckhuysen, B. M. In Situ Spectroscopic Investigation of the Cobalt-Catalyzed Oxidation of Lignin Model Compounds in Ionic Liquids. *Green Chem.* **2011,** *13* (3), 671–680.

Zare, A.; Hasaninejad, A.; Zare, A. R. M.; Parhami, A.; Sharghi, H.; Khalafi-Nezhad, A. Zinc Oxide as a New, Highly Efficient, Green, and Reusable Catalyst for Microwave-Assisted Michael Addition of Sulfonamides to α, β-Unsaturated Esters in Ionic Liquids. *Canad. J. Chem.* **2007,** *85* (6), 438–444.

Zeng, Y.; Wang, Y.; Xu, Y.; Song, Y.; Jiang, J.; Jin, Z. Pd Nanoparticles in the Thermoregulated Ionic Liquid and Organic Biphasic System: An Efficient and Recyclable Catalyst for Heck Reaction. *Catal. Lett.* **2013,** *143* (2), 200–205.

Zhao, G.; Jiang, T.; Gao, H.; Han, B.; Huang, J.; Sun, D. Mannich Reaction Using Acidic Ionic Liquids as Catalysts and Solvents. *Green Chem.* **2004,** *6* (2), 75–77.

Zhu, A.; Li, L.; Wang, J.; Zhuo, K. Direct Nucleophilic Substitution Reaction of Alcohols Mediated by a Zinc-Based Ionic Liquid. *Green Chem.* **2011,** *13* (5), 1244–1250.

Zulfiqar, F.; Kitazume, T. Lewis Acid-Catalysed Sequential Reaction in Ionic Liquids. *Green Chem.* **2000,** *2* (6), 296–297.

CHAPTER 6

Supercritical Solvents

RAVI CHANGWAL[1], ABHILASHA JAIN[2], SHIKHA PANCHAL[1], SHWETA SHARMA[1], and RAMESHWAR AMETA[1]

[1]*Department of Chemistry, PAHER University, Udaipur, India*

[2]*Department of Chemistry, St. Xavier's College, Mumbai, India*

ABSTRACT

In a supercritical fluid, distinct liquid and gas phases do not exist at a temperature and pressure above its critical point. It can effuse like a gas in porous solids and dissolve some materials like a liquid. As such, supercritical fluids are suitable as an alternative for organic solvents in a wide range of laboratory as well as industrial processes. Carbon dioxide and water are the most commonly used supercritical fluids. These solvents have been used in a number of chemical reactions such as oxidation, reduction, coupling, alkylation/acylation, esterification, hydrolysis, decarboxylation, extraction, rearrangements, polymerization, etc. Use of supercritical water and carbon dioxide as solvent in various chemical reactions has been summarized.

6.1 INTRODUCTION

A supercritical fluid is a substance at a temperature and pressure above its critical point, where distinct liquid and gas phases do not exist. It can effuse through solids like a gas and dissolve materials like a liquid. In addition, close to the critical point, small changes in pressure or temperature result in large changes in density, allowing many properties of a supercritical fluid to

Green Chemistry, 2nd Edition: Fundamentals and Applications. Suresh C. Ameta & Rakshit Ameta (Eds.)

be "fine-tuned." Supercritical fluids are suitable as a substitute for organic solvents in a range of industrial and laboratory processes. Carbon dioxide and water are the most commonly used supercritical fluids, being used for decaffeination and power generation, respectively.

6.2 SOME MAJOR SUPERCRITICAL FLUIDS

6.2.1 SUPERCRITICAL WATER

Water is described as superheated water, subcritical water, or pressurized hot water between 100°C and its supercritical point is 370°C. Supercritical water has been used in synthetic organic chemistry because of having some unique properties different from those of ambient water (Kajmoto, 1999; Savage, 1999; Broli et al., 1999; Siskin and Katrinzky, 2001). Water has similar properties to an organic solvent such as methanol. The near critical water (NCW) region is described as 250–300°C at 100–180 bar (Raner et al., 1995; Stadler et al., 2003). It has some unique properties also and these characteristics are:

i) It has lower viscosity as compared to water that results in faster diffusion of the compound.
ii) It has lower surface tension than water.
iii) More solubility of two polar compounds due to less hydrogen bonding.
iv) Its heat capacity is 2–6 times of liquid water, which improves transfer of heat.
v) Single homogeneous phase results in it with no interfacial mass transfer limitation.
vi) Dramatic changes were observed in density and in ionic product of water, near critical temperature.

All these properties result in an increased rate of reaction and extraction.

There are many types of reactions that occur in water at high temperature, for example.

i) Polyester and other condensation polymers are cleaved to their starting materials at 573 K.
ii) Rubber tires are converted into an oil with 44% yield at 673 K.
iii) Cellulose base waste can be converted into sugars.
iv) It is a very effective medium for acid and base catalyzed reactions.

v) Gasification of biomass can be done in it, which includes food waste, organic waste from industries, and waste from forestry or agriculture.
vi) Oxidation reactions can be used for waste water treatment. Supercritical water oxidation (SCWO) technique is environmental friendly because no organic solvent or additives are required. In this technique, oxygen, air, or hydrogen peroxide can be used as an oxidant. In SCWO, theoretically all organic compounds (pollutant) can be converted mainly into carbon dioxide and water. It is a cost-effective technique.

The use of superheated water as a suitable solvent for the extraction of nonpolar compounds from an environmental sample was proposed by Hawthrone et al. (1994). It is one of the best options as a solvent for chromatography because it is a green solvent having minimum cost, safe, nontoxic, noninflammable, and recyclable. The reduction in viscosity with temperature enables high flow rates. Moreover, it can be used with a wide range of detectors, where water has virtually no background signal. It is also compatible with flame ionization detection. So, it can be concluded that pressurized hot water and supercritical water is a great promise for the future green chemical technology.

6.2.2 ORGANIC REACTIONS IN SUPERCRITICAL WATER (SCW)

- Diel-Alder cycloaddition reactions have been carried out in SCW (Korzenski and Kolis., 1997).
- Heck coupling reaction was conducted in SCW (Reardon et al., 1995; Diminnie et al., 1995).
- Aldol condensation reaction could be conducted in SCW (An et al., 1997).
- Claisen, Rupe, Meyer-Schuster, and pinacol-pinacolone rearrangements have also been carried out in SCW (Kuhlmann et al., 1994a; Crittendon and Parsons, 1994).
- Dehydration of alcohol (Crittendon and Parsons, 1994; Kuhlmann et al, 1994b; Xu and Antal, 1994, 1997) and decarboxylation of carboxylic acids (Carlsson et al., 1994; Yu and Savage, 1998) have been reported in SCW.
- In SCW, carbon–silicon bond cleavage of organosilicon reaction is also reported (Itami et al., 2003). At 390°C, the reaction of arylsilanes in SCW causes C-Si bond cleavage.

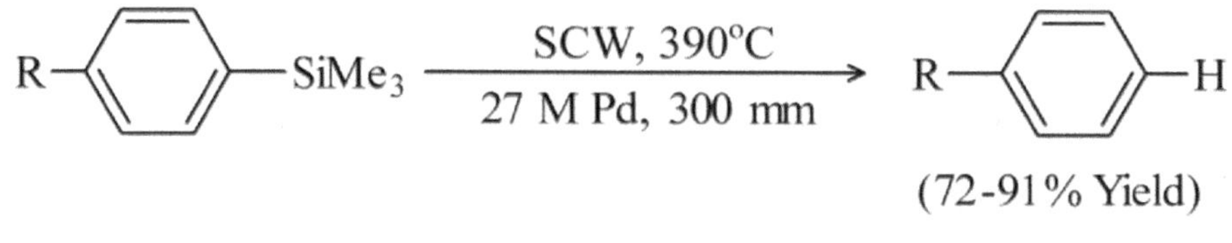

6.2.3 *ORGANIC REACTION IN NEAR CRITICAL WATER (NCW) REGION*

i) **Diels-Alder cycloaddition:** Korzenski and Kolisa (1997) prepared a series of Diels-Alder reactions using cyclopentadiene (diene) and diethyl fumarate (dienophiles) in supercritical water as a reaction medium. It leads to high yields of clean products (adducts) without added catalysts.

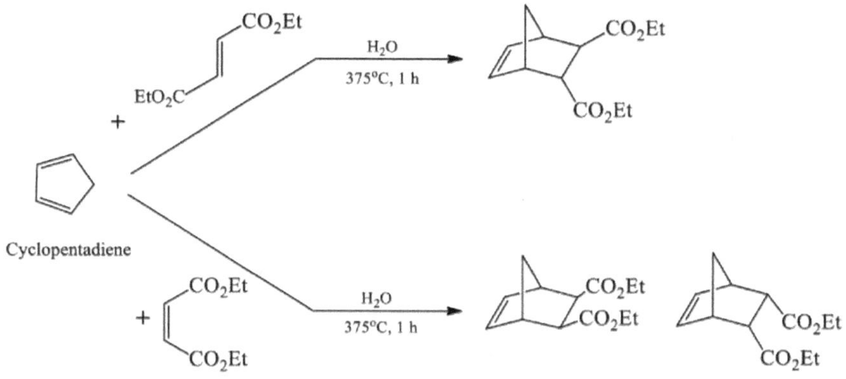

ii) **Claisen rearrangement:** It was reported in near critical water at 240°C in a MW oven, which gives 84% yield (Raner et al., 1995; An et al., 1997).

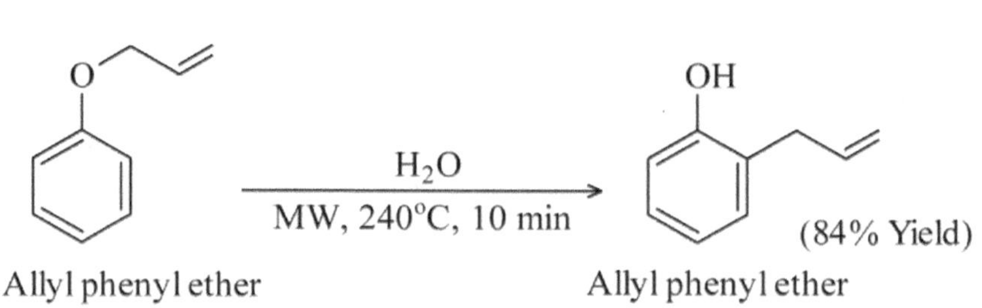

Allyl phenyl ether Allyl phenyl ether

iii) **Pinacol-Pinacolone rearrangement:** It has also been carried out in near critical water at 270°C (Kremsner and Kappe, 2005).

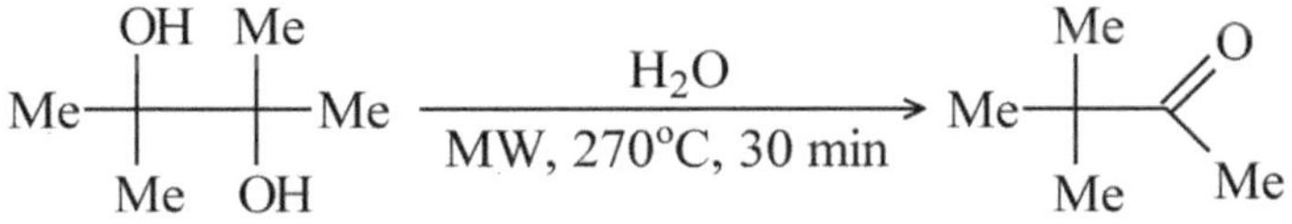

Allyl phenyl ether Pinacolone

iv) **Fischer indole synthesis:** When this reaction was carried out by heating with water in a MW oven at 270°C, then 64% yield of indole derivative has been obtained (Strauss and Trainor, 1995; Strauss, 1999).

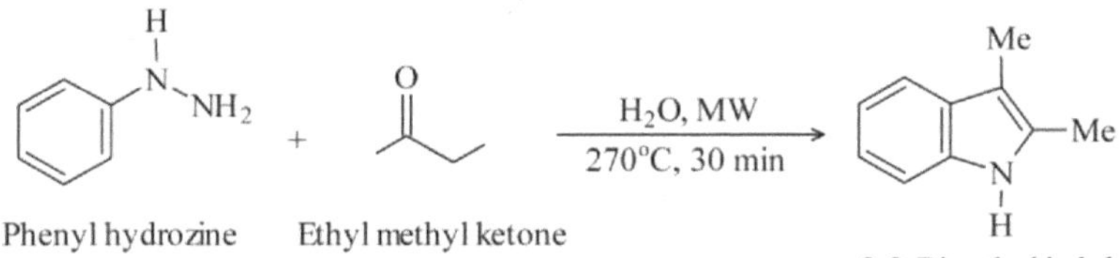

Phenyl hydrozine Ethyl methyl ketone 2,3-Dimethyl indole

v) **Decarboxylation of carboxylic acids:** Decarboxylation of indole -2-carboxylic acid by heating with water at near critical region gives indole with 100% yield (An et al., 1997). It requires 255°C temperature and 20 min. for the completion of reaction.

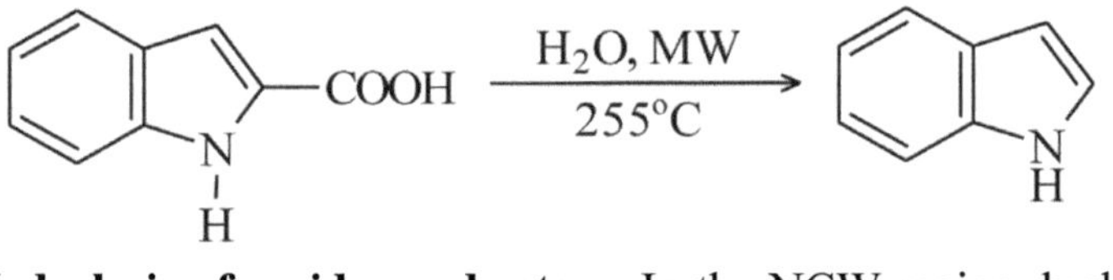

vi) **Hydrolysis of amides and esters:** In the NCW region, hydrolysis of amides and esters does not require addition of acid/base or any other catalyst. Ethyl benzoate and benzamide are hydrolyzed to benzoic acid in MW at 295°C for 2 and 4 h, respectively without addition of any catalyst (Kremsner and Kappe, 2005). It gives 92–95% yield of product.

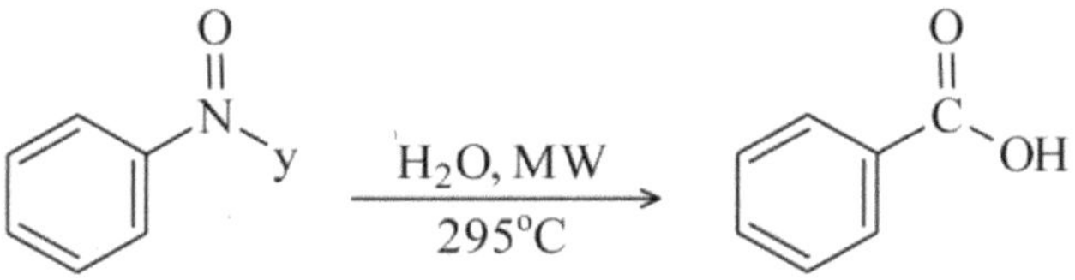

y = $-OC_2H_5$ or $-NH_2$

vii) Some other reactions: It was observed that the reactivity of some organic compounds in near critical region or in high temperature water is enhanced by some water-soluble compounds, for example, mineral acid, ammonia, which are produced by hydrolysis of ester or amines and can act as an acid or base catalyst (Akiya and Savage, 2002).

Chakinala et al. (2013) used a homologous series of carboxylic acids with a linear chain of C1C8 and alcohols and observed the effect of structure of the molecules on the gasification at two different concentrations in supercritical water. It was observed that carboxylic acids concentration had a strong effect on the gasification efficiency (GE), while no such effect was there in alcohol series. It was also revealed that alcohols were easily gasified than the organic acids, but the carbon GE order of alcohols was just reverse as compared to that of acids, particularly for short chain compounds. An interesting oscillatory behavior of methane and CO_2 product yield was there with an increasing chain length.

Guo et al. (2013) reported supercritical water gasification (SCWG) of glycerol and analyzed the liquid products. It was suggested that the main sources of hydrogen were pyrolysis of glycerol and steam reforming of intermediates, but yield of hydrogen is very small from watergas shift reaction (WGSR). Guo et al. (2014) also developed a model for gasification and homogeneous decomposition of indole in supercritical water. Toluene, aniline, and benzene were intermediate products, and H_2, CH_4, and CO_2 were found as terminal gaseous products. The fastest step is ring-opening of indole to afford aniline in a first few minutes, then a large variety of gasifiable products were depleted, which was found to be dominating in the range of 5–55 min., followed by water gas shift as the main H2-producing reaction in a longer period.

Freitas and Guirardello (2015) studied the effect of addition of carbon dioxide (as co-reactant) in supercritical water gasification (SCWG) using sugarcane bagasse as a raw material (waste). It was observed that syngas with a H2/CO molar ratio close to 2 was obtained, when CO2 was added as a co-reactant at 35 wt.%. This ratio of H2/CO is useful in FischerTropsch synthesis. Cantero et al. (2015) hydrolyzed cellulose in supercritical water to obtain a sugar yield more than 95 wt.% in only 0.02 s, whereas the yield of 5-hydroxymethylfurfural (5-HMF) was less than 0.01 wt.%. It was observed that the main product was glycolaldehyde (60 wt.%) on increasing reaction time to 1 s. Different biomasses (glucose, cellulose, fructose, and wheat bran) were also used. The dehydration to 5-HMF and isomerization of glucose to fructose was found to be strongly dependent on concentration.

The supercritical water upgrading (SCWU) of heavy oils was found to decrease the sulfur content as well as average molecular weight without rejecting any carbon (as coke products) (Timko et al., 2015). It was observed that sulfide decomposition during SCWU followed a radical chain reaction pathway. It was revealed that sulfide decomposition processes through aldehyde and thioaldehyde intermediates and water played important roles as a catalyst in aldehyde decarbonylation and as a reactant thioaldehyde hydrolysis. It was also found that ZnO had a capacity to increase sulfur removal during this SCWU process, without addition of molecular hydrogen.

Čolnik et al. (2021) presented the chemical recycling of polyethylene terephthalate (PET) waste using sub- and supercritical water (SubCW and SCW). Two types of PET waste were selected for hydrolytic depolymerization: (i) Colorless and (ii) Colored bottles. It was reported that the highest yield of terephthalic acid (TPA) was observed at 300°C in 30 min giving 90.0 ± 0.4% yield from colorless PET waste and 85.0 ± 0.2% from green PET waste.

Zhao et al. (2022) reported that the carbon can be extracted from organic waste and then converted into syngas with calorific value. They studied gasification of polystyrene (PS) in supercritical water with CO_2 in the temperature range (400−700°C) and time 0–30 min. The carbon conversion efficiency of PS plastic could reach at 47.6% at 700°C in supercritical water within only 20 min under feedstock condition.

6.3 CARBON DIOXIDE

Gore (1861) gave the process of preparing liquid CO_2 for the first time. Carbon dioxide exists in three phases, that is, solid, liquid, and gas. Solid phase of CO_2 is called ‘dry ice’ and it is used for cooling. Gas phase is well known and the solid transforms to gas without liquification at atmospheric temperature and pressure. Only in certain specified conditions, it can be liquefied. With the increasing pressure on gas or heating of solid CO_2, liquid phase can be achieved. The critical temperature of CO_2 is 31°C. All three phases of carbon dioxide exist simultaneously at the temperature –56°C and 5.1 atm. It exists as a supercritical fluid at 31°C and 73 atm. In this condition, it has unique properties. It has viscosity similar to the gas phase and density similar to the liquid phase.

Some advantages of supercritical carbon dioxide ($scCO_2$) are as follows:

i) A high diffusion rate offers potential for increased reaction rates.

ii) It has high compressibility. Large changes in solvent properties with relatively small change in pressure have been reported, by which infinite range of solvent properties can be achieved. A small amount of co-solvent can further modify solvent properties.
iii) Due to the high solubility of light gases, some catalysts and substrates bring all compounds together in single homogeneous phase so that it has a potential for homogeneous catalytic processes.
iv) It is an excellent medium for oxidation and reduction reactions.

Thus, supercritical carbon dioxide can be used as a solvent in various ways, for example, for dry cleaning, extracting natural products and solvent for organic reactions.

In dry cleaning, the most commonly used solvent is PERC (perchloroethylene), which is a carcinogenic compound and also one of the responsible factors for ozone layer depletion. This problem can be overcome by using CO_2 as a solvent for dry cleaning. It dissolves nonpolar substances and with the addition of some surfactants, the solubility of oils and greases in carbon dioxide can be increased. Nowadays, the micelle technologies have produced dry cleaning machines using liquid CO_2 and a surfactant to dry clean clothes.

Supercritical CO_2 can be used as a solvent for extraction because

- No hydrolysis (which generally occurs in steam distillation).
- No loss of volatile components.
- No thermal degradation products.
- Free of inorganic salts or heavy metals.
- High concentration of valuable ingredients and high extraction yield.
- Free of any microbial life.
- Environmentally benign solvent.
- High solubility toward hydrocarbons, ethers, esters, etc. whereas polar compounds, for example, sugars, tannins, glycosides, etc. are insoluble.

As $scCO_2$ has all these advantages and hence, there are numerous applications, when it is used for extraction of natural products:

- Essential oils from turmeric, coriander, ginger, ajowan, etc. have been extracted with $scCO_2$.
- It is extensively used for natural coffee decaffeination.
- Various food products, for example, pepper, cumin, cardamom, cloves, etc. can be isolated.

Vegetables tannin materials have been isolated using $scCO_2$.

Supercritical CO_2 can be used as a solvent for organic synthesis because it is environmentally benign and nontoxic solvent (Anatas et al., 1998). Since the characteristics of $scCO_2$ are intermediate between that of liquid and gas, so with the manipulation of temperature and pressure, its solvent properties can be changed dramatically (Mchugh and Krukonis, 1986; Johnston, 1989). Some applications of $scCO_2$ in organic synthesis are

i) Diels–Alder reaction

Reaction between 2-t-butyl-1,3-butadiene and methyl acrylate gives addition product with 54% yield, when carried out in $scCO_2$ (Renslo et al., 1997).

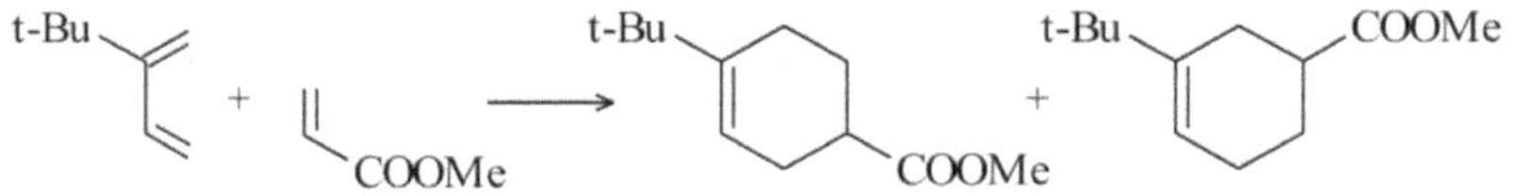

An aza–Diels-Alder reaction of Danishefsky's diene with an imine in the presence of scandium perfluoroalkane sulfonate in $scCO_2$ gives the adduct with 99% yield (Matsuo et al., 2000).

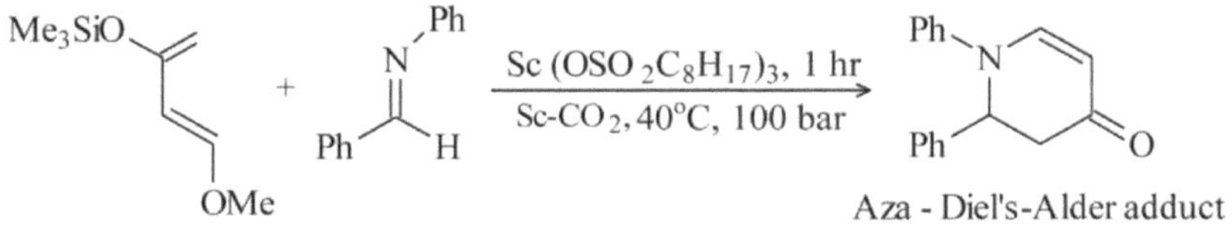

Cott et al. (2005) carried out Diels–Alder reactions between maleic anhydride and furan derivatives in supercritical CO_2.

ii) Freidel–Crafts reaction

The following transformation has been achieved by using the Freidel-Crafts reaction in $scCO_2$ (Burk et al., 1995).

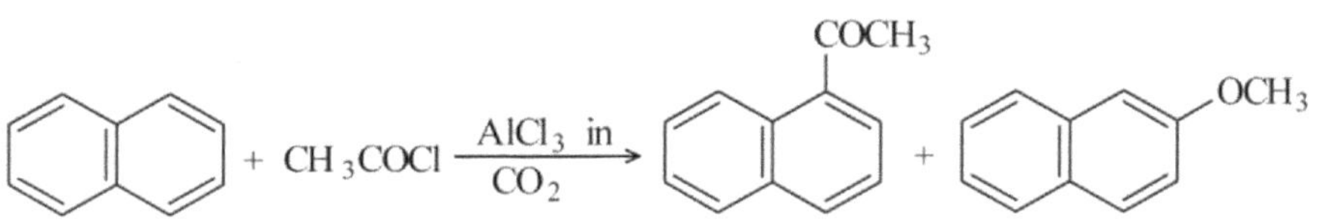

Continuous Friedal–Crafts alkylations in supercritical carbon dioxide using solid acid catalysts are described by Amandi et al. (2005).

iii) Supercritical polymerization

The $scCO_2$ has been used for the polymerization, where it is mixed with a surfactant to enhance the solubility for hydrocarbon-based molecules. Different types of polymers have been synthesized in $scCO_2$ (Cooper, 2000).

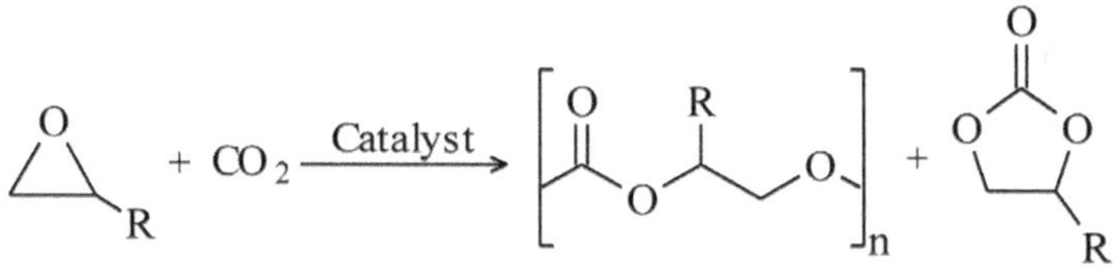

It has been reported that use of water in $scCO_2$ enhances the solubility of organics (Johnston et al., 1996). The following polymerization reaction has also been accomplished using $scCO_2$.

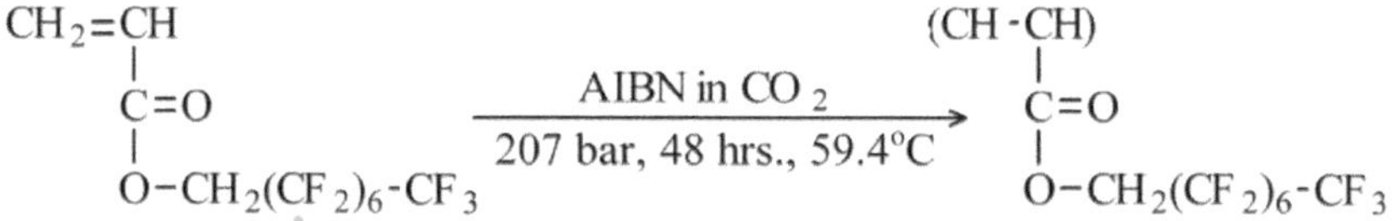

Lee et al. (2012) used supercritical carbon dioxide as an alternative solvent for the synthesis and purification of poly(L-lactic acid). This benign one-pot process replaced the use of toxic organic solvents and produced polymers with tunable molecular weight. Yang et al. (2013) fabricated well controlled porous foams of graphene oxide modified poly(propylene carbonate) using supercritical carbon dioxide. Poly(propylene carbonate) is a new amorphous, biodegradable, and biocompatible aliphatic polyester. It has a potentially wide range of applications like packing and biomedical materials. The fabricated porous materials were non-cytotoxic and therefore these are promising materials for tissue engineering applications.

Polymerization of 3-undecylbithiophene and preparation of poly(3-undecylbithiophene)/ polystyrene composites in supercritical carbon dioxide was studied by Abbett et al. (2003). Poly(methyl methacrylate-ran-perfluoroalkyl ethyl methacrylate) copolymers having varying perfluoroalkyl ethyl methacrylate ester (Zonyl-TM) contents were synthesized by Cengiz et al. (2011) in supercritical CO_2.

iv) Oxidation reactions

Oxidations have also been carried out in liquid CO_2 as the solvent (Pesiri et al., 1998). Sharpless asymmetric oxidation of allyic alcohol has been carried out in liquid CO_2.

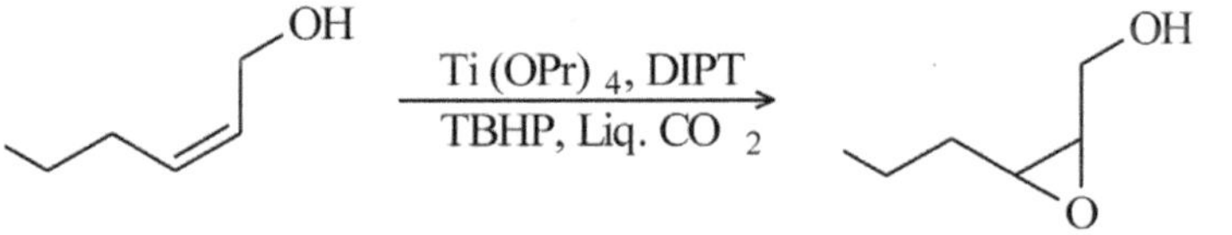

Diastereoselective sulfoxidation of cystein derivative in $scCO_2$ was achieved giving the anti – diastereomer as the sole product of this reaction (Oakes et al., 1999).

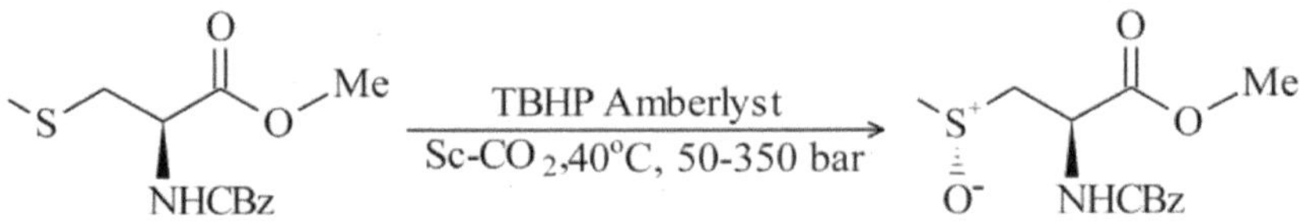

It has been observed that oxidation of alcohols to carbonyl compounds in $scCO_2$ can be carried out with high selectivity as well as at a high rate (Jerner et al., 2000). Mello et al. (2009) carried out Baeyer–Villiger oxidation of ketones in supercritical carbon dioxide at 250 bar and 40° C under flow conditions with a silica-supported peracid. The reagent can be recycled by treatment with acidic 70% hydrogen peroxide at 0°C.

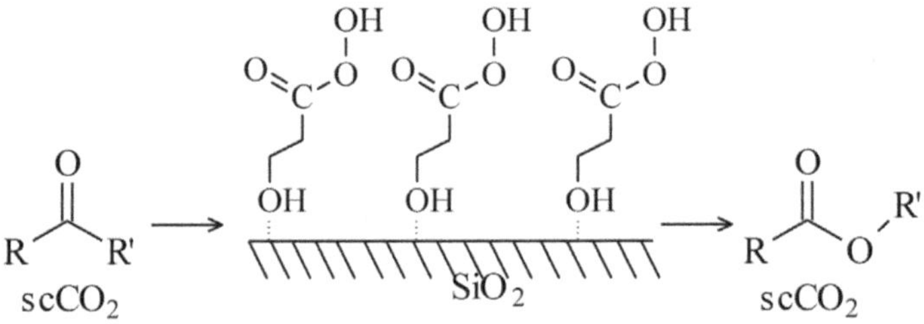

Chapman et al. (2010) reported that primary and secondary alcohols can be oxidized selectively with O_2 in continuous flow in supercritical CO_2. Selective oxidation of styrene to acetophenone has been investigated over Pd–Au catalysts with H_2O_2 in supercritical carbon dioxide medium (Wang et al., 2007).

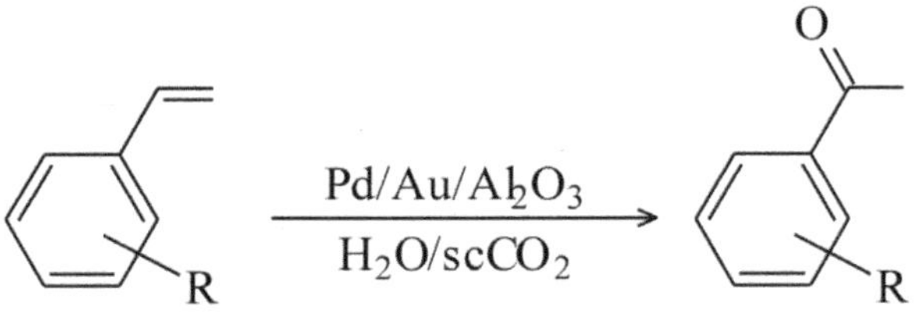

R = H, 4-Me, 3-Cl, 3-NO_2

A continuous stream of pentachlorophenol (PCP, 10–20 mg min^{-1}) in supercritical carbon dioxide was dechlorinated efficiently by a heated column (25 × 1 cm diameter) of a zero-valent silver–iron (Ag^0/Fe^0) bimetallic mixture (Kabir and Marshall, 2001). Mello et al. (2012) carried out epoxidation of olefins with a silica-supported peracid in supercritical carbon dioxide under Flow conditions.

v) Hydrogenation reactions

The hydrogenation of alkene in $scCO_2$ has immense importance in industry (Jessop et al., 1999; Jessop and Leitner, 1999; Baiker, 1999), because hydrogen is soluble in $scCO_2$ and it is able to bring together hydrogen substrates and catalysts in a single homogeneous reaction. γ–Butyrolactone, which is one of the most valuable alternatives of chlorinated solvent, can be synthesized by hydrogenation of maleic anhydride with hydrogen in the presence of Pd/Al_2O_3 using $scCO_2$ at 200°C (Pillai and Sahle-Demessi, 2002).

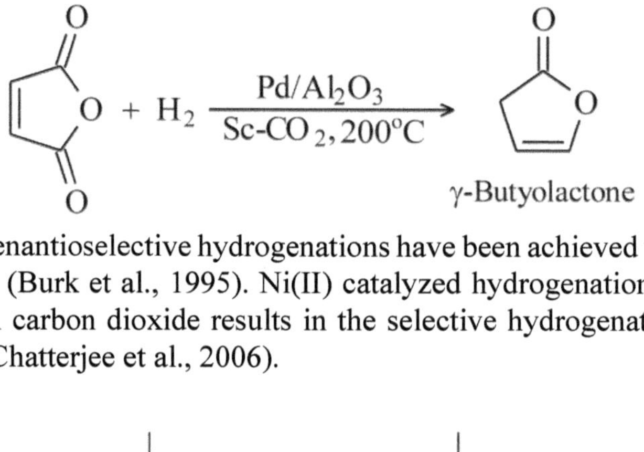

Various enantioselective hydrogenations have been achieved using $scCO_2$ as a solvent (Burk et al., 1995). Ni(II) catalyzed hydrogenation of citral in supercritical carbon dioxide results in the selective hydrogenation of C=O over C=C (Chatterjee et al., 2006).

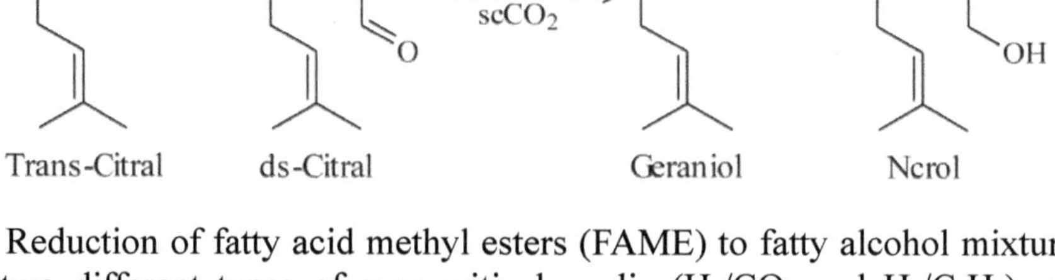

Reduction of fatty acid methyl esters (FAME) to fatty alcohol mixtures in two different types of supercritical media (H_2/CO_2 and H_2/C_3H_8) was compared using two different hydrogenation catalysts (Andersson et al., 2000). Li et al. (2001) reported that $Zn–H_2O–CO_2$ is an excellent reducing reagent for the reduction of aldehydes in supercritical carbon dioxide. Selective hydrogenation of α,β-unsaturated aldehydes to saturated aldehydes has been achieved using Pd/C catalyst in the $scCO_2$ medium by Zhao et al. (2003a). Hydrogenation of 2-butyne-1,4-diol to butane-1,4-diol has also been successfully conducted in supercritical carbon dioxide by Zhao et al. (2003b) at 323 K with a high selectivity of 84% for butane-1,4-diol at 100% conversion.

The hydrogenation of benzonitrile to benzylamine with high conversion (90.2%) and selectivity (90.9%) was achieved in supercritical carbon dioxide without using any additive over Pd/MCM-41 catalyst (Chatterjee et al., 2010). The heterogeneously catalyzed hydrogenation of carvone with green solvent supercritical carbon dioxide was reported by Melo et al. (2011). Two Pd(II) complexes with the tridentate ligands (SOO/SOS donor atom) have been used as catalyst in homogeneous hydrogenation of olefins with molecular hydrogen in super critical carbon dioxide (Yılmaz et al., 2010).

vi) Heck reaction

It has been reported that the reaction between iodobenezene and methyl acrylate catalyzed by $Pd(OAc)_2$ in the presence of fluorous ligand using $scCO_2$ gave 92% yield of methyl cinnamate.

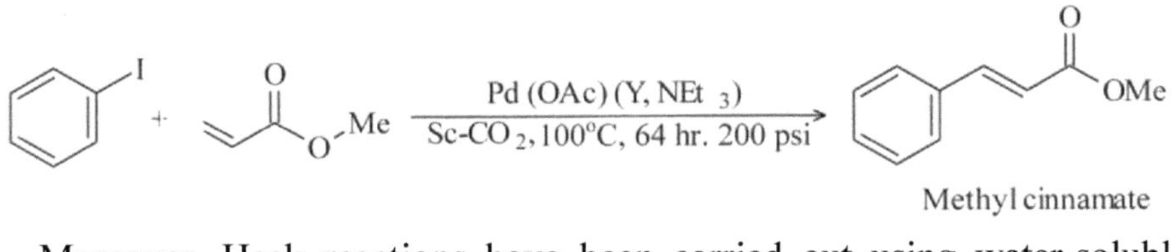

Moreover, Heck reactions have been carried out using water-soluble catalysts in the $scCO_2$/water biphasic system (Bhanage et al., 1999).

vii) Coupling reaction

Pd catalyzed biaryl formation by homocoupling of iodobenezene in $scCO_2$ has been carried out (Shezad et al., 2002).

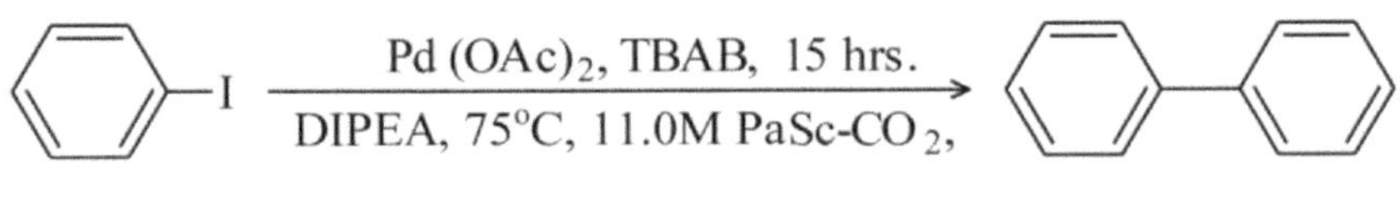

The homocoupling of iodoarenes catalyzed by $Pd(OCOCF_3)_2$/P(2-furyl)$_3$ also occurs best in $scCO_2$ and under solventless reaction conditions (Shezad et al., 2002).

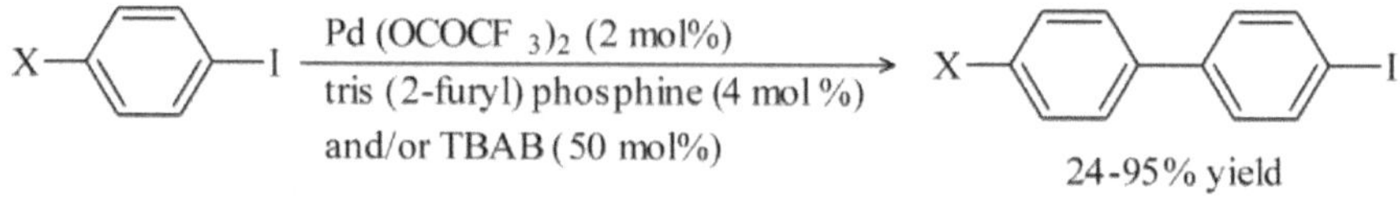

viii) Photochemical reaction

Synthesis of 2-acyl-1,4-hydroquimone by photoinduced addition of aldehydes to α, β-unsaturated carboxyl compounds is an environmentally benign method, which can be further improved by the use of $scCO_2$ in place of benzene as a solvent (Pacut et al., 2001).

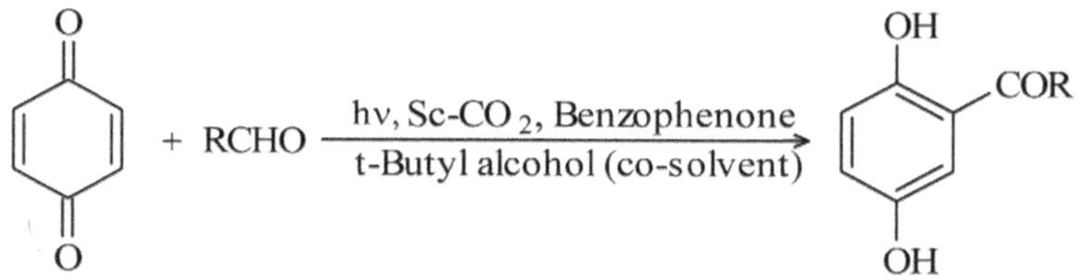

ix) Alkylation/acylation

The acetalization of terminal olefins with electron withdrawing groups was carried out smoothly in supercritical carbon dioxide under oxygen atmosphere using polystyrene supported benzoquinone as cocatalyst with palladium chloride (Wang et al., 2005). Amandi et al. (2007) used supercritical carbon dioxide as a solvent for continuous alkylation of phenol using a solid acid catalyst, γ-Al_2O_3, with either cyclohexene or cyclohexanol as alkylating agents.

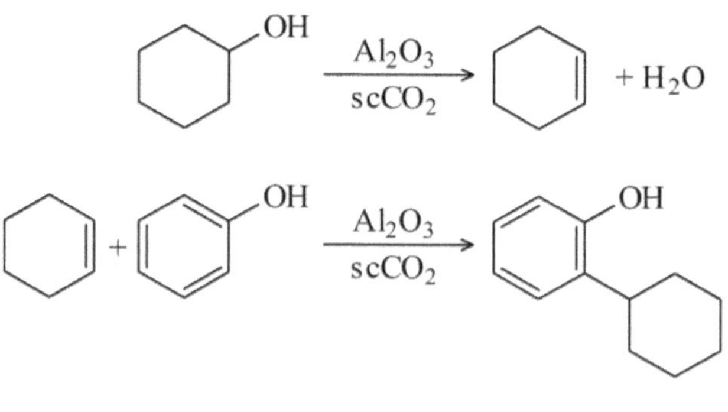

x) Esterification

Esterification of 2-ethyl-1-hexanol with 2-ethylhexanoic acid to produce 2-ethylhexyl-2-ethylhexanoate has been investigated in supercritical CO_2 by Ghaziaskar et al. (2006).

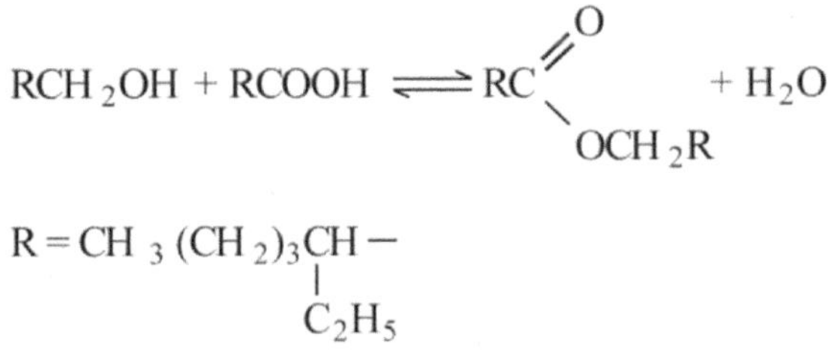

xi) Extraction

Supercritical CO_2 extraction of phenolic compounds from *Zostera marina* residues was optimized by developing a mathematical model based on mass transfer balances (Pilavtepe et al., 2012). Zarena et al. (2012) used supercritical fluid technology to extract the active constituents from *mangosteen pericarp*. Xanthons were extracted and their antioxidant activity was measured by 2,2-diphenyl-1-picrylhydrazyl (DPPH) radical scavenging assay. Supercritical CO_2 extraction of oleic sunflower seeds was shown to be effective and it yielded a product very similar to pure refined oil (Kiriamiti et al., 2002).

Vedaraman et al. (2004) have investigated the conventional methods of cholesterol extraction from cattle brain using supercritical carbon dioxide and with different cosolvents while carotenoids have been extracted from rosehip fruit using supercritical CO_2 in various extraction conditions by Machmudah et al. (2008). A two-step process was used for extraction of alkannin derivatives from *Alkanna tinctoria* with supercritical CO_2 followed by alkaline hydrolysis of alkannin derivatives. The highest total alkannins (1.47%) were extracted with $scCO_2$, which was higher than conventional hexane extraction (1.24%) providing a solvent-free alternative for industrial production (Akgun et al., 2011).

Soh and Zimmerman (2011) reported lipid extraction from wet algae using $scCO_2$, which was optimized for biodiesel production potential in terms of FAME yield and selectivity. Roop et al. (1989) determined the method of extraction of phenol from water with supercritical carbon dioxide, while supercritical carbon dioxide has been found to be a good solvent for the extraction of flavor from milk fat (Haan et al., 1990).

Supercritical-CO_2 was used by Baldino et al. (2018) for extraction of rotenoids from *Derris elliptica* roots. The final product was obtained with a concentration of 93% w/w of rotenone and rotenoids. A very minute quantity of wax was also found (up to 0.05% w/w).

Patil et al. (2018) used supercritical carbon dioxide as a clean and green solvent for extraction of bio-oil/lipids from algae (*Nannochloropsis salina*).

It was observed that lipid extraction efficiency was improved and extraction energy consumption was reduced on a microwave pretreatment before supercritical extraction.

Ouédraogo et al. (2018) extracted flavonoids from *Odontonema strictum* leaves using supercritical carbon dioxide with ethanol. It was revealed that antioxidant activity of the extracts was found to be 49.21% under optimum conditions, but conventional solvent (CSE) extract was 37.05% only.

Rosa rugosa Thunb. seed oil (RR) was extracted by Grajzer et al. (2021) using supercritical CO_2. Higher antioxidant activity of RR was there (2.1 mM kg^{-1} Trolox equivalent) and good oxidative stability.

xii) Catalysis

McCarthy et al. (2002) reported a new procedure for multiphase catalysis, where the organometallic catalyst is immobilized in $scCO_2$ phase and water is used as the mobile phase for polar substrates and products. In the catalyst Ph-SBA-15-PPh_3-Pd, the Pd species were anchored inside the mesoporous material, and act as nanoreactors for Suzuki reaction in supercritical carbon dioxide (Feng et al., 2010). Ciftci and Saldaña (2012) studied enzymatic synthesis of phenolic lipids from flaxseed oil and ferulic acid in supercritical carbon dioxide media using immobilized lipase from *Candida Antarctica.*

Enzymatic synthesis of phenolic lipids from flaxseed oil and ferulic acid was also studied in supercritical carbon dioxide media using immobilized lipase from *Candida Antarctica* (Ghoreishi and Heidari, 2012). Enantioselective enzymatic hydrolysis of benzoyl benzoin catalyzed by *Candida cylindracea* (CCL) lipase was carried out in supercritical carbon dioxide (Celebi et al., 2007), while the lipase catalyzed butanolysis of triolein was carried out in an ionic liquid and selective extraction of product was done using supercritical carbon dioxide by Miyawaki and Tatsuno (2008).

xiii) Solubility

Kazemi et al. (2012) measured the solubilities of ferrocene and acetylferrocene in supercritical carbon dioxide using an analytical method in a quasi-flow apparatus, while the solubilities of benzene derivatives in supercritical carbon dioxide were also determined by the saturation method over the pressure range (9.5 to 14.5) MPa (Reddy and Madras, 2011). Marceneiro et al. (2011) measured the solid solubilities of 1,4-naphthoquinone and 5-hydroxy-2-methyl-1,4-naphthoquinone (also known as plumbagin) in supercritical carbon dioxide using a static analytical method at 308.2, 318.2, and 328.2 K, and pressures between 9.1 and 24.3 MPa.

Rajasekhar et al. (2010) used the saturation method for the determination of the solubility of a drug, *n*-(4-ethoxyphenyl)ethanamide (phenacetin), in supercritical carbon dioxide at 308, 318, and 328 K and 9 to 19 MPa. Solid drug's solubility in supercritical fluids (SCFs) was also experimentally determined by Hojjati et al. (2007). Lee et al. (1999) measured the solubility of disperse dyes in supercritical carbon dioxide An enhancement in solubility of poly(vinyl ester) in supercritical carbon dioxide can be achieved by decreasing the strength of the polymer–polymer interactions (Girard et al., 2012). Solubility measurements of zopiclone and nimodipine in supercritical carbon dioxide were also reported by Medina and Bueno (2001).

xiv) Synthesis

Rohr et al. (2001) reported the solvent-free synthesis of β-[(diethylcarbamoyl) oxy]styrene from phenylacetylene and diethylamine in supercritical carbon dioxide is greatly accelerated for a series of ruthenium catalysts compared to the same reaction in toluene. The O_2 and the mixed solvent (MeOH–supercritical carbon dioxide) can control the chemoselectivity of the palladium catalyzed carbonylation of amines. Du et al. (2005) reported organic solvent-free process for the synthesis of propylene carbonate from supercritical carbon dioxide and propylene oxide catalyzed by insoluble ion exchange resins. The process represents a simple, ecologically safer, and cost-effective route to cyclic carbonates with high product quality, as well as an easy product recovery and catalyst recycling.

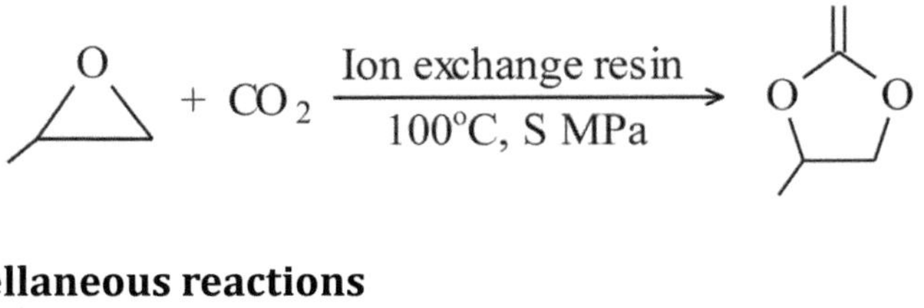

xv) Miscellaneous reactions

Besides these reactions, there are a number of reactions, which have been reported in $scCO_2$ with good yields and faster speed, for example, Pausom-Khand reaction (Jeong et al., 1997), Baylis-Hillman reaction (Rose et al., 2002), hydroboration of styrene (Carter et al., 2000), carbamate synthesis (Yoshida et al., 2000), 2-pyrone synthesis (Reetz et al., 1993), and polymerization reactions (Beckman, 2004).

It can now be concluded that $scCO_2$ is environmentally benign medium. It has been used in many reactions for example hydrogenation, hydroformylation, coupling reactions, oxidation reaction, photochemical reaction, as well as for stereochemical control in reaction (Luzzio, 2001).

α-Olefins undergo highly selective self-metathesis catalyzed by supported Re-oxide in the presence of supercritical carbon dioxide (Selva et al., 2009).

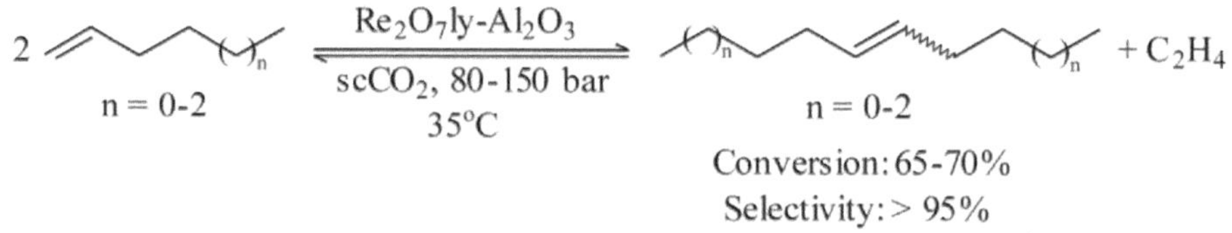

The oxybromination of aniline and phenol derivatives was performed by Ganchegui and Leitner (2007) with high conversions and selectivities in $H_2O/scCO_2$ exploiting the intrinsic reactivity of the biphasic system.

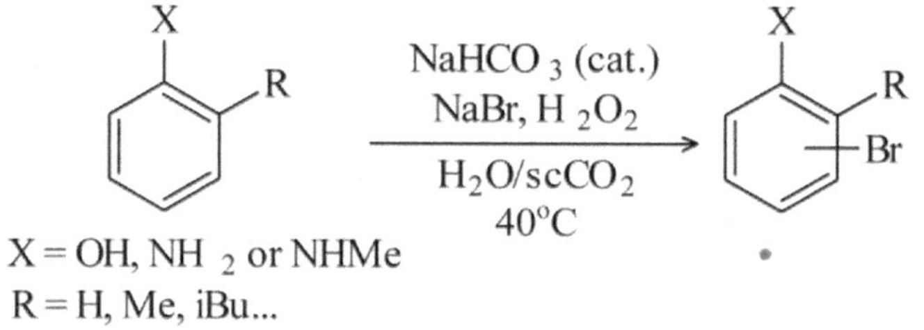

Sulfur-cured natural rubber was devulcanized by using supercritical CO_2 as a swelling solvent (Kojima et al., 2004). Wacker reaction was carried out smoothly in supercritical carbon dioxide or ROH/supercritical carbon dioxide. The results show that both $scCO_2$ and co-solvent can remarkably affect the selectivity toward methyl ketone and the presence of ROH accelerates this reaction (Jiang et al., 2000). Viguera et al. (2013) have used supercritical carbon dioxide for the removal of lubricating oils from metallic contacts, while Weinstein et al. (2005) explored the use of liquid and supercritical carbon dioxide for the blending of poly(vinyl acetate) and citric acid.

Methanol-wetted zinc borates produced either from borax decahydrate and zinc nitrate hexahydrate ($2ZnO \cdot 3B_2O_3 \cdot 7H_2O$) or from zinc oxide and boric acid ($2ZnO \cdot 3B_2O_3 \cdot 3H_2O$) were dried by both conventional and supercritical carbon dioxide drying methods (Gonen et al., 2009). Pizarro et al. (2009) have determined the binary diffusion coefficients of 2-ethyltoluene, 3-ethyltoluene, and 4-ethyltoluene in supercritical carbon in a pressure range of 15.0–35.0 MPa and temperatures 313, 323, and 333 K by means of the Taylor–Aris dispersion technique. Moderately thermophilic bacterial strain CC-HSB-11[T] (*Muricauda lutaonensis*) has been identified to produce zeaxanthin as a predominant xanthophyll by liquid chromatography−tandem mass spectrometry (LC−MS/MS). Micronization of zeaxanthin was achieved through the supercritical carbon dioxide antisolvent precipitation method (Hameed et al., 2011).

Yuan et al. (2003) reported detoxification of aryl-organochlorine compounds by catalytic reduction in supercritical carbon dioxide. An automated continuous flow reactor has been built and used to study acid catalyzed etherification reactions in $scCO_2$ (Walsh et al., 2005). Cid et al. (2005) carried out an excellent dye fixation on cotton in supercritical carbon dioxide using fluorotriazine reactive dyes. As a result, water usage and dye containing waste are eliminated and energy consumption can also be reduced. Catalytic and selective hydroxylation of phenol was carried out by Baldi et al. (2010) under environmentally benign reaction conditions in the presence of a Fe(III)-EPS (EPS = exopolysaccharide) catalyst. A novel method was developed by Long et al. (2012) for the dyeing of cotton fabric with a vinylsulfone reactive dispersed dye in supercritical carbon dioxide.

Xie et al. (2012) have reported a new method of postmodifying the particle size and morphology of $LiFePO_4$ via supercritical carbon dioxide. Enantiodifferentiating photocyclization of 5-hydroxy-1,1-diphenyl-1-pentene sensitized by bis(1,2,4,5-di-*O*-isopropylidene-α-fructopyranosyl) 1,4-naphthalenedicarboxylate was performed in near critical and supercritical carbon dioxide media containing organic entrainers to obtain a chiral tetrahydrofuran derivative in enantiomeric excess (ee) (Nishiyama et al., 2012).

Zinc sulfide nanoparticles have been deposited on carbon nanotubes (CNTs) from a single-source diethyldithiocarbamate precurse (Casciato et al., 2012), while nanoporous silica materials have been prepared using an activated carbon as a mold and supercritical carbon dioxide as a solvent by Wakayama and Fukushima (2000). Kuo et al. (2011) investigated anti-inflammatory effects of supercritical carbon dioxide extract and its isolated carnosic acid from *Rosmarinus officinalis* leaves. Colloidal carbon spheres were synthesized by the carbonization of squalane, a nonvolatile hydrocarbon solvent in supercritical carbon dioxide (Barrett et al., 2011). Household steel wool was used as a reducing agent for the reductive dechlorination of polychlorinated biphenyls (PCBs) in supercritical carbon dioxide (Chen et al., 2012).

Sun et al. (2014) reported a novel coal extraction method with high efficiency, separability, diffusibility, and environmental friendliness using supercritical carbon dioxide (scCO2)/1-methyl-2-pyrrolidone (NMP) mixed solvent. They used eight bituminous coals, all with higher extraction yields, but bituminous coal with 85.12% carbon content gave observed extraction yield.

The production of eugenyl acetate was investigated by Santos et al. (2016) via esterification of eugenol and acetic anhydride in $scCO_2$. They used two commercial lipases (Novozym 435 and Lipozyme 435) as catalysts. It was reported that the use of Novozym 435 could achieve higher conversion and specific productivity of eugenyl acetate as compared to Lipozyme 435.

The effectiveness of supercritical carbon dioxide for the recovery of cobalt in lithium-ion batteries (LIBs) was evaluated by Bertuol et al. (2016). It was reported that extraction of more than 95 wt.% of the cobalt was possible with supercritical CO_2 in 60 min.

Zheng et al. (2017) investigated dyeing feasibility of wool with different dyes such as disperse red 54, disperse yellow 82, disperse red 92, disperse blue 148, disperse red 153, and disperse blue 183 in supercritical carbon dioxide. Higher fixation and K/S values rates were obtained with two disperse dyes (disperse blue 148 and dispersed red 153) than that of other disperse dyes. These were also found to improved gradually from 70 to 110°C, 15 to 25 MPa, 60 to 120 min, as well as mesh number (40 to 100) for disperse red 153, and for disperse blue 148 from 80 to 120°C, 15 to 25 MPa, 60 to 120 min, as well as mesh number (40 to 100) during the dyeing procedure using supercritical carbon dioxide. They could obtain excellent permeability and color fastness along with slightly reduction in tensile strength and breaking force of dyed wool fibers.

A new approach was developed by Hu et al. (2021) making use of supercritical carbon dioxide to anchor Fe_3O_4 nanoparticles (NPs) efficiently on graphene foam (GF) without using any surfactants. They obtained moderately spaced Fe_3O_4 NPs arrays on the surface of GF. The particle size of Fe_3O_4 NPs exhibited a narrow distribution (11 ± 4 nm in diameter) and this composite could deliver a high capacity (1200 mAh g^{-1}) up to 500 cycles at 1 C. It reaches higher value Fe_3O_4, which was used as anode material for lithium-ion batteries.

Golzary and Abdoli (2020) reported copper recovery from printed circuit boards (PCBs) waste using supercritical CO_2 extraction. Maximum extraction efficiency could reach 97% with high purity (98.7%) at 60°C in 30 min., and 1 m L min^{-1} CO_2 flow rate.

Solvents were always a problematic issue for organic synthesis and there was a regular search for an alternate for organic solvents. Water and carbon dioxide can serve as solvents, but in critical conditions only. They are called supercritical fluids. Many organic syntheses have been already worked out in these solvents and many more are feasible in the future.

6.4 RECENT DEVELOPMENTS

Liao et al. (2022) developed a green method for the synthesis of Pd nanoparticles/graphene composites using a choline chloride–oxalic acid deep eutectic solvent (DES). They did not use any surfactant or reducing agent. The DES are normally composed of hydrogen-bond donors and halide salts, and many of them are biodegradable and biocompatible. It was reported that size and dispersion of Pd particles was found to be improved, when supercritical carbon dioxide ($scCO_2$) was used as it has gaslike near-zero surface tension and diffusivity. It was revealed that as-prepared sc-Pd NPs/GR/SPCE exhibited excellent activity toward oxidation of glycerol as compared to the composites, which were not fabricated by $scCO_2$ processes.

Conceição et al. (2022) reported the preparation of Zn(II) coordination polymer $[Zn(L1)(NMeF)]_n \cdot n(NMeF)$ (Zn-CP **1**) via the solvothermal reaction of $Zn(NO_3)_2 \cdot 6H_2O$ with 5-{(pyren-4-ylmethyl)amino}isophthalic acid (H_2L1) in *N*-methylformamide (NMeF). They also evaluated its potential as a catalyst in Knoevenagel condensation of malononitrile and benzaldehyde in $scCO_2$ medium under mild conditions (Figure 6.1). It was observed that yield of reaction increases as aprotic (tetrahydrofurane) to protic (ethanol and water) polar co-solvents. Full conversion could be achieved in case of water. It was reported that $scCO_2$ is not the most suitable medium for this reaction in the absence of a protic co-solvent. It was also revealed that this catalyst can be recycled without any significant loss of activity.

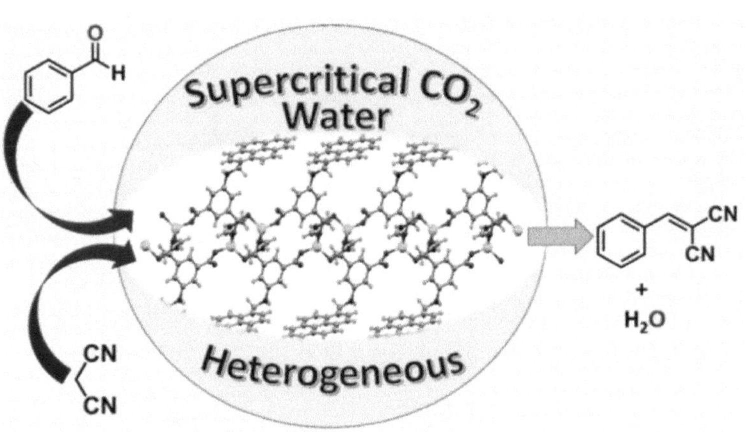

FIGURE 6.1 Knoevenagel condensation in $scCO_2$.

Source: Reprinted with permission from Conceição et al., 2022. © 2022 Elsevier.

López et al. (2022) evaluated different homogeneous catalysts (copper bromide, and copper(II) acetate) and heterogeneous catalysts (Cu wire, pre-treated Cu wire, Cu plate, pre-treated Cu Plate and Cu/β-SiC) for alkyne–azide cycloaddition reaction. Toluene was used as an organic solvent while supercritical CO_2 ($scCO_2$) was used as green solvent. The effect of various parameters such as reaction time, catalyst loading, reusability and leaching of the catalysts were studied to achieve optional conditions for this reaction in $scCO_2$. It was observed that use of pre-treated copper plate revealed 57% increase in reaction yield as compared to the non–pre-treated copper plate. It was also revealed that Cu Wire (without pre-treatment) yielded 94.2% product even on reusing it for five more consecutive cycles.

The e-waste is a rich source of polymers, metals, ceramics, and glass fibers. Preetam et al. (2022) investigated a subcritical to supercritical methanol environment to pretreat e-waste and they recovered nonmetallic fractions as well as concentrate metals from e-waste. They carried out experiments at 150 to 300°C with initial atmospheric pressure. They studied depolymerization by changing reaction time from 30 min to 240 min and solid to liquid ratio between 1:10 and 1:30 g mL^{-1} in the presence of N_2 environment. It was reported that high metal concentrations (> 90%) could be obtained after supercritical treatment such as nickel, copper, gold, zinc, and silver. A greener and efficient approach using supercritical solvent has been proposed for this e-waste recycling (Figure 6.2).

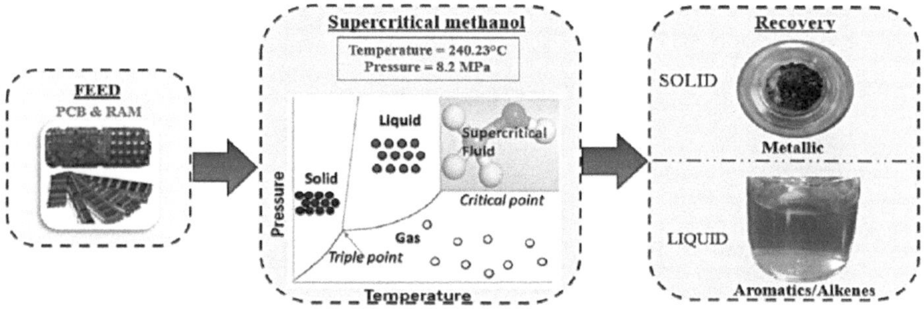

FIGURE 6.2 *Recovery of metals and nonmetallic fractions by treatment of e-waste.*

Source: Reprinted with permission from Preetam et al., 2022. © 2022 Elsevier.

Demirkol et al. (2022) carried out extraction of coumarin, alkaloids, and furanocoumarin from *Ruta chalepensis* roots. They used supercritical CO_2, subcritical water, and subcritical ethyl alcohol. The effect of operating conditions for sub- and supercritical extractions was determined by varying the extraction temperature (100–160°C) and extraction pressure (60 and 80 atm). It was observed that maximum extraction efficiency (40.6%) was there, when subcritical water was used. They identified five classes of components in these extractions, out of which furanocoumarins were the most abundant components with 31.0% yield with subcritical water as the solvent.

Pérez et al. (2022) depolymerized lignins from hardwoods, softwoods, or crop residues and obtained by-products of various processes such as biorefining or pulping in alkaline supercritical water. They could get high value-added compounds with low molecular weight in only 300 ms. It was observed that char formation was lower in most of the cases, but the proportion of light and heavy oils were dependent on starting material used.

KEYWORDS

- **supercritical fluids**
- **supercritical water**
- **supercritical carbon dioxide**
- **supercritical polymerization**

REFERENCES

Abbett, K. F.; Teja, A. S.; Kowalik, J.; Tolbert, L. Polymerization of 3-Undecylbithiophene and Preparation of Poly (3-Undecylbithiophene)/Polystyrene Composites in Supercritical Carbon Dioxide. *Macromole.* **2003,** *36* (9) 3015–3019.

Akgun, I. H.; Erkucuk, A.; Pilavtepe, M.; Celiktas, O. Y. Optimization of Total Alkannin Yields of Alkanna Tinctoria by Using Sub-and Supercritical Carbon Dioxide Extraction. *J. Supercrit. Fluids* **2011,** *57* (1), 31–37.

Akiya, N.; Savage, P. E. Roles of Water for Chemical Reactions in High-Temperature Water. *Chem. Rev.* **2002,** *102* (8), 2725–2750.

Amandi, R.; Hyde, J. R.; Ross, S. K; Lotz, T. J.; Poliakoff, M. Continuous Reactions in Supercritical Fluids: A Cleaner, More Selective Synthesis of Thymol in Supercritical CO_2. *Green Chem.* **2005,** *7*, 288–293.

Amandi, R.; Scovell, K.; Licence, P.; Lotz, T. J.; Poliakoff, M. The Synthesis of O-Cyclohexylphenol in Supercritical Carbon Dioxide: Towards a Continuous Two-Step Reaction. *Green Chem.* **2007,** *9*, 797–801.

An, J.; Bagnell, L.; Cablewski, T.; Strauss, C. R.; Trainor, R. W. Applications of High-Temperature Aqueous Media for Synthetic Organic Reactions. *J. Org. Chem.* **1997,** *62*, 2505–2511.

Anatas, P. T.; Warner, P. T.; Warner, J. C. *Green Chemistry, Theory and Practice*; Oxford University Press: Oxford, 1998.

Andersson, M. B. O.; King, J. W.; Blomberg, L. G. Synthesis of Fatty Alcohol Mixtures from Oleochemicals in Supercritical Fluids. *Green Chem.* **2000,** *2*, 230–234.

Baiker, A. Supercritical Fluids in Heterogeneous Catalysis. *Chem. Rev.* **1999,** *99* (2), 453–473.

Baldi, F.; Marchetto, D.; Zanchettin, D.; Sartorato, E.; Paganelli, S.; Piccolo, O. A Bio-Generated Fe (III)-Binding Exopolysaccharide Used as New Catalyst for Phenol Hydroxylation. *Green Chem.* **2010,** *12*, 1405–1409.

Baldino, L.; Scognamiglio, M.; Reverchon, E. Extraction of Rotenoids from Derris Elliptica Using Supercritical CO_2. *J. Chem. Technol. Biotechnol.* **2018,** *93* (12), 3656–3660.

Barrett, C. A.; Singh, A.; Murphy, J. A.; O'Sullivan, C.; Buckley, D. N.; Ryan, K. M. Complete Synthesis of Germanium Nanocrystal Encrusted Carbon Colloids in Supercritical CO_2 and Their Superhydrophobic Properties. *Langmuir* **2011,** *27* (17), 11166–11173.

Beckman, E. J. Supercritical and Near-Critical CO_2 in Green Chemical Synthesis and Processing. *J. Supercrit. Fluids* **2004,** *28* (2–3), 121–191.

Bertuol, D. A.; Machado, C. M.; Silva, M. L.; Calgaro, C. O.; Dotto, G. L.; Tanabe, E. H. Recovery of Cobalt from Spent Lithium-Ion Batteries Using Supercritical Carbon Dioxide Extraction. *Waste Manag.* **2016,** *51*, 245–251.

Bhanage, B. M.; Ikushima, Y.; Shirai, M.; Arai, M. Heck Reactions Using Water-Soluble Metal Complexes in Supercritical Carbon Dioxide. *Tetrahedron Lett.* **1999,** *40* (35), 6427–6430.

Broli, D.; Kaul, C.; Kramer, A.; Krammer, P.; Richter, T.; Jung, M. et al. Chemistry in Supercritical Water. *Angew. Chem. Int. Ed.* **1999,** *38* (20), 2998–3014.

Burk, M. J.; Feng, S.; Gross, M. F.; Tumas, W. Asymmetric Catalytic Hydrogenation Reactions in Supercritical Carbon Dioxide. *J. Am. Chem. Soc.* **1995,** *117*, 8277–8278.

Cantero, D.; Bermejo, M.; Cocero, M. Governing Chemistry of Cellulose Hydrolysis in Supercritical Water. *ChemSusChem*, **2015,** *8* (6), 1026–1033.

Carlsson, M.; Habenicht, C.; Kam, L. C.; Antal, M.; Bian, N.; Cunninghom, R. J. et al. Study of the Sequential Conversion of Citric to Itaconic to Methacrylic Acid In Near-Critical and Supercritical Water. *Ind. Eng. Chem. Res.* **1994,** *33* (8), 1989–1996.

Carter, C. A. G.; Baker, R. T.; Nolan, S. P.; Tumas, W. Enhanced Regioselectivity of Rhodium-Catalysed Alkenehydroboration in Supercritical Carbon Dioxide. *Chem. Commun.* **2000,** *5*, 347–348.

Casciato, M. J.; Levitin, G.; Hess, D. W.; Grover, M. A. Synthesis of Optically Active ZnS–Carbon Nanotube Nanocomposites in Supercritical Carbon Dioxide via a Single Source Diethyldithiocarbamate Precursor. *Ind. Eng. Chem. Res.* **2012,** *51*, 11710–11716.

Celebi, N.; Yildiz, N.; Demir, A. S.; Calimli A. Enzymatic Synthesis of Benzoin in Supercritical Carbon Dioxide. *J. Supercrit. Fluids* **2007,** *41* (3), 386–390.

Cengiz, U.; Gengec, N. A.; Kaya, N. U.; Erbil, H. Y.; Sarac, A. S. Mechanical and Thermal Properties of Perfluoroalkyl Ethyl Methacrylate–Methyl Methacrylate Statistical Copolymers Synthesized in Supercritical Carbon Dioxide. *J. Fluor. Chem.* **2011,** *132* (5), 348–355.

Chakinala, A.; Kumar, S.; Kruse, A.; Kersten, S.; van Swaaij, W.; Brilman, D. Supercritical Water Gasification of Organic Acids and Alcohols: The Effect of Chain Length. *J. Supercrit. Fluids*, **2013,** *74*, 8–21.

Chapman, A. O.; Akien, G. R.; Arrowsmith, N. J.; Licence, P.; Poliakoff, M. Continuous Heterogeneous Catalytic Oxidation of Primary and Secondary Alcohols in $scCO_2$. *Green Chem.* **2010,** *12*, 310–315.

Chatterjee, M.; Chatterjee, A.; Raveendran, P.; Ikushima, Y. Hydrogenation of Citral in Supercritical CO_2 Using a Heterogeneous Ni (II) Catalyst. *Green Chem.* **2006,** *8*, 445–449.

Chatterjee, M.; Kawanami, H.; Sato, M.; Ishizaka, T.; Yokoyama, T.; Suzuki, T. Production of Linear Alkanevia Hydrogenative Ring Opening of a Furfural-Derived Compound in Supercritical Carbon Dioxide. *Green Chem.* **2010,** *12*, 87–93.

Chen, Y.; Liao, W.; Yak, H. Complete Reduction of Polychlorinated Biphenyls by Acid-Etched Steel Wool in Supercritical Carbon Dioxide. *Ind. Eng. Chem. Res.* **2012,** *51*, 6625–6630.

Cid, M. V. F.; Spronsen, J. V.; Kraan, M. V. D.; Veugelers, W. J. T.; Woerlee, G. F.; Witkamp, G. J. Excellent Dye Fixation on Cotton Dyed in Supercritical Carbon Dioxide Using Fluorotriazine Reactive Dyes. *Green Chem.* **2005,** *7*, 609–616.

Ciftci, D.; Saldaña, M. D. A. Enzymatic Synthesis of Phenolic Lipids Using Flaxseed Oil and Ferulic Acid in Supercritical Carbon Dioxide Media. *J. Supercrit. Fluids* **2012,** *72*, 255–262.

Čolnik, M.; Knez, Ž.; Škerget, M. Sub-and Supercritical Water for Chemical Recycling of Polyethylene Terephthalate Waste. *Chem. Eng. Sci.*, **2021,** *233*. DOI: 10.1016/j.ces.2020.116389.

Conceição, N. R.; Nobre, B. P.; Karmakar, A.; Palavra, A. M.; Mahmudov, K. T.; da Silva, M. F. C. G. et al. Knoevenagel Condensation Reaction in Supercritical Carbon Dioxide Medium Using a Zn (II) Coordination Polymer As Catalyst. *Inorganica Chimica Acta* **2022,** *538*. DOI: 10.1016/j.ica.2022.120981.

Cooper, A. L. Polymer Synthesis and Processing Using Supercritical Carbon Dioxide. *J. Mater. Chem.* **2000,** *10*, 207. DOI: 10.1039/A906486I

Cott, D. J.; Ziegler, K. J.; Owens, V. P.; Glennon, J. D.; Graham, A. E.; Holmes, J. D. Diels–Alder Reactions Between Maleic Anhydride and Furan Derivatives in Supercritical CO_2. *Green Chem.* **2005,** *7*, 105–110.

Crittendon, R. C.; Parsons, E. J. Transformations of Cyclohexane Derivatives in Supercritical Water. *Organometallics* **1994,** *13* (7), 2587–2591.

Demirkol, O.; Erşatır, M.; Giray, E. S.; Kırıcı, S. Comparison of the Effects of Green and Sustainable Extraction Methods on the Extraction Yield and Chemical Composition of Ruta Chalepensis Roots. *Sustain. Chem. Pharm. 29*, **2022. DOI:** 10.1016/j.scp.2022.100750.

Diminnie, J.; Metts, S.; Parsons, E. J. In Situ Generation and Heck Coupling of Alkenes in Superheated Water. *Organometallics* **1995,** *14* (8), 4023–4025.

Du, Y.; Cai, F.; Kong, D.; He, L. Organic Solvent-Free Process for the Synthesis of Propylene Carbonate from Supercritical Carbon Dioxide and Propylene Oxide Catalyzed By Insoluble Ion Exchange Resins. *Green Chem.* **2005,** *7*, 518–523.

Feng, X.; Yan, M.; Zhang, T.; Liu, Y.; Bao, M., Preparation and Application of SBA-15-Supported Palladium Catalyst for SUZUKI REACTION in Supercritical Carbon Dioxide. *Green Chem.* **2010,** *12*, 1758–1766.

Freitas, A.; Guirardello, R. Use of CO_2 as a Co-Reactant to Promote Syngas Production in Supercritical Water Gasification of Sugarcane Bagasse. *J. CO_2 Util.* **2015,** *9*, 66–73.

Ganchegui, B.; Leitner, W. Oxybromination of Phenol and Aniline Derivatives in $H_2O/scCO_2$ Biphasic Media. *Green Chem.* **2007,** *9*, 26–29.

Ghaziaskar, H. S.; Daneshfar, A.; Calvo, L. Continuous Esterification or Dehydration in Supercritical Carbon Dioxide. *Green Chem.* **2006,** *8*, 576–581.

Ghoreishi, S. M.; Heidari, E. Extraction of Epigallocatechin Gallate from Green Tea via Modified Supercritical CO_2: Experimental, Modeling and Optimization. *J. Supercrit. Fluids* **2012,** *72*, 36–45.

Girard, E.; Tassaing, T.; Camy, S.; Condoret, J.; Marty, J.; Destarac, M. Enhancement of Poly (Vinyl Ester) Solubility in Supercritical CO_2 by Partial Fluorination: The Key Role of Polymer–Polymer Interactions. *J. Am. Chem. Soc.* **2012,** *134* (29), 11920–11923.

Golzary, A.; Abdoli, M. A. Recycling of Copper from Waste Printed Circuit Boards by Modified Supercritical Carbon Dioxide Combined with Supercritical Water Pre-Treatment. *J. CO_2 Util.* **2020,** *41*. DOI: 10.1016/j.jcou.2020.101265

Gönen, M.; Balköse, D.; Gupta, R. B.; Ülkü S. Supercritical Carbon Dioxide Drying of Methanol–Zinc Borate Mixtures. *Ind. Eng. Chem. Res.* **2009,** *48* (14), 6869–6876.

Gore, G. *II.* On the Properties of Liquid Carbonic Acid. *Proc. R. Soc. London* **1861,** *11*, 85–86.

Grajzer, M.; Wiatrak, B.; Gębarowski, T.; Matkowski, A.; Grajeta, H.; Rój, E. et al. Chemistry, Oxidative Stability and Bioactivity of Oil Extracted from Rosa rugosa (Thunb.) Seeds by Supercritical Carbon Dioxide. *Food Chem.* **2021,** *335*. DOI: 10.1016/j.foodchem.2020.127649

Guo, S.; Guo, L.; Yin, J.; Jin, H. Supercritical Water Gasification of Glycerol: Intermediates and Kinetics. *J. Supercrit. Fluids* **2013,** *78*, 95–102.

Guo, Y.; Wang, S.; Huelsman, C.; Savage, P. Kinetic Model for Reactions of Indole Under Supercritical Water Gasification Conditions. *Chem. Eng. J.* **2014,** *241*, 327–335.

Haan, A. B. de; Graauw, J. de; Schaap, J. E.; Badings, H. T. Extraction of Flavors from Milk Fat with Supercritical Carbon Dioxide. *J. Supercrit. Fluids* **1990,** *3* (1), 15–19.

Hameed, A.; Arun, A. B.; Ho, H.-P.; Chang, C.-M. J.; Rekha, P. D.; Maw- Lee, R. et al. C.-C. Supercritical Carbon Dioxide Micronization of Zeaxanthin from Moderately Thermophilic Bacteria Muricauda Lutaonensis CC-HSB-11T. *J. Agric. Food Chem.* **2011,** *59* (8), 4119–4124.

Hawthrone, S. B.; Yang, Y.; Miller, D. Extraction of Organic Pollutants from Environmental Solids with Sub-and Supercritical Water. J. *Anal. Chem.* **1994,** *66* (18), 2912–2920.

Hojjati, M., Yamini, Y.; Khajeh, M.; Vatanara, A. Solubility of Some Statin Drugs in Supercritical Carbon Dioxide and Representing the Solute Solubility Data with Several Density-Based Correlations. *J. Supercrit. Fluids* **2007,** *41* (2), 187–194.

Hu, X.; Ma, M.; Zeng, M.; Sun, Y.; Chen, L.; Xue, Y. et al. Supercritical Carbon Dioxide Anchored Fe_3O_4 Nanoparticles on Graphene Foam and Lithium Battery Performance. *ACS Appl. Mater. Interf.* **2021,** *6* (24), 22527–22533. doi:10.1021/am5066255

Itami, K.; Terakawa, K.; Yoshida, J. I.; Kajimoto, O. Efficient and Rapid C–Si Bond Cleavage in Supercritical Water. *J. Am. Chem. Soc.* **2003,** *125* (20), 6058–6059.

Jeong, N.; Hwang, S. H.; Lee, Y. W.; Lim, J. S., Catalytic Pauson–Khand Reaction in Super Critical Fluids. *J. Am. Chem. Soc.* **1997,** *119* (43), 10549–10550.

Jerner, G.; Sueur, D.; Mallat, T.; Baiker, A. Partial Oxidation of Alcohols in Supercritical Carbon Dioxide. *Chem. Commun.* **2000,** 2247–2248.

Jessop, P. G.; Ikaviya, T.; Noyori, R. Homogeneous Catalysis in Supercritical Fluids. *Chem. Rev.* **1999a,** *99*, 475–493.

Jessop, P. G.; Leitner, W. *Chemical Synthesis Using Supercritical Fluids*; Wiley-VCH: Washington, 1999b.

Jiang, H.; Jia, L.; Li, J. Wacker Reaction in Supercritical Carbon Dioxide. *Green Chem.* **2000,** *2*, 161–164.

Johnston, K. P. Supercritical Fluid Science and Technology; Johnston, K. P., Penninger, J. M. L. (Eds.), *American Chem. Soc. Washington, DC*, **1989,** *109* (11), 1–12.

Johnston, K. P.; Harrison, K. I.; Clarke, M. J.; Howdle, S. M.; Heitz, M. P.; Bright, F. V. et al. Water-in-Carbon Dioxide Microemulsions: An Environment for Hydrophiles Including Proteins. *Sciences* **1996,** *271* (5249), 624–626.

Kabir, A.; Marshall, W. D. Dechlorination of Pentachlorophenol in Supercritical Carbon Dioxide with a Zero-Valent Silver–Iron Bimetallic Mixture. *Green Chem.* **2001,** *3*, 47–51.

Kajmoto, O. Solvation in Supercritical Fluids: Its Effects on Energy Transfer and Chemical Reactions. *Chem. Rev.* **1999,** *99* (2), 355–390.

Kazemi, S.; Belandria, V.; Janssen, N.; Richon, D.; Peters, C. J.; Kroon, M. C. Solubilities of Ferrocene and Acetylferrocene in Supercritical Carbon Dioxide. *J. Supercrit. Fluids* **2012,** *72*, 320–325.

Kiriamiti, H. K.; Rascol, E.; Marty, A.; Condoret, J. S. Extraction Rates of Oil from High Oleic Sunflower Seeds with Supercritical Carbon Dioxide. *Chem. Eng. Proce.* **2002,** *41* (8), 711–718.

Kojima, M.; Tosaka, M.; Ikeda, Y. Chemical Recycling of Sulfur-Cured Natural Rubber Using Supercritical Carbon Dioxide. *Green Chem.* **2004,** *6*, 84–89.

Korzenski, M. B.; Kolis, J. W. Diels-Alder Reactions Using Supercritical Water as an Aqueous Solvent Medium. *Tetrahedron Lett.* **1997,** *38* (32), 5611–5614.

Kremsner, J. M.; Kappe, C. O. Microwave-Assisted Organic Synthesis in Near-Critical Water at 300°C–A Proof-of-Concept Study. *Eur. J. Org. Chem.* **2005,** *2005* (17), 3672–3679.

Kuhlmann, B.; Arnett, E. M.; Siskin, M. HD Exchange in Pinacolone by Deuterium Oxide at High Temperature and Pressure. *J. Org. Chem.* **1994a,** *59* (18), 5377–5380.

Kuhlmann, B.; Arnett, E. M.; Siskin, M. Classical Organic Reactions in Pure Superheated Water. *J. Org. Chem.* **1994b,** *59* (11), 3098–3101.

Kuo, C-F.; Su, J-D.; Chiu, C-H.; Peng, C-C.; Chang, C-H.; Sung, T-Y. et al. Anti-Inflammatory Effects of Supercritical Carbon Dioxide Extract and Its Isolated Carnosic Acid from Rosmarinus Officinalis Leaves. *J. Agric. Food Chem.* **2011,** *59* (8), 3674–3685.

Lee, J. W.; Min, J. M.; Bae, H. K. Solubility Measurement of Disperse Dyes in Supercritical Carbon Dioxide. *J. Chem. Eng. Data* **1999,** *44* (4), 684–687.

Lee, S. Y.; Valtchev, P.; Dehghani, F. Synthesis and Purification of Poly (L-Lactic Acid) Using a One Step Benign Process. *Green Chem.* **2012,** *14*, 1357–1366.

Li, G., Jiang, H., Li, J. Use of Water as a Direct Hydrogen Donor in Supercritical Carbon Dioxide: A Novel and Efficient Zn–H_2O–CO_2 System for Selective Reduction of Aldehydes to Alcohols. *Green Chem.* **2001,** *3* (5), 250–251.

Li, J.; Jiang, H.; Chen, M. Respective Conversion of N-Butylamine to Methyl Nn-Butylcarbamate and Oxalbutyline in Supercritical Carbon Dioxide. *Green Chem.* **2001,** *3*, 137–139.

Liao, C. H.; Chen, J. Y.; Liu, G. Y.; Xu, Z. R.; Lee, S.; Chiang, C. K. et al. Supercritical Fluid-Assisted Fabrication of Pd Nanoparticles/Graphene Using a Choline Chloride–Oxalic Acid Deep Eutectic Solvent for Enhancing the Electrochemical Oxidation of Glycerol. *ACS Omega*. **2022,** *7* (23), 19930–19938.

Long, J.; Xiao, G.; Xu, H.; Wang, L.; Cui, C.; Liu, J. et al. Dyeing of Cotton Fabric with a Reactive Disperse Dye in Supercritical Carbon Dioxide. *J. Supercrit. Fluids* **2012,** *69*, 13–20.

López, S.; García-Vargas, J. M.; García, M. T.; Rodríguez, J. F.; Gracia, I.; Ramos, M. J. Copper-Containing Catalysts for Azide–Alkyne Cycloaddition in Supercritical CO_2. *Catalysts*, **2022,** *12* (2). DOI: 10.3390/catal12020194.

Luzzio, F. The Henry Reaction: Recent Examples. *Tetrahedron* **2001,** *57*, 915–945.

Machmudah, S.; Kawahito, Y.; Sasaki, M.; Goto, M. Process Optimization and Extraction Rate Analysis of Carotenoids Extraction from Rosehip Fruit Using Supercritical CO_2. *J. Supercrit. Fluids* **2008,** *44* (3), 308–314.

Marceneiro, S.; Braga, M. E. M.; Dias, A. M. A.; Sousa, H. C. De. Measurement and Correlation of 1, 4-Naphthoquinone and of Plumbagin Solubilities in Supercritical Carbon Dioxide. *J. Chem. Eng. Data* **2011,** *56* (11), 4173–4182.

Matsuo, J.; Tsuchige, T.; Odashima, K.; Kobayashi, S. Lewis Acid Catalysis in Supercritical Carbon Dioxide. Use of Scandium Tris (Heptadecafluorooctanesulfonate) as a Lewis Acid Catalyst in Diels-Alder and Aza Diels-Alder Reactions. *Chem. Lett.* **2000,** *29* (2), 178–179.

McCarthy, M.; Stemmer, H.; Leitner, W. Catalysis in Inverted Supercritical CO_2/Aqueous Biphasic Media. *Green Chem.* **2002,** *4*, 501–504.

Mchugh, M.; Krukonis, V. *Supercritical Fluid Extraction, Principles and Practice*; Butterworths: Boston, 1986; pp 1–11.

Medina, I.; Bueno, J. L. Solubilities of Zopiclone and Nimodipine in Supercritical Carbon Dioxide. *J. Chem. Eng. Data* **2001,** *46* (5), 1211–1214.

Mello, R.; Alcalde-Aragonés, A.; Olmos, A.; González-Núñez, M. E.; Asensio, G. Epoxidation of Olefins with a Silica-Supported Peracid in Supercritical Carbon Dioxide Under Flow Conditions. *J. Org. Chem.* **2012,** *77* (10), 4706–4710.

Mello, R.; Olmos, A.; Parra-Carbonell, J.; González-Núñez, M. E.; Asensio, G. Baeyer–Villiger Oxidation of Ketones with a Silica-Supported Peracid in Supercritical Carbon Dioxide Under Flow Conditions. *Green Chem.* **2009,** *11*, 994–999.

Melo, C. I.; Bogel-Łukasik, R.; Silva, M. G. D.; Bogel-Łukasik, E. Advantageous Heterogeneously Catalysed Hydrogenation of Carvone with Supercritical Carbon Dioxide. *Green Chem.* **2011,** *13*, 2825–2830.

Miyawaki, O.; Tatsuno, M. Lipase-Catalyzed Butanolysis of Triolein in Ionic Liquid and Selective Extraction of Product Using Supercritical Carbon Dioxide. *J. Biosci. Bioeng.* **2008,** *105* (1), 61–64.

Nishiyama, Y.; Wada, T.; Kakiuchi, K.; Inoue, Y. Entrainer Effects on Enantiodifferentiating Photocyclization of 5-Hydroxy-1, 1-Diphenylpentene in Near-Critical and Supercritical Carbon Dioxide. *J. Org. Chem.* **2012,** *77* (13), 5681–5686.

Oakes, R. S.; Clifford, A. A.; Bartle, K. D.; Thornton-Pett, M.; Rayner, C. M. Sulfur Oxidation in Supercritical Carbon Dioxide: Dramatic Pressure Dependant Enhancement of Diastereoselectivity for Sulfoxidation of Cysteine Derivatives. *Chem. Commun.* **1999,** 247–248.

Ouédraogo, J. C. W.; Dicko, C.; Kini, F. B.; Bonzi-Coulibaly, Y. L.; Dey, E. S. Enhanced Extraction of Flavonoids from Odontonema Strictum Leaves with Antioxidant Activity Using Supercritical Carbon Dioxide Fluid Combined with Ethanol. *J. Supercrit. Fluids* **2018,** *131*, 66–71.

Pacut, R.; Grimn, M. L.; Kraus, G. A.; Tanko, J. M. Photochemistry in Supercritical Carbon Dioxide: The Benzophenone-Mediated Addition of Aldehydes to α, β-Unsaturated Carbonyl Compounds. *Tetrahedron Lett.* **2001,** *42* (8), 1415–1418.

Patil, P.; Dandamudi, K.; Wang, J.; Deng, Q.; Deng, S. Extraction of Bio-Oils from Algae with Supercritical Carbon Dioxide and Co-Solvents. *J. Supercrit. Fluids* **2018,** *135*, 60–68.

Pérez, E.; Abad-Fernández, N.; Lourençon, T.; Balakshin, M.; Sixta, H.; Cocero, M. J. Base-Catalysed Depolymerization of Lignins in Supercritical Water: Influence of Lignin Nature and Valorisation of Pulping and Biorefinery by-Products. *Biomass Bioenergy.* **2022,** *163.* DOI: 10.1016/j.biombioe.2022.106536.

Pesiri, D. R.; Morita, D. K.; Glaze, W.; Tumas, W. Selective Epoxidation in Dense Phase Carbon Dioxide. *Chem. Commun.* **1998,** *88*, 1015–1016.

Pilavtepe, M.; Yucel, M.; Helvaci, S. S.; Demircioglu, M.; Celiktas, O. Y. Optimization and Mathematical Modeling of Mass Transfer Between Zostera Marina Residues and Supercritical CO_2 Modified with Ethanol. *J. Supercrit. Fluids* **2012,** *68*, 87–93.

Pillai, U. R.; Sahle-Demessi, E. Selective Hydrogenation of Maleic Anhydride to γ-Butyrolactone Over Pd/Al_2O_3 Catalyst Using Supercritical CO_2 as Solvent. *Chem. Commun.* **2002,** 422–423.

Pizarro, C.; Iglesias, O. S.; Medina, I.; Bueno, J. L. Binary Diffusion Coefficients of 2-Ethyltoluene, 3-Ethyltoluene, and 4-Ethyltoluene in Supercritical Carbon Dioxide. *J. Chem. Eng. Data* **2009,** *54* (5), 1467–1471.

Preetam, A.; Mishra, S.; Naik, S. N.; Pant, K. K.; Kumar, V. A Sustainable Approach for Material and Metal Recovery from E-Waste Using Subcritical to Supercritical Methanol. *Waste Manag.* **2022,** *145*, 29–37.

Rajasekhar, Ch., Garlapati, C.; Madras, G. Solubility of n- (4-Ethoxyphenyl) Ethanamide in Supercritical Carbon Dioxide. *J. Chem. Eng. Data* **2010,** *55* (3), 1437–1440.

Raner, K. D.; Strauss, C. R.; Trainer, R. W. A New Microwave Reactor for Batchwise Organic Synthesis. *J. Org. Chem.* **1995,** *60* (8), 2456–2460.

Reardon, P.; Metts, S.; Crittendon, C.; Daugherity, P. Parson, E. J. Palladium-Catalyzed Coupling Reactions in Superheated Water. *Organometrallics* **1995,** *14* (8), 3810–3816.

Reddy, S. N.; Madras, G. Solubilities of Benzene Derivatives in Supercritical Carbon Dioxide. *J. Chem. Eng. Data* **2011,** *56* (4), 1695–1699.

Reetz, M. T.; Konen, W.; Strack, T. Supercritical Carbon Dioxide as a Reaction Medium and Reaction Partner. *Chimica* **1993,** *47*, 493–493.

Renslo, A. R.; Weinstien, R. D.; Tester, J. W.; Danherical, R. L. Concerning the Regiochemical Course of the Diels-Alder Reaction in Supercritical Carbon Dioxide. *J. Org. Chem.* **1997,** *62* (13), 4530–4533.

Rohr, M.; Geyer, C.; Wandeler, R.; Schneider, M. S.; Murphy, E. F.; Baiker, A. Solvent-Free Ruthenium-Catalysed Vinylcarbamate Synthesis from Phenylacetylene and Diethylamine in 'Supercritical' Carbon Dioxide. *Green Chem.* **2001,** *3*, 123–125.

Roop, R. K.; Akgerman, A.; Dexter, B. J.; Irvin, T. R. Extraction of Phenol from Water with Supercritical Carbon Dioxide. *J. Supercrit. Fluids* **1989,** *2* (2–3), 51–56.

Rose, P. M.; Clifford, A. A.; Rayner, C. M. The Baylis–Hillman Reaction in Supercritical Carbon Dioxide: Enhanced Reaction Rates, Unprecedented Ether Formation, and a Novel Phase-Dependent 3-Component Coupling. *Chem. Commun.* **2002,** 968–969.

Santos, P.; Zabot, G.; Meireles, M.; Mazutti, M.; Martínez, J. Synthesis of Eugenyl Acetate by Enzymatic Reactions in Supercritical Carbon Dioxide. *Biochem. Eng. J.*, **2016,** *114*, 1–9. DOI: 10.1016/j.bej.2016.06.018

Savage, P. E. Organic Chemical Reactions in Supercritical Water. *Chem. Rev.* **1999,** *99*, 603–622.

Selva, M.; Perosa, A.; Fabris, M.; Canton, P. The Metathesis of α-Olefins Over Supported Re-Catalysts in Supercritical CO_2. *Green Chem.* **2009,** *11*, 229–238.

Shezad, N.; Clifford, A. A.; Rayner, C. M. Pd-Catalysed Coupling Reactions in Supercritical Carbon Dioxide and Under Solventless Conditions. *Green Chem.* **2002,** *4*, 64–67.

Siskin, M.; Katrinzky, M. Reactivity of Organic Compounds in Superheated Water: General Background. *Chem. Rev.* **2001,** *101* (4), 825–835.

Soh, L.; Zimmerman, J. Biodiesel Production: The Potential of Algal Lipids Extracted with Supercritical Carbon Dioxide. *Green Chem.* **2011,** *13*, 1422–1429.

Stadler, A.; Yousti, B. H.; Dallinger, D.; Walla, P.; Vander Eycken, E.; Kaval, N. et al. Scalability of Microwave-Assisted Organic Synthesis from Single-Mode to Multimode Parallel Batch Reactors. *Org. Process. Res. Dev.* **2003,** *7* (5), 707–716.

Strauss, C. R.; Invited Review. A Combinatorial Approach to the Development of Environmentally Benign Organic Chemical Preparations. *Aust. J. Chem.* **1999,** *52* (2), 83–96.

Strauss, C. R.; Trainor, K. W. Developments in Microwave-Assisted Organic Chemistry. *Aust. J. Chem.* **1995,** *48* (10), 1665–1692.

Sun, Y.; Wang, X.; Feng, T.; Yu, G.; Wang, F. Evaluation of Coal Extraction with Supercritical Carbon Dioxide/1-Methyl-2-Pyrrolidone Mixed Solvent. *Energy Fuels* **2014,** *28* (2), 816–824.

Timko, M.; Ghoniem, A.; Green, W. Upgrading and Desulfurization of Heavy Oils by Supercritical Water. *J. Supercrit. Fluids* **2015,** *96*, 114–123.

Vedaraman, N.; Brunner G.; Kannan, C. S.; Muralidharan, C.; Rao, P. G.; Raghavan, K. V. Extraction of Cholesterol from Cattle Brain Using Supercritical Carbon Dioxide. *J. Supercrit. Fluids* **2004,** *32* (1–3), 231–242.

Viguera, M.; Gómez-Salazar, J. M.; Barrena, M. I.; Calvo, L. The Removal of Lubricating Oils from Metallic Contacts with Supercritical CO_2. *J. Supercrit. Fluids* **2013,** *73*, 51–56.

Wakayama, H.; Fukushima, Y. Nanoporous Silica Prepared with Activated Carbon Molds Using Supercritical CO_2. *Chem. Mater.* **2000,** *12* (3), 756–761.

Walsh, B.; Hyde, J. R.; Licence, P.; Poliakoff, M. The Automation of Continuous Reactions in Supercritical CO_2: The Acid-Catalysed Etherification of Short Chain Alcohols. *Green Chem.* **2005,** *7*, 456–463.

Wang, X.; Venkataramanan, N. S.; Kawanami, H.; Ikushima, Y. Selective Oxidation of Styrene to Acetophenone Over Supported Au–Pd Catalyst with Hydrogen Peroxide in Supercritical Carbon Dioxide. *Green Chem.* **2007,** *9*, 1352–1355.

Wang, Z.; Jiang, H.; Qi, C.; Wang, Y.; Dong, Y.; Liu, H., PS–BQ: An Efficient Polymer-Supported Cocatalyst for the Wacker Reaction in Supercritical Carbon Dioxide. *Green Chem.* **2005,** *7*, 582–585.

Weinstein, R. D.; Gribbin; J. J.; Najjar, D. D. Liquid and Supercritical Carbon Dioxide Assisted Blending of Poly (Vinyl Acetate) and Citric Acid. *Ind. Eng. Chem. Res.* **2005,** *44* (10), 3480–3484.

Xie, M.; Zhang, X.; Laakso, J.; Wang, H.; Levänen, E. New Method of Postmodifying the Particle Size and Morphology of $LiFePO_4$ via Supercritical Carbon Dioxide. Cryst. Growth *Des.* **2012,** *12* (5), 2166–2168.

Xu, X.; Antal, M. J. Kinetics and Mechanism of Isobutene Formation from T-Butanol in Hot Liquid Water. *AlchE. J.* **1994,** *40* (9), 1524–1534.

Xu, X.; Antal, M. J. Mechanism and Temperature-Dependent Kinetics of the Dehydration of Tert-Butyl Alcohol in Hot Compressed Liquid Water. *Ind. Eng. Chem. Res.* **1997,** *36* (1), 23–41.

Yang, G.; Su, J.; Gao, J.; Hu, X.; Geng, C.; Fu, Q. Fabrication of Well-Controlled Porous Foams of Graphene Oxide Modified Poly (Propylene-Carbonate) Using Supercritical Carbon Dioxide and Its Potential Tissue Engineering Applications. *J. Supercrit. Fluids* **2013,** *73*, 1–9.

Yılmaz, F.; Mutlu, A.; Ünver, H.; Kurtça, M.; Kani, İ. Hydrogenation of Olefins Catalyzed by Pd (II) Complexes Containing a Perfluoroalkylated S, O-Chelating Ligand in Supercritical CO_2 and Organic Solvents. *J. Supercrit. Fluids* **2010,** *54* (2), 202–209.

Yoshida, M.; Hava, N.; Okuyama, S. Catalytic Production of Urethanes from Amines and Alkyl Halides in Supercritical Carbon Dioxide. *Chem. Commun.* **2000,** 151–152.

Yu, J.; Savage, P. E. Decomposition of Formic Acid Under Hydrothermal Conditions. *Ind. Eng. Chem. Res.* **1998,** *37* (1), 2–10.

Yuan, T.; Majid, A.; Marshall, W. D. Detoxification of Aryl-Organochlorine Compounds by Catalytic Reduction in Supercritical Carbon Dioxide. *Green Chem.* **2003,** *5*, 25–29.

Zarena, A. S.; Sachindra, N. M.; Sankar, K. U. Optimisation of Ethanol Modified Supercritical Carbon Dioxide on the Extract Yield and Antioxidant Activity from *Garcinia mangostana* L. *Food Chem.* **2012,** *130* (1), 203–208.

Zhao F.; Ikushima, Y.; Arai, M. Hydrogenation of 2-Butyne-1, 4-Diol to Butane-1, 4-Diol in Supercritical Carbon Dioxide. *Green Chem.* **2003a,** *5*, 656–658.

Zhao, F.; Ikushima, Y.; Chatterjee, M.; Shirai, M.; Arai, M. An Effective and Recyclable Catalyst for Hydrogenation of α, β-Unsaturated Aldehydes into Saturated Aldehydes in Supercritical Carbon Dioxide. *Green Chem.* **2003b,** *5*, 76–79.

Zhao, S.; Wang, C.; Bai, B.; Jin, H.; Wei, W. Study on the Polystyrene Plastic Degradation in Supercritical Water/CO_2 Mixed Environment and Carbon Fixation of Polystyrene Plastic in CO_2 Environment. *J. Hazard. Mater.* **2022,** *421*, 126763, DOI: 10.1016/j.jhazmat.2021.126763.

Zheng, H.; Xu, Y.; Zhang, J.; Xiong, X.; Yan, J.; Zheng, L. An Ecofriendly Dyeing of Wool with Supercritical Carbon Dioxide Fluid. *J. Clean. Prod.* **2017,** *143*, 269–277.

CHAPTER 7

Other Green Solvents

NEHA GODHA[1], ABHILASHA JAIN[2], RITU VYAS[3], AARTI AMETA[4], and P. B. PUNJABI[5]

[1]*Department of Chemistry, PAHER University, Udaipur, India*

[2]*Department of Chemistry, St. Xavier's College, Mumbai, India*

[3]*Department of Chemistry, Pacific Institute of Technology, Udaipur, India*

[4]*Department of Chemistry, Guru Nanak Girls' P.G. College, Udaipur, India.*

[5]*Department of Chemistry, M. L. Sukhadia University, Udaipur, India.*

ABSTRACT

Water is accespted as a universal solvent and it is green in nature also. Apart from it, there are some other green solvents such as polyethylene glycol, glycerol, cyclopentylmethyl ether, 2-methyltetrahydrofuran, ethyl lactate, perfluorinated (fluorous) solvent, p-cymene, limonene, gamma-valerolactone, etc. These solvents have been used successfully for a wide range of organic reactions. These reactions including oxidation, reduction, photochemical reaction, hydroxylation, dehalogenation, decarboxylation, phosphorylation, cycloaddition, condensation, coupling, polymerization, etc. have been discussed.

7.1 INTRODUCTION

Most of the chemical transformations required solvents to have a better contact between catalysts and reagents. They also ensured an easy isolation process of obtained products from the reaction mixture. As different organic solvents are normally used in organic syntheses, therefore these are the main

Green Chemistry, 2nd Edition: Fundamentals and Applications. Suresh C. Ameta & Rakshit Ameta (Eds.)

sources of generated wastes. Commonly used solvents are flammable, volatile, and hazardous to environment and human beings.

There are two main approaches to solve the problems caused due to these solvents. In the first approach, the solvent is not used altogether. Many reactions can be carried out under the neat conditions. But in solvent-free techniques, many reactions are not possible and many reagents and intermediates are not stable outside the solution. The second approach is the replacement of organic solvents with greener solvents. Nowadays, ionic liquids and supercritical fluids have been used quite commonly and the use of these has already been discussed. There are some more compounds, which can serve as green solvents in organic syntheses.

The most popular among these green solvents are water, polyethylene glycol, glycerol, cyclopentylmethyl ether, 2-methyltetrahydrofuran, ethyl lactate, perfluorinated solvent, γ-valero lactone, p-cymene, etc.

7.2 SOME MAJOR GREEN SOLVENTS

7.2.1 WATER

Water is one of the greener solvents. It is readily available, least expensive, safe, and nonhazardous to environment. Water is a "universal solvent" in nature. Living cells represent the most complex chemical reactions (termed as biochemical reactions) and all such reactions occur in the environment with > 90% water. Inorganic reactions are also carried out using water as a solvent.

Water is the best solvent among all the green solvents because it has many advantages like:

- Environmental benefits
- Cost/economic factors
- Safety
- Synthetic efficiency
- Simple operation and
- Have great potential for new synthetic methodologies.

- **Characteristics properties of water**

It exists in three forms—solid (ice), liquid, and vapor.

Boiling point and freezing point: Boiling point of water is 373.15 K, whereas freezing point is 273.15 K at 1 atm pressure.

Viscosity: The viscosity of water is 1.01 $mm^2\ s^{-1}$ at 20°C. Its viscosity affects the movement of solute in water as well as sedimentation rate of suspended particles.

Surface tension and density: The surface tension of water is 72.75 mN m^{-2}, whereas density is 1.0000 kg^{-3} at 4°C.

Heat capacity: Heat capacity of water is 4182 J kg^{-1} at 20°C. Due to high heat capacity, the rapid exchange in temperature results in low changes in water temperature. So in large- scale processes, temperature of endo- and exothermic reactions can be controlled.

Dielectric constant: The value of dielectric constant of water is 80. Due to its high value, polar compounds are readily soluble in water.

Hydrogen bonding: One molecule of water can form four hydrogen bonds with other water molecules. Hydrogen bonding in water endows a unique characteristic, which is known as dual activator property. Water can activate both nucleophiles and electrophiles and accelerates polar reactions, which involve polar transition state or intermediates.

Water is a versatile solvent in nature and it is used in synthetic organic chemistry. Even with less solubility of most of the organic compounds in it, the entire scenario has changed by the pioneer discovery of Rideout and Breslow (1980) and Grieco et al. (1983), who have used water as a solvent in organic synthesis for the first time. Water has many physicochemical characteristics, which make it a unique solvent and these characteristics can be modified as per requirement of the reaction by using various catalysts and additives like surfactants (Lindstrom, 2002).

Most of the important reactions in organic synthesis have been tried using water as a solvent or one of the components in the solvent mixture; of course, with some modifications in the conventional methodologies.

Bhat et al. (2020) used water as a green solvent with reactants such as active methylene compound and substituted aromatic aldehydes under microwave exposure. The products were obtained with 84–91% yields.

- **Applications**

i) Diels-Alder reaction—It is a [4 + 2] cycloaddition reaction between diene and a dienophile.

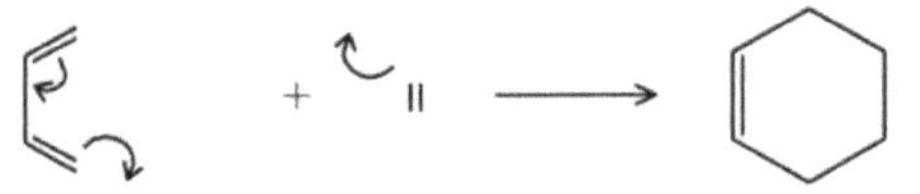

Butadiene (Diene) + Ethene (Dienophile) → Cyclohexane (Adduct)

The solvent affects the stereoselectivity of some reaction (Breslow et al., 1983; Breslow and Maitra, 1984). For example, cyclopentadiene reacts with butanone to give a product, where the ratio of endo/exo products was 21.4, when these are stirred at 0.15 M concentration in water, whereas the ratio is only 8.5 in ethanol. The stereochemical changes could be explained by the need to minimize the transition state surface area in water solution favoring the more compact endo-structure (Berson et al., 1962; Samil et al., 1985).

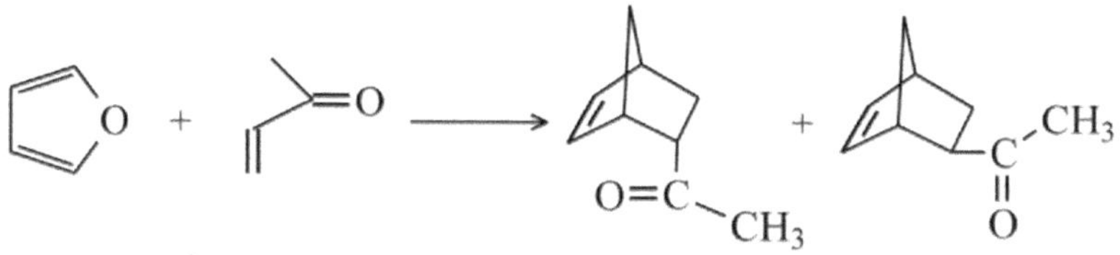

- ***Hetero Diels-Alder reaction***

Hetero Diels-Alder reaction was reported in aqueous medium by Larsen and Grieco (1985). In this reaction, simple iminium salts generated *in situ* under Mannich-like conditions reacted with dienes in water to give aza-Diels-Alder reaction products. Retro Diels-Alder reaction and Aza-Diels-Alder reactions also occurred readily in water (Grieco et al., 1987). It has also been reported that the rate of given Diels-Alder reaction increases more than 700 times when carried out in water.

ii) Claisen rearrangement

This reaction involves [3, 3]—sigmatropic shift.

In Claisen rearrangement of chorismic acid, pure water is used to promote the reaction.

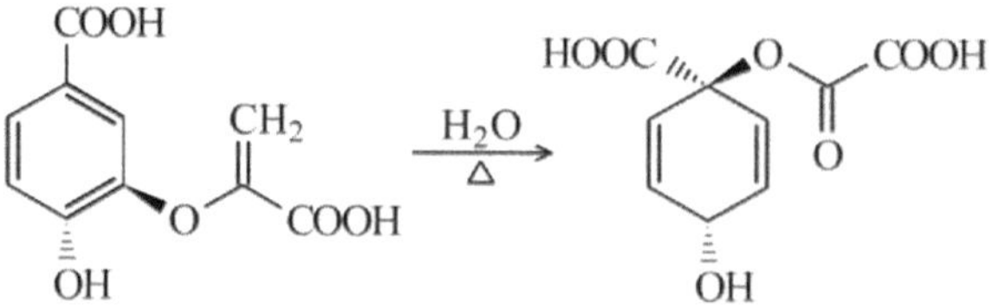

Chorismic acid

It was observed that the rate of this reaction is increased with the use of water as a polar solvent as compared to nonpolar solvent (White and

Wolfarth, 1970; Ponaras, 1983). It has been reported that Claisen rearrangement of allyl vinyl ether in water gives the aldehyde with 82% yield (Grieco et al., 1989).

iii) 2 σ + 2 σ + 2 π Cycloaddition

Quadricyclane with dimethyl acetylenedicarboxylate (DMAD) gives an additional product. The rate of reaction was found maximum in water (Domingo et al., 2008).

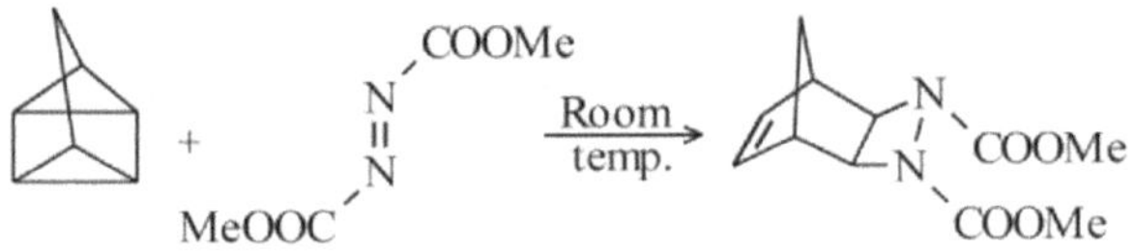

iv) Aldol condensation

The aldol condensation involves self-addition of aldehydes containing an α-hydrogen atom, giving β-hydroxyaldehyde, which undergoes dehydration resulting in α, β-unsaturated aldehyde. Aldol reaction of silyl enol ethers with aldehydes was reported, which was promoted by water (Lubineau, 1986; Lubineau and Meyer, 1988).

v) Benzoin condensation

The reaction of aromatic aldehydes in the presence of sodium or potassium cyanide in aqueous ethanolic solution gives α-hydroxy ketones (Lapworth, 1903, 1904; Ide and Buck, 1984).

$$2\,\mathrm{Ph}-\overset{\mathrm{H}}{\overset{|}{\mathrm{C}}}=\mathrm{O} \xrightarrow[\mathrm{EtOH},\ \mathrm{H_2O}]{\mathrm{CN}^{\ominus},\ \Delta} \mathrm{Ph}-\overset{\mathrm{O}}{\overset{\|}{\mathrm{C}}}-\underset{\mathrm{H}}{\underset{|}{\overset{\mathrm{OH}}{\overset{|}{\mathrm{C}}}}}-\mathrm{Ph}$$

The benzoin condensation in the aqueous medium using inorganic salts is about 200 times faster than in ethanol (Kool and Breslow, 1988).

vi) Claisen-Schmidt condensation

This reaction involves the condensation of aromatic aldehydes, which do not have α-hydrogen, with an aliphatic aldehyde or ketone having α-hydrogen in the presence of a strong base to form α, β-unsaturated aldehyde or ketone (Claisen and Claparede, 1881, Schmidt and Ber, 1881).

$$C_6H_5CHO + CH_3CHO \xrightarrow{NaOH} HCH_5C_6{=}CH.CHO$$

Cinnamaldehyde

The reaction of cyclohexanone with benzaldehyde in water gives high yield.

vii) Heck reaction

It involves the coupling of an alkene with a halide in the presence of Pd (0) catalyst to form a new alkene.

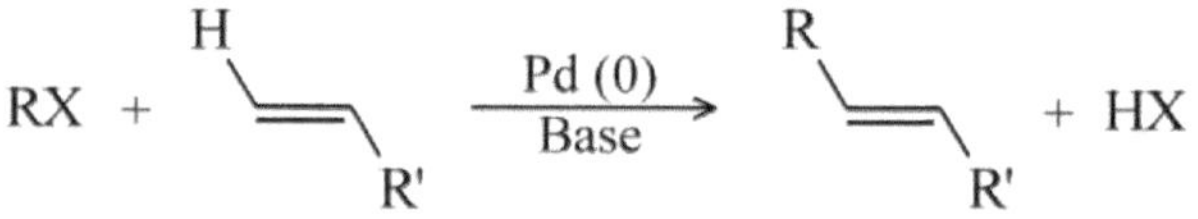

Heck reaction can proceed very well in water. Water plays an important role in the transformation of catalyst precursor into Pd (0) species and to generate zero valent Pd species. It can be performed under mild conditions in the presence of water and acetate ion.

viii) Knoevenagel reaction

This is a condensation reaction of an aldehyde or a ketone with the compound having active $-CH_2-$ group in the presence of a weak base (Knoevenagel, 1898; Johnson, 1942).

$$RCHO + CH_2(COOH)_2 \xrightarrow{Base} R{-}CH{=}C(COOH)_2$$

$$\xrightarrow{\Delta,\ -CO_2} R{-}CH = CH{-}COOH$$

Crotonic acid

This reaction has been carried out between aldehydes and acetonitrile in water.

Knoevenagel-type addition product can be obtained by the reaction of acryclic derivative in the presence of water and DABCO [1,4-diazabicyclo [2,2.2] octane with 90–98% yield (Auge et al., 1994).

ix) Oxidation

Although many of the oxidation reactions are carried out in the aqueous medium, some more innovative oxidation reactions have been reported in aqueous medium.

Oxidation of alkenes and alkynes—Oxidation of alkene using aqueous solution of $KMnO_4$ and in the presence of a phase transfer catalyst or 18-crown-6 gives carboxylic acid with good yield.

$$\underset{\text{Alkene}}{R{-}(CH_2)_n{-}CH{=}CH_2} \xrightarrow[\text{PTC or 18-Crown-6}]{\text{aq. }KM_nO_4} \underset{\text{Carboxylic acid}}{R{-}(CH_2)_n{-}CH_2{-}COOH}$$

Similarly, alkynes can also be oxidized and they give a mixture of carboxylic acids.

$$R{-}C{\equiv}C{-}R' + 4\,[O] \xrightarrow[\text{aq. medium}]{KMnO_4} R\,COOH + R'\,COOH$$

Oxidation of aldehydes and ketones—Oxidation of aromatic aldehydes by aqueous performic acid has been reported at low temperature (0–4°C). Baeyer-Villiger oxidation of ketone was also carried out in aqueous heterogeneous medium using m-chlordroxybenzole acid (MCPBA) at room temperature (Fringuell et al., 1989).

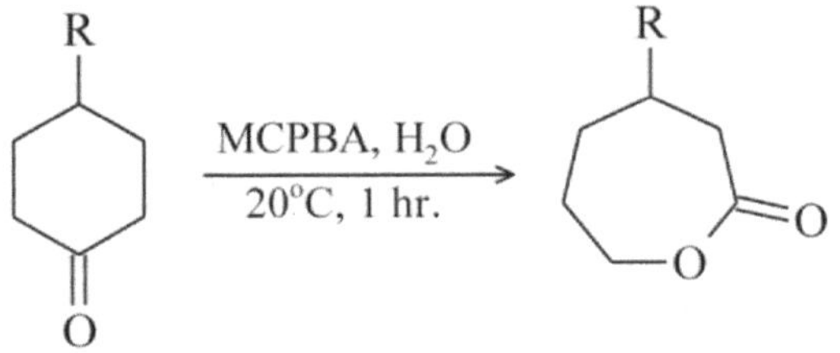

Oxidation of amines—Oxidation of aromatic amines having –OH or –COOH group to nitro compound by oxone in 20–50% aqueous acetone gives 73–84% yield (Webb and Seneviratrie, 1995). It is possible to synthesize N-oxide from aminopyridine directly (without protection of amino group) by using oxone in water under neutral or basic condition in good yields (Robke and Behrman, 1993).

Oxidation of sulfides—Oxidation of sulfides to sulphone can be carried out by using sodium perborate (SPB) in aqueous methanolic sodium hydroxide (Mckillop and Tarbin, 1987), whereas oxidation of sulfides to sulphoxides can be done in 70% aqueous tetrabutyl phosphonium hydroxide (TBPH) in water in heterogeneous phase at low temperature (20–70°C) (Fringuelli et al., 1993).

Oxidation of phenols and anilines—The oxidative bromination of phenols and anilines at ambient temperatures using molecular bromine was reported by Ghorpade et al. (2018). It was catalyzed by graphene oxide

(GO) and transformations had 100% atom economy and high selectivities for the tribromoanilines and -phenols. This method can also be tried with N-bromosuccinimide (NBS) effectively as brominating reagent.

x) Reduction

Reduction of many compounds in aqueous medium has been reported with high yields.

Reduction of alkene and alkyne—The reduction of carbon–carbon double bond of α, β-unsaturated carbonyl compounds using $Zn/NiCl_2$ in methoxy-ethanol–water system has been reported (Petrier and Luche, 1987).

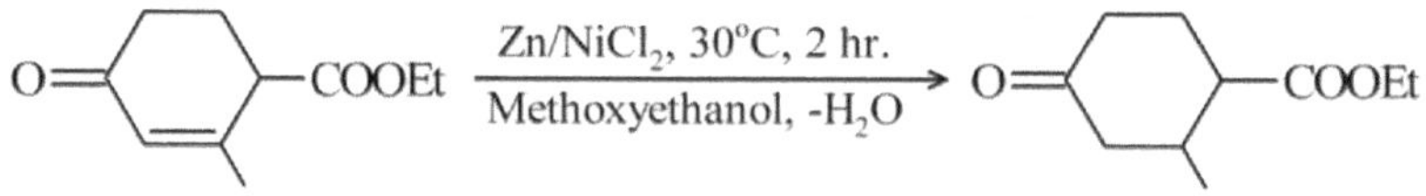

Reduction of alkynes can be carried out with water-soluble monosulphonated and trisulphonated triphenylphosphine (Larpent and Meignan, 1993).

Reduction of aldehydes and ketones—The reduction of carbonyl compound in aqueous media has been carried out by sodium borohydride at room temperature. Samarium iodide in aqueous THF and cadmium chloride-magnesium in the H_2O-THF system can also be employed for the reduction of carbonyl compounds (Hasegawa and Curran, 1993; Bordoloi, 1993).

Reduction of aromatic ring—The heterocyclic compounds can be reduced by SmI_2-H_2O system with good yield at 0°C (Hanessian and Girand, 1994). Aromatic compound can also be reduced in aqueous medium with ruthenium chloride/trioctylamine (TOA) at room temperature (Fache et al., 1995).

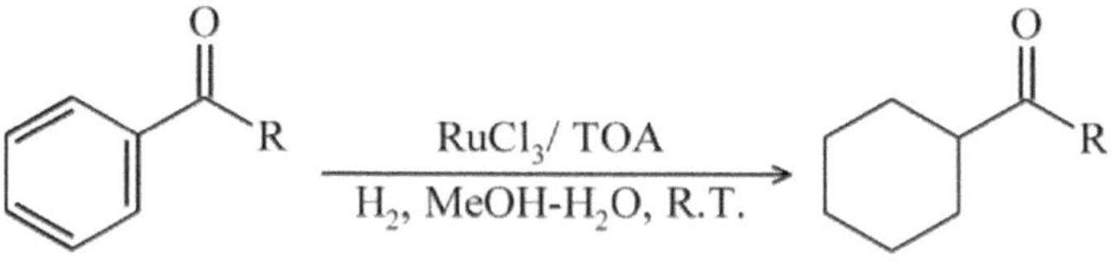

xi) Photochemical reactions

Photodimerization of thymine and uracil has been reported in aqueous medium (Ramamurthy, 1986). Organic compounds like stilbenes, alkyl cinnamates, coumarin, etc. can also be dimerized in water with good yields.

Some more examples of photochemical reactions in aqueous medium are

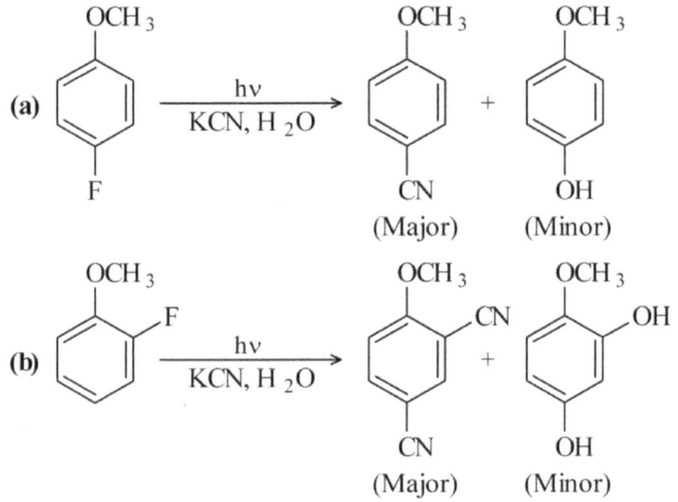

xii) Reaction of carbanion equivalents

Reactions of carbonyl compounds and imines with allyl halides allowed use of water as a solvent along with no requirement of low temperature unlike the organolithium and organomagnesium reagents, where metals such as indium, tin, and zinc are used as metal mediators (Keh et al., 2003).

xiii) *Barbier–Grignard type carbonyl alkylation*

Barbier-type allylation—This reaction is highly regio- and stereoselective. In this reaction, the use of water as a solvent has very high selectivity for α-adduct as a product (100%) as compared to dimethyl sulfide (DMS) with 65% yield. Here, water is required for the formation of oxonium ion intermediate, which furnishes α-adduct (Tan et al, 2003).

xiv) Reaction of radicals

Formation of lactone—It has been reported that yield of lactone formation is 78% in water while it is 0% in benzene and hexane (Yorimitsu et al., 2000).

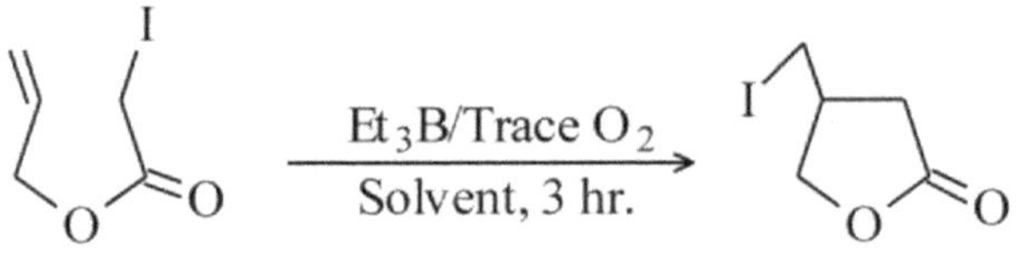

xv) Dehalogenation

Xiao et al. (2019) carried out dehalogenation of some benzyl halides (chlorides, bromides, or iodides) in the presence of H_3PO_3 system. Benzyl halides afforded a wide range of diarylmethanes (good yields) through electrophilic substitution reactions with arenes (electron-rich).

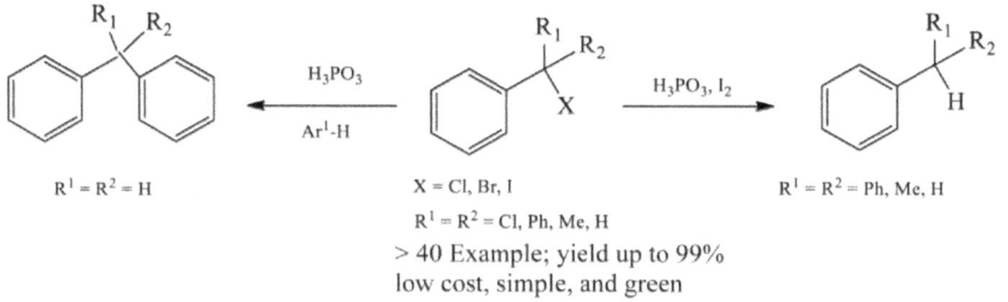

xvi) Multicomponent reactions

Passerini reaction—The rate of this reaction is increased 300 times, when it is carried out in water (Pirrung and Sarma, 2004).

Ugi reaction—There is an acceleration in the rate of this reaction as well as high yields are obtained, when water is used as a solvent (Pirrung and Sarma, 2004).

xii) Sulphonation

Xie et al. (2018) reported an ecofriendly method for the synthesis of various sulfonylated N-heteroaromatics in water under metal-free, organic-solvent-free, neutral, and mild reaction conditions. The advantages of this reaction are readily available reagents, wide substrate range, high regioselectivity with chemoselectivity, and make this protocol very significant. It was reported that pure products can be easily obtained via filtration, washing (alcohol) without any extraction and recrystallization.

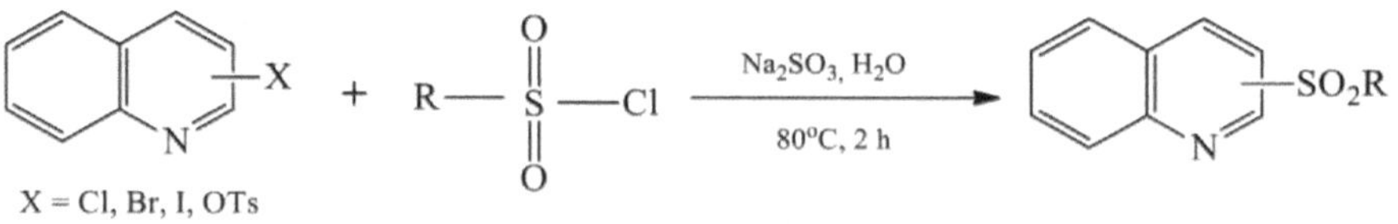

An efficient, ligand-free, and additive-free Suzuki–Miyaura coupling was reported by Chinthakindi et al. (2016). It was found compatible with the aromatic sulfonyl fluoride functional group and provided a wide range of biaryl sulfonyl fluorides as bioorthogonal “click” reagents at room temperature.

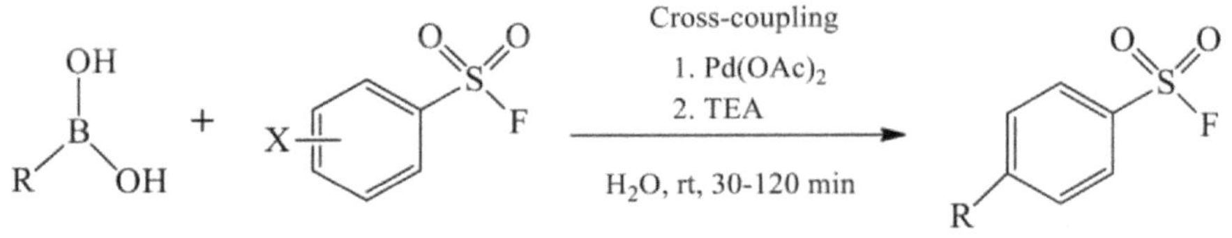

where R = Aryl or heteroaryl
X = -o, m, p-Br/ p-I

Chemoselective
Ligand free
Green protocol
Open flask chemistry
Up to 97% yield

xiii) Phosphorylation

A metal-free 2,4,5,6-tetra(9H-carbazol-9-yl)isophthalonitrile (4CzIPN)-catalyzed phosphorylation/cyclization reaction (visible-light-induced) was observed by Yuan et al. (2020) in water at room temperature. The final products were 3-phosphorylated benzothiophenes. A series of 3-phosphorylated benzothiophenes with different various functional groups were synthesized by this procedure in good to excellent yields.

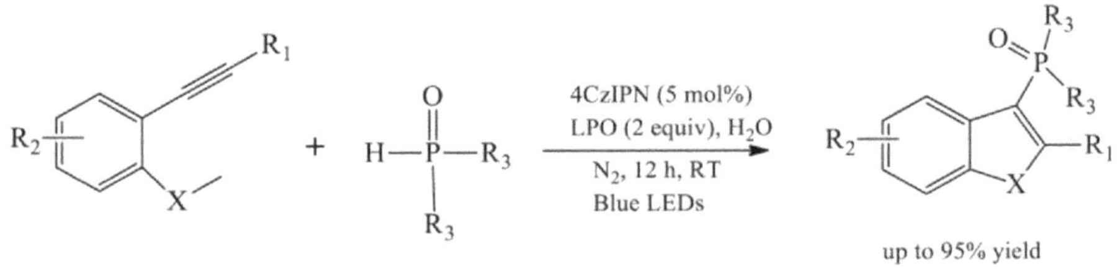

- Acid-free, easy scalability
- Metal free; Visible light promoted
- Water as solvent; room temperature

Liu et al. (2020) developed a metal-free visible-light-induced radical cascade cyclization and prepared 2-phosphorylated thioflavones from phosphine oxides and methylthiolatedalkynones in water as a green solvent.

xiv) Synthesis of heterocycle

An eco-friendly methodology was developed by Xie et al. (2017) for synthesis of different functionalized quinolin-2(1H)-ones in water at ambient temperature. They did not use any organic solvent base. This protocol gives excellent yield with shorter reaction time (1–8 min). It was revealed that this methodology has an E-factor of 1.6.

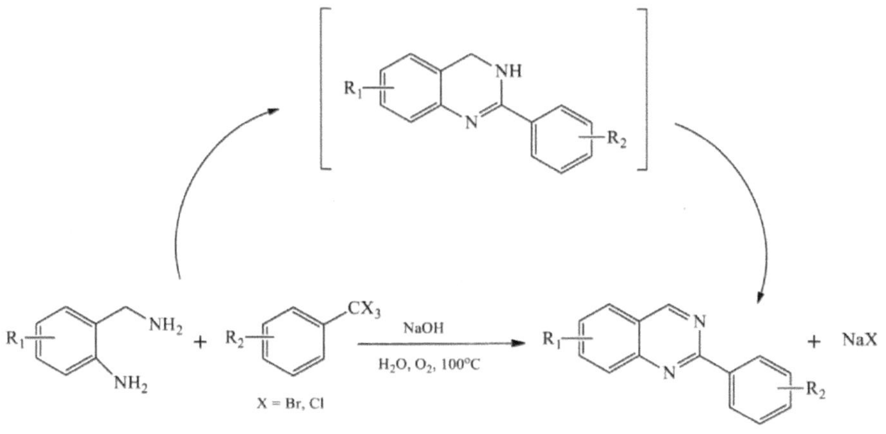

- Transition-metal free protocol
- Use of water and molecular oxygen
- Formation of being by-products
- Isolation of products by simple crystallization

A transition-metal-free method was developed by Chatterjee et al. (2018) for the synthesis of quinazolines. They used 2-aminobenzylamines and α,α,α-trihalotoluenes as starting materials. The reaction was carried out in the presence of sodium hydroxide and molecular oxygen in water at 100°C.

xv) Hydroxylation

4-Dodecylbenzenesulfonic acid (DBSA) catalyzed hydrolysis reaction of *β*-ethylthio-*β*-indolyl-*α*, *β*-unsaturated ketones was carried out in water by Hu et al. (2019). It afforded 3-ethanoyl/aroylacetylindoles as the product. It was observed that this reaction proceeds in the presence of 10 mol% DBSA in water under reflux with excellent yield of the product.

xvi) Others

A highly efficient, one-pot, and three-component reaction of amine and carbon disulfide with alkyl vinyl ether via Markovnikov addition reaction was carried out in water under a mild and green procedure with excellent yield and complete regiospecificity (Halimehjani et al., 2010).

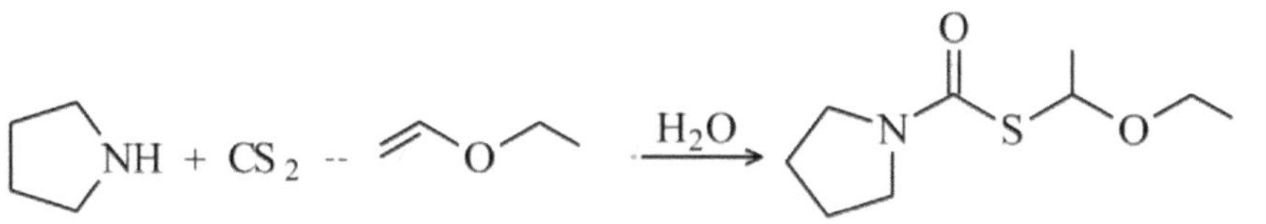

Water was used as a promoter for the synthesis of 2-aminobenzothiazoles or 2-aminobenzoxazoles in tandem reaction by Zhang et al. (2011).

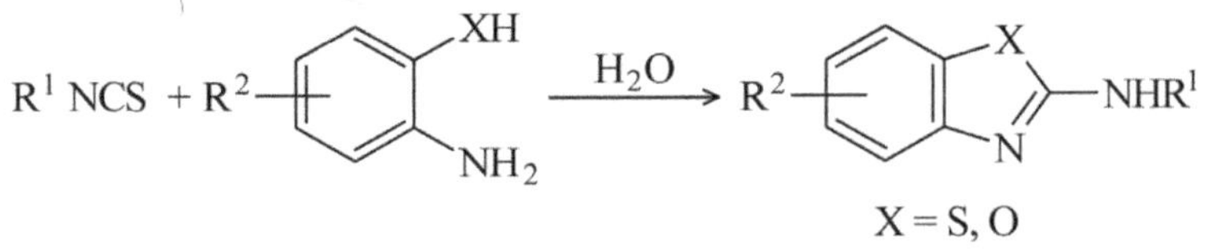

Peng et al. (2010) used stilbazo as a promoter for the synthesis of biaryl compounds by a ligand-free Suzuki–Miyaura reaction in water at room temperature.

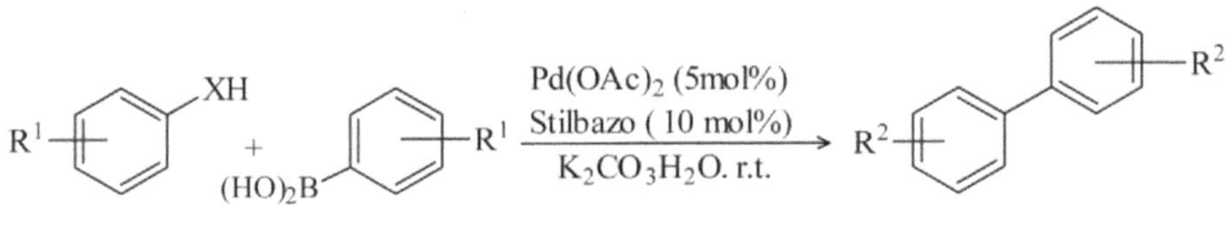

Addition of ethylene to aniline to generate N-ethylaniline was reported out by Dub et al. (2010) in the presence of a catalyst $PtBr_2/Br^-$ in aqueous medium. It was found to be an atom economical addition.

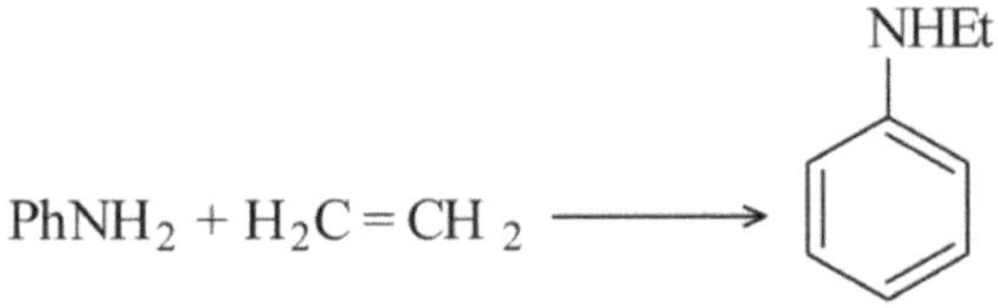

A catalyst-free amination of 2-mercaptobenzoxazoles in water was developed by Tankam et al. (2018) under microwave irradiation. The product was obtained by direct amination with various amines with moderate to high yields. The synthesis of Suvorexant, a medicine for the treatment of insomnia, was carried out using this amination process, which was completed in 1 h at 100–150°C under microwave irradiation without using any external catalyst or additive.

Huang et al. (2018) developed a new aerobic cross-dehydrogenative coupling of C–H and S–H to get aryl sulfides in water. They utilized CoPcS as the catalyst and O_2 as the oxidant.

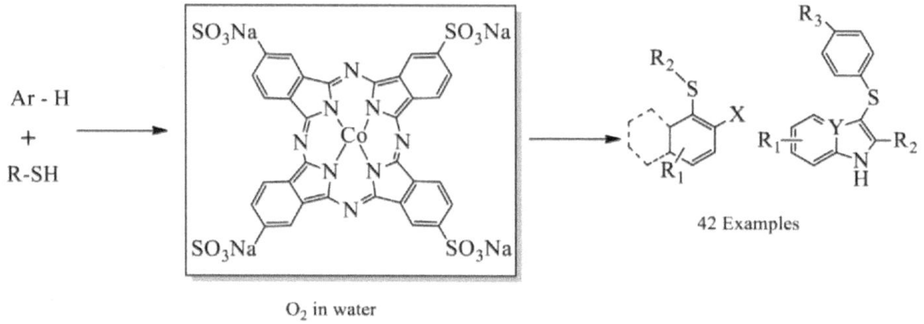

The radical cyclizations of enynes/dienes with alcohols in water have been carried out by Wang et al. (2020). This methodology was designed to work in mild reaction conditions and that too without any catalyst. The $K_2S_2O_8$ was used as a green oxidant and water as a green solvent.

7.2.2 POLYETHYLENE GLYCOL (PEG)

Polyethylene glycol is a linear polymer formed from the polymerization of ethylene oxide. It is available in a variety of molecular weights. The numerical designations of PEG indicates the average molecular weight, for example, PEG-200, PEG-400, PEG-2000 etc. Low molecular weight PEGs are liquid and completely miscible in water whereas PEGs having high molecular weight are waxy white solids and highly soluble in water.

PEG is inexpensive, recoverable, biologically compatible, non-toxic, thermally stable and biodegradable (Harris, 1992a, b; Harris and Zalipsky, 1997). In addition to this, PEG and its monomethyl ethers have low vapor pressure and nonflammable so that these compounds present simple workup procedure and can be recovered as well as recycled. Thus, it can be considered as an environmentally benign solvent.

PEG is employed as a support for the various transformations (Dickerson et al., 2002). It is a biologically acceptable polymer (PEG), which has immense importance in drug delivery (Kolate et al.2014) bioconjugates (Harris, 1992, b), and bioseparations (Albertsson, 1986). According to US FDA, it is considered safe and approved for internal consumption (Harris, 1992a, b; Herold et al., 1989; Molineux, 2002). It can also be used as an efficient medium for phase transfer catalyst (PTC). Now a days, it is used as a solvent for many organic reactions. PEG is commercially available at low cost. Generally, low molecular weights (< 2000) PEGs have been used because of low melting point or their existence as liquid at the room temperature. PEGs are stable to acid, base and high temperatures (Guo et al., 2002; Chen et al., 2004a, 2004b; Naik and Doraiswamy, 1998). These are not affected by oxygen, hydrogen peroxide or other oxidation systems (Haimov and Neumann, 2002).

A number of organic compounds and metallic complexes are soluble in polyethylene glycol (PEG). Apart from it, PEG is nonvolatile, biodegradable, thermally stable, non-toxic, recyclable and inexpensive. As a result PEG can be successfully used in lot of organic reactions, as green medium such as carbon–carbon coupling reaction multicomponent reaction, addition reaction, carbon-hetero coupling reaction, condensation reaction, oxidation reaction, substitution reaction, reductive reaction, and so on (Figure 7.1).

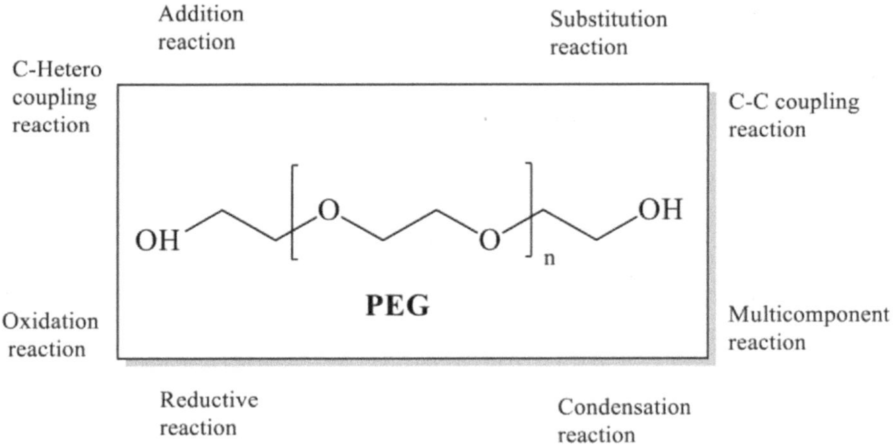

FIGURE 7.1 Different reaction polyethylene glycol.

i) Oxidation

In the oxidation of olefins to dihydroxy compounds, PEG-400 is used as a solvent and OsO_4 acts as a catalyst for the reaction (Chandrasekhar et al., 2003).

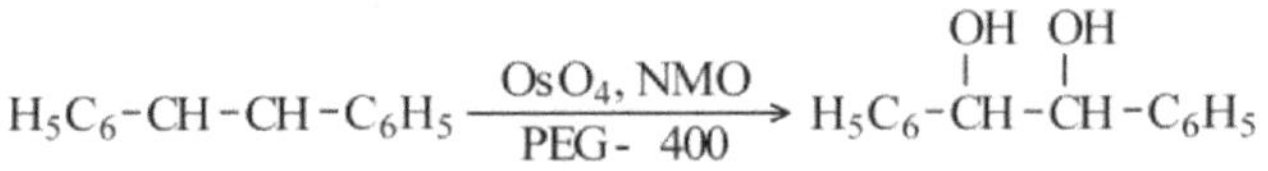

After the extraction of the product, PEG—400 can be reused.

$K_2Cr_2O_7$ has been used (which is soluble in PEG—400) for the oxidation of benzyl bromide to benzaldehyde with good yield (Santaniello et al., 1979).

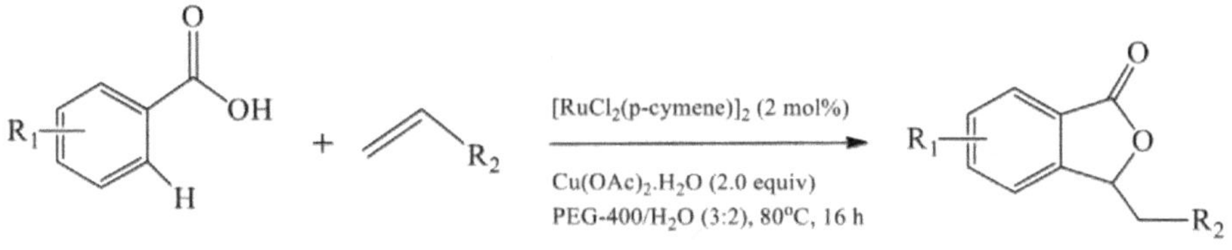

$RuCl_2$(p-cymene)]$_2$ in a mixture of poly(ethylene glycol) (PEG-400) and water is reported to be an highly efficient catalyst reaction between alkenes and benzoic acids involving cross-dehydrogenative C–H bond alkenylation (Zhao et al., 2015) The reaction afforded a number of derivatives (phthalide) in excellent yields at 80°C, $Cu(OAc)_{2.}H_2O$ was used as an oxidant.

ii) Reduction

It has been reported that the reduction of alkyl and acyl esters to the corresponding alcohols by sodium borohydride was enhanced in PEG-400.

$$\text{R'COOR} \xrightarrow[\text{PEG- 400}]{\text{NaBH}_4} \text{R'-CH}_2\text{-OH}$$

Carbonyl compounds can be reduced by $NaBH_4$ more easily and efficiently, when PEG-400 was used as a solvent rather than THF. The reduction of halides as well as acyl chloride can also be accomplished by $NaBH_4$ in PEG-400 conveniently (Santaniello et al., 1983). An efficient, safe reduction of alkenes using $NaBH_4$–$NiCl_2{\cdot}6H_2O$ and in ethanol/PEG-400 under mild conditions was reported in Li et al. (2018) using the one-pot convenient catalytic system.

iii) Substitution reaction

In PEG-300, the reaction of tert-butyl chloride with water gives corresponding alcohol (Leininger et al., 2002).

$$\text{(H}_3\text{C)}_3\text{-C-CL} + \text{H}_2\text{O} \xrightarrow{\text{PEG- 400}} \text{(CH}_3\text{)}_3\text{-CH-OH}$$

The reaction of alkyl halides (RCH_2Br) with acetate, iodide, and cyanide gives the corresponding substituted products in PEG-400 (Santanielto, 1984). Potassium thioacetate in PEG-400 has been used as a nucleophilic reagent for the alkyl halides (R—CHXR'), which gives 92–98% yield of product (Ferravoski et al., 1987).

iv) Diels–Alder reaction

Diels–Alder reaction of 2,3-dimethyl-1,3-butadiene with acrolein in PEG-300 gives the addition product in good yield.

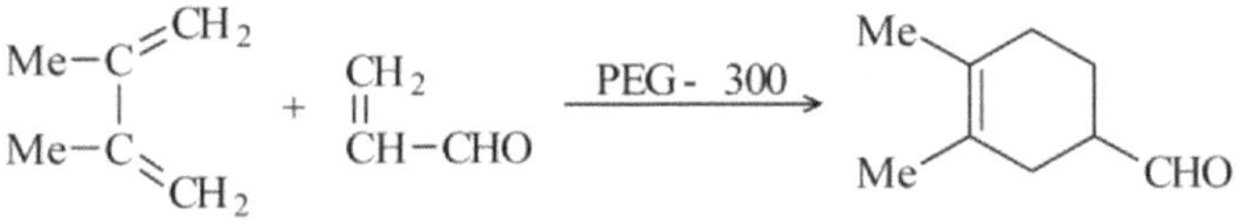

v) Heck coupling reaction

It has been reported that the Heck coupling reaction can be carried out in molten liquid PEG-2000 at 80°C with good yield (Chandrasekhar et al., 2002). After the extraction of product, PEG-2000 and $Pd(OAc)_2$ could be reused.

$$R{-}C_6H_4{-}Br + CH_2{-}CH{-}X \xrightarrow[Pd\,(OA)_2]{PEG\text{-}\ 400} R{-}C_6H_4{-}CH{=}CH{-}X$$

$X = \text{-COOEt}; C_6H_5\text{-}$

vi) Baylis–Hillman reaction

Unreactive aldehydes and activated olefins react, when PEG-400 is used as a recyclable solvent for the Baylis–Hillman reaction (Chandrasekhar et al., 2004).

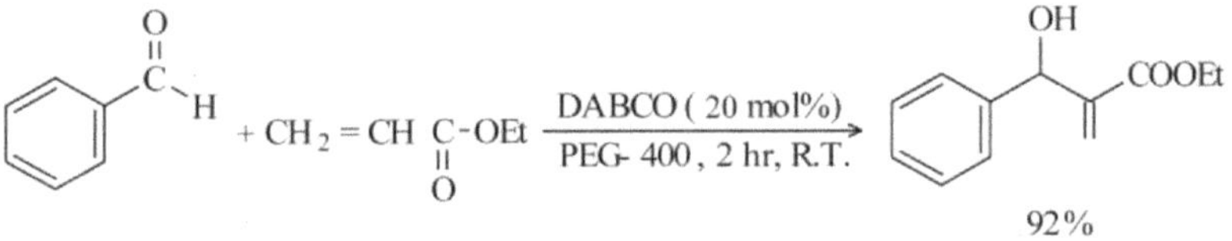

vii) Suzuki cross-coupling reaction

Suzuki cross-coupling reaction is a Pd catalyzed C-C coupling reaction of organoboron compounds with aromatic aldehydes in the presence of a base using PEG-400 as a reaction medium. The yield of product biaryl was found to be 55–81% (Namboodiri and Varma, 2001).

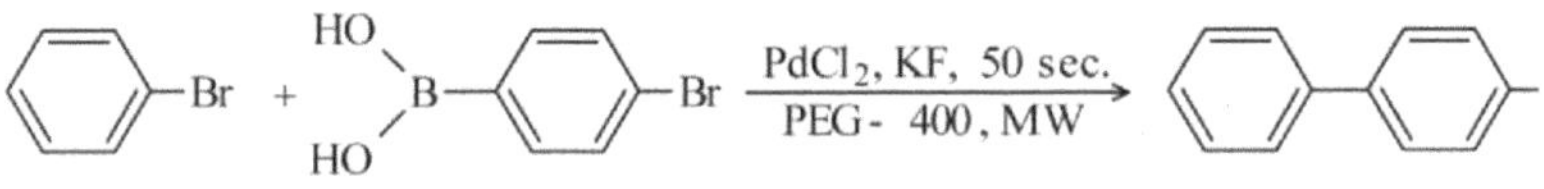

viii) Decarboxylation of cinnamic acid

The decarboxylation of substituted α-phenyl cinnamic acid derivative has been achieved by Kumar et al. (2007) using catalytic amount of methylimidazole and aqueous $NaHCO_3$ in PEG under microwave irradiation.

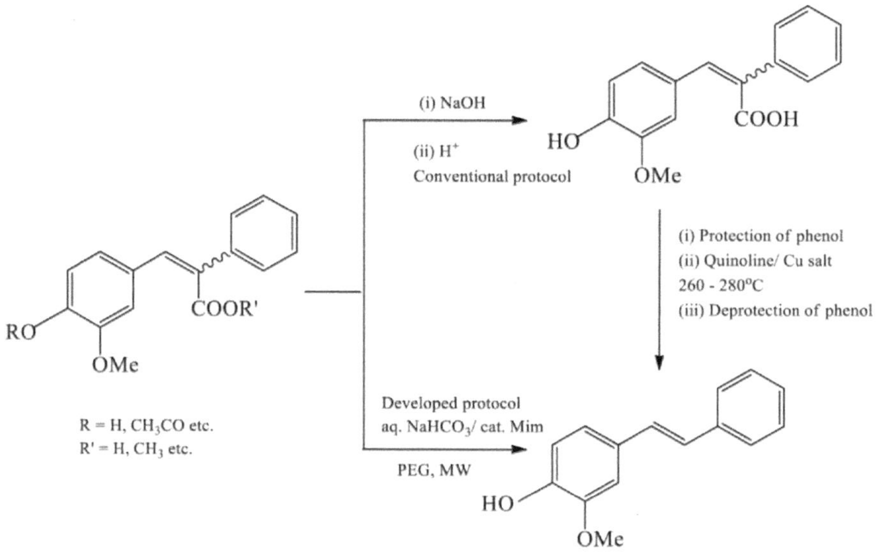

ix) Synthesis of 2-amino-2 chromones

The condensation reaction of aldehyde, malononitrile, and α-napthnol in PEG-H_2O gave 2-amino-2-chromones in high yield at room temperature. The reaction was catalyzed by nano-sized MgO (Kumar et al., 2007).

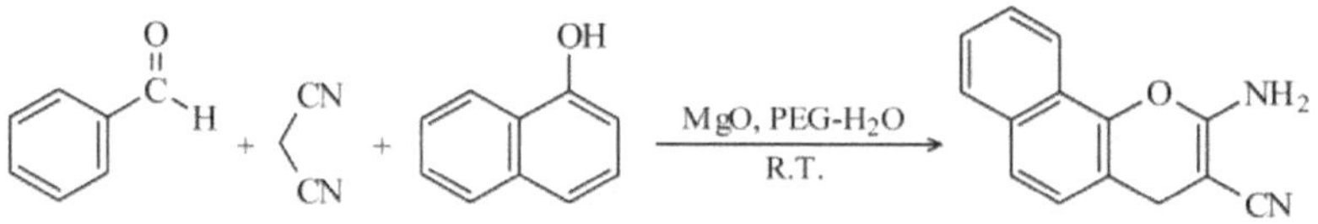

This reaction has a very simple experimental procedure, and cost effectiveness while recyclability of catalyst and use of environmental friendly solvent are another beneficial features of this reaction. On the basis of these applications and its characteristics, it can be concluded that PEG and its aqueous solution can be one of the best alternatives of organic solvents. It can be used as a phase transfer catalyst also. The use of PEG in many enzymatic transformations has also been reported (Tjerneld et al., 1985; Persson et al., 1991; Mandenius et al., 1988; Kondo et al., 1994)

x) Sonogashira coupling reaction

Zhao et al. (2014) used as an extremely active catalyst $PdCl_2(PPh_3)_2$ in a mixture of water and poly(ethylene glycol) (PEG-2000) for Sonogashira coupling reaction (carbonylative) of terminal alkynes with aryl iodides. This reaction was used to prepare flavones from *o*-iodophenol and terminal alkynes. This catalytic system can be reused 6 times without any major loss in its catalytic activity.

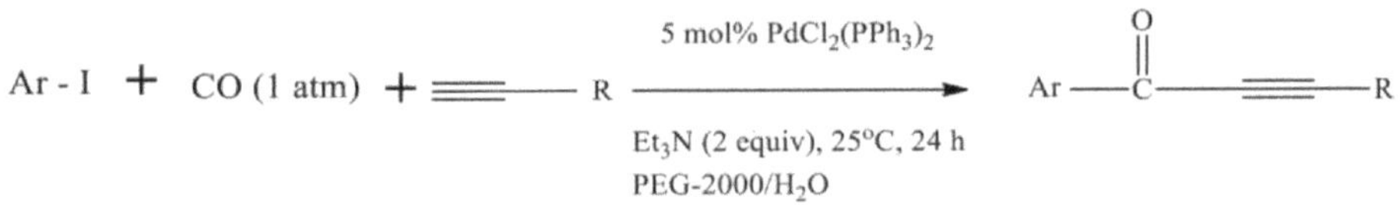

xi) Addition

Michael addition of aldehydes (stereoselective and organocatalytic) was observed by Feu et al. (2014) using PEG as a solvent. The *trans*-β-nitrostyrenes was the final product and it was obtained in good yields and stereoselectivities.

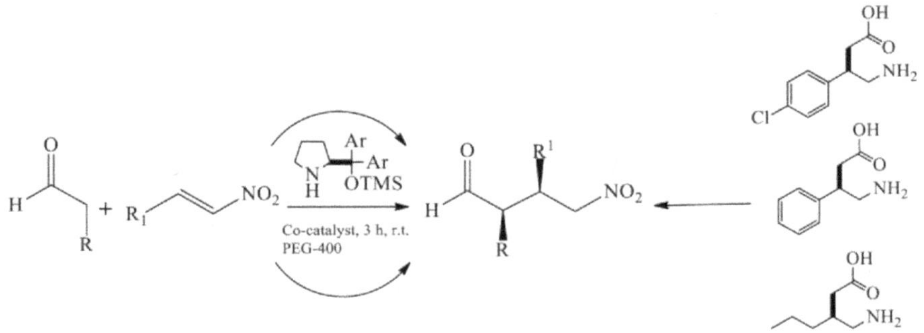

xii) Oxidative cyclization

Tiwari and Bhanage (2016) synthesized 1,3,5-triazines from amidines with benzylamines and N-substituted benzylamines. They used PEG-600 as a green solvent. This protocol is a green chemical route as it is transition-metal phosphine ligand free and molecular oxygen is utilized (O2) as an oxidant. They could obtain a series of 1,3,5-triazines derivatives with excellent yields in a limited time.

xiii) Synthesis of heterocycles

A cascade synthesis of [1,3] chromeno[1,3]oxazin and oxazino quinoline derivatives has been reported using PEG as a green solvent by Yadav et al. (2020) via three component reaction of various aromatic amines 4-methylumbelliferone/4-hydroxyquinoline-2(1H)-one and formaldehyde.

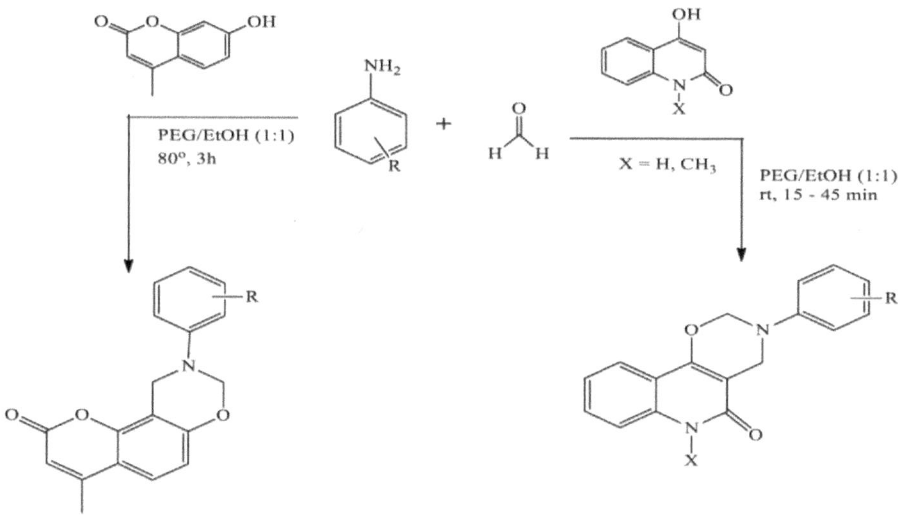

9 Examples
Up to 90% yields

17 Examples
Up to 96% yields

- Catalyst free reaction
- Environment friendly and cost effective
- Simple operation
- Mild reaction conditions

A green, as well as efficient methodology for the synthesis of a variety of isoxazolyl chromeno[2,3-*b*]pyridine-3-carboxylate derivatives have been reported by Ponduri et al.(2018) via one-pot reaction of isoxazolyl enamino esters and 3-formylchromones. They used water as a reaction medium and polyethylene glycol-400 (PEG-400) as the green promoter.

Khan et al. (2015) developed a green one-pot protocol for synthesizing 2-amino selenopyridine derivatives using ultrasound-assisted reactions of malononitrile, aldehydes, and benzeneselenol. They used polyethylene glycol (PEG-400) as a green solvent. It was observed that sterically hindered o,o-disubstituted aromatic aldehydes could be obtained from corresponding functionalized seleno dihydropyridines. Here, four new bonds are formed in one pot and these are two CC, one CN, and one CSe bonds.

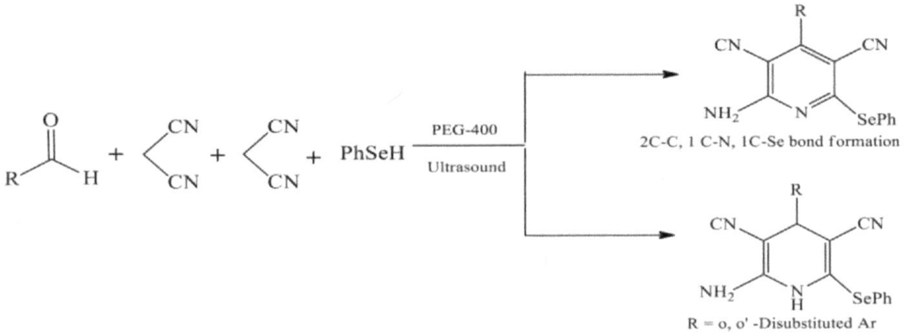

It was reported by Kidwai et al. (2010) that ceric ammonium nitrate (CAN) efficiently catalyzes the synthesis of benzimidazole derivatives from o-phenylenediamine and aldehydes in PEG. This seems to be a facile route for the synthesis of benzimidazoles with good yields and that too with little catalyst loading.

xiv) Friedel–Craft reaction

A one-pot multicomponent aza-Friedel–Crafts reaction was carried out by Bosica and Abdilla (2017) under neat, heterogeneous, and green conditions. This reaction proceeds in the presence of 1.25 mmol% of 30% w/w silico-tungstic acid, which was supported on Amberlyst 15 beads. It was reported that primary and secondary anilines afforded 3-substituted indole as product in good to excellent yields at room temperature.

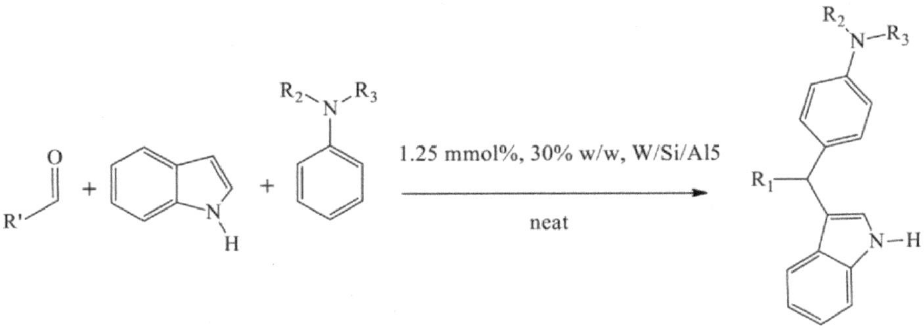

xv) Others

A sustainable protocol has been developed by Gaikwad and Bhanage (2019) for the synthesis of aromatic esters by a carbonylative method using di-μ-chlorobis [5-hydroxy-2-[1-(hydroxyimino-κN) ethyl] phenyl-κC] palladium (II) dimer catalyst in PEG-400 (greener and recyclable solvent). This reaction was carried out using CO in a balloon at room temperature and direct insertion of CO was found to lead to step and high atom economy. It was claimed that Pd/PEG-400 system could be used again for five consecutive cycles without any major loss of activity and selectivity

7.2.3 GLYCEROL

Glycerol is **not only a green solvent** but it can also serve as an **effective replacement for many other solvents.** Alkylaminophenols, 2-substituted pyridines, and 2*H*-chromenes were prepared by Rosholm et al. (2015) using multicomponent Petasis borono–Mannich (PBM) reaction in reasonable to good yields. Glycerol was used as a solvent and it provided higher yield in some reactions.

- ***Advantages of Glycerol as a Solvent***

Polarity: Glycerol is a polar molecule, and it is capable of dissolving many polar organic compounds as well as hydrophobic substrates such as ethers and hydrocarbons.

Volatility and boiling point: Boiling point of glycerol is 290°C and it has nonvolatile behavior so it could easily be separated from a solute using distillation. Its high boiling point enables glycerol to be used at higher temperatures.

High yields: In certain reactions, use of glycerol results in high yields due to possibility of hydrogen bonding by the glycerol molecule. The reaction selectivity is more in glycerol for various reasons including its polarity, structure, and solubility properties resulting in higher yields.

Microwave heating: It has been reported that glycerol can tolerate the heating, even when microwaves are used. This is used in many microwave-assisted organic synthesis and results in a cleaner reaction and less reaction time.

Apart from these, there are some key properties of glycerol that make it a green solvent. These are low toxicity, low vapor pressure, low environmental impact, availability, easy handling, and storage. It can also be used with catalysts.

i) Organic synthesis

It has been reported that some organic synthesis reactions, for example, Heck reactions, Suzuki reaction, and hydrogenation reactions are carried out with high yields in glycerol. Aza-Michael reaction between p-anisidine and n-butyl acrylate can proceed smoothly under catalyst-free conditions in glycerol as a solvent, whereas many other solvents were found to be ineffective (Gu and Jerome, 2010).

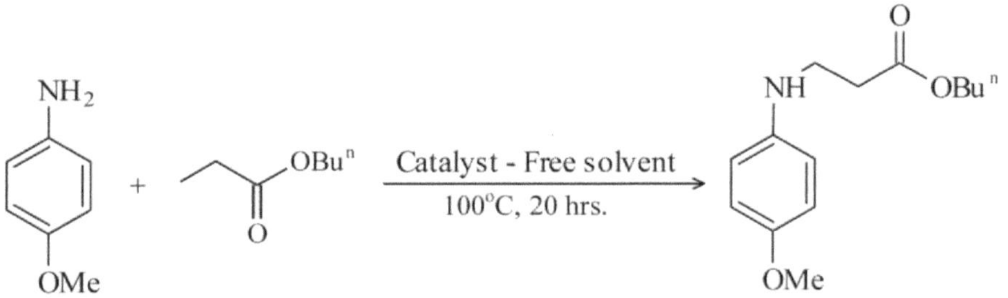

Glycerol has been applied as an efficient, biodegradable, reusable, and green promoting medium for the one-pot three- component diversity-oriented synthesis of 4H-pyrans (Safaei et al., 2012).

Thurow et al. (2013) reported synthesis of 2-organylselanyl pyridines by reactions of organylselenols with 2-chloropyridines. They used hypophosphorous acid as a reducing agent and glycerol as a solvent. A wide range of selenium substituted pyridines were obtained in good to high yields.

Chahdoura et al. (2015) used palladium nanoparticles (immobilized in a glycerol phase) in multi-step syntheses of heterocycles. It was reported that two- and three-component carbonylative couplings were followed by an intramolecular cyclization leading to tetrahydroisoquinolin-1,3-diones N-substituted (na)phthalimides, and isoindole-1-ones in high yields. Similarly, 2-benzofurans and dihydrobenzofurans could be also obtained by Cu-free Sonogashira coupling/hetero-cyclization tandem processes. It was revealed that the Pd-based catalytic glycerol phase can be recycled up to ten times maintaining its activity and selectivity.

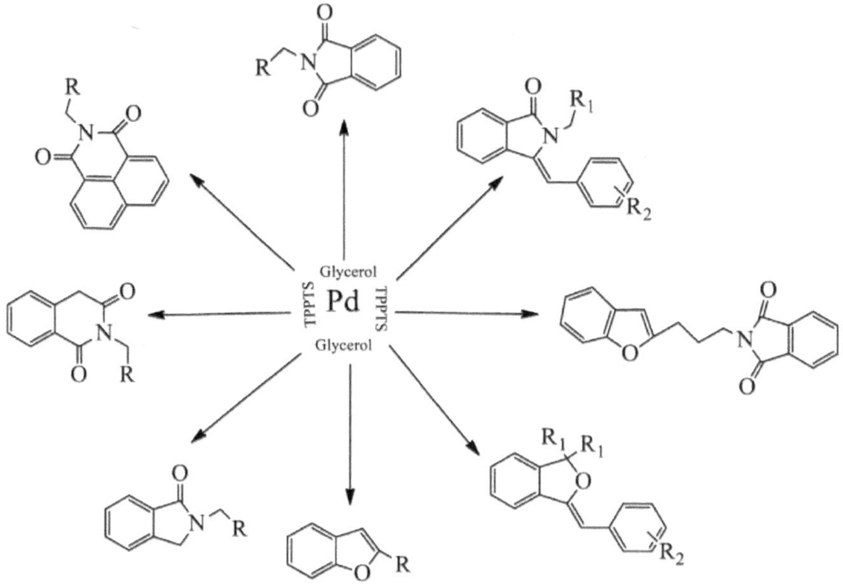

- One-pot procedure, high yield
- Broad substrate scope, recycling

Taduri et al. (2014) developed an efficient, mild, as well as green methodology for the synthesis of some 2-hetero/styryl-benzimidazoles. These were synthesized by the condensation of cinnamic acids with o -phenylenediamine using glycerol containing triacetylborate (10–20mol%) at 150–180°C as the medium. Condensation of aromatic aldehydes with 2-methylbenzimidazole derivatives was also carried out by them using glycerol containing triacetylborate (10–20 mol%) as the reaction medium. A three-component synthesis of triazolo[1,2-a]indazole-triones in one pot was investigated by Shekouhy et al. (2015). They carried out this synthesis under catalyst-free conditions using glycerol as a benign solvent. A wide range of carbonyl compounds are condensed with aromatic aldehydes with a reactive -methylene group and urazole derivatives. Reaction was completed in short times with good to excellent yields of products.

ii) Enhancing reaction selectivity

The selectivity of certain reactions has been improved on using glycerol as a solvent, to produce higher product yields, for example, ring opening reaction of p-anisidine with styrene oxide can be performed without any catalyst, if glycerol is used as a solvent. This reaction demonstrates a better regioselectivity. It was also reported that reaction involving styrene, paraformaldehyde, and dimedone shows an extremely higher yield in glycerol. The rate of the hetero-Diels-Alder reaction was also increased in glycerol because it is a polar protic solvent.

Overall, the structure of glycerol, its polarity, and the intermolecular forces can make a reaction more selective. Thus, the product yield obtained depends on the way in which a solvent interacts with the starting material. Moreover, glycerol has been used uniquely for its high selectivity to give desired products in one-pot two-step reactions. By using a single reactor, the yield of the product increases on one hand, while separation processes and purification of intermediate compounds are avoided on the other hand, making this process green, for example, (i) Reaction involving arylhydrazines, β-ketone esters, formaldehyde, and styrenes (ii) Reaction involving indoles, arylhydrazine, β-ketone, and esters, and (iii) One-pot sequential reaction involving phenylhydrazine, ethyl 4-methoxybenzoylacetate, α-methylstyrene, and paraformaldehdye in glycerol.

iii) Solvent for biocatalysis

Glycerol can be used as a solvent giving high yield with the use of natural catalysts due to its low toxicity and high affinity for hydrophilic compounds, for example, Baker's yeast catalyzed reduction of ketones and bioreduction of 2'-chloroacetophenone using glycerol as a co-solvent to obtain high yields.

iv) Catalyst design and recycling

In the process of homogeneous catalyst recovery, solvents are used to immobilize and recycle the catalyst. It reduces waste and recyles the catalyst. Glycerol has a better solubility with the homogeneous catalyst with ionic compounds and hence, it can be used for catalyst recycling. It has been observed that in the catalytic formation of bisindolylmethane in glycerol over $CeCl_3$/Lewis acid, the reaction products can be selectively extracted from the glycerol/$CeCl_3$ mixture by liquid phase extraction with ethyl acetate Silveira et al. (2009) and therefore, allowing a convenient recycling of both; $CeCl_3$ and glycerol.

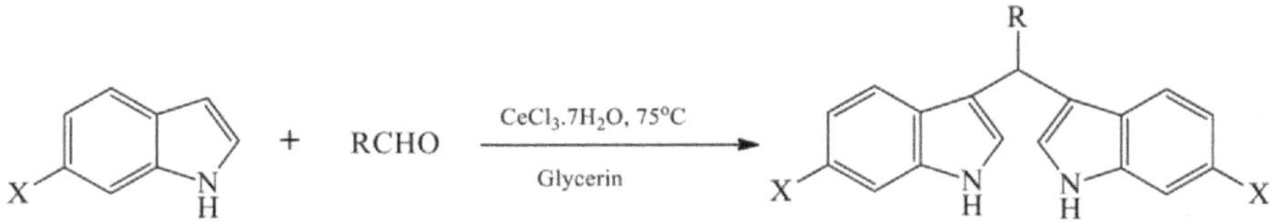

X= H and Br

R= C_6H_5, 4-$NO_2C_6H_4$, 3-ClC_6H_4, 4-$CH_3OC_6H_4$, 2-CLC_6H_4, 2-$CH_3C_6H_4$, C_6H_5CH=CH 3,4,-OCH_2O-C_6H_3, 2-furyl and C_4H_9

Glycerol can also dissolve organometallics complexes and thus allow nonionic catalysts to be recovered, for example, the hydrogenation reaction catalyzed by the [Ru(p-cumene)Cl_2]$_2$ complex using glycerol. This Ru complex catalyst is not ionized in glycerol, and is recycled after the extraction of the reaction products with diethyl ether (Wolfson et al., 2009).

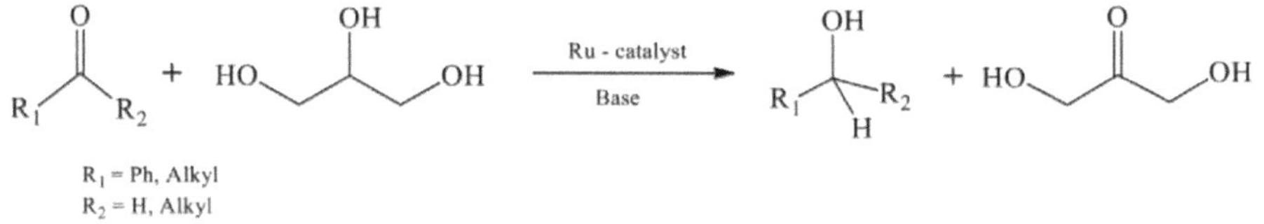

v) Solvent for separation

Bioethanol is one of the important biofuel and it is a green alternative to gasoline. The purification of bioethanol by extractive distillation was achieved by glycerol. It was found that the separation process of ethanol was more effective, when glycerol was used. Here, the ethanol, water, and glycerol were all recovered with more than 99% purity.

vi) Use in materials chemistry

Glycerol shows promising properties in the preparation of materials. Its high boiling point and low vapor pressure allow it to be used for the reactions at high temperatures. Good solubility of inorganic and organic compounds in glycerol is another added advantage.

vii) Nanoparticles

One simple method to prepare metal particles was suggested by Sinha and Sharma (2002) through heating a metallic salt and then reducing it. Glycerol has a high boiling point and it can also act as a reducing reagent. Some copper particles were successfully obtained with a purity of greater than 99%, when $Cu(OH)_2$, CuO, and $Cu(OAc)_2$ were heated under ambient conditions in glycerol at a temperature less than 240°C.

They also reported that silver nitrate produces silver particles with a high yield and uniformity using glycerol as a solvent as well as a reducing reagent using a similar process (Sinha and Sharma, 2005). The size of the silver particles can be changed by changing the amount of $AgNO_3$.

Chahdoura et al. (2014) observed synthesis of copper(I) oxide nanoparticles, while they were stabilized by poly(vinylpyrrolidone) in neat glycerol under hydrogen atmosphere. Then as-prepared material was applied in C-heteroatom couplings and azide–alkyne cycloadditions with desired products selectively with yields > 90 %.

viii) Coupling

Nickel catalyzed Suzuki reactions of aryl diazonium salts with phenyl boronic acids were reported by Bhojane et al. (2016) in glycerol medium. It was revealed that different aryl diazonium salts reacted with aryl boronic acids effectively affording the corresponding diaryl compounds (good to excellent yields).

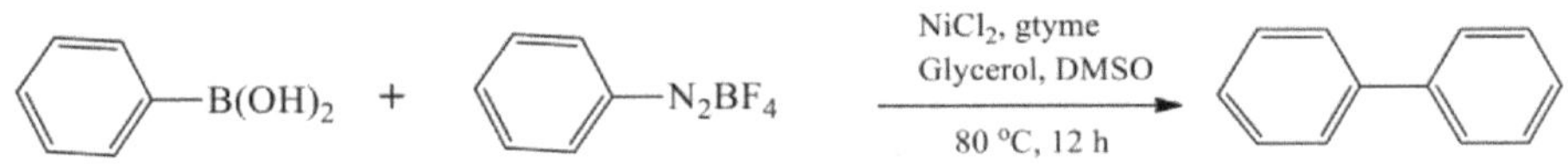

ix) Condensation

Nascimento et al. (2015) developed a simple method to synthesize 4-arylselanylpyrazoles by reaction of α-arylselanyl-1,3-diketones with arylhydrazines at 60°C under N_2 atmosphere using glycerol as solvent. This is a direct cyclocondensation reaction and it can be performed with α-arylselanyl-1,3-diketones and arylhydrazines bearing electron-withdrawing as well as electron donating groups. The corresponding 4-arylselanylpyrazoles were obtained in moderate to good yields.

x) Cycloaddition

Metal-free intermolecular Huisgen cycloadditions was successful carried out by Rodríguez-Rodríguez et al. (2015) using nonactivated internal alkynes in neat glycerol under both, thermal as well as microwave dielectric heating. It was observed that no reaction occurred in other protic solvents, such as ethanol, water, or diols. Trujillo et al. (2019) investigated copper(I)-catalyzed azide-alkyne cycloaddition (CuAAC) reactions in glycerol. They studied four synthetic protocols to have a comparison of time required for the, purity and yield of 1*H*-1,2,3-triazole products. The compounds synthesized were non-fluorinated triazoles along with some new fluorinated triazoles.

xi) Other

An efficient method has been developed by Nemati et al. (2016) for the synthesis of 1,2,4,5-tetraaryl and 2,4,5-triaryl imidazole derivatives. They used glycerol as a green solvent and it is catalyst-free route also. The yields of products were found comparable to or better than that obtained by conventional method. The use of green solvent in this reaction makes it cost-effective and safe.

The click reaction (catalyzed by CuI) was carried out by Pasupuleti and Bez (2019). They did not need an inert atmosphere, when they used CuI/l-proline system and glycerol as solvent. The protocol exhibited good stability particularly towards sensitive functional groups such as 1,2,4-trioxanes and acetonides.

7.2.4 CYCLOPENTYL METHYL ETHER (CPME)

Cyclopentyl methyl ether (CPME) is new hydrophobic ether solvent. Unlike other common ether solvents, CPME has unique excellent properties and it has many other properties that make it a greener, easy to use and a more cost effective solvent for many types of synthesis. Some properties of CPME are -

High hydrophobicity: It is easily separated and recovered from water, reduces emissions and waste water. It is widely used as a reaction, extraction, and crystallization solvent, in simple and one-pot syntheses.

Wide liquidity range: It has a wide range of applications from lower to higher temperatures for accelerating reaction rate.

Low heat of vaporization: The heat of vaporization of CPME is 69.2 Kcal kg^{-1} and due to this low value, it saves energy for distillation and recovery.

Resist peroxide formation: Ethereal solvents have some explosive nature arising from peroxide generation. The ether radical is more stable and more peroxide is accumulated during the storage. The ether radical of CPME is unstable compared to other ethereal solvents. Thus, it is concluded that the peroxide formation from CPME is very slow as compared to other ethers.

Relatively stable to acids and bases: It has good stability under acidic and basic conditions.

It has limited solubility in water, easy drying, narrow explosion area, and high boiling point. Azeotropes with water are formed. It makes CPME a good alternative to other solvents such as dioxane (carcinogenic), tetrahydrofuran (THF), 1,2-dimethoxyethane (DME), and 2-methyl tetrahydrofuran (2-MeTHF).

These unique properties of CPME give high recovery rate (> 90%). All these properties collectively contribute to benign chemistry, as it reduces total amount of solvents used, waste water, carbon dioxide emissions, and waste solvent. This solvent can also help in process innovation, where it can provide saving in process time, simplified total process, shortening work-up time, and as a result, costs are also reduced. The melting point of CPME is –140°C, whereas boiling point is 106°C. The viscosity of CPME is found to be 0.55 cP and its surface tension is 25.17 mN m^{-1}. It has a low specific heat value (0.4346 Kcal kg^{-1} K) and its dielectric constant is very less compared to water (4.76). It has low density (0.86 g cm^{-3}).

Azzena et al. (2019) evened cyclopentyl methyl ether as the green solvent. It has a high boiling point, low toxicity, low melting point, chemical stability with a wide range of conditions hydrophobicity stability toward the abstraction of hydrogen atoms, relatively low latent heat of vaporization easy recovery, and recyclability.

CPME is a better alternative for other ethereal solvents. As a solvent, it can be used in the following processes

- Extraction
- Crystallization
- Polymerization
- Coating

i) *Reactions with alkylating agents*

Methylation of alcohol with methyl triflate (Me-OTf) (which is a powerful methylating agent) has been carried out in CPME (Watanabe et al., 2007).

ii) *Friedel-Crafts type reactions*

In the Friedel-Crafts reaction, Lewis acid catalyst and halogenated solvents are required. As CPME can attain anhydrous conditions, it is useful to keep anhydrous reaction media for metal triflates or prevent the decomposition. But CPME cannot be compatible with a strong Lewis-acid catalyst ($AlCl_3$). Due to this, some Ti Lewis acids are recommended with CPME (Watanabe et al., 2007).

iii) *Grignard-type reactions*

CPME is a preferred solvent for Grignard-type reactions because it can maintain anhydrous conditions without any particular precautions. Magnesium turnings were placed on a flask for the preparation of Grignard reagent. Then it was covered with CPME and a small piece of iodine was added. A solution of alkyl bromide was also added while heating. After the completion of the addition, the mixture was heated for a while. A small amount of magnesium still remained in the flask. The Grignard reagent thus prepared was cooled to 0°C. The solution became cloudy with the precipitation of the Grignard reagent. After that a solution of carbonyl compound was added to the Grignard reagent and allowed to attain room temperature. The reaction achieved completion at this stage (Watanabe et al., 2007).

Cyclopentyl methyl ether (CPME) can be used as a green solvent for Grignard reactions. Here, diisobutylaluminum hydride was used as an activator of magnesium and a series of Grignard reagents could be prepared using CPME as the solvent. It can be efficiently recycled without any adverse effect in the yield of consecutive experiment. It was reported that some of the Grignard reagents prepared in CPME were found stable for several months.

(iv) Reactions with transition metal catalysts

The ethereal solvents are used for transformations with transition metal catalysts (Pd, Ni, Rh, Ir). High reaction temperature is preferable and anhydrous conditions are also useful for multicomponent couplings for such catalytic reactions. Like other ethers and toluene, CPME can also take part in the Pd-catalyzed asymmetric allylic alkylation. Due to high boiling point, CPME is successfully employed as the solvent for such coupling reactions for the completion of reaction within a short time. High boiling points and easy workup made the overall reaction sequence very effective and convenient (Watanabe et al., 2007).

Recently, Kawatsura et al. (2010) reported a Suzuki-Miyaura coupling reaction catalyzed by a ruthenium complex. In this reaction, substituted halobenzene were coupled with $ArB(OH)_2$ to give diaryl compounds. Arylation is also reported with 5- member heterocyclic rings using aryl halide in the presence of Pd-base and CPME (Beydoun and Doucet, 2011).

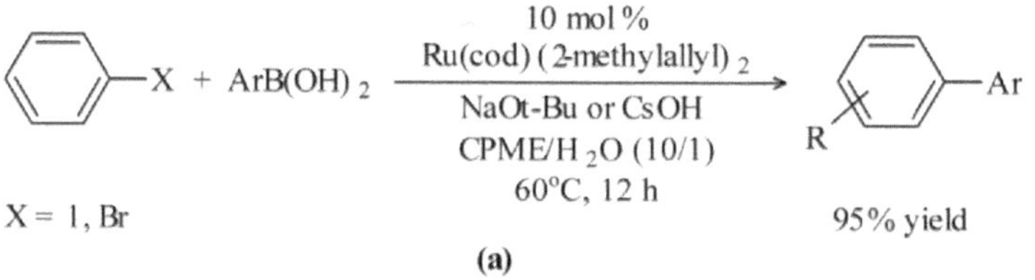

(a)

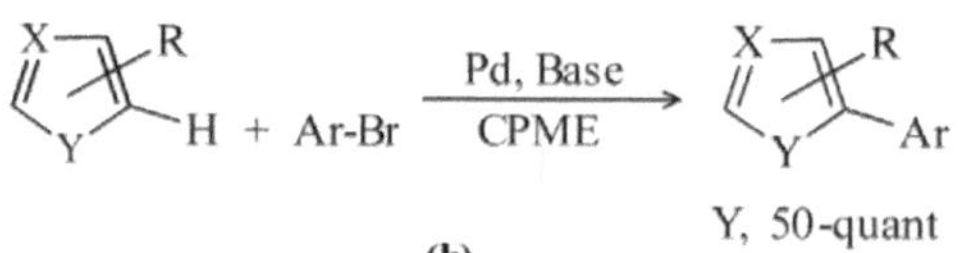

(b)

v) *Palladium catalyzed direct arylation of heteroaromatics*

Cyclopentyl methyl ether can be employed for the palladium catalyzed direct arylation of heteroaromatics. The direct 5-arylation of thiazoles, thiophenes, or furans was carried out by using aryl bromides (a coupling partner) in the presence of CPME and 0.5–1 mol% of palladium catalysts with moderate to high yields (Watanabe et al., 2007).

vi) ***Condensation reactions***

Aldol and Claisen condensations require basic media; thus, it is compatible with CPME. Claisen-Schmidt condensation with a pyridine derivative is feasible in CPME with the proper base catalysis. These base catalyzed reactions indicate the usefulness of CPME for general condensation reactions. In addition to this advantage, the desired product can be isolated by the simple extraction from CPME, leaving the polar starting materials in aqueous solution. Thus, use of CPME makes the workup very easy (Watanabe et al., 2007).

Some secondary amines (symmetric) were synthesized by Wang et al. (2018). They used self-condensation of primary amines for this purpose and a photocatalyst (palladium-loaded titanium dioxide) in CPME. It was observed that these reactions can yield a series of secondary amines with moderate to excellent yields and at ambient temperature, 30°C.

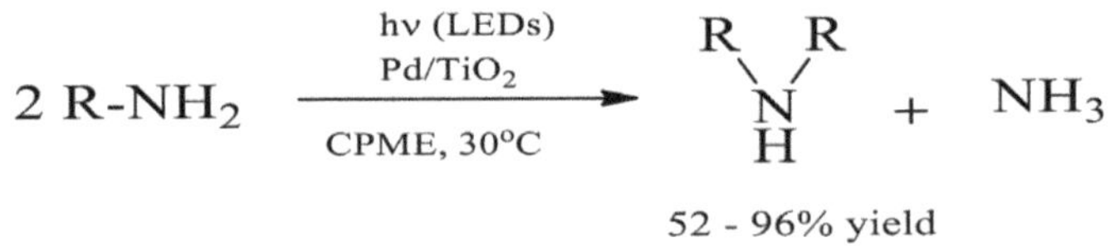

vii) Enolate chemistry

An interesting solvent effect of CPME emerged during enolate formation. An asymmetric methylation (chirality transfer reaction), which might be manifested through the formation of a rigid enolate from *O*-methyl mandelic acid, in the medium of CPME was reported by Kawabata's group. Aggregated enolate formation is controlled by the interference of CPME with base or enolate itself.

viii) ***Transformations***

Classical transformations, which have been utilized in the pharmaceutical industries, are renewed by the use of CPME as a solvent. It demonstrates the application of CPME in the manipulation of the heterocyclic intermediates. It provides a green solution for improving chemical process by minimizing the solvent waste stream and improving laboratory safety due to CPME's unique composition, which resists the formation of peroxides. It is stable than THF and 2-MeTHF (stabilizer required). Therefore, the frequency of peroxide testing is reduced.

It is a novel hydrophobic ether solvent and used in many organometallic reactions, where it provides better yield and higher selectivity than THF. It forms an azeotrope rich with water and can be more easily dried as compared to THF and 2-MeTHF. It has limited miscibility in water (1.1 g/100 g at 23°C), easily separated and recovered from water reducing the waste stream. CPME has a higher boiling point (106°C) as compared to THF and 2-MeTHF and higher reaction temperature reduces overall reaction time. It has low heat of vaporization and therefore, less solvent is lost during reflux.

ix) Oxidation

Oxidation of an alcohol to ketone in the presence of MnO2 was carried out by Watanabe et al. (2007). They could achieve 90% conversion using CPME as a green solvent.

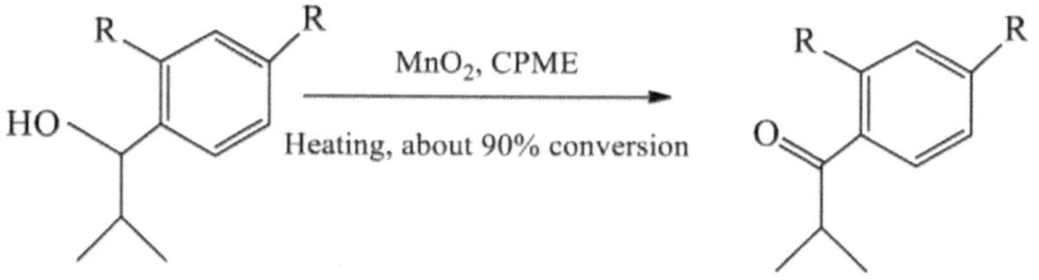

x) Reduction

Kobayashi et al. (2013) evaluated cyclopentyl methyl ether as a green solvent for radical reactions. They carried out hydrostannation, hydrothiolation, and hydrosilylation. It was also observed that CPME degraded slightly into cyclopentanol, cyclopentanone, methyl pentanoate, and 2-cyclopenten-1-ol.

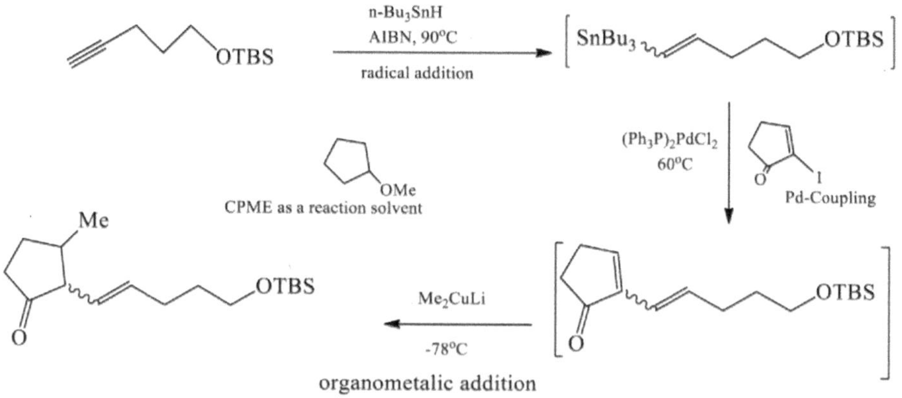

The reductive cleavage of methyl glucoside acetals was reported by Gozlan et al. (2014) using hydrogen as a reducing agent at 120°C. This reaction is palladium-catalyzed and cyclopentyl methyl ether (CPME) was used as a solvent without acid co-catalyst. It was reported that the respective methyl glucoside monoethers could be obtained with good isolated yields (3781%), but high selectivities (8699%).

xi) Alkylation

CPME is also used for O-methylation of pyridinol (Watanabe et al., 2007).

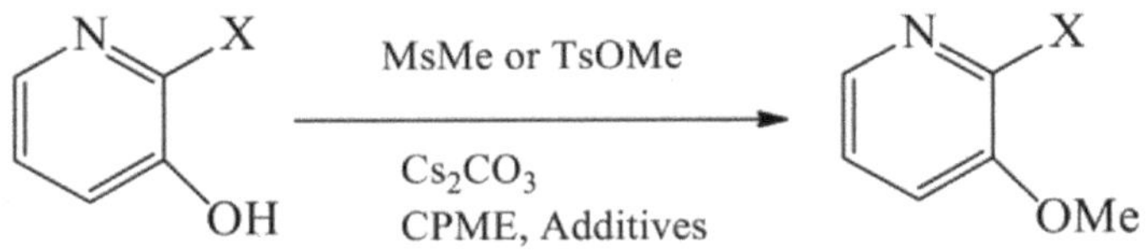

xii) Reaction in the presence of acid

Watanabe et al. (2009) observed Pinner reaction in CPME, where cyanide is converted to methoxy imine.

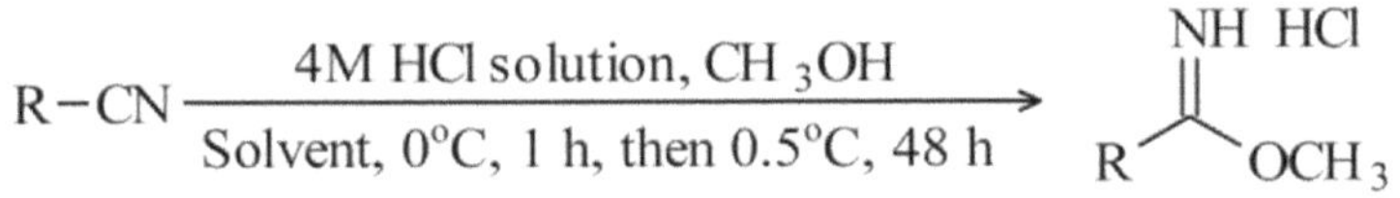

xiii) Polymerization

Cross coupling reaction of p-dibromobenzene and m-dibromobenzene takes place in the presence of nickel complex and magnesium (Yamamoto, 2003).

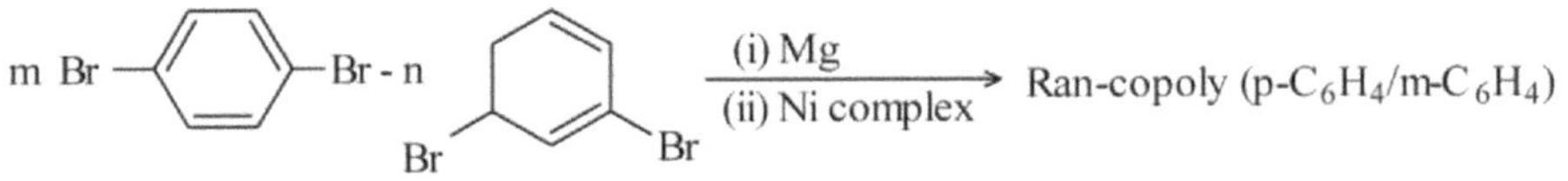

xiv) Nanoparticles

Nanoparticles of gold were obtained by Sugie et al. (2008) treating $HAuCl_4$ with a thiol.

$$HAuCl_4 \text{ - } 4H_2O + HS \text{ - } nC_{12}H_{25} \xrightarrow[\text{CPME}]{Et_3SiH} \text{Gold nanoparticles}$$

xv) Others

Lenstra et al. (2018) observed a highly efficient catalytic Staudinger reduction to convert organic azides to amines using CPME and obtained an excellent yield. It was observed that the reaction exhibited excellent functional group tolerance toward reduction, such as alkene, esters, amides, ketones, nitriles, sulfones, and benzyl ethers.

$$R \text{ - } N_3 \xrightarrow[\text{CPME, reflux, 20h}]{3\text{ mol\% } Ph_3P\text{, PMHS}} R \text{ - } NH_2$$

- Up to 99% isolated yield
- No column chromatography
- Excellent functional group tolerance
- Amenable to gram-scale synthesis

A novel approach was developed by Liu et al. (2018), synthesize furfural from xylose via sulfonic acid-functionalized metal–organic frameworks (MOFs) (MIL-101(Cr)-SO3H) using a cyclopentyl methyl ether (CPME)/ H_2O-NaCl (biphasic) solvent system. It was reported that xylose high yield of furfural (70.8%) with 97.8% xylose conversion could be achieved under optimum conditions; were discussed in detail. It was reported that furfural under following conditions: xylose (0.4 g), catalyst (0.27 g), NaCl (26 wt.%), mixed solution of CPME/H_2O (mass ratio 2:1) (12 g), at 443 K for 180 min. The catalyst can be recycled five times.

The recovery of formic acid (FA), acetic acid (AA), and propionic acid (PA) from aqueous solutions was reported by Türk et al. (2020) in CPME. They used tributyl phosphate (TBP) as an extractant, cyclopentyl methyl ether (CPME) and 2-methyltetrahydrofuran (2-MeTHF) have been utilized as green solvents and diluents. The loading factor (Z) distribution coefficient (D), and extraction efficiency (E%) values were determined to be 22.41–78.41, 0.289–4.003, and 0.198–2.218%, respectively. It was revealed that extraction efficiency in both the diluents increased in the following order:

$$PA > AA \geq FA.$$

Coeck and De Vos (2020) reported one-pot reductive amination of carboxylic acids to amines using ruthenium-tungsten bimetallic catalyst. They used H_2 and NH_3 as the reactants. This reaction can be performed with relatively low cost and green solvent, cyclopentyl methyl ether (CPME). It afforded 99% selectivity for the primary amine.

xvi) Solar cells

Li et al. (2019) observed that the performance of all polymer cells can be further enhanced to a higher value of 11% by adjusting the bulk heterojunction (BHC) morphology using a green solvent cyclopentyl methyl ether (CPME), which is otherwise limited to 8–10%

Farahat et al. (2017) could achieve a power conversion efficiency (PCE) of 3.13% using CPME, where donor molecule was two-dimensional conjugated small molecule and [6,6]-phenyl-C61-butyric acid methyl ester (PC61BM) (acceptor) to form (BHJ) in organic photovoltaic (OPV) device. This low PCE may be due to low solubility of PC61BM in CPME. The PCE was increased to 7% on using a mixture CPME: Toluene (60:40).

7.2.5 2-METHYLTETRAHYDROFURAN (2-METHF)

The 2-MeTHF is derived from renewable resources such as corn cobs and sugar cane bagasse. The 2-MeTHF offers both economical and environmentally friendly advantages over tetrahydrofuran, when is used as an organometallic solvent. It is an aprotic solvent, which resembles toluene in its physical properties. Bromo and iodo Grignard reagents tend to be more soluble in 2-MeTHF, while chloro Grignard reagent tends to be less soluble in it. It forms an azeotrope rich with water, which can be more easily dried than dichloromethane or tetrahydrofuran. It is a truly green alternative to dichloromethane and tetrahydrofuran. It has limited miscibility in water (14 g/100 g at 23°C). The 2-MeTHF is easy to separate and recovering it from water reduces its quantity in the waste stream. Its boiling point is 80°C, which is higher as compared to tetrahydrofuran (66°C); thus, higher reaction temperature reduces overall reaction time. It has low heat of vaporization, so less solvent is lost during reflux and therefore, it saves energy during distillation and recovery of solvent.

It is used as an alternative to tetrahydrofuran for organometallic reactions.

- Grignard
- Reformatskii (reformatsky)

- Lithiation
- Hydride reduction and
- Metal catalyzed coupling (Heck, Still, and Suzuki)

It is used as an alternative to dichloromethane for biphasic reactions, for example, alkylation, amidation, and nucleophilic substitution reactions.

The 2-MeTHF has an added advantage of having some beneficial physical and chemical properties. Corn cobs and sugar cane are renewable resources for furfural, which on hydrogenation yields 2-MeTHF while THF is obtained from 1,4-butandiol, which is an oil-derived substance. The 2-MeTHF is the only aprotic solvent similar to THF derived from renewable resources and it is industrially available.

The incineration of solvents in most of the chemical industries adds to greenhouse gases causing greenhouse effect, but incineration of 2-MeTHF does not contribute to it as it returns the carbon dioxide back to the atmosphere, which was captured by previous years' crop.

2-MeTHF is not miscible with water and it is comparable or even better than THF in terms of its chemical properties. However, 2-MeTHF resembles toluene in terms of its physical properties. It provides an easy and clean phase separation during work-up. These advantages make the process simpler and more robust, translating into higher through-put and reduced cost.

2-MeTHF reduces the solvent and energy variable costs. Thus, it has better extractive properties than the classic THF/toluene mixture. This means that the number of extraction steps can be reduced by using it and simultaneously, the recovery of the product is also increased. The 2-MeTHF solution of crude product can be dried through a simple distillation at atmospheric pressure. The water-rich 2-MeTHF azeotrope will create rapidly an anhydrous solution providing the option to add a new reagent without product isolation, for example, the classic reaction sequence from carbonyl to alcohol followed by alcohol to ester is particularly adequate by cutting almost 50% and it is improved by avoiding isolation of the intermediate alcohol. It is much easier to recycle and be dried as compared to THF.

The 2-MeTHF requires only simple distillation at atmospheric pressure, whereas THF is recycled and dried using swing distillation. The recycling and drying of 2-MeTHF is cost effective. The 2-MeTHF is a versatile solvent covering a wide range of applications (Comanita, 2006). It has comparatively more stability than THF even in acidic reactions (David, 2007). The hydrolysis of 2-MeTHF is also much slower because of its immiscibility with water. Lithium aluminum hydride ($LiAlH_4$) is more soluble in 2-MeTHF (about 10%). $LiAlH_4$ can reduce aldehydes, esters, and acids in 2-MeTHF and the products are similar as obtained in THF.

A one-pot dehydrobromination of a 2-bromoacrylate ester with lithium 2,2,6,6-tetramethylpiperidide (LTMP) and aldehydes in 2-MeTHF has been carried out by Pace et al. (2012).

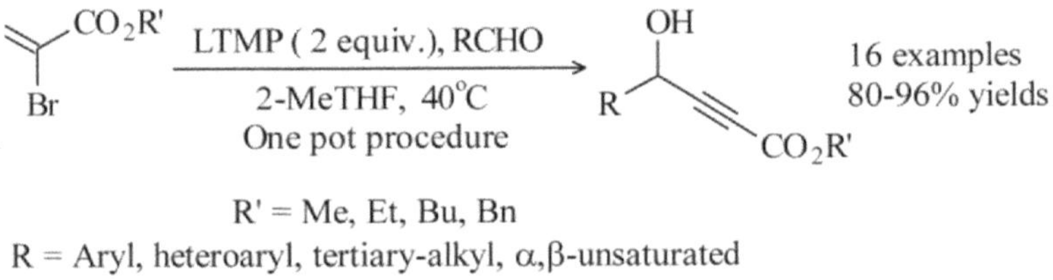

Some esterifications were also carried out in 2-methyltetrahydrofuran (as an excellent substitute for THF) in biocatalyzed processes in organic media. This application for this green solvent is a proof of opening a new field utilizing MeTHF in biotransformations (Simeó et al., 2009).

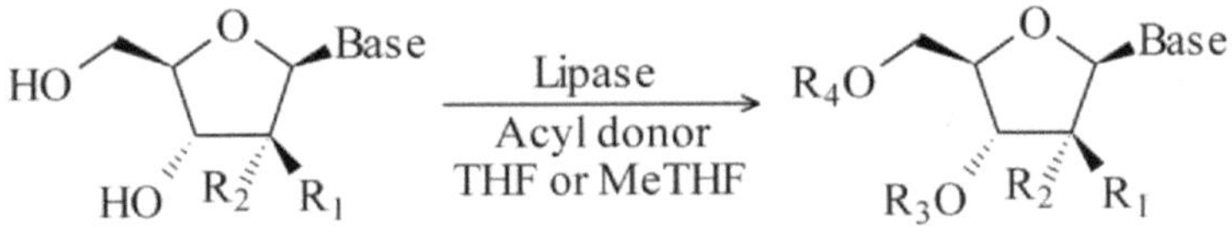

Milton and Clarke (2010) reported the cross-coupling of Grignard reagents at 5M in methyltetrahydrofuran with no added reaction solvents. The 2-MeTHF was also used in an improved microwave accelerated synthesis of the [Pd(L)Cl_2] precatalysts from sodium tetrachloropalladate in very high yield.

2-Methyltetrahydrofuran is a useful bio-based co-solvent for benzaldehyde catalyzed reactions (Shanmuganathan et al., 2010).

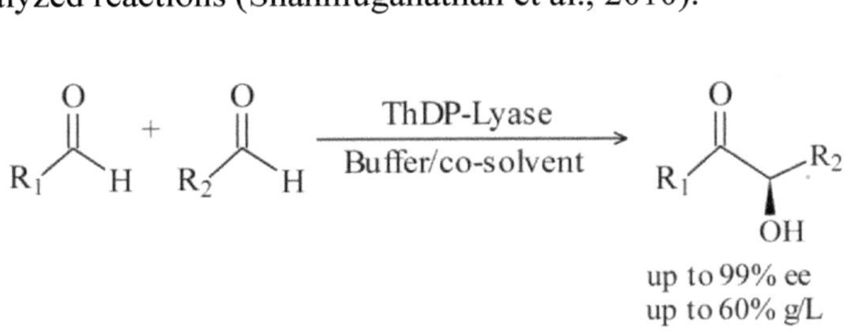

i) Coupling reactions

The formation of C-C bonds involved iron-catalyzed cross-coupling reactions using cost-effective, earth-abundant base-metal catalysis in complex areas of natural product, pharmaceutical development, and synthesis of fine chemical. This process can have good sustainability, if proper green and renewable solvents are identified that can be used in place of conventional solvents. It still remains a challenge. Bisz and Szostak (2018) reported that the iron-catalyzed cross-coupling of tosylates and aryl chlorides provides good to excellent yields using eco-friendly solvent 2-methyltetrahydrofuran (2-MeTHF). This solvent has an excellent functional group tolerance and that too with very mild conditions. This protocol exhibited a great potential of 2-MeTHF as a green solvent for sustainable iron-catalyzed cross-coupling reactions.

The palladium-NHC-catalyzed (NHC=N-heterocyclic carbene) Suzuki-Miyaura cross-coupling of esters and amides via highly chemoselective NC(O) and OC(O) cleavage with aryl boronic acids using eco-friendly, green, and sustainable and 2-methyltetrahydrofuran (2-MeTHF) (solvent) was reported by Lei et al. (2019). Air- and moisture-stable Pd(II) NHC precatalysts for cross-coupling of a variety of amides and aryl esters with aryl boronic acids are possible with excellent yields.

ii) Reduction

A simple and eco-friendly catalytic procedure has been developed by Nardi et al. (2015) for selective reduction of different α,β-unsaturated ketones in 2-MeTHF. The corresponding allylic alcohols were obtained as the product with high chemo- and diastereoselectivity, when $Er(OTf)_3$ and $NaBH_4$ were used. It was claimed that this method reduces the amount of catalyst and $NaBH_4$ as compared to traditional procedures.

7.2.6 ETHYL LACTATE

It is also known as lactic acid ethyl ester. Lactate esters are commonly used solvents in the paint and coating industries. These have numerous attractive advantages like 100% biodegradability, easy to recycle, noncorrosive, and noncarcinogenic. Ethyl lactate has replaced solvents such as NMP, toluene, acetone, and xylene, which has resulted in making the work place relatively safer. Ethyl lactate has been used as a green solvent because it has low VOC, high solvency power for resin and polymer, and a high boiling point. It has almost eliminated the common use of chlorinated solvents.

Ethyl lactate is a bio-based solvent, which is eco-friendly as well as economically viable and it is as effective as compared to other petroleum-based solvents. It has been used for multi-component reactions, asymmetric induction, light-induced synthesis, ligand-free coupling reactions, cycloaddition reaction, stereoselective synthesis, etc.

Pereira et al. (2011) reviewed the field of use of ethyl lactate as a green solvent because ethyl lactate is formed by the esterification of lactic acid by ethanol, which are biomass-derived compounds.

Wan et al. (2012) developed a sustainable catalyst system consisting of H_2O/ethyl lactate, $Pd(OAc)_2$, and K_2CO_3. They used it for Suzuki–Miyaura reactions using various aryl bromides and iodides to incorporate arylboronic acids under ligand-free conditions.

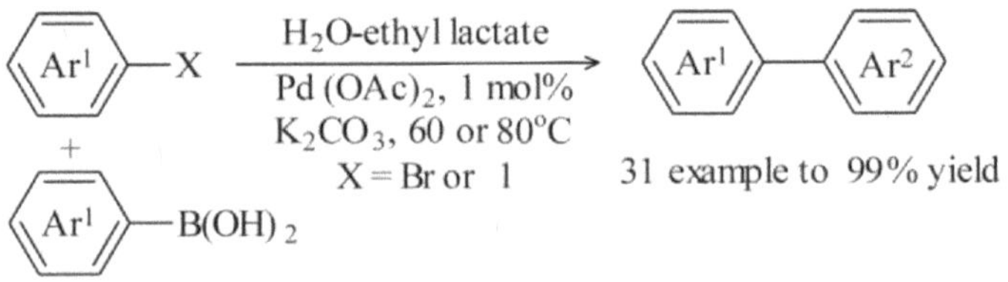

Bennett et al. (2009) used ethyl lactate as a tunable solvent for the synthesis of aryl aldimines

The bio-based chemical ethyl lactate (EL) was used as an excellent medium by Wan et al. (2014) for the Glaser-type homo- and cross-coupling reactions of terminal alkynes. It was reported that good to excellent yields of conjugate diynes were obtained and that too under ligand-free and mild heating conditions in the presence of CuI and molecular oxygen.

Jilva et al. (2019) used ultrasound assisted for extracting lycopene from tomato waste using an eco-friendly solvent mixture, which contains ethyl lactate and ethyl acetate. Erbium (III) chloride in ethyl lactate has been used as an environmentally friendly system for the reaction of furfural and amines. It afforded different N,N-substituted trans-4,5-diaminocyclopent-2-enones diastereoselective, which can be further used as synthetic intermediates to form functionalized derivatives.

7.2.7 PERFLUORINATED (FLUOROUS) SOLVENTS

Horvath and Rabai (1994) introduced this term, which has analogy with aqueous medium. These compounds were defined by Gladysz and Curran (2002) as being compounds that are highly fluorinated and based upon sp^3 hybridized carbon. Perfluorinated hydrocarbons are found to be unique solvents. These compounds are immiscible with water and most of the common organic solvents and form third liquid phase. These are chemically benign and environment friendly because these are nontoxic, nonflammable, thermally stable, recyclable and having high ability to dissolve oxygen. Fluorous fluids have high density, low intermolecular interaction, low surface tension, low dielectric constant, and high stability.

Perfluorous liquids, for example, perfluoroethers, perfluoroalkanes, perfluroamines, etc. exhibited unique characteristics, which make them suitable alternative to most of the common organic solvents. The boiling points of these liquids depend on their molar mass and it is lower than the corresponding alkanes. The density of perfluorous alkanes is higher than water and other organic molecules. Oxygen, carbon dioxide, and hydrogen-like gases are highly soluble in perfluorocarbons. Thus, these perfluorinated hydrocarbons permit some selective and efficient oxidation reaction under mild conditions.

Melting point and boiling point of some perfluorinated solvents are given in Table 7.1.

TABLE 7.1 Melting and boiling points of perfluorinated solvents

Compound	Formula	M. P. (°C)	B.P. (°C)
Perfluorohexane	C_6F_{14}	−87°C	75°C
Perfluoroheptane	C_7F_{16}	−78°C	82°C
Perfluorodecalin	$C_{10}F_{18}$	−10°C	142°C
Perfluoromethylcyclohexane	C_7F_{14}	−45°C	72°C
Perfluorotributyl amine	$C_{12}F_{27}N$	−50°C	173°C

Perfluorocarbons are non-ozone depleting compounds. Newly introduced fluoroiodocarbons, which are nonflammable, noncorrosive, and non-ozone depleting, are also emerging as possible replacements for CFCs.

Fluorous phase technique has a different green approach. Although these are solvents but are not the only solvent replacements. Because of their extremely nonpolar characteristics, these are not suitable for organic reactions and are used in conjunction with organic solvent to form biphasic system (Clarke et al., 2004). In this technique, reagents or catalysts, which are soluble in fluorous fluids, remain in fluorous phase whereas starting materials or substrates are dissolved in organic solvents or water, which are immiscible with fluorous fluids. These two distinct layers become homogeneous. On heating, reactant and substrate come in contact with each other and thus, the reaction takes place. These layers are separated again, when temperature is lowered down. As products remain in organic layer while unused reactant and catalyst remain in fluorous phase, which leads to an easy separation of products as well as recycling of the catalysts. Thus, the use of organic solvent for extraction can be avoided with this technique.

i) Oxidation reactions

In the oxidation of various aliphatic and benzylic alcohols, fluorous biphasic system (FBS) has been used by Nishimura et al. (2000). The formation of epoxides has been reported in the oxidation of alcohols and alkenols by Maayan et al. (2003) by using FBS.

ii) Asymmetric allylic alkylation

In the asymmetric palladium catalyzed alkylation of the allylic substrate, the use of fluorous chiral bisoxazolines was reported by Bayardon and Sinou (2003). Here, the high yield of the product has been obtained.

iii) Chlorination and bromination of alcohol

Chlorination and bromination of alcohol were achieved by Nakamura et al. (2003) by using phase vanishing reactions.

iv) Stille coupling

This coupling reaction has been reported with perfluoro-tagged tin compounds (Hoshino et al., 1997; Still, 1986).

v) Miscellaneous reactions

Friedel-Craft acylation, Mitsunobu reaction, and palladium–catalyzed C-C cross coupling reactions such as Suzuki reaction, Heck reaction, etc. have been carried out under fluorous conditions (Barrett et al., 2000; Mikami et al., 2001; Dandapani and Curran, 2002; Monineau et al., 1999; Betzemeier and Knochel, 1997; Yilmakz et al., 2020).

Thus, it can be concluded that fluorous fluids are greener solvent in organic synthesis.

7.2.8 p-CYMEME

The *p*-cymene is a biorenewable solvent and can be used for the metathesis of various substrates (Granato et al., 2017). It is a nontoxic compound, which can be obtained in large amounts as a side product of the cellulose and citrus industry. It was reported that this solvent prevents the consecutive double-bond isomerization of the product and affords the best yields in cross-metathesis of estragole with methyl acrylate.

7.2.9 LIMONENE

Molecular components are normally deposited from solution by spin-coating in the fabrication of thin-film electronic devices such as solar cells. As-ecofriendly toxic chlorinated solvents are commonly used in, and therefore, ecofriendly alternatives are required. These alternative solvents should be renewable sources derived, able to dissolve typical molecular electronic materials and also inexpensive. A survey was made by the deposition of layers of [6,6]-phenyl-C61-butyric acid methyl ester (PCBM) and poly(3-hexylthiophene) (P3HT) examined, which are widely used in solar cells as electron acceptor and donor, respectively. Lamarche et al. (2017) found that limonene is a green solvent produced by citrus fruits and other plants and it is effective for depositing PCBM on P3HT to create bilayer architectures.

7.2.10 GAMMA-VALEROLACTONE

Gamma-valerolactone (GVL) is an excellent solvent (Tang et al., 2020; Kerkel et al., 2021). It is a biomass-derived chemical.

γ-valerolactone is a bioderived solvent and it can be used in solid-phase peptide synthesis incorporating amino acid onto p-alkoxybenzyl alcohol resin (Al Musaimi et al., 2018).

Presently, cathode manufacturing for lithium-ion batteries utilizes N-methyl-2- pyrrolidone (NMP) as a coating solvent, as it has a petrochemical origin and undesirable toxicological properties. There is an urgent demand for other alternatives. Ravikumar et al. (2021) evaluated γ-valerolactone, which is a promising green candidate.

Besides these green solvents, there are some more compounds, which can be considered green solvents in the future. Some of these are N, N-dimethylpropyleneurea (DMPU), 1, 3-propanediol, 1, 3-dioxolan, etc.

Water is a universal as well as a green solvent. Cyclopentyl methyl ether, 2-methyl tetrahydrofuran, polyethylene glycol, ethyl lactate, glycerol, p-cymene, limonene, γ-valerolactone, etc. are among other green solvents. Search is still on for new solvents, which are more greener in nature and can replace traditional toxic solvents.

7.3 RECENT DEVELOPMENTS

The demand for organic electronic devices is regularly increasing with necessity of development of the scalability and green solvent processibility of some π-conjugated materials. Perylene diimide (PDI)-based materials are promising materials as interlayers for electronic devices, because of their high charge mobility in films and low E_{LUMO} energy. Pettipas et al. (2022) developed a green synthesis for a series of PDI-based electronically active materials, which are slot-die-coated into uniform thin films in air using green solvents. They reported that these materials on slot-die coating from ethanol may serve as cathode interlayers air-processed organic photovoltaics. It was also observed that power conversion efficiencies of device could reach 10% with PM6:Y6 active layer.

Dimethyl carbonate was used as a green solvent in synthesis in the presence of the 55 aroylated heterocycles with 95% yield, including modification of pharmaceuticals and natural products such as thioflavones, indolo[2,1-*a*]isoquinolin-6(5*H*)ones, quinoxalin-2(1*H*)-ones, benzimidazo[2,1-*a*]isoquinolin-6(5*H*)-ones, quaternary 3,3-dialkyl 2-oxindoles, and benzo[*e*][1,2,3]oxathiazine 2,2-dioxides in ambient conditions in the presence of air-visible-light-induced catalyst-/additive-free strategy (Zeng et al., 2022). The acylation reaction can be carried out using 4-acyl-1,4-dihydropyridines (acyl-DHPs) as acylating reagents with additional oxidants, without catalysts, and mild conditions.

There is still an existing challenge to search for an effective solvent system, which can dissolve cellulose and lignin in biomass residues simultaneously so as to fabricate lignocellulose hydrogels (LHs). Zheng et al. (2022) pretreated corncob residues from furfural production with alkaline peroxide to regulate the contents of lignin. Then lignin/cellulose composites with different lignin contents were dissolved and regenerated again by a green $ZnCl_2/CaCl_2$ solvent system. It was reported that such inorganic salt solvents served as linkers to get flexible LHs. It was revealed that material containing 10.75% lignin exhibited best compressive stress (76.71 kPa). These hydrogels have superior ionic conductivity and these were assembled into a solid-state electrolyte to be used in a zinc-ion hybrid supercapacitor.

An aldol reaction of α-ketoamide with α,β-unsaturated ketone was reported for the construction of α-hydroxy amides derivatives with quaternary carbon substitution in aqueous medium by KOH catalyzed up to 95% yield. Qin et al. (2022) reported an aldol reaction of a range of α,β-unsaturated ketone with α-ketoamide via the activation of potassium hydroxide in the aqueous medium. The approach provided an atom-efficient strategy for synthesizing a wide variety of multifunctional α-hydroxy amides. The reaction can be carried out under mild reaction conditions in the presence of green solvent. It was found that this method can be applied to substrates with different substituents and substitution patterns, affording the product yields up to 95%.

Shi et al. (2022) used dimethyl carbonate for an efficient and eco-friendly carbon nitride nanosheet (NM-g-C_3N_4)-catalyzed decarboxylative coupling reaction of imidazo[1,2-*a*]pyridines, benzo[*d*]imidazo[2,1-*b*]thiazole) with *N*-phenylglycines. The NM-g-C_3N_4 could be easily recovered by centrifugation. It was reported that it can be recycled and reused at least seven times without any significant decrease in its catalytic activity. Thus, one can easily avoid external oxidants, toxic solvents, and harsh reaction conditions with sustainable protocol.

Harikrishna et al. (2022) synthesized an efficient RuO_2/MWCNT nanocatalyst. They used this catalyst in synthesis of sulfonyl-quinoline derivatives using NH_4OAc, substituted aldehydes, dimedone, and phenylsulfonyl acetonitrile with ethanol as a green solvent. The conversion was reported to be good under optimal conditions with higher yields (91–98%) in 15 min. It was also revealed that the catalyst could be separated very easily and it can be reused eight times without any considerable loss in its catalytic efficiency.

KEYWORDS

- **cyclopentyl methyl ether**
- **glycerol**
- **polyethylene glycol**
- **2-methyltetrahydrofuran**
- **ethyl lactate**
- **perfluorinated (fluorous) solvent**

REFERENCES

Abdullah, M. A.; Hussein, H. A.; Alshajrawi, O. M. S. Ethyl Lactate as a Green Solvent in the Pharmaceutical Industry. In *Green Sustainable Process for Chemical and Environmental Engineering and Science*; Elsevier: Amsterdam, 2021; pp 185–194.

Abreu, C. M.; Maximiano, P.; Guliashvili, T.; Nicolas, J.; Serra, A. C.; Coelho, J. F. Cyclopentyl Methyl Ether as a Green Solvent for Reversible-Addition Fragmentation Chain Transfer and Nitroxide-Mediated Polymerizations. *RSC Adv.* **2016,** *6* (9), 7495–7503.

Akiya, N.; Savage, P. E. Roles of Water for Chemical Reactions in High-Temperature Water. *Chem. Rev.* **2002,** *102* (8), 2725–2750.

Al Musaimi, O.; El-Faham, A.; Almarhoon, Z.; Basso, A.; de la Torre, B. G.; Albericio, F. Bypassing Osmotic Shock Dilemma in a Polystyrene Resin Using the Green Solvent Cyclopentyl Methyl Ether (CPME): A Morphological Perspective. *Polymers* **2019a,** *11* (5). DOI: 10.3390/polym11050874

Al Musaimi, O.; El-Faham, A.; Basso, A.; Beatriz, G.; Albericio, F. γ-Valerolactone (GVL): An Eco-Friendly Anchoring Solvent for Solid-Phase Peptide Synthesis. *Tetrahedron Lett.* **2019b,** *60* (38). DOI: 10.1016/j.tetlet.2019.151058.

Albertsson, P. A. *Partion of Cell Particales and Macromolecules, Partition of Cell Particles and Macromolecules: Separation and Purification of Biomolecules, Cell Organelles, Membranes and Cells in Aqueous Polymer Two Phase Systems and Their Use in Biochemical Analysis and Biotechnology*, 3rd ed.; Wiley: New York, 1986. DOI: 10.1002/cbf.290050311

An, J.; Bagnell, L; Cablewski, T.; Strauss, C. R.; Trainor, R. W. Applications of High-Temperature Aqueous Media for Synthetic Organic Reactions. *J. Org. Chem.* **1997,** *62* (8), 2505–2511.

Anatas, P. T.; Warner P. T.; Warner, J. C. *Green Chemistry, Theory and Practice*; Oxford University Press: Oxford, 1998.

Auge, A.; Lubin, M.; Lubineau, A. Acceleration in Water of the Baylis-Hillman Reaction. *Tetrahedron Lett.* **1994,** *35* (43), 7947–7948.

Azzena, U.; Carraro, M.; Pisano, L.; Monticelli, S.; Bartolotta, R.; Pace, V. Cyclopentyl Methyl Ether: A Elective Ecofriendly Ethereal Solvent in Classical and Modern Organic Chemistry. *ChemSusChem* **2019,** *12* (1), *40*. DOI: 10.1002/cssc.201801768

Baiker, A. Supercritical Fluids in Heterogeneous Catalysis. *Chem. Rev.* **1999,** *99* (2), 453–473.

Barrett, A. G. M.; Braddock, D. C.; Catterick, D.; Chadwick, D.; Henschke, K. P.; Mckinnell, R. M. Fluorous Biphase Catalytic Friedel-Crafts Acylation: Ytterbium Tris (Perfluoroalkanesulfonyl)methide catalysts. *Syn. Lett.* **2000,** *2000* (6) 847–849.

Bayardon, J.; Sinou, D. Fluorous Chiral Bisoxazolines: Synthesis and Applications to an Asymmetric Allylic Alkylation. *Tetrahedron Lett.* **2003,** *44* (7), 1449–1451.

Beckman, E. J. J. Supercritical and Near-Critical CO_2 in Green Chemical Synthesis and Processing, Supercritical Fluids. *J. Supercrit. Fluids* **2004,** *28* (2–3), 121–191.

Bennett, J. S.; Charles, K. L.; Miner, M. R.; Heuberger, C. F.; Spina, E. J.; Bartels, M. F.; Foreman, T. Ethyl Lactate as a Tunable Solvent for the Synthesis of Aryl Aldimines. *Green Chem.* **2009,** *11* (2), 166–168.

Berson, J. A.; Hamlet, Z.; Muller, W. A. The Correlation of Solvent Effects on the Stereoselectivities of Diels-Alder Reactions by Means of Linear Free Energy Relationships: A New Empirical Measure of Solvent Polarity. *J. An. Chem. Soc.* **1962,** *84*, 297–304.

Betzemeier, B.; Knochel, P. Palladium-Catalyzed Cross-Coupling of Organozinc Bromides with Aryl Iodides in Perfluorinated Solvents. *Angew. Chem. Int. Edu.* **1997,** *36*, 2623–2624.

Beydoun, K.; Doucet, H. Cyclopentyl Methyl Ether: An Alternative Solvent for Palladium-Catalyzed Direct Arylation of Heteroaromatics. *Chem. Sus. Chem.* **2011,** *4* (4), 526–534.

Bhanage, B. M.; Ikushima, Y.; Shirai, M.; Arai, M. Heck Reactions Using Water-Soluble Metal Complexes in Supercritical Carbon Dioxide. *Tetrahedron Lett.* **1999,** *40* (35), 6427–6430.

Bhat, A. R.; Najar, M. H.; Dongre, R. S.; Akhter, M. S. Microwave Assisted Synthesis of Knoevenagel Derivatives Using Water as Green Solvent. *Curr. Res.Green Sustain.Chem.* **2020,** *3*. DOI: 10.1016/j.crgsc.2020.06.001.

Bhojane, J. M.; Sarode, S. A.; Nagarkar, J. M. Nickel–Glycerol: A Efficient, Recyclable Catalysis System for Suzuki Cross Coupling Reactions Using Aryl Diazonium Salts. *New J. Chem.* 2016, *40* (2), 1564–1570.

Bisz, E.; Szostak, M. 2-Methyltetrahydrofuran: A Green Solvent for Iron-Catalyzed Cross-Coupling Reactions. *ChemSusChem* **2018,** *11* (8).

Bordoloi, M. Cadmium Chloride-Magnesium-Water: A New System for Reduction of Various Organic Functionalities, *Tetrahedron Lett.* **1993,** *34* (10), 1681–1684.

Bosica, G.; Abdilla, R. A Regioselective One-Pot Aza-Friedel–Crafts Reaction for Primary, Secondary and Tertiary Anilines Using a Heterogeneous Catalyst. *Green Chem.* **2017,** *19* (23), 5683–5690.

Breslow, R.; Maitra, U. On the Origin of Product Selectivity in Aqueous Diels-Alder Reactions. *Tetrahedron Lett.* **1984,** *25* (12), 1239–1240.

Breslow, R.; Maitra, U.; Rideout, D. Selective Diels-Alder Reactions in Aqueous Solutions and Suspensions. *Tetrahedron Lett.* **1983,** *24* (18), 1901–1904.

Broli, D.; Kaul, C.; Kramer, A.; Krammer, P.; Richter, T.; Jung, M.; Vogel, H.; Zehner, P. Chemistry in Supercritical Water. *Angew. Chem. Int. Ed.* **1999,** *38* (20), 2998–3014.

Burk, M. J.; Feng, S.; Gross, M. F.; Tumas, W. Asymmetric Catalytic Hydrogenation Reactions in Supercritical Carbon Dioxide. *J. Am. Chem. Soc.* **1995,** *117* (31), 8277–8278.

Carlsson, M.; Habenicht, C.; Kam, L. C.; Antal, M.; Bian, N.; Cunninghom, R. J.; Jones, M. Study of the Sequential Conversion of Citric to Itaconic to Methacrylic Acid in Near-Critical and Supercritical Water. *Ind. Eng. Chem. Res.* **1994,** *33* (8), 1989–1996.

Carter, C. A. G.; Baker, R. T.; Nolan, S. P.; Tumas W. Enhanced Regioselectivity of Rhodium-Catalysedalkenehydroboration in Supercritical Carbon Dioxide. *Chem. Commun.* **2000,** *2000*, 347–348.

Chahdoura, F.; Mallet-Ladeira, S.; Gómez, M. Palladium Nanoparticles in Glycerol: A Clear-Cut Catalyst for One-Pot Multi-Step Processes Applied in the Synthesis of Heterocyclic Compounds. *Org. Chem. Front.* **2015,** *2* (4), 312–318.

Chahdoura, F.; Pradel, C.; Gómez, M. Copper (I) Oxide Nanoparticles in Glycerol: A Convenient Catalyst for Cross-Coupling and Azide–Alkyne Cycloaddition Processes. *ChemCatChem*, **2014,** *6* (10), 2929–2936.

Chandrasekhar, S.; Narsihmulu, Ch.; Saritha, B.; Sultana, S. S. Poly (Ethyleneglycol) (PEG): A Rapid and Recyclable Reaction Medium for the DABCO-Catalyzed Baylis–Hillman Reaction. *Tetrahedron Lett.* **2004,** *45* (30), 5865–5867.

Chandrasekhar, S.; Narsihmulu, Ch.; Sultana, S. S.; Reddy, N. R. Osmium Tetroxide in Poly (Ethylene Glycol) (PEG): A Recyclable Reaction Medium for Rapid Asymmetric Dihydroxylation Under Sharpless Conditions. *Chem. Commun.* **2003,** 1716–1717.

Chandrasekhar, S.; Narsihmulu, Ch.; Sultana, S. S.; Reddy, N. R. Poly (Ethylene Glycol) (PEG) as a Reusable Solvent Medium for Organic Synthesis. Application in the Heck Reaction. *Org. Lett.* **2002,** *4* (25), 4399–4401.

Chatterjee, T.; Kim, D. I.; Cho, E. J. Base-Promoted Synthesis of 2-Aryl Quinazolines from 2-Aminobenzylamines in Water. *Org.chem.* **2018,** *83* (14), 7423–7430.

Chen, J.; Spear, S. K.; Huddleston, J. G.; Holbrey, J. H.; Rogers, R. D. Application of Poly (Ethylene Glycol)-Based Aqueous Biphasic Systems as Reaction and Reactive Extraction Media. *Ind. Eng. Chem. Res.* **2004a,** *43* (17), 5358–5364.

Chen, J.; Spear, S. K.; Huddleston, J. G.; Holbrey, J. H.; Rogers, R. D. Application of Polyethylene Glycol-Based Aqueous Biphasic Reactive Extraction to the Catalytic Oxidation of Cyclic Olefins. *J. Chromatogr. B. Biomed. Appl.* **2004b,** *807* (1), 145–149.

Chinthakindi, P. K.; Kruger, H. G.; Govender, T.; Naicker, T.; Arvidsson, P. I. On-Water Synthesis of Biaryl Sulfonyl Fluorides. *J. Org. Chemi.* **2016,** *81* (6), 2618–2623.

Claisen, L.; Claparede, A. Condensationen von KetonenmitAldehyden. *Ber. Deutsch. Chem. Ges.* **1881,** *14* (2), 2460–2468.

Clarke, D.; Ali, M. A.; Clifford, A. A.; Parratt, A.; Rose, P.; Schwinn, D.; Bannwarth, W. C.; Rayner, M. Reactions in Unusual Media. *Curr. Top. Med. Chem.* **2004,** *4* (7), 729–771.

Coates, R. M.; Rogers, B. D.; Hobbs, S. J.; Peck, D. R.; Curran, D. P. Synthesis and Claisen Rearrangement of Alkoxyallyl Enol Ethers: Evidence for a Dipolar Transition State. *J. Am. Chem. Soc.* **1987,** *109* (4), 1160–1170.

Coeck, R.; De Vos, D. E. One-Pot Reductive Amination of Carboxylic Acids: A Sustainable Method for Primary Amine Synthesis. *Green Chem.* **2020,** *22* (15), 5105–5114.

Comanita, B. *Specialty Chemicals Magazine, Oct.* 2006, 23. http://www.pennakem.com/pdfs/methfpenngreenche mistry.pdf (accessed January 26, 2015).

Cooper, A. L. Polymer Synthesis and Processing Using Supercritical Carbon Dioxide. *J. Meter. Chem.* **2000,** *10*, 207–234.

Crittendon, R. C.; Parsons, E. Transformations of Cyclohexane Derivatives in Supercritical Water. *J. Organometallics* **1994,** *13* (7), 2587–2591.

Dandapani, S.; Curran, D. P. Fluorous Mitsunobu Reagents and Reactions, *Tetrahedron* **2002,** *58*, 3855–3864.

David. F. Solvent Applications of 2-Methyltetrahydrofuran in Organometallic and Biphasic Reactions. *Org. Proc. Res. Dev.* **2007,** *11* (1), 156–159.

Dickerson, T. J.; Reed, N. N.; Janda, K. D. Soluble Polymers as Scaffolds for Recoverable Catalysts and Reagents. *Chem. Rev.* **2002,** *102* (10), 3325–3344.

Diels, O.; Alder, K. Synthesen in der hydroaromatischenReihe. XI. Mitteilung. (Dien-synthesen of cyclopentadiens, cyclo-hexadiens and butadiensmitacetylen-dicarbonsäure und ihrenEstern. *Liebigs, Ann.* **1931,** *490* (1), 243–257.

Domingo, L. R.; Saéz, J. A.; Zaragozá, R. J.; Arnó, M. Understanding the Participation of Quadricyclane as Nucleophile in Polar [2σ+ 2σ+ 2π] Cycloadditions Toward Electrophilic π Molecules. *J. Org. Chem.* **2008,** *73* (22), 8791–8799.

Dub, P. A.; Rodriguez-Zubiri, M.; Baudequin, C.; Poli, R. Hydroamination of Ethylene by Aniline: Catalysis in Water. *Green Chem.* **2010,** *12* (8), 1392–1396.

Fache, F.; Lehueds, S.; Lemaine, M. A Catalytic Stereo- and Chemo-Selective Method for the Reduction of Substituted Aromatics. *Tetrahedron Lett.* **1995,** *36* (6), 885–888.

Farahat, M. E.; Perumal, P.; Budiawan, W.; Chen, Y. F.; Lee, C. H.; Chu, C. W. Efficient Molecular Solar Cells Processed from Green Solvent Mixtures. *J. Mater. Chem. A*, **2017,** *5* (2), 571–582.

Ferravoski, P.; Fiecchi, A.; Grisenti, P.; Santaniello, E.; Trave, S. Polyethylene Glycols as Solvents for Anionic Activation: Synthesis of Thioacetates by Means of Potassium Thioacetate in Polyethylene Glycol 400. *Synth. Commun.* **1987,** *17* (13), 1569–1575.

Feu, K. S.; Alexander, F.; Silva, S.; de Moraes Junior, M. A.; Corrêa, A. G.; Paixão, M. W. Polyethylene Glycol (PEG) as a Reusable Solvent Medium for an Asymmetric Organocatalytic Michael Addition. Application to the Synthesis of Bioactive Compounds. *Green Chem.* **2014,** *16* (6), 3169–3174.

Fringuell, F.; Pellegrino, R.; Piermattic, O.; Pizzo, F. Facile and Selective Oxidation of Sulfides to Sulfoxides by T-Butyl Hydroperoxide in Aqueous Medium. *Synth. Commun.* **1993,** *23* (22), 3157.

Fringuelli, F.; Germani, R.; Pizzo, F.; Savelli, G. Baeyer-Villiger Reaction in Water. *Gazz. Chem. Ital.* **1989,** *119*, 249–249.

Gaikwad, V. V.; Bhanage, B. M. Oxime Palladacycle in PEG as a Highly Efficient and Recyclable Catalytic System for Phenoxycarbonylation of Aryl Iodides with Phenols. *Appl. Organometallic Chem.* **2019,** *33* (4). DOI: 10.1002/aoc.4741.

Ghorpade, P. V.; Pethsangave, D. A.; Some, S.; Shankarling, G. S. Graphene Oxide Promoted Oxidative Bromination of Anilines and Phenols in Water. *Org. Chem.* **2018,** *83* (14), 7388–7397.

Gladysz, J. A.; Curran, D. P. Fluorous Chemistry: From Biphasic Catalysis to a Parallel Chemical Universe and Beyond. *Tetrahedron* **2002,** *58* (20), 3823–3825.

Gore, G. II. On the Properties of Liquid Carbonic Acid. *Proc. R. Soc. Landon* **1861,** *11*, 85–86.

Gozlan, C.; Lafon, R.; Duguet, N.; Redl, A.; Lemaire, M. Catalytic Reductive Cleavage of Methyl α-D-Glucoside Acetals to Ethers Using Hydrogen as a Clean Reductant. *RSC Adv.* **2014,** *4* (92), 50653–50661.

Granato, A. V.; Santos, A. G.; Dos Santos, E. N. p-Cymene as Solvent for Olefin Metathesis: Matching Efficiency and Sustainability. *ChemSusChem.* **2017,** *10* (8), 1832–1837.

Grieco, P. A.; Rrandes, E. B.; Mecann, S.; Clark, J. D. Water as a Solvent for the Claisen Rearrangement: Practical Implications for Synthetic Organic Chemistry. *J. Org. Chem.* **1989,** *54* (25), 5849–5851.

Grieco, P. P.; Garner, A.; He, Z. "Micellar" Catalysis in the Aqueous Intermolecular Diels-Alder Reaction: Rate Acceleration and Enhanced Selectivity. *Tetrahedron Lett.* **1983,** *24* (18), 1897–1900.

Grieco, P.; Parker, D. T.; Cornwell, M.; Ruckle, R. Retro Aza Diels-Alder Reactions: Acid-Catalyzed Heterocycloreversion of 2-Azanorbornenes in Water at Ambient Temperature. *J. Am. Chem. Soc.* **1987,** *109* (19), 5859–5861.

Gu, Y.; Jerome, F. Glycerol as a Sustainable Solvent for Green Chemistry. *Green Chem.* **2010,** *12*, 1127–1138.

Guo, Z.; Li, M.; Willauer, H. D.; Huddleston, J. G.; April, G. C.; Rogers, R. D. Evaluation of Polymer-Based Aqueous Biphasic Systems as Improvement for the Hardwood Alkaline Pulping Process. *Ind. Eng. Chem. Res.* **2002,** *41* (10), 2535–2542.

Haimov, A.; Neumann, R. Polyethylene Glycol as a Non-Ionic Liquid Solvent for Polyoxometalate Catalyzed Aerobic Oxidation. *Chem. Commun.* **2002,** 867–876.

Halimehjani, A. Z.; Marjani, K.; Ashouri, A. Synthesis of Dithiocarbamate by Markovnikov Addition Reaction in Aqueous Medium. *Green Chem.* **2010,** *12* (7), 1306–1310.

Hanessian, S.; Girand, C. One Step α-Deoxygenation of Unprotected Aldonolactones Using Samarium Diiodide-THF/H_2O System—A New Synthesis of 2-Deoxy-D-Ribose. *Syn. Lett.* **1994,** 1994 (10) 861–862.

Harikrishna, S.; Gangu, K. K.; Robert, A. R.; Ganja, H.; Kerru, N.; Maddila, S. et al., An Ecofriendly and Reusable Catalyst RuO_2/MWCNT in the Green Synthesis of Sulfonyl-Quinolines. *Process Saf. Environ. Prot.* **2022,** *159*, 911–917.

Harris, J. M. *Polyethylene Glycol Chemistry, Biochemical and Biomedical Applications*; Ed. Haris, J. M.; Plenium Press: New York and London, 1992a; p 7.

Harris, J. M. *Polyethylene Glycol Chemistry, Biotechnological and Biomedium Applications*; Plenum Press: New York, 1992b; p 3.

Harris, J. M.; Zalipsky, S. Introduction to Chemistry and Biological Applications of Poly (Ethylene Glycol); In *Poly (Ethylene Glycol), Chemistry and Biological Applications*; ACS Books: Washington, DC, 1997.

Hasegawa, C.; Curran, D. P. Additive and Solvent Effects on Samarium Diiodide Reductions: The Effects of Water and DMPU. *J. Org. Chem.* **1993,** *58* (18), 5008–5010.

Hawthrone, S. B.; Yang, Y.; Miller, D. Extraction of Organic Pollutants from Environmental Solids with Sub- and Supercritical Water. *J. Anal. Chem.* **1994,** *66* (18), 2912–2920.

Herold, D. A.; Keil, K.; Bruns, D. E. Oxidation of Polyethylene Glycols by Alcohol Dehydrogenase. *Biochem. Pharmacol.* **1989,** *38* (1), 73–76.

Horvath, I. T.; Rabai, J. Facile Catalyst Separation Without Water: Fluorous Biphase Hydroformylation of Olefins. *Science* **1994,** *266* (5182), 72–75.

Hoshino, M.; Degenkolb, P.; Curran, D. P. Palladium-catalyzed stille couplings with fluorous tin reactants. *J. Org. Chem.* **1997,** *62* (24), 8341–8349.

Hu, X.; Yu, H.; Wang, W.; Jiang, S.; Liu, Q.; He, J. Synthesis of 3-ethanoyl/aroylacetylindoles in Water. *Chinese J. Org. Chem.* **2019,** *39* (11), 3183–3189.

Huang, X.; Chen, Y.; Zhen, S.; Song, L.; Gao, M.; Zhang, P. et al., Cobalt-Catalyzed Aerobic Cross-Dehydrogenative Coupling of C–H and Thiols in Water for C–S Formation. *Org. Chem.* **2018,** *83* (14), 7331–7340.

Itami, K.; Terakawa, K.; Yoshida, J. I.; Kajimoto, O.; Efficient and Rapid C−Si Bond Cleavage in Supercritical Water. *J. Am. Chem. Soc.* **2003,** *125* (20), 6058–6059.

Ito, H.; Nishiyama, J.; Adschiri, T.; Arai, K. Synthesis of ε-Caprolactam from ε-Caprolactone and Ammonia in Supercritical Water, *KobunshiRobunshu* **2001,** *58* (12), 679.

Jad, Y. E.; Govender, T.; Kruger, H. G.; El-Faham, A.; de la Torre, B. G.; Albericio, F. Green Solid-Phase Peptide Synthesis (GSPPS) 3. Green Solvents for Fmoc Removal in Peptide Chemistry. *Org. Process Res. Dev.* **2017,** *21* (3), 365–369.

Johnson, J. R. Perkin Reaction and Related Reactions. *Org. Reactions* **1942,** *1*, 210–265.

Johnston, K. P. In Supercritical Fluid Science and Technology; Johnston, K. P., Penninger, J. M. L. Eds.; *Am. Chem. Soc: Washington D.C.* **1989,** 1–12.

Johnston, K. P.; Harrison, K. I.; Clarke, M. J.; Howdle, S. M.; Heitz, M. P.; Bright, F. V.; Carlier, C.; Randolph, T. W. Water-in-Carbon Dioxide Microemulsions: An Environment for Hydrophiles Including Proteins. *Science* **1996,** *271* (5249), 624–626.

Kajmoto, O. Solvation in Supercritical Fluids: Its Effects on Energy Transfer and Chemical Reactions. *Chem. Rev.* **1999,** *99* (2), 355–390.

Kawabata, T.; Öztürk, O.; Chen, J.; Fuji, K. Chirality Transfer During Alkylation of Chiral Amides. *Chem. Commun.* **2003,** *1*, 162–163.

Kawatsura, M.; Kamesaki, K.; Yamamoto, M.; Hayase, S.; Itoh, T. Homogeneous Ruthenium Precatalyst for Suzuki–Miyaura Coupling Reaction. *Chem. Lett.* **2010,** *39* (10), 1050–1051.

Keh, C. C. K.; Wei, C.; Li, C. J. The Barbier−Grignard-Type Carbonyl Alkylation Using Unactivated Alkyl Halides in Water. *J. Am. Chem. Soc.* **2003,** *125* (14), 4062–4063.

Kerkel, F.; Markiewicz, M.; Stolte, S.; Müller, E.; Kunz, W. The Green Platform Molecule Gamma-Valerolactone–Ecotoxicity, Biodegradability, Solvent Properties, and Potential Applications. *Green Chem.* **2021,** *23* (8), 2962–2976.

Khan, M. N.; Karamthulla, S.; Choudhury, L. H.; Faizi, M. S. H. Ultrasound Assisted Multicomponent Reactions: A Green Method for the Synthesis of Highly Functionalized Selenopyridines Using Reusable Polyethylene Glycol as Reaction Medium. *RSC Adv.* **2015,** *5* (28), 22168–22172.

Kidwai, M.; Jahan, A.; Bhatnagar, D. Polyethylene Glycol: A Recyclable Solvent System for the Synthesis of Benzimidazole Derivatives Using CAN as Catalyst. *J. Chem. Sci.* **2010,** *122* (4), 607–612.

Knoevenagel, F. Condensation of Malonic Acid with Aromatic Aldehydes by Ammonia and Amines. *Ber. Dtsch. Chem. Ges.* **1898,** *31*, 2596–2619.

Kobayashi, S.; Kuroda, H.; Ohtsuka, Y.; Kashihara, T.; Masuyama, A.; Watanabe, K. Evaluation of Cyclopentyl Methyl Ether (CPME) as a Solvent for Radical Reactions. *Tetrahedron* **2013,** *69* (10), 2251–2259.

Kobayashi, S.; Shibukawa, K.; Miyaguchi, Y.; Masuyama, A. Grignard Reactions in Cyclopentyl Methyl Ether. *Asian J. Org. Chem.* **2016,** *5* (5), 636–645.

Kolate, A.; Baradia, D.; Patil, S.; Vhora, I.; Kore, G.; Misra, A. PEG—A Versatile Conjugating Ligand for Drugs and Drug Delivery Systems. *J. Controlled Release* **2014,** *192*, 67–81.

Kondo, A.; Urabe, T.; Higashitani, K. J. Bioconversions in an Aqueous Two-Phase System Using Enzymes Immobilized on Ultrafine Silica Particles. *J. Ferment Bioeng.* **1994,** *77* (6), 700–703.

Kool, E. T.; Breslow, R. Dichotomous Salt Effects in the Hydrophobic Acceleration of the Benzoin Condensation. *J. Am. Chem. Soc.* **1988,** *110* (5), 1596–1597.

Korzenski, M. B.; Kolis, J. W. Diels-Alder Reactions Using Supercritical Water as an Aqueous Solvent Medium. *Tetrahedron Lett.* **1997,** *38* (32), 5611–5614.

Kremsner, J. M.; Kappe, C. O. Microwave-Assisted Organic Synthesis in Near-Critical Water at 300°C—A Proof-of-Concept Study. *Eur. J. Org. Chem.* **2005,** *2005* (17), 3672–3679.

Kua, Y. L.; Gan, S.; Morris, A.; Ng, H. K. Ethyl Lactate as a Potential Green Solvent to Extract Hydrophilic (Polar) and Lipophilic (Non-Polar) Phytonutrients Simultaneously from Fruit and Vegetable by-Products. *Sustain. Chem. Pharm.* **2016,** *4*, 21–31.

Kuhlmann, B.; Arnett, E. M.; Siskin, M. H-D Exchange in Pinacolone by Deuterium Oxide at High Temperature and Pressure. *J. Org. Chem.* **1994,** *59* (18), 5377–5380.

Kuhlmann, B.; Arnettm, E. M.; Siskin, M. Classical Organic Reactions in Pure Superheated Water. *J. Org. Chem.* **1994,** *59* (11), 3098–3101.

Kumar, D.; Reddy, V. B.; Mishra, B. G.; Rana, R. K.; Nadagouda, M. N.; Varma, K. S. Nanosized Magnesium Oxide as Catalyst for the Rapid and Green Synthesis of Substituted 2-Amino-2-Chromenes. *Tetrahedron* **2007,** *63* (15), 3093–3097.

Kumar, V.; Sharma, A.; Sharma, A.; Sinha, A. K. Remarkable Synergism in Methylimidazole-Promoted Decarboxylation of Substituted Cinnamic Acid Derivatives in Basic Water Medium Under Microwave Irradiation: A Clean Synthesis of Hydroxylated (E)-Stilbenes. *Tetrahedron* **2007,** *63* (32), 7640–7646.

Lamarche, M.; Dang, M. T.; Lefebvre, J.; Wuest, J. D.; Roorda, S. Limonene as a Green Solvent for Depositing Thin Layers of Molecular Electronic Materials with Controlled Interdiffusion. *ACS Sustain. Chem. Eng.* **2017,** *5* (7), 5994–5998.

Lapworth, A. J. CXXII.—Reactions Involving the Addition of Hydrogen Cyanide to Carbon Compounds, Part II: Cyanohydrins Regarded as Complex Acids, *J. Chem. Soc.* **1904,** *85*, 1206–1214.

Lapworth, A. J.; XCVI.—Reactions Involving the Addition of Hydrogen Cyanide to Carbon Compounds, *J. Chem. Soc.* **1903,** *83*, 995–1005.

Larpent, C.; Meignan, G. Chemistry in Water-Part VI Catalytic Isomerization and Stereochemistry of Reduction of Acetylenics Mediated by Water-Soluble Phosphines. *Tetrahedron Lett.* **1993,** *34* (27), 4331–4334.

Larsen, S. D.; Grieco, P. A.; Aza Diels-Alder Reactions in Aqueous Solution: Cyclocondensation of Dienes with Simple Iminium Salts Generated Under Mannich Conditions, *J. Am. Chem. Soc.* **1985,** *107* (6), 1968–1769.

Lei, P.; Ling, Y.; An, J.; Nolan, S. P.; Szostak, M. 2-Methyltetrahydrofuran (2-MeTHF): A Green Solvent for Pd–NHC-Catalyzed Amide and Ester Suzuki-Miyaura Cross-Coupling by N– C/O– C Cleavage. *Adv. Synth. Catal.* **2019,** *361* (24), 5654–5660.

Leininger, N. F.; Clontz, R.; Gainer, J. L.; Kirwan, D. V. In *Clean Solvents, Alternative Media for Chemical Reactions and Processing*; Abraham, M. A., Moens, L.; ACS Sympossium Series 819; American Chemical Society; Washington, DC, 2002; pp 208–223.

Lenstra, D. C.; Lenting, P. E.; Mecinović, J. Sustainable Organophosphorus-Catalysed Staudinger Reduction. *Green Chem.* **2018,** *20* (19), 4418–4422.

Li, K.; Liu, C.; Wang, K.; Ren, Y.; Li, F. Sodium Borohydride-Nickel Chloride Hexahydrate in EtOH/PEG-400 as an Efficient and Recyclable Catalytic System for the Reduction of Alkenes. *RSC Adv.* 2018, *8* (14), 7761–7764.

Li, Z.; Ying, L.; Zhu, P.; Zhong, W.; Li, N.; Liu, F.; Cao, Y. A Generic Green Solvent Concept Boosting the Power Conversion Efficiency of All-Polymer Solar Cells to 11%. *Energy Environ. Sci.* **2019,** *12* (1), 157–163.

Lindstrom, U. M. Stereoselective Organic Reactions in Water. *Chem. Rev.* **2002,** *102* (8), 2751–2772.

Liu, X. C.; Chen, X. L.; Liu, Y.; Sun, K.; Peng, Y. Y.; Qu, L. B. et al., Visible-Light-Induced Metal-Free Synthesis of 2-Phosphorylated Thioflavones in Water. *ChemSusChem* **2020,** *13* (2), 298–303.

Liu, Y.; Ma, C.; Huang, C.; Fu, Y.; Chang, J. Efficient Conversion of Xylose into Furfural Using Sulfonic Acid-Functionalized Metal–Organic Frameworks in a Biphasic System. *Ind. Eng. Chem. Res.* **2018,** *57* (49), 16628–16634.

Liwei, X.; Fucai, D.; Zheng, L.; Xuemin, J.; Jie, K.; Guangxian, L. Polyethylene Glycol: A New Medium for Green Organic Synthesis. *Chinese J. Org. Chem.* **2019,** *39* (3), 648–660.

Lubineau, A. Water-Promoted Organic Reactions: Aldol Reaction Under Neutral Conditions. *J. Org. Chem.* **1986,** *51* (11), 2142–2144.

Lubineau, A.; Meyer, E. Water-Promoted Organic Reactions. Aldol Reaction of Silyl Enol Ethers with Carbonyl Compounds Under Atmospheric Pressure and Neutral Conditions. *Tetrahedron* **1988,** *44* (19), 6065–6070.

Maayan, G.; Fish, R. H.; Neuman, R. Polyfluorinated Quaternary Ammonium Salts of Polyoxometalate Anions: Fluorous Biphasic Oxidation Catalysis with and Without Fluorous Solvents. *Org. Lett.* **2003,** *5* (20), 3547–3550.

Majdanski, T. C.; Vitz, J.; Meier, A.; Brunzel, M.; Schubert, S.; Nischang, I.; Schubert, U. S. "Green" Ethers as Solvent Alternatives for Anionic Ring-Opening Polymerizations of Ethylene Oxide (EO): In-Situ Kinetic and Advanced Characterization Studies. *Polymer* **2018,** *159*, 86–94.

Mandenius, C. F.; Nilsson, B.; Persson, I.; Tjerneld, F. Kinetic Models for Enzymic Cellulose Degradation in Aqueous Two-Phase Systems. *Biotechnol. Bioeng.* **1988,** *31*, 203–207.

Matsuo, J.; Tsuchige, T.; Odashima, K.; Kobayashi, S. Lewis Acid Catalysis in Supercritical Carbon Dioxide. Use of Scandium Tris (Heptadecafluorooctanesulfonate) as a Lewis Acid Catalyst in Diels-Alder and Aza Diels-Alder Reactions. *Chem. Lett.* **2000,** *29* (2) 178–179.

Maximiano, P.; Mendes, J. P.; Mendonça, P. V.; Abreu, C. M.; Guliashvili, T.; Serra, A. C. et al., Cyclopentyl Methyl Ether: A New Green Co-Solvent for Supplemental Activator and Reducing Agent Atom Transfer Radical Polymerization. *J. Polym. Sci. A Polym. Chem.* **2010,** *53* (23), 2722–2729.

Mchugh, M.; Krukonis, V. *Supercritical Fluid Extraction, Principles and Practice*; Butterworths: Boston, 1986; pp 1–11.

Mckillop, A.; Tarbin, J. A. Functional Group Oxidation Using Sodium Perborated1. *Tetrahedron* **1987,** *43* (8), 1753–1758.

Mikami, K.; Mikami, Y.; Matsumota, Y.; Nishikido, J.; Yamamoto, F.; Nakajima, H. Lewis Acid Catalysis by Lanthanide Complexes with Tris (Perfluorooctanesulfonyl) Methide Ponytails in Fluorous Recyclable Phase. *Tetrahedran Lett.* **2001,** *42* (2), 289–292.

Milton, E. J.; Clarke, M. L. Palladium-Catalysed Grignard Cross-Coupling Using Highly Concentrated Grignards in Methyl-Tetrahydrofuran. *Green Chem.* **2010,** *12* (3), 381–383.

Molineux, G. Pegylation: Engineering Improved Pharmaceuticals for Enhanced Therapy. *Cancer Treat. Rev.* **2002,** *28*, 13–16.

Monineau, J.; Pozzi, G.; Quici, S.; Sinou, Palladium-Catalyzed Heck Reaction in Perfluorinated Solvents. *Tetrahedron Lett.* **1999,** *40* (43), 7683–7686.

Mouret, A.; Leclercq, L.; Mühlbauer, A.; Nardello-Rataj, V. Eco-Friendly Solvents and Amphiphilic Catalytic Polyoxometalate Nanoparticles: A Winning Combination for Olefin Epoxidation. *Green Chem.* **2014,** *16* (1), 269–278.

Naik, S. D.; Doraiswamy, L. K. Phase Transfer Catalysis: Chemistry and Engineering. *Alch E-J.* **1998,** *44* (3), 612–646.

Nakamura, H.; Usui, T.; Kudora, H.; Ryu, I.; Matsubara, H.; Yasuda, S.; Curran, D. P. Fluorous Solvent as a New Phase-Screen Medium Between Reagents and Reactants in the Bromination and Chlorination of Alcohols. *Org. Lett.* **2003,** *5* (8), 1167–1169.

Namboodiri, V. V.; Varma, R. S. Microwave-Accelerated Suzuki Cross-Coupling Reaction in Polyethylene Glycol (PEG). *Green Chem.* **2001,** *3*, 146–148.

Narayan, S.; Muldoon, H.; Finn, M. G.; Fokin, V. V.; Kolb, H. C.; Sharpless, K. B. "On Water": Unique Reactivity of Organic Compounds in Aqueous Suspension, *Angew. Chem. Int. Ed.* **2005,** *44* (21), 3275–3279.

Nardi, M.; Sindona, G.; Costanzo, P.; Oliverio, M.; Procopio, A. Eco-Friendly Stereoselective Reduction of α, β-Unsaturated Carbonyl Compounds by Er (OTf) 3/$NaBH_4$ in 2-MeTHF. *Tetrahedron* **2015,** *71* (7), 1132–1135.

Nascimento, J. E.; Oliveira, D. H. D.; Abib, P. B.; Alves, D.; Perin, G.. Jacob, R. G. Synthesis of 4-Arylselanylpyrazoles Through Cyclocondensation Reaction Using Glycerol as Solvent. *J. Braz. Chem. Soc.* **2015,** *26*, 1533–1541.

Nemati, F.; Hosseini, M. M.; Kiani, H. Glycerol as a Green Solvent for Efficient, One-Pot and Catalyst Free Synthesis of 2, *4*, 5-triaryl and 1, *2*, 4, 5-Tetraaryl Imidazole Derivatives. *J. Saudi Chem. Soc.* **2016,** *20*, S503–S508.

Nishimura,T.; Maeda, Y.; Kakiuchi, N.; Uemura, S. Palladium (II)-Catalysed Oxidation of Alcohols Under an Oxygen Atmosphere in a Fluorous Biphase System (FBS). *J. Chem. Soc. Perkin Trans.* **2000,** *1*, 4301–4305.

Oakes, R. S.; Clifford, A. A.; Bartle, K. D.; Thornton-Pett, M.; Rayner, C. M.; Sulfur Oxidation in Supercritical Carbon Dioxide: Dramatic Pressure Dependant Enhancement of Diastereoselectivity for Sulfoxidation of Cysteine Derivatives, *J. Chem. Soc. Chem. Commun.* **1999,** 247–248.

Pace, V.; Castoldi, L.; Alcantara, A. R.; Holzer, W. Robust Eco-Friendly Protocol for the Preparation of γ-hydroxy-α, β-acetylenic Esters by Sequential One-Pot Elimination–Addition of 2-Bromoacrylates to Aldehydes Promoted by LTMP in 2-MeTHF. *Green Chem.* **2012,** *14* (7), 1859–1863.

Pacut, R.; Grimn, M. L.; Kraus, G. A.; Tanko, J. M. Photochemistry in Supercritical Carbon Dioxide: The Benzophenone-Mediated Addition of Aldehydes to α, β-Unsaturated Carbonyl Compounds. *Tetrahedron Lett.* **2001,** *42* (8), 1415–1418.

Pasupuleti, B. G.; Bez, G. CuI/L-Proline Catalyzed Click Reaction in Glycerol for the Synthesis of 1, *2*, 3-Triazoles. *Tetrahedron Lett.* **2019,** *60* (2), 142–146.

Paul, S.; Pradhan, K.; R Das, A. Ethyl Lactate as a Green Solvent: A Promising Bio-Compatible Media for Organic Synthesis. *Curr. Green Chem.* **2016,** *3* (1), 111–118.

Peng, Y. Y.; Liu, J.; Lei, X.; Yin, Z. Room-Temperature Highly Efficient Suzuki–Miyaura Reactions in Water in the Presence of Stilbazo. *Green Chem.* **2010,** *12* (6), 1072–1075.

Pereira, C. S.; Silva, V. M.; Rodrigues, A. E. Ethyl Lactate as a Solvent: Properties, Applications and Production Processes–A Review. *Green Chem.* **2011,** *13* (10), 2658–2671.

Persson, I.; Stalbrand, H.; Tjerneld, F.; Hahn-Hagerdal, B.; Semicontinuous Production of Cellulolytic Enzymes with Trichoderma Reesei Rutgers C30 in an Aqueous Two-Phase System. *Appl. Biochem. Biotechnol.* **1991,** *27*, 27–36.

Pesiri, D. R.; Morita, D. K.; Glaze, W.; Tumas, W. Selective Epoxidation in Dense Phase Carbon Dioxide. *J. Chem. Soc. Chem. Commun.* **1998,** *9*, 1015–1016.

Petrier, C.; Luche, J. L. Ultrasonically Improved Reductive Properties of an Aqueous Zn-NiCl2 System-1 Selective Reduction of α, β-Unsaturated Carbonyl Compounds. *Tetrahedron Lett.* **1987,** *28* (21), 2347–2350.

Pettipas, R. D.; Hoff, A.; Gelfand, B. S.; Welch, G. C. Green Solvent-Processible n–h-Functionalized Perylene Diimide Materials for Scalable Organic Photovoltaics. *ACS Appl. Mater. Interf.* **2022,** *14* (2), 3103–3110.

Pillai, U. R.; Sahle-Demessi, E. Selective Hydrogenation of Maleic Anhydride to γ-Butyrolactone Over Pd/Al_2O_3 Catalyst Using Supercritical CO_2 as Solvent. *Chem. Commun.* **2002,** 422–423.

Pirrung, M. C.; Sarma, K. D.; Multicomponent Reactions Are Accelerated in Water. *J. Am. Chem. Soc.* **2004,** *126* (2), 444–445.

Ponaras, A. A.;A new variant of the Claisen Rearrangement Capable of Creating the Bond Between Two Quaternary Centers. *J. Org. Chem.* **1983,** *48* (21), 3866–3868.

Ponduri, R.; Kumar, P.; Vadali, L. R.; Modugu, N. R. Water-PEG-400 Mediated an Efficient One-Pot Eco-Friendly Synthesis of Functionalized Isoxazole Substituted Chromeno [2, 3-b] Pyridine-3-Carboxylate Derivatives. *ChemistrySelect* 2018, *3* (27), 7766–7770.

Prache, N.; Abreu, S.; Sassiat, P.; Thiébaut, D.; Chaminade, P. Alternative Solvents for Improving the Greenness of Normal Phase Liquid Chromatography of Lipid Classes. *J. Chromatogr. A* **2016,** *1464*, 55–63.

Qin, X.; Wu, C.; Lu, F.; Wang, Z. Y.; Jiang, J.; Liu, H. The Aldol Reaction of α-Ketoamide with α, β-Unsaturated Ketone in KOH Aqueous Medium. *ChemistrySelect* **2022,** *7* (4). DOI: 10.1002/slct.202104146.

Ramamurthy, R. Organic Photochemistry in Organized Media. *Tetrahedron Lett.* **1986,** *42* (21), 5753–5839.

Raner, K. D.; Strauss C. R.; Trainor, R. W. A New Microwave Reactor for Batchwise Organic Synthesis. *J. Org. Chem.* **1995,** *60* (8), 2456–2460.

Ravikumar, V. R.; Schröder, A.; Köhler, S.; Çetinel, F. A.; Schmitt, M.; Kondrakov, A.; B. γ-Valerolactone: An Alternative Solvent for Manufacturing of Lithium-Ion Battery Electrodes. *ACS Appl. Energy Mater.* **2021,** *4* (1), 696–703.

Reardon, P.; Metts, S.; Crittendon, C.; Daugherity, P.; Parson, E. Palladium-Catalyzed Coupling Reactions in Superheated Water. *J. Organometrallics.* **1995,** *14* (8), 3810–3816.

Renslo, A. R.; Weinstien, R. D.; Tester, J. W.; Danherical, R. L.; Concerning the Regiochemical Course of the Diels-Alder Reaction in Supercritical Carbon Dioxide. *J. Org. Chem.* **1997,** *62* (13), 4530–4533.

Rideout, D. C.; Breslow, R.; Hydrophobic Acceleration of Diels-Alder Reactions. *J. Am. Chem. Soc.* **1980,** *102* (26), 7816–7817.

Robke, G. D.; Behrman, E. Conversion of Aminopyridines into N-Oxides by Caro's Acid Anion (Peroxymonosulfate). *J. Chem. Res.* (S) **1993,** 412–413.

Rodríguez-Rodríguez, M.; Gras, E.; Pericàs, M. A.; Gómez, M. Metal-Free Intermolecular Azide–Alkyne Cycloaddition Promoted by Glycerol. *Chem. Eur. J.* **2015,** *21,* 18706–18710.

Rose, P. M.; Clifford, A. A.; Rayner, C. M. The Baylis–Hillman Reaction in Supercritical Carbon Dioxide: Enhanced Reaction Rates, Unprecedented Ether Formation, and a Novel Phase-Dependent 3-Component Coupling. *Chem. Commun.* **2002,** 968–969.

Rosholm, T.; Gois, P. M.; Franzen, R.; Candeias, N. R. Glycerol as an Efficient Medium for the Petasis Borono–Mannich Reaction. *ChemistryOpen* 2015, *4* (1), 39.

Safaei, H. R.; Shekouhy, M.; Rahmanpur, S.; Shirinfeshan, A. Glycerol as a Biodegradable and Reusable Promoting Medium for the Catalyst-Free One-Pot Three Component Synthesis of 4H-Pyrans. *Green Chem.* **2012,** *14* (6), 1696–1704.

Samil, A. A. Z.; Desavignac, A.; Rico, I.; *A.* Latters, Waterless Microemulsions-IV: Diels-Alder Reaction of Cyclopentadiene and Methylacrylate as a Probe of Formamide Microemulsions. *Tetrahedron.* **1985,** *41* (18), 3683–3688.

Santaniello, E.; Manzocchi, A.; Sozzani, P. Polyetylene Glycols as Host Solvents: Applications to Organic Synthesis. *Tetrahedron Lett.* **1979,** *20* (47), 4581–4582.

Santaniello, E.; Fiecchi, A.; Manzocchi, A.; Ferraboschi, P. Reductions of Esters, Acyl Halides, Alkyl Halides, and Sulfonate Esters with Sodium Borohydride in Polyethylene Glycols: Scope and Limitation of the Reaction. *J. Org. Chem.* **1983,** *48* (18), 3074–3077.

Santanielto, E. In *Crown Ethers and Phase Transfer Catalysis in Polymer Science*; Mathias, L. J., Carrater, C. E., Jr., Eds.; Plenum: New York, 1984; p 397.

Savage, P. E. Organic Chemical Reactions in Supercritical Water. *Chem. Rev.* **1999,** *99,* 603–622.

Schmidt, J. G.; Ueber die Einwirkung von Aceton auf Furfurol und auf Bittermandelöl-beiGegenwart von Alkalilauge. *Ber.* **1881,** *14,* 1459–1461.

Shanmuganathan, S.; Natalia, D.; van den Wittenboer, A.; Kohlmann, C.; Greiner, L.; de María, P. D. Enzyme-Catalyzed C–C Bond Formation Using 2-Methyltetrahydrofuran (2-MTHF) as (Co) Solvent: Efficient and Bio-Based Alternative to DMSO and MTBE. *Green Chem.* **2010,** *12* (12), 2240–2245.

Shekouhy, M.; Sarvestani, A. M.; Khajeh, S.; Khalafi-Nezhad, A. Glycerol: A More Benign and Biodegradable Promoting Medium for Catalyst-Free One-Pot Multi-Component Synthesis of Triazolo [1, 2-a] Indazole-Triones. *RSC Adv.* **2015,** *5* (78), 63705–63710.

Shezad, N.; Clifford, A. A.; Rayner, C. M.; Pd-Catalysed Coupling Reactions in Supercritical Carbon Dioxide and Under Solventless Conditions. *Green Chem.* **2002,** *4,* 64–67.

Shi, T.; Liu, Y. T.; Wang, S. S.; Lv, Q. Y.; Yu, B. Recyclable Carbon Nitride Nanosheet-Photocatalyzed Aminomethylation of Imidazo [1, 2-a] Pyridines in Green Solvent. *Chin. J. Chem.* **2022,** *40* (1), 97–103.

Silva, Y. P.; Ferreira, T. A.; Celli, G. B.; Brooks, M. S. Optimization of Lycopene Extraction from Tomato Processing Waste Using an Eco-Friendly Ethyl Lactate–Ethyl Acetate Solvent: A Green Valorization Approach. *Waste Biomass Valori.* **2019,** *10* (10), 2851–2861.

Silveira, C. C.; Mendes, S. R.; Líbero, F. M.; Lenardão, E. J.; Perin, G. Glycerin and $CeCl_3$ $7H_2O$: A New and Efficient Recyclable Medium for the Synthesis of Bis (Indolyl) Methanes. *Tetrahedron Lett.* **2009,** *50* (44), 6060–6063.

Simeó, Y.; Sinisterra, J. V.; Alcántara, A. R. Regioselective Enzymatic Acylation of Pharmacologically Interesting Nucleosides in 2-Methyltetrahydrofuran, a Greener Substitute for THF. *Green Chem.* **2009,** *11* (6), 855–862.

Sinha, A.; Sharma, B. P. Preparation of Copper Powder by Glycerol Process. *Mater. Res. Bull.* **2002,** *37* (3), 407–416.

Sinha, A.; Sharma, B. P. Preparation of Silver Powder Through Glycerol Process. *Bull. Mater. Sci.* **2005,** *28* (3), 213–217.

Siskin, M.; Katrinzky, M.; Reactivity of Organic Compounds in Superheated Water: General Background. *Chem. Rev.* **2001,** *101* (4), 825–835.

Song, B.; Yu, Y.; Wu, H. Solvent Effect of Gamma-Valerolactone (GVL) on Cellulose and Biomass Hydrolysis in Hot-Compressed GVL/Water Mixtures. *Fuel.* **2018,** *232*, 317–322.

Stadler, A.; Yousti, B. H.; Dallinger, D.; Walla, P.; Eycken, E. V.; Kaval, N.; Kappe, C. O.; Scalability of Microwave-Assisted Organic Synthesis: From Single-Mode to Multimode Parallel Batch Reactors. *Org. Process. Res. Dev*. **2003,** *7* (5), 707–716.

Still, J. K. The Palladium-Catalyzed Cross-Coupling Reactions of Organotin Reagents with Organic Electrophiles [New Synthetic Methods (58). *Angew. Chem. Int. Ed. Engl.* **1986,** *25*, 508–524.

Strauss, C. R. Trainor, K. W. Developments in Microwave-Assisted Organic Chemistry. *Aust. J. Chem*. **1995,** *48* (10), 1665–1692.

Strauss, C. R.; Invited Review: A Combinatorial Approach to the Development of Environmentally Benign Organic Chemical Preparations. *Aust. J. Chem*. **1999,** *52* (2), 83–96.

Sugie, A.; Somete, T.; Kanie, K.; Muramatsu, A.; Mori, A. Triethylsilane as a Mild and Efficient Reducing Agent for the Preparation of Alkanethiol-Capped Gold Nanoparticles. *Chem. Commun.* **2008,** *2008*, 3882–3884.

Taduri, A. K.; Babu, P. N.; Devi, B. R. Glycerol Containing Triacetylborate Mediated Syntheses of Novel 2-Heterostyryl Benzimidazole Derivatives: A Green Approach. *Org. Chem. Int.* **2014,** *2014*. DOI: org/10.1155/2014/260726

Tan, K. T.; Chng, S. S.; Cheng, H. S.; Loh, T. P.; Development of a Highly α-Regioselective Metal-Mediated Allylation Reaction in Aqueous Media: New Mechanistic Proposal for the Origin of α-Homoallylic Alcohols. *J. Am. Chem. Soc*. **2003,** *125* (10), 2958–2963.

Tang, X.; Sun, Y.; Zeng, X.; Lei, T.; Li, H.; Lin, L. γ-Valerolactone—An Excellent Solvent and a Promising Building Block. In *Biomass, Biofuels, Biochemicals, Saravanamurugan, S.*; Li, H., Riisager, A., Pandey, A., Eds.; Elsevier: Amsterdam, 2020; pp 199–226.

Tankam, T.; Srisa, J.; Sukwattanasinitt, M.; Wacharasindhu, S. Microwave-Enhanced On-Water Amination of 2-Mercaptobenzoxazoles to Prepare 2-Aminobenzoxazoles. *Org. Chem.* **2018,** *83* (19), 11936–11943.

Thurow, S.; Webber, R.; Perin, G.; Lenardão, E. J.; Alves, D. Glycerol/hypophosphorous acid: A Efficient System Solvent-Reducing Agent for the Synthesis of 2-Organylselanyl Pyridines. *Tetrahedron Lett.* **2013,** *54* (24), 3215–3218.

Tiwari, A. R.; Bhanage, B. M. Polythene glycol (PEG) as a Reusable Solvent System for the Synthesis of 1, *3*, 5-Triazines via Aerobic Oxidative Tandem Cyclization of Benzylamines and N-Substituted Benzylamines with Amidines Under Transition Metal-Free Conditions. *Green Chem.* **2016,** *18* (1), 144–149.

Tjerneld, T.; Persson, I.; Albertsson, P. A.; Hagerdal, B. H.; Enzymatic Hydrolysis of Cellulose in Aqueous Two-Phase Systems. I. Partition of Cellulases from Trichoderma Reesei. *Biotechnol. Bio. Eng.* **1985,** *27* (7), 1036–1043.

Trujillo, M.; Hull-Crew, C.; Outlaw, A.; Stewart, K.; Taylor, L.; George, L. et al., Green Methodologies for Copper (I)-Catalyzed Azide-Alkyne Cycloadditions: A Comparative Study. *Molecules* **2019,** *24* (5). DOI: 10.3390/molecules24050973

Türk, F. N.; Çehreli, S.; Baylan, N. Reactive Extraction of Monocarboxylic Acids (Formic, Acetic, and Propionic) Using Tributyl Phosphate in Green Solvents (Cyclopentyl Methyl Ether and 2-Methyltetrahydrofuran). *J. Chem. Eng. Data.* **2020,** *66* (1), 130–137.

Vafaeezadeh, M.; Hashemi, M. M. Polyethylene Glycol (PEG) as a Green Solvent for Carbon–Carbon Bond Formation Reactions. *J. Mole. Liq.* **2015,** *207*, 73–79.

Wan, J. P.; Cao, S.; Jing, Y. Copper-Catalyzed Homo-and Cross-Coupling Reactions of Terminal Alkynes in Ethyl Lactate. *Appl. Organometallic Chem.* **2014,** *28* (8), 631–634.

Wan, J. P.; Wang, C.; Zhou, R.; Liu, Y. Sustainable H 2 O/Ethyl Lactate System for Ligand-Free Suzuki–Miyaura Reaction. *RSC Adv.* **2012,** *2* (23), 8789–8792.

Wang, D. K.; Fang, Y. L.; Zhang, J.; Guan, Y. T.; Huang, X. J.; Zhang, J. et al., Radical Cyclizations of Enynes/Dienes with Alcohols in Water Using a Green Oxidant. *Org. Biomol. Chem.* **2020,** *18* (41), 8491–8495.

Wang, L. M.; Kobayashi, K.; Arisawa, M.; Saito, S.; Naka, H. Pd/TiO_2-Photocatalyzed Elf-Condensation of Primary Amines to Afford Secondary Amines at Ambient Temperature. *Organic Lett.* **2018,** *21* (2), 341–344.

Watanabe, K.; Kogoshi, N.; Miki, H.; Torisawa, Y. Improved Pinner Reaction with CPME as a Solvent. *Synth. Commun.* **2009,** *39* (11), 2008–2013.

Watanabe, K.; Yamagiwa, N.; Torisawa, Y. Cyclopentyl Methyl Ether as a New and Alternative Process Solvent. *Org. Proc Res. Develop.* **2007,** *11* (2), 251–258.

Webb, K. S.; Seneviratrie, V. A Mild Oxidation of Aromatic Amines. *Tetrahedron Lett.* **1995,** *36* (14), 2377–2378.

White, W. N.; Wolfarth, E. F. The o-Claisen Rearrangement. VIII. Solvent Effects. *J. Org. Chem.* **1970,** *35* (7), 2196–2199.

Wolfson, A.; Dlugy, C.; Shotland, Y.; Tavor, D. Glycerol as Solvent and Hydrogen Donor in Transfer Hydrogenation–Dehydrogenation Reactions. *Tetrahedron Lett.* **2009,** *50* (43), 5951–5953.

Xiao, J.; Ma, Y.; Wu, X.; Gao, J.; Tang, Z.; Han, L. B. Phosphonic Acid Mediated Practical Dehalogenation and Benzylation with Benzyl Halides. *RSC Adv.* **2019,** *9* (39), 22343–22347.

Xie, L. Y.; Duan, Y.; Lu, L. H.; Li, Y. J.; Peng, S.; Wu, C. et al., Fast, Base-Free and Aqueous Synthesis of Quinolin-2 (1H)-Ones Under Ambient Conditions. *ACS Sustain. Chem. Eng.* 2017, *5* (11), 10407–10412.

Xie, L. Y.; Peng, S.; Tan, J. X.; Sun, R. X.; Yu, X.; Dai, N. N. Waste-Minimized Protocol for the Synthesis of Sulfonylated N-Heteroaromatics in Water. *ACS Sustain. Chem. Eng.* 2018, *6* (12), 16976–16981.

Xu, X.; Antal, M. J. Kinetics and Mechanism of Isobutene Formation from T-Butanol in Hot Liquid Water. *Alch E. J.* **1994,** *40* (9), 1524.

Xu, X.; Antal, M. Mechanism and Temperature-Dependent Kinetics of the Dehydration of Tert-Butyl Alcohol in Hot Compressed Liquid Water. *J. Ind. Eng. Chem. Res.* **1997,** *36* (1), 23–41.

Yadav, M. B.; Balwe, S. G.; Kim, J. T.; Cho, B. G.; Jeong, Y. T. PEG-Assisted One-Pot Three-Component Synthesis of 1, 3-Oxazino Quinoline and Chromeno 1, 3-Oxazin Derivatives Under Catalyst Free Condition. *Synth. Commun.* **2020,** *50* (10), 1456–1467.

Yorimitsu, H.; Nakamurat, T.; Shinokubo, H.; Oshime, K.; Omoto, K.; Fujimoto, H.; Powerful Solvent Effect of Water in Radical Reaction: Triethylborane-Induced Atom-Transfer Radical Cyclization in Water. *J. Am. Soc.* **2000,** *122* (45), 11041–11047.

Yu, J.; Savage, P. E.; Decomposition of Formic Acid Under Hydrothermal Conditions. *Ind. Eng. Chem. Res.* **1998,** *37* (1), 2–10.

Yuan, X. Y.; Zeng, F. L.; Zhu, H. L.; Liu, Y.; Lv, Q. Y.; Chen, X. L. et al., A Metal-Free Visible-Light-Promoted Phosphorylation/Cyclization Reaction in Water Towards 3-Phosphorylated Benzothiophenes. *Org. Chem. Front.* **2020,** *7* (14), 1884–1889.

Zeng, F. L.; Xie, K. C.; Liu, Y. T.; Wang, H.; Yin, P. C.; Qu, L. B. et al., Visible-Light-Promoted Catalyst-/Additive-Free Synthesis of Aroylated Heterocycles in a Sustainable Solvent. *Green Chem.* **2022,** *24* (4), 1732–1737.

Zhang, X.; Jia, X.; Wang, J.; Fan, X. An Economically and Environmentally Sustainable Synthesis of 2-Aminobenzothiazoles and 2-Aminobenzoxazoles Promoted by Water. *Green Chem.* **2011,** *13* (2), 413–418.

Zhao, H.; Cheng, M.; Zhang, J.; Cai, M. Recyclable and Reusable $PdCl_2(PPh_3)_2$/PEG-2000/H_2O System for the Carbonylative Sonogashira Coupling Reaction of Aryl Iodides with Alkynes. *Green Chem.* **2014,** *16* (5), 2515–2522.

Zhao, H.; Zhang, T.; Yan, T.; Cai, M. Recyclable and Reusable [$RuCl_2$ (p-cymene)]$_2$/Cu $(OAc)_2$/PEG-400/H_2O System for Oxidative C–H Bond Alkenylations: Green Synth. of Phthalides. *J. Org. Chem.* **2015,** *80* (17), 8849–8855.

Zheng, T.; Yang, L.; Li, J.; Cao, M.; Shu, L.; Yang, L. et al., Lignocellulose Hydrogels Fabricated from Corncob Residues Through a Green Solvent System. *Int. J. Biol. Macromol.* **2022,** *217*, 428–434.

CHAPTER 8

Photocatalysis: An Emerging Technology

SHUBANG VYAS[1], INDU BHATI[2], PARAS TAK[1], H. S. SHARMA[3], and RAKSHIT AMETA[4]

[1]*Department of Chemistry, PAHER University Udaipur, India*

[2]*Department of Chemistry, M. L. Sukhadia University, Udaipur, India*

[3]*Department of Chemistry, Govt. P.G. College, Kota, India*

[4]*Department of Chemistry, J. R. N. Rajasthan Vidhyapeeth, (Deemed to be University), Udaipur, India*

ABSTRACT

Photocatalysis is an emerging area of research. It falls under the arena of advanced oxidation processes (AOPs). Mostly, hydroxyl radicals are generated by this route, which are considered strongest oxidizing agent next to fluorine. As such, it can oxidize many toxic, stable and recalcitrant molecules to harmless smaller fragments. It is an eco-friendly approach for the treatment of wastewaters. Semiconducting materials are mostly used as photocatalysts. Photcatalytic materials have been utilized in naïve and various modified forms (sensitized, doped, co-catalyzed, composite, etc.) to degrade dyes, pesticides, pharmaceutical drugs, phenols, nitro compounds, detergents, etc. with reasonably good efficiency. This is being described here.

8.1 INTRODUCTION

Although 70% area of Earth is covered with water, inspite of that it is well known that only 1% of this water can be used for human consumption and other purposes. Men started industrialization of a number of products for the

Green Chemistry, 2nd Edition: Fundamentals and Applications. Suresh C. Ameta & Rakshit Ameta (Eds.)

development. As a result, there are a huge number of small and big industries all over the world. A number of heavy metals, inorganic compounds, organic compounds like phenols, pesticides, fertilizers, surfactants, drugs, detergents, insecticides, dyes, and other chemical products are disposed directly into the water resources by these industries, without any effective treatment strategy. All these substances are toxic and harmful to human beings, animals, and plants. Many of these compounds can be readily absorbed through the skin and create health problems.

Innovative steps should be taken for purification and recycling of wastewater of various industries, as water has now become a key symbol of protest all around the world. Color removal from the textile wastewaters has become an issue of interest during the last few years because of the toxicity of the dyes and more often, the colored wastewater from the textile industries also decreases the visibility of the receiving waters affecting the animal kingdom adversely.

Photocatalysis is gaining popularity in recent years and it is quite promising also for the treatment of resilient pollutants.

8.2 PHOTOCATALYSIS

Photocatalysis includes such reactions, which utilize light to activate a substance (particularly a semiconductor), which modifies the rate of a chemical reaction without being involved itself. The definition of "photocatalysis" accepted by IUPAC after long debate is a catalytic reaction involving light absorption by a substrate (1988). This substrate, which is a semiconductor, absorbs light and acts as a catalyst for that chemical reaction, is known as photocatalyst. Extensive reviews of this field have been made from time to time by various researchers (Gratzel, 1983; Pelizzetti and Serpone, 1986; Ameta et al., 1992, 2010a, Kamat, 1993; Ameta and Ameta 2017).

8.2.1 PHOTOCATALYTIC REACTIONS

The photocatalytic reactions can be classified into two categories on the basis of physical state/appearance of reactants.

- **Homogeneous Photocatalysis**

When the catalyst and reactant; both are in the same phases, that is, gas, solid or liquid, then the photocatalytic reaction is called homogeneous photocatalysis. Different dyes/organic substances and colored coordination compounds are best examples of homogeneous photocatalysts.

- **Heterogeneous Photocatalysis**

When the catalyst and reactant; both are in different phases, then the photocatalytic reaction is called heterogeneous photocatalysis. The common example of this kind is a solid photocatalyst in contact with either liquid or a gas phase.

8.2.2 PHOTOCATALYSTS

All the photocatalysts are normally semiconductors, but all semiconductors are not necessarily photocatalysts. Semiconductor is a substance, where the energy gap between conduction band (lowest unoccupied molecular orbital, LUMO) and valence band (highest occupied molecular orbital, HOMO), ranges from 1.5 to 3.0 eV.

8.2.3 BAND GAP

The energy difference between the valence band and the conduction band is known as the band gap (E_g). On the basis of this band gap, the materials are classified in three categories (i) E_g < 1.0 eV, metal or conductor, (ii) E_g > 5.0 eV, insulator or nonconductor and (iii) E_g ~ 1.5–3.0 eV, semiconductor. This classification is shown in Figure 8.1.

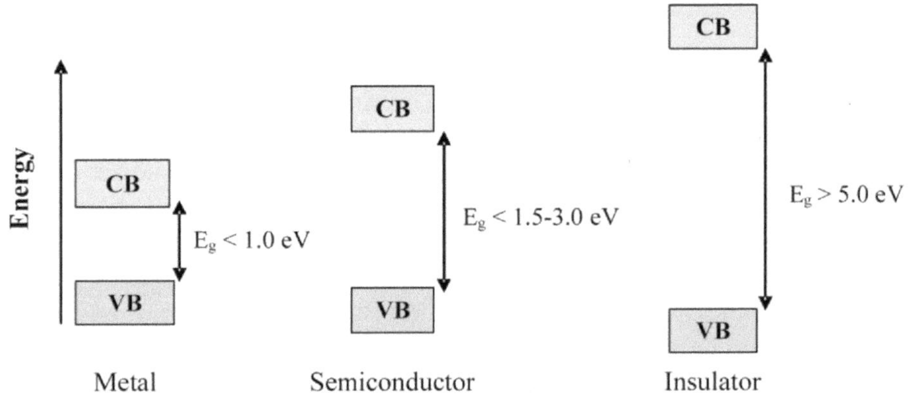

FIGURE 8.1 Band gap and nature of materials.

The band gaps of various semiconductors are given in Table 8.1.

TABLE 8.1 Band Gaps of Different Semiconductors.

Semiconductor	Band gap (eV) at 300 K
ZnS (wurtzite)	3.91
ZnS (zinc blende)	3.54
TiO_2	3.20
ZnO	3.03
WO_3	2.60
CdS	2.42
CdSe	1.70

When the sufficient energy (equal to or more than the band gap) is provided to a semiconductor, e^- from the valence band absorbs this energy and jumps to the conduction band leaving behind a hole in valence band. Therefore, a semiconductor, which is an insulator at normal temperature and pressure, may conduct some electric current on exposure to light containing energy equal to or more than the corresponding band gap of semiconductor. Hence, semiconductors are capable of conducting electricity even at room temperature in the presence of light and that is why, they work as photocatalyst.

8.2.4 MECHANISM OF PHOTOCATALYSIS

When the photocatalyst is illuminated by light, the energy of photons is utilized by the e^- of valence band promoting it to the conduction band while a hole (h^+) is created in valence band by removal of this electron. This process creates "photoexcitation state." Thus, a photocatalyst absorbs appropriate radiation (Vis/UV) from sunlight or illuminated light source (fluorescent lamps), producing a pair of electron (e^-) and hole (h^+). This excited e^- can be used for reducing an acceptor substrate whereas the h^+ may be utilized for oxidation of donor molecules (Figure 8.2).

In photogenerated catalysis, the photocatalytic activity depends on the ability of the catalyst to create e^-– h^+ pairs. But in the absence of any other influence or process, the excited state may diffuse down the concentration gradient of e^-– h^+ pair, as there is an equal possibility for e^- to lose the energy and get back to its position in valence band, where hole was present.

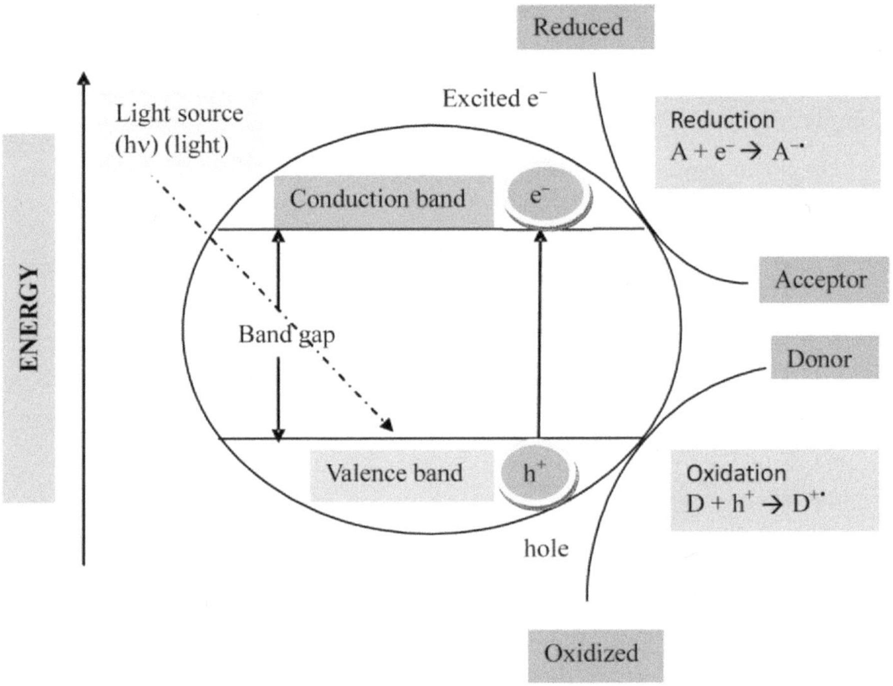

FIGURE 8.2 Meachnism of photocatalysis (e^-–h^+ pair generation).

8.2.5 FATE OF EXCITED ELECTRON–HOLE PAIR

The fate of excited electron and hole is decided by the relative positions of the conduction and valence bands of semiconductor and the redox levels of the substrate (Fig. 8.3). The importance of photocatalysis lies in the fact that by choosing a photocatalyst (semiconductor) of the desired band gap, one can drive a reaction in a desired direction, as photocatalyst provides an oxidation as well as reduction environment, simultaneously.

The field of heterogeneous photocatalysis has already grown out of its infancy and now emerged as a major field of research. The recent development of ultrafine semiconductor particles with many interesting photocatalytic properties has added some newer dimensions. The ability of such semiconductors to carry out redox processes with greater efficiency and selectivity than in the homogeneous solutions has made them potential candidates for the conversion and storage of solar energy and in the mineralization of chemical pollutants. Photocatalysis has a wide scope of its uses in variety of applications, which includes wastewater treatment, conservation and storage of energy, self cleaning, deodorization, sterilization, air purification, antifouling, antifogging, nitration, sulphonation, halogenation, etc.

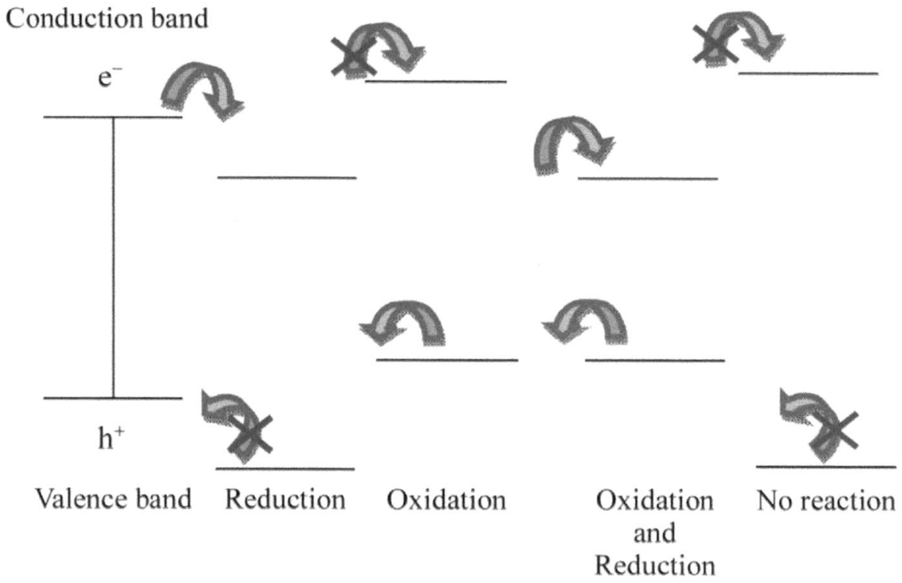

FIGURE 8.3 Various possibilities of reactions.

8.3 WASTEWATER TREATMENT PROCESSES

A typical system for surface water treatment generally consists of presettling, coagulation/flocculation (Montgomery, 1985) (sediment removal), granular filtration (Faust and Aly, 1999) (sediment removal), corrosion control (USEPA, 1989) (pH adjustment or addition of corrosion inhibitors), and disinfection. These methods suffer from the one or other drawbacks. However, photocatalysis provides more cleaner and environment healthy/benign method for wastewater treatment (Ameta, et al., 2003). Mainly, the responsible chemicals for water contamination are dyes, pesticides, surfactants, phenols, halo-organics, herbicides, organic compounds, pharmaceuticals, etc.

8.3.1 DYES

Many of the dyeing, textile, and printing industries discharge their colored effluents in the nearby water resources without proper treatment, which pollute the water. This colored water can not be used for any useful purpose due to its hazardous nature. Photocatalysis provides an eco-friendly method for removal of these toxic dye molecules from effluents. Various semiconductors have been used by researchers for this purpose. Tanaka et al. (2000) photocatalytically degrade some commercially used azo dyes in TiO_2 suspension and proposed that the disappearance of azo dyes proceeds through both; oxidation and reduction processes, while TiO_2/UV photocatalytic degradation of methylene blue has been investigated in aqueous heterogeneous suspensions (Tatsuma et al., 1999; Houas et al., 2001).

Arabatzis et al. (2003) synthesized and charaterized Au/TiO_2 thin films, which was used for the degradation of azo dye, while Wei et al. (2007) used V-doped TiO_2 for photocatalytic degradation of methyl orange. Cerium-doped TiO_2 catalyst was prepared by controlled hydrolysis of titanium alkoxide based on esterification reaction by Tang et al. (2007).

Ameta et al. (2000) studied the photocatalytic degradation of orange–G over ZnO in the presence of a surfactant, whereas Pare et al. (2009) studied the photocatalytic degradation of lissamine fast yellow using ZnO in the aqueous suspension. ZnO was also used for photodegradation of methylene blue and eosin by Chakrabarti and Dutta (2004). Photocatalytic bleaching of rhodamine-B and rhodamine-6G over zinc oxide and lead oxide was studied by Mansoori et al. (2004), whereas photobleaching of amaranth dye in the presence of ZnO photocatalyst was reported by Kothari et al. (2004). Ameta et al. (2002, 2006) have done an extensive work on photocatalytic bleaching of some dyes. Kothari et al. (2007) observed the photocatalytic bleaching of Evans blue over zinc oxide while Vyas et al. (2005) investigated the photocatalytic bleaching of eosin using zinc oxide and effect of surface charge. Punjabi et al. (2005) reported the photoreduction of Congo red by ascorbic acid and EDTA as reductant and cadmium sulphide as photocatalyst.

Photodegradation of rhodamine-B (RB) catalyzed by TiO_2 films was demonstrated by Ma and Yao (1998). The mechanism of this reaction was proposed as

$$RB \xrightarrow{h\nu} RB^*$$

$$RB^* + TiO_2 \longrightarrow RB^{+\bullet} + TiO_2\,(e^-)$$

$$RB^* + O_2 \longrightarrow RB^+ + O_2^{-\bullet}$$

$$TiO_2\,(e^-) + O_2 \longrightarrow TiO_2 + O_2^{-\bullet}$$

$$RB^{+\bullet} \xrightarrow{O_2} \text{Products}$$

Doping of semiconductor with metals or nonmetals enhances its activity. Vaya et al. (2008) reported the effect of transition metal ions doping on the photocatalytic activity of ZnS whereas Li et al. (2008a) prepared ZnS polymer composites, which exhibited high efficiency for degrading methyl orange, methylene blue, and eosin B. Puretedal et al. (2009) used nanoparticle (NP) zinc sulfide doped with manganese, nickel, and copper as a nanophotocatalyst for the degradation of organic dyes; methylene blue and safranin. The photocatalysts like TiO_2, ZnO, CdS, etc., can oxidize organic pollutants like dyes and other harmful wastes into nontoxic and less harmful materials (Poulis et al., 2003; Bilgi and Demir, 2005; Papic et al., 2006; Behnajady et al., 2006).

A new method for doping metal ions on to the TiO_2 surface has been developed by Xu et al. (2004). It was observed that the surface-doped TiO_2 exhibited higher photocatalytic activity than pure TiO_2 for the degradation of methyl orange in water. Fe^{3+}-doped TiO_2 catalyst was prepared by Tong et al. (2008) and used for the degradation of methyl orange. Preparation and characterization and use of nanosilver-doped mesoporous titania photocatalyst for dye degradation was investigated by Binitha et al. (2009).

Sol–gel synthesis of TiO_2 nanoparticles and photocatalytic degradation of methyl orange in aqueous TiO_2 suspensions was investigated by Yang et al. (2006). Tian et al. (2008) synthesized Au/TiO_2 catalyst from Au(I)–thiosulphate complex and observed its photocatalytic activity for the degradation of methyl orange while Wang et al. (2009) characterized and reported the photocatalytic activity of poly(3-hexylthiophene)-modified TiO_2 for the degradation of methyl orange under visible light.

Li et al. (2008b) TiO_2 prepared immobilized nanoparticles of TiO_2 supported on natural porous mineral and used it for photocatalytic degradation of azo dyes. However, Magalhaes and Lago (2009) grafted TiO_2 on expanded polystyrene (PS) beads, which were used for the solar degradation of dyes. UV/titanium dioxide degradation of two xanthene dyes, erythrosine B and eosin Y was studied by Pereira et al. (2013) in a photocatalytic reactor, while Mahadwad et al. (2011) carried out photocatalytic degradation of reactive black-5 (RB-5) dye using supported TiO_2 photocatalyst-based adsorbent as a semiconductor photocatalyst in a batch reactor. Santhanalakshmi and Komalavalli (2012) investigated kinetics of degradation of reactive yellow, acid blue, methyl orange, and acid green with TiO_2 in H_2O_2 aqueous solution.

Visible light-induced degradation of methylene blue in the presence of photocatalytic ZnS and CdS nanoparticles has been carried out by Soltani et al. (2012). Shankar et al. (2011) studied photocatalytic degradation of methylene blue using nano ZnO. Ameta et al. (2011) used Sb_2S_3 semiconductor for photocatalytic degradation of naphthol green B. They proposed that the dye molecules were oxidized by hydroxyl radicals, which were generated due to the reaction between hydroxyl anion and hole of the semiconductor. The role of hydroxyl radicals as an active oxidizing species was confirmed by carrying out the reaction in the presence of hydroxyl-radicals scavenger, that is, 2-propanol, where the rate was drastically reduced. They have proposed the following mechanism for the degradation of the naphthol green B:

$$SC \xrightarrow{h\nu} h^+(VB) + e^-(CB)$$

$$h^+ + OH^- \rightarrow {}^{\bullet}OH$$

$$^{\bullet}OH + Dye \rightarrow Products$$

Shanthi and Kuzhalosai (2012) studied photocatalytic degradation of acid red 27 azo dye in aqueous solution using nanosized ZnO, while Madhusudhana et al. (2011) studied photocatalytic degradation of coralene dark red 2B azo dye using calcium zincate nanoparticles in the presence of natural sunlight. Yang and Luan (2012) synthesized and characterized a composite polymer polyaniline/Bi_2SnTiO_7 and studied its photocatalytic activity using methylene blue system, while Kaur et al. (2011) synthesized, characterized, and studied the photocatalytic activity of La_2CoO_4 using azure B as the model system. Municipal wastewater and brilliant blue dye were photocatalytically degraded by Parvin et al. (2012) using hydrothermally synthesized surface-modified silver-doped ZnO. Sm_2FeTaO_7 photocatalyst was used for the degradation of indigo carmine dye under solar light irradiation by Torres-Martínez et al. (2012). Ameta et al. (2010) used nanosized chromium-doped TiO_2 supported on zeolite for methylene blue degradation, while Sacco et al. (2012) carried out photocatalytic degradation of methylene blue under visible light on N-doped TiO_2 photocatalysts.

Moafi et al. (2013) synthesized W-doped ZnO nanocomposites with different W contents. The photocatalytic activity of undoped ZnO and W-doped ZnO was observed in photodegradation of methylene blue. It was observed that it gave exhibited maximium activity, when doping W was found to be 4 mol%.

The WO_3-loaded Ag–ZnO (WO_3–Ag–ZnO) was prepared by Subash et al. (2013). As-prepared WO_3–Ag–ZnO was used for the degradation of acid black 1 (AB 1) under UV-A light. They confirmed mineralization of AB 1 by COD measurements. Kulkarni and Bhanage (2014) prepared hybrid Ag@AgCl plasmonic NPs (average size 37 nm) using sugar cane juice. This method avoids the use of any external reducing or capping agents, solvents, and external halide ion sources. As-prepared photocatalyst exhibited good photocatalytic ability for the degradation of methylene blue and methyl orange under visible light.

Magnesium-doped titanium dioxide (Mg^{2+}-TiO_2) was synthesized by Avasarala et al. (2016) with different magnesium weight percentages (0.25–1.0 wt.%). The photocatalytic efficiency of as-synthesized catalysts was investigated for the degradation of methyl orange in the presence of visible light irradiation. One pot synthesis of ZnO nanoflowers (ZnO-NFs) has been reported by Mohammed et al. (2016) using asymmetric Zn(II) dimeric complex, [Zn(hmp-H)$_2$(H_2O)(μ-Cl)Zn(μ-Cl)(Cl)$_3$],(A), where hmp-H=2-(2-hydroxymethyl)pyridine). These ZnO-NFs were used as photocatalyst for the degradation of different dyes (Congo red, and eosin B, Chicago sky blue, methyl orange). It was revealed that improved photocatalytic activity toward methyl orange degradation (99.46%) was there as compared to other dyes.

BiOI nanoplates were deposited on a film of TiO_2 nanoparticles by Odling and Robertson (2017) using sequential ionic layer adsorption and reaction (SILAR) method. It was reported that modified film with five SILAR cycles were optimal in the photocatalytic degradation of rhodamine B. CdSe quantum dots (QDs) hybridized with graphene oxide (GO) were synthesized by Thirugnanam et al. (2017) via chemical precipitation method. Using the brilliant green dye, it was revealed that photocatalytic efficiency is approximately 81.9 and 95.5%, respectively, when CdSe QDs and the CdSe/GO nanocomposite were used under sunlight irradiation in 90 min. This enhanced photocatalytic activity of nanocomposite may be due to the reduction of electron–hole pair recombination and high specific surface area due to introduction of graphene oxide.

Chen et al. (2017) synthesized zinc oxide via sol–gel method. As-prepared ZnO exhibited 99.70% removal rate for methyl orange. Kuvarega et al. (2018) embedded N, Pd co-doped TiO_2 (different proportions) in a polysulfone and used it for the degradation eosin yellow under visible light irradiation. They could achieve 92% dye degradation with the 7% N,Pd TiO_2/PSf nanocomposite membrane in 3 h.

Undoped and Ca-doped CeO_2 quantum dots were synthesized by Ramasamy et al. (2018) via sol–gel method. It was observed that photocatalytic activity of CeO_2 quantum dots increases as the concentration of Ca doping was increased. The rate of photodegradation of methylene blue over Ca-doped CeO_2 quantum dots was found to be relatively high (84%) as compared to bare and other doped samples under sunlight.

A synthesis strategy was developed by Cheng et al. (2019) for synthesis of fluorescent carbon quantum dots (CQDs). They used aqua mesophase pitch as the carbon source using hydrothermal method. As-obtained CQDs were modified with thionyl chloride and ammonia, which exhibited an excellent photocatalytic activity for the degradation of rhodamine B, methyl blue, and indigo carmine. It was revealed that for the degradation percentage of N-CQDs on RhB could achieve 97% degradation under natural light in 4 h and retained 93% activity even after using five times.

A TiO_2-coated Tunisian clay (TiO_2-clay) was synthesized by Hadjltaief et al. (2019). They could observe almost complete discoloration of reactive blue in 20 min in the presence of UV irradiation. Abhilash et al. (2019) synthesized Fe_2O_3/Cu_2O nanocomposite through facile hydrothermal technique. It was confirmed that hexagonal rod-shaped bare Cu_2O, rhombohedral-shaped Fe_2O_3 and was there in composite assembly. Rhodamine-B and Janus green were selected as model pollutants. As-synthesized materials exhibited high stability; and apart from this, they were active against *Pseudomonas aeruginosa*, *Escherichia coli*, *Staphylococcus aureus*, and *Bacillus subtilis*, but have less toxicity against the Musmusculus skin melanoma cells (B16-F10).

Novel $FeVO_4/Bi_7O_9I_3$ nanocomposites were fabricated with different weight percentages by Chachvalvutikul et al. (2019). They used cyclic microwave irradiation, which was followed by wet impregnation. The 6.25%wt-$FeVO_4/Bi_7O_9I_3$ nanocomposite showed better photocatalytic degradation of dyes (rhodamine B, methylene blue, and methyl orange) with efficiencies of decolorization as 98.9, 81.3, and 94.9% within 3 h, respectively. It was also reported that nanocomposite exhibited excellent stability and reusability. They attributed this enhancement in the photocatalytic activity of nanocomposite to synergistic effects of a favorable type-II heterojunction and visible-light absorption.

The photocatalytic degradation of fast green was studied by Jat et al. (2019) under visible light using SnO_2–TiO_2. This composite was prepared via hydrothermal method using $SnCl_4$ (hydrate) as a precursor for SnO_2 quantum dots. It was revealed that there is higher photocatalytic activity as compared to titania nanopowder for the degradation of fast green. The influence of operational parameters on the rate of degradation was also studied, such as concentration of dye, pH, light intensity, and amount of catalyst.

Wang et al. (2020) prepared new ternary *n*-$BaTiO_3$/Ag/p-AgBr heterostructured photocatalysts by coupling p-type AgBr particles and n-type $BaTiO_3$ onto Ag nanowires. The photodegradation performances of as-prepared photocatalysts were evaluated by using simulated sunlight and rhodamine B as the model system. It was reported that *n*-$BaTiO_3$/Ag/p-AgBr ternary composite photocatalysts exhibited excellent photodegradation as-compared to bare $BaTiO_3$ and AgBr particles. It was observed that

20%$BaTiO_3$/1.2%Ag/AgBr composite can degrade 99.3% rhodamine B in 12 min. It exhibited higher photodegradation activity, which is almost 3.1 times to bare AgBr particles.

8.3.2 PESTICIDES

Pesticides and herbicides have their own importance to save crops from insects, but due to their bioaccumulation nature, these create problems. These pollutants are often toxic and cause adverse effects on human and animal lives, when present even in low concentrations. Organo chemicals such as dicofol, BHC, cypermethrin, and organophosphorous pesticides are commonly used pesticides and insecticides. Among all, photocatalytic oxidation is one of the best and suitable technologies for the elimination of these organic pollutants (Kerzhentsev et al., 1996; Doong and Chang, 1997). Photocatalytic oxidation processes have been successfully utilized for the removal/degradation of pesticides and herbicides from water.

Photocatalytic degradation of pesticide pirimiphosmethyl has been carried out by Hermann et al. (1999). Methoxychor and *p,p'*-DDT have successfully been degraded by TiO_2 in aqueous suspension (Zaleska et al., 2000). A TiO_2 film developed by Choi et al. (2000) has been used to degrade polychlorinated dibenzo-*p*-dioxine under UV irradiation, while Muszkat et al. (2002) studied the photocatalytic degradation of pesticides and biomolecules in water. TiO_2 nanoparticles have been used for transformation of chlorinated volatile organic compounds by Liu et al. (2002).

Tamini et al. (2006) studied the degradation of pesticide methomyl in aqueous solution by ultraviolet irradiation in the presence of TiO_2, whereas Konstantinou and Albanis (2003) used photocatalytic process for the transformation of pesticides in aqueous titania suspension using artificial and solar light. Yu et al. (2007) carried out photocatalytic degradation of organochlorine pesticides on a nano-TiO_2 coated film, while Senthilnathan and Philip (2010a) degraded mixed pesticides. Re^{3+}-doped nano-TiO_2 was used by Zhang et al. (2010) for photocatalytic degradation of carbofuran, whereas nitrogen-doped titania with different nitrogen-containing organic compounds was used under UV and visible light for the degradation of lindane (Senthilnathan and Philip, 2010b).

Machuca-Martínez and Colina-Márquez (2011) studied the effect of pH and the catalyst concentration on TiO_2-based photocatalytic degradation of three commercial pesticides, that is, 2,4-D, diuron, ametryne, while Verma et al. (2013) studied titanium dioxide mediated photocatalytic degradation of malathion in aqueous suspension. Zinc oxide has also been used for photocatalytic oxidation of organophosphorous pesticides by Dehghani and Fadaei (2012). Miguel et al. (2012) studied photocatalytic degradation of 44 organic pesticides with titanium dioxide in natural water.

The TiO_2-mediated photocatalytic degradation of quinalphos was studied by Kaur and Sud (2012). It was revealed that complete degradation of quinalphos could be obtained in 24 h. Synthesis, characterization, and application of nanophotocatalyst, heterostructured ZnO/$TiO_2$$TiO_2$ photocatalyst (Z9TZ9T), was prepared by Kaur et al. (2013). The particle size of as-synthesized Z9TZ9T was found to be 21.48 nm. The pesticides (quinalphos and monocrotophos) were degraded in the presence of nanophotocatalysts UV light.

Hollow anatase titania spheres were synthesized by Baharvand et al. (2014) using fructose and tetrabutyl titanate. It was revealed that photocatalytic activity of as-synthesized TiO_2 for the photodecomposition of chlorpyrifos was 18.67% higher than that with commercial TiO_2. Ananpattarachai and Kajitvichyanukul (2015) prepared N-doped TiO_2 nanopowders. It was observed that it can degrade *p,p*′-DDT effectively under both; UV and visible lights. Almost 100% degradation of *p,p*′-DDT could be achieved on using N-doped TiO_2 catalyst and reaction rate was six fold higher than UV light.

Composites were synthesized by Abdennouri et al. (2016) via immobilizing TiO_2 onto surfactant-pillared clay using cetyltrimethyl ammonium bromide. The adsorption performance and photocatalytic activites of these samples were evaluated using 2,4-dichlorophenoxypropionic acid (2,4-DP) and 2,4-dichlorophenoxyacetic acid (2,4-D) as model pollutants. It was observed that removal efficiency increases on increasing Ti content in the pillared clay. Lannoy et al. (2017) prepared Au/TiO_2 photocatalysts and then used for the degradation of the herbicide phenoxyacetic acid in water. Different cyclodextrins (CD) were used to drive this reaction. It was observed that randomly methylated β-CD afforded Au/TiO_2 composites with high crystallinity, large surface area, and optimum Au particle size.

The In,S-TiO_2@rGO nanocomposite was prepared by Khavar et al. (2018) using an ultrasonic-assisted solvothermal method. When TiO_2 was co-doped with S and In elements and its composite was prepared with rGO, then it was found to have increased photocatalytic activity under visible light. The complete degradation and 95.5% mineralization of 20 mg L^{-1} atrazine was attained with 3.0 mol.% of In, 1 mol.% of S and 5 wt.% of rGO within 20 min. As-prepared In,S co-doped TiO_2@rGO nanocomposite could retain its catalytic activity even after four consecutive recycles.

Vela et al. (2018) used two commercial TiO_2 nanopowders (Kronos vlp 7000 and Degussa P25) as photocatalysts to degrade quinalphos, malathion, fenitrothion,vinclozoline, fenarimol, and dimethoate. All these pesticides exhibited endocrine disrupting activity. The time required for 90% degradation of these pesticides was in the range of 79–1270 min. Hanh et al. (2019) synthesized Cu-doped ZnO materials and used for photocatalytic degradation of monocrotophos pesticide (MCP). The highest MCP degradation efficiency was obtained with 3 wt.% Cu/ZnO among all synthesized Cu–ZnO.

Kushniarou et al. (2019) investigated photocatalyzed degradation of pesticides commonly used on vegetables, citrus, vines, and fruit crops in aqueous suspensions of TiO_2 with $Na_2S_2O_8$ under natural sunlight. It was observed that time required for disappearance of 90% pesticides was as follows:

Cyproconazole (4.9 h) > Metalaxil (6.1 h) > Propyzamide (7.9 h) > Cyproconazole (8.9 h)

The Fe-doped TiO_2 was deposited on bentonite by Phuong et al. (2019). As-synthesized materials were used for removal of diazinon under both; visible light and dark conditions. It was indicated that the Fe–TiO_2/Bent-Fe exhibited high photocatalytic activity for removal of diazinon even under visible light. The optimal pH and photocatalyst dosage were 4.5 and 0.5 g L^{-1}, respectively. Ghodsi et al. (2020) synthesized g-C_3N_4/Fe_3O_4/Ag nanocomposite via a hydrothermal method. It was observed that this nanocomposite successfully removed diazinon from aqueous solutions by photocatalytic process. The complete removal of diazinon could be achieved after an hour.

Tang et al. (2020) prepared a dual Z-scheme photocatalyst AgI/Ag_3PO_4/g-C_3N_4. This photocatalyst exhibited higher activity for the degradation of nitenpyram than its individual components. It was reported that rate constant for the degradation of nitenpyram was about 2.4, 2.9 and 16.2, times than that of Ag_3PO_4, AgI and g-C_3N_4, respectively. The GO-doped metal ferrites (GO-Fe_3O_4 and GO-$CoFe_2O_4$) were prepared by Tabasum et al. (2021). They evaluated photocatalytic potentials of these catalysts for the degradation of acetamiprid. It was revealed that about 97 and 90% degradation of the acetamiprid could be achieved by using GO-Fe_3O_4 and GO-$CoFe_2O_4$, respectively, during the first hour only under UV radiations.

8.3.3 SURFACTANTS

Surfactants are extensively used in many fields such as pharmacy, cosmetics, textile, industry, agriculture, biotechnology, and also in daily life due to their favorable physicochemical properties. A large quantity of surfactant is released to the aquatic environment after its use. It causes serious threat to aquatic environment because of their high foaming and lower oxygenation potential, which leads to death of water-borne organism. Surfactants cause short term as well as long term changes in the ecosystem. 0.5–1.0 mg L^{-1} limit has been fixed by many environmental and public health regulatory authorities. Surfactants are classified into ionic and nonionic types according to their chemical nature.

Singhal et al. (1997) carried out the degradation of cetylpyridinium chloride, while Rao and Dubey (1996) studied photocatalytic degradation of binary and ternary mixture of three surfactants, viz., dodocylbenzene sulfonic acid, sodium salt (DBS), cetylpyridinium chloride (CPC), and triton X-100 (TX-100) using TiO_2 as a photocatalyst. Hidaka et al. (1990) degraded various kinds of cationic (*e.g.*, C_{16}-HTAB and C_{12}-BDDAC), anionic (e.g., C_{12}-DBS, C_{12}-DS, C_{12}-LES-3, C_{14}-AOS, and C_{12}-DG) and non-ionic (e.g., NPE*n* (*n* = 7, 9, 17, 50), C_{18}-PEA-15, $C_{12–14}$-BHA, $C_{12–14}$-PAE-10, $C_{12–14}$-NOE and $C_{14–16}$-NOE) surfactants by TiO_2 semiconductor particles under UV illumination. These surfactants are converted completely into CO_2.

Solar photodegradation of two commercial surfacatants, sodium dodecyl sulfate (SDS) and DBS has been studied by Amat et al. (2004) whereas Kimura et al. (2004) investigated photocatalytic degradation of nonionic surfactant polyoxyethylene alkyl ether in water using immobilized TiO_2 photocatalyst. The photocatalytic degradation of two industrial grade surfactants, sodium lauryl sulphate and sodium dodecylbenzenesulfonate was achieved using TiO_2 immobilized on glass (Lizama et al., 2005). TiO_2–CO_2 composite oxide was used for photocatalytic degradation of dodecylbenzenesulfonate (DBS) under visible irradiation (Han et al., 2009). Bardos et al. (2011) have carried out titanium dioxide-mediated photocatalyzed degradation of benzenesulfonate.

Zhang et al. (2011) carried out photocatalytic degradation of phenanthrene (PHE) over TiO_2 in aqueous solution containing nonionic surfactant micelles while photocatalytic degradation of anionic surfactant linear alkylbenzene sulfonate was reported by Giahi et al. (2012) using ZnO nanoparticles on irradiation with UV light. The ZnO nanocatalyst was synthesized by Samadi et al. (2017) and used for UV-induced removal of SDS from aquatic solutions. It was observed that about 98% of surfactant was removed at 40 min.

Aoudjit et al. (2019) synthesized Zn_2Al-LDH and TiO_2/Zn_2Al-LDH materials via coprecipitation. The photocatalytic degradation of SDS was investigated over these materials. It was indicated that TiO_2 3.6/Zn_2Al-LDH exhibited better photocatalytic activity in comparision to Zn_2Al-LDH sample while Huszla et al. (2021) evaluated the activity of zinc oxide nanoparticles (photocatalyst) for degrading two nonionic surfactants (C12E10 and Triton X-100). It was reported that use of as-obtained ZnO nanoparticles could achieve photodegradation of both the surfactant Triton X-100 (82%) and C12E10 (92%) in 1 h of UV exposure.

8.3.4 HYDROCARBONS

Platinum-loaded TiO_2 has been used for photocatalytic hydroxylation of aromatic ring by using water as an oxidant (Park and Choi, 2005). Lair et al. (2007) studied photocatalytic degradation of naphthalene over TiO_2 in the presence of inorganic anions. Tang et al. (2008) carried out the synthesis of different sized cuprous oxide nanocrystallites and studied their photocatalytic activity, while Fuerte et al. (2002) synthesized nanosized Ti-W oxide and observed the effect of doping level in the photocatalytic degradation of toluene under sunlight excitation.

Marci et al. (2003) reported the photocatalytic oxidation of toluene on irradiated TiO_2, while Zhang et al. (2003) carried out a comparative study on the decomposition of gaseous toluene by O_3/UV, TiO_2/UV and O_3/TiO_2/UV. Photocatalytic degradation of 1,4-dioxane in TiO_2 suspension has been observed by Yamazaki et al. (2007) while Garcia et al. (2005) investigated oxidation of light alkanes over titania supported palladium/vanadium catalyst. Bai et al. (2017) fabricated a range of TiO_2–graphene composites (P25-GR). They studied removal of phenanthrene, fluoranthene, and benzo[a]pyrene in the presence of these components. It was observed that 80% of these polycyclic hydrocarbons were removed after 2 h.

The incorporation of metal-oxide nanoparticles into polymers has become important due to their numerous applications. The Fe-doped ZnO nanoparticles were incorporated into PVA nanofibers by Sekar et al. (2018). They observed the photocatalytic activity of nanofibers for the degradation of naphthalene. It was reported that there was 96 and 81% efficiency for calcined and un-calcined nanofibers, respectively. They observed the photocatalytic activity of nanofibers for the degradation of naphthalene. It was reported that there was 96 and 81% efficiency for calcined and un-calcined nanofibers, respectively. Multimetal oxides nanocomposite photocatalyst such as $Gd_2O_2CO_3 \cdot ZnO \cdot CuO$ was prepared by Mukwevho et al. (2019) via co-precipitation method. It was reported that $Gd_2O_2CO_3 \cdot ZnO \cdot CuO$ composite show a higher efficiency for removing phenanthrene and related polycyclic aromatic hydrocarbons as compared to Cu–CuO/ZnO and CuO nanoparticles.

Sheikholeslami et al. (2019) synthesized γ-Fe_2O_3 nanoparticles via co-precipitation method. The best removal efficiency (95%) could be achieved in the presence of maghemite nanoparticles in 5 days and 97% in 90 min under visible light and UV light, respectively for photocatalytic degradation of BTEX. Lu et al. (2020) prepared $CeVO_4$ photocatalysts via ethylene glycol (EG)-aided hydrothermal method. The photodegradation performance of naphthalene in the presence of $CeVo_4$ was observed in natural seawater.

8.3.5 ALCOHOLS

The V-doped TiO_2 photocatalyst was prepared by Klosek and Raftery (2001), which was used for photoxidation of ethanol in visible light. It is believed that under visible irradiation, the vanadium center donates an electron to the TiO_2 conduction band, which allows the oxidation of surface-absorbed molecules. Mohamed et al. (2002) studied the photocatalytic oxidation of some selected aryl alcohols in acetonitrile whereas heterogeneous photocatalytic dehydrogenation of ethanol over TiO_2 has been reported by Kawai and Sakata (1980). Kirchnerova et al. (2005) performed oxidation of *n*-butanol over commercial TiO_2 under flourescent visible light lamp.

Zhang et al. (2012) reported the use of gold nanoparticles supported on zeolite supports (Au/zeolite) as photocatalysts for oxidation of benzyl alcohol with a high selectivity (99%) under visible light. It was reported by Tapley et al. (2013) that surface plasmon excitation of these supported gold nanoparticles with H_2O_2 is responsible for oxidation of benzyl alcohol and sec-phenethyl alcohol to acetophenone and benzaldehyde, respectively. They used ZnO, Al_2O_3 and hydrotalcite as solid supports. It was reported that hydrotalcite-derived nanocomposites afforded 90% yield of acetophenone in 40 min of LED irradiation.

Bellardita et al. (2018) prepared bare and P-doped graphitic carbon nitride (g-C_3N_4) photocatalysts. The photocatalytic activity of these photocatalysts for the oxidation of benzyl alcohol, 4-methoxy benzyl alcohol, and piperonyl alcohol was evaluated in water suspension under both; UV as well as visible light irradiation. Xiao et al. (2018) carried out oxidation of benzyl alcohol into benzaldehyde in the presence of $Bi_{24}O_{31}Br_{10}$. It was reported that as-synthesized $Bi_{24}O_{31}Br_{10}$ exhibited higher activity for this oxidation of benzyl alcohol into benzaldehyde (with >99% conversion and selectivity) in the presence of blue LED irradiation.

The $H_2Ti_3O_7$ nanowires (NWs) supported by Au, Ag, and Pd monometallic nanoparticles (NPs) and Au–Pd bimetallic NPs were prepared by Du et al. (2019) and used for oxidation of benzyl alcohol. As-prepared Pd/$H_2Ti_3O_7$ NWs catalyst exhibited increased photocatalytic performance under light irradiation. It was revealed that there was no decline in activity and selectivity for this catalyst even after five recycles. Feng et al. (2020a) decorated bi-metal nanoparticles Au–Pd on $ZnIn_2S_4$ nanosheets. This photocatalytic system was used for photocatalytic oxidation of aromatic alcohols. It was revealed that 0.5 wt.% Au–Pd/$ZnIn_2S_4$ photocatalysts exhibited the highest photocatalytic activity.

8.3.6 PHENOLS

Phenolic compounds are widely used in industrial and daily life, but due to their stability and carcinogenic character, they threat the human health and water ecosystem. Hatipoglu et al. (2004) studied the photocatalytic degradation of *m*-cresol while Kang et al. (2000) reported the photocatalytic degradation of 4-chlorophenol in water over TiO_2 and TiO_2/CdS powder. San et al. (2001) investigated photodegradation of 3-aminophenol and Chen et al. (2002) examined the photooxidation of phenol and benzene using TiO_2. The influence of codoping of Zn(II) + Fe(III) on the photocatalytic activity of TiO_2 for phenol degradation was studied by Yuan et al. (2002), whereas Sakthivel and Kisch (2003) reported that C/TiO_2 photocatalyst was active under artificial solar light and could efficiently decompose tetrachlorophenol.

Semiconductor promoted photooxidation of phenol was carried out by Okamoto et al. (1985). They proposed a mechanism for the reaction, which involved stepwise hydroxylation, via di-, tri- and tetra-hydroxybenzenes (mixtures of the various isomers), leading to formic acid, which was finally oxidized to carbon dioxide.

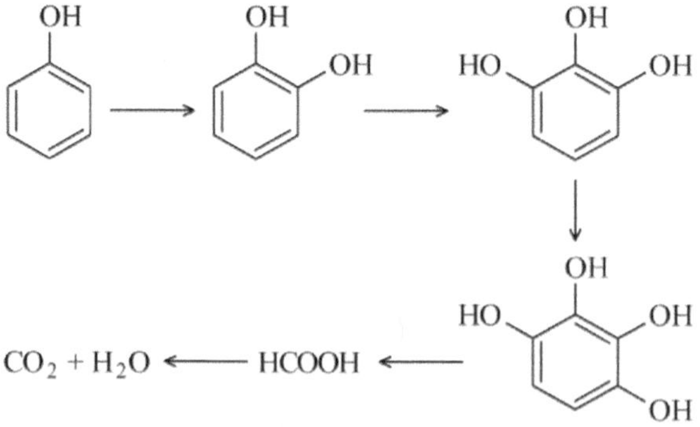

McManamon et al. (2011) improved photocatalytic degradation rates of phenol using novel porous ZrO_2-doped TiO_2 nanoparticulate powders, while photocatalytic degradation of phenol in natural seawater using visible light active carbon modified (CM)-*n*-TiO_2) nanoparticles under UV light and natural sunlight illuminations was carried out by Shaban et al. (2012). Heltina et al. (2015) synthesized titania nanotube (TNT)-carbon nanotube (CNT) composite under acidic conditions and ultrasonic treatment. This composite was used to degrade phenol.

Yadav et al. (2017)-doped TiO_2 with tungsten and formed a nanocomposite with reduced graphene oxide (rGO). This composite rGO/1W-TiO_2 exhibited the highest activity in photodegradation of *p*-nitro phenol (87%) as compared to rGO/TiO_2, 1W-TiO_2 and pure TiO_2 in 3 h.

Shen et al. (2018) constructed g-C_3N_4/Ag/Ag_3PO_4 composites, which exhibited much improved photocatalytic activity. It can decompose phenol at a rate, which was almost 60 and 2.5 times higher than g-C_3N_4 and Ag/Ag_3PO_4, respectively. They also investigated its photocatalytic activity for the degradation of gaseous isopropanol (IPA) and could achieve 63% degradation in 4 h.

Three bimetallic metal oxide nanoparticles were synthesized by Rani and Shanker (2018); Ni–Cu oxide, Cu–Cr oxide, and Ni–Cr oxide using *Aegle marmelos* leaf extract. These have different morphologies like Ni–Cu oxide (nanorods); Ni–Cr oxide (nanospheres) and Cu–Cr oxide (nanoflowers). They also evaluated photocatalytic potential of these nanoparticles for removal of various phenols (phenol, 3-aminophenol, and 2, 4-dinitrophenol). The order of their removal is summarized in Table 8.2:

TABLE 8.2 Removal Efficiency of Different Phenols.

Substrate	Cu–Cr oxide (%)	Ni–Cr oxide (%)	Ni–Cu oxide (%)
Phenol	89	92	95
3-Aminophenol	92	95	97
2,4-Dinitrophenol	87	89	91

Moradi et al. (2019) prepared Fe-doped TiO_2 nanoparticles. It was observed that photocatalytic degradation of phenol solution increased significantly in the presence of $Fe_{0.5}$-TiO_2 from 33 to 57% in 90 min visible light irradiation.

The preparation of ZnO composites sensitized by CQDs was reported by Liang et al. (2020) and then used in photocatalytic degradation of phenol under visible-light exposure. It was revealed that photocatalytic performance of ZnO composites sensitized by CQDs was found to be 60% higher than that of pure ZnO and it has almost no significant change after 10 cycles. The

ZnO/Ag porous nanorod (NR), $ZnO/Ag/Ag_2O$ photocatalysts were prepared by Feng et al. (2020) and photocatalytic degradation of phenol was investigated. A complete degradation of phenol could be achieved in only 90 min.

8.3.7 HALO COMPOUNDS

The halo-organic compounds are stable compounds; thus, they show low chemical reactivity. These properties make them long lasting and nonbiodegradable. The halo-organic compounds can be used as solvents, agrochemicals, dyes, drugs, etc. These compounds represent a major class of environmental pollutants as these compounds are passed to the aquatic plants and animals in different ways, which are consumed as food by human beings and other animals becoming a part of food chain.

Various methods like extraction, incineration, chemical degradation, electrochemical treatment, sonochemical destruction, and photochemical process are in use for the treatment of halo-organics pollutants. Among all these methods, photochemical processes have been found useful for the degradation of halo-compounds into less toxic compounds in water effluents (Ollis, 1982; Cesareo et al., 1986; Ollis et al., 1991). The photocatalytic degradation of $CHCl_3$, $CHBr_3$, CCl_4, and $CCl_3CO_2^-$ has been investigated by Choi and Hoffmann (1997) in aqueous TiO_2 suspension.

Krysova et al. (1998) used TiO_2 for the photocatalytic degradation of diuron, whereas Jirkovsky et al. (1997) studied a direct photolysis of diuron under various conditions at 254 nm. Bhatkhande et al. (2004) carried out photocatalytic degradation of chlorobenzene, while Pandiyan et al. (2002) used photochemical methods for the destruction and dehalogenation of chlorophenols.

The effectiveness of ZnO-assisted photocatalytic degradation of trihalomethanes (THMs), triclosan (TCS), and triclocarban (TCC), was reported by Hwangbo et al. (2019) under low intensity of UV exposure. An enhanced dehalogenation of TCS and TCC was observed with tetrapodal zinc oxide (T-ZnO). The visible-light-active graphitic carbon nitride ($g\text{-}C_3N_4$) samples were synthesized by Cheng et al. (2019). Then it was used to degrade trihalomethanes (THMs) and haloacetonitriles (HANs).

8.3.8 CARBONYL COMPOUNDS

Yanagida et al. (1989) reported the photoreduction of aldehyde and related derivatives by ZnS. Identification of intermediates and reaction mechanism of the photomineralization of polycarboxylic benzoic acid (trimellitic acid) in UV-irradiation aqueous suspensions of titania have been studied by Assabbane et al. (2000). The photoassisted decomposition of salicylic acid on TiO_2 and Pd/TiO_2 films was reported by Sukhaser et al. (1995). A comparative study of the degradation of benzamide and acetic acid on TiO_2 was also made by Heintz et al. (2000).

Fukui et al. (2017) reported that a titanium(IV) oxide photocatalyst is effective to carry out heterogeneous Meerwein-Ponndorf-Verley (MPV)-type reduction of different benzaldehydes under a metal-free condition. Morawski et al. (2017) prepared TiO_2-reduced graphene oxide photocatalysts using different amounts of rGO in the presence of 1-butanol. The maximum photodegradation rate of acetic acid was observed for TiO_2 decorated with 0.5 wt.% of rGO.

The preparation of active carbon (AC-titania) composites has been reported by Pauová et al. (2020). They also observed photocatalytic activity of composites using toluene in the gaseous phase and benzoic acid in the aqueous phase. It was revealed that composites prepared from milled active carbon for the removal of benzoic acid was more than 2.5 times better as compared with non-milled AC. A sustainable oxygen-dependent route for oxidizing different aromatic and aliphatic aldehydes to carboxylic acids was reported Hajimohammadi et al. (2021) under visible light and sunlight. They used cobalt phthalocyanine tetrasulfonic acid (CoPcS) supported on reduced graphene oxide (RGO) for this purpose. They could achieve products with 100% selectivity and (81–100%) conversion.

8.3.9 NITROGEN-CONTAINING COMPOUNDS

Nitrogen-containing compounds comprise different kinds of compounds such as amino acids, proteins, nitro, drugs, herbicides, pesticides, dyes, and many more. Some of them cause serious environmental threat because of their stability and toxicity. There is a pressing demand for such technologies for removal of these pollutants to get clean and pure water, which should be economic and environmentally benign.

During the past 15 years, Jing et al. (2011) reviewed this field with a focus on the progress of nitrogen-containing organic compounds removal in wastewater by TiO_2-mediated photocatalytic activity. They have summarized the important factors affecting the relevant photocatalytic activity. They have also proposed that different rates of photocatalytic degradation of N-containing compounds are the result of a strong interaction among pollutants structure, TiO_2 properties, and photocatalytic reaction conditions. Nitroaromatics are toxic compounds, which are insoluble in nature and therefore, these are not easily biodegraded.

The oxidation of dissolved nitrogen in natural water has been observed by Takeda and Fujiwara (1996) using titanium dioxide and platinized titanium dioxide. Ciping et al. (1993) investigated the free-radical intermediates generated in photocatalytic oxidation of some organic compounds containing nitrogen atom, whereas photocatalytic degradation of a series of primary, secondary, tertiary amines, and other N-containing organic compounds over UV-illuminated film of TiO_2 has been studied by Low et al. (1991). The photocatalytic mineralization of nitrobenzene, nitrosobenzene, phenylhydroxylamine, aniline, and 4-nitrosophenol has been investigated by Piccinini et al. (1997) in the presence of TiO_2, while Tayade and Key (2010) carried out photocatalytic degradation of nitrobenzene with TiO_2 nanotubes.

Alberici et al. (2001) used TiO_2/UV–Vis photocatalytic process in the destruction of nitrogen-containing organic compounds. Pyridine (C_5H_5N), propylamine ($C_3H_7NH_2$), and diethylamine ($C_4H_{10}NH$) were used by them as the substrates for the photodegradation in the presence and the absence of oxygen. Aqueous TiO_2 suspension has also been used for photocatalytic degradation of six membered heteroatomic compounds such as pyridazine, pyrimidine and pyrazine. A complete mineralization has been achieved in this case.

The oxides of nitrogen [NO and NO_2 (NO_x)] have variety of negative impact on environment and human health. Photocatalysis has been proved to be a successful method for removal of NO_x from the atmosphere (Devahasdin et al., 2003; Wang et al., 2007). Dillert et al. (1995, 1996) studied photocatalytic degradation of trinitrotoluene and other nitro aromatic compounds, whereas Augugliaro et al. (1991) studied photocatalytic degradation of nitrophenols in aqueous titanium dioxide dispersion.

Nahen et al. (1997) observed the photocatalytic degradation of trinitrotoluene, trinitrobenzene, and dinitrobenzene on exposure to UV light using titanium dioxide as catalyst and found that degradation follows oxidative and reductive pathways; both, while Schmelling and Gray (1995) compared the transformation and mineralization of TNT (trinitrotoluene) under photocatalytic and direct photolytic reactions. The 5-nitro-1,2,4-triazol-3-one (NTO) is a powerful explosive present in industrial wastewater, which was completely mineralized in 3 h with TiO_2 by Campion et al. (1999).

Wang et al. (1999) used photocatalytic technique to degrade 2-nitrophenol in aqueous solution in the presence of titanium dioxide, whereas the effect of nano-TiO_2 on photocatalytic degradation of nitrobenzene was investigated by Makarova et al. (2000) for the removal of nitro aromatic compounds from contaminated water streams. Vohra and Tanaka (2002) employed TiO_2-mediated photocatalytic degradation process to treat aqueous 2-, 3- and 4-nitrotoluene pollutants. The mineralization products included NH_4^+, NO_3^- and CO_2.

Selvam and Swaminatham (2012) prepared *N*-doped TiO_2 and used for the selective one-pot synthesis of quinaldines from nitrobenzenes under UV and visible light.

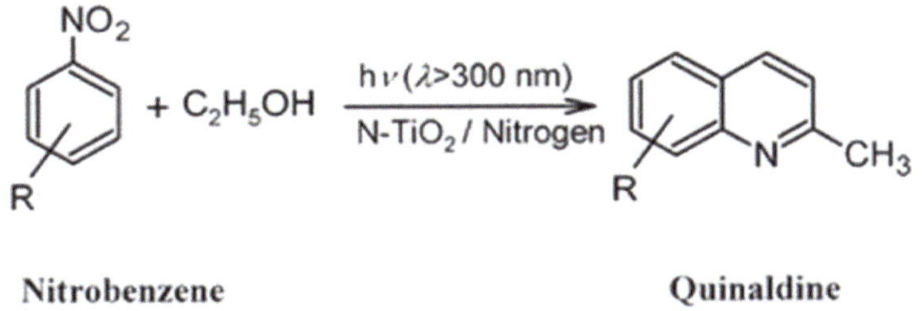

Degradation of *p*-nitrophenol (PNP) was carried out by photocatalytic process using magnetic titania nanoparticles. It was reported that up to 90% PNP was degraded under the optimum conditions. Nickel oxide nanoparticles were grafted on reduced graphene oxide by Al-Nafiey et al. (2017). The (rGO/NiO) was found to be effective visible light active photocatalyst for reducing nitroaromatic compounds. It was revealed that it can be easily recovered by using simple external magnet. It could also be reused for six cycles without any loss in its activity.

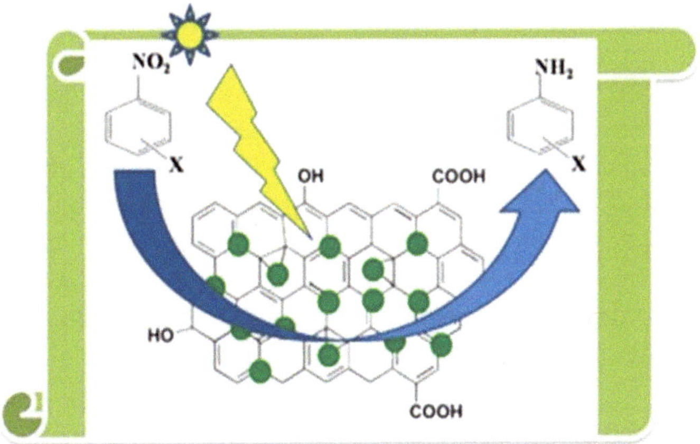

FIGURE 8.4 Reduction of nitrobenzenes to aniline.

Source: Reprinted with permission from Al-Nafiey et al., 2017. © 2017 Elsevier.

Eskandari and Kazemi (2018) synthesized magnetic $Fe_3O_4/SiO_2/CdS$ nanocomposites. As-prepared photocatalysts were used for the photocatalytic reduction of nitro compounds under visible LED irradiation under mild conditions. Iridium/cerium dioxide (Ir/CeO_2) photocatalysts with different Ir contents (0.5, 1.0 and 2.0 wt.%) were prepared by Castañeda et al. (2019) and used in photoreduction of 4-nitrophenol under UV irradiation and Na_2SO_3.

8.3.10 HYDROGEN PRODUCTION

World is facing energy scarcity for the last few years. Hydrogen is advocated as a fuel of future provided it is generated from water in the presence of sunlight. Here, photocatalysis can play significant role in producing hydrogen through photosplitting of water, which is abundantly available on Earth.

The synthesis of self-assembled aligned hexagonal prismatic Cu-doped ZnO nanoparticles has been reported by Kanade et al. (2007) with average particle size was in range of 40–85 nm. They could achieve maximum hydrogen production rate as 1932 μmol h^{-1} in the presence of visible light irradiation. It was observed that photocatalytic activity of Cu–ZnO synthesized in organic media was higher as compared to prepared in aqueous medium.

The effect of copper doping onto TiO_2 (photocatalyst) was investigated Yoong et al. (2009) on hydrogen generation under visible light. They used wet impregnation and complex precipitation methods for preparation for the copper-doped catalyst as the starting material. It was observed that highest yield of hydrogen could be obtained with 10 wt.%, Cu/TiO_2, which was calcined for half an hour at 300°C.

Peng et al. (2011) synthesized multiwalled carbon nanotubes (MWCNTs)/CdS nanocomposites via hydrothermal method and direct growth of CdS nanoparticles on the functionalized MWCNT surface. It was observed that 10 wt.% MWCNTs/CdS exhibited much higher photocatalytic hydrogen production efficiency as well as photostability than the pure CdS nanoparticles.

MWCNTs and $ZnIn_2S_4$ composites were prepared by Chai et al. (2012) via a facile hydrothermal method. The highest photocatalytic hydrogen production efficiency with high apparent quantum efficiency (23.3%) was observed with 3 wt.% MWCNTs/$ZnIn_2S_4$ composite.

$ZnIn_2S_4$ microspheres (ZIS MSs) were decorated with CQDs and platinum nanoparticles (NPs) by Li et al. (2014) as dual co-catalysts. It was reported that as-prepared ZIS MSs co-loaded with CQDs and Pt exhibited significantly higher photocatalytic H_2 production rate of 1032.2 μmol h^{-1} g^{-1} in triethanolamine aqueous solution in the presence of visible-light irradiation.

This enhancement was attributed to an integrative effect of enhanced light absorption, good crystallization, and high electrical conductivity of CQDs.

Chouhan et al. (2016) synthesized Co-doped ZnO nanorods (Co–ZnO NRs) through hydrothermal method using cetyltrimethylammonium bromide. It was observed that nano-sized sensitizer CdS onto the Co–ZnO NRs' surface reduces its band gap. This hetero-assembly (1.5% Pt/CdS/ Co–ZnO NRs) exhibited excellent photocatalytic responses such as quantum efficiency (1.98%) and hydrogen generation capacity (67.20 mmol H_2 g^{-1}) under 1 Sun light exposure.

Pan et al. (2018) prepared CQDs modified porous g-C_3N_4/TiO_2 two-dimensional (2D) nano-heterojunctions. It was observed that hydrogen production was enhanced to almost double as compared to that of unmodified samples. Such an enhancement was attributed to dual function of CQDs with high catalytic activity of H_2O_2 decomposition as well as unique up-conversion photoluminescence.

Cadmium sulfide fibers were synthesized by Quiroz-Cardoso et al. (2019) and then modified with graphene oxide sheets (GO) and nickel nanoparticles. The concentrations of 1, 3, and 10 wt.% were evaluated to find an optimal content of GO. The photoactivity of these composites was evaluated using four LED lamps (4 W each) $\lambda = 450$ nm using ethanol solutions (50/50 vol%) as sacrificial molecule. The activity of Ni/GO-CdS 1% composite was found to be increased by 6.3 times (8866 μmol $h^{-1}g^{-1}$) than the photoactivity reported for the only semiconductor CdS (1410 μmol h^{-1} g^{-1}).

Ismael (2019) synthesized bare TiO_2 and ruthenium (Ru)-doped TiO_2 nanoparticles via precipitation method. They evaluated photocatalytic activities of bare TiO_2 and Ru-doped (0.05–0.2 mol.%) TiO_2 nanoparticles for hydrogen production. The maximum hydrogen production rate was found to be 3400 μmol h^{-1} in aqueous methanol, which is almost more than double as compared to bare TiO_2 (1500 μmol h^{-1}).

Cao et al. (2021) used NiCoP nanoparticles for photocatalytic dehydrogenation of formic acid, when these NPs were anchored on CdS nanorods (NiCoP@CdS NRs). They could achieve H_2 production rate of ~354 mol mg^{-1} h^{-1} in the presence of NiCoP@CdS nanorods under visible light.

8.3.11 REDUCTION OF CARBON DIOXIDE

Photocatalytic conversion of carbon dioxide to reduced products (formic acid, formaldehyde, methanol, and methane) is a convinent method to harvest solar energy and reduce ever increasing environmental pollution and solve both the problems; the scarcity of energy and deteriorating quality of environment, simultaneously. The conversion of CO_2 to value-added chemicals through visible light photocatalysis has a great potential.

Hsu et al. (2013) carried out photocatalytic conversion of CO_2 to methanol. They used graphene oxides (GOs) as photocatalysts. It was reported that this photocatalytic conversion rate on modified graphene oxide (GO-3) was 0.172 mol g^{-1} cat. h^{-1} under visible light, which is almost six times higher than the pure TiO_2.

Ag-loaded TiO_2 (Ag/TiO_2) nanocomposites were fabricated by Liu et al. (2014) via microwave-assisted chemical reduction method. It was reported that 2.5% Ag/TiO_2 exhibited the highest activity with 405.2 mol g^{-1} cat. methanol yield, which was about 9.4 times more than that with pure TiO_2.

Graphene oxide (GO)-tethered Co(II) phthalocyanine complex CoPcGO was synthesized by Kumar et al. (2014) and used as photocatalyst (recyclable) for reducing carbon dioxide into methanol (main product). They used triethylamine as the sacrificial donor. The methanol yield was determined as 3781.8881 μmol/g^{-1}cat. after 48 h.

A new graphene oxide-doped-oxygen-rich TiO_2 (GO–OTiO_2) hybrid heterostructure was synthesized by Tan et al. (2015) via facile wet chemical impregnation. It was reported that photostability of the resulting GO–OTiO_2 composite was enhanced, with a reactivity of 95.8% even after 6 h of light irradiation. It was revealed that 5GO–OTiO_2 with 5 wt.% GO loading, exhibited the highest photoactivity, and could achieve a CH_4 yield up to 1.718 μmol^{-1} g^{-1}cat. after 6 h of reaction. The yield obtained was found to be 14-folds higher as compared to Degussa P25.

The Ti-KIT-6 and Ti-SBA-15-spherical with different Si/Ti ratios (photo-catalysts) were synthesized by Hussain et al. (2015). It was reported that Ti-KIT-6 (Si/Ti = 100) showed better CH_4 production rate (4.15 mol g^{-1}cat. h^{-1}) as compared to Ti-KIT-6-dried (2.63 mol g^{-1}cat. h^{-1}) and the Ti-SBA-15-dried/calcined (3.45, 1.85 mol g^{-1}cat. h^{-1}, respectively). The CH_3OH, CO, and H_2 were other main products apart from CH_4.

Ong et al. (2015) deposited noble-metal Pt nanoparticles (2.5 nm) on graphitic carbon nitride (g-C_3N_4). It was observed that Pt-loaded g-C_3N_4 (Pt/CN) exhibited a significant enhancement in the photoreduction of carbon dioxide to methanol under visible light in the presence of water vapor. The 2 wt.% Pt-loaded g-C_3N_4 (2Pt/CN) nanocomposites gave optimum yield of CH_4 (13.02 mol g1cat.) in 10 h, which was about five times as compared to pure g-C_3N_4 (2.55 mol g^{-1}cat.).

Visible-light-active silver halides AgX (where X = Cl and Br) were deposited on the protonated graphitic carbon nitride (pCN) by Ong et al. (2016) via a sonication-assisted deposition-precipitation. The photocatalytic performance of as-prepared AgX/pCN was evaluated for reduction of carbon dioxide to methanol in water. The highest photocatalytic activity achieved was 10.92 μmol g^{-1}cat. with 30AgBr/pCN, which was almost 4.2 and 34.1 times higher as compared to pCN and single-phase AgBr, respectively.

Sharma and Lee (2017) immobilized nanocomposite (NC) of nickel-loaded TiO_2 photocatalysts on activated carbon fibers and used for photocatalytic reduction of CO_2 to methanol under both; UV and visible light irradiation. The photocatalytic activity of as-prepared nickel-loaded TiO_2 was enhanced and the yield of methanol in 2 h could reach 755.1 μmol g^{-1} and 986.3 μmol g^{-1} under UV and visible light irradiation, respectively.

Zhang et al. (2017) synthesized a range of microsized nanoporous titanium dioxide (TiO_2), which was decorated with metallic copper and loaded it with different amounts of Cu. It exhibited good photocatalytic activity for reduction of CO_2 with water. A loading of Cu (≈ 0.4 wt%) exhibited the best photocatalytic activity for reducing CO_2 to CH_4, which was 21-fold higher than that of Degussa P25 TiO_2.

Tan et al. (2018) designed a photocatalyst of bimetal (Ag/Pd) nanoalloys, which were supported on nitrogen-doped TiO_2 nanosheet. It was observed that CO_2 could be efficiently reduced to CH_4 under mild conditions in aqueous solution. They could achieve maximum production rate of CH_4 (79.0 μmol g^{-1} h^{-1}) with high selectivity.

A series of direct Z-scheme composites of β-$AgVO_3$ nanoribbons and $InVO_4$ nanoparticles, ($InVO_4$/β-$AgVO_3$) were prepared by Yang et al. (2019). It was observed that CO evolution rate of 12.61 μmol g^{-1} h^{-1} could be achieved with 20% In–Ag without using any cocatalyst or sacrificial agent. This was 11 times higher than that with pure $InVO_4$ (1.12 μmol·g^{-1}·h^{-1}).

Rangappa et al. (2020) prepared a nanohybrid system, which consists of photodeposited Pt nanoparticles on a coreshell structure of CdS/ZnS. The photocatalytic reduction of CO_2 in aqueous solution afforded CO and methane using a hole scavenger (triethanolamine). The production of carbon monoxide and methane by CdS/ZnS/Pt photocatalyst was higher than that with CdS.

A novel catalyst was developed by He et al. (2021) as a combination of NiB amorphous alloy with In_2O_3 semiconductor. It was reported that very high HCOOH yield (5158.0 μmol g^{-1} h^{-1}) was obtained under sunlight irradiation.

Li et al. (2021) designed $Cu_2O/g\text{-}C_3N_4$ heterojunction and used for photocatalytic CO_2 reduction to ethanol. The production rate of ethanol reached 0.71 mmol g^{-1} h^{-1}, which was about 1.89 and 7.05 times of photocatalysis and thermal catalysis, respectively.

8.3.12 OTHERS

The influence of TiO_2 particle size and morphology has been addressed with the aim of increasing the photocatalytic efficiency of powdered materials. Titania nanoflowers were synthesized by oxidizing pure titanium with hydrogen peroxide solutions containing hexamethylenetetramine and nitric acid at a low temperature of 353 K (Wu et al., 2006). These nanomaterials are widely utilized as photocatalyst to treat various wastewaters (Balasubramaniam et al., 2004), degrade organic pollutants, and also for cleaning environment (Bhatkhande et al., 2001).

TiO_2-coated surfaces are increasingly studied for their ability to inactivate microorganisms. The activity of glass coated with thin films of TiO_2, CuO, and hybrid TiO_2/CuO prepared by atmospheric chemical vapor decomposition (Ap–CVD) was studied by Ditta et al. (2008). Visible-light-induced photocatalysis by titania particles for the reduction of environmental toxins on a global scale has been widely studied. A comparative study of TiO_2 supported on alumina and glass beads for the catalytic decomposition of leather was made by Sakthivel et al. (2002).

The photodegradibility of polymer like polyvinyl chloride (Kim et al., 2006), polythene (PE) and PS (Zan et al., 2004) was carried out by using TiO_2 nanoparticles, which is an eco-friendly method and also the need of the day, whereas Carlos et al. (2000) reported the toxicity of lignin and Kraft effluents toward *E. coli* was completely removed by using Ag–ZnO catalyst.

Jain and Ameta (2008) reported the photocatalytic oxidation of arabinose and glucose over cadmium sulfide. Formaldehyde formed was further oxidized in the presence of hydroxyl radicals into acid and water.

$$CdS + h\nu \rightarrow e^-_{cb} + h^+_{vb}$$

$$H_2O + h^+ \rightarrow {}^{\bullet}OH + H^+$$

$$8\ {}^{\bullet}OH + C_5H_{10}O_4\ \text{(Arabinose)} \rightarrow HCHO + 4HCOOH + 4H_2O$$

$$10\ {}^{\bullet}OH + C_6H_{12}O_6\ \text{(Glucose)} \rightarrow HCHO + 5HCOOH + 5H_2O$$

Cellulose was photocatalytically degraded on supported TiO_2 and ZnO by Yeber et al. (2000). Cellulose bleaching effluent was completely decolorized and the total phenol content was reduced by 85% after 120 min treatment with both the catalysts. Villasenor et al. (2002) carried out the photocatalytic degradation of organic matter dissolved in Kraft Black liquor, which is an important effluent from pulp and paper industries.

Production of hydrogen from water splitting using different photocatalysts has gained attention of researchers from all over the world, as the hydrogen is considered as a rich fuel source in future. Sun et al. (2013) observed photocatalytic generation of hydrogen from water. They used molecular and semiconductor nanowire photosensitizers and combination with cobalt pentapyridine complex.

Mangrulkar et al. (2012) reported hydrogen production via water splitting using nano-ferrites. They used Pt as co-catalyst and ethanol as the sacrificial donor. Nanoferrite is developing as a photocatalyst with evolution rate of hydrogen as 8.275 mol h^{-1} under visible light as compared to 0.0046 mol h^{-1} on using commercial iron oxide.

The wastewater from different industries is left as such without any proper treatment. This pollutes the nearby water resources. There is an urgent need to find out effective ways for the treatment of this polluted water. Although a number of treatment methods are available, but these are associated with some or the other demerits. Photocatalysis has emerged as a promising technology for the treatment of wastewater and time is not far off, when this technology may surpass other alternative techniques.

8.4 RECENT DEVELOPMENTS

A green synthesis of Nb–ZnO was reported by Nguyen et al. (2022) using *Vernonia amygdalina* leaves extract. It was observed that Nb–ZnO samples exhibited higher photocatalytic activity as compared to ZnO. The highest tetracycline degradation efficiency was reached to 93.2% in 3 h using Nb:Zn molar ratio of 1:1. Nabi et al. (2022) synthesized TiO_2 nanoparticles using lemon peel extract. The TiO_2 particles were of spherical shape with particle size ranging between 80 and 140 nm. Photocatalytic performance of as-prepared TiO_2 was also evaluated for the degradation of rhodamine B (70%), and it was found to be more efficient as compared to commercial TiO_2.

Sebuso et al. (2022) used ZnO and graphene nanostructures for fabricating multilayer graphene/zinc oxide (MLG/ZnO) nanocomposites, with different ratios of MLG to ZnO (1:1, 1:2, 1:3). It was reported that photodegradation of brilliant black under sunlight by MLG, MLG/ZnO_1, ZnO, MLG/ZnO_2 and MLG/ZnO_3 were found to be 7, 39, 63, 81, and 93%, respectively. Iqbal et al. (2022) used leaf extract of *Citrus Medica Linn.* to synthesize pure and lanthanum (La)-(1, 2, and 3 wt.%)-doped copper oxide nanoparticles (CuO-NPs). The synthesized material was used as photocatalyst for the degradation of methylene blue dye. It was observed that band gap of material shifted toward visible region (3.03–2.71 eV) and maximum degradation efficiency of methylene blue (84%) could be achieved in 150 min on using 2% La-doped CuO-NPs.

Ansari et al. (2022) prepared titania using *Acorus calamus* (*A. calamus*) leaf extract as a capping as well as reducing agent. As-prepared nanoparticles were found to be globular (average size of 15–40 nm). It was reported that these TiO_2 nanoparticles exhibited higher photocatalytic activity, degrading rhodamine B up to 96.59%. It was also revealed that this biosynthesized TiO_2 showed significant antimicrobial activity against gram-positive staining (*S. aureus, B. subtilis*) over Gram-negative (*E. coli, P. aeruginosa,*) bacteria as compared toTiO_2.

Panchal et al. (2022) used *Aloe Vera* plant extract for synthesizing MgO, Ag NPs, and Ag/MgO-nanocomposites. These NPs and composites were applied for photocatalytic degradation of methylene blue and phenol and it was observed that Ag/MgO–NCs showed 90.18 dye and 80.67% phenol degradation in 2 h under sunlight. Razavi et al. (2022) synthesized strontium hexaferrite nanoparticles used by three different methods: Microwave, sol–gel Pechini, and sol–gel auto-combustion, and ultrasound-assisted auto-combustion to prepare $SrFe_{12}O_{19}$@Ag, $SrFe_{12}O_{19}$@Au core-shell in the presence of beetroot juice. Then these nanoproducts were applied as photocatalysts to degrade eosin with maximum degradation by $SrFe_{12}O_{19}$@Ag and $SrFe_{12}O_{19}$@Au, as 95.9, and 93.88%, respectively.

KEYWORDS

- **photocatalysis**
- **degradation**
- **dyes**
- **pesticides**
- **photooxidation**
- **photocatalytic**

REFERENCES

Abdennouri, M.; Baâlala, M.; Galadi, A.; El Makhfouk, M.; Bensitel, M.; Nohair, K. Photocatalytic Degradation of Pesticides by Titanium Dioxide and Titanium Pillared Purified Clays. *Arab. J. Chem.* **2016,** *9*, S313–S318.

Abhilash, M. R.; Akshatha, G.; Srikantaswamy, S. Photocatalytic Dye Degradation and Biological Activities of the Fe_2O_3/Cu_2O Nanocomposite. *RSC Adv.* **2019,** *9* (15), 8557–8568.

Alberici, R. M.; Canela, M. C.; Eberlin, M. N.; Jardim, W. F. Catalyst Deactivation in the Gas Phase Destruction of Nitrogen-Containing Organic Compounds Using TiO_2/UV–VIS. *Appl. Catal. B: Environ.* **2001,** *30*, 389–397.

Al-Nafiey, A.; Kumar, A.; Kumar, M.; Addad, A.; Sieber, B.; Szunerits, S. et al. Nickel Oxide Nanoparticles Grafted on Reduced Graphene Oxide (rGO/NiO) as Efficient Photocatalyst for Reduction of Nitroaromatics Under Visible Light Irradiation. *J. Photochem. Photobiol. A Chem.* **2017,** *336*, 198–207.

Amat, A. M.; Arques, A.; Miranda, M. A.; Sequi, S. Photo-Fenton Reaction for the Abatement of Commercial Surfactants in a Solar Pilot Plant. *Solar Energy* **2004,** *77*, 559–566.

Ameta, A.; Bhati, I.; Ameta, R.; Ameta, S. C. Use of Nanosized Chromium Doped TiO_2 Supported on Zeolite for Methylene Blue Degradation. *Indo. J. Chem.* **2010a,** *10*, 20–25.

Ameta, S. C.; Punjabi, P. B.; Ameta, R.; Bhati, I. *Recent Progress in Chemistry and Chemical Engineering Research*; Haghi, A. K., Ed.; Nova Science Publisher: USA, 2010b; Chapter 11, pp 173–204.

Ameta, R.; Ameta, S. C., Ed. *Photocatalysis: Principles and Applications*; CRC Press: Boca Raton, 2017.

Ameta, R.; Jain, S.; Bhatt, C. V.; Ameta, S. C. Photocatalytic Degradation of Orange-G on ZnO Powder in the Presence of a Surfactant. *Rev. Roum. Chim.* **2000,** *45*, 49–56.

Ameta, R.; Vardia, J.; Punjabi, P. B.; Ameta, S. C. Use of Semiconducting Iron (III) Oxide in Photocatalytic Bleaching of Some Dyes. *Indian J. Chem. Tech.* **2006,** *13*, 114–118.

Ameta, R.; Punjabi, P. B.; Ameta, S. C. Photodegradation of Naphthol Green B in the Presence of Semiconducting Antimony Trisulphide. *J. Serbian Chem. Soc.* **2011,** *76*, 1049–1055.

Ameta, S. C.; Ameta, R.; Punjabi, P. B.; Sharma, B. K.; Lodha, A. Heterogeneous and Homogeneous Photocatalysis, Part II: Organic Substrates. *Asian J. Chem. Revs.* **1992,** *3*, 1–11.

Ameta, S. C.; Choudhary, R.; Ameta, R.; Vardia, J. Photocatalysis: A Promising Technology for Wastewater Treatment. *J. Indian Chem. Soc.* **2003,** *80*, 257–265.

Ameta, S. C.; Sharma, A.; Sharma, R.; Rathore, S. S. Use of Zinc Oxide Particulate System as a Photocatalyst: Photobleaching of Rose Bengal. *J. Indian Chem. Soc.* **2002,** *79*, 929–931.

Ananpattarachai, J.; Kajitvichyanukul, P. Photocatalytic Degradation of p,p'-DDT Under UV and Visible Light Using Interstitial N-Doped TiO_2. *J. Environ. Sci. Health B.* **2015,** *50*, 247–260.

Ansari, A.; Siddiqui, V. U.; Rehman, W. U.; Akram, M. K.; Siddiqi, W. A.; Alosaimi, A. M. et al. Green Synthesis of TiO_2 Nanoparticles Using *Acorus calamus* Leaf Extract and Evaluating Its Photocatalytic and In Vitro Antimicrobial Activity. *Catalysts* **2022,** *12* (2). DOI: 10.3390/catal12020181.

Aoudjit, F.; Cherifi, O.; Halliche, D. Simultaneously Efficient Adsorption and Photocatalytic Degradation of Sodium Dodecyl Sulfate Surfactant by One-Pot Synthesized TiO_2/Layered Double Hydroxide Materials. *Separ. Sci. Technol.* **2019,** *54* (7), 1095–1105.

Arabatzis, M.; Stergiopoulos, T.; Andreeva, D.; Kitova, S.; Neophytides, S. G.; Falaras, P. Characterization and Photocatalytic Activity of Au/TiO_2 Thin Films for Azo-Dye Degradation. *J. Catal.* **2003,** *220*, 127–135.

Assabbane, A.; Ichou, Y. A.; Tahiri, H.; Guillard, C.; Herrmann, J. N. Photocatalytic Degradation of Polycarboxylic Benzoic Acids in UV-Irradiated Aqueous Suspensions of Titania: Identification of Intermediates and Reaction Pathway of the Photomineralization of Trimellitic Acid (1, *2*, 4-Benzene Tricarboxylic Acid). *Appl. Catal.* **2000,** *24*, 71–87.

Augugliaro, V.; Palmisano, L.; Schiavello, M.; Sclafani, A. Photocatalytic Degradation of Nitrophenols in Aqueous Titanium Dioxide Dispersion. *Appl. Catal.* **1991,** *69*, 323–340.

Avasarala, B. K.; Tirukkovalluri, S. R.; Bojja, S. Magnesium Doped Titania for Photocatalytic Degradation of Dyes in Visible Light. *J. Environ. Anal. Toxicol.* **2016,** *6* (1). DOI: 10.4172/2161–0525.1000358.

Baharvand, A.; Ali, R.; Yusof, A. M.; Ibrahim, A. N.; Chandren, S.; Nur, H. Preparation of Anatase Hollow TiO_2 Spheres and Their Photocatalytic Activity in the Photodegradation of Chlorpyrifos. *J. Chinese Chem. Soc.* **2014,** *61* (11), 1211–1216.

Bai, H.; Zhou, J.; Zhang, H.; Tang, G. Enhanced Adsorbability and Photocatalytic Activity of TiO2-Graphene Composite for Polycyclic Aromatic Hydrocarbons Removal in Aqueous Phase. *Colloids Surf. B.* **2017,** *150*, 68–77.

Balasubramaniam, G.; Dionysiou, D. D.; Suidan, M. T.; Baudin, I.; Laine, J. M. Evaluating the Activities of Immobilized TiO_2 Powder Films for the Photocatalytic Degradation of Organic Contaminants in Water. *Appl. Catal.* **2004,** *47*, 73–84.

Bardos, E. S.; Markovics, O.; Horvath, O.; Toro, N.; Kiss, G. Photocatalytic Degradation of Benzenesulfonate on Colloidal Titanium Dioxide. *Water Res.* **2011,** *45*, 1617–1628.

Behnajady, M. A.; Modirshala, N.; Hamzavi, R. Kinetic Study on Photocatalytic Degradation of CI Acid Yellow 23 by ZnO Photocatalyst. *J. Hazard. Mater.* **2006,** *133*, 226–232.

Bellardita, M.; García-López, E. I.; Marcì, G.; Krivtsov, I.; García, J. R.; Palmisano, L. Selective Photocatalytic Oxidation of Aromatic Alcohols in Water by Using P-Doped g-C_3N_4. *Appl. Catal. B: Environ.* **2018,** *220*, 222–233.

Bhatkhande, D. S.; Sawant, S. B.; Schouten, J. C.; Pangarkar, V. G. Photocatalytic Degradation of Chlorobenzene Using Solar and Artificial UV Radiation. *J. Chem. Technol. Biotechnol.* **2004,** *79*, 354–360.

Bhatkhande, S.; Pangarkar, V. G.; Beenackers, A. A. C. M. Photocatalytic Degradation for Environmental Applications–A Review. *J. Chem. Technol. Biotechnol.* **2001,** *77*, 102–116.

Bilgi, S.; Demir, C. Identification of Photooxidation Degradation Products of CI Reactive Orange 16 Dye by Gas Chromatography-Mass Spectrometry. *Dyes Pigments.* **2005,** *66*, 69–76.

Binitha, N. N.; Yaakob, Z.; Reshmi, M. R.; Sugunan, S.; Ambili, V. K.; Zetty, A. A. Preparation and Characterization of Nano Silver-Doped Mesoporous Titania Photocatalysts for Dye Degradation. *Catal. Today* **2009,** *147*, 76–80.

Campion, L. L.; Giannotti, C.; Ouazzani, J. Photocatalytic Degradation of 5-Nitro-1, *2*, 4-Triazol-3-One NTO in Aqueous Suspention of TiO_2. *Chemosphere* **1999,** *38*, 1561–1570.

Cao, H.; Wang, T.; Minja, A. C.; Jiang, D.; Du, P. NiCoP Nanoparticles Anchored on CdS Nanorods for Enhanced Hydrogen Production by Visible Light-Driven Formic Acid Dehydrogenation. *Int. J. Hydrogen Energy* **2021,** *46* (64), 32435–32444.

Carlos, A.; Gouvea, K.; Wypych, F.; Moraes, S. G.; Duran, N.; Zamora, R. Semiconductor-Assisted Photocatalytic Degradation of Reactive Dyes in Aqueous Solution. *Chemosphere* **2000,** *40*, 427–432.

Castañeda, C.; Alvarado, I.; Martínez, J. J.; Brijaldo, M. H.; Passos, F. B.; Rojas, H. Enhanced Photocatalytic Reduction of 4-Nitrophenol Over Ir/CeO_2 Photocatalysts Under UV Irradiation. *J. Chem. Technol. Biotechnol.* **2019,** *94* (8), 2630–2639.

Cesareo, D.; Di, D. A.; Marchini, S.; Passerini, L.; Tosato, M. L. Environmental Photochemistry of Chlorinated Aromatics in Aqueous Media: A Review of Data. *Homo.-Hetero. Photocatal.* **1986,** *174*, 593–627.

Chachvalvutikul, A.; Jakmunee, J.; Thongtem, S.; Kittiwachana, S.; Kaowphong, S. Novel $FeVO_4/Bi_7O_9I_3$ Nanocomposite with Enhanced Photocatalytic Dye Degradation and Photoelectrochemical Properties. *Appl. Surf. Sci.* **2019,** *475*, 175–184.

Chai, B.; Peng, T.; Zeng, P.; Zhang, X. Preparation of a MWCNTs/$ZnIn_2S_4$ Composite and Its Enhanced Photocatalytic Hydrogen Production Under Visible-Light Irradiation. *Dalton Trans.* **2012,** *41* (4), 1179–1186.

Chakrabarti, S.; Dutta, B. K. Photocatalytic Degradation of Model Textile Dyes in Wastewater Using ZnO as Semiconductor Catalyst. *J. Hazard. Mater.* **2004,** *112*, 269–278.

Chang, X.; Yao, X.; Ding, N.; Yin, X.; Zheng, Q.; Lu, S. et al. Photocatalytic Degradation of Trihalomethanes and Haloacetonitriles on Graphitic Carbon Nitride Under Visible Light Irradiation. *Sci. Total Environ.* **2019,** *682*, 200–207.

Chen, J.; Eberlein, L.; Cooper, C. H.; Langford, H. Pathways of Phenol and Benzene Photooxidation Using TiO_2 Supported on a Zeolite. *J. Photochem. Photobiol.* **2002,** *148*, 183–189.

Chen, X.; Wu, Z.; Liu, D.; Gao, Z. Preparation of ZnO Photocatalyst for the efficient and Rapid Photocatalytic Degradation of Azo Dyes. *Nanoscale Res. Lett.* **2017,** *12* (1). DOI: 10.1186/s11671–017–1904–4.

Cheng, Y.; Bai, M.; Su, J.; Fang, C.; Li, H.; Chen, J.; Jiao, J. Synthesis of Fluorescent Carbon Quantum Dots from Aqua Mesophase Pitch and Their Photocatalytic Degradation Activity of Organic Dyes. *J. Mater. Sci. Technol.* **2019,** *35* (8), 1515–1522.

Choi, W.; Hoffmann, M. R. Novel Photocatalytic Mechanisms for $CHCl_3$, $CHBr_3$, and CCl_3CO_2-Degradation and the Fate of Photogenerated Trihalomethyl Radicals on TiO_2. *Environ Sci. Technol.* **1997,** *31*, 89–95.

Choi, W.; Hong, S. J.; Chang, Y. S.; Cho, Y. Photocatalytic Degradation of Polychlorinated Dibenzo-P-Dioxins on TiO_2 Film Under UV or Solar Light Irradiation. *Environ. Sci. Technol.* **2000,** *34*, 4810–4815.

Chouhan, N.; Ameta, R.; Meena, R. K.; Mandawat, N.; Ghildiyal, R. Visible Light Harvesting Pt/CdS/Co-Doped ZnO Nanorods Molecular Device for Hydrogen Generation. *Int. J. Hydrogen Energy* **2016,** *41* (4), 2298–2306.

Ciping, C.; Daohui, L.; Guangzhi, X. Free Radicals Generated in Photocatalytic Oxidation of Some Amines and Diamines. *J. Environ. Sci.* **1993,** *5*, 464–469.

Dehghani, M. H.; Fadaei, A. M. Potoccatalytic Oxidation of Oganophosphorus Pesticides Using Zinc Oxide. *Res. J. Chem. Environ.* **2012,** *16, 3876–3881.*

Devahasdin, S.; Fan, C.; Li, K.; Chen, D. H. TiO_2 Photocatalytic Oxidation of Nitric Oxide: Transient Behavior and Reaction Kinetics. *J. Photochem. Photobiol.* **2003,** *156*, 161–170.

Dillert, R.; Brandt, M.; Fornefett, I.; Siebers, U.; Bahnemann, D. Photocatalytic Degradation of Trinitrotoluene and Other Nitroaromatic Compounds. *Chemosphere* **1995,** *30*, 2333–2341.

Dillert, R.; Fornefett, I.; Siebers, U.; Bahnemann, D. Photocatalytic Degradation of Trinitrotoluene and Trinitrobenzene: Influence of Hydrogen Peroxide. *J. Photochem. Photobiol.* **1996,** *94*, 231–236.

Ditta, I. B.; Steele, A.; Liptrot, C.; Tobin, J. Tyler, H. Yates, H. M. et al. Photocatalytic Antimicrobial Activity of Thin Surface Films of TiO_2, CuO and TiO_2/CuO Dual Layers on *Escherichia coli* and Bacteriophage T4. *Appl. Microbiol. Biotechnol.* **2008,** *79*, 127–133.

Doong, R.; Chang, W. Photoassisted Titanium Dioxide Mediated Degradation of Organophosphorus Pesticides by Hydrogen Peroxide. *J. Photochem. Photobiol.* **1997,** *107*, 239–244.

Du, M.; Zeng, G.; Huang, J.; Sun, D.; Li, Q.; Wang, G.; Li, X. Green Photocatalytic Oxidation of Benzyl Alcohol Over Noble-Metal-Modified $H_2Ti_3O_7$ Nanowires. *ACS Sustain. Chem. Eng.* **2019,** *7* (10), 9717–9726.

Eskandari, P.; Kazemi, F. Preparation and Characterization of Fe_3O_4/SiO_2/CdS Nanocomposites as Efficient Magnetic Photocatalysts for the Reduction of Nitro Compounds Under Visible LED Irradiation. *J. Photochem. Photobiol. A: Chem.* **2018,** *364*, 233–239.

Faust, S. D.; Aly, O. M. *Chemistry of Water Treatment*, 2nd ed.; Lewis Publishers: Boca Raton, 1999; pp 1–581.

Feng, C.; Chen, Z.; Jing, J.; Hou, J. The Photocatalytic Phenol Degradation Mechanism of Ag-Modified ZnO Nanorods. *J. Mater. Chem. C*, **2020,** *8* (9), 3000–3009.

Feng, C.; Yang, X.; Sun, Z.; Xue, J.; Sun, L.; Wang, J. Dual Interfacial Synergism in Au-Pd/ZnIn2S4 for Promoting Photocatalytic Selective Oxidation of Aromatic Alcohol. *Appl. Surf. Sci.* **2020a,** 501. DOI: 10.1016/j.apsusc.2019.144018.

Fuerte, A.; Hernandez–Alonso, M. D.; Maria, J.; Martinez–Arias, A.; Fernandez–Garcia, M. F.; Conesa, J. C. et al. Nanosize Ti–W Mixed Oxides: Effect of Doping Level in the Photocatalytic Degradation of Toluene Using Sunlight-Type Excitation. *J. Catal.* **2002,** *212*, 1–9.

Fukui, M.; Kouda, H.; Tanaka, A.; Hashimoto, K.; Kominami, H. Heterogeneous Meerwein-Ponndorf-Verley–Type Reduction of Aromatic Aldehydes Having Other Reducible Functional Groups Over a TiO_2 Photocatalyst. *ChemistrySelect* **2017,** *2* (7), 2293–2299.

Garcia, T.; Solsona, B.; Murphy, D. M.; Antcliff, K. L.; Taylor, S. H. Deep Oxidation of Light Alkanes Over Titania-Supported Palladium/Vanadium Catalysts. *J. Catal.* **2005,** *229*, 1–11.

Ghodsi, S.; Esrafili, A.; Kalantary, R. R.; Gholami, M.; Sobhi, H. R. Synthesis and Evaluation of the Performance of g-C_3N_4/Fe_3O_4/Ag Photocatalyst for the Efficient Removal of Diazinon: Kinetic Studies. *J. Photochem. Photobiol. A: Chem.* **2020,** *389*. DOI: 10.1016/j.jphotochem.2019.112279.

Giahi M.; Habibi S.; Toutounchi S.; Khavei M. Photocatalytic Degradation of Anionic Surfactant Using Zinc Oxide Nanoparticles. *Russ. J. Phys. Chem.* **2012,** *86*, 689–693.

Gratzel, M. *Energy Resources Through Photochemistry and Catalysis*; A. P.: New York, 1983.

Hadjltaief, H. B.; Gálvez, M. E.; Zina, M. B.; Da Costa, P. TiO_2/Clay as a Heterogeneous Catalyst in Photocatalytic/Photochemical Oxidation of Anionic Reactive Blue 19. *Arab. J. Chem.* **2019,** *12* (7), 1454–1462.

Hajimohammadi, M.; Azizi, N.; Tollabimazraeno, S.; Tuna, A.; Duchoslav, J.; Knör, G. Cobalt (II) Phthalocyanine Sulfonate Supported on Reduced Graphene Oxide (RGO) as a Recyclable Photocatalyst for the Oxidation of Aldehydes to Carboxylic Acids. *Catal. Lett.* **2021,** *151* (1), 36–44.

Han, C.; Li, Z.; Shen, J. Photocatalytic Degradation of Dodecyl-Benzenesulfonate Over TiO_2–Cu_2O Under Visible Irradiation. *J. Hazard Mater.* **2009,** *168*, 215–219.

Hanh, N. T.; Tri, N. L. M.; Van Thuan, D.; Tung, M. H. T.; Pham, T. D.; Minh, T. D. et al. Monocrotophos Pesticide Effectively Removed by Novel Visible Light Driven Cu Doped ZnO Photocatalyst. *J. Photochem. Photobiol. A Chem.* **2019,** *382*. DOI: 10.1016/j.jphotochem.2019.111923.

Hatipoglu, A.; San, N.; Cinar, Z. An Experimental and Theoretical Investigation of the Photocatalytic Degradation of Meta-Cresol in TiO_2 Suspensions: A Model for the Product Distribution. *J. Photochem. Photobiol.* **2004,** *165*, 119–129.

He, J.; Lyu, P.; Jiang, B.; Chang, S.; Du, H.; Zhu, J.; Li, H. A Novel Amorphous Alloy Photocatalyst (NiB/In_2O_3) Composite for Sunlight-Induced CO_2 Hydrogenation to HCOOH. *Appl. Catal. B Environ.* **2021,** *298*. DOI: 10.1016/j.apcatb.2021.120603.

Heintz, O.; Robert, D.; Weber, J. V. Comparison of the Degradation of Benzamide and Acetic Acid on Different TiO_2 Photocatalysts. *J. Photochem. Photobiol.* **2000,** *135*, 77–80.

Heltina, D.; Wulan, P. P. D. K. Synthesis and Characterization of Titania Nanotube-Carbon Nanotube Composite for Degradation of Phenol. *Int. J. Technol.* **2015,** *6* (7), 1137–1145.

Hermann, J. M.; Guillard, C.; Arguello, M.; Aguera, A.; Tejedor, A.; Piedra, L. et al.. Photocatalytic Degradation of Pesticide Pirimiphos-Methyl: Determination of the Reaction Pathway and Identification of Intermediate Products by Various Analytical Methods. *Catal Today* **1999,** *54*, 353–367.

Hidaka, H.; Yamada, S.; Suenaga, S.; Zhao, J.; Serpone, N.; Pelizzetti, E. Photodegradation of Surfactants: Part VI Complete Photocatalytic Degradation of Anionic, Cationic and Nonionic Surfactants in Aqueous Semiconductor Dispersions. *J. Mol. Catal.* **1990,** *59*, 279–290.

Houas, A.; Lachheb, H.; Ksibi, N.; Elaloui, E.; Guillard, C. Hermann, Photocatalytic Degradation Pathway of Methylene Blue in Water. J. M. *Appl. Catal. B: Environ.* **2001,** *31*, 145–157.

Hsu, H. C.; Shown, I.; Wei, H. Y.; Chang, Y. C.; Du, H. Y.; Lin, Y. G. et al. Graphene Oxide as a Promising Photocatalyst for CO_2 to Methanol Conversion. *Nanoscale* **2013,** *5* (1), 262–268.

Hussain, M.; Akhter, P.; Russo, N.; Saracco, G. New Optimized Mesoporous Silica Incorporated Isolated Ti Materials Towards Improved Photocatalytic Reduction of Carbon Dioxide to Renewable Fuels. *Chem. Eng. J.* **2015,** *278*, 279–292.

Huszla, K.; Wysokowski, M.; Zgoła-Grześkowiak, A.; Staszak, M.; Janczarek, M.; Jesionowski, T.; Wyrwas, B. UV-Light Photocatalytic Degradation of Non-Ionic Surfactants Using ZnO Nanoparticles. *Int. J. Environ. Sci. Technol.* **2021**. DOI: 10.1007/s13762–021–03160–1.

Hwangbo, M.; Claycomb, E. C.; Liu, Y.; Alivio, T. E.; Banerjee, S.; Chu, K. H. Effectiveness of Zinc Oxide-Assisted Photocatalysis for Concerned Constituents in Reclaimed Wastewater: 1, 4-Dioxane, Trihalomethanes, Antibiotics, Antibiotic Resistant Bacteria (ARB), and Antibiotic Resistance Genes (ARGs). *Sci. Total Environ.* **2019,** *649*, 1189–1197.

Iqbal, T.; Masood, A.; Khalid, N. R.; Tahir, M. B.; Asiri, A. M.; Alrobei, H. Green Synthesis of Novel Lanthanum Doped Copper Oxide Nanoparticles for Photocatalytic Application: Correlation Between Experiment and COMSOL Simulation. *Ceram. Int.* **2022,** *48* (10), 13420–13430.

Ismael, M. Highly Effective Ruthenium-Doped TiO_2 Nanoparticles Photocatalyst for Visible-Light-Driven Photocatalytic Hydrogen Production. *New J. Chem.* **2019,** *43* (24), 9596–9605.

IUPAC Glossary of Terms Used in Photochemistry, *Pure Appl. Chem.* **1988,** *60*, 1055–1106.

Jain, S.; Ameta, S. C. Photocatalytic Oxidation of Arabinose and Glucose Over Cadmium Sulphide. *Res. J. Chem. Environ.* **2008,** *12*, 61–64.

Jat, K. K.; Bhatt, J.; Ameta, S. C. Photodegradation of Fast Green by Using SnO_2 Quantum Dots/TiO_2 Nanoparticles Composite. *J. Appl. Chem.* **2019,** *8* (1), 139–145.

Jing, J.; Liu, M.; Colvin, V. L.; Li, W.; Yu, W. W. Photocatalytic Degradation of Nitrogen-Containing Organic Compounds Over TiO_2. *J. Mol. Catal. A: Chem.* **2011,** *351*, 17–28.

Jirkovsky, J.; Faure, V.; Boule, P. Photolysis of diuron. *Pestic Sci.* **1997,** *50*, 42–52.

Kamat, P. V. Photochemistry on Nonreactive and Reactive (Semiconductor) Surfaces. *Chem. Revs.* **1993,** *93*, 267–300.

Kanade, K. G.; Kale, B. B.; Baeg, J. O.; Lee, S. M.; Lee, C. W.; Moon, S. J. et al. Self-Assembled Aligned Cu Doped ZnO Nanoparticles for Photocatalytic Hydrogen Production Under Visible Light Irradiation. *Mater. Chem. Phys.* **2007,** *102* (1), 98–104.

Kang, M. G.; Jung, H. S.; Kim, K. J. Effect of Chloride Ions on 4-Chlorophenol Photodegradation in the Absence and Presence of Titanium Silicalite-2. *J. Photochem. Photobiol.* **2000,** *136*, 117–123.

Kaur, P.; Bansal, P.; Sud, D. Heterostructured Nanophotocatalysts for Degradation of Organophosphate Pesticides from Aqueous Streams. *J. Korean Chem. Soc.* **2013,** *57* (3), 382–388.

Kaur, P.; Khant, A.; Khandelwal, R. C. Synthesis and Characterization of La_2CoO_4 and Its Use as a Photocatalyst. *Int. J. Chem. Sci.* **2011,** *9*, 980–988.

Kaur, P.; Sud, D. Photocatalytic Degradation of Quinalphos in Aqueous TiO_2 Suspension: Reaction Pathway and Identification of Intermediates by GC/MS. *J. Mol. Catal. A Chem.* **2012,** *365*, 32–38.

Kawai, T.; Sakata, T. Photocatalytic Hydrogen Production from Liquid Methanol and Water. *J. Chem. Soc. Chem. Commun.* **1980,** *15*, 694–695.

Kerzhentsev, M.; Guillard, C.; Herrmann, J. M.; Pichat, P. Photocatalytic Pollutant Removal in Water at Room Temperature: Case Study of the Total Degradation of the Insecticide Fenitrothion (Phosphorothioic Acid O, O-Dimethyl-O- (3-Methyl-4-Nitro-Phenyl) Ester). *Catal. Today* **1996,** *27*, 215–220.

Khavar, A. H. C.; Moussavi, G.; Mahjoub, A. R.; Satari, M.; Abdolmaleki, P. Synthesis and Visible-Light Photocatalytic Activity of In, S-TiO_2@ rGO Nanocomposite for Degradation and detoxification of Pesticide Atrazine in Water. *Chem. Eng. J.* **2018,** *345*, 300–311.

Kim, S. H.; Kwak, S. Y.; Suzuki, T. Photocatalytic Degradation of Flexible PVC/TiO_2 Nanohybrid as an Eco-Friendly Alternative to the Current Waste Landfill and Dioxin-Emitting Incineration of Post-Use PVC. *Polymer.* **2006,** *47*, 3005–3016.

Kimura, T.; Yoshikawa, N.; Matsumura, N.; Kawase, Y. Photocatalytic Degradation of Nonionic Surfactants with Immobilized TiO_2 in an Airlift Reactor. *J. Environ. Sci. Health A Tox. Hazard Subst. Environ. Eng.* **2004,** *39*, 2867–2881.

Kirchnerova, J.; Cohen, M. L.; Guy, C.; Klvana, D. Photocatalytic Oxidation of N-Butanol Under Fluorescent Visible Light Lamp Over Commercial TiO_2 (Hombicat UV100 and Degussa P25). *Appl. Catal.* **2005,** *282*, 321–332.

Klosek, S.; Raftery, D. Visible Light Driven V-Doped TiO_2 Photocatalyst and Its Photooxidation of Ethanol. *J. Phys. Chem. B.* **2001,** *105*, 2815–2819.

Konstantinou, I. K.; Albanis, T. A. Photocatalytic Transformation of Pesticides in Aqueous Titanium Dioxide Suspensions Using Artificial and Solar Light: Intermediates and Degradation Pathways. *Appl. Catal. B: Environ.* **2003,** *42*, 319–335.

Kothari, S.; Ameta, P.; Ameta, R. Photocatalytic Bleaching of Evans Blue Over Zinc Oxide Particulate System. *Indian J. Chem.* **2007,** *46*, 432–435.

Kothari, S.; Jain, N.; Ameta, R. Photocatalytic Bleaching of Amaranth Dye Over ZnO Powder. *Indian J. Chem. Tech.* **2004,** *11*, 423–426.

Krysova, H.; Krysa, J.; Macounova, K.; Jirkovsky, J. Photocatalytic Degradation of Diuron [3- (3, 4-Dichlorophenyl)-1, 1-Dimethylurea] on the Layer of TiO_2 Particles in the Batch Mode Plate Film Reactor. *J. Chem. Technol. Biotechnol.* **1998,** *72*, 169–175.

Kulkarni, A. A.; Bhanage, B. M.; Ag@AgCl Nanomaterials Synthesis Using Sugar Cane Juice and Its Application in Degradation of Azo Dyes. *ACS Sustain. Chem. Eng.* **2014,** *2*, 1007–1013.

Kumar, P.; Kumar, A.; Sreedhar, B.; Sain, B.; Ray, S. S.; Jain, S. L. Cobalt Phthalocyanine Immobilized on Graphene Oxide: An Efficient Visible-Active Catalyst for the Photoreduction of Carbondioxide. *Chem. Eur. J.* **2014,** *20*, 6154–6161.

Kushniarou, A.; Garrido, I.; Fenoll, J.; Vela, N.; Flores, P.; Navarro, G.; Solar Photocatalytic Reclamation of Agro-Waste Water Polluted with Twelve Pesticides for Agricultural Reuse. *Chemosphere* **2019,** *214*, 839–845.

Kuvarega, A. T.; Khumalo, N.; Dlamini, D.; Mamba, B. B. Polysulfone/N, Pd co-doped TiO_2 Composite Membranes for Photocatalytic Dye Degradation. *Sep. Purif. Technol.* **2018,** *191*, 122–133.

Lair, A.; Ferronato, C.; Chovelon, J. M.; Herrmann, J. M. Naphthalene Degradation in Water by Heterogeneous Photocatalysis: An Investigation of the Influence of Inorganic Anions. *J. Photochem. Photobiol.* **2007,** *193*, 193–203.

Lannoy, A.; Bleta, R.; Machut-Binkowski, C.; Addad, A.; Monflier, E.; Ponchel, A. Cyclodextrin-Directed Synthesis of Gold-Modified TiO_2 Materials and Evaluation of Their Photocatalytic Activity in the Removal of a Pesticide from Water: Effect of Porosity and Particle Size. *ACS Sustain. Chem. Eng.* **2017,** *5* (5), 3623–3630.

Li, J. H.; Lu, A. H.; Liu, F.; Fan, L. Z. Synthesis of ZnS/Dravite Composite and Its Photocatalytic Activity on Degradation of Methylene Blue. *Solid State Ion.* **2008a,** *179*, 1387–1390.

Li, F.; Sun, S.; Jiang, Y.; Xia, M.; Sun, M.; Xue, B. Photodegradation of an Azo Dye Using Immobilized Nanoparticles of TiO_2 Supported by Natural Porous Mineral. *J. Hazard. Mater.* **2008b,** *152*, 1037–1044.

Li, P.; Liu, L.; An, W.; Wang, H.; Cui, W. Efficient Photothermal Catalytic CO_2 Reduction to CH_3CH_2OH Over $Cu_2O/g\text{-}C_3N_4$ Assisted by Ionic Liquids. *Appl. Surf. Sci.* **2021,** *565*. DOI: 10.1016/j.apsusc.2021.150448.

Li, Q.; Cui, C.; Meng, H.; Yu, J. Visible-Light Photocatalytic Hydrogen Production Activity of $ZnIn_2S_4$ Microspheres Using Carbon Quantum Dots and Platinum as Dual Co-Catalysts. *Chem. Asian J.* **2014,** *9* (7), 1766–1770.

Liang, H.; Tai, X.; Du, Z.; Yin, Y. Enhanced Photocatalytic Activity of ZnO Sensitized by Carbon Quantum Dots and Application in Phenol Wastewater. *Opti. Mater.* **2020,** *100*, 109674.

Liu, E.; Kang, L.; Wu, F.; Sun, T.; Hu, X.; Yang, Y.; Liu, H.; Fan, J. Photocatalytic Reduction of CO_2 into Methanol Over Ag/TiO_2 Nanocomposites Enhanced by Surface Plasmon Resonance. *Plasmonics* **2014,** *9*, 61–70.

Liu, G. H.; Zhu, Y. F.; Zhang, X. R.; Xu, B. Q. Chemiluminescence Determination of Chlorinated Volatile Organic Compounds by Conversion on Nanometer TiO_2. *Anal. Chem.* **2002,** *24*, 6279–6284.

Lizama, C.; Bravo, C.; Caneo, C.; Ollino, M. Photocatalytic Degradation of Surfactants with Immobilized TiO_2: Comparing Two Reaction Systems. *Environ. Technol.* **2005,** *26*, 909–914.

Low, G. K. C.; McEvoy, R.; Mathews, R. W. Formation of Nitrate and Ammonium Ions in Titanium Dioxide Mediated Photocatalytic Degradation of Organic Compounds Containing Nitrogen Atoms. *Environ. Sci. Technol.* **1991,** *25*, 460–467.

Lu, G.; Song, B.; Li, Z.; Liang, H.; Zou, X. Photocatalytic Degradation of Naphthalene on $CeVO_4$ Nanoparticles Under Visible Light. *Chem. Eng. J.* **2020,** *402*. DOI: 10.1016/j.cej. 2020.125645.

Ma, Y.; Yao, J. N. Photodegradation of Rhodamine B Catalyzed by TiO_2 Thin Films. *J. Photochem. Photobiol.* **1998,** *116A*, 167–170.

Machuca-Martínez, F.; Colina-Márquez, J. Á. Effect of the Initial pH and the Catalyst Concentration on TiO_2-Based Photocatalytic Degradation of Three Commercial Pesticides. *Ing. Desarro.* **2011,** *29*, 84–100.

Madhusudhana, N.; Yogendra, K.; Mahadevan, K. M.; Naik, S. Photocatalytic Degradation of Coralene Dark Red 2B Azo Dye Using Calcium Zincate Nanoparticle in Presence of Natural Sunlight: An Aid to Environmental Remediation. *Int. J. Chem. Engg. Appl.* **2011,** *2*, 294–298.

Magalhaes, F.; Lago, R. M. Floating Photocatalysts Based on TiO_2 Grafted on Expanded Polystyrene Beads for the Solar Degradation of Dyes. *Solar Energy* **2009,** *83*, 1521–1526.

Mahadwad, O. K.; Parikh, P. A.; Jasra, R. V.; Patil, C. Photocatalytic Degradation of Reactive Black-5 Dye Using TiO_2 Impregnated ZSM-5. *Bull. Mater. Sci.* **2011,** *34*, 551–556.

Makarova, O. V.; Rajh, T.; Thurnauer, M. C. Surface Modification of TiO_2 Nanoparticles for Photochemical Reduction of Nitrobenzene. *Environ. Sci. Technol.* **2000,** *34*, 4797–4803.

Mangrulkar, P. A.; Polshettiwar, V.; Labhsetwar, N. K.; Varma, R. S.; Rayalu, S. S. Nano-Ferrites for Water Splitting: Unprecedented High Photocatalytic Hydrogen Production Under Visible Light. *Nanoscale* **2012,** *4*, 5202–5209.

Mansoori, R. A.; Kothari, S.; Ameta, R. Use of ZnO as a Photocatalyst in Photocatalytic Bleaching of Rhodamine-B. *J. Indian Chem. Soc.* **2004,** *81*, 335–337.

Marci, G.; Addano, M.; Augugliaro, V.; Coluccia, S.; Garcia–Lopez, E.; Laddo, V.; Mastra, G. Photocatalytic Oxidation of Toluene on Irradiated TiO_2: Comparison of Degradation Performance in Humidified Air, in Water and in Water Containing a Zwitterionic Surfactant. *J. Photochem. Photobiol.* **2003,** *160*, 105–114.

McManamon, C.; Holmes, J. D.; Morris, M. A. Improved Photocatalytic Degradation Rates of Phenol Achieved Using Novel Porous ZrO_2-Doped TiO_2 Nanoparticulate Powders. *J. Hazard. Mater.* **2011,** *193*, 120–127.

Miguel, N.; Ormad, M. P.; Mosteo, R.; Ovelleiro, J. L. Photocatalytic Degradation of Pesticides in Natural Water: Effect of Hydrogen Peroxide. *Int. J. Photoenergy* **2012**. DOI: 10.1155/2012/371714

Moafi, H. F.; Zanjanchi, M. A.; Fallah, S. A. Tungsten-Doped ZnO Nanocomposite: Synthesis, Characterization, and Highly Active Photocatalyst Toward Dye Photodegradation. *Mater. Chem. Phys.* **2013,** *139*, 856–864.

Mohamed, O. S.; Gaber, A. M.; Abdel-Wahab, A. A. Photocatalytic Oxidation of Selected Aryl Alcohols in Acetonitrile. *J. Photochem. Photobiol.* **2002,** *148*, 205–210.

Mohammad, A.; Kapoor, K.; Mobin, S. M. Improved Photocatalytic Degradation of Organic Dyes by ZnO-Nanoflowers. *ChemistrySelect* **2016,** *1* (13), 3483–3490.

Montgomery, J. M.; *Consulting Engineers, Water Treatment: Design and Design*; John Wiley: New York, 1985; pp 1–696.

Moradi, V.; Ahmed, F.; Jun, M. B.; Blackburn, A.; Herring, R. A. Acid-Treated Fe-Doped TiO_2 as a High Performance Photocatalyst Used for Degradation of Phenol Under Visible Light Irradiation. *J. Environ. Sci.* **2019,** *83*, 183–194.

Morawski, A. W.; Kusiak-Nejman, E.; Wanag, A.; Kapica-Kozar, J.; Wróbel, R. J.; Ohtani, B. et al. Photocatalytic Degradation of Acetic Acid in the Presence of Visible Light-Active TiO_2-Reduced Graphene Oxide Photocatalysts. *Catal. Today*, **2017,** *280*, 108–113.

Mukwevho, N.; Fosso-Kankeu, E.; Waanders, F.; Kumar, N.; Ray, S. S.; Mbianda, X. Y. Photocatalytic Activity of $Gd_2O_2CO_3 \cdot ZnO \cdot CuO$ Nanocomposite Used for the Degradation of Phenanthrene. *SN Appl. Sci.* **2019,** *1* (1). DOI: 10.1007/s42452–018–0012–0.

Muszkat, L.; Feigelson, L.; Bir, L.; Muszkat, K. A. Photocatalytic Degradation of Pesticides and Bio-Molecules in Water. *Pest Manage. Sci.* **2002,** *58*, 1143–1148.

Nabi, G.; Ain, Q. U.; Tahir, M. B.; Nadeem Riaz, K.; Iqbal, T.; Rafique, M. Green Synthesis of TiO_2 Nanoparticles Using Lemon Peel Extract: Their Optical and Photocatalytic Properties. *Int. J. Environ. Anal. Chem.* **2022,** *102* (2), 434–442.

Nahen, M.; Bahnemann, D.; Dillert, R.; Fels, G. Photocatalytic Degradation of Trinitrotoluene: Reductive and Oxidative Pathways. *J. Photochem. Photobiol.* **1997,** *110*, 191–199.

Nguyen, T. H. A.; Doan, V. D.; Tran, A. V.; Nguyen, V. C.; Nguyen, A. T.; Vasseghian, Y. Green Synthesis of Nb-doped ZnO Nanocomposite for Photocatalytic Degradation of Tetracycline Antibiotic Under Visible Light. *Mater. Lett.* **2022,** *308*. DOI: 10.1016/j.matlet.2021.131129.

Odling, G.; Robertson, N. SILAR BiOI-Sensitized TiO_2 Films for Visible-Light Photocatalytic Degradation of Rhodamine B and 4-Chlorophenol. *Chem. Phys. Chem.* **2017,** *18* (7), 728–735.

Okamoto, K.; Yamanoto, Y.; Tanaka, H.; Itaya, A. Heterogeneous Photocatalytic Decomposition of Phenol Over TiO_2 Powder. *Bull. Chem. Soc. Jpn.* **1985,** *58*, 2023–2028.

Ollis, D. F. Pelizzetti, E.; Serpone, N. Photocatalyzed Destruction of Water Contaminants. *Environ. Sci. Technol.* **1991,** *25*, 1522–1529.

Ollis, D. F. Contaminant Degradation in Water. *Environ. Sci. Technol.* **1982,** *19*, 480–484.

Ong, W. J.; Putri, L. K.; Tan, L. L.; Chai, S. P.; Yong, S. T. Heterostructured AgX/g-C_3N_4 (X= Cl and Br) Nanocomposites via a Sonication-Assisted Deposition-Precipitation Approach: Emerging Role of Halide Ions in the Synergistic Photocatalytic Reduction of Carbon Dioxide. *Appl. Catal. B Environ.* **2016,** *180*, 530–543.

Ong, W. J.; Tan, L. L.; Chai, S. P.; Yong, S. T. Heterojunction Engineering of Graphitic Carbon Nitride (gC 3 N 4) via Pt Loading with Improved Daylight-Induced Photocatalytic Reduction of Carbon Dioxide to Methane. *Dalton Trans.* **2015,** *44* (3), 1249–1257.

Pan, J.; You, M.; Chi, C.; Dong, Z.; Wang, B.; Zhu, M. et al. The Two Dimension Carbon Quantum Dots Modified Porous g-C_3N_4/TiO_2 Nano-Heterojunctions for Visible Light Hydrogen Production Enhancement. *Int. J. Hydrogen Energy*, **2018,** *43* (13), 6586–6593.

Panchal, P.; Paul, D. R.; Gautam, S.; Meena, P.; Nehra, S. P.; Maken, S.; Sharma, A. Photocatalytic and Antibacterial Activities of Green Synthesized Ag Doped MgO Nanocomposites Towards Environmental Sustainability. *Chemosphere* **2022,** *297.* DOI: 10.1016/j.chemosphere.2022.134182

Pandiyan, T.; Rivas, O. M.; Martinez, J. O.; Amezcua, G. B.; Martinez-Carrillo, M. A. Comparison of Methods for the Photochemical Degradation of Chlorophenols. *J. Photochem. Photobiol.* **2002,** *146*, 149–155.

Papic, S.; Koprivanac, N.; Bozic, A. L. Advanced Oxidation Processes in Azo Dye Wastewater Treatment. *Water Environ. Res.* **2006,** *78*, 572–579.

Pare, B.; Singh, P.; Jannalgadda, S. B.; Artificial Light Assisted Photocatalytic Degradation of Lissamine Fast Yellow Dye in ZnO Suspension in a Slurry Batch Reactor. *Indian J. Chem.* **2009,** *48*, 1364–1369.

Park, H.; Choi, W. Photocatalytic Conversion of Benzene to Phenol Using Modified TiO_2 and Polyoxometalates. *Catal. Today* **2005,** *101*, 291–297.

Parvin, T.; Keerthiraj, N.; Ibrahim, I. A.; Phanichphant, S.; Byrappa, K. Photocatalytic Degradation of Municipal Wastewater and Brilliant Blue Dye Using Hydrothermally Synthesized Surface-Modified Silver-Doped ZnO Designer Particles. *Int. J. Photoenergy* **2012,** *44*, 1056–1059.

Paušová, Š.; Pacileo, L.; Baudys, M.; Hrubantová, A.; Neumann-Spallart, M.; Dvoranová, D. et al. Active Carbon/TiO_2 Composites for Photocatalytic Decomposition of Benzoic Acid in Water and Toluene in Air. *Catal. Today*, **2020,** 2020. DOI: 10.1016/j.cattod.2020.06.048.

Pelizzetti, E.; Serpone, N. *Homogeneous and Heterogeneous Photocatalysis*; D. Reidel Pub. Co.: Holland, 1986.

Peng, T.; Zeng, P.; Ke, D.; Liu, X.; Zhang, X. Hydrothermal Preparation of Multiwalled Carbon Nanotubes (MWCNTs)/CdS Nanocomposite and Its Efficient Photocatalytic Hydrogen Production Under Visible Light Irradiation. *Energy Fuels*, **2011,** *25* (5), 2203–2210.

Pereira, L.; Pereira, R.; Oliveira, C. S.; Apostol, L.; Pons, M. G. M.-N.; Zahraa, O.; Alves, M. M. UV/TiO_2 photocatalytic degradation of xanthene dyes. *J. Photochem. Photobiol.* **2013,** *89*, 33–39.

Phuong, N. M.; Chu, N. C.; Van Thuan, D.; Ha, M. N.; Hanh, N. T.; Viet, H. D. T. et al. Novel Removal of Diazinon Pesticide by Adsorption and Photocatalytic Degradation of Visible Light-Driven Fe-TiO_2/Bent-Fe Photocatalyst. *J. Chem.* **2019,** *2019*. DOI: 10.1155/2019/2678927.

Piccinini, P.; Minero, C.; Vincenti, M.; Pelizzetti, E. Photocatalytic Mineralization of Nitrogen-Containing Benzene Derivatives. *Catal. Today* **1997,** *39*, 187–195.

Poulis, I.; Micropoulou, E.; Panou, R.; Kostopolou, E. Photooxidation of Eosin Y in the Presence of Semiconducting Oxides. *Appl. Catal.* **2003,** *41*, 345–355.

Punjabi, P. B.; Ameta, R.; Vyas, R.; Kothari, S. Photoreduction of Congo Red by Ascorbic Acid and EDTA Over Cadmium Sulphide as Photocatalyst. *Indian J. Chem.* **2005,** *44*, 2266–2269.

Puretedal, H. R.; Norozi, A.; Keshararz, M. H.; Semnani, A. Nanoparticles of Zinc Sulfide Doped with Manganese, Nickel and Copper as Nanophotocatalyst in the Degradation of Organic Dyes. *J. Hazard. Mater.* **2009,** *162*, 674–681.

Quiroz-Cardoso, O.; Oros-Ruiz, S.; Solís-Gómez, A.; López, R.; Gómez, R. Enhanced Photocatalytic Hydrogen Production by CdS Nanofibers Modified with Graphene Oxide and Nickel Nanoparticles Under Visible Light. *Fuel*, **2019,** *237*, 227–235.

Ramasamy, V.; Mohana, V.; Rajendran, V. Characterization of Ca doped CeO_2 Quantum Dots and Their Applications in Photocatalytic Degradation. *OpenNano*, **2018,** *3*, 38–47.

Rangappa, A. P.; Praveen Kumar, D.; Hong, Y.; Jeong, S.; Reddy, D. A.; Song, J. K. et al. Construction of a Highly Efficient and Durable 1D Ternary CdS/ZnS/Pt Nanohybrid Catalyst for Photocatalytic CO_2 Reduction into Chemical Fuels Under Solar Light Irradiation. *ACS Appl. Energy Mater.* **2020,** *3* (11), 10533–10540.

Rani, M.; Shanker, U. Photocatalytic Degradation of Toxic Phenols from Water Using Bimetallic Metal Oxide Nanostructures. *Colloids Surf. A Physicochem. Eng. Aspects* **2018,** *553*, 546–561.

Rao, N.; Dubey, S. Photocatalytic Degradation of Mixed Surfactants and Some Commercial Soap/Detergent Products Using Suspended TiO_2 Catalysts. *J. Mol. Catal. A: Chem.* **1996,** *104*, 197–199.

Razavi, F. S.; Ghanbari, D.; Salavati-Niasari, M. Green Synthesis of $SrFe_{12}O_{19}$@Ag and SrFe12O19@ Au as Magnetic Plasmonic Nanocomposites with High Photocatalytic Performance for Degradation of Organic. *Chemosphere*, 291 (2022), Article 132741.

Sacco, O.; Stoller, M.; Vaiano, V.; Ciambelli, P.; Chianese, A.; Sannino, D. Photocatalytic Degradation of Organic Dyes Under Visible Light on N-Doped TiO_2 Photocatalysts. *Int. J. Photoenergy* **2012**. DOI: 10.1155/2012/626759.

Sakthivel, S.; Kisch, H. Daylight Photocatalysis by Carbon-Modified Titanium Dioxide. *Angew. Chem. Int. Ed.* **2003,** *42*, 4908–4911.

Sakthivel, S.; Shanker, M. V.; Palanichamy, M.; Arbindoo, B.; Murugesan, V. Photocatalytic Decomposition of Leather Dye: Comparative Study of TiO_2 Supported on Alumina and Glass Beads. *J. Photochem. Photobiol.* **2002,** *148*, 153–159.

Samadi, M. T.; Dorraji, M. S. S.; Atashi, Z. Photo Catalytic Removal of Sodium Dodecyl Sulfate from Aquatic Solutions with prepared ZnO Nanocrystals and UV Irradiation. *Avicenna J. Environ. Health Eng.* **2017,** *1* (1), 17–23.

San, N.; Hathipoglu, V.; Kocturk, G.; Cinar, Z. Prediction of Primary Intermediates and the Photodegradation Kinetics of 3-Aminophenol in Aqueous TiO_2 Suspensions. *J. Photochem. Photobiol.* **2001,** *139*, 225–232.

Santhanalakshmi, J.; Komalavalli, R. Nano TiO_2 Assisted Degradation of Textile Dyes in H_2O_2 Aqueous Solution: Kinetic Studies with pH and Mass Effects. *Chem Sci Trans.* **2012,** *1*, 522–529.

Schmelling, D. C.; Gray, K. A. Photocatalytic Transformation and Mineralization of 2, *4*, 6-Trinitrotoluene (TNT) in TiO_2 Slurries. *Water Res.* **1995,** *29*, 2651–2662.

Sebuso, D. P.; Kuvarega, A. T.; Lefatshe, K.; King'ondu, C. K.; Numan, N.; Maaza, M.; Muiva, C. M. Green Synthesis of Multilayer Graphene/ZnO Nanocomposite for Photocatalytic Applications. *J. Alloys Compd.* **2022,** *900*. DOI: 10.1016/j.jallcom.2021.163526.

Sekar, A. D.; Muthukumar, H.; Chandrasekaran, N. I.; Matheswaran, M. Photocatalytic Degradation of Naphthalene Using Calcined FeZnO/PVA Nanofibers. *Chemosphere* **2018,** *205*, 610–617.

Selvam, K.; Swaminathan, M. Nano N-TiO_2 Mediated Selective Photocatalytic Synthesis of Quinaldines from Nitrobenzenes. *RSC Adv.* **2012,** *2* (7), 2848–2855.

Senthilnathan, J.; Philip, L. Removal of Mixed Pesticides from Drinking Water System Using Surfactant-Assisted Nano-TiO_2. *Water, Air Soil Poll.* **2010a,** *210*, 143–154.

Senthilnathan, J.; Philip, L. Photocatalytic Degradation of Lindane Under UV and Visible Light Using N-Doped TiO_2. *Chem. Engg. J.* **2010b,** *161*, 83–92.

Shaban, Y. A.; El-Sayed, M. A.; El-Maradny, A. A.; Al-Farawati, R. K.; Al-Zobidi, M. I. Photocatalytic Degradation of Phenol in Natural Seawater Using Visible Light Active Carbon Modified (CM)-n-TiO_2 Nanoparticles Under UV Light and Natural Sunlight Illuminations. *Chemosphere* **2012,** *91*, 307–313.

Shankar, H.; Saravanan, R.; Rajasudha, G.; Stephen, A.; Narayanan, V. Photocatalytic Degradation of Aqueous Methyl Orange Using Nanotitania Loaded Mo-MCM-41. *Int. J. Nanosci.* **2011,** *10*, 1131–1135.

Shanthi M.; Kuzhalosai V. Photocatalytic Degradation of an Azo Dye, Acid Red 27, in Aqueous Solution Using Nano ZnO. *Indian J. Chem. A* **2012,** *51*, 428–434.

Sharma, A.; Lee, B. K. Photocatalytic Reduction of Carbon Dioxide to Methanol Using Nickel-Loaded TiO_2 Supported on Activated Carbon Fiber. *Catal. Today*, **2017,** *298*, 158–167.

Sheikholeslami, Z.; Kebria, D. Y.; Qaderi, F. Investigation of Photocatalytic Degradation of BTEX in Produced Water Using γ-Fe_2O_3 Nanoparticle. *J. Ther. Anal. Calorim.* **2019,** *135* (3), 1617–1627.

Shen, Y.; Zhu, Z.; Wang, X.; Gong, J.; Zhang, Y. Synthesis of Z-scheme g-C_3N_4/Ag/Ag_3PO_4 Composite for Enhanced Photocatalytic Degradation of Phenol and Selective Oxidation of Gaseous Isopropanol. *Mater. Res. Bull.* **2018,** *107*, 407–415.

Singhal, B.; Porwal, A.; Sharma, A.; Ameta, R.; Ameta, S. C. Photocatalytic Degradation of Cetylpyridinium Chloride Over Titanium Dioxide Powder. *J. Photochem. Photobiol.* **1997,** *108*, 85–88.

Soltani, N.; Saion, E.; Hussein, M. Z.; Erfani, M.; Abedini, A.; Bahmanrokh, G. et al. Visible Light-Induced Degradation of Methylene Blue in the Presence of Photocatalytic ZnS and CdS Nanoparticles. *Int. J. Mol. Sci.* **2012,** *13*, 12242–12258.

Subash, B.; Krishnakumar, B.; Swaminathan, M.; Shanthi, M. Enhanced Photocatalytic Performance of WO_3 Loaded Ag–ZnO for Acid Black 1 Degradation by UV–A Light. *J. Mol. Catal. A Chem.* **2013,** *366*, 54–63.

Sukhaser, J.; Wold, A.; Gao, Y. M.; Dwight, K. Photoassisted Decomposition of Salicylic Acid on TiO_2 and Pd/TiO_2 Films. *J. Solid State Chem.* **1995,** *119*, 339–343.

Sun, Y.; Sun, J.; Jeffrey, R. L.; Yang, P.; Chang, C. Photocatalytic Generation of Hydrogen from Water Using a Cobalt Pentapyridine Complex in Combination with Molecular and Semiconductor Nanowire Photosensitizers. *J. Chem. Sci.* **2013,** *4*, 118–124.

Tabasum, A.; Alghuthaymi, M.; Qazi, U. Y.; Shahid, I.; Abbas, Q.; Javaid, R. et al. UV-Accelerated Photocatalytic Degradation of Pesticide Over Magnetite and Cobalt Ferrite Decorated Graphene Oxide Composite. *Plants*, **2021,** *10* (1). DOI: 10.3390/plants10010006.

Takeda, K.; Fujiwara, K. Characteristics on the Determination of Dissolved Organic Nitrogen Compounds in Natural Waters Using Titanium Dioxide and Platinized Titanium Dioxide Mediated Photocatalytic Degradation. *Water Res.* **1996,** *30*, 323–330.

Tamini, M.; Qourzal, S.; Assabane, A.; Chovelon, J. M.; Ferronato, C.; Ait-Ichou, Y. Photocatalytic Degradation of Pesticide Methomyl: Determination of the Reaction Pathway and Identification of Intermediate Products. *Photochem. Photobiol. Sci.* **2006,** *5*, 477–482.

Tan, D.; Zhang, J.; Shi, J.; Li, S.; Zhang, B.; Tan, X. et al. Photocatalytic CO_2 Transformation to CH_4 by Ag/Pd Bimetals Supported on N-doped TiO_2 Nanosheet. *ACS Appl. Mater. Interf.* **2018,** *10* (29), 24516–24522.

Tan, L. L.; Ong, W. J.; Chai, S. P.; Goh, B. T.; Mohamed, A. R. Visible-Light-Active Oxygen-Rich TiO_2 Decorated 2D Graphene Oxide with Enhanced Photocatalytic Activity Toward Carbon Dioxide Reduction. *Appl. Catal. B Environ.* **2015,** *179*, 160–170.

Tanaka, K.; Padermpole, K.; Hisanaga, T. Photocatalytic Degradation of Commercial Azo Dyes. *Water Res.* **2000,** *34*, 327–333.

Tang, A.; Xiao, Y.; Ouyang, J.; Nie, S. Preparation, Photo-Catalytic Activity of Cuprous Oxide Nano-Crystallites with Different Sizes. *J. Alloys Comp.* **2008,** *457*, 447–451.

Tang, M.; Ao, Y.; Wang, C.; Wang, P. Facile synthesis of dual Z-scheme g-C_3N_4/Ag_3PO_4/AgI Composite Photocatalysts with Enhanced Performance for the Degradation of a Typical Neonicotinoid Pesticide. *Appl. Catal. Environ.* **2020,** *268*. DOI: 10.1016/j.apcatb.2019.118395.

Tang, T.; Zhang, J.; Tian, B.; Chen, F.; He, D.; Anpo, M. Preparation of Ce–TiO_2 catalysts by Controlled Hydrolysis of Titanium Alkoxide Based on Esterification Reaction and Study on Its Photocatalytic Activity. *J. Colloid Interface Sci.* **2007,** *315*, 382–388.

Tapley, G. L. H.; Silvero, M. J.; Alejo, C. J. B.; Bejar, M. G.; McTiernan, C. D.; Grenier, M. et al. Supported Gold Nanoparticles as Efficient Catalysts in the Solventless Plasmon Mediated Oxidation of Sec-Phenethyl and Benzyl Alcohol. *J. Phys. Chem.* C **2013,** *117*, 12279–12288.

Tatsuma, T.; Tachibana, S.; Miwa, T.; Tryk, D. A. Fujishima, A. Remote Bleaching of Methylene Blue by UV-Irradiated TiO_2 in the Gas Phase. *J. Phys. Chem. B.* **1999,** *103*, 8033–8035.

Tayade, R. J.; Key, D. L. Synthesis and Characterization of Titanium Dioxide Nanotubes for Photocatalytic Degradation of Aqueous Nitrobenzene in the Presence of Sunlight. *Mater. Sci. Forum.* **2010,** *657*, 62–74.

Thirugnanam, N.; Song, H.; Wu, Y. Photocatalytic Degradation of Brilliant Green Dye Using CdSe Quantum Dots Hybridized with Graphene Oxide Under Sunlight Irradiation. *Chinese J. Catal.* **2017,** *38* (12), 2150–2159.

Tian, B.; Zhang, J.; Tong, T.; Chen, F. Improving the Visible Light Photocatalytic Activity of Mesoporous TiO_2 via the Synergetic Effects of B Doping and Ag Loading. *Appl. Catal. B, Environ.* **2008,** *79*, 394–401.

Tong, T.; Zhang, J. Tian, B.; Chen, F.; He, D. Preparation of Fe^{3+}-Doped TiO_2 Catalysts by Controlled Hydrolysis of Titanium Alkoxide and Study on Their Photocatalytic Activity for Methyl Orange Degradation. *J. Hazard. Mater.* **2008,** *155*, 572–579.

Torres-Martínez, L. M.; Ruiz-Gómez, M. A.; Figueroa-Torres, M. Z.; Juárez-Ramírez, I.; Moctezuma, E. Sm_2FeTaO_7 Photocatalyst for Degradation of Indigo Carmine Dye Under Solar Light Irradiation. *Int. J. Photoenergy* **2012,** *2012*. DOI: 10.1155/2012/939608.

USEPA, *Technologies for Upgrading Existing or Designing New Drinking Water Treatment Facilities*, EPA/625/4–89/023, 1989.

Vaya, D.; Benjamin, S.; Sharma, V. K.; Ameta, S. C. Effect of Transition Metal Ions Doping on ZnS for Eosin Y Degradation. *Bull. Catal. Soc. India.* **2008,** *7*, 56–59.

Vela, N.; Calín, M.; Yáñez-Gascón, M. J.; Garrido, I.; Pérez-Lucas, G.; Fenoll, J.; Navarro, S. Photocatalytic Oxidation of Six Pesticides Listed as Endocrine Disruptor Chemicals from Wastewater Using Two Different TiO_2 Samples at Pilot Plant Scale Under Sunlight Irradiation. *J. Photochem. Photobiol. A: Chem.* **2018,** *353*, 271–278.

Verma, A.; Sheoran, M.; Toor, A. P. Titanium Dioxide Mediated Photocatalytic Degradation of Malathion in Aqueous Phase. *Indian J. Chem. Technol.* **2013,** *20*, 46–51.

Villasenor, J.; Duran, N.; Mansilla, H. D. Photocatalyzed Mineralization of Kraft Black Liquor on ZnO/Fe_2O_3 Coupled Semiconductor. *J. Environ. Technol.* **2002,** *23*, 955–959.

Vohra, M. S.; Tanaka, K. Photocatalytic Degradation of Nitrotoluene in Aqueous TiO_2 Suspension. *Water Res.* **2002,** *36*, 59–64.

Vyas, R.; Swarnkar, H.; Ameta, S. C. The Photocatalytic Bleaching of Eosin Using ZnO: Effect of Surface Charge. *Chem. Environ. Res.* **2005,** *14*, 71–78.

Wang, D.; Zhang, J.; Luo, Q.; Li, X.; Duan, Y.; An, J. Characterization and Photocatalytic Activity of Poly (3-Hexylthiophene)-Modified TiO_2 for Degradation of Methyl Orange Under Visible Light. *J. Hazard. Mater.* **2009,** *169*, 546–550.

Wang, H.; Wu, Z.; Zhao, W. Guan, B. Photocatalytic Oxidation of Nitrogen Oxides Using TiO_2 Loading on Woven Glass Fabric. *Chemosphere* **2007,** *66*, 185–190.

Wang, K.-H.; Hsieh, Y.-H.; Chou, M.-Y.; Chang, C.-Y. Photocatalytic Degradation of 2-Chloro and 2-Nitrophenol by Titanium Dioxide Suspensions in Aqueous Solution. *Appl. Catal. B: Environ.* **1999,** *21*, 1–8.

Wang, Y.; Yang, H.; Sun, X.; Zhang, H.; Xian, T. Preparation and Photocatalytic Application of Ternary N-$BaTiO_3$/Ag/p-AgBr Heterostructured Photocatalysts for Dye Degradation. *Mater. Res. Bull.* **2020,** *124*, 110754.

Wei, H.; Tang, X. H.; Liang, J. R.; Tan, S. Y. Preparation, Characterization and Photocatalytic Activities of Boron-and Cerium-Codoped TiO_2. *J. Environ. Sci. (China)* **2007,** *19*, 90–96.

Wu, M.; Huang, B.; Wang, M.; Osaka, A. Titania Nanoflowers with High Photocatalytic Activity. *J. Am. Ceram. Soc.* **2006,** *89*, 2660–2663.

Xiao, X.; Zheng, C.; Lu, M.; Zhang, L.; Liu, F.; Zuo, X.; Nan, J. Deficient $Bi_{24}O_{31}Br_{10}$ as a Highly Efficient Photocatalyst for Selective Oxidation of Benzyl Alcohol Into Benzaldehyde Under Blue LED Irradiation. *Appl. Catal. B: Environ.* **2018,** *228*, 142–151.

Xu, J. C.; Shi, Y. L.; Huang, J. E.; Wang, B. Li, H. L. Doping Metal Ions Only Onto the Catalyst Surface. *J. Mol. Catal. A.* **2004,** *219*, 351–355.

Yadav, M.; Yadav, A.; Fernandes, R.; Popat, Y.; Orlandi, M.; Dashora, A. et al. Tungsten-Doped TiO_2/Reduced Graphene Oxide Nano-Composite Photocatalyst for Degradation of Phenol: A System to Reduce Surface and Bulk Electron-Hole Recombination. *J. Environ. Manage.* **2017,** *203*, 364–374.

Yamazaki, S.; Yamabe, N.; Nagano, S.; Fukuda, A. Adsorption and Photocatalytic Degradation of 1, 4-Dioxane on TiO_2. *J. Photochem. Photobiol.* **2007,** *185*, 150–155.

Yanagida, S.; Ishimarru, Y.; Mujake, Y.; Shiragami, T.; Hashimoto K.; Sakata, T. Semiconductor Photocatalysis. 8. Zinc Sulfide-Catalyzed Photoreduction of Aldehydes and Related Derivatives: Two-Electron-Transfer Reduction and Relationship with Spectroscopic Properties. *J. Phys. Chem.* **1989,** *93*, 2576–2682.

Yang, H.; Zhang, K.; Shi, R.; Li, X.; Dong, X.; Yu, Y. Sol–Gel Synthesis of TiO_2 Nanoparticles and Photocatalytic Degradation of Methyl Orange in Aqueous TiO_2 Suspensions. *J. Alloys Comp.* **2006,** *413*, 302–306.

Yang, J.; Hao, J.; Xu, S.; Wang, Q.; Dai, J.; Zhang, A.; Pang, X. $InVO_4$/β-$AgVO_3$ Nanocomposite as a Direct Z-Scheme Photocatalyst Toward Efficient and Selective Visible-Light-Driven CO_2 Reduction. *ACS Appl. Mater. Interf.* **2019,** *11* (35), 32025–32037.

Yang, Y.; Luan, Synthesis, Property Characterization and Photocatalytic Activity of the Novel Composite Polymer Polyaniline/Bi_2SnTiO_7. *J. Mol.* **2012,** *17*, 2752–2772.

Yeber, M. C.; Rodriguez, J.; Freer, J.; Duran, N.; Mansilla, H. D. Photocatalytic Degradation of Cellulose Bleaching Effluent by Supported TiO_2 and ZnO. *Chemosphere* **2000,** *41*, 1193–1197.

Yoong, L. S.; Chong, F. K.; Dutta, B. K. Development of Copper-Doped TiO_2 Photocatalyst for Hydrogen Production Under Visible Light. *Energy*, **2009,** *34* (10), 1652–1661.

Yu, B.; Zeng, J.; Gong, L.; Zhang, M.; Zhang, L.; Chen, X. Investigation of the Photocatalytic Degradation of Organochlorine Pesticides on a Nano-TiO_2 Coated Film. *Talanta* **2007,** *72*, 1667–1674.

Yuan, Z. H.; Jia, J. H.; Zhang, L. D. Influence of Co-Doping of Zn $(II)^+$ Fe (III) on the Photocatalytic Activity of TiO_2 for Phenol Degradation. *Mater. Chem. Phys.* **2002,** *73*, 323–326.

Zaleska, A.; Hupka, J.; Wiergowski, M.; Biziuk, M. Photocatalytic Degradation of Lindane, p, p′-DDT and Methoxychlor in an Aqueous Environment. *J. Photochem. Photobiol.* **2000,** *135*, 213–220.

Zan, L.; Tian, L.; Liu, Z.; Peng, Z. A New Polystyrene–TiO_2 Nanocomposite Film and Its Photocatalytic Degradation. *Appl. Catal.* **2004,** *264*, 237–242.

Zhang, P.; Liang, F.; Yu, G.; Chen, Q.; Zhu, W. A Comparative Study on Decomposition of Gaseous Toluene by O_3/UV, TiO_2/UV and O_3/TiO_2/UV. *J. Photochem. Photobiol.* **2003,** *156*, 189–194.

Zhang, R.; Wang, J.; Choi, J.; Hu, L.; Mu, K. Photocatalytic Degradation of Pesticide Residues with RE^{3+}-Doped Nano-TiO_2. *J. Rare Earths* **2010,** *28*, 353–356.

Zhang, T.; Low, J.; Huang, X.; Al-Sharab, J. F.; Yu, J.; Asefa, T. Copper-Decorated Microsized Nanoporous Titanium Dioxide Photocatalysts for Carbon Dioxide Reduction by Water. *ChemCatChem*, **2017,** *9* (15), 3054–3062.

Zhang, X.; Ke, X.; Zhu, H.; Zeolite-Supported Gold Nanoparticles for Selective Photooxidation of Aromatic Alcohols Under Visible-Light Irradiation. *Chem. Eur. J.* **2012,** *18*, 8048–8056.

Zhang, Y.; Wong, J. W.; Liu, P.; Yuan, M. Heterogeneous Photocatalytic Degradation of Phenanthrene in Surfactant Solution Containing TiO_2 Particles. *J. Hazard. Mater.* **2011,** *191*, 136–143.

CHAPTER 9

Photo-Fenton Reactions: A Green Chemical Route

MEGHAVI GUPTA[1], NOOPUR AMETA[2], SURBHI BENJAMIN[3], and P. B. PUNJABI[2]

[1]*Department of Chemistry, PAHER University, Udaipur, India*

[2]*Department of Chemistry, M. L. Sukhadia University, Udaipur, India*

[3]*Department of Chemistry, J. R. N. Rajasthan Vidhyapeeth, (Deemed to be University), Udaipur, India*

ABSTRACT

Organic compounds are easily oxidized by Fenton's reagent, but this method has certain demerits that sludge of ferric ions is formed, which is to be separated from time to time and fresh amount of ferrous salt is to be added. This can be improved upon by exposing the reaction mixture with UV or visible light; thus, converting this reaction to photo-Fenton reaction. It is a cyclic process, where ferrous ions are regenerated removing both these limits of Fenton's reaction. As it generated hydroxyl radicals, it is also an advanced oxidation process. It has been used to degrade various organic contaminants such as hydrocarbons, carboxylic acids, phenols, nitro compounds, chloro compounds, pesticides, dyes, drugs, etc. This green chemical route has been summarized with different organic pollutants.

9.1 INTRODUCTION

There is practically no human activity that does not produce waste products and in addition, there is a direct relationship between the standard of living

Green Chemistry, 2nd Edition: Fundamentals and Applications. Suresh C. Ameta & Rakshit Ameta (Eds.)

in a society or country and the amount of waste products produced. These days, the criteria of determining the status of a country are well defined as developed, developing, and under developed based on the production of waste products by that particular country. Until recently, discharging waste in the environment was one of the major ways of eliminating them, until the auto purifying capacity of the environment becomes insufficient. The main problem stems from the waste coming from industry and agriculture, despite the fact that the population also plays an important role in environmental contamination, through anthropological activities.

Numerous physical, chemical, and biological methods have been investigated to control organic contaminants (Gupta et al., 2009). Conventional processes used to treat wastewater from textile industry includes chemical precipitation with alum or ferrous sulfate, which suffers from drawbacks such as generation of a large volume of sludge leading to the disposal problem, the contamination of chemical substances in the treated wastewater, etc. Moreover, these processes are inefficient in completely oxidizing organic compounds of complex structure. To overcome these problems, advanced oxidation processes (AOPs) have been developed to generate hydroxyl free radicals by different techniques.

Nowadays, scientists are very much concerned with AOPs (Legrini et al., 1993) based on the intermediacy of hydroxyl and other radicals to oxidize recalcitrant, toxic, and nonbiodegradable compounds to various by products and eventually to inert less harmful or almost harmful end products. Hydrogen peroxide is increasingly favored as an environmentally acceptable bleaching agent in both; domestic and industrial situations (Gould et al., 2001; Costa et al., 2004). Common AOPs involve photocatalysis, Fenton processes, photo-Fenton process, ozonation, photochemical, and electrochemical oxidation.

Among all the AOPs, the photo-Fenton oxidation has emerged as a very promising technology because of its high efficiency and cost-effectiveness. Photo-Fenton reaction is one of the very common AOPs used frequently for the degradation of organic pollutants. It is low cost, environment friendly, less time consuming, and easy to handle.

Phenols, pesticides, fertilizers, detergents, dyes, chlororganic, and many other chemical products are disposed directly into the environment, without being properly treated (discharging controlled or uncontrolled) into nearby water resources. Aromatic nitro compounds are commonly used in industrial processes (manufacture of pesticides, dyes, and explosives) and as a consequence, they appear as contaminants in every kind of water (especially in surface waters) and industrial waste waters. These substances produce a high toxicity, provoking serious health problems like blood infections, eye and skin irritation, affecting the central nervous system and sometimes, these are carcinogenic also. Several studies have shown the presence of these substances in surface waters (Howard, 1991) and ground waters (Duguet et al., 1990). Textile mills are major consumers of water and consequently, they are one of the largest groups of industries causing intense water pollution.

The extensive use of chemicals and water resulted in generation of large quantities of highly polluted waste water. The dye-containing colored water is almost of no use, but if this colored solution is bleached to give colorless water, it may be used for washing, cooling, irrigation, and cleaning purposes. Thus, photochemical bleaching may provide a low-cost method to solve the problem of water pollution to some extent.

Glaze et al. (1987) defined AOPs as water treatment processes at ambient temperature and pressure, which involve the generation of highly reactive radicals (especially hydroxyl radicals) in sufficient quantity to decontaminate polluted water. These treatment processes are considered as very promising methods for the remediation of contaminated ground, surface, and waste waters containing nonbiodegradable organic pollutants. Hydroxyl radicals are extraordinarily reactive species that attack most of the organic molecules. The kinetics of reaction is generally first order with respect to the concentration of hydroxyl radicals and to the concentration of the species to be oxidized. Hydroxyl radicals are also characterized by a little selectivity of attack, an attractive feature of an oxidant to be used in waste water treatment. Several organic compounds are susceptible to degradation by means of hydroxyl radicals. The attack by hydroxyl radical, in the presence of oxygen, initiates a complex cascade of oxidative reactions leading to mineralization.

Fenton's reagent is one of the most powerful inorganic oxidizing agents. Fenton (1894) discovered that the addition of ferrous salt to hydrogen peroxide initiated the rapid decomposition of α-hydroxy acids, such as tartaric acid and α-glycols, but this reagent did not receive wide attention for a long period. Forty years later, the mechanism was postulated, which revealed that the effective oxidative agent in the Fenton reaction was the hydroxyl radical (Haber and Weirs, 1934). The Fenton reactions can be outlined as

$$M^{n+} + H_2O_2 \rightarrow M^{(n+1)} + OH^- + {}^{\bullet}OH$$

where M is a transition metal such as Fe or Cu.

9.2 THE FENTON'S REAGENT

The well-known reactions of hydrogen peroxide with Fe^{2+} generates $^{\bullet}OH$ radicals. The ferrous ion reacts with hydrogen peroxide (Fenton's reagent) stoichiometrically to give $^{\bullet}OH$.

$$Fe^{2+} + H_2O_2 \rightarrow Fe^{3+} + OH^- + {}^{\bullet}OH$$

Reactions that generate hydroxyl radical in solution at low temperature have attracted interest for destruction of toxic organic compounds in waste waters.

De Laat and Gallard (1999, 2000) reported that in the absence of light and complexing ligands other than water, the most accepted mechanism of H_2O_2 decomposition in acidic homogeneous aqueous solution involves the formation of hydroxyperoxyl or superoxide ion radicals ($HO_2^{\bullet}/O_2^{\bullet-}$) and hydroxyl ($^{\bullet}OH$) radicals. The OH radical in solution attacks almost every organic contaminant and the regeneration of metal ion can follow different paths.

It was observed that the Fenton reaction rates were increased by irradiation with UV/Visible light (Ruppert et al., 1993; Pignatello and Sun, 1993). During the Fenton reaction, the Fe^{3+} ions are accumulated in the system and after all the Fe^{2+} ions are consumed, the reaction practically stops. Faust and Hoigne (1990) proposed the photochemical regeneration of Fe^{2+} ions by photoreduction of Fe^{3+} ions. The newly generated ferrous ions react with H_2O_2 generating a second $^{\bullet}OH$ radical and ferric ion and this cycle continues.

$$Fe^{3+} + H_2O + h\nu \rightarrow Fe^{2+} + {}^{\bullet}OH + H^+$$

Fenton and photo-Fenton reactions depend not only on H_2O_2 concentration and iron added, but also on the operating pH value. The stoichiometric coefficient for the Fenton reaction was approximately 0.5 mol of organic compound/mol of H_2O_2. The process was found to eliminate the toxic substances and to increase the biodegradability of the treated water. Some work about the treatment of textile water by means of Fenton and photo-Fenton process has been carried out and most of them showed their effectiveness for color removal and COD reduction (Balanosky et al., 1999; Kang et al., 2000; Perez et al., 2002a).

Zepp et al. (1992) studied the reaction between Fe (II) and H_2O_2 in the wider pH range (3–8) for hydroxyl radical formation. Recently, two new electrochemical procedures for the detoxification of acidic waste waters (the so called electro-Fenton and photoelectro-Fenton processes, where H_2O_2 is electrogenerated) have been developed and shown their good efficiencies for the mineralization of aniline and 4-chlorophenol (Brillas et al., 1998a, 1998b).

9.3 PHOTO-FENTON REACTION

About two decades ago, it was found that the irradiation of Fenton reaction systems with UV/visible light strongly accelerated the rate of degradation of a variety of pollutants (Huston and Pignatello, 1999). This rate enhancement of Fenton oxidation has been named as photo-Fenton oxidation. Nowadays, the photo-Fenton oxidation has emerged as a very promising technology because of its high efficiency and cost-effectiveness compared with other AOPs.

This behavior upon irradiation is principally due to the photochemical reduction of Fe(III) to Fe(II) in aqueous medium. Secondly, the Fe(II) so formed is then oxidized to Fe(III) on reaction with H_2O_2 as in dark Fenton process. Thus, process becomes cyclic generating 2$^{\bullet}$OH radicals per Fe-atom utilized. In addition, in the presence of light, some other reactions that produce hydroxyl radical or increase the production rate of hydroxyl radical can also occur.

$$Fe^{3+} + H_2O \xrightarrow{\text{Visible light}} Fe^{2+} + {}^{\bullet}OH + H^+$$

$$Fe^{2+} + H_2O_2 \longrightarrow Fe^{3+} + {}^{\bullet}OH + {}^{-}OH$$

$$H_2O_2 + h\nu \longrightarrow 2{}^{\bullet}OH$$

$$Fe(OH)^{2+} + h\nu \longrightarrow Fe^{2+} + {}^{\bullet}OH$$

$$Fe^{3+} + H_2O_2 \xrightarrow{\text{Visible light}} Fe^{2+} + HO_2^{\bullet} + H^+$$

$${}^{\bullet}OH + H_2O_2 \longrightarrow HO_2^{\bullet} + H_2O$$

$$Fe^{2+} + {}^{\bullet}OH \longrightarrow Fe^{3+} + {}^{-}OH$$

$$Fe^{3+} + HO_2^{\bullet} \longrightarrow Fe^{2+} + O_2 + H^+$$

$$Dye + {}^{\bullet}OH \longrightarrow \text{Smaller products}$$

Such systems are often called the photo-Fenton systems.

The rate of reaction in the photo-Fenton process is faster than the conventional thermal Fenton process. On the basis of these reactions, it can be explained that rate of Fenton reaction is increased by irradiation because the ferrous ion required in Fenton reactions are produced during the process, which reacts with hydrogen peroxide in excess to produce more and more hydroxyl radicals.

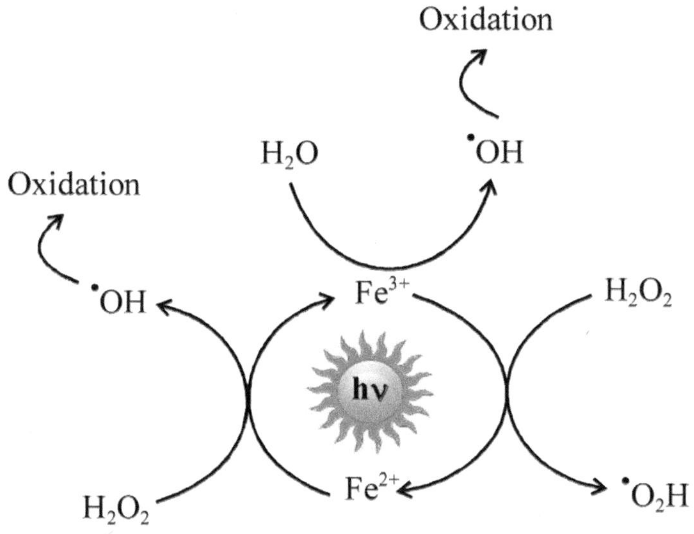

FIGURE 9.1 Schematic diagram of photo-Fenton process

Thus, photo-Fenton processes are better than dark Fenton processes, because photo-Fenton process consumes less H_2O_2 and requires only catalytic amounts of Fe(II). Any residual hydrogen peroxide that is not consumed in the process will spontaneously decompose into water and molecular oxygen and is thus, a "clean" reagent in itself. These features make photo-Fenton-based AOP, a leading candidate for cost efficient, environmental friendly treatment of industrial effluents on a small to moderate scale (Pignatello et al., 2006).

An early example of an industrial-scale application of the photo-Fenton process was the decontamination of 500 L batches of an industrial effluent containing 2,4-dimethylaniline in a photochemical reactor fitted with a 10 kW medium-pressure mercury lamp (Oliveros et al., 1997).

Photo-Fenton reactions are cyclic in nature and therefore, addition of H_2O_2 only will keep this process continuous to generate $^{\bullet}OH$ radicals while in Fenton process, reaction stops after all the Fe^{2+} ions are consumed.

In aqueous solution, Fe^{2+} and Fe^{3+} states are commonly present in the form of octahedral complexes. In photo-Fenton process, the ferric iron is most critical iron species as its hydroxide precipitates at lower pH than those of ferrous iron. The main compounds, absorbing light in the Fenton system, are ferric ion complexes, for example, $[Fe(OH)]^{2+}$ and $[Fe(RCO_2)]^{2+}$, which are highly photoreactive at pH below 3. These complexes give ferrous ions by undergoing ligand-to-metal charge transfer excitation as

$$Fe^{3+}(L^-)_n \longrightarrow Fe^{2+}(L)_{n-1} + Lox^{\bullet}$$

Moreover, the oxidation of ligand may lead to further degradation of target compounds. Ferric complexes also undergo photoreduction. These complexes have different absorption properties and thus, photoreduction may takes place at different wavelengths and with different quantum yields. These complexes of ferric ions are commonly present with acids because acids are general by products generated during the mineralization process.

Depending upon the phase, the photo-Fenton reaction may be carried out under homogeneous or heterogeneous conditions. Under homogeneous conditions, the catalyst Fe^{2+}/H_2O_2 remains soluble in aqueous acidic medium, whereas under heterogeneous conditions, the metal ions like Fe^{2+}, Fe^{3+}, Cu^{2+}, etc. gets anchored on some low-cost carriers like bentonite, zeolite, Fe_2O_3, silica, etc.

9.4 PHOTODEGRADATION OF SOME ORGANIC COMPOUNDS

9.4.1 HYDROCARBONS

The kinetics and mechanism of degradation of benzene derivatives with Fenton's reagent in aqueous medium were studied by Augusti et al. (1998). The active role of some transient intermediate species during the Fenton's mediated degradation of quinoline in oxidative media was suggested by Nedoloujko and Kiwi (1997). The composition of the products is considerably dependent on the presence of dioxygen, and pH (Loef and Stein, 1963). In the presence of dioxygen, the main product formed in a neutral medium is 2, 4-hexadiendial, while in an acidic medium, it does not arise at all. According to them, in the absence of dioxygen, the predominant product should be phenolic compounds. Dillert et al. (1996) reported the influence of H_2O_2 on photocatalytic degradation of trinitrotoluene and trinitrobenzene. Kuznetsova et al. (1996) studied the catalytic properties of heteropoly complexes containing Fe(III) ion in oxidation of benzene by H_2O_2.

The mechanism of reaction between benzene and H_2O_2 in the presence of dioxygen with added Fe^{3+} or Cu^{2+} ions was investigated by Jacob et al. (1977). Three reaction products; phenol, 2-hydroxy-2,4-hexadiendial, and 3-hydroxyl-2,4-hexadiendial have been obtained. Their ratio depends on the intensity of photolytic radiations.

9.4.2 CARBOXYLIC ACIDS

The hydroxyl radicals are produced in photo-Fenton reaction and this radical attacks the aromatic ring of carboxylic acids resulting in the formation of hydroxyl derivatives. The subsequent process involves further oxidation accompanied by opening of the aromatic ring and the formation of aliphatic carboxylic acids. The formation of phenol and benzene is probably a consequence of splitting of hydrogen from the carboxylic group, which is followed by decarboxylation.

The degradation of gallic acid (3, 4, 5-trihydroxy benzoic acid) in aqueous solution by UV/H_2O_2 treatment via Fenton's reagent and the photo-Fenton system was reported by Benitez et al. (2005). Nogueira et al. (2002) carried out the solar photodegradation of dichloroacetic acid and 2,4-dichlorophenol (DCP) using an enhanced photo-Fenton process. The photo-assisted Fenton degradation of salicylic acid, using strongly acidic ion exchange resin (SAIER) exchanged with Fe ions as catalyst, in the presence of UV light and H_2O_2 was studied by Feng et al. (2004a). Quici et al. (2005) studied the destruction of oxalic acid by combined heterogeneous photocatalysis and photo-Fenton reaction using UV/Fe/H_2O_2 and the UV/TiO_2/Fe/H_2O_2.

The inhibiting and chemiluminescent properties of benzoic acid and acetyl salicylic acid in the presence of Fenton reagent was observed by Zakhrov and Kumpan (1996). The molecular and structural characteristic of humic acid during photo-Fenton processes were studied by Fukushima et al. (2001). Varghese et al. (2007) studied the degradation of cyanuric acid with a combination of gamma radiolysis and Fenton reaction.

Acero et al. (2001) reported studies on the degradation of p-hydroxyphenylacetic acid by photo-assisted Fenton reaction. The pronounced photocatalytic efferts of $FeCl_3$ and Na_2 [Fe $(CN)_5$ NO] on the hydroxylation of some carboxylic acids initiated by H_2O_2 /UV radiation was reported by Sedlak et al. (1989). It was concluded that the reaction proceeds via several parallel mechanisms.

9.4.3 PHENOLS AND ITS DERIVATIVES

Martinez et al. (2007) reported the incorporation of iron species over different silica supports for the heterogeneous photo-Fenton oxidation of phenol whereas Sykora et al. (1997) have studied the influence of metal ions including Fenton and photo-Fenton reactions on homogeneous photoxidation of phenol. Fenton and photo-Fenton reactions were used for phenol removal from high salinity effluent by Maciel et al. (2004). Martínez et al. (2005) studied the heterogeneous photo-Fenton degradation of phenolic aqueous solution over iron-containing SBA—15 catalysts.

Kušić et al. (2006) investigated the application of AOPs, dark Fenton, and photo-assisted Fenton type processes; Fe^{2+}/ H_2O_2, Fe^{3+}/H_2O_2, UV/ Fe^{2+}/ H_2O_2, UV/Fe^{3+}/H_2O_2, and UV/ Fe^0/H_2O_2, for the degradation of phenol present in waste water. A mechanism for the formation of phenoxy radicals during photo-oxidation of phenol in the presence of Fe^{3+} was suggested by Nadtochenko and Kiwi (1998).

The investigation of the mechanism of phenol decomposition of Fe-C–TiO_2 and Fe–TiO_2 photocatalysts via photo-Fenton process was carried out by Tryba et al. (2006a) while carbon coating of Fe-C–TiO_2 photocatalyst was used on phenol decomposition under UV irradiation via photo-Fenton process (Tryba et al., 2006b). Li et al. (2007) and Katsumata et al. (2004) investigated the photo-Fenton photodegradation of bisphenol A in aqueous solution with iron oxides. Du et al. (2006) investigated the role of intermediate in the degradation of 4-chlorophenol, 4-nitrophenol, and phenol by Fenton process.

Benitez et al. (2005) reported the degradation of gallic acid in aqueous solution by UV/H_2O_2 treatment, Fenton's reagent, and photo-Fenton system. The kinetic modeling and reaction pathway of 2, 4-DCP transformation by photo-Fenton oxidation has been reported by Chu et al. (2005), while Kavitha and Palanivelu (2003) made a comparative study of the degradation of 2- chlorophenol by Fenton and photo-Fenton processes.

Feng and Le-Cheng (2004) reported the degradation, kinetics, and mechanisms of phenol in photo-Fenton process while Arana et al. (2001) carried out treatment of the highly concentrated phenolic wastewater by the photo-Fenton reaction and reported the mechanism of reaction. The photocatalytic degradation of resorcinol over titanium dioxide using photo-Fenton's related reagents was investigated by Ameta et al. (2006). The role of ferrous ion in Fenton and photo-Fenton processes for the degradation of phenol was suggested by Kavitha and Palanivelu (2004). Ghaly et al. (2001)

reported studies on the photochemical oxidation of p-chlorophenol by UV/H_2O_2 and photo-Fenton process.

Phenol degradation in water through a heterogeneous photo-Fenton process catalyzed by Fe-treated laponite was studied by Iurascu et al. (2009). Martinez et al. (2005) studied heterogeneous photo-Fenton degradation of phenolic aqueous solutions over iron-containing SBA-15 catalyst. Parida and Pradhan (2010) reported that Fe/meso-Al_2O_3 is an efficient photo-Fenton catalyst for the adsorptive degradation of phenol.

9.4.4 HALO COMPOUNDS

Kwon et al. (1999) and Du et al. (2007) investigated the role of oxygen in the degradation pathway of 4-chlorophenol by Fenton system. An attempt was made by Saritha et al. (2007) to degrade 4-chloro-2-nitrophenol (4C-2-NP), which is widely available in bulk drug and pesticides wastes. Sabhi and Kiwi (2001) reported the degradation of 2,4-DCP on Nafion-Fe (1.78%) under visible light irradiation in the presence of H_2O_2.

The degradation of a phototypical halogenated aromatic pollutant, 4-chloroaniline, which was photoinduced by Fe(III) species in acidic aqueous solutions (pH 2–4) of $Fe(ClO_4)_3$, was reported by Mailhot et al. (2004). Yeh et al. (2002) investigated the role of soil organic matter during the oxidation of chlorophenols with Fe^{2+}-catalyzed H_2O_2 (Fenton oxidation) system. Homogeneous degradation of 1,2,9,10-tetrachlorodecane in aqueous solution using hydrogen peroxide, iron, and UV light was observed by El-Morsi et al. (2002).

9.4.5 NITRO COMPOUNDS

Kiwi et al. (1997) reported the combined Fenton and biological flow reactor degradation of p-nitrotoluene–ortho-sulphonic acid (p-NTS). Fenton reagent, UV/H_2O_2, and UV/Fenton's regents were used to mineralize dinitrotoluenes and trinitrotoluene of spent acid in toluene nitration process. Sun et al. (2007) reported the kinetic study of the degradation of p-nitroaniline by Fenton oxidation process. Fenton regent has also been used in the oxidation of nitroaromatic explosives, namely, 2,4,6-trinitrophenol ammonium picrate, 2,4-dinitrotoluenes, 2,4,6-trinitrotoluene (TNT) and hexahydro-1,3-5-trinitro-1,3,5-triazine (RDX). Liou et al. (2004) performed a series of photo-Fenton reactions for the degradation of 2, 4, 6-TNT. Optimization of the solar photo-Fenton process in the treatment of contaminated water has been reported by Rodriguez et al. (2005). Ahmadimoghaddam et al. (2010) observed degradation of 2,4-dinitrophenol by photo-Fenton process. A comparative study of hydrogen peroxide photolysis, Fenton reagent, and photo-Fenton for the degradation of nitrophenols has been carried out by Goi and Trapido (2002). Aggarwal et al. (2017) used TiO_2-coated clay beads and cement in photodegradation of 2-chloro-4-nitrophenol. It was claimed that 97.19% degradation of 2-chloro-4-nitrophenol could be achieved in 105 min using solar photo-Fenton process.

9.4.6 PESTICIDES

The photo-Fenton process can potentially be integrated into waste water treatment process to enhance the organic compound removal. It can operate at low concentrations of contaminant and can mineralize the compound or convert it into less toxic form. The efficiency of the process is found to be maximum at pH 2.8; however, it has been found that with addition of suitable complexing agent for Fe^{3+}, the process can be operated close to neutral pH. In this study, citric acid was used as a complexing agent and 2, 4-DCP as model contaminant, where pH (5–8.89) was observed to be the feasible pH range. Concludingly, it can be said that the photo-Fenton process may be used practically as treatment option for waters contaminated with pesticides and other organic compounds that are poorly biodegradable. The influence of pH on the degradation of the heribicide tebuthuron using in situ generated Fe (III)–citrate complexes by photo-Fenton process under solar irradiation was investigated by Silva et al. (2007).

The photocatalytic removal of fenitrothion in pure and natural waters by photo-Fenton reaction was reported by Derbalah et al. (2004). The photo-Fenton degradation of the heribicide tebuthiuron, diuron, and 2,4-D of heribicides in aqueous solution using ferrioxalate complex (FeO_x) as a source of Fe^{2+} under irradiation was reported by Paterlini and Nogueira (2005). Maldonado et al. (2007) investigated the photocatalytic degradation of pesticides using TiO_2 and Fenton as well as photo-Fenton processes. Gromboni et al. (2007) observed the microwave-assisted photo-Fenton decomposition of chlorfenvinphos and cypermethrin in residual water, while waste water containing five common pesticides methomyl, dimethoate, oxamyl, cymoxanil, and pyrimethanil has been mineralized by solar AOPs—biological coupled system (Oller et al. 2007).

The photo-Fenton process was successfully applied to a mixture of ten commercial available pesticides that served as a model for a proposed recycling plant for pesticide bottles. A solution of mixed pesticides (alachor, atrazine, chlorfenvinphos, diuron, and isoproturon) was considered for degradation using photo-Fenton as a preliminary step before biotreatment (Lapertot et al., 2006; 2007; Malato et al., 2002). Badawy et al. (2006) used the combination of the Fenton reaction, UV/H_2O_2, and the photo-Fenton process in the degradation of organophosphorous-containing substrates such as fenitrothion, diazinon, and profenofos. The photocatalytic degradation of two selected insecticides (dimethoate and methyl parathion) using photo-Fenton reaction was also studied by Evgenidou et al. (2007).

Fallman et al. (1999) showed the applicability of the photo-Fenton method for treating water containing pesticides. Coupling of photo-Fenton and biological treatment for the removal of diuron and linuron from water was done by Farre et al. (2006). Identification of the intermediates generated during the degradation of diuron and linuron herbicides by the photo-Fenton reaction was also carried out done by Farre et al. (2007). Flox et al. (2007) used photoelectron-Fenton with UV-4 and solar light for mineralization of herbicide mecoprop.

The photo-Fenton and biological integrated process for degradation of a mixture of pesticides has been reported by Lapertot et al. (2007). Katsumata et al. (2005) observed the degradation of linuron in aqueous solution by the photo-Fenton reaction while kinetics and products of photo-Fenton degradation of trizophos was studied by Lin et al. (2004). A comparative investigation about the oxidation of the herbicide, 2, 4-dichlorophenoxyacetic acid (2,4-D) by $Fe^{2+}/H_2O_2/$ UV and ferrous oxalate/ H_2O_2/UV processes was carried out by Kwan and Chu (2004).

Katsumata et al. (2006) carried out the photodegradation of alachor (which is one of the acetanilide herbicides) in the presence of the Fenton reagent and citrate. The degradation of fungicide carbendazim via photo-Fenton reactions has been studied by Costa et al. (2019) in photoreactors, which was solar irradiated. It was reported that 96% removal of carbendazim could be achieved via this process.

Abdelhaleem and Chu (2020) developed a photo-Fenton like reagent (FeIII impregnated N-doped TiO_2 (FeNT)/H_2O_2) and used it for degradation of carbofuran, a frequently used pesticide. The photo-Fenton degradation of organophosphate pesticide diazinon (DZN) was studied by Zekkaoui et al. (2021). The highest yield of degradation (96%) was obtained using photo-Fenton process.

9.4.7 DYES

Carneiro et al. (2007) investigated the homogeneous photodegradation of reactive blue-4 using a photo-Fenton process under artificial and solar irradiation. Use of photo-Fenton's reagent for the photochemical bleaching of metanil yellow has been reported by Kumar et al. (2008a). He et al. (2002) investigated the photo-Fenton degradation of an azo dye at neutral pH. Muruganandham and Swaminathan (2004) reported the decolorization of reactive orange-4 by Fenton and photo-Fenton oxidation technology. Decolorization of azo dye reactive black 5 by Fenton and Photo-Fenton oxidation has been done by Lucas and Peres (2006). Degradation and sludge production of textile dyes by Fenton and photo-Fenton processes has been studied by Liu et al. (2007).

Decolorization and mineralization of acid yellow 23 by Fenton and photo-Fenton processes has been reported by Modirshala et al. (2007) while the degradation of naphthol green B by photo-Fenton reagent was observed by Kumar et al. (2008b). A study of kinetic parameters related to the decolorization and mineralization of reactive dyes from textile dyeing industries using Fenton and photo-Fenton processes has been made by Nunez et al. (2007). Solar photocatalytic degradation of azo dyes by photo-Fenton process was carried out by Chacon et al. (2006). Wu et al. (1999) reported the photodegradation of malachite green in the presence of Fe^{2+}/H_2O_2 under visible irradiation. Selvam et al. (2005) studied the combined homogeneous and heterogeneous photocatalytic decolorization and degradation of a chlorotriazine reactive dye, reactive orange 4 using ferrous sulfate/ferrioxalate with H_2O_2 and TiO_2-P25 particles.

Nitampegliotis et al. (2006) studied the decolorization kinetics of procion dye from textile dyeing industry using photo-Fenton reaction. The influence of alizarin violet-3B dye on the Fenton reaction of organic compounds under visible irradiation was examined by Ma et al. (2005). The degradation of acid orange-7 dye by three different photochemical processes; photoperoxidation, Fenton, and photo-Fenton was observed by Scheeren et al. (2002).

Mineralization of C.I. acid red 14 azo dye by UV/Fe-ZSM 5/ H_2O_2 process has been investigated by Kasiri et al. (2010). Decolorization of reactive dyes by modified photo-Fenton process under irradiation with sunlight has been studied by Chaudhuri and Wei (2009). Bacardit et al. (2007) observed effect of salinity on the photo-Fenton process. Degradation of organic pollutants by the photo-Fenton process has been investigated by Kim and Vogelpohl (1998). The photo-Fenton reaction and the TiO_2/UV process for waste water treatment have been suggested by Bauer et al. (1999).

Factors affecting the kinetic parameters related to the degradation of direct yellow 50 by Fenton and photo-Fenton processes has been observed by Mahmoud and Ismail (2011). Performance of the photo-Fenton process in the degradation of a model azo dye mixture has been studied by Macias-Sanchez et al. (2011). Heterogeneous catalytic treatment of synthetic dyes in aqueous media using Fenton and photo-assisted Fenton process has been studied by Soon and Hameed (2011). Heterogeneous photo-Fenton oxidation of reactive azo dye solutions using iron-exchanged zeolite as a catalyst was observed by Tekbas et al. (2011).

Heterogeneous photo-Fenton degradation of reactive brilliant orange X-GN over iron-pillared montmorillonite under visible irradiation was observed by Chen et al. (2009). Yang et al. (2009) observed degradation of methylene blue by heterogeneous Fenton reaction using titanomagnetite at neutral pH values. A study of catalytic behavior of aromatic additives on the photo-Fenton degradation of phenol red was examined by Jain et al. (2009). Fenton- and photo-Fenton-like degradation of a textile dye by heterogeneous processes with Fe/ZSM-5 zeolite has been reported by Duarte and Madeira (2010).

Rasoulifard et al. (2011) suggested photoassisted hetero-Fenton decolorization of azo dye from contaminated water by Fe–Si mixed oxide nanocomposite. Kumar et al. (2011) carried out comparative studies of degradation of dye intermediate (H-acid) using TiO_2/UV/H_2O_2 and photo-Fenton process. Zhang et al. (2011a) suggested application of heterogenous catalyst of tris (1, 10)-phenanthroline iron (II) loaded on zeolite for the photo-Fenton degradation of methylene blue. Rapid decolorization of rhodamine-B by UV/Fe (III)-penicillamine process under neutral pH has been studied by Xue et al. (2011), while Saatci (2010) reported decolorization and mineralization of remazol red F3B by Fenton and photo-Fenton processes.

Fe(III) supported ceria as an effective catalyst for the heterogeneous photo-oxidation of basic orange 2 in aqueous solution under sunlight has been reported by Martinez et al. (2011). Photoassisted degradation of azo dyes over FeO_xH_{2x-3}/Fe^0 in the presence of H_2O_2 at neutral pH values has been investigated by Nie et al. (2007). Fernandez et al. (1999) observed the photo-assisted Fenton degradation of nonbiodegradable azo dye (orange II) in iron-free solution, mediated by cation transfer membranes. The combined homogeneous and heterogeneous photocatalytic decolorization and degradation of a chlorotriazine reactive azo dye, reactive orange 4 using ferrous sulfate/ferrioxalate with H_2O_2 and TiO_2-P25 particles was studied by Swaminathan et al. (2006).

The photochemical degradation of azo dyes, namely red MX-5B, reactive black-5, and orange-G using low iron concentrations in Fenton and Fenton-like systems was studied by Hsueh et al. (2005). The decolorization of solution containing a common textile and leather dye, acid red 14, at pH 3 using Fenton, UV/ H_2O_2/O_2, UV/H_2O_2/Fe^{2+}, UV/H_2O_2/Fe^{3+}, and UV/H_2O_2/Fe^{3+}/oxalate processes was carried out by Daneshvar and Khataee (2006). Different workers have investigated the use of photo-Fenton process for color removal from textile waste waters of textile industries (Lloyd et al., 1997; Kang et al., 2002; Kwan and Chu, 2003; Zheng et al., 2004). Laponite and bentonite clay-based Fe nanocomposite have been developed as suspended photo-Fenton catalysts for the degradation of organic dyes (Serp et al., 2003).

Amorim et al. (2013) used blast furnace dust as a catalyst to degrade reactive red 195 (RR195), an azo dye by photo-Fenton-like process. This dust contains iron sources hematite, magnetite and maghemite. It was observed that addition of catalyst considerably increased the reaction rates by more than five times.

Hernández-Rodríguez et al. (2014) investigated mineralization as well as decolorization of wastewaters from wool dyeing industries using photo-Fenton process. They selected red, yellow, and blue dyebaths with anthraquinone dyes and additives as colored effluents. It was revealed that 100% of color removal could be obtained under optimal conditions.

Two composite photocatalysts, (Fe_3O_4 or $NiFe_2O_4$) with MIL-53(Fe) were fabricated by Nguyen et al. (2018) and used as catalysts for the degrading rhodamine B under visible light. It was observed that $NiFe_2O_4$-doped MIL-53(Fe) sample exhibited a higher photocatalytic activity for degradation of this dye than that of Fe_3O_4/MIL-53(Fe) sample. A Mexican natural zeolite (MNZ) was impregnated by Domenzain-Gonzalez (2019) with Fe at different concentrations (5 and 10 mg $FeCl_3$ g^{-1} MNZ) (MNZ/Fe) for photo-Fenton degradation of reactive black 5 (RB5). The best result for this photo-Fenton degradation was achieved at 91% discoloration and 68.5% removal of chemical oxygen demand.

The degradation of Ponceau S in aqueous medium was observed by Laftani et al. (2019) using the photo-Fenton process. It was reported that more than 94.3% degradation of Ponceau S could be achieved. A comparative study was conducted by Sajjala and Sairam (2020) between TiO_2-mediated solar photo-Fenton treatment and solar photo-Fenton in decolorizing Sudan IV dye. It was reported that although TiO_2 mediated solar photo-Fenton process was better than solar photo-Fenton process due to operating pH and time of irradiation period, but it is expensive due to high cost of TiO_2.

Heterogeneous photo-Fenton degradation of procion red dye in presence of visible irradiation and bismuth ferrite catalyst was investigated by Severo et al. (2020). It was reported that $BiFeO_3$ catalyst prepared with EDTA exhibited superior catalytic activity, and it could reach 99% degradation in 2 h, as compared to $BiFeO_3$ prepared without EDTA (70% degradation).

Manganese ferrite was synthesized by Morais et al. (2020) via two routes: (i) Using reagents commercial ($MnFe_2O_4$-R) and (ii) Manganese received from cathodes of spent $ZnMnO_2$ batteries ($MnFe_2O_4$-B). These manganese ferrites were used as catalysts for the degradation (photo-Fenton) of methylene blue. The $MnFe_2O_4$-R and $MnFe_2O_4$-B achieved 98% and 92% decolorization in 2 h of reaction, respectively.

Nadeem et al. (2020) prepared heterogeneous photo-Fenton catalysts ($ZnFe_2O_4$ and graphene oxide (GO) based $ZnFe_2O_4$ composite) and used for degradation of synzol red reactive dye. It was reported that 57% and 94% dye degradation could be obtained with $ZnFe_2O_4$ and GO-$ZnFe_2O_4$, respectively, in 1 h. It was found that the composite (GO-$ZnFe_2O_4$) exhibited higher degradation of dye as compared to $ZnFe_2O_4$. Nanosheets of Fe-chitosan/montmorillonite (Fe-CS/MMTNS) were prepared by Zhao et al. (2020) and used for degradation of methylene blue under visible light in the presence of H_2O_2. Zhang et al. (2020) prepared potassium ferrite crystals with different sizes. They used it as heterogeneous Fenton catalyst to degrade methylene blue and crystal violet in the presence of H_2O_2 under visible-light irradiation. It was revealed that 100% methylene blue and 92% crystal violet were degraded within 35 min and catalyst was found to be stable after four cycles.

9.4.8 DRUGS

Experimental design of Fenton and photo-Fenton reactions for the treatment of ampicillin solutions has been observed by Rozas et al. (2010). Developing a reusable and durable catalyst for heterogeneous photo-Fenton process is of great importance for the practical application in the environment remediation.

Flumequine is an antimicrobial agent (broad-spectrum) of quinolone class. Rodrigues-Silva et al. (2013) degraded water samples containing flumequine using the Fenton and photo-Fenton processes. It was found that maximum degradation efficiency of flumequine was more than 94% in an hour. Rad et al. (2015) applied photo-Fenton process for the treatment of phenol and paracetamol in a binary system. They used cobalt ferrite nanoparticles in this photo-Fenton process. Maximum degradation efficiencies of phenol and paracetamol were found to be 95% and 85%, respectively.

The acidification of the metoprolol (MET) was avoided by using resorcinol in photo-Fenton process by Romero et al. (2016). It was claimed that complete MET degradation could be obtained in 3 min using photo-Fenton process. The photo-Fenton mineralization of metronidazole (MTZ) has been studied by Ammar et al. (2016). It was reported that solar photo-Fenton system could achieve 96% of COD removal under optimal experimental conditions in 12 min. The ornidazole (ORZ) and ofloxacin (OFX) was degraded in presence of naturally available soil (Changotra et al. 2017). Soil contains different iron oxides (hematite, goethite, magnetite, wustite, and pyrite). It was reported that about 95% and 92% of OFX and ORZ were removed, respectively.

Parabens are endocrine-disrupting chemicals. The solar photo-Fenton technology has been used by Zúñiga-Benítez et al. (2018) for removal of two parabens in aqueous solutions. Bansal et al. (2019) utilized automobile shredder residue (ASR) fly ash as a source of iron to degrade tetracycline (TC). It was observed that degradation proceeds at neutral pH through visible-light-driven photo-Fenton-like process. These beads retained activity and exhibited stability after ten recycling tests and as a result real-time applications of ASR fly ash was possible in wastewater treatment. A layered hollow $Fe_3O_4/Fe_{1-x}S@MoS_2$ composite was prepared by Li et al. (2020). The maximum adsorption capacity of removal of TC hydrochloride by $Fe_3O_4/Fe_{1-x}S$ $@MoS_2$-50% sample was found to be 748.9 mg L^{-1}.

9.4.9 OTHERS

The cyano complexes raise the quantum yields by as much as three folds of magnitude. Lei et al. (1998) investigated the oxidative degradation of polyvinyl alcohols by photo-Fenton process. The oxidative degradation of polyvinyl alcohol by the photo-Fenton process was also studied by Guardani et al. (2006).

Ormad et al. (2006) studied the treatment of winery waste waters by the photo-Fenton reaction in homogeneous phase, which are difficult to treat by conventional biological process. The oxidative degradation of poly(ethylene glycol) by Fenton and photo-Fenton reactions was investigated by Prousek and Duriskova (1998). Degradation of formaldehyde in the presence of methanol by photo-Fenton process was carried out by Kajitvichyanukul et al. (2008).

Kim et al. (1997) reported the landfill leachate treatment by a photo-assisted Fenton reaction while Gulyas (1997) carried out the removal of recalcitrant organics from industrial waste waters using AOPs. The intermediates in the Fenton-type reagents were investigated by Walling (1998). Macfaul et al. (1998) provided a radical account of oxygenated Fenton chemistry, while Gallard et al. (1998) observed the effect of pH on the oxidation rate of organic compounds by Fe^{2+}/ H_2O_2. Nogueira et al. (1999) investigated the photolytic degradation of two common water pollutants, phenol and trichloroethylene using Fenton's reagent/UV/ferrioxalate/H_2O_2/UV, and TiO_2/UV. Sarria et al. (2003) observed the effectiveness of photo-Fenton treatment of a biorecalcitrant waste water generated in textile activities and the biodegradability of the photo-treated solution.

The superior biodegradability of waste waters mediated by immobilized Fe-fabrics as compared to Fenton homogeneous reactions was reported by Bozzi et al. (2003). Safarzadeh-Amiri et al. (1997) investigated the UV–Visible photolysis of ferrioxalate in the presence of hydrogen peroxide. A comparison of different AOPs for phenol degradation was made by Esplugas et al. (2002). Parra et al. (2004) reported the synthesis, testing, and characterization of a novel Nafion membrane with superior performance in photo-assisted immobilized Fenton catalysis. The role of UV and natural sunlight in photodegradation of parathion in aqueous TiO_2 and Fe (0) solution in presence of H_2O_2 was observed by Doong and Chang (1998) whereas Luong and Lin (2000) observed the control of Fenton reactions for soil remediation. Arslon-Alaton and Gurses (2004) investigated the degradation of procaine penicillin-G formulation effluent by Fenton-like (Fe^{3+}/H_2O_2) and UV–A Light assisted Fenton-like (Fe^{3+}/H_2O_2) UV–A processes.

The photo-Fenton process in heterogeneous phase as an alternative methodology for the treatment of winery waste waters was presented by Mosteo et al. (2006). Photo-Fenton assisted ozonation of p-coumaric acid present in olive mill wastewater was investigated by Monteagudo et al. (2005). Oxidation of some explosives by Fenton and photo-Fenton processes has been reported by Liou et al. (2003). Galvao et al. (2006) reported the application of the photo-Fenton process for the treatment of wastewater contaminated with diesel. Removal of organic contaminants by Fenton and photo-Fenton processes from paper pulp effluents was reported by Perez et al. (2002b).

Tokumara et al. (2007) studied photo-Fenton process for excess sludge disintegration. Experimental design of Fenton and photo-Fenton reactions for the treatment of cellulose bleaching effluents has been reported by Torrades et al. (2003). White et al. (2003) studied the role of the photo-Fenton reaction in the production of hydroxyl radicals and photo-bleaching of colored dissolved organic matter in a coastal river of the Southeastern United States. Multivariable approach to the photo-Fenton process as applied to the degradation winery wastewater was suggested by Ormad et al. (2006).

Park et al. (2006) made a comparison of Fenton and photo-Fenton processes for livestock wastewater treatment. Degradation of polyethylene glycol in aqueous solution by photo-Fenton and H_2O_2/UV processes has been reported by Giroto et al. (2010). Mohajeri et al. (2010) studied the influence of Fenton reagent on oxidative mineralization and decolorization of municipal landfill leachate while Herney-Ramirez et al. (2010) studied heterogeneous photo-Fenton oxidation with pillared clay-based catalysts for wastewater treatment.

Fe_2O_3-pillared rectorite has been investigated by Zhang et al. (2010) as an efficient and stable Fenton-like heterogeneous catalyst for photodegradation of organic contaminants. Wastewater treatment by catalytic wet oxidation using H_2O_2 and pillared clays containing iron as heterogeneous catalyst has been widely investigated (Guelou et al., 2003; Guo and Al-Dahaan, 2003). Highly active S-modified $ZnFe_2O_4$ heterogeneous catalyst and its photo-Fenton behavior under UV–visible irradiation was studied by Liu et al. (2011). Yip et al. (2005) carried out novel heterogeneous acid-activated clay supported copper catalyzed photobleaching and degradation of textile organic pollutant using photo-Fenton-like reaction.

Elmolla and Chaudhari (2010) studied effect of photo-Fenton operating conditions on the performance of photo-Fenton-sequencing batch reactor (SBR) process for recalcitrant wastewater treatment. Fenton and photo-Fenton processes coupled to upflow anaerobic sludge blanket reactor to treat coffee pulping wastewater has been reported by Kondo et al. (2010). Zelmanov and Semiat (2008) studied iron (III) oxide-based nanoparticles as catalysts in advanced organic aqueous oxidation of pollutants. Microstructure and photo-Fenton performance of trinuclear iron cluster intercalated montmorillonite catalyst has been investigated by Zhang et al. (2011b). Ju et al. (2011) investigated sol–gel synthesis and photo-Fenton-like catalytic activity of $EuFeO_3$ nanoparticles. Landfill leachate treatment by Fenton, photo-Fenton processes and their modifications have been suggested by Krzysztoszek and Naumczyk (2012).

Synthesis, characterization, and visible light photo-Fenton catalytic activity of hydroxy Fe/Al intercalated montmorillonite has been investigated by Li et al. (2011). Vermilyea and Voelker (2009) observed photo-Fenton reaction at near neutral pH in which oxidation of photoproduced ferrous iron by hydrogen peroxide, produces reactive oxidants that may be important for degradation of biologically and chemically recalcitrant organic compounds in surface waters.

Hydroxyl radical production via the photo-Fenton reaction in the presence of fulvic acid has been studied by Southworth and Voelker (2003). Vilar et al.

(2012) suggested application of Fenton and solar photo-Fenton processes for the treatment of a sanitary landfill leachate in a pilot plant with CPCs.

Evaluation of performance of a photo-Fenton process for pollutant removal from textile effluents has been made in a batch system by Modenes et al. (2012). Lucas et al. (2012) treated pulp mill wastewater by both; Fenton (Fe^{2+}/H_2O_2) and solar photo-Fenton ($Fe^{2+}/H_2O_2/UV$) processes. It was revealed that solar photo-Fenton could reach 90% DOC degradation in less time. The treatment efficiency of aniline wastewater by biological oxidation, photo-Fenton, and combined from (photo-Fenton and biological oxidation) were evaluated by Yu et al. (2012). It was revealed that about 62.5% of H_2O_2 can be saved by combining photo-Fenton and biological oxidation processes.

Photo-Fenton oxidation and its combination with aerobic SBR were selected to degrade real textile wastewater by Blanco et al. (2014) they could achieve 97% and 95%, COD and total organic carbon (TOC) reduction, respectively. It was reported that treated water can be useful for internal reuse. Mahdad et al. (2016) applied AOPs for the treatment of composting leachate of municipal solid by an activated sludge process. About 69.64% and 84.09% COD and color removal were observed in a photo-Fenton process, respectively.

The efficiency of 2-mercaptobenzothiazole (MBT) degradation was evaluated by Redouane-Salah et al. (2018) using heterogeneous photo-Fenton process with local natural clay powder (NCP). It was observed that 42.5% and 62% of MBT was degraded in the presence of H_2O_2 and under sunlight irradiation in 180 min (using NCP), respectively. Wei et al. (2018) prepared bentonite-supported Fe(II)/phosphotungstic acid composite and used in photo-Fenton degradation of ethyl xanthate in presence of visible light. It was revealed that it had an excellent stability and it can retain catalytic activity for four cycles.

Sharma et al. (2020) prepared CuO/g-C_3N_4 (CuO/GCN) photocatalyst with H_2O_2. The catalytic process (photo-Fenton) was utilized for degradation of 2, 4-dimethylphenol (DMP). It was observed that coupling of CuO/GCN system with H_2O_2 significantly improved its activity. As-prepared 4%CuO/g-C_3N_4/H_2O_2 system proved to be the best and it exhibited much higher photoremoval efficiency (99% degradation of DMP) in 2 h than CuO/g-C_3N_4 system only.

Almeida et al. (2021) applied Fenton, photo-Fenton, UV/H_2O_2, and UV/Fe^{3+} processes for degradation of caffeine. It was observed that degradation of caffeine with artificial radiation was superior than with natural radiation.

Du et al. (2021) constructed sodium alginate (SA)/hydrogel beads/poly(vinyl alcohol) (PVA). They carried out by photo-Fenton degradation of

TC under visible-light irradiation using these PVA/SA-FeC13 beads. It was indicated that the removal rate of TC could reach 90.5%. In addition, PVA/SA-6% FeC13 beads could be recovered easily and used further in water purification.

Fenton reaction is well known for oxidation. Photo-Fenton reagent is a recent development in this field, which takes care of environment because there is no sludge formation because photo-Fenton reaction is reversible and does not require any addition of ferrous ions. Side-wise, it will generate two hydroxyl radicals without a reasonable change of pH of the medium as compared to Fenton reaction, where only one hydroxyl ion is generated along with an increase with pH.

9.5 RECENT DEVELOPMENTS

Endocrine disruptors are recalcitrant pollutants in aquatic ecosystems. Gamarra-Güere et al. (2022) reported degradation of methylparaben (MeP) via photo-Fenton process, which was completed in 16 min. under optimal conditions: Fe^{2+} (4 mg L^{-1}) and H2O2 (52 mg L^{-1})and UV lamp (4 W). Ding et al. (2022) synthesized mesoporous Fe-g-C_3N_4 via thermal shrinkage polymerization. They used ferrous oxalate as iron source as well as pore-forming agent. It was reported that as-prepared Fe-g-C_3N_4 system exhibited an excellent photo-Fenton performance for removal of TC in 2 h as well as good recyclability.

Zhang et al. (2022) fabricated iron-doped graphitic carbon nitride on surface of diatomite via a single-step polymerization (FGD-x), where x represents Fe loading content. It was observed that FGD-x efficiently degrades TC hydrochloride by photo-Fenton process. It was reported that FGD-3 ($FeCl_3$ loading 75 mg) exhibited best catalytic performance in degradation of TC with H_2O_2 (1 mmol L^{-1}) in a wide pH range from 2.0 to 7.0. The FGD-3 could degrade about 98.3% of 20.0 mg L^{-1} drug at pH 4.0 within 100 min in presence of visible light.

Bashir et al. (2022) prepared fibrous biochar zerovalent iron nanoparticles (nZVI) particles and used for photo-Fenton degradation of methylene blue and catalytic reduction of 4-nitrophenol. It was reported that as-prepared nZVI-FBC exhibited excellent stability and reusability for a number of consecutive cycles. Ortega-Moreno et al. (2022) synthesized and, activated MOF MIL-53(Fe). It was then used for photo-Fenton degradation of sulfamethoxazole (SMX) assisted by natural sunlight and UVA LED irradiation. It was reported that around 96% SMX could be degraded in 2 h under almost-neutral conditions.

Ghazzaf et al. (2022) synthesized magnetic nanoparticles from iron rust (waste). Then they used it as a catalyst for photo-Fenton degradation of acid red 97 (AR97). It was reported that as-prepared Fe_3O_4 nanoparticles (spherical-like shape) exhibited higher catalytic activity as compared to α-Fe_2O_3, octahedral-shaped Fe_3O_4, and α-FeOOH. It was reported that about 81% of TOC was removed in 3 h. Ju et al. (2022) used Z-type heterojunction MIL-101(Fe)/g−C_3N_4 as photocatalysts, which could achieve an in situ H_2O_2 production rate of 4370 μmol h^{-1}, which was 5 times higher than the diphase control. It also degrades methyl orange (99%) in 130 min, while degradation in presence of in diphase control is much less (21%).

KEYWORDS

- **Fenton reaction**
- **photo-Fenton reaction**
- **Dyes**
- **halo compounds**
- **nitro compounds**
- **pesticides**

REFERENCES

Abdelhaleem, A.; Chu, W. Prediction of Carbofuran Degradation Based on the Hydroxyl Radical's Generation Using the FeIII Impregnated N Doped-TiO_2/H_2O_2/Visible LED Photo-Fenton-Like Process. *Chem. Eng. J.* **2020,** *382*. DOI: 10.1016/j.cej.2019.122930.

Acero, J. L.; Benitez, F. J.; Real, F. J.; Leal, A. I. Degradation of p-Hydroxyphenylacetic Acid by Photoassisted Fenton Reaction. *Water Sci. Technol.* **2001,** *44* (5), 31–38.

Aggarwal, K.; Verma, A.; Bansal, P.; Singh, A. Heterogeneous Photo-Fenton and Photocatalytic Degradation Studies of 2-Chloro-4-Nitrophenol (2Cl4NP) Using Foundry Sand and TiO2 Coated Cement/Clay Beads. *Braz. J. Chem. Eng.* **2017,** *34*, 747–758.

Ahmadimoghaddam, M.; Mesdaghinia, A.; Naddafi, K.; Nasseri, S.; Mahvi, A. H.; Vaezi, F. et al. Degradation of 2,4-Dinitrophenol by Photo Fenton Process. *Asian J. Chem.* **2010,** *22* (2), 1009–1016.

Almeida, L. N. B.; Josué, T. G.; Fidelis, M. Z.; Abreu, E.; Bechlin, M. A.; dos Santos, O. A. A. et al. Process Comparison for Caffeine Degradation: Fenton, Photo-Fenton, UV/H_2O_2 and UV/Fe^{3+}. *Water Air Soil Pollut.* **2021,** *232* (4). DOI: 10.1007/s11270–021–05115–1.

Ameta, S. C.; Punjabi, P. B.; Kumari, C. A Comparative Study of Fenton's, Photo-Fenton's and Other Related Reagents: Resorcinol-TiO_2 System. *J. Indian Chem. Soc.* **2006,** *83* (1), 42–48.

Ammar, H. B.; Brahim, M. B.; Abdelhédi, R.; Samet, Y. Enhanced Degradation of Metronidazole by Sunlight via Photo-Fenton Process Under Gradual Addition of Hydrogen Peroxide. *J. Molecul. Catal. A: Chem.* 2016, *420*, 222–227.

Amorim, C. C.; Leão, M. M.; Moreira, R. F.; Fabris, J. D.; Henriques, A. B. Performance of Blast Furnace Waste for Azo Dye Degradation Through Photo-Fenton-Like Processes. *Chem. Eng. J.* **2013,** *224*, 59–66.

Arana, J.; Rendón, E. T.; Rodríguez, J. D.; Melián, J. H.; Díaz, O. G.; Peña, J. P. Highly Concentrated Phenolic Wastewater Treatment by the Photo-Fenton Reaction, Mechanism Study by FTIR-ATR. *Chemosphere* **2001,** *44* (5), 1017–1023.

Arslan-Alaton, I.; Gurses, F. Photo-Fenton-Like and Photo-Fenton-Like Oxidation of Procaine Penicillin G Formulation Effluent. *J. Photochem. Photobiol. A Chem.* **2004,** *165* (1–3), 165–175.

Augusti, R.; Dias, A. O.; Rocha, L. L.; Lago, R. M. Kinetics and Mechanism of Benzene Derivative Degradation with Fenton's Reagent in Aqueous Medium Studied by MIMS. *J. Phys. Chem. A* **1998,** *102* (52), 10723–10727.

Bacardit, J.; Stötzner, J.; Chamarro, E.; Esplugas, S. Effect of Salinity on the Photo-Fenton Process. *Ind. Eng. Chem. Res.* **2007,** *46* (23), 7615–7619.

Badawy, M. I.; Ghaly, M. Y.; Gad-Allah, T. A. Advanced Oxidation Processes for the Removal of Organophosphorus Pesticides from Wastewater. *Desalination* **2006,** *194* (1–3), 166–175.

Balanosky, E.; Fernandez, J.; Kiwi, J.; Lopez, A. Degradation of Membrane Concentrates of the Textile Industry by Fenton Like Reactions in Iron-Free Solutions at Biocompatible pH Values (pH≈ 7–8). *Water Sci. Technol.* **1999,** *40* (4–5), 417–424.

Bali, U.; Çatalkaya, E. Ç.; Şengül, F. Photochemical Degradation and Mineralization of Phenol: A Comparative Study. *J. Environ. Sci. Health A* **2003,** *38* (10), 2259–2275.

Bandara, J.; Pulgarin, C.; Peringer, P.; Kiwi, J. Chemical (Photo-Activated) Coupled Biological Homogeneous Degradation of P-Nitro-O-Toluene-Sulfonic Acid in a Flow Reactor. *J. Photochem. Photobiol. A Chem.* **1997,** *111* (1–3), 253–263.

Bansal, P.; Bui, T. S.; Lee, B. K. Potential Applications of ASR Fly Ash in Photo-Fenton Like Process for the Degradation of Tetracycline at Neutral pH: Fixed-Bed Approach. *Chem. Eng. J.* **2020,** *391*. DOI: 10.1016/j.cej.2019.123509.

Bashir, A.; Pandith, A. H.; Qureashi, A.; Malik, L. A.; Gani, M.; Perez, J. M. Catalytic Propensity of Biochar Decorated with Core-Shell nZVI@ Fe_3O_4: A Sustainable Photo-Fenton Catalysis of Methylene Blue Dye and Reduction of 4-Nitrophenol. *J. Environ. Chem. Eng.* 2022, 10 (3). DOI: 10.1016/j.jece.2022.107401.

Bauer, R.; Waldner, G.; Fallmann, H.; Hager, S.; Klare, M.; Krutzler, T. et al. The Photo-Fenton Reaction and the TiO_2/UV Process for Waste Water Treatment–Novel Developments. *257 Catal. Today* **1999,** *53*, 131–144.

Benitez, F. J.; Real, F. J.; Acero, J. L.; Leal, A. I.; Garcia, C. Gallic Acid Degradation in Aqueous Solutions by UV/H_2O_2 Treatment, Fenton's Reagent and the Photo-Fenton System. *J. Hazard. Mater.* **2005,** *126* (1–3), 31–39.

Blanco, J.; Torrades, F.; Morón, M.; Brouta-Agnésa, M.; García-Montaño, J. Photo-Fenton and Sequencing Batch Reactor Coupled to Photo-Fenton Processes for Textile Wastewater Reclamation: Feasibility of Reuse in Dyeing Processes. *Chem. Eng. J.* **2014,** *240*, 469–475.

Bozzi, A.; Yuranova, T.; Mielczarski, E.; Mielczarski, J.; Buffat, P. A.; Lais, P. et al. Superior Biodegradability Mediated by Immobilized Fe-Fabrics of Waste Waters Compared to Fenton Homogeneous Reactions. *Appl. Catal. B: Environ.* **2003,** *42* (3), 289–303.

Brillas, E.; Mur, E.; Sauleda, R.; Sanchez, L.; Peral, J.; Domènech, X.; Casado, J. Aniline Mineralization by AOP's: Anodic Oxidation, Photocatalysis, Electro-Fenton and Photoelectro-Fenton Processes. *Appl. Catal. B: Environ.* **1998a,** *16* (1), 31–42.

Brillas, E.; Sauleda, R.; Casado, J. Degradation of 4-Chlorophenol by Anodic Oxidation, Electro-Fenton, Photoelectron-Fenton, and Peroxi-Coagulation Processes. *J. Electrochem. Soc.* **1998b,** *145* (3), 759.

Carneiro, P. A.; Nogueira, R. F. P.; Zanoni, M. V. B. Homogeneous Photodegradation of CI Reactive Blue 4 Using a Photo-Fenton Process Under Artificial and Solar Irradiation. *Dyes Pigm.* **2007,** *74* (1), 127–132.

Chacón, J. M.; Leal, M. T.; Sánchez, M.; Bandala, E. R. Solar Photocatalytic Degradation of Azo-Dyes by Photo-Fenton Process. *Dyes Pigm.* **2006,** *69* (3), 144–150.

Changotra, R.; Rajput, H.; Dhir, A. Natural Soil Mediated Photo Fenton-Like Processes in Treatment of Pharmaceuticals: Batch and Continuous Approach. *Chemosphere* 2017, *188*, 345–353.

Chaudhuri, M.; Wei, T. Y. Decolourisation of Reactive Dyes by Modified Photo-Fenton Process Under Irradiation with Sunlight. *Nat. Environ. Pollut. Technol.* **2009,** *8* (2), 359–363.

Chen, Q.; Wu, P.; Li, Y.; Zhu, N.; Dang, Z. Heterogeneous Photo-Fenton Photodegradation of Reactive Brilliant Orange X-GN Over Iron-Pillared Montmorillonite Under Visible Irradiation. *J. Hazard. Mater.* **2009,** *168* (2–3), 901–908.

Chu, W.; Kwan, C. Y.; Chan, K. H.; Kam, S. K. A Study of Kinetic Modelling and Reaction Pathway of 2, 4-Dichlorophenol Transformation by Photo-Fenton-Like Oxidation. *J. Hazard. Mater.* **2005,** *121* (1–3), 119–126.

Costa, E. P.; Bottrel, S. E. C.; Starling, M. C. V.; Leão, M. M.; Amorim, C. C. Degradation of Carbendazim in Water via Photo-Fenton in Raceway Pond Reactor: Assessment of Acute Toxicity and Transformation Products. *Environ. Sci. Pollut. Res.* 2019, *26* (5), 4324–4336.

Costa, F. A.; dos Reis, E. M.; Azevedo, J. C.; Nozaki, J. Bleaching and Photodegradation of Textile Dyes by H_2O_2 and Solar or Ultraviolet Radiation. *Solar Energy* **2004,** *77* (1), 29–35.

Daneshvar, N.; Khataee, A. R. Removal of Azo Dye CI Acid Red 14 from Contaminated Water Using Fenton, UV/H_2O_2, UV/H_2O_2/Fe (II), UV/H_2O_2/Fe (III) and UV/H_2O_2/Fe (III)/Oxalate Processes: A Comparative Study. *J. Env. Sci. Health A* **2006,** *41* (3), 315–328.

De Laat, J.; Gallard, H. Catalytic Decomposition of Hydrogen Peroxide by Fe (III) in Homogeneous Aqueous Solution: Mechanism and Kinetic Modeling. *Environ. Sci. Technol.* **1999,** *33* (16), 2726–2732.

Derbalah, A. S.; Nakatani, N.; Sakugawa, H. Photocatalytic Removal of Fenitrothion in Pure and Natural Waters by Photo-Fenton Reaction. *Chemosphere* **2004,** *57* (7), 635–644.

Dillert, R.; Fornefett, I.; Siebers, U.; Bahnemann, D. Photocatalytic Degradation of Trinitrotoluene and Trinitrobenzene: Influence of Hydrogen Peroxide. *J. Photochem. Photobiol. A Chem.* **1996,** *94* (2–3), 231–236.

Ding, C.; Kang, S.; Li, W.; Gao, W.; Zhang, Z.; Zheng, L. et al. Mesoporous Structure and Amorphous Fe-N Sites Regulation in Fe-g-C_3N_4 for Boosted Visible-Light-Driven Photo-Fenton Reaction. *J. Colloid Interface Sci.* **2022,** *608* (3), 2515–2528.

Domenzain-Gonzalez, J.; Castro-Arellano, J. J.; Galicia-Luna, L. A.; Lartundo-Rojas, L. Photo-Fenton Degradation of RB5 Dye in Aqueous Solution Using Fe Supported on Mexican Natural Zeolite. *Int. J. Photoenergy* **2019,** *2019*. DOI: 10.1155/2019/4981631.

Doong, R. A.; Chang, W. H. Photodegradation of Parathion in Aqueous Titanium Dioxide and Zero Valent Iron Solutions in the Presence of Hydrogen Peroxide. *J. Photochem. Photobiol. A: Chem.* **1998,** *116* (3), 221–228.

Du, Y.; Zhou, M.; Lei, L. Role of the Intermediates in the Degradation of Phenolic Compounds by Fenton-Like Process. *J. Hazard. Mater.* **2006,** *136* (3), 859–865.

Du, Z.; Liu, F.; Xiao, C.; Dan, Y.; Jiang, L. Fabrication of Poly (Vinyl Alcohol)/Sodium Alginate Hydrogel Beads and Its Application in Photo-Fenton Degradation of Tetracycline. *J. Mater. Sci.* **2021,** *56* (1), 913–926.

Duarte, F.; Madeira, L. M. Fenton-and Photo-Fenton-Like Degradation of a Textile Dye by Heterogeneous Processes with Fe/ZSM-5 Zeolite. *Sep. Sci. Technol.* **2010,** *45* (11), 1512–1520.

Duguet, J. P.; Anselme, C.; Mazounie, P.; Mallevialle, J. Application of Combined Ozone–Hydrogen Peroxide for the Removal of Aromatic Compounds from a Groundwater. *J. Int. Ozone Assoc.* **1990,** *12*, 281–294.

Elmolla, E. S.; Chaudhuri, M. Effect of Photo-Fenton Operating Conditions on the Performance of Photo-Fenton-SBR. *J. Appl. Sci.* **2010,** *10* (24), 3236–3242.

El-Morsi, T. M.; Emara, M. M.; Abd El Bary, H. M.; Abd-El-Aziz, A. S.; Friesen, K. J. Homogeneous Degradation of 1, 2, 9, 10-Tetrachlorodecane in Aqueous Solutions Using Hydrogen Peroxide, Iron and UV Light. *Chemosphere* **2002,** *47* (3), 343–348.

Esplugas, S.; Gimenez, J.; Contreras, S.; Pascual, E.; Rodríguez, M. Comparison of Different Advanced Oxidation Processes for Phenol Degradation. *Water Res.* **2002,** *36* (4), 1034–1042.

Evgenidou, E.; Konstantinou, I.; Fytianos, K.; Poulios, I. Oxidation of Two Organophosphorous Insecticides by the Photo-Assisted Fenton Reaction. *Water Res.* **2007,** *41* (9), 2015–2027.

Farré, M. J.; Brosillon, S.; Domènech, X.; Peral, J. Evaluation of the Intermediates Generated During the Degradation of Diuron and Linuron Herbicides by the Photo-Fenton Reaction. *J. Photochem. Photobiol. A: Chem.* **2007,** *189* (2–3), 364–373.

Farré, M. J.; Doménech, X.; Peral, J. Assessment of Photo-Fenton and Biological Treatment Coupling for Diuron and Linuron Removal from Water. *Water Res.* **2006,** *40* (13), 2533–2540.

Faust, B. C.; Hoigné, J. Photolysis of Fe (III)-Hydroxy Complexes as Sources of OH Radicals in Clouds, Fog and Rain. *Atmos. Environ. A. Gen. Top.* **1990,** *24* (1), 79–89.

Feng, H. E.; Le-Cheng, L. E. I. Degradation Kinetics and Mechanisms of Phenol in Photo-Fenton Process. *J. Zhejiang Univ. Sci. A* **2004,** *5* (2), 198–205.

Feng, J.; Hu, X.; Yue, P. L. Degradation of Salicylic Acid by Photo-Assisted Fenton Reaction Using Fe Ions on Strongly Acidic Ion Exchange Resin as Catalyst. *Chem. Eng. J.* **2004,** *100* (1–3), 159–165.

Feng, J.; Hu, X.; Yue, P. L. Discoloration and Mineralization of Orange II by Using a Bentonite Clay-Based Fe Nanocomposite Film as a Heterogeneous Photo-Fenton Catalyst. *Water Res.* **2005,** *39* (1), 89–96.

Feng, J.; Hu, X.; Yue, P. L.; Zhu, H. Y.; Lu, G. Q. Discoloration and Mineralization of Reactive Red HE-3B by Heterogeneous Photo-Fenton Reaction. *Water Res.* **2003,** *37* (15), 3776–3784.

Fenton, H. J. H. LXXIII.–Oxidation of Tartaric Acid in Presence of Iron. *J. Chem. Soc. Trans.* **1894,** *65*, 899–910.

Fernandez, J.; Bandara, J.; Lopez, A.; Buffat, P.; Kiwi, J. Photoassisted Fenton Degradation of Nonbiodegradable Azo Dye (Orange II) in Fe-Free Solutions Mediated by Cation Transfer Membranes. *Langmuir* **1999,** *15* (1), 185–192.

Flox, C.; Garrido, J. A.; Rodríguez, R. M.; Cabot, P. L.; Centellas, F.; Arias, C. et al. Mineralization of Herbicide Mecoprop by Photoelectro-Fenton with UVA and Solar Light. *Catal. Today* **2007,** *129* (1–2), 29–36.

Fukushima, M.; Tatsumi, K.; Nagao, S. Degradation Characteristics of Humic Acid During Photo-Fenton Processes. *Environ. Sci. Technol.* **2001,** *35* (18), 3683–3690.

Gallard, H.; De Laat, J. Kinetic Modelling of Fe (III)/H_2O_2 Oxidation Reactions in Dilute Aqueous Solution Using Atrazine as a Model Organic Compound. *Water Res.* **2000,** *34* (12), 3107–3116.

Gallard, H.; de Laat, J.; Legube, B. Effect of pH on the Oxidation Rate of Organic Compounds by Fe-II/H_2O_2 Mechanisms and Simulation. *New J. Chem.* **1998,** *22* (3), 263–268.

Galvão, S. A. O.; Mota, A. L.; Silva, D. N.; Moraes, J. E. F.; Nascimento, C. A.; Chiavone-Filho, O. Application of the Photo-Fenton Process to the Treatment of Wastewaters Contaminated with Diesel. *Sci. Total Environ.* **2006,** *367* (1), 42–49.

Gamarra-Güere, C. D.; Dionisio, D.; Santos, G. O. S.; Lanza, M. R. V.; de Jesus Motheo, A. Application of Fenton, Photo-Fenton and Electro-Fenton Processes for the Methylparaben Degradation: A Comparative Study. *J. Environ. Chem. Eng.* **2022,** *10* (1). DOI: 10.1016/j.jece.2021.106992.

García-Montaño, J.; Torrades, F.; García-Hortal, J. A.; Domenech, X.; Peral, J. Degradation of Procion Red H-E7B Reactive Dye by Coupling a Photo-Fenton System with a Sequencing Batch Reactor. *J. Hazard. Mater.* **2006,** *134* (1–3), 220–229.

Ghaly, M. Y.; Härtel, G.; Mayer, R.; Haseneder, R. Photochemical Oxidation of p-Chlorophenol by UV/H_2O_2 and Photo-Fenton Process: A Comparative Study. *Waste Manag.* **2001,** *21* (1), 41–47.

Ghazzaf, H.; Nechchadi, B.; Jouali, A.; Salhi, A.; El Krati, M.; Tahiri, S. Synthesis of Heterogeneous Photo-Fenton Catalyst from Iron Rust and Its Application to Degradation of Acid Red 97 in Aqueous Medium. *J. Environ. Chem. Eng.* **2022,** *10* (3). DOI: 10.1016/j.jece.2022.107570.

Giroto, J. A.; Guardani, R.; Teixeira, A. C. S. C.; Nascimento, C. Study on the Photo-Fenton Degradation of Polyvinyl Alcohol in Aqueous Solution. *Chem. Eng. Process.* **2006,** *45* (7), 523–532.

Giroto, J. A.; Teixeira, A. C.; Nascimento, C. A.; Guardani, R. Degradation of Poly (Ethylene Glycol) in Aqueous Solution by Photo-Fenton and H_2O_2/UV Processes. *Ind. Eng. Chem. Res.* **2010,** *49* (7), 3200–3206.

Glaze, W. H.; Kang, J. W.; Chapin, D. H. The Chemistry of Water Treatment Processes Involving Ozone, Hydrogen Peroxide and Ultraviolet Radiation. *J. Int. Ozone Assoc.* **1987,** *9*, 335–352.

Goi, A.; Trapido, M. Hydrogen Peroxide Photolysis, Fenton Reagent and Photo-Fenton for the Degradation of Nitrophenols: A Comparative Study. *Chemosphere* **2002,** *46* (6), 913–922.

Gould, D. M.; Griffith, W. P.; Spiro, M. Polyoxometalate Catalysis of Dye Bleaching by Hydrogen Peroxide[J]. *J. Mol. Catal. A Chem.* **2001,** *175*, 289–291.

Gromboni, C. F.; Kamogawa, M. Y.; Ferreira, A. G.; Nóbrega, J. A.; Nogueira, A. R. A. Microwave-Assisted Photo-Fenton Decomposition of Chlorfenvinphos and Cypermethrin in Residual Water. *J. Photochem. Photobiol. A Chem.* **2007,** *185* (1), 32–37.

Guélou, E.; Barrault, J.; Fournier, J.; Tatibouët, J. M. Active Iron Species in the Catalytic Wet Peroxide Oxidation of Phenol Over Pillared Clays Containing Iron. *Appl. Catal. B: Environ.* **2003,** *44* (1). DOI: 10.1016/S0926–3373 (03)00003–1.

Gulyas, H. Processes for the Removal of Recalcitrant Organics from Industrial Wastewaters. *Water Sci. Technol.* **1997,** *36* (2–3), 9–16.

Guo, J.; Al-Dahhan, M. Catalytic Wet Oxidation of Phenol by Hydrogen Peroxide Over Pillared Clay Catalyst. *Ind. Eng. Chem. Res.* **2003,** *42* (12), 2450–2460.

Gupta, V. K.; Carrott, P. J. M.; Ribeiro Carrott, M. M. L.; Suhas. Low-Cost Adsorbents: Growing Approach to Wastewater Treatment—A Review. *Crit. Rev. Environ. Sci. Technol.* **2009,** *39* (10), 783–842.

Haber, F.; Weiss, J. The Catalytic Decomposition of Hydrogen Peroxide by Iron Salts. *Proc. R. Soc. Lond. A Math. Phys. Sci.* **1934,** *147* (861), 332–351.

He, J.; Tao, X.; Ma, W.; Zhao, J. Heterogeneous Photo-Fenton Degradation of an Azo Dye in Aqueous H_2O_2/Iron Oxide Dispersions at Neutral pHs. *Chem. Lett.* **2002,** *31* (1), 86–87.

Hernández-Rodríguez, M. J.; Fernández-Rodríguez, C.; Doña-Rodríguez, J. M.; González-Díaz, O. M.; Zerbani, D.; Peña, J. P. Treatment of Effluents from Wool Dyeing Process by Photo-Fenton at Solar Pilot Plant. *J. Environ. Chem. Eng.* 2014, *2* (1), 163–171.

Herney-Ramirez, J.; Vicente, M. A.; Madeira, L. M. Heterogeneous Photo-Fenton Oxidation with Pillared Clay-Based Catalysts for Wastewater Treatment: A Review. *Appl. Catal. B: Environ.* **2010,** *98* (1–2), 10–26.

Howard, P. H.; Michalenko, E. M.; Jarvis, W. F.; Basu, D. K.; Sage, G. W.; Meylan, W. M.; Beauman, J. A.; Anthony Gray, D., Eds. *Handbook of Environmental Fate and Exposure Data: For Organic Chemicals*, 3rd ed.; Routledge, 1991. DOI: 10.1201/9780203719305

Hsueh, C. L.; Huang, Y. H.; Wang, C. C.; Chen, C. Y. Degradation of Azo Dyes Using Low Iron Concentration of Fenton and Fenton-Like System. *Chemosphere* **2005,** *58* (10), 1409–1414.

Huston, P. L.; Pignatello, J. J. Degradation of Selected Pesticide Active Ingredients and Commercial Formulations in Water by the Photo-Assisted Fenton Reaction. *Water Res.* **1999,** *33* (5), 1238–1246.

Iurascu, B.; Siminiceanu, I.; Vione, D.; Vicente, M. A.; Gil, A. Phenol Degradation in Water Through a Heterogeneous Photo-Fenton Process Catalyzed by Fe-Treated Laponite. *Water Res.* **2009,** *43* (5), 1313–1322.

Jacob, N.; Balakrishnan, I.; Reddy, M. P. Characterization of the Hydroxyl Radical in Some Photochemical Reactions. *J. Phys. Chem.* **1977,** *81* (1), 17–22.

Jain, A.; Lodha, S.; Punjabi, P. B.; Sharma, V. K.; Ameta, S. C. A Study of Catalytic Behaviour of Aromatic Additives on the Photo-Fenton Degradation of Phenol Red. *J. Chem. Sci.* **2009,** *121* (6), 1027–1034.

Ju, L.; Chen, Z.; Fang, L.; Dong, W.; Zheng, F.; Shen, M. Sol–Gel Synthesis and Photo-Fenton-Like Catalytic Activity of $EuFeO_3$ Nanoparticles. *J. Am. Ceramic Soc.* **2011,** *94* (10), 3418–3424.

Ju, Y.; Li, H.; Wang, Z.; Liu, H.; Huo, S.; Jiang, S. et al. Solar-Driven on-Site H_2O_2 Generation and Tandem Photo-Fenton Reaction on a Triphase Interface for Rapid Organic Pollutant Degradation. *Chem. Eng. J.* **2022,** *430* (4). DOI: 10.1016/j.cej.2021.133168.

Kajitvichyanukul, P.; Lu, M. C.; Jamroensan, A. Formaldehyde Degradation in the Presence of Methanol by Photo-Fenton Process. *J. Environ. Manag.* **2008,** *86* (3), 545–553.

Kang, S. F.; Liao, C. H.; Chen, M. C. Pre-Oxidation and Coagulation of Textile Wastewater by the Fenton Process. *Chemosphere* **2002,** *46* (6), 923–928.

Kang, S. F.; Liao, C. H.; Po, S. T. Decolorization of Textile Wastewater by Photo-Fenton Oxidation Technology. *Chemosphere* **2000,** *41* (8), 1287–1294.

Kasiri, M. B.; Aleboyeh, H.; Aleboyeh, A. Mineralization of CI Acid Red 14 Azo Dye by UV/Fe-ZSM5/H_2O_2 Process. *Environ. Technol.* **2010,** *31* (2), 165–173.

Katsumata, H.; Kaneco, S.; Suzuki, T.; Ohta, K.; Yobiko, Y. Degradation of Linuron in Aqueous Solution by the Photo-Fenton Reaction. *Chem. Eng. J.* **2005,** *108* (3), 269–276.

Katsumata, H.; Kaneco, S.; Suzuki, T.; Ohta, K.; Yobiko, Y. Photo-Fenton Degradation of Alachlor in the Presence of Citrate Solution. *J. Photochem. Photobiol. A Chem.* **2006,** *180* (1–2), 38–45.

Katsumata, H.; Kawabe, S.; Kaneco, S.; Suzuki, T.; Ohta, K. Degradation of Bisphenol A in Water by the Photo-Fenton Reaction. *J. Photochem. Photobiol. A Chem.* **2004,** *162* (2–3), 297–305.

Kavitha, V.; Palanivelu, K. Degradation of 2-Chlorophenol by Fenton and Photo-Fenton Processes—A Comparative Study. *J. Environ. Sci. Health A* **2003,** *38* (7), 1215–1231.
Kavitha, V.; Palanivelu, K. The Role of Ferrous Ion in Fenton and Photo-Fenton Processes for the Degradation of Phenol. *Chemosphere* **2004,** *55* (9), 1235–1243.
Kim, S. M.; Geissen, S. U.; Vogelpohl, A. Landfill Leachate Treatment by a Photoassisted Fenton Reaction. *Water Sci. Technol.* **1997,** *35* (4), 239–248.
Kim, S. M.; Vogelpohl, A. Degradation of Organic Pollutants by the Photo-Fenton-Process. *Chem. Eng. Technol.* **1998,** *21* (2), 187–191.
Kondo, M. M.; Leite, K. U.; Silva, M. R.; Reis, A. D. Fenton and Photo-Fenton Processes Coupled to UASB to Treat Coffee Pulping Wastewater. *Sep. Sci. Technol.* **2010,** *45* (11), 1506–1511.
Krzysztoszek, A.; Naumczyk, J. Landfill Leachate Treatment by Fenton, Photo-Fenton Processes and Their Modification. *J. Adv. Oxid. Technol.* **2012,** *15* (1), 53–63.
Kumar, A.; Paliwal, M.; Ameta, R.; Ameta, S. C. A Novel Route for Waste Water Treatment: Photo-Assisted Fenton Degradation of Naphthol Green B. *Collect. Czechoslov. Chem. Commun.* **2008,** *73* (5), 679–689.
Kumar, A.; Paliwal, M.; Ameta, R.; Ameta, S. C. Use of Photo-Fenton's Reagent for the Photochemical Bleaching of Metanil Yellow. *Proc. Nat. Acad. Sci. India Phys. Sci.* **2008,** *78*, 123–128.
Kumar, B. N.; Anjaneyulu, Y.; Himabindu, V. Comparative Studies of Degradation of Dye Intermediate (H-Acid) Using TiO_2/UV/H_2O_2 and Photo-Fenton Process. *J. Chem. Pharma. Res.* **2011,** *3* (2), 718–731.
Kušić, H.; Koprivanac, N.; Božić, A. L.; Selanec, I. Photo-Assisted Fenton Type Processes for the Degradation of Phenol: A Kinetic Study. *J. Hazard. Mater.* **2006,** *136* (3), 632–644.
Kuznetsova, L. I.; Detusheva, L. G.; Fedotov, M. A.; Likholobov, V. A. Catalytic Properties of Heteropoly Complexes Containing Fe (III) Ions in Benzene Oxidation by Hydrogen Peroxide. *J. Mol. Catal. A Chem.* **1996,** *111* (1–2), 81–90.
Kwan, C. Y.; Chu, W. A Study of the Reaction Mechanisms of the Degradation of 2,4-Dichlorophenoxyacetic Acid by Oxalate-Mediated Photooxidation. *Water Res.* **2004,** *38* (19), 4213–4221.
Kwan, C. Y.; Chu, W. Photodegradation of 2,4-Dichlorophenoxyacetic Acid in Various Iron-Mediated Oxidation Systems. *Water Res.* **2003,** *37* (18), 4405–4412.
Kwon, B. G.; Lee, D. S.; Kang, N.; Yoon, J. Characteristics of p-Chlorophenol Oxidation by Fenton's Reagent. *Water Res.* **1999,** *33* (9), 2110–2118.
Laftani, Y.; Boussaoud, A.; Chatib, B.; El Makhfouk, M.; Hachkar, M.; Khayar, M. Comparison of Advanced Oxidation Processes for Degrading Ponceau S Dye. Application of Photo-Fenton Process. *Maced. J. Chem. Chem. Eng.* **2019,** *38* (2), 197–205.
Lapertot, M.; Ebrahimi, S.; Dazio, S.; Rubinelli, A.; Pulgarin, C. Photo-Fenton and Biological Integrated Process for Degradation of a Mixture of Pesticides. *J. Photochem. Photobiol. A Chem.* **2007,** *186* (1), 34–40.
Lapertot, M.; Pulgarín, C.; Fernández-Ibáñez, P.; Maldonado, M. I.; Pérez-Estrada, L.; Oller, I. et al. Enhancing Biodegradability of Priority Substances (Pesticides) by Solar Photo-Fenton. *Water Res.* **2006,** *40* (5), 1086–1094.
Legrini, O.; Oliveros, E.; Braun, A. M. Photochemical Processes for Water Treatment. *Chem. Rev.* **1993,** *93* (2), 671–698.
Lei, L.; Hu, X.; Yue, P. L.; Bossmann, S. H.; Göb, S.; Braun, A. M. Oxidative Degradation of Poly Vinyl Alcohol by the Photochemically Enhanced Fenton Reaction. *J. Photochem. Photobiol. A Chem.* **1998,** *116* (2), 159–166.

Li, F. B.; Li, X. Z.; Li, X. M.; Liu, T. X.; Dong, J. Heterogeneous Photodegradation of Bisphenol A with Iron Oxides and Oxalate in Aqueous Solution. *J. Colloid Interf. Sci.* **2007,** *311* (2), 481–490.

Li, H.; Wu, P.; Dang, Z.; Zhu, N.; Li, P.; Wu, J. Synthesis, Characterization, and Visible-Light Photo-Fenton Catalytic Activity of Hydroxy Fe/Al-Intercalated Montmorillonite. *Clays Clay Miner.* **2011,** *59* (5), 466–477.

Li, J.; Zhang, X.; Wang, T.; Zhao, Y.; Song, T.; Zhang, L.; Cheng, X. Construction of Layered Hollow $Fe_3O_4/Fe_{1-x}S@MoS_2$ Composite with Enhanced Photo-Fenton and Adsorption Performance. *J. Environ. Chem. Eng.* **2020,** *8* (3). DOI: 10.1016/j.jece.2020.103762.

Lin, K.; Yuan, D.; Chen, M.; Deng, Y. Kinetics and Products of Photo-Fenton Degradation of Triazophos. *J. Agric. Food Chem.* **2004,** *52* (25), 7614–7620.

Liou, M. J.; Lu, M. C.; Chen, J. N. Oxidation of Explosives by Fenton and Photo-Fenton Processes. *Water Res.* **2003,** *37* (13), 3172–3179.

Liou, M. J.; Lu, M. C.; Chen, J. N. Oxidation of TNT by Photo-Fenton Process. *Chemosphere* **2004,** *57* (9), 1107–1114.

Liu, L.; Zhang, G.; Wang, L.; Huang, T.; Qin, L. Highly Active S-Modified $ZnFe_2O_4$ Heterogeneous Catalyst and Its Photo-Fenton Behavior Under UV–Visible Irradiation. *Ind. Eng. Chem. Res.* **2011,** *50* (12), 7219–7227.

Liu, L.; Zhang, G.; Wang, L.; Huang, T.; Qin, L. Highly Active S-Modified $ZnFe_2O_4$ Heterogeneous Catalyst and Its Photo-Fenton Behavior Under UV–Visible Irradiation. *Ind. Eng. Chem. Res.* **2011,** *50* (12), 7219–7227.

Liu, Q. Y.; Liu, Y. X.; Lu, X. J. Combined Photo-Fenton and Biological Oxidation for the Treatment of Aniline Wastewater. *Procedia Environ. Sci.* **2012,** *12*, 341–348.

Liu, R.; Chiu, H. M.; Shiau, C. S.; Yeh, R. Y. L.; Hung, Y. T. Degradation and Sludge Production of Textile Dyes by Fenton and Photo-Fenton Processes. *Dyes Pigments* **2007,** *73* (1), 1–6.

Lloyd, R. V.; Hanna, P. M.; Mason, R. P. The Origin of the Hydroxyl Radical Oxygen in the Fenton Reaction. *Free Radic. Biol. Med.* **1997,** *22* (5), 885–888.

Loeff, I.; Stein, G. The Radiation and Photochemistry of Aqueous Solutions of Benzene. *J. Chem. Soc.* **1963,** *1963*, 2623–2633.

Lucas, M. S.; Peres, J. A. Decolorization of the Azo Dye Reactive Black 5 by Fenton and Photo-Fenton Oxidation. *Dyes Pigm.* **2006,** *71* (3), 236–244.

Lucas, M. S.; Peres, J. A.; Amor, C.; Prieto-Rodríguez, L.; Maldonado, M. I.; Malato, S. Tertiary Treatment of Pulp Mill Wastewater by Solar Photo-Fenton. *J. Hazard. Mater.* **2012,** *225*, 173–181.

Luong, H. V.; Lin, H. K.; Controlling the Fenton Reaction for Soil Remediation. *Anal. Lett.* **2000,** *33*, 3051–3065.

Ma, J.; Song, W.; Chen, C.; Ma, W.; Zhao, J.; Tang, Y. Fenton Degradation of Organic Compounds Promoted by Dyes Under Visible Irradiation. *Environ. Sci. Technol.* **2005,** *39* (15), 5810–5815.

MacFaul, P. A.; Wayner, D. D. M.; Ingold, K. U. A Radical Account of "Oxygenated Fenton Chemistry". *Acc. Chem. Res.* **1998,** *31* (4), 159–162.

Macías-Sánchez, J.; Hinojosa-Reyes, L.; Guzmán-Mar, J. L.; Peralta-Hernández, J. M.; Hernández-Ramírez, A. Performance of the Photo-Fenton Process in the Degradation of a Model Azo Dye Mixture. *Photochem. Photobiol. Sci.* **2011,** *10* (3), 332–337.

Maciel, R.; Sant'Anna Jr, G. L.; Dezotti, M. Phenol Removal from High Salinity Effluents Using Fenton's Reagent and Photo-Fenton Reactions. *Chemosphere* **2004,** *57* (7), 711–719.

Mahdad, F.; Younesi, H.; Bahramifar, N.; Hadavifar, M. Optimization of Fenton and Photo-Fenton-Based Advanced Oxidation Processes for Post-Treatment of Composting Leachate of Municipal Solid Waste by an Activated Sludge Process. *KSCE J. Civ. Eng.* 2016, *20* (6), 2177–2188.

Mahmoud, G. E. A.; Ismail, L. F. M. Factors Affecting the Kinetic Parameters Related to the Degradation of Direct Yellow 50 by Fenton and Photo-Fenton Processes. *J. Basic Appl. Chem.* **2011,** *1* (8), 70–79.

Mailhot, G.; Hykrdová, L.; Jirkovský, J.; Lemr, K.; Grabner, G.; Bolte, M. Iron (III)-Photoinduced Degradation of 4-Chloroaniline in Aqueous Solution. *Appl. Catal. B: Environ.* **2004,** *50* (1), 25–35.

Malato, S.; Blanco, J.; Cáceres, J.; Fernández-Alba, A. R.; Agüera, A.; Rodrıguez, A. Photocatalytic Treatment of Water-Soluble Pesticides by Photo-Fenton and TiO_2 Using Solar Energy. *Catal. Today* **2002,** *76* (2–4), 209–220.

Maldonado, M. I.; Passarinho, P. C.; Oller, I.; Gernjak, W.; Fernández, P.; Blanco, J. et al. Photocatalytic Degradation of EU Priority Substances: A Comparison Between TiO_2 and Fenton Plus Photo-Fenton in a Solar Pilot Plant. *J. Photochem. Photobiol. A Chem.* **2007,** *185* (2–3), 354–363.

Martínez, F.; Calleja, G.; Melero, J. A.; Molina, R. Heterogeneous Photo-Fenton Degradation of Phenolic Aqueous Solutions Over Iron-Containing SBA-15 Catalyst. *Appl. Catal. Environ.* **2005,** *60* (3–4), 181–190.

Martínez, F.; Calleja, G.; Melero, J. A.; Molina, R. Iron Species Incorporated Over Different Silica Supports for the Heterogeneous Photo-Fenton Oxidation of Phenol. *Appl. Catal. B : Environ.* **2007,** *70* (1–4), 452–460.

Martínez, S. S.; Sanchez, J. V.; Estrada, J. R. M.; Velasquez, R. F. FeIII Supported on Ceria as Effective Catalyst for the Heterogeneous Photo-Oxidation of Basic Orange 2 in Aqueous Solution with Sunlight. *Sol. Energy Mater. Sol. Cells* **2011,** *95* (8), 2010–2017.

Módenes, A. N.; Espinoza-Quiñones, F. R.; Manenti, D. R.; Borba, F. H.; Palácio, S. M.; Colombo, A. Performance Evaluation of a Photo-Fenton Process Applied to Pollutant Removal from Textile Effluents in a Batch System. *J. Environ. Manag.* **2012,** *104*. DOI: 10.1016/j.jenvman.2012.03.032.

Mohajeri, S.; Aziz, H. A.; Isa, M. H.; Bashir, M. J.; Mohajeri, L.; Adlan, M. N. Influence of Fenton Reagent Oxidation on Mineralization and Decolorization of Municipal Landfill Leachate. *J. Environ. Sci. Health Part A* **2010,** *45* (6), 692–698.

Monteagudo, J. M.; Carmona, M.; Duran, A. Photo-Fenton-Assisted Ozonation of p-Coumaric Acid in Aqueous Solution. *Chemosphere* **2005,** *60* (8), 1103–1110.

Morais, V. S.; Barrada, R. V.; Moura, M. N.; Almeida, J. R.; Moreira, T. F. M.; Gonçalves, G. R. et al. Synthesis of Manganese Ferrite from Spent Zn–MnO_2 Batteries and Its Application as a Catalyst in Heterogeneous Photo-Fenton Processes. *J. Environ. Chem. Eng.* **2020,** *8*. DOI: 10.1016/j.jece.2020.103716.

Mosteo, R.; Ormad, P.; Mozas, E.; Sarasa, J.; Ovelleiro, J. L. Factorial Experimental Design of Winery Wastewaters Treatment by Heterogeneous Photo-Fenton Process. *Water Res.* **2006,** *40* (8), 1561–1568.

Muruganandham, M.; Swaminathan, M. Decolourisation of Reactive Orange 4 by Fenton and Photo-Fenton Oxidation Technology. *Dyes Pigments* **2004,** *63* (3), 315–321.

Nadeem, N.; Zahid, M.; Tabasum, A.; Mansha, A.; Jilani, A.; Bhatti, I. A.; Bhatti, H. N. Degradation of Reactive Dye Using Heterogeneous Photo-Fenton Catalysts: $ZnFe_2O_4$ and GO-$ZnFe_2O_4$ Composite. *Mater. Res. Express* **2020,** *7* (1). DOI: 10.1088/2053–1591/ab66ee.

Nadtochenko, V. N.; Kiwi, J. Mechanism of Formation of Phenoxyl Radicals During the Photo-Oxidation of Phenol in the Presence of Fe III. *J. Chem. Soc. Perkin Trans.* **1998,** *2*, (6), 1303–1306.

Nedoloujko, A.; Kiwi, J. Transient Intermediate Species Active During the Fenton-Mediated Degradation of Quinoline in Oxidative Media: Pulsed Laser Spectroscopy. *J. Photochem. Photobiol. A: Chem.* **1997,** *110* (2), 141–148.

Nguyen, V. H.; Bach, L. G.; Bui, Q. T. P.; Nguyen, T. D.; Vo, D. V. N.; Vu, H. T. et al. Composite Photocatalysts Containing MIL-53 (Fe) as a Heterogeneous Photo-Fenton Catalyst for the Decolorization of Rhodamine B Under Visible Light Irradiation. *J. Environ. Chem. Eng.* **2018,** *6* (6), 7434–7441.

Nie, Y.; Hu, C.; Qu, J.; Zhou, L.; Hu, X. Photoassisted Degradation of Azodyes Over FeO_xH_{2x-3}/Fe_0 in the Presence of H_2O_2 at Neutral pH Values. *Environ. Sci. Technol.* **2007,** *41* (13), 4715–4719.

Nogueira, R. F. P.; Alberici, R. M.; Mendes, M. A.; Jardim, W. F.; Eberlin, M. N. Photocatalytic Degradation of Phenol and Trichloroethylene: On-Line and Real-Time Monitoring via Membrane Introduction Mass Spectrometry. *Ind. Eng. Chem. Res.* **1999,** *38* (5), 1754–1758.

Nogueira, R. F. P.; Trovó, A. G.; Modé, D. F. Solar Photodegradation of Dichloroacetic Acid and 2,4-Dichlorophenol Using an Enhanced Photo-Fenton Process. *Chemosphere* **2002,** *48* (4), 385–391.

Ntampegliotis, K.; Riga, A.; Karayannis, V.; Bontozoglou, V.; Papapolymerou, G. Decolorization kinetics of Procion H-Exl Dyes from Textile Dyeing Using Fenton-Like Reactions. *J. Hazard. Mater.* **2006,** *136* (1), 75–84.

Núñez, L.; García-Hortal, J. A.; Torrades, F. Study of Kinetic Parameters Related to the Decolourization and Mineralization of Reactive Dyes from Textile Dyeing Using Fenton and Photo-Fenton Processes. *Dyes Pigments* **2007,** *75* (3), 647–652.

Oliveros, E.; Legrini, O.; Hohl, M.; Müller, T.; Braun, A. M. Industrial Waste Water Treatment: Large Scale Development of a Light-Enhanced Fenton Reaction. *Chem. Eng. Process.* **1997,** *36* (5), 397–405.

Oller, I.; Malato, S.; Sánchez-Pérez, J. A.; Maldonado, M. I.; Gassó, R. Detoxification of Wastewater Containing Five Common Pesticides by Solar AOPs–Biological Coupled System. *Catal. Today* **2007,** *129* (1–2), 69–78.

Ormad, M. P.; Mosteo, R.; Ibarz, C.; Ovelleiro, J. L. Multivariate Approach to the Photo-Fenton Process Applied to the Degradation of Winery Wastewaters. *Appl. Catal. B Environ.* **2006,** *66* (1–2), 58–63.

Ortega-Moreno, G. A.; Ayala-Durán, S. C.; Barbero, B. P.; Narda, G. E.; Bernini, M. C.; Nogueira, R. F. P. Photo-Fenton Degradation of Sulfamethoxazole Using MIL-53 (Fe) Under UVA LED Irradiation and Natural Sunlight. *J. Environ. Chem. Eng.* **2022,** *10* (3). DOI: 10.1016/j.jece.2022.107678.

Parida, K. M.; Pradhan, A. C. Fe/meso-Al_2O_3: An Efficient Photo-Fenton Catalyst for the Adsorptive Degradation of Phenol. *Ind. Eng. Chem. Res.* **2010,** *49* (18), 8310–8318.

Park, J. H.; Cho, I. H.; Chang, S. W. Comparison of Fenton and Photo-Fenton Processes for Livestock Wastewater Treatment. *J. Env. Sci. Health B* **2006,** *41* (2), 109–120.

Parra, S.; Henao, L.; Mielczarski, E.; Mielczarski, J.; Albers, P.; Suvorova, E. et al. Synthesis, Testing, and Characterization of a Novel Nafion Membrane with Superior Performance in Photoassisted Immobilized Fenton Catalysis. *Langmuir* **2004,** *20* (13), 5621–5629.

Paterlini, W. C.; Nogueira, R. F. P. Multivariate Analysis of Photo-Fenton Degradation of the Herbicides Tebuthiuron, Diuron and 2, 4-D. *Chemosphere* **2005,** *58* (8), 1107–1116.

Pérez, M.; Torrades, F.; Domènech, X.; Peral, J. Fenton and Photo-Fenton Oxidation of Textile Effluents. *Water Res.* **2002,** *36* (11), 2703–2710.

Perez, M.; Torrades, F.; García-Hortal, J. A.; Domènech, X.; Peral, J. Removal of Organic Contaminants in Paper Pulp Treatment Effluents Under Fenton and Photo-Fenton Conditions. *Appl. Catal. B Environ.* **2002,** *36* (1), 63–74.

Pignatello, J. J.; Oliveros, E.; MacKay, A. Advanced Oxidation Processes for Organic Contaminant Destruction Based on the Fenton Reaction and Related Chemistry. *Crit. Rev. Environ. Sci. Technol.* **2006,** *36* (1), 1–84.

Pignatello, J. J.; Sun, Y. Photo-Assisted Mineralization of Herbicide Wastes by Ferric Ion Catalyzed Hydrogen Peroxide. **1993.** DOI: 10.1021/bk-1993–0518.ch004

Prousek, J.; duriskova, I. Oxidacni degradace polyethylenglykolu (peg) za pouziti fentonovy a fotochemicke fentonovy reakce. *Chem. Listy* **1998,** *92* (3), 218–220.

Quici, N.; Morgada, M. E.; Piperata, G.; Babay, P.; Gettar, R. T.; Litter, M. I. Oxalic Acid Destruction at High Concentrations by Combined Heterogeneous Photocatalysis and Photo-Fenton Processes. *Catal. Today* **2005,** *101* (3–4), 253–260.

Rad, L. R.; Irani, M.; Pourahmad, H.; Sayyafan, M. S.; Haririan, I. Simultaneous Degradation of Phenol and Paracetamol During Photo-Fenton Process: Design and Optimization. *J. Taiwan Inst. Chem. Eng.* 2015, *47*, 190–196.

Rasoulifard, M. H.; Hosseini Monfared, H.; Masoudian, S. Photo-Assisted Hetero-Fenton Decolorization of Azo Dye from Contaminated Water by Fe–Si Mixed Oxide Nanocomposite. *Environ. Technol.* **2011,** *32* (14), 1627–1635.

Redouane-Salah, Z.; Malouki, M. A.; Khennaoui, B.; Santaballa, J. A.; Canle, M. Simulated Sunlight Photodegradation of 2-Mercaptobenzothiazole by Heterogeneous Photo-Fenton Using a Natural Clay Powder. *J. Environ. Chem. Eng.* **2018,** *6* (2), 1783–1793.

Rodrigues-Silva, C.; Maniero, M. G.; Rath, S.; Guimarães, J. R. Degradation of Flumequine by the Fenton and Photo-Fenton Processes: Evaluation of Residual Antimicrobial Activity. *Sci. Total Environ.* **2013,** *445*, 337–346.

Rodríguez, M.; Malato, S.; Pulgarin, C.; Contreras, S.; Curcó, D.; Giménez, J.; Esplugas, S. Optimizing the Solar Photo-Fenton Process in the Treatment of Contaminated Water. Determination of Intrinsic Kinetic Constants for Scale-Up. *Solar Energy* **2005,** *79* (4), 360–368.

Rodriguez, M.; Sarria, V.; Esplugas, S.; Pulgarin, C. Photo-Fenton Treatment of a Biorecalcitrant Wastewater Generated in Textile Activities: Biodegradability of the Photo-Treated Solution. *J. Photochem. Photobiol. A Chem.* **2002,** *151* (1–3), 129–135.

Romero, V.; Acevedo, S.; Marco, P.; Giménez, J.; Esplugas, S. Enhancement of Fenton and Photo-Fenton Processes at Initial Circumneutral pH for the Degradation of the β-Blocker Metoprolol. *Water Res.* **2016,** *88*, 449–457.

Rozas, O.; Contreras, D.; Mondaca, M. A.; Pérez-Moya, M.; Mansilla, H. D. Experimental Design of Fenton and Photo-Fenton Reactions for the Treatment of Ampicillin Solutions. *J. Hazard. Mater.* **2010,** *177* (1–3), 1025–1030.

Ruppert, G.; Bauer, R.; Heisler, G. The Photo-Fenton Reaction-Wastewater Treatment Process an Effective Photochemical. *J. Photochem. Photobiol. A. Chem.* **1993,** *73* (1), 75–78.

Saatci, Y. Decolorization and Mineralization of Remazol Red F3B by Fenton and Photo-Fenton Processes. *J. Environ. Eng.* **2010,** *136* (9), 1000–1005.

Sabhi, S.; Kiwi, J. Degradation of 2,4-Dichlorophenol by Immobilized Iron Catalysts. *Water Res.* **2001,** *35* (8), 1994–2002.

Safarzadeh-Amiri, A.; Bolton, J. R.; Cater, S. R. Ferrioxalate-Mediated Photodegradation of Organic Pollutants in Contaminated Water. *Water Res.* **1997,** *31* (4), 787–798.

Sajjala, S. R.; Sairam, V. De-Colorization of Sudan IV Dye Solution by Solar Photo Fenton and TiO_2 Mediated Solar Photo Fenton Processes: A Comparative Study. *Int. J. Adv. Res. Eng. Technol.* **2020,** *11* (6), 774–785.

Saritha, P.; Aparna, C.; Himabindu, V.; Anjaneyulu, Y. Comparison of Various Advanced Oxidation Processes for the Degradation of 4-Chloro-2 Nitrophenol. *J. Hazard. Mater.* **2007,** *149* (3), 609–614.

Sarria, V.; Deront, M.; Péringer, P.; Pulgarin, C. Degradation of a Biorecalcitrant Dye Precursor Present in Industrial Wastewaters by a New Integrated Iron (III) Photoassisted–Biological Treatment. *Appl. Catal. B Environ.* **2003,** *40* (3), 231–246.

Scheeren, C. W.; Paniz, J. N. G.; Martins, A. F. Comparison of Advanced Processes on the Oxidation of Acid Orange 7 Dye. *J. Environ. Sci. Health A* **2002,** *37* (7), 1253–1261.

Sedlak, P.; Luňak, S.; Brodilová, J.; Lederer, P. Photocatalytic Effects of Fe (III) Compounds on the Hydroxylation of Benzoic Acid by Hydrogen Peroxide Initiated by UV-Radiation. *React. Kinet. Catal. Lett.* **1989,** *39* (2), 249–253.

Selvam, K.; Muruganandham, M.; Swaminathan, M. Enhanced Heterogenous Ferrioxalate Photo-Fenton Degradation of Reactive Orange-4 by Solar Light. *Solar Energy Materi. Solar Cells* **2005,** *89*, 61–74.

Serp, P.; Corrias, M.; Kalck, P. Carbon Nanotubes and Nanofibers in Catalysis. *Appl. Catal. A Gen.* **2003,** *253* (2), 337–358.

Severo, C. E.; Dotto, G. L.; Silvestri, S.; Nunes, I. S.; Salla, S. J.; Martinez-de, C. A. et al. Improved Catalytic Activity of EDTA–Modified $BiFeO_3$ Powders for Remarkable Degradation of Procion Red by Heterogeneous Photo–Fenton Process. *J. Environ. Chem. Eng.* 2020, *8* (4). DOI: 10.1016/j.jece.2020.103853.

Sharma, K.; Raizada, P.; Hosseini-Bandegharaei, A.; Thakur, P.; Kumar, R.; Thakur, V. K. et al. Fabrication of Efficient CuO/Graphitic Carbon Nitride Based Heterogeneous Photo-Fenton Like Catalyst for Degradation of 2, 4-Dimethylphenol. *Process Saf. Environ.* **2020,** *142*, 63–75.

Silva, M. R. A.; Trovó, A. G.; Nogueira, R. F. P. Degradation of the Herbicide Tebuthiuron Using Solar Photo-Fenton Process and Ferric Citrate Complex at Circumneutral pH. *J. Photochem. Photobio. A: Chem.* **2007,** *191* (2–3), 187–192.

Soon, A. N.; Hameed, B. H. Heterogeneous Catalytic Treatment of Synthetic Dyes in Aqueous Media Using Fenton and Photo-Assisted Fenton Process. *Desalination* **2011,** *269* (1–3), 1–16.

Southworth, B. A.; Voelker, B. M. Hydroxyl Radical Production via the Photo-Fenton Reaction in the Presence of Fulvic Acid. *Environ. Sci. Technol.* **2003,** *37* (6), 1130–1136.

Sun, J. H.; Sun, S. P.; Fan, M. H.; Guo, H. Q.; Qiao, L. P.; Sun, R. X. A Kinetic Study on the Degradation of p-Nitroaniline by Fenton Oxidation Process. *J. Hazard. Mater.* **2007,** *148* (1–2), 172–177.

Sýkora, J.; Pado, M.; Tatarko, M.; Izakovič, M. Homogeneous Photo-Oxidation of Phenols: Influence of Metals. *J. Photochem. Photobiol. A Chem.* **1997,** *110* (2), 167–175.

Torrades, F.; Pérez, M.; Mansilla, H. D.; Peral, J. Experimental Design of Fenton and Photo-Fenton Reactions for the Treatment of Cellulose Bleaching Effluents. *Chemosphere* **2003,** *53* (10), 1211–1220.

Tryba, B.; Morawski, A. W.; Inagaki, M.; Toyoda, M. Mechanism of Phenol Decomposition on $FeCTiO_2$ and $FeTiO_2$ Photocatalysts via Photo-Fenton Process. *J. Photochem. Photobiol. A Chem.* **2006a,** *179* (1–2), 224–228.

Tryba, B.; Morawski, A. W.; Inagaki, M.; Toyoda, M. Effect of the Carbon Coating in Fe–C–TiO_2 Photocatalyst on Phenol Decomposition Under UV Irradiation via Photo-Fenton Process. *Chemosphere* **2006b,** *64* (7), 1225–1232.

Varghese, R.; Aravind, U. K.; Aravindakumar, C. T. Fenton-Enhanced γ-Radiolysis of Cyanuric Acid. *J. Hazard. Mater.* **2007,** *142* (1–2), 555–558.

Vermilyea, A. W.; Voelker, B. M. Photo-Fenton Reaction at Near Neutral pH. *Environmental Sci. Technol.* **2009,** *43* (18), 6927–6933.

Vilar, V. J.; Moreira, J. M.; Fonseca, A.; Saraiva, I.; Boaventura, R. A. Application of Fenton and Solar Photo-Fenton Processes to the Treatment of a Sanitary Landfill Leachate in a Pilot Plant with CPCs. *J. Adv. Oxid. Technol.* **2012,** *15* (1), 107–116.

Walling, C. Intermediates in the Reactions of Fenton Type Reagents. *Acc. Chem. Res.* **1998,** *31*, 155–157.

Wei, G.; Li, Y.; Cai, S.; Li, Z.; Mo, J.; Zhang, L. Photo-Fenton Degradation of Ethyl Xanthate Catalyzed by Bentonite-Supported Fe (II)/Phosphotungstic Acid Under Visible Light Irradiation. *Water Sci. Technol.* **2018,** 2017 (2), 473–480.

White, E. M.; Vaughan, P. P.; Zepp, R. G. Role of the Photo-Fenton Reaction in the Production of Hydroxyl Radicals and Photobleaching of Colored Dissolved Organic Matter in a Coastal River of the Southeastern United States. *Aquat. Sci.* **2003,** *65* (4), 402–414.

Wu, K.; Xie, Y.; Zhao, J.; Hidaka, H. Photo-Fenton Degradation of a Dye Under Visible Light Irradiation. *J. Mol. Catal. Chem.* **1999,** *144* (1), 77–84.

Xue, X. F.; Liu, Y. X.; Shao, Y. Q.; Deng, N. S. Rapid Decolorization of Rhodamine B by UV/Fe (III)-Penicillamine Process Under Neutral pH: Compared with UV/Fe (III)-Oxalate. *Adv. Mater. Res.* **2011,** *183*, 130–134.

Yang, S.; He, H.; Wu, D.; Chen, D.; Ma, Y.; Li, X. et al. Degradation of Methylene Blue by Heterogeneous Fenton Reaction Using Titanomagnetite at Neutral pH Values: Process and Affecting Factors. *Ind. Eng. Chem. Res.* **2009,** *48* (22), 9915–9921.

Yeh, C. K. J.; Kao, Y. A.; Cheng, C. P. Oxidation of Chlorophenols in Soil at Natural pH by Catalyzed Hydrogen Peroxide: The Effect of Soil Organic Matter. *Chemosphere* **2002,** *46* (1), 67–73.

Yip, A. C. K.; Lam, F. L. Y.; Hu, X. A Novel Heterogeneous Acid-Activated Clay Supported Copper Catalyst for the Photobleaching and Degradation of Textile Organic Pollutant Using Photo-Fenton-Like Reaction. *Chem. Commun.* **2005,** *25*, 3218–3220.

Zakharov, I. V.; Kumpan, Y. V. Inhibiting and Chemiluminescent Properties of Benzoic Acid and Acetylsalicylic Acid in the Fenton Reagent System. *Kinetics Catal.* **1996,** *37* (2), 174–178.

Zekkaoui, C.; Berrama, T.; Dumoulin, D.; Billon, G.; Kadmi, Y. Optimal Degradation of Organophosphorus Pesticide at Low Levels in Water Using Fenton and Photo-Fenton Processes and Identification of By-Products by GC-MS/MS. *Chemosphere* 2021, *279*. DOI: 10.1016/j.chemosphere.2021.130544.

Zelmanov, G.; Semiat, R. Iron (3) Oxide-Based Nanoparticles As Catalysts in Advanced Organic Aqueous Oxidation. *Water Res.* **2008,** *42* (1–2), 492–498.

Zepp, R. G.; Faust, B. C.; Hoigne, J. Hydroxyl Radical Formation in Aqueous Reactions (pH 3–8) of Iron (II) with Hydrogen Peroxide: The Photo-Fenton Reaction. *Environ. Sci. Technol.* **1992,** *26* (2), 313–319.

Zhang, G.; Gao, Y.; Zhang, Y.; Guo, Y. Fe_2O_3-Pillared Rectorite as an Efficient and Stable Fenton-Like Heterogeneous Catalyst for Photodegradation of Organic Contaminants. *Environ. Sci. Technol.* **2010,** *44* (16), 6384–6389.

Zhang, S.; Liang, S.; Wang, X.; Long, J.; Li, Z.; Wu, L. Trinuclear Iron Cluster Intercalated Montmorillonite Catalyst: Microstructure and Photo-Fenton Performance. *Catal. Today* **2011b,** *175* (1), 362–369.

Zhang, J.; Hu, F. T.; Liu, Q. Q.; Zhao, X.; Liu, S. Q. Application of Heterogenous Catalyst of Tris (1, 10)-Phenanthroline Iron (II) Loaded on Zeolite for the Photo-Fenton Degradation of Methylene Blue. *React. Kinet. Mech. Catal.* **2011b,** *103* (2), 299–310.

Zhang, X.; Geng, Z.; Jian, J.; He, Y.; Lv, Z.; Liu, X. et al. Potassium Ferrite as Heterogeneous Photo-Fenton Catalyst for Highly Efficient Dye Degradation. *Catal.* **2020,** *10* (3). DOI: 10.3390/catal10030293.

Zhang, Y.; Chen, X.; Cui, M. S.; Guo, Z.; Chen, Y. H.; Cui, K. P. et al. Binding Fe-Doped g-C_3N_4 on the Porous Diatomite for Efficient Degradation of Tetracycline via Photo-Fenton Process. *J. Environ. Chem. Eng.* **2022,** *10* (3). DOI: 10.1016/j.jece.2022.107406.

Zhao, Y.; Kang, S.; Qin, L.; Wang, W.; Zhang, T.; Song, S. et al. Self-Assembled Gels of Fe-Chitosan/Montmorillonite Nanosheets: Dye Degradation by the Synergistic Effect of Adsorption and Photo-Fenton reaction. *Chem. Eng. J.* **2020,** *379*. DOI: 10.1016/j.cej.2019.122322.

Zheng, H. L.; Xiang, X. Y. Photo-Fenton Oxidation Processes Used in the Degradation of Rhodamine B. *Guang pu xue yu Guang pu fen xi= Guang pu* **2004,** *24* (6), 726–729.

Zúñiga-Benítez, H.; Muñoz-Calderón, A.; Peñuela, G. A. Removal of a Mix of Benzophenones and Parabens Using Solar Photo-Fenton and a Cylinder Parabolic Collector in Aqueous Solutions. *J. Environ. Chem. Eng.* **2018,** *6* (6), 7347–7357.

CHAPTER 10

Sonochemistry: A Pollution-Free Pathway

NEHA KAPOOR[1], GARIMA AMETA[2], SURBHI BENJAMIN[3], VIKAS SHARMA[4], and SHIPRA BHARDWAJ[5]

[1]*Department of Chemistry, PAHER University, Udaipur, India*

[2]*Department of Chemistry, M. L. Sukhadia University, Udaipur, India*

[3]*Department of Chemistry, J. R. N. Rajasthan Vidhyapeeth, (Deemed to be University), Udaipur, India*

[4]*Jain Mandir Road, Kota Jn., Kota, India*

[5]*Department of Chemistry, Govt. Meera Girls' College, Udaipur, India*

ABSTRACT

Ultrasound can be used as a source of energy to drive certain organic reactions of synthetic importance. Its use is relatively pollution free as it does not utilize any chemical; rather its exposure leads to a particular desired reaction. It is also used in polymer synthesis and heterogeneous catalysis. The ultrasound has been used in carrying out a number of reactions such as Reformatsky reaction, Ullmann reaction, Blaise reaction, Sonogashira coupling, Curtius rearrangement, and synthesis of dihydropyrimidinones, hydantoins, spiro-oxindoles, dihyropyridines and polyhydroquinolines, etc. The use of ultrasound in chemical reactions can be considered a green chemical pathway for preparing various useful compounds.

10.1 INTRODUCTION

Different forms of energy, such as heat, light can drive some chemical reactions. But in the past few decades, ultrasound has emerged as a potential

Green Chemistry, 2nd Edition: Fundamentals and Applications. Suresh C. Ameta & Rakshit Ameta (Eds.)

source to enhance the chemical reactivity, besides being used in non-chemical situations, such as the medical diagnosis, navigation by bats, cleaning and drilling of teeth, SONAR, and material testing. It has been found that when ultrasound in low frequency range is passed through a chemical system, it influences its chemical reactivity. The study of effects of ultrasound on chemical reactivity is termed as sonochemistry.

The chemical effect of ultrasound was first reported by Richards and Loomis (1927). The basis for the present-day generation of ultrasound was established as far back as 1880 with the discovery of piezoelectric effect by the Curie and Curie (1880, 1881). Crystalline materials showing this effect are known as piezoelectric materials. Ultrasonic devices consist of transducers (energy converters), which are composed of these piezoelectric materials. An inverse piezoelectric effect is used in transducers, that is, a rapidly alternating potential is placed across the faces of piezoelectric crystal, which generates dimensional change, and thus converts electrical energy into sound energy. The first ultrasonic transducer was a whistle developed by Galton (1983), who was then investigating the frequency of human hearing.

10.2 ULTRASOUND

10.2.1 CLASSIFICATION

Scientifically, sound is the transmission of energy through the generation of acoustic pressure waves in the medium. Sound is said to be a mechanical wave, that is, it requires a medium and the particles in the medium are vibrating to transfer the waves. The frequency of the wave determines its regime and it is classified in to different kinds as given below:

Infrasound–Sound waves having frequency less than 20 Hz.
Audible sound–Sound waves having frequency between 20 Hz–20 KHz.
Ultrasound–Sound waves with frequency more than 20 KHz.
Hypersound–Sound waves with frequency higher than 10 GHz.

10.2.2 PRINCIPLES OF SONOCHEMISTRY

Sonochemical effects are due to the phenomenon of acoustic cavitation, that is, the creation, growth, and implosive collapse of gas-filled bubbles in a liquid in response to an applied ultrasonic field. Cavitation was first identified and reported by Thornycraft and Barnaby (1895). When a liquid is irradiated by ultrasound, microbubbles will appear, grow, and oscillate extreme quickly and even collapse near a solid surface generating microjets and shock waves. Moreover, in the liquid phase surrounding the particles, high micro-mixing will increase the heat and mass transfer and even the diffusion of species inside the pores of the solid.

10.2.3 PHENOMENON OF CAVITATION

Sound waves are basically pressure waves. Sound wave consists of alternating compression and expansion cycles. When an acoustic field is applied to a liquid, the sonic vibrations create an acoustic pressure (P_a) at any time (t), which is given by the equation:

$$P_a = P_A \sin 2\pi ft$$

where P_A is the maximum pressure amplitude of the wave and f is the frequency of the sound wave.

When ultrasound is passed through a liquid, the total pressure (P) in the liquid is given by

$$P = P_a - P_h$$

where P_h is the hydrostatic pressure

During the negative cycle of the wave, the distance between the molecules of the liquid will vary (oscillate) about a mean position. If the distance between the molecules exceeds the critical molecular distance, R (e.g., for water, the value of R is 10^{-8} cm), then the liquid will breakdown and voids will be created, that is, formation of cavitation bubbles. During the positive cycle of the wave, the bubbles grew in size due to the positive acoustic pressure and then finally collapse, leading to the formation of new nuclei for the next cavitation.

This is the mechanism for a cavity formation. If attempts are made to remove contaminants, particulates, precipitates, etc. from a solution, these are generally small dust moles or crystalline materials present in it. These irregular surfaces allow gas to be trapped. Acoustic waves consist of alternating compression and rarefaction waves. Upon rarefaction, the gas, which is trapped in the dust mole, is pulled out to become a free-standing bubble.

The occurrence of these collapses near a solid surface will generate microjets and shock wave (Suslick and Casadonte, 1987). Moreover, in the liquid phase surrounding the particles, high micromixing will increase the heat and mass transfer and even the diffusion of species inside the pores of the solid.

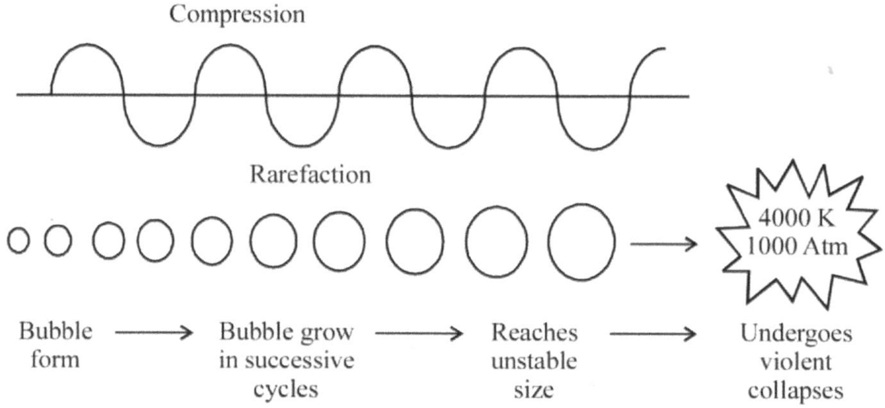

FIGURE 10.1 Impulsive collapse of bubbles.

10.2.4 FACTORS AFFECTING CAVITATION

i) **Ultrasound frequency:** Mostly, sonochemical work involves frequencies between 20 and 50 KHz. It has been observed that sonochemical effect is limited at higher frequencies. It is due to the reason that the bubble has less time to grow and collapse at higher frequencies.

ii) **Presence of gas:** Dissolved gases act as nucleation sites for cavitation. As gases are removed from the reaction mixture because of the implosion of the cavitation bubbles, initiation of new cavitational events becomes increasingly difficult. Bubbling gases through the mixture facilitates the production of cavitation bubbles.

iii) **Effect of external pressure:** On increasing the external pressure, P_h, it requires the application of greater acoustic pressure, P_a and hence, the system requires higher ultrasonic intensities to generate cavitation bubbles.

iv) **Temperature:** Sonication proceeds more efficiently at lower temperatures. It is due to the effect on the solvent properties, such as densities, surface tensions, viscosities.

10.2.5 SOURCES OF ULTRASOUND IN SONOCHEMISTRY

i) **Ultrasonic cleaning bath:** The simple ultrasonic bath is economical and is an easily available source of ultrasonic irradiation in the chemical laboratories. It has several drawbacks, such as the ultrasonic intensity is limited for every ultrasonic cleaning bath, temperature control is also difficult and reproducibility of results is relatively poor because inter-laboratory comparisons are difficult. Different baths have different frequencies and power outputs.

ii) **Ultrasonic probe system:** In this type of system, a metal "probe" is used, which is attached to the transducer introduced directly into a reaction system itself for increasing the amount of ultrasonic power.

The horn is part of an assembly, which amplifies small vibrations of the piezoelectric crystal to a larger amplitude. This probe system has several advantages over a cleaning bath.

The ultrasonic probe system is of two types:–

- Cup-horn system
- Flow-cell system

iii) **Submersible transducer:** In this type of system, transducer is directly immersed into the reaction. It is an alternative to the cleaning bath and also used for batch sonication.

iv) **Whistle reactor:** In this system, ultrasonic intensities are very low. Generally, this equipment is used for emulsification, polymerization and phase-transfer reactions. Liquids are pumped at a high rate through a narrow gap onto a thin metal blade. This sets the blade into vibration with a sufficiently high frequency to cause cavitation, which causes the reaction to proceed.

v) **Tube reactor:** Tube is surrounded by a radical transducer. Reaction mixture flows through this tube. The ultrasonic energy is focused toward the middle of the tube with much lower powers at the inner surface. In this way, erosion problems are reduced.

10.3 ORGANIC SYNTHESIS

The first report about the effect of ultrasound to chemical reactions is by Richards and Loomis (1927) involving rate studies on the hydrolysis of dimethyl sulphate and the iodine "clock" reaction (the reduction of potassium iodate by sulfurous acid) Later, Porter and Young (1938) reported that ultrasound increases the rate of the Curtius rearrangement.

The driving force for developments in the field of use of ultrasound in organic synthesis has many facets. The increasing requirement for environmental technology, which minimizes of the productions of waste at source (Cains et al., 1998). Ultrasound may offer cleaner reactions by improving product yields and selectivities, enhancing product recovery and quality through application to crystallization, product recovery and purification processes. Ultrasound enhances the rate particularly of those reactions which involve free radical intermediates (Singh et al., 2001). Sonication allows the use of non-activated and crude reagents as well as an aqueous solvent system, and therefore, it is eco-friendly and nontoxic. Ultrasound is widely used for improving the conditions, reducing long reaction times, avoiding high temperatures, unsatisfactory yield and incompatibility with other functional groups (Yadav et al., 2001).

Han and Boudjouk (1982) found significant increased yields and rates of Reformatsky reactions by ultrasound.

$$CH_3(CH_2)_6CHO + BrCH_2COOEt \xrightarrow[\bullet)))]{\text{Zn, Dioxane}} CH_3(CH_2)_6CHOH\ CH_2COOEt$$

It was observed that ultrasound promoted synthesis gave 95% yield within 5 min, whereas the conventional method gives only 61% yield with 12 h at 80°C on stirring.

Ceric ammonium nitrate (CAN) effectively catalyzes the three-component condensation of an aldehyde, β-ketoester, and urea in methanol to afford the corresponding dihydropyrimidinones in excellent yields under sonication. The reaction of benzaldehyde, ethyl acetoacetate, and urea for 3.5 h sonication resulted in the formation of 3,4-dihydropyrimidine-2(1H)-one (92% yield), and it is believed to proceed through a single electron transfer with initial formation of a β-ketoester radical that adds to the imine intermediate (Li et al., 1998).

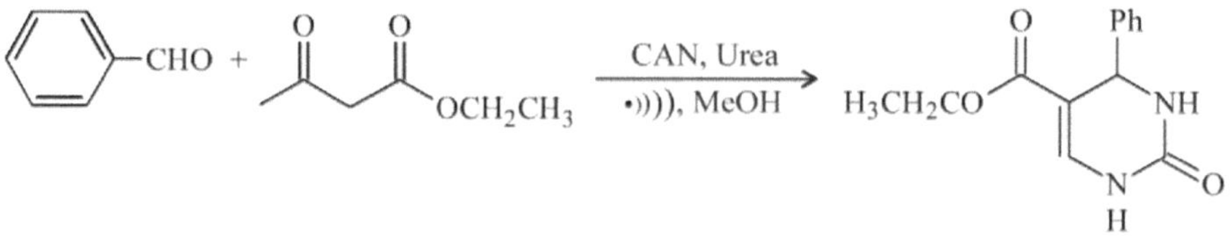

Li et al. (1999) improved Bucherer-Bergs method for the synthesis of 5,5-disubstituted hydantoins using ultrasound and Claisen-Schmidt condensation for cycloalkanenes and chalcones.

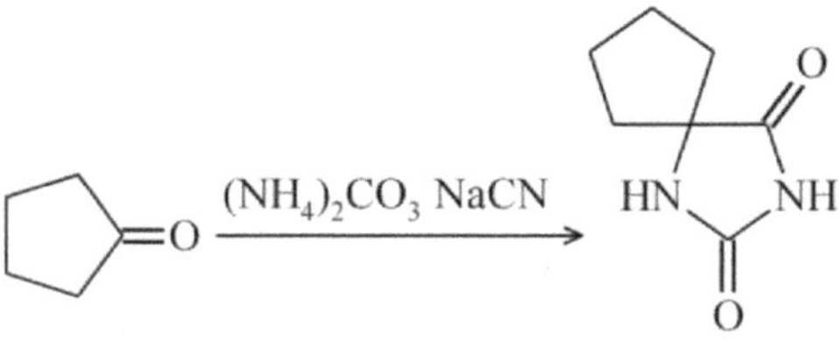

Hydantoin

	Traditional	US
Time (h)	4	3.5
T (°C)	150 (Autoclave)	45
Yield (%)	70	89

Sonication methods were used to synthesize selenium heterocycles and anhydrides by Hu et al. (1997) and Wang and Zhao (1996). Ultrasound increased 5–10% yield of cis- and trans-2,6-diphenyl-1,4-diselenofulvenes. Traditionally, the anhydrides were produced under phase-transfer condition at −10°C using aqueous NaOH as non-organic phase. The anhydrides can be prepared in a single organic phase even at 45°C using ultrasound. The synthesis of indoles was also improved under sonication (Koulocheri and Harautounain, 2001). Ultrasound increased the yield by about 10–15% and reduced the reaction time to one-fifth under a polyphosphoric acid (PPA) condition.

Ultrasound can seriously affect photocatalytic ring opening of α-epoxyketones by 1-benzyl-2,4,6-triphenylpyridinium tetrafluoroborate (NBTPT) as photocatalyst in methanol because of the efficient mass transfer of the reactions and the excited state of NBTPT (Memarian and Saffar-Teluri, 2007). The higher yields and shorter reaction times are the main advantages of this method.

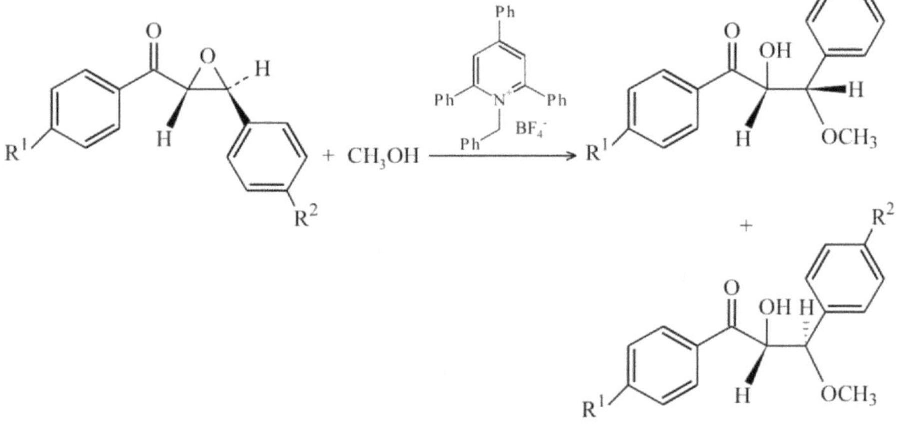

A multicomponent synthesis of spiro-oxindoles was carried out in the presence of a catalytic amount of p-TSA as an inexpensive and available catalyst in EtOH under ultrasound irradiation (Dabiri et al., 2011). The method is quite simple. It starts from readily accessible commercial starting materials, and provides biologically interesting products in good yields and shorter reaction times.

When ultrasound is applied to an Ullmann reaction that normally requires a 10-fold excess of copper and 48 h of reaction time, it can be reduced to only four-fold excess of copper and a reaction time of hardly 10 h. The particle size of the copper shrinks from 87 to 25 μm, but the increase in the surface area cannot fully explain the increase in reactivity. It was suggested that sonication also assists in the breakdown of intermediates and desorption of the products form the surface (Mason, 1999).

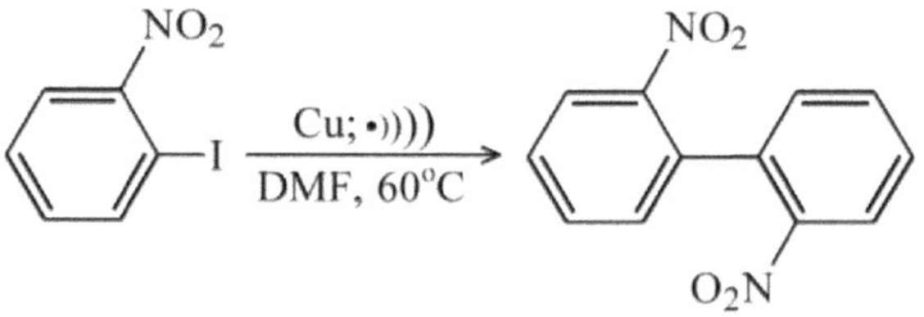

Typically, ionic reactions are accelerated by physical effects like better mass transport, which is also called "False sonochemistry." If the extreme conditions within the bubbles lead to totally new reaction pathways, for example, via radicals generated in the vapor phase that would only have a transient existence in the bulk liquid, one speaks about "sonochemical switching." Such a switching has been observed in the following Kornbulum-Russel reaction, where sonication favors a single electron transfer SET pathway (Mason, 1999).

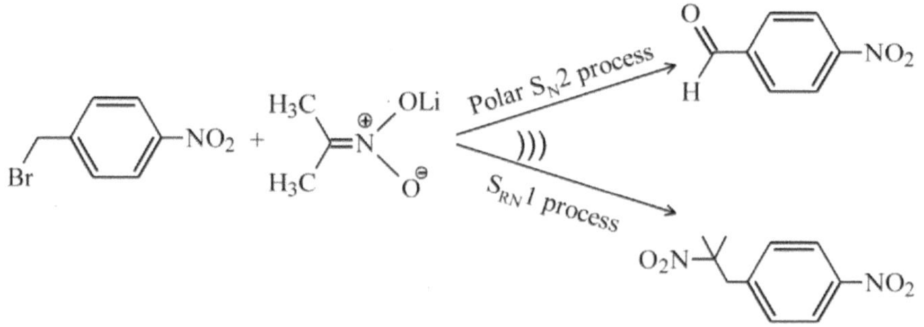

β-Amino-α,β-unsaturated esters are produced by a sonochemical reaction. Blaise reaction of nitriles, zinc powder, zinc oxide, and ethyl bromoacetate in THF was reported in a commercial ultrasonic cleaning bath (Lee and Chang, 1997).

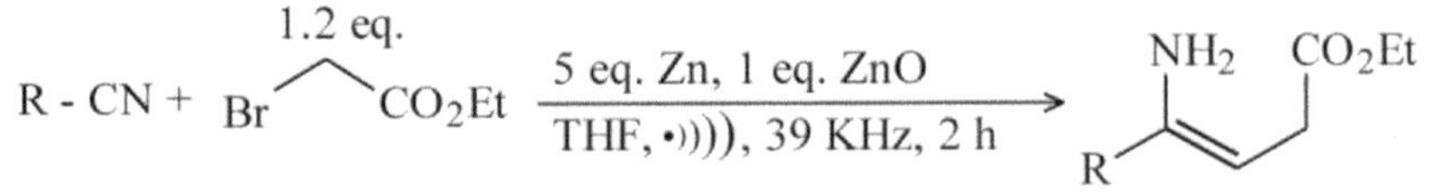

A palladium catalyzed and ultrasonic promoted Sonogashira coupling (1,3-dipolar cycloaddition of acid chlorides, terminal acetylenes, and sodium azide) in one pot enables an efficient synthesis of 4,5-disubstituted-1,2,3-(NH)-triozoles in excellent yield (Li et al., 2009).

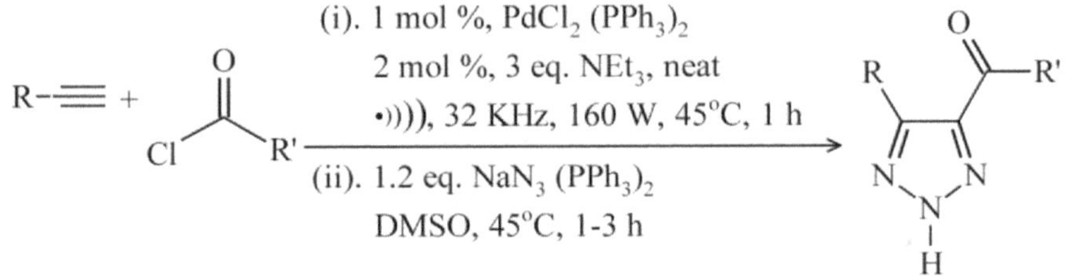

A facile one-pot procedure for the synthesis of urea linked peptide mimetics and neoglycopeptides under Curtius rearrangement conditions employing Deoxo-fluor and $TMSN_3$ is efficient under ultrasonication and it circumvents the isolation of acryl azide and isocyanate intermediates (Hemantha et al., 2009).

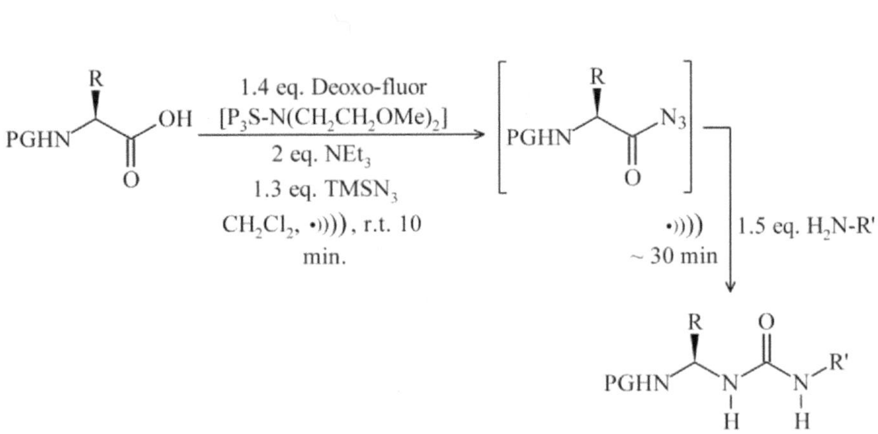

Structurally and functionally diverse N-carbamoylamino acids were obtained through the alkylation of monosubstituted parabanic acid followed by hydrolysis of the intermediate products in very good yields and excellent purity (Bogolubsky et al., 2008).

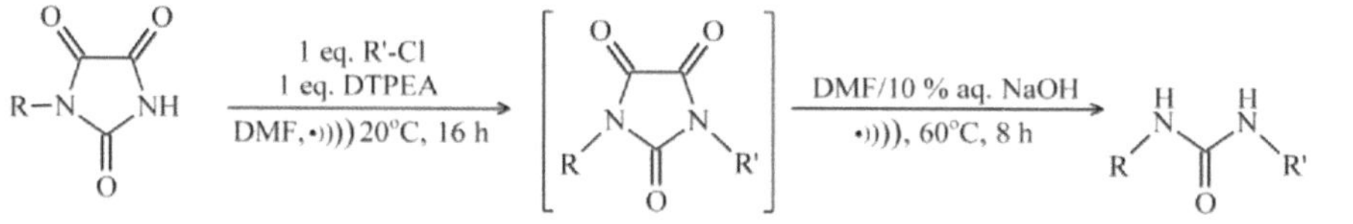

Parabanic acid N-Carbamoylamino acid

Hantzsch 1,4-dihyropyridine and polyhydroquinoline derivatives were synthesized in excellent yields in aqueous micelles. The reaction is catalyzed by P-TSA and strongly accelerated by ultrasonic irradiation (Kumar and Maurya, 2008).

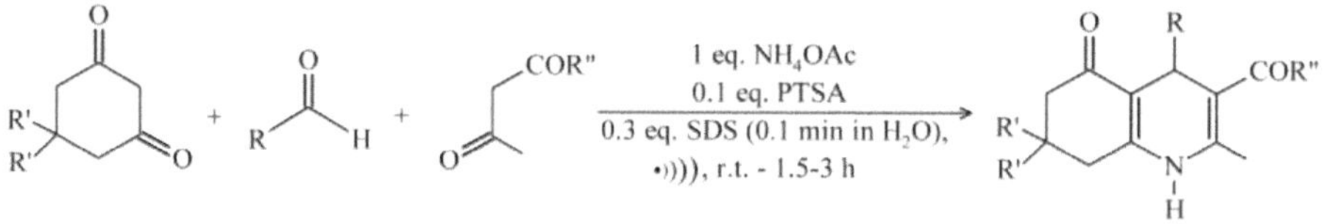

Application of ultrasound accelerates the conversion of hydroxamic acid from carboxylic acids in the presence of 1-proponephosphonic acid cyclic anhydride (Vasantha et al., 2010). Further, T3P has also been employed to activate the hydroxamates, leading to isocyanates via Lossen rearrangement. Trapping with suitable nucleophile affords the corresponding ureas and carbonates.

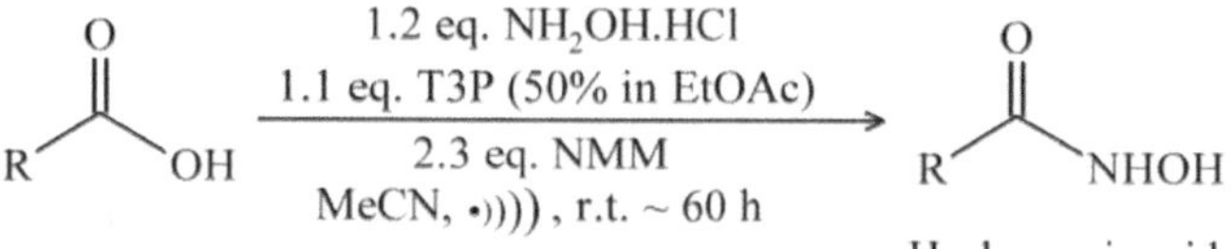

Hydroxamic acid

10.4 BIOLOGICAL REACTIONS

Enzymes are being used for enhancing the reactivity of biological materials or organic synthesis because of the regio- and stereospecific nature of their reactions. Yeast cells can have the cuticle removed and enzyme released in active form by the use of ultrasound (Bujans et al., 1998). Attaching enzymes to hydrophobic surfaces of polymers or macromolecules presents a problem because reactants in aqueous solution will encounter the enzyme. Stirring can solve this problem but not very efficiently. The application of low-intensity ultrasound (7.6 MHz, 1.5 KW m^{-2}) speeds up the reactions of the enzymes α-amylase and glucoamylase, which have been immobilized on polystyrene (Schmidt et al., 1987).

One of the most well-known uses of ultrasound in biotechnology is for microbial cell disruption (Suslick, 1998). When a cellular material is placed in an ultrasonic field, the shock waves produced by surrounding cavitational events are capable of causing mechanical damage to the surrounding cellular materials. Other important use of ultrasound is to synthesize N-acetylamino acids from the amino acids and acetic anhydride without racemization (Reddy and Ravindranath, 1992a). This reaction was later advantageously incorporated as a synthesis step in the production of α, β- and cyclic spaglumic acids (Reddy and Ravindranath, 1992b).

Lee et al. (1989, 1990) used the ultrasound promoted cycloadditions to synthesize series of natural compounds. The cycloaddition of diene **(1)** with enone **(2)** gave a 76% yield of cycloadducts with the desired regiomers, in the ratio of 5: 1 under sonication for 2 h at 45°C as compared with a 15% yield of cycloadducts in a ratio of 1: 1 when refluxed in benzene for 8 h. Deprotection of **(3)** yields natural compound tanshindiol B.

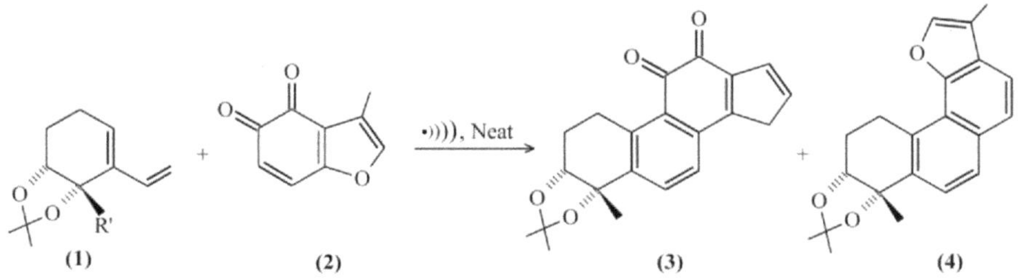

•)))), 2 h, 45°C, 76%, **(3)/(4)** = 5 : 1, Reflux in C_6H_6, 8 h, 15%, **(3)/(4)** = 1:1

High-intensity ultrasound was used to prepare an aqueous suspension of proteinaceous microspheres (long-lived) filled with either air, or water-insoluble liquids for application in the filled of medical science. Emulsification proved to be insufficient in producing these microspheres (long lived) and chemical reactions are also proved to be in critical in forming them, which required oxygen. Particularly, hydrogen atoms are produced from sonolysis of water, which will react with oxygen to generate superoxide.

Curcumin is a natural yellow colorant of turmeric (a spice used primarily as a food colorant and also to flavor several foods). It is a nutraceutical compound used worldwide for medicinal as well as food purposes. It has attracted special attention due to its potential pharmacological activities such as to protect cells from β-amyloid in Alzheimer's disease and cancer preventive properties. Curcumin is insoluble in water at acidic and neutral pH. In order to make it water soluble, it is envisaged that attachment of a polar group and molecule would enhance the hydrophilicity of the molecule. This can be achieved by making its suitable sugar derivatives.

Ultrasound brought about acceleration and increases the yields of the curcumin glucosides in the Koenigs-Knorr type reaction of 2,3,4,6-tetra-O-acetyl–1-bromoglucose with the potassium salt of curcumin [bis-1,7-(3'-methoxy-4'-hydroxy) phenyl-5-hydroxy-1,4,6-heptatrien-3-one] under the biphasic reaction conditions in the presence of benzyltributyl ammonium chloride as a phase-transfer catalyst (Parvathy and Srinivas, 2008). The reaction was stereoselective leading to the preponderant formation of either mono or di-glucoside tetraacetates of curcumin under controlled conditions in mono- and biphasic reactions, respectively.

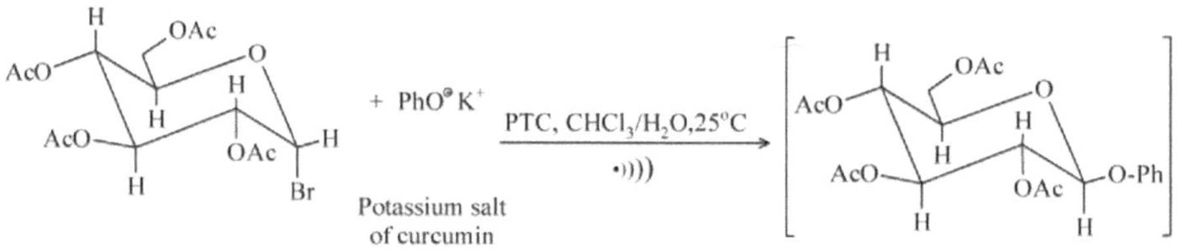

α-Tetra-O-acetyl-α-D-glucopyranosyl bromide

Curcumin-di-β-glucoside tetraacetate

2-Methoxy-6-alkyl-1,4-benzoquinones occur widely in nature, particularly in plants and most of them show potent biological activities, such as anticancer activity, radiosensitization activity, and 5- lipoxygenase inhibitory activity. Ultrasound-assisted Witting reaction of alkyltriphenyl phosphonium bromides with O-vanillin in basic aqueous conditions followed by reduction with Na/n-BuOH gave 2-methoxy-6-alkylphenols. Oxidation of 2-methoxy-6-alkyl phenols with Fremy's salt produced 2-methoxy-6-alkyl-1,4-benzoquinenes (Wu et al., 2009).

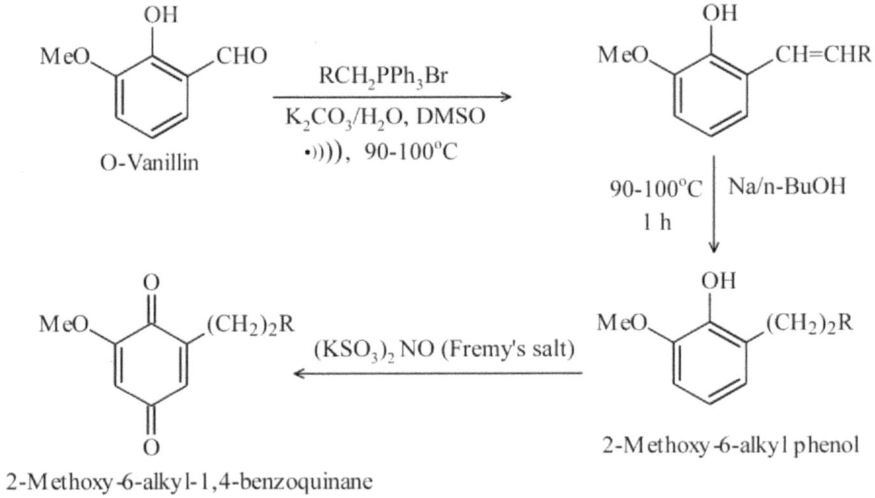

N-Arylhyroxylamines are an important class of compound frequently used as key intermediates in the synthesis of fine chemical, natural products, and some promising biologically active compounds. They also display a wide range of physiological and pharmacological activities. The Zn/$HCOONH_4$/CH_3CN system is used for preparing N-arylhydroxylamines by the reduction of the corresponding nitroarenes under ultrasound (Qi-Xun et al., 2009). This method is quite efficient, environmentally benign, highly chemoselective, especially simple, and most practical.

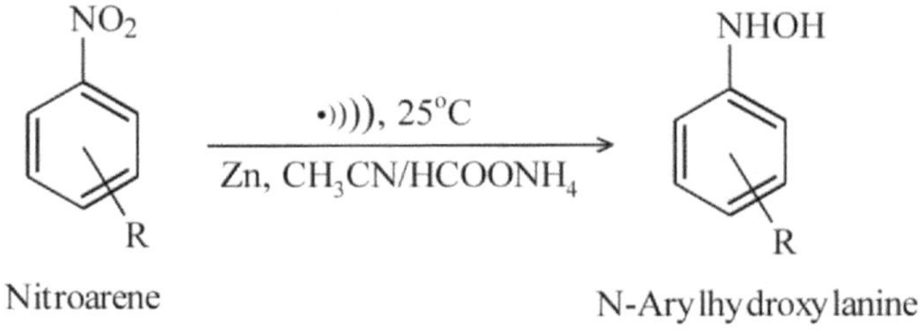

Carbazole and especially heterocycle-containing carbazole derivatives are embodied in many naturally occurring products and possess a broad spectrum of useful biological activities, such as antitumor, antimitotic, and antioxidative activities. They are also widely used as building blocks for new organic materials and play a very important role in electroactive and photoactive devices. On the other hand, the benzofuran derivatives also show some important biological properties, such as antimicrobial, anticonvulsant, anti-inflammatory, antitumor, and antifungal activities.

A facile synthesis of 3-(2-benzofuroyl) carbozols (9), 3,6-bis(2-benzofuroyl) carbozoles (10) and naphtha[2,1-6]furoylcarbozoles (11) and (12) has been reported (Gao et al., 2011). The synthesis mainly relies on the ultrasound-assisted Rap-Stoermer reaction of 3-chloroacetyl (5) or 3,6-dichloroacetyl-9-ethyl-9H-carbozole (6) with various salicyladehydes (7) as well as 2-hydroxy-1-naphthaldehyde (8) in CH_3CN with the presence of PEG 400 as catalyst. Using this method, it is easy to obtain benzofuroylcarbazoles in lesser time and that too with good yields.

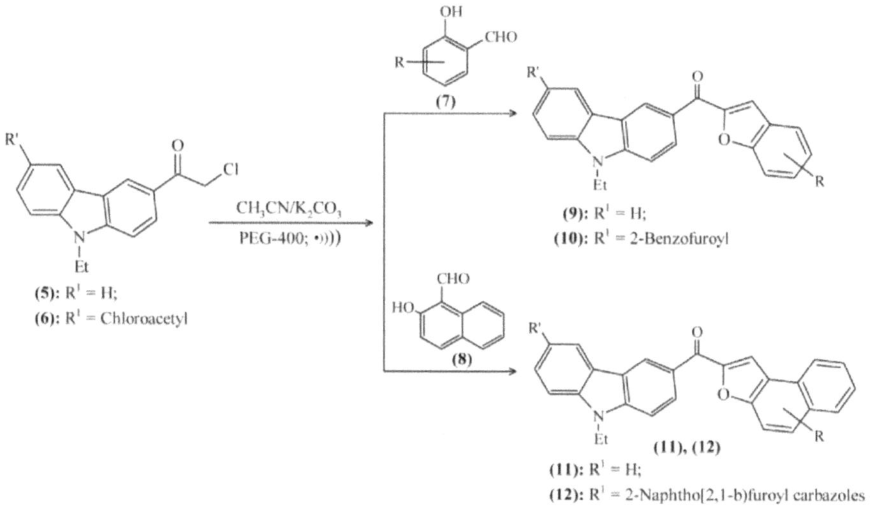

A number of biologically important compounds contain an aryl sulfone moiety, which exhibited antibacterial and antitumor properties as well as inhibitors for several enzymes, such as cyclooxygenase-2 (COX-2), HIV-1 reverse transcriptase, integrin VLA-4, and the ATPase. New compounds incorporating aryl sulphone moieties may display biological activities. Some homoallylic alcohol derivatives were synthesized by conventional and ultrasound-assisted methods in aqueous solution. Their derivatives are linked to sulfonyl dibenzene (Mady et al., 2013).

Larpkiattaworn et al. (2010) reported that biodiesel can be produced by using ultrasound to enhance the transesterification reaction for both homogeneous and heterogeneous systems. Ultrasonic irradiation can provide the properties of biodiesel, such as viscosity and mono-, di-, and tri-glyceride contents, which are within the limit of EN standard. Moreover, this method could reduce the transesterification reaction time using K/Al_2O_3 catalyst.

10.5 POLYMER SYNTHESIS

Ultrasound has been successfully applied to a wide range of polymer synthesis. High frequency ultrasound in the range of 1–10 MHz has been applied for the determination of structure and conformation of polymers (Pethrick, 1991). The chemical effects of ultrasound arise from cavitation, that is, the collapse of microscopic bubbles in a liquid. Upon implosion of a cavity, extreme conditions in the bubble occur (5000 K and 200 bar) and high-strain rates are generated outside the bubble (10^{-1} s^{-1}). Monomer molecules are dissociated by the high temperature inside the hot spot, whereas polymer chains are fractured by the high-strain rates outside the cavitation bubble. These two reactions lead to the formation of radicals, which can initiate a free radical polymerization. The majority of the radicals are generated by scission of polymer chains (Kuijpers, 2004).

An important parameter in ultrasound-induced bulk polymerizations is the viscosity. The polymerization reactions yielded high molecular weight polymers. In a pen reaction, the long chains formed cause a drastic increase in the viscosity. A high viscosity hinders cavitation and consequently reduces the production rate of radicals. Emulsion and precipitation polymerizations provide a potential solution to this problem. Ultrasound-induced bulk polymerization is usually performed at room temperature. This low temperature is chosen because radical formation induced by ultrasound is more efficient at lower temperatures. In emulsion polymerizations, a heterogeneous reaction system is involved. The polymers are insoluble in the continuous aqueous phase and therefore, the viscosity of the water phase does not increase upon reaction. Ultrasound-induced emulsion polymerization is a well studied system, where indeed high conversions can be obtained (Cooper et al., 1996; Zhang et al., 2002). During precipitation polymerizations, the polymer precipitates from the reaction mixture, resulting in a constant viscosity, and thus, a constant radical formation rate by ultrasound. In this perspective, high pressure carbon dioxide (CO_2) is an interesting medium as most monomers have a high solubility in CO_2, whereas it exhibits an antisolvent effect for most polymers.

10.5.1 RADICAL POLYMERIZATION

Generally, free radical polymerization consists of four elementary steps: initiation, propagation, chain transfer, and termination (Rudin, 1990). When ultrasound is used to initiate polymerization, radicals can be formed from monomer and from polymer molecules both (Feldman, 1995). This implies that due to radical formation by polymer scission, additional elementary step is introduced in ultrasound-induced polymerization as:

Initiation

$M + \text{Cavitation} \longrightarrow 2\,R^{\bullet}$ $\qquad R_i = 2\,k_{dmon}\,[M]$

Propagation

$M_n^{\bullet} + M \longrightarrow M_{n+1}^{\bullet}$ $\qquad R_p = k_p\,[M]\,[M_n^{\bullet}]$

Chain transfer

• to monomer

$M_n^{\bullet} + M \longrightarrow M_n + M^{\bullet}$ $\qquad R_{ctm} = k_{ctm}\,[M]\,[M_n^{\bullet}]$

• to polymer

$M_n^{\bullet} + M_m \longrightarrow M_n + M_m^{\bullet}$ $\qquad R_{ctp} = k_{ctp}\,[M_n^{\bullet}]\,[M_m]$

Termination

• by combination

$M_n^{\bullet} + M_m^{\bullet} \longrightarrow M_{n+m}$ $\qquad R_{tc} = k_{tc}\,[M_n^{\bullet}]\,[M_m^{\bullet}]$

• by disproportionation

$M_n^{\bullet} + M_m^{\bullet} \longrightarrow M_n + M_m$ $\qquad R_{td} = k_{td}\,[M_n^{\bullet}]\,[M_m^{\bullet}]$

Polymer scission

$M_n + \text{Cavitation} \longrightarrow M_m^{\bullet} + M_{n-m}^{\bullet}$ $\qquad R_d = 2\,k_{dpol}\,[M_n]$

10.5.2 POLYMER SCISSION

Fracture of the polymer chain occurs at a random site in the degradation of polymers. An alternative method is ultrasound-induced polymer scission, which involves a much better controlled, non-random process (Niezette and Linkens, 1978). Ultrasound-induced polymer breakage is a direct consequence of cavitation, because no degradation was observed under conditions that suppress cavitation. In this non-random scission process, the polymer is fractured at the center of the chain (Glyn et al., 1972; Vander Hoff and Gall, 1977; Madras et al., 2000).

One of the applications of ultrasound-induced polymer scission is the production of block copolymers. Copolymers are used in many applications, where different polymers act as compatibilizing agents between immiscible polymers. The synthesis of block copolymers by ultrasound starts with the dissolution of a homopolymer in a different monomer. Subsequently, ultrasonic scission of the polymer chains generates polymeric radicals, which initiate the polymerization reaction with the monomer present. In this way, ultrasound provides the controlled formation of block copolymers. Solution of two different polymers in a nonreactive solvent can also lead to the formation of block copolymers. In this case, the generated polymeric radicals have to undergo termination by cross-combination.

Polysilanes were made under sonication by Kim and Matyjaszewski (1988). The polydispersities of these polymers could be low, $\bar{M}w/\bar{M}n < 1.2$. They believed that three phenomena might be related to the formation of monomodal polymers.

a) The preferential contribution of the type of intermediate in the sonochemical reductive coupling.
b) The formation of high-quality sodium dispersion, which is continuously regenerated during the coupling process.
c) The selective degradation of polysilanes with higher molecular weight.

During the synthesis of alkyl silicon network polymers, the poly (alkylsilynes), $(SiR)_n$, was greatly facilitated by the sonochemically generated NaK emulsions in hydrocarbon solvent (Bianconi et al., 1989). By preventing the passivation of the reductant (NaK) with salt and growing polymer, sonication initiates the reductive condensations of alkyltrichlorosilanes in inert saturated hydrocarbon solvents, thereby, preventing the complications and side reactions often associated with ethereal solvent and electron transfer reagents.

The preparation of polyurethanes from a number of diisocyanides and diols under sonication was also reported (Price et al., 2002). The sonication made the reactions faster at the early stages and led to higher molecular weights in all cases. In the process of searching for clean and low emission polymerization techniques, supercritical fluid technology, sonochemistry, and microemulsion techniques have attracted more and more interest because of their unique advantages over conventional techniques.

Zhang et al. (2009) reported sonochemical preparation of polymer nanocomposites. When some volatile organometallic precursors were sonochemically decomposed in low volatility solvents, then they produced some nanostructured materials. These can have catalytic activities. These nanostructured metals, oxides, carbides, sulfides, and alloys, nanometric colloids, and supported on catalysts can be prepared by this general route (Abedini and Mousavi, 2010). Another important application of sonochemistry in materials preparation includes the modification of polymers and polymer surfaces.

Nanostructures of two new Pb(II) three-dimensional coordinate polymers (where [Pb2(4-pyc)2I2(H_2O)]n and [Pb(3-pyc)I]n and 3-Hpyc = 3-pyridine carboxylic acid) were synthesized by sonochemical method. PbO block structures were obtained by calcinations of the nanostructures of compounds at 400°C (Sadeghzadeh et al., 2010).

Patra et al. (2012) synthesized PMMA/clay nanocomposites by ultrasound-assisted polymerization technique, which does not require emulsifier ultrasound (different frequencies and power). It was reported that dispersion of the clay layers with polymer matrix. Oxygen permeability of the samples was studied and it was found that the oxygen flow rate was reduced by the combined effect of clay loading and ultrasound. The flame-retardant property of the nanocomposites due to clay dispersion was also investigated by the measurement of limiting oxygen index (LOI).

10.6 HETEROGENEOUS CATALYSIS

10.6.1 LIQUID–LIQUID SYSTEMS

Ultrasound forms very fine emulsions in systems with two immiscible liquids, which is very beneficial when working with phase-transfer catalyzed or biphasic systems. When very fine emulsions are formed, the surface area available for reaction between the two phases is significantly increased; thus, increasing the rate of the reaction. This aspect of ultrasound has also been used for coal, oil, and water mixtures to increase the efficiency of combustion, as well as to decrease the amount of pollutants produced during the combustion process (Dooher et al., 1980).

10.6.2 LIQUID-SOLID SYSTEMS

The most pertinent effects of ultrasound on liquid–solid systems are mechanical, and these are attributed to symmetric and asymmetric cavitations. When a bubble is able to collapse symmetrically, localized area of high temperatures and pressures are generated in the fluid. In addition, shock waves are produced, which have the potential of creating microscopic turbulence within interfacial films surrounding nearby solid particles, also referred to as microstreaming (Elder, 1959). This phenomenon increases the transfer coefficient, as well as possibly the thin film. Hagenson and Doraiswamy (1998) obtained the evidence of a two-fold increase in the intrinsic mass transfer coefficient when modeling experimental data were obtained for the synthesis of dibenzyl sulfide in the presence and absence of ultrasound.

The shock waves produced by cavitation bubbles increase the momentum of solid powders in solution, so that they can collide with a great force. When the solid is inorganic, the particles are fractured upon collision, leading to an overall decrease in the average particle size. When the bubble is collapsing near a solid surface, it is several orders of magnitude larger than the cavitating bubble, that is, a surface greater than 200 μm in diameter when operating at 20 KHz (Suslick, 1990). Symmetric cavitation is hindered and collapse occurs asymmetrically (Neppiras, 1980). As the bubble collapses, microjets of solvent are formed perpendicular to the solid surface. These microjets have an estimated speed of 100 M s^{-1} (aqueous solution) and lead to pitting and erosion of the surface, in addition to the well-known cleaning effects associated with ultrasound. This behavior leads to enhancement in some heterogeneous reactions.

The allylation of ketones and aldehydes by allylic alcohols has been improved using ultrasonic irradiation of a palladium–tin dichloride catalyst in less polar solvents. Inverted regioselectivity was observed as compared with homogeneous carbonyl allylation in polar solvents.

$$H_3OCH = CHCH_2OH + H_5C_6CHO \xrightarrow[\text{Pd/SnCl}_2]{\bullet)))} H_3CCH = CHCH_2CH\,(OH)\,C_6H_5$$

Carbon sonogels have been produced by Tonanon's group (2005a, 2005b). Irradiation with high-intensity ultrasound promotes the sol–gel polycondensation of resorcinol and formaldehyde, typical precursors for carbon supports. As with the oxides, gelation occurs more rapidly when exposed to ultrasonic irradiation. The resulting carbon gels have increased mesoporosity as compared with those prepared without ultrasound and moderate surface areas (500–800 m^2g^{-1}). Such materials could be ideal supports for Pt-based fuel cell catalysts.

A nanostructured bifunctional catalyst, Mo_2C/ZSM-5, was prepared by irradiation of $Mo(CO)_6$ and HZSM-5 in a slurry with hexadecane (Dantsin and Suslick, 2000). As the event responsible for the formation of metal clusters is in the gas phase of the collapsing bubbles. Egg shell catalysts are formed and ~ 2 nm Mo_2C particles decorate the surface of the ZSM-5 support. Studies on the dehydroaromatization of methane to benzene were also performed.

Several supported hydrodesulfurization (HDS) catalysts have been prepared using ultrasound. Dhas et al. (2001) prepared Co and Ni promoted MoS_2 supported on alumina through high-intensity ultrasonic irradiation of isobutene slurries containing $Mo(CO)_6$, $Co_2(CO)_8$, elemental sulfur, and Al_2O_3 or Ni-Al_2O_3 under argon flow. The sonochemically prepared catalysts are extremely active catalysts for the HDS of thiophene and dibenzothiophene with activities several times those of comparable catalysts under identical conditions. Lee et al. (2008) have also prepared MoS_2/Al_2O_3. Higher loadings of Mo can be achieved through the use of ultrasound, resulting in a more active catalyst. A CoMoS/Al_2O_3 catalyst has also been prepared by combining sonochemical and CVD techniques (Lee et al., 2005).

The reactivity of p-phenyl substituted β-enamino compounds using the acidic clay montmorillonite, K-10, as a solid support under sonication was investigated (Valduga and Santis, 1997; 1998). The results indicated the influence of the K-10 support on the regiochemistry of these reactions. Because of steric reasons in the first step of this reaction, the interaction of K-10 with the nitrogen of the amino group makes the carboxylic carbon more electrophilic and the addition of the methyl hydrazine occurs by initial addition of the unsubstituted nitrogen followed by cyclization to give pyrazole. However, the regiochemistry was inverted. It was believed that this is due to a stronger interaction between K-10 and the nitro group than that between K-10 and the nitrogen or oxygen atoms of enamino ketone, thereby, moving the reaction path to the more conventional one.

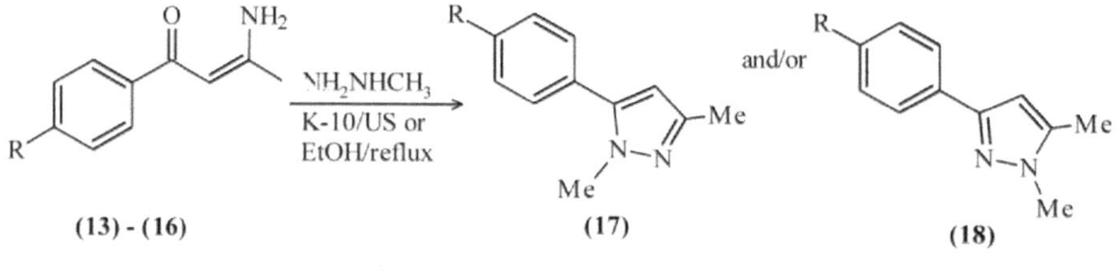

	R	Ratio of isomers EtOH/reflux		(17) and (18) K-10/US	
		(17)	**(18)**	**(17)**	**(18)**
(13)	H	0	100	100	0
(14)	Me	55	45	93	7
(15)	OMe	45	55	94	6
(16)	NO_2	0	100	20	80

A comparison between conventional and ultrasound-assisted heterogeneous catalysis, that is, hydrogenation of 3-buten-1-ol in aqueous solution was reported by Disselkamp et al. (2004). Iron-containing SBA-15 material (Fe-SBA-15) is a promising catalyst for the treatment of phenolic aqueous solutions by coupling ultrasound with heterogeneous catalytic wet peroxide oxidation (Molina et al., 2006). The optimal hydrogen peroxide concentration was two times the stoichiometric amount for TOC degradation ranging from 30 to 40%, which represents a low oxidant dosage. The catalyst loading plays more important role than the oxidant concentration in the activity of the sono-Fenton catalytic system. The remarkable stability (less than 4 ppm loss for the best reaction conditions) of the Fe-SBA-15 heterogonous catalyst under ultrasonic irradiation is particularly noteworthy.

A diverse set of applications of ultrasound have been explored in the synthesis of nanostructured materials, including both direct sonochemical reactions and ultrasonic spray pyrolyses (Bang and Suslick, 2010). The usefulness of sonochemistry as a synthetic tool resides in its versatilily. With a simple modification in reaction conditions, various forms of nanostructured materials can be synthesized. It has been further extended to the preparation of carbons, polymers, and biomaterials.

In ultrasonic spray pyrolysis, ultrasound does not induce chemical reactions of itself. Instead, the ultrasound serves to nebulize precursor solutions producing micron-sized droplets that confine chemical reactions within their interiors. Such phase-isolated microreactors allow for the facile control over chemical composition and complete retention of the bulk chemical composition on the micron size scale. One can produce spherical nanoparticles with narrow particle size distribution. Final particle sizes depend on generated precursor droplet size distribution, if one assumes that every droplet undergoes same process steps and transform to a particle.

10.7 ORGANOMETALLIC PROCESSES

The use of high-intensity ultrasound to enhance the reactivity of metals as stoichiometric reagents have became an important synthetic technique for many heterogeneous organometallic reactions, especially those involving reactive metals, such as magnesium, lithium, and zinc. Grignard reagent was formed under the influence of ultrasonic waves.

$$R-X+Mg \xrightarrow{\bullet)))} R-Mg-X$$

These effects are quite general and apply to reactive inorganic salts as well. Reactivity rate enhancements of more than 10-fold are common. Yields are often substantially improved and by-products are avoided. A few examples of the sonochemical reactions are:

$$C_6H_5Br + Li \xrightarrow{\bullet)))} C_6H_5Li + LiBr$$

$$R - Br + Li + R_2'\,NCHO \xrightarrow[H_2O]{\bullet)))} RCHO + R_2'\,NH$$

$$2\ o\text{-}C_6H_4\,(NO_2)\,I + Cu \xrightarrow{\bullet)))} o\text{-}(O_2N)\,H_4C_6{-}C_6H_4\,(NO_2) + 2\ CuI$$

$$R\,R'HC - OH + KMnO_4\ (s) \xrightarrow{\bullet)))} R\,R'C = O$$

$$C_6H_5CH_2Br + KCN \xrightarrow[Al_2O_3]{\bullet)))} C_6H_5CH_2CN$$

$$MCl_5 + Na + CO \xrightarrow{\bullet)))} M\,(CO)_6\ (M = V, Nb, Ta)$$

The mechanism of the rate enhancements in reactions of metals has been unveiled by monitoring the effect of ultrasonic irradiation on the kinetics of the chemical reactivity of the solids, examining the effects of irradiation on surface structure and size distributions of powders and solids and determining the depth profiles of the surface elemental composition. The power of this three-pronged approach has been proved in studies of the sonochemistry of transition metal powders. Suslick and Doktyoz (1989) found that ultrasonic irradiation of liquid, zinc powders, lead to dramatic changes in structure. The high-velocity interparticle collisions produced in such slurries cause smoothing of individual particles and agglomeration of particles into extended aggregates. Surface composition was probed by Auger electron spectroscopy and mass spectrometry to generate depth profiles of these powders. They revealed that ultrasonic irradiation effectively removed the inactive surface oxide coating. The removal of such passivating coating dramatically improves the reaction rates.

Another application of sonochemistry involves the preparation of amorphous metals. If one can cool a molten metal alloy quickly enough, it can be frozen into a solid before it has a chance to crystallize. Such amorphous metallic alloys lack long-range crystalline order and have unique electronic, magnetic, and corrosion resistant properties. The production of amorphous metals, however, is difficult because extremely rapid cooling of molten metals is necessary to prevent crystallization. Cooling rates (10^7 to 10^5 K s^{-1}) are required while plunging red hot steel into water, which produces cooling at relative lower rate of cooling (~2500 K s^{-1}). The ultrasound has also been used to synthesize amorphous metal powders by the decomposition of some volatile organometallic compounds (Suslick and Doktycz, 1989).

This exciting discovery opens new applications of ultrasound for the low-temperature synthesis of unusual phases, that is, the sonolysis of iron pentacarbonyl produces nearly pure amorphous iron, which was characterized by a variety of techniques to prove its lack of long-range order. Scanning electron micrographs show conchoidal fractures (those with smoothly curved surfaces, which are typical of an amorphous material) and at higher magnification, it reveals a coral like porosity coming from the agglomeration of small clusters of iron.

Surprisingly, the optimal solvent was found to be dioxane. Dioxane is not a suitable solvent for common Reformatsky reaction, because it has a tendency to promote enolization. Neither ether nor benzene, the common Reformatsky solvents produced high yields even after several hours of sonication. An insonated Reformatsky reaction for the preparation of a β-lactam from Zn, ethylbromoacetate, and a diaryl Schiff base gave a 95% yield with 4 h at room temperature, whereas the conventional method of refluxing in toluene gives only 60% yield (Bose et al., 1984). An ultrasound promoted synthesis of hydroesters by Reformatsky reactions using indium metals was reported by Lee et al. (2001). THF is the best solvent. The reaction of benzaldehyde with ethyl bromoacetate in the presence of indium in THF afforded ethyl 3-hydroxy-3-phenylpropanoate in 97% yield in 2 h as compared with 70% yield in 17 h with stirring.

Luche et al. (1982) investigated the use of sonication in a variety of organometallic reactions. Sonication plays an essential role in the formation of a mixture of n-butyl bromide, lithium, and 2-cyclohexanone in the presence of copper (I) iodide, on stirred or low-desity irradiation, thus yielding mostly 1-n-butyl-2-cyclohexan-1-ol. They investigated the sonicated preparation of organozinc reagents and their conjugate addition to α-enones (Luche et al., 1983; Petrier et al., 1984). Later, they found that these organometallic reagents (prepared by sonication) gave rise to clean and selective conjugate additions to α,β-unsaturated aldehydes and ketones in the presence of catalytic amounts of nickel acetylacetonate (Petrier et al., 1985). Einhorn et al. (1986) and Einhorn and Luche (1988) also investigated the Bowault reactions. They discovered that Bouveault reaction intermediate prepared under sonochemical conditions easily undergo ortho-directed ithiation (Einhorn and Luche 1987). The use of tetrahydropyran as solvent dramatically increases the rates and yields of metallation, which can be accomplished with an in situ generated alkyl lithium.

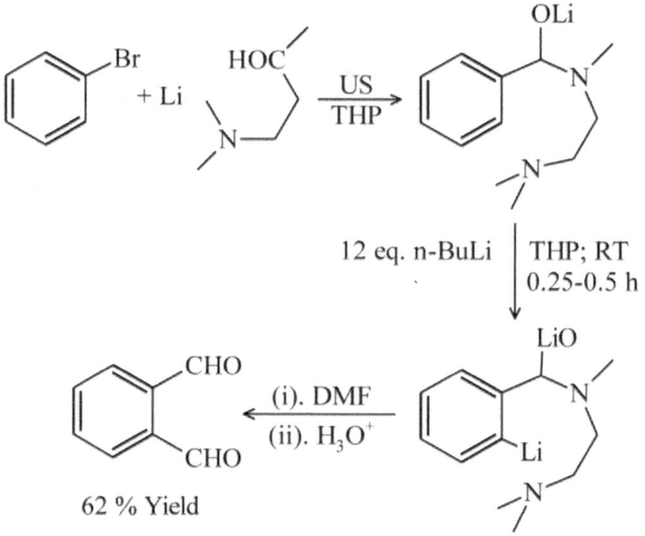

Another impressive result is in situ generation and the use of butyl-ithium reagents under sonochemical conditions (Einhorn and Luche, 1987; De Souza-Barboza et al., 1988). Lithium diisopropylamide (LDA) can be prepared without much effort from diisopropylamine, lithium, and butyl halides by sonicating the mixture in dry THF at 15–18°C. Much better yield was obtained from butyl chloride (92%) than bromide (40%).

The Simmons-Smith cyclopropanation of alkenes with diiodomethane and Zn under sonication has been investigated by Repič and Vogt (1982). Methyl oleate gave a 99% yield of cyclopropanated adduct after 2 h sonication as compared with 50% yield without sonication, while (–)-α-pinene gave a 90% yield under 4 h insonication and 12% without sonication.

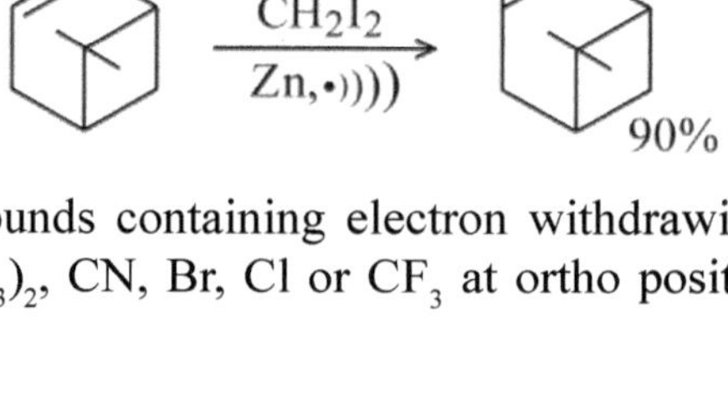

Arylzinc compounds containing electron withdrawing groups, such as CO_2CH_3, $CON(CH_3)_2$, CN, Br, Cl or CF_3 at ortho position prepared under sonication were applied to palladium (O)-catalyzed cross-coupling with aryl halides (Takagi, 1993). Methyl 2-iodobenzoate reacted with Zn under sonication giving 87% arylzinc compound in 1,1,3,3-tetramethylurea, which was coupled with ethyl 2-bromobenzoate in 100% yield. However, electron donating groups diminished the reactivities of aryl iodides greatly. So the reaction did not reach to completion under sonication even at 50°C.

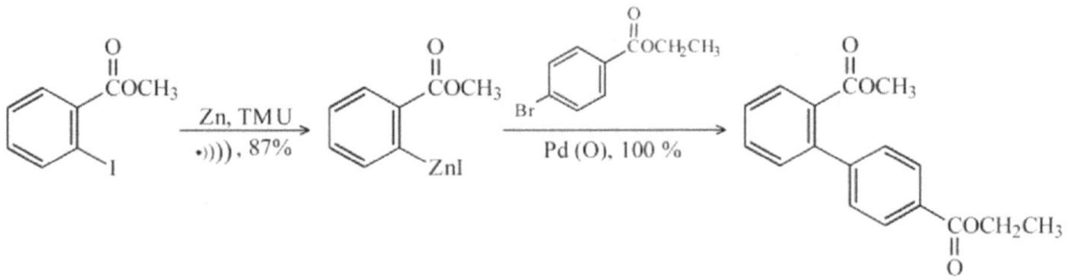

Phenylenedizinc(II) compounds were also prepared in N,N,N,N'-tetramethy-lethylenediamine and 1,1,3,3-tetramethylurea under sonication (Takagi et al., 1994), which can be used for Pd(0) catalyzed synthesis of symmetrically 1,2-disubstituted benzenes. However, the reaction did not succeed when coupled with non-aryl group.

The ultrasound was used for the conjugate addition to α-enones in the presence of zinc–copper couple. Ultrasound increased the yields from poor (20–30%) to good or excellent (Luche and Allavena, 1988; Luche et al., 1988). The ultrasound irradiation was effective in enhancing the reactivity of organomagnesium reagents toward ethylene ketals of α,β-unsaturated aldehydes and it is an efficient alternative to traditional heating or using Lewis acid catalysis (Lu et al., 1998).

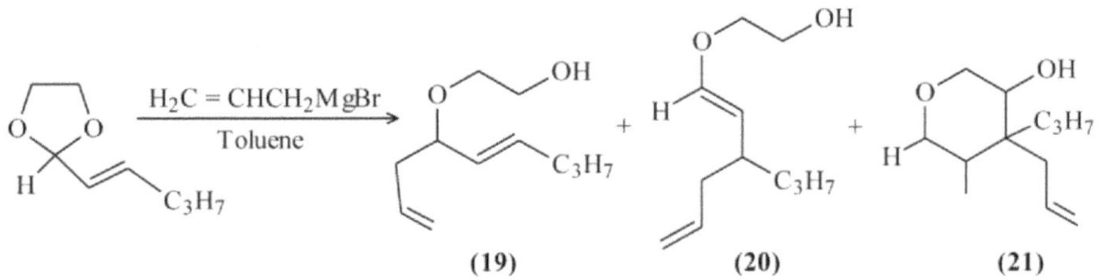

(19) **(20)** **(21)**

Yield	•)))) : 100%	Stirred : 17%
ratio **19** : **(20)** + **(21)**	85 : 15	84 : 16

10.8 SCALE-UP CONSIDERATION

To scale-up a reaction, which is accelerated by ultrasound, there are many factors to be considered. First, it is important to know, what is the role of ultrasound in the rate enhancement? Whether the effects are truly chemical (i.e., is the enhancement due to the formation of free radicals) or they are primarily physical? If they are physical one, which effects are most important for the enhancement of the reaction? If particle degradation is the main critical role of ultrasound, then a sonochemical reactor may not be necessary. Instead, the solids can be sonicated before these are placed within a conventional reactor. However, if another physical effect of ultrasound are playing important role such as surface renewal and/or the enhanced rate of mass transfer, then sonication will be required over due course of the reaction.

In some cases, the ultrasound may generate a reaction intermediate, which catalyzes the reaction, as in the case of the Diels-Alder cycloaddition reaction between cyclopentadiene and methyl vinyl ketone. However, if this is the only role of ultrasound, it may prove to be more cost-effective to add physically the necessary intermediate rather than generating it by the ultrasound. Once the conclusion is reached that ultrasound is required to obtain the desired reaction enhancement, several factors need to be considered before proposing scale-up procedure. The properties of the fluid and dissolved gases are extremely important to the type and amount of sonication required. In addition, the presence of solids, their nature, size, and structure will also affect the selection of a reactor.

In addition to the kinetics of the reaction and nature of reaction mixture, one should also have knowledge of the ultrasonic conditions and optimum system, such as pressure, dissipated power, frequency, ambient reaction, temperature, ultrasonic field, and their interactions. Addition of equipment within a reactor (i.e., baffles, stirrers, and cooling coils) affects the distribution of ultrasonic energy because of wave reflection. All these scale-up.

10.9 OTHER APPLICATIONS

10.9.1 HYDROGEN PRODUCTION

CdS nanocrystals embedded in MoO_3–CdS core–shell nanospheres were prepared by Shen et al. (2012) via a sonochemical method without any templates or surfactants. It was reported that the rate of photocatalytic hydrogen yield with MoO_3–CdS core–shell nanospheres could reach 5.25 mmol without using noble metal co-catalysts.

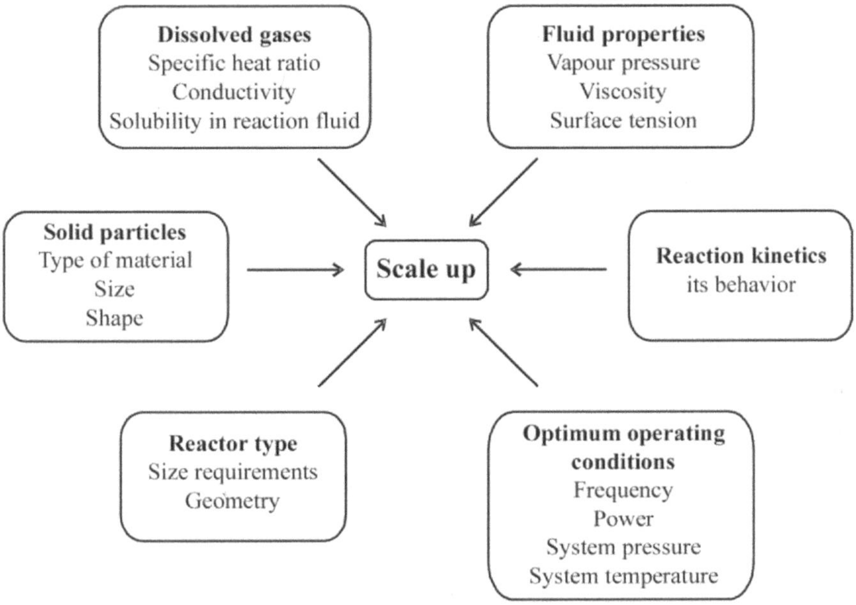

FIGURE 10.2 Scale-up consideration.

Hydrogen is advocated as a fuel of the future. The Bi_2S_3/CdS nanocrystal composites were synthesized by Hao et al. (2014) via ultrasound-assisted method. An enhanced photocatalytic activity could be achieved using this heterostructure in the presence of visible light. An initial rate of H_2 evolution up to 5.5 mmol h^{-1} g^{-1} was observed without resorting to any co-catalysts.

10.9.2 EXTRACTION

Dey and Rathod (2013) reported ultrasound-assisted extraction (UAE) of β-carotene from *Spirulina platensis*. They observed optimal conditions as 1.5 g *Spirulina* (2 min pre-soaked in methanol) in 50 mL n-heptane at 30°C temperature, 167 W cm^{-2} electrical acoustic intensity and 61.5% duty cycle for 8 min. It was reported that maximum extraction was 47.10% under these conditions.

González-Centeno et al. (2015) carried out aqueous extraction of total phenolic content from grape pomace by-products and evaluated their antioxidant capacity. It was reported that about three, four, and eight times lesser time is required on using acoustic process.

Sonochemical synthesis of metal–organic framework nanocubes (MOF-5-NCs) was carried out by Pezeshkpour et al. (2018). Phenolic acids were extracted in prepared broccoli extracts. They also evaluated antimicrobial activity of broccoli and MOF-5NCs extract against *Pseudomonas aeruginosa* (gram-negative bacteria). It was observed that MIC of MOF-5-NCs and broccoli extract on these strains were found to be 13 and 7.81 mg mL^{-1}, respectively.

10.9.3 NANOPARTICLES

Nanoparticles of NiO were prepared by Ghobadifard et al. (2015) via sonochemical route without any calcination step and surfactants. The gas sensing devices for NO_2 and CO were fabricated with nanocrystals onto alumina substrates at lower working temperature. These sensors (based on NiO NPs) were quite selective to nitrogen oxide at 200°C while high responses were obtained for carbon monoxide at 350°C. Wang et al. (2015) synthesized magnetite (Fe_3O_4) and $Zn_{0.5}Fe_{0.5}Fe_2O_4$ nanoparticles (NPs) via sonochemical coprecipitation reactions. These NPs prepared in the presence of ultrasonic irradiation exhibited improved magnetic properties.

Silver nanoparticles were prepared by Yakoot and Selem (2016). As-prepared nanoparticles were found suitable as preservative in some cosmetic items and can be used for the production of nanoparticles of other metals as it is a very simple method without using any toxic reagents. Ultrasound synthesis of Eu^{3+} ions-doped ZrO_2 nanoparticles was reported by Yadav et al. (2019) using *Alove vera* as a fuel. It was observed that as-prepared phosphor material is red emitting and hence, it can be used in display applications.

10.9.4 WATER TREATMENT

Khataee et al. (2015) synthesized undoped and Pr-doped ZnO nanoparticles using sonochemical method. They investigated the sonochemical activity for the degradation of acid red 17 (AR17) in the presence of ultrasonic irradiation. It was reported that decolorization efficiency of as-prepared ultrasound-assisted undoped ZnO and Pr-doped ZnO (5%) was 46 and 100% within 70 min, respectively. It was also observed that degradation efficiency was enhanced from 79 to 93 and 85% in 50 min on addition of hydrogen peroxide and peroxydisulfate as enhancer, respectively.

Muzakki et al. (2016) synthesized ZnO/CuO and TiO_2/CuO nanocomposites with different Ti/Cu and Zn/Cu ratios using sol–gel method. They evaluated their sonocatalytic, photocatalytic, and photosonocatalytic activities for degradation of methylene blue as a model textile dye. The rapid degradation of methylene blue revealed that the incorporation of CuO in TiO_2/CuO and ZnO/CuO nanocomposites was responsible for their higher photocatalytic activity. The Bi_2O_3 nanoplates were synthesized by Sánchez-Martínezn et al. (2016) via an ultrasound-assisted precipitation method. The photocatalytic activity of Bi_2O_3 samples was evaluated with respect to the degradation of rhodamine B, indigo carmine, and tetracycline hydrochloride in aqueous solution.

A magnetic $CoFe_2O_4$@ZnS core–shell nanocomposite was fabricated by Farhadi et al. (2017) via one-step hydrothermal method. It was reported that average crystallite size of $CoFe_2O_4$@ZnS nanocomposite was 18 nm. The band gap of this nanocomposite was found to be 3.4 eV, which is suitable for its use in sono-/photocatalytic processes. As-prepared nanocomposite was utilized as sonocatalyst for degrading different organic pollutants. Complete degradation of methylene blue (25 mg L^{-1}) was observed within 70 min in the presence of $CoFe_2O_2$@ZnS nanocomposite and H_2O_2 (4 mM).

The photolysis (UV-C) and sonication were used by Joseph et al. (2017) in effectively degrading reactive black 5 (RB5) simultaneously. It was reported that sonophotolysis rate was high in alkaline medium (67.7%) as compared with acidic one (46.9%). Taghipour et al. (2018) synthesized nanosized rod-like metal organic polymer (MOP) at room temperature in 30 min. It was reported that the reaction of zinc salt with 1,4-phenylenedioxy diacetic acid gave $[Zn(C_{10}H_8O_6)(H_2O)_4]$. Then, it was loaded by sonication on activated carbon.

The [Zn(BDC)(DMF)] crystal (metal organic framework, MOF) was synthesized by Samuel et al. (2018) via ultrasonic irradiation. The catalytic activity of this MOF for degradation of 4-nitrophenol (4-NP) was evaluated under direct sunlight. It exhibited higher photocatalytic activity in the presence of $NaBH_4$ under sunlight irradiation to reduce 4-NP to 4-aminophenol (4-AP) within 10 min. A facile synthesis of SnO_2 2D nanoflakes (two-dimensional) has been reported by Yashas et al. (2020). They used these for sonophotocatalytic of degradation tetracycline hydrochloride. SnO_2 2D nanoflakes showed excellent photocatalytic degradation under visible light. The optimum degradation was found to be 88.82% and that too in 135 min. Hayati et al. (2020) synthesized MgO/CNT nanocomposites (MCs) through an ultrasound-assisted hydrothermal method. The ability of MCs/UV/ultrasound system for sulfadiazine (SDZ) degradation was evaluated. It was reported that complete degradation of SDZ could be achieved in 80 min.

Wang et al. (2021) synthesized copper-doped graphitic carbon nitride ($Cu-C_3N_4$) and used it as an effective Fenton-like catalyst for the degradation of methylene blue. It was observed that 96% methylene blue degradation could be achieved within 30 min.

10.9.5 ORGANIC SYNTHESIS

Kumar et al. (2014) synthesized silver nanoparticles using starch under sonication. Silver nanoparticles were obtained with diameter 23–97 nm with average particle size of 45.6 nm. These silver nanoparticles can be used for the synthesis of 2-aryl substituted benzimidazoles. The 3-selanylindoles were prepared by Vieira et al. (2015) via the direct selanylation of indoles. They used diorganyldiselenides as a source of selenium presence of ultrasound DMSO as the solvent irradiation using catalyst, CuI (20 mol%).

Silva et al. (2016) prepared 12 isatin derivatives, 5'-(4-alkyl/aryl-1*H*-1,2,3-triazoles), using 5-azido-spiro[1,3-dioxolane-2,3'-indol]-2'(1'*H*)-one in the presence of ultasound irradiation and various alkynes under acidic conditions. It was reported that yields increased to 78–98%, and reaction time was reduced to 5 min compared with the conventional methods.

Sonochemical [3 + 2] cycloadditions of organocatalytic enamine-azide and β-oxoamides with a series of aryl azides was reported by Xavier et al. (2017). These sonochemical-assisted reactions were amenable to a wide range of -oxo amides or aryl azides, and N-aryl-1,2,3-triazoyl carboxamides affording good to excellent yields in a short period. The chalcone and bis-chalcone derivatives were synthesized by Polo et al. (2019) under sonication via Claisen-Schmidt condensation. The chemical, antioxidative properties, and chemical reactivity for acetylcholinesterase (AChE) inhibition were reported.

A multicomponent synthesis of 12 2-(*N*-heterocycle) substituted 1,3,4-oxadiazoles was reported by Bhatt et al. (2020) under ultrasound exposure. Out of all, 5-bromo-1-((4-chlorophenyl)((5-(4-hydroxyphenyl)-1,3,4-oxadiazol-2-yl)amino)methyl) indoline-2,3-dione exhibited significant cytotoxicity against all the three human cancer cell lines (liver cancer cell line Hep G2, colorectal cancer cell line HT-29) and breast cancer cell line MCF-7. Sultan et al. (2020) reported that condensation of aminotrizoles with a number of aromatic aldehydes afforded Schiff bases in good yields in only 3–5 min on exposure to ultrasound. As-synthesized compounds were found active against gram-negative (*Escherichia coli, Pseudomonas aeruginosa, Shigella sonnei,* and *Salmonella typhi*) and two gram-positive (*Bacillus subtilis and Staphylococcus aureus)* strains.

10.9.6 SOLAR CELLS

The copper oxide nanoparticles were prepared by Zhang et al. (2015) via sonochemistry. They could obtain colloidal crystalline CuO nanoparticles (diameter of 3–12 nm). These nanoparticles were then applied as an anode buffer layer in a PCDTBT:PC71BM bulk heterojunction solar cell. It replaced the commonly used hygroscopic PEDOT:PSS. It was observed that UV–ozone (UVO) treatment enhanced the work function of CuO buffer layer from − 4.7 to − 5.4 eV and photovoltaic performance from 6.00 to 6.44% as compared with PEDOT:PSS-based solar cells.

Yuan et al. (2019) reported ultrasound-assisted synthesis of hybrid black/red phosphorus quantum dots (BP/RPQDs). They used as-prepared BP/RPQDs in dye-sensitized solar cells (DSSCs). It was observed that hybrid BP/RPQDs have prolonged recombination lifetime of photogenerated carriers and light-response in a wider range ultraviolet to visible and even near-infrared region upon assembly into liquid-junction DSSCs. It was reported that BP/RPQDs-N719 co-sensitizer could achieve efficiency of 8.02% as compared with 7.60% for N719-only solar cell.

The exposure of a chemical reaction to ultrasound creates bubbles, which will explode creating very high temperature and pressure. This drives a chemical reaction and yields some products. The ultrasound can be used to synthesize some organic compounds, polymerized some monomers as well as destroy some of the organic contaminants. Sonochemistry has a lot of possibilities, but it has not been investigated in detail so far.

10.10 RECENT DEVELOPMENTS

George (2022) synthesized two samples of Ag_2CrO_4 nanoparticles by ultrasound-assisted route without calcinations at elevated temperature. It was observed that the as-prepared Ag_2CrO_4 was orthorhombic Ag_2CrO_4 nanoparticles with band gap as 1.45 and 1.95 eV, respectively. Ayoub et al. (2022) developed a catalyst-free process for the synthesis of maleic acid from furfural in the presence of ultrasound irradiations (high frequency). It was reported that 70% selectivity of maleic acid with 92% of furfural conversion could be achieved without using any catalyst and that too under mild conditions. They used hydrogen peroxide as the oxidant for this purpose.

Ahmady et al. (2022) developed a eco-friendly method for the synthesis of fenitrothion using sonochemical-assisted approach. They obtained fenitrothion from the reaction of 3-methyl-4-nitrophenol as an intermediate. First, m-cresol was nitrated by ferric nitrate (nitrating agent) in the presence of triethylammonium nitrate (ionic liquid) in the presence of ultrasonic irradiation. It required shorter reaction time with high yields. It was also reported that ionic liquid can be reused for five consecutive runs. Talebzadeh et al. (2022) also used an eco-friendly sonochemical process to prepare $Gd_2Sn_2O_7$ (GSO) nanoparticles. They used lysine (amino acid) as a green capping agent while ethylenediamine was used as an alkaline agent. They also compared the photodegradation performance of $Gd_2Sn_2O_7/SnO_2/Gd_2O_3$ nanocomposites for the degradation of erythrosine (98%), which was more than $Gd_2Sn_2O_7$ and $Gd_2Sn_2O_7/SnO_2$.

Dehane et al. (2022) reported sonochemical production of hydrogen in the presence of rare gases (He, Ar, and Xe). It was found that the yield of hydrogen follows the order: Xe > Ar > He. This order was attributed to the thermal conductivity of these gases, in reverse order (He > Ar > Xe). Mahdavi et al. (2022) fabricated photocatalytic nanocomposites (Dy_2O_3-SiO_2) via sonochemical method with tetraethylenepentamine (Tetrene). They could obtain a porous, sphere-shaped nanocomposites with particle size in the range of 20–60 nm by using ultrasonic power (400 W) for 10 min. The photocatalytic performance of as-prepared sample of nanocomposite was evaluated for photodecomposition of seven dyes (methyl orange, malachite green, rhodamine B, erythrosine, thymol blue, acid red 14, and eriochrome black T). It was reported that this nanophotocatalyst could degrade 92.9% erythrosine.

KEYWORDS

- **sonochemistry**
- **ultrasound**
- **polymer synthesis**
- **hetergeneous catalysis**
- **organic synthesis**
- **organometallic processes**

REFERENCES

Abedini, R.; Mousavi, S. M. Preparation and Enhancing of Materials Using Ultrasound Technique: Polymers, Catalysts and Nanostructure Particles. *Petroleum Coal* **2010,** *52* (2), 81–98.

Ahmady Zarei, A.; Javani, M.; Saghanezhad, S. J. Green Sonochemical Synthesis of Technical Fenitrothion Using Protic Ionic Liquid. *Iran. J. Chem. Chem. Eng.* **2022,** *24*. DOI: 10.30492/IJCCE.2022.543432.5029.

Ameta, G.; Benjamin, S.; Sharma, V.; Bhardwaj, S. Sonochemistry: A Pollution Free Pathway. *Green Chem.* 2013.

Ayoub, N.; Toufaily, J.; Guénin, E.; Enderlin, G. Catalyst-Free Process for Oxidation of Furfural to Maleic Acid by High Frequency Ultrasonic Activation. *Green Chem.* **2022,** *24,* 4164–4173.

Bang, J. H.; Suslick, K. S.; Applications of Ultrasound to the Synthesis of Nanostructured Materials. *Adv. Mater*. 2010, *22*, 1039–1059.

Bhatt, P.; Sen, A.; Jha, A. Design and Ultrasound Assisted Synthesis of Novel 1, 3, 4-Oxadiazole Drugs for Anti-Cancer Activity. *ChemistrySelect*, **2020,** *5* (11), 3347–3354.

Bianconi, P. A.; Schilling, F. C.; Weidman, T. W. Ultrasound-Mediated Reductive Condensation Synthesis of Silicon-Silicon-Bonded Network Polymers. *Macromolecules* **1989,** *22* (4), 1697–1704.

Bogolubsky, A. V.; Ryabukhin, S. V.; Pakhomo, G. G.; Ostapchuk, E. N.; Shivanyuk, A. N.; Tolmachev, A. A. Synlett., A Facile Synthesis of *N*-Carbamoylamino Acids 2008; pp. 2279–2282.

Bose, A. K.; Gupta, K.; Manhas, M. S. β-Lactam Formation by Ultrasound-Promoted Reformatsky Type Reaction. *J. Chem. Soc. Chem. Commun.* 1984, 86–87.

Bujans, J.; Guajordo, R.; Loler, K. S.; Enantioselective Enzymic Sterol Synthesis by Ultrasonically Stimulated Bakers' Yeast. *J. Am. Chem. Soc.* **1998,** *110*, 604–606.

Cains, P. W.; Martin, P. D.; Price, C. J. The Use of Ultrasound in Industrial Chemical Synthesis and Crystallization. 1. Applications to Synthetic Chemistry. *Org. Proc. Res. Develop.* 1998, *2*, 34–48.

Cooper, G.; Grieser, F.; Biggs, S.; Butyl Acrylate/Vinyl Acetate Copolymer Latex Synthesis Using Ultrasound As an Initiator. *J. Colloid. Interface Sci.* 1996, *184*, 52–63.

Curie, J.; Curie, P. Contractions et dilatations produites par des tensions électriques dans les cristaux hémièdres à faces inclinées. *Compt. Rend.* **1881,** *93*, 1137–1140.

Curie, J.; Curie, P. Développement par compression de l'électricité polaire dans les cristaux hémièdres à faces inclinées. *Bulletin Minéral.* **1880,** *3* (4), 90–93.

Dabiri, M.; Tisseh, Z. N.; Bahramnejad, M.; Bazgir, A. Sonochemical Multi-Component Synthesis of Spirooxindoles. *Ultrason. Sonochem.* **2011,** *18* (5), 1153–1159.

Dantsin, G.; Suslick, K. S. Sonochemical Preparation of a Nanostructured Bifunctional Catalyst. *J. Am. Chem. Soc.* **2000,** *122* (21), 5214–5215.

De Souza-Barboza, J. C.; Petrier, C.; Luche, J. L. Ultrasound in Organic Synthesis. 13. Some Fundamental Aspects of the Sonochemical Barbier Reaction. *J. Org. Chem.* **1988,** *53* (6), 1212–1218.

Dehane, A.; Merouani, S.; Chibani, A.; Hamdaoui, O. Clean Hydrogen Production by Ultrasound (Sonochemistry): The Effect of Noble Gases. *Curr. Res. Green Sustain. Chem.* **2022,** *5*. DOI: 10.1016/j.crgsc.2022.100288.

Dey, S.; Rathod, V. K. Ultrasound Assisted Extraction of β-Carotene from Spirulina Platensis. *Ultrason. Sonochem.* **2013,** *20* (1), 271–276.

Dhas, N. A.; Ekhtiarzadeh, A.; Suslick, K. S. Sonochemical Preparation of Supported Hydrodesulfurization Catalysts. *J. Am. Chem. Soc.;* **2001,** *123* (34), 8310–8316.

Disselkamp, R. S.; Judd, K. M.; Hart, T. R.; Peden, C. H.; Posakony, G. J.; Bond, L. J. A Comparison Between Conventional and Ultrasound-Mediated Heterogeneous Catalysis: Hydrogenation of 3-buten-1-ol Aqueous Solutions. *J. Catal.* **2004,** *221* (2), 347–353.

Dooher, J.; Genberg, R.; Moon, S.; Gilmartin, B.; Jakatt, S.; Skura, J. et al. Combustion Studies of Water/Oil Emulsion on a Commercial Boiler Using No. 2 Oil and Low and High Sulphur No. 6 Oil. *Fuel,* **1980,** *59* (12), 883–892.

Einhorn, J.; Einhorn, C.; Luche, J. L. Ultrasound in Organic Synthesis 15. Radical Cyclisation of o-Allyl Benzamides via the Sonochemically Generated Radical Anions. *Tetrahedron Lett.* **1988,** *29* (18), 2183–2184.

Einhorn, J.; Luche, J. L. Ultrasound in Organic Synthesis 10 Selective Ortho-Lithiation of the Bouveault Reaction Intermediate. *Tetrahedron Lett.* **1986,** *27* (16), 1793–1796.

Einhorn, J.; Luche, J. L. Ultrasound in Organic Synthesis. 12. In Situ Generation and Uses of Butyllithium Reagents in Several Synthetic Reactions. *J. Org. Chem.* **1987,** *52* (18), 4124–4126.

Elder, S. A. Cavitation Microstreaming. *J. Acoust. Soc. Am.* **1959,** *31* (1), 54–64.

Farhadi, S.; Siadatnasab, F.; Khataee, A. Ultrasound-Assisted Degradation of Organic Dyes Over Magnetic $CoFe_2O_4$@ZnS Core-Shell Nanocomposite. *Ultrason. Sonochem.* **2017,** *37*, 298–309.

Feldman, D. Ultrasound Applications in Polymer Science. *Polym. News* **1995,** *20* (5), 138–144.

Galton, F. *Inquiries Into Human Faculty and Its Development*; Macmillan: London, 1983.

Gao, W.; Zheng, M.; Li, Y. A Novel and Facile Synthesis of 3- (2-benzofuroyl)-and 3, 6-bis (2-Benzofuroyl) Carbazole Derivatives. *Beilstein Jo. Org. Chem.* **2011,** 7 (1), 1533–1540.

George, P. P. Sonochemical Synthesis and Characterization of Ag_2CrO_4 Nanoparticles, a Green Chemistry Approach. *Asian Basic Appl. Res. J.* **2022,** *6* (1), 25–33.

Ghobadifard, M.; Mahmoudi, M.; Khelghati, M.; Maleki, G.; Farhadi, S.; Aslani, A. Sono-Chemical Synthesis, Characterization and Gas Sensing Properties of NiO Nanoparticles. *J. Nano. Adv. Mat*, **2015,** *3*, 107–114.

Glyn, P. A. R.; Vander Hoff, B. M. E.; Rully, P. M. A General Model for Prediction of Molecular Weight Distributions of Degraded Polymers. Development and Comparison with Ultrasonic Degradation Experiments. *J. Macromol. Sci. Chem.* **1972,** *A6*, 1653–1664.

González-Centeno, M. R.; Comas-Serra, F.; Femenia, A.; Rosselló, C.; Simal, S. Effect of Power Ultrasound Application on Aqueous Extraction of Phenolic Compounds and Antioxidant Capacity from Grape Pomace (*Vitis vinifera L.*): Experimental Kinetics and Modeling. *Ultrason. Sonochem.* **2015,** *22*, 506–514.

Hagenson, L. C.; Doraiswamy, L. K. Comparison of the Effects of Ultrasound and Mechanical Agitation on a Reacting Solid-Liquid System. *Chem. Eng. Sci.* **1988,** *53* (1), 131–148.

Han, B. H.; Boudjouk, P. Organic Sonochemistry: Sonic Acceleration of the Reformatsky Reaction. *J. Org. Chem.* **1982,** *47* (25), 5030–5032.

Hao, L. X.; Chen, G.; Yu, Y. G.; Zhou, Y. S.; Han, Z. H.; Liu, Y. Sonochemistry Synthesis of Bi_2S_3/CdS Heterostructure with Enhanced Performance for Photocatalytic Hydrogen Evolution. *Int. J. Hydrogen Energy* **2014,** 39, 14479–14486.

Hayati, F.; Isari, A. A.; Anvaripour, B.; Fattahi, M.; Kakavandi, B. Ultrasound-Assisted Photocatalytic Degradation of Sulfadiazine Using MgO@ CNT Heterojunction Composite: Effective Factors, Pathway and Biodegradability Studies. *Chem. Eng. J.* **2020,** 381. DOI: 10.1016/j.cej.2019.122636.

Hemantha, H. P.; Chennakrishnareddy, G.; Vishwanatha, T. M.; Sureshbabu, V. V. One-Pot Synthesis of Ureido Peptides and Urea-Tethered Glycosylated Amino Acids Employing Deoxo-Fluor and TMSN3. *Synlett* **2009,** *2009*, 407–410.

Hu, Y.; Wang, J. X.; Li, S. Synthesis of Anhydrides from Acyl Chlorides Under Ultrasound Condition. *Synth. Commun.* **1997,** *27* (2), 243–248.

Joseph, C. G.; Taufiq-Yap, Y. H.; Krishnan, V. Ultrasonic Assisted Photolytic Degradation of Reactive Black 5 (RB5) Simulated Wastewater. *ASEAN J. Chem. Eng.* **2017,** *17* (2), 37–50.

Khataee, A.; Karimi, A.; Arefi-Oskoui, S.; Soltani, R. D. C.; Hanifehpour, Y.; Soltani, B.; Joo, S. W. Sonochemical Synthesis of Pr-Doped ZnO Nanoparticles for Sonocatalytic Degradation of Acid Red 17. *Ultrason. Sonochem.* **2015,** *22*, 371–381.

Kim, H. K.; Matyjaszewski, K. Preparation of Polysilanes in the Presence of Ultrasound. *J. Am. Chem. Soc.* **1988,** *110*, 3321–3323.

Koulocheri, S. D.; Harautounain, S. A.; Ultrasound-Promoted Synthesis of 2, 3-Bis (4-Hydroxyphenyl) Indole Derivatives as Inherently Fluorescent Ligands for the Estrogen Receptor. *Eur. J. Org. Chem.* **2001,** 1723–1729.

Kuijpers, M. W. A. *Ultrasound-Induced Polymer Reaction Engineering in High-Pressure Fluids*; Technische Universiteit Eindhoven; Eindhoven, 2004.

Kumar, A.; Maurya, R. A. Efficient Synthesis of Hantzsch Esters and Polyhydroquinoline Derivatives in Aqueous Micelles. *Synlett* **2008,** *2008*, 883–885.

Kumar, B.; Smita, K.; Cumbal, L.; Debut, A.; Pathak, R. N. Sonochemical Synthesis of Silver Nanoparticles Using Starch: A Comparison. *Bioinorg. Chem. Appl.* **2014,** *2014*. DOI: 10.1155/2014/784268.

Larpkiattaworn, S.; Jeerapan, C.; Tongpan, R.; Tongon, S. Ultrasonic on Transesterification Reaction for Biodiesel Production. In *7th Biomass Asia Workshop, Jakarta*, 2010.

Lee, A. S. Y.; Chang, R. Y. A Simple and Highly Efficient Synthesis of β-Amino-α, β-Unsaturated Ester via Sonochemical Blaise Reaction. *Tetrahedran Lett.* **1997,** *38*, 443–446.

Lee, J. J.; Kim, H.; Koh, J. H.; Jo, A.; Moon, S. H. Performance of $CoMoS/Al_2O_3$ Prepared by Sonochemical and Chemical Vapor Deposition Methods in the Hydrodesulfurization of Dibenzothiophene and 4,6-Dimethyldibenzothiophene. *Appl. Catal. Environ.* **2005,** *58* (1–2), 89–95.

Lee, J. J.; Kim, H.; Moon, S. H. Preparation of Highly Loaded, Dispersed MoS_2/Al_2O_3 Catalysts for the Deep Hydrodesulfurization of Dibenzothiophenes. *Appl. Catal. Environ.* **2008,** *41* (1–2), 171–180.

Lee, J.; Mei, H. S.; Snyder, J. K. Synthesis of Miltirone by an Ultrasound-Promoted Cycloaddition. *J. Org. Chem.* **1990,** *55* (17), 5013–5016.

Lee, P. H.; Bang, K.; Lee, K.; Sung, S. Y.; Chang, S. Ultrasound Promoted Synthesis of β-Hydroxyesters by Reformatsky Reaction Using Indium Metal. *Synth. Commun.* **2001,** *31* (24), 3781–3789.

Li, J. T.; Chen, G. F.; Wang, J. X.; Li, T. S. Ultrasound promoted synthesis of α, α'-bis (substituted furfurylidene) cycloalkanones and chalcones. *Synth. Commun.* **1999,** *29* (6), 965–971.

Li, J.; Wang, D.; Zhang, Y.; Li, J.; Chen, B. Facile One-Pot Synthesis of 4, 5-Disubstituted 1, 2, 3- (NH)-Triazoles Through Sonogashira Coupling/1, 3-Dipolar Cycloaddition of Acid Chlorides, Terminal Acetylenes, and Sodium Azide. *Org. Lett.* **2009,** *11* (14), 3024–3027.

Lu, T. J.; Cheng, S. M.; Sheu, L. J. Ultrasound Accelerated Coupling Reaction of Grignard Reagents with 1, 3-Dioxolanes of α, β-Unsaturated Aldehydes. *J. Org. Chem.* **1998,** *63* (8), 2738–2741.

Luche, J. L.; Allavena, C.; Petrier, C.; Dupuy, C. Ultrasound in Organic Synthesis 17 Mechanistic Aspects of the Conjugate Additions to α-Enones in Aqueous Media. *Tetrahedron Lett.* **1988,** *29* (42), 5373–5374.
Luche, J. L.; Petrier, C.; Gemal, A. L.; Zikra, N. Ultrasound in Organic Synthesis. 2. Formation and Reaction of Organocopper Reagents. *J. Org. Chem.* **1982,** *47* (19), 3805–3806.
Luche, J. L.; Petrier, C.; Lansard, J. P.; Greene, A. E. Ultrasound in Organic Synthesis. 4. A Simplified Preparation of Diarylzinc Reagents and Their Conjugate Addition to. Alpha.-enones. *J. Org. Chem.* **1983,** *48* (21), 3837–3839.
Luche, J. L.; Allavena, C. Ultrasound in Organic Synthesis 16. Optimisation of the Conjugate Additions to α, β-Unsaturated Carbonyl Compounds in Aqueous Media. *Tetrahedron Lett.* **1988,** *29*, 5369–5372.
Madras, G.; Kumar, S.; Chattopadhyay, S. Continuous Distribution Kinetics for Ultrasonic Degradation of Polymers. *Polym. Degrad. Stab.* **2000,** *69* (1), 73–78.
Mady, M. F.; El. Kateb, A. A.; Zeid, I. F.; Jorgensen, K. B.; Comparative Studies on Conventional and Ultrasound-Assisted Synthesis of Novel Homoallylic Alcohol Derivatives Linked to Sulfonyl Dibenzene Moiety in Aqueous Media. *J. Chem.* **2013,** *2013*. DOI: 10.1155/2013/364036.
Mahdavi, K.; Zinatloo-Ajabshir, S.; Yousif, Q. A.; Salavati-Niasari, M. Enhanced Photocatalytic Degradation of Toxic Contaminants Using Dy_2O_3-SiO_2 Ceramic Nanostructured Materials Fabricated by a New, Simple and Rapid Sonochemical Approach. *Ultrason. Sonochem.* **2022,** *82*. DOI: 10.1016/j.ultsonch.2021.105892.
Mason, T. J. *Sonochemistry*; Oxford University Press, 1999.
Memarian, H. R.; Saffar-Teluri, A. Photosonochemical Catalytic Ring Opening of α-Epoxyketones. *Beilstein J. Org. Chem.* **2007,** *3* (1), 2.
Molina, R.; Martínez, F.; Melero, J. A.; Bremner, D. H.; Chakinala, A. G. Mineralization of Phenol by a Heterogeneous Ultrasound/Fe-SBA-15/H2O2 Process: Multivariate Study by Factorial Design of Experiments. *Appl. Catal. Environ.* **2006,** *66* (3–4), 198–207.
Muzakki, A.; Shabrany, H.; Saleh, R. Synthesis of ZnO/CuO and TiO_2/CuO Nanocomposites for Light and Ultrasound Assisted Degradation of a Textile Dye in Aqueous Solution. *AIP Conf. Proc.* 2016, *1725*. DOI: 10.1063/1.4945505.
Neppiras, E. A. Acoustic Cavitation. *Phys. Rep.* **1980,** *61* (3), 159–251.
Niezette, J.; Linkens, A. Contribution to the Ultrasonic Degradation of Polystyrene Solutions. *Polymer* **1978,** *19* (8), 939–942.
Parvathy, K. S.; Srinivas, P. Ultrasound-Assisted Reaction of 2,3,4,6-tetra-O-acetyl-alpha-D-Glucopyranosyl Bromide with Potassium Salt of Curcumin Under PTC Conditions. *Ultrson. Sonochem.* **2008,** *55*, 571–577.
Patra, S. K.; Prusty, G.; Swain, S. K. Ultrasound Assisted Synthesis of PMMA/Clay Nanocomposites: Study of Oxygen Permeation and Flame Retardant Properties. *Bul. Mater. Sci.* **2012,** *35* (1), 27–32.
Pethrick, R. A. Ultrasonic Studies of Polymeric Solids and Solutions. *Adv. Sonochem.* **1991,** *2*, 65–134.
Petrier, C.; De Souza Barbosa, J.; Dupuy, C.; Luche, J. L. Ultrasound in Organic Synthesis. 7. Preparation of Organozinc Reagents and Their Nickel-Catalyzed Reactions with. Alpha., Beta.-Unsaturated Carbonyl Compounds. *J. Org. Chem.* **1985,** *50* (26), 5761–5765.
Petrier, C.; Luche, J. L.; Dupuy, C. Ultrasound in Organic Synthesis 6. An Easy Preparation of Organozinc Reagents and Their Conjugate Addition to α-Enones. *Tetrahedron Lett.* **1984,** *25* (32), 3463–3466.

Pezeshkpour, V.; Khosravani, S. A.; Ghaedi, M.; Dashtian, K.; Zare, F.; Sharifi, A. et al. Ultrasound Assisted Extraction of Phenolic Acids from Broccoli Vegetable and Using Sonochemistry for Preparation of MOF-5 Nanocubes: Comparative Study Based on Micro-Dilution Broth and Plate Count Method for Synergism Antibacterial Effect. *Ultrason. Sonochem.* **2018,** *40*, 1031–1038.

Polo, E.; Ibarra-Arellano, N.; Prent-Peñaloza, L.; Morales-Bayuelo, A.; Henao, J.; Galdámez, A. et al. Ultrasound-Assisted Synthesis of Novel Chalcone, Heterochalcone and Bis-Chalcone Derivatives and the Evaluation of Their Antioxidant Properties and as Acetylcholinesterase Inhibitors. *Bioorg Chem.* **2019,** *90*. DOI: 10.1016/j.bioorg.2019.103034.

Porter, C. W.; Young, L. A Molecular Rearrangement Induced by Ultrasonic Waves. *J. Am. Chem. Soc.;* **1938,** *60* (6), 1497–1500.

Price, G. J.; Lenz, V; Ansell, C. W. G.; The Effect of High Intensity Ultrasound on the Synthesis of Some Polyurethanes. *Eur. Polym. J.* **2002,** *38*, 1531–1536.

Qi-Xun, S.; Rong-Wen, L.; Xin-Yu, H.; Lian-Hai, L.; Shu-Fen, Z. Ultrasound-Assisted Highly Efficient Reduction of Nitroarenes to Corresponding N-Arylhydroxylamines. *Chem. Res. Chinese Univ.* **2009,** *25* (2), 183–188.

Reddy, A. V.; Ravindranath, B. Acetylation Under Ultrasonic Conditions: Convenient Preparation of N-Acetylamino Acids. *Synthetic Commun.* **1992a,** *22* (2), 257–264.

Reddy, A. V.; Ravindranath, B. Synthesis of α-, β-and Cyclic Spaglumic Acids. *Int. J. Pept. Protein Res.* **1992b,** *40* (5), 472–476.

Repič, O.; Vogt, S. Ultrasound in Organic Synthesis: Cyclopropanation of Olefins with Zinc-Diiodomethane. *Tetrahedron Lett.* **1982,** *23* (27), 2729–2732.

Richards, W. T.; Loomis, A. L. The Chemical Effects of High Frequency Sound Waves I. A Preliminary Survey. *J. Am. Chem. Soc.* **1927,** *49* (12), 3086–3100.

Rudin, A. *The Elements of Polymer Science and Engineering*; Academic Press: San Diego, 1990.

Sadeghzadeh, H.; Morsali, A.; Retailleau, P. Ultrasonic-Assisted Synthesis of Two New Nano-Structured 3D Lead (II) Coordination Polymers: Precursors for Preparation of PbO Nano-Structures. *Polyhedron* **2010,** *29* (2), 925–933.

Samuel, M. S.; Bhattacharya, J.; Parthiban, C.; Viswanathan, G.; Singh, N. P. Ultrasound-Assisted Synthesis of Metal Organic Framework for the Photocatalytic Reduction of 4-Nitrophenol Under Direct Sunlight. *Ultrason. Sonochem.* **2018,** *49*, 215–221.

Sánchez-Martínez, D.; Juárez-Ramírez, I.; Torres-Martínez, L. M.; de León-Abarte, I. Photocatalytic Properties of Bi_2O_3 Powders Obtained by an Ultrasound-Assisted Precipitation Method. *Ceram. Int.* **2016,** *42* (1), 2013–2020.

Schmidt, P.; Rosenfield, E.; Milner, R.; Czerner, R.; Schellenberger, A. Theoretical and Experimental Studies on the Influence of Ultrasound on Immobilized Enzymes. *Biotechnol. Bioinorg.* **1987,** *30*, 928–935.

Shen, Z.; Chen, G.; Yu, Y.; Wang, Q.; Zhou, C.; Hao, L. Sonochemistry Synthesis of Nanocrystals Embedded in a MoO_3–CdS Core–Shell Photocatalyst with Enhanced Hydrogen Production and Photodegradation. *J. Mater. Chem.* **2012,** *22*, 19646–19651

Silva, B. N.; Pinto, A. C.; Silva, F. C.; Ferreira, V. F.; Silva, B. V. Ultrasound-Assisted Synthesis of Isatin-Type 5'- (4-Alkyl/Aryl-1H-1, 2, 3-Triazoles) via 1, 3-Dipolar Cycloaddition Reactions. *J. Braz. Chem. Soc.* **2016,** *27*, 2378–2382.

Singh, J.; Kaur, J.; Nayyar, S.; Bhandari, M.; Kad, G. L. Ultrasound Mediated Synthesis of a Few Naturally Occurring Compounds, **2001,** *40B*, 386–390.

Sultan, A.; Ur Rehman, M. H.; Sajjad, N.; Irfan, A.; Ullah, I.; Mustaqeem, M.; et al. A Facile Sonochemical Protocol for Synthesis of 3-Amino- and 4-Amino-1,2,4-Triazole Derived

Schiff Bases as Potential Antibacterial Agents. *Plos One.* **2020,** *15* (6). doi.org/10.1371/journal.pone.0229891.

Suslick, K. S. Sonochemistry. *Sciences* **1990,** *247* (4949), 1439–1445.

Suslick, K. S. *Ultrasound: Its Chemical, Physical, and Biological Effects*; VCH Publishers, 1998.

Suslick, K. S.; Casadonte, D. J. Heterogeneous Sonocatalysis with Nickel Powder. *J. Am. Chem. Soc.* **1987,** *109* (11), 3459–3461.

Suslick, K. S.; Doktycz, S. J. The Sonochemistry of Zinc Powder. *J. Am. Chem. Soc.* **1989,** *111* (6), 2342–2344.

Taghipour, T.; Karimipour, G.; Ghaedi, M.; Asfaram, A. Mild Synthesis of a Zn (II) Metal Organic Polymer and Its Hybrid with Activated Carbon: Application as Antibacterial Agent and in Water Treatment by Using Sonochemistry: Optimization, Kinetic and Isotherm Study. *Ultrason. Sonochem.* **2018,** *41*, 389–396.

Takagi, K. Ultrasound-Promoted Synthesis of Arylzinc Compounds Using Zinc Powder and Their Application to Palladium (0)-Catalyzed Synthesis of Multifunctional Biaryls. *Chem. Lett.* **1993,** *22* (3), 469–472.

Takagi, K.; Shimoishi, Y.; Sasaki, K. Ortho-Phenylenedizinc (II) Compounds Ultrasound-Promoted Synthesis from o-Diiodobenzene and Zinc Powder and Its Synthetic Application. *Chem. Lett.* **1994,** *23* (11), 2055–2058.

Talebzadeh, Z.; Masjedi-Arani, M.; Amiri, O.; Salavati-Niasari, M. Green Sonochemistry Fabrication of Pure $Gd_2Sn_2O_7$ Nanoparticles with Advanced Photocatalytic Efficiency for Elimination of Dye Pollutions. *Int. J. Hydrog. Energy* **2022,** *47* (8), 5269–5280.

Thornycraft, J.; Barnaby, S. W.; Tospedo Boot Destroyers. *Proc. Inst. Civil Eng.* **1895,** *122*, 51–103.

Tonanon, N.; Siyasukh, A.; Tanthapanichakoon, W.; Nishihara, H.; Mukai, S. R.; Tamon, H. Improvement of Mesoporosity of Carbon Cryogels by Ultrasonic Irradiation. *Carbon* **2005a,** *43* (3), 525–531.

Tonanon, N.; Siyasukh, A.; Wareenin, Y.; Charinpanitkul, T.; Tanthapanichakoon, W.; Nishihara, H.; et al. 3D Interconnected Macroporous Carbon Monoliths Prepared by Ultrasonic Irradiation. *Carbon* **2005b,** *43* (13), 2808–2811.

Valduga, C. J.; Santis, D. B. *J. Heterocyclic Chem.* **1998,** *36*, 505; [Links] Valduga, C. J.; Braibante, H. S.; Braibante, M. E. F. *J. Heterocyclic Chem.* **1997,** *34*, 1453.

Van der Hoff, B. M. E.; Gall, C. E. A Method for Following Changes in Molecular Weight Distributions of Polymers on Degradation: Development and Comparison with Ultrasonic Degradation Experiments. *J. Macromol. Sci. Chem.* **1977,** *11* (9), 1739–1758.

Vasantha, B.; Hemantha, H. P.; Sureshbabu, V. V. 1-Propanephosphonic Acid Cyclic Anhydride (T3P) as an Efficient Promoter for the Lossen Rearrangement: Application to the Synthesis of Urea and Carbamate Derivatives. *Synthesis* **2010,** *2010* (17), 2990–2996.

Vieira, B. M.; Thurow, S.; Brito, J. S.; Perin, G.; Alves, D.; Jacob, R. G. et al. Sonochemistry: An Efficient Alternative to the Synthesis of 3-Selanylindoles Using CuI as Catalyst. *Ultrason. Sonochem.* **2015,** *27*, 192–199.

Wang, C.; Huang, R.; Sun, R.; Wang, H. Ultrasound Assisted Fenton-Like Degradation of Dyes Using Copper Doped Graphitic Carbon Nitride. *Water Sci. Technol.* **2021,** *84* (5), 1146–1158.

Wang, J. X.; Zhao, K. Synthesis of cis and Trans-2, 6-diphenyl-1, 4-diselena-fulvenes from Phenyiacetylene with Selenium and Base Under PTC-Ultrasound Conditions. *Synthetic Commun.* **1996,** *26* (8), 1617–1622.

Wang, Y.; Nkurikiyimfura, I.; Pan, Z. Sonochemical Synthesis of Magnetic Nanoparticles. *Chem. Eng. Commun.* **2015,** *202* (5), 616–621.

Wu, L. Q.; Yang, C. G.; Yang, L. M.; Yang, L. J. Ultrasound-Assisted Wittig Reaction: A Short, Efficient Synthesis of 2-Methoxy-6-Alkyl-1, 4-Benzoquinones. *J. Chinese Chem. Soc.* **2009,** *56* (1), 47–50.

Xavier, D. M.; Goldani, B. S.; Seus, N.; Jacob, R. G.; Barcellos, T.; Paixao, M. W. et al. Sonochemistry in Organocatalytic Enamine-Azide [3+ 2] Cycloadditions: A Rapid Alternative for the Synthesis of 1, 2, 3-Triazoyl Carboxamides. *Ultrason. Sonochem.* **2017,** *34,* 107–114.

Yadav, H. A.; Eraiah, B.; Nagabhushana, H. Ultrasound Assisted Sonochemical Synthesis of ZrO_2: Eu^{3+} Nanophosphor. *AIP Conf. Proceed.* **2019,** *2115*. DOI: org/10.1063/1.5112939.

Yadav, S. J.; Reddy, B. V. S.; Reddy, K. B.; Raj, K. R.; Prasad, A. R.; Ultrasound-Accelerated Synthesis of 3,4-Dihydropyrimidin-2 (1*H*)-Ones with Ceric Ammonium Nitrate. *J. Chem. Soc. Perkin, Trans.* **2001,** *1*, 1939–1941.

Yakoot, S. M.; Salem, N. A. A Sonochemical-Assisted Simple and Green Synthesis of Silver Nanoparticles and Its Use in Cosmetics. *Int. J. Pharmacol.* **2016,** *12*, 572–575.

Yashas, S.R.; Shivaraju, H.P.; Thinley, T. et al. Facile Synthesis of SnO_2 2D Nanoflakes for Ultrasound-Assisted Photodegradation of Tetracycline Hydrochloride. *Int. J. Environ. Sci. Technol.* **2020,** *17*, 2593–2604.

Yuan, H.; Zhao, Y.; Wang, Y.; Duan, J.; He, B.; Tang, Q. Sonochemistry-Assisted Black/Red Phosphorus Hybrid Quantum Dots for Dye-Sensitized Solar Cells. *J. Power Sources* **2019,** *410*, 53–58.

Zhang, C.; Wang, Q.; Xia, H.; Qiu, G. Ultrasonically Induced Microemulsion Polymerization of Styrene. *Eur. Polym. J.* **2002,** *38* (9), 1769–1776.

Zhang, J.; Wang, J.; Fu, Y.; Zhang, B.; Xie, Z. Sonochemistry-Synthesized CuO Nanoparticles as an Anode Interfacial Material for Efficient and Stable Polymer Solar Cells. *RSC Adv.* **2015,** *5* (36), 28786–28793.

Zhang, K.; Park, B. J.; Fang, F. F.; Choi, H. J. Sonochemical Preparation of Polymer Nanocomposites. *Molecules* **2009,** *14* (6), 2095–2110.

CHAPTER 11

Microwave-Assisted Organic Synthesis: A Need of the Day

SEEMA KOTHARI[1], CHETNA AMETA[2], K. L. AMETA[3], B. K. SHARMA[4], RAJAT AMETA[5], and RAKSHIT AMETA[6]

[1]*Department of Chemistry, PAHER University, Udaipur, India*

[2]*Department of Chemistry, M. L. Sukhadia University, Udaipur, India*

[3]*Department of Chemistry, Modi University, Lakshmangarh, India*

[4]*Department of Chemistry, G. G. Govt. P.G. College, Banswara, India*

[5]*Zyfine Cadila, Ahmedabad, India*

[6]*Department of Chemistry, J. R. N. Rajasthan Vidhyapeeth, (Deemed to be University), Udaipur, India*

ABSTRACT

Many reactions can be carried out in presence of microwaves. Here, microwaves are the energy source. It is used for the heating with certain benefits such as efficient, uniform and selective heating as compared to conventional heating. It provides better yields and purity and that too in reduced time. Microwaves are used in different reactions such as oxidation, reduction, alkylation, esterification, rearrangement, cycloaddition, condensation, protection, coupling, and so on. All such reactions in presence of microwaves have been discussed here.

11.1 INTRODUCTION

After almost one and a half century of the first chemical revolution, a new kind of chemical revolution has come up that is "Green Chemistry." The

Green Chemistry, 2nd Edition: Fundamentals and Applications. Suresh C. Ameta & Rakshit Ameta (Eds.)

fundamental idea of green chemistry is that the manufacturer of any chemical needs to consider what will be the fate of human life after this particular chemical is generated and used in society?

The 12 principles of green chemistry can be used to access a particular synthetic protocol's greenness (Anastas and Warner, 2000). These principles address some basic aspects, such as the use of various solvents, the amount of chemical waste produced, the use of catalyst and reagents (quantity and reusability), the amount of chemical step (energy efficiency) and atom economy, and the use of safer chemicals and reaction conditions. It is very difficult for a new synthetic protocol to satisfy all the 12 principles, which is not expected also, but if most of the principles are satisfied by a protocol, then the developed process will be considered as a green route.

There are two alternative ways to categorize different approaches from green chemistry point of view.

- Synthesis via an environment-friendly synthetic pathways or processes.
- To develop some new benign replacements which are capable of achieving the desired performance without any adverse ecological impacts.

One can achieve greener protocol through a proper choice of starting materials (feed stock), atom economic methodologies with minimum chemical steps, the use of appropriate greener solvents and reagents, and efficient strategies for the isolation of product and its purification. Thus, a major goal of this protocol should be to maximize the efficient use of safer raw materials on the one hand and reduce the wastes produced in a particular process on the other, simultaneously.

Therefore, there is need to find out solutions to problems faced in developing green and sustainable synthetic methods in the fields of healthcare and fine chemicals like posing significant challenges to the synthetic organic chemists, the pressure to produce these substances expeditiously and that too in an environmentally benign fashion. It is worthwhile to note that rapid development of "Green Organic Chemistry" is due to the recognition that eco-friendly products and processes will be economical in the long term as these do not require the treatment of "end-of-the-pipe" pollutants, and by-products so commonly generated by conventional synthetic procedures and one such technology is microwave-assisted organic synthesis (MAOS) (Kingston and Haswell, 1997; Lidstrom and Tiernery, 2005).

Microwave heating under controlled conditions is an invaluable technology because it not only dramatically reduces the reaction time, typically from days or hours to minutes or even seconds, that is to speed up the reaction, but it also fulfills the aim of green chemistry by reducing side reactions increasing yields and improves reproducibility (Hayes, 2002). This approach has now become a central tool in this rapid paced, time-sensitive field and it has also blossomed into a useful technique for a variety of applications in organic synthesis, where high-yielding protocols and facility of purification are highly desirable (Bradley, 2001; Kulmert, 2002; Dudley and Stiegman, 2018). Furthermore, this technique is energy efficient and the possibilities for application in combinatorial, parallel, and automated environmentally benign chemistry are obvious.

Ortiz et al. (2018) observed the influence of microwave exposure in MAOS. They used computational chemistry and developed a predictive model. Different related parameters were determined such as nonthermal and thermal effects. It has a with clear added advantages on experimental methods, where the effect cannot be separated to the extent (almost impossible).

11.2 MICROWAVE-ASSISTED CHEMISTRY

11.2.1 MICROWAVES AS ENERGY SOURCE

Microwave radiations are electromagnetic radiations, which are widely used as a source of heating in organic synthesis. Microwaves have enough momentum to activate reaction mixture to cross-energy barrier and complete the reaction in lesser time. A microwave oven consists of a magnetron, a wave guide feed and an oven cavity. A magnetron is a thermionic diode that works on the principle of dielectric heating by converting part of the electric power into electromagnetic energy and the rest of it into heat energy. Microwaves occupy a place in the electromagnetic spectrum between infrared waves and radio waves, ranging in wavelengths between 0.01 and 1 m, and operate in a frequency range between 0.3 and 30 GHz. The typical bands for industrial applications are 915 ± 15 and 2450 ± 50 MHz. The wavelength between 1 and 25 cm are extensively used for RADAR transmissions and the remaining wavelength range is used for telecommunications. The entire microwave region is therefore not available for heating applications and the equipment operating at 2.45 GHz, corresponding to a wavelength of 12.2 cm, is quite commonly used. The energy carried by microwave at 2.45 GHz is 1 Joule per mole of quanta, which is relatively very small energy.

11.2.2 MICROWAVES AS A TOOL FOR SYNTHETIC CHEMISTRY

The earliest description of the magnetron (the high-power generator of microwave power), a diode with a cylindrical anode was reported by Hull (1921a, 1921b). The potential of microwave heating for organic synthesis has been explored in last three and half decades after the first reports appeared in 1986 (Gedye et al., 1986; Giguere et al., 1986). Initially, reactions were performed in domestic microwave ovens using appropriate solvents. After that, several groups started investigating reactions in solvent-free conditions including "dry media" usually with open vessels (Stadler and Kappe, 2000; Vidal et al., 2000). The use of microwave units specially designed for the purpose of synthesis are expensive and it becomes rather difficult at times. Thus, unmodified home microwave units are suitable in some cases. However, simple modifications (e.g., a reflux condenser) can enhance the safety factor. High-pressure chemistry should only be carried out in special reactors with a microwave oven specifically designed for this purpose. A further point in favor of using the more expensive apparatus is the question of reproducibility, as only these specialized machines can achieve good field homogeneity, and in some cases, these can even be directed on the reaction vessel.

It has long been known that molecules undergo excitation with electromagnetic radiation. This effect is utilized in household microwave ovens to heat up food. However, chemists have been using microwaves only as a reaction methodology for a few years. Some of the first examples gave amazing results, which led to a flood of interest in microwave-accelerated synthesis (Bose et al., 1991, 1994; Banik et al., 1992, 1993).

The MW heating has not been restricted to organic chemistry only, but its application to various aspects of inorganic chemistry and polymer chemistry has also been investigated with several advantage of an eco-friendly approach. In the past few decades, this technique has found a valuable place in the synthetic chemist's tool box, which is evident from a large number of publications (Lidstrom et al., 2001; Watkins, 2002; Perreux and Loupy, 2001; Strauss, 2002; Loupy, 2006; Ortiz et al., 2018; Priecel and Lopez-Sanchez, 2019; Kumar et al., 2020), particularly acylation reaction (Moghaddam and Sharifi, 1995; Monteil-Rivera and Paquet, 2015; Naeimi et al., 2017; Rajbongshi et al., 2020), addition reaction, elimination reaction (Mogilaiah et al., 2003; Wu et al,. 2016; Blanco-Vega et al., 2017), alkylation reaction (Abramovitch et al., 1995; Apsunde and Trudell, 2014), alkynes metathesis (Miljanic et al., 2003; Kugelgen et al., 2019; Grau and Tsogoeva, 2020), alkylation reaction (Lourenço et al., 2013; Rubiño et al., 2014), allylation reaction (Motorina et al., 1996; Liu et al., 2019; Novanna et al., 2020), amination reaction (Mccarroll et al., 2003; Tankam et al., 2018; Seubert et al., 2020), aromatic nucleophillic substitution reaction (Bozinovic et al., 2016; Fuchibe et al,. 2017), arylation reaction (Wali et al., 1995; Mehra et al., 2016; Sawant et al., 2018; Liu et al., 2020), carbonylation reaction (Yamazaki and Kondo, 2002; Maddocks et al., 2020), combinatorial reaction (Al-obeidi et al., 2003), condensation reaction (Ameta et al., 2010; Viola et al., 2014; Yuan et al., 2020), coupling reaction (Burton et al., 2003; Krömer et al., 2014), cyanation reaction (Arevela et al., 2003; Sawant and Bhanage, 2014), cyclization reaction (Shrimali et al., 2009; Cochrane et al., 2021), cycloaddition reaction (Lerestif et al., 1995; Chakraborty et al., 2013; Yıldız et al., 2016), deacetylation reaction (Kumar et al., 2003), dehalogenation reaction (Calinescu et al., 2003; Goswami et al., 2017; Murata et al., 2017), Diel's-Alder reaction (Mavoral et al., 1995; Liu et al., 2017; Simin et al., 2020), dimerization reaction (Santagada et al., 2003; Xue et al., 2018; Marx and Ndabab, 2021), trans-esterification reaction (Roy and Gupta, 2003; Shinde and Yadav, 2014), enantioselective reaction (Diaz-Ortiz et al., 2003; Han and Gong, 2019; Singh et al., 2021), halogenation reaction (Inagaki et al., 2003; Iwasaki et al., 2019; Das and Basak, 2020), synthesis of heterocyclic compounds (Pathak et al., 2019; Gupta et al., 2021), hydrolysis reaction (Chen et al., 2015; Portero-Barahona et al., 2019; Chen et al., 2019), Mannich reaction (Sitha et al., 2010; Aljohani et al., 2019; Peipei et al., 2019; Jin et al., 2021), synthesis of metal organic frame work (MOF) (Liang and D'Alessandro, 2013; Babu et al., 2017), oxidation reaction (Kiasat et al., 2003; Hosseinpour et al, 2014, Sutradhar et al., 2019), phosphorylation synthesis (Gospondinova et al., 2002; Artem'ev et al., 2015; Yin et al., 2018; Karu and Gedu, 2019), polymerization reaction (Vu et al., 2003; Monterde et al., 2019; Loudy et al., 2020), protection reaction (Liu et al., 2015), rearrangement reaction (Srikrishna and Kumar, 1995; Horikoshi et al., 2015; González-Liste et al., 2015; Schultze and Schmidt, 2019; Dev et al., 2021), multicomponent reactions (Karamthulla et al., 2014; Fardood et al., 2019), solvent-free synthesis (Shah and Mohanraj, 2014; Borade al., 2020) and so on.

The initial slow development of this technology in the last 1980s and early 1990s has been attributed to lack of its controllability and reproducibility coupled with detail understanding of the basics of MW dielectric heating.

11.3 PRINCIPLE

The basic principle behind heating by microwaves is the interaction of charged particle of the reaction material with electromagnetic wavelength of a particular frequency. The phenomenon of producing heat by electromagnetic irradiation involves either collision or conduction and sometimes both. Two basic principles are involved in heating the materials by microwaves.

11.3.1 DIPOLAR POLARIZATION

Dipolar polarization is the phenomenon responsible for the majority of microwave heating. It depends upon the polarity of solvent and compound. In polar molecules, different electronegativities of individual atoms result in a permanent electric dipole, which is sensitive to external electric fields and will attempt to align with them by rotation. This realignment is rapid for a free molecule, but in liquid, the instantaneous alignment is prohibited by the presence of other molecules. A limit is, therefore, placed on the ability of the dipole to respond to an electric field, which affects the behavior of the molecule with different frequencies of electric field, for example, under low-frequency irradiation, the dipole may react by aligning itself in phase with the electric field. Molecules will polarize uniformly and thus, no random motion results.

Under high frequency irradiation, the polar molecule will attempt to follow the field, but intermolecular inertia stops any significant motion before the field has reversed, in this case, the dipole do not have sufficient time to respond the field, and it does not rotate. As no motion is induced in the molecules, no energy transfer will take place, and therefore, no heating results. In the case of intermediate frequency, the field will be such that the molecule is almost (but not quite) able to keep in phase with the field polarity. The microwave frequency is low enough so that the dipoles have enough time to respond to the alternating field, and therefore to rotate, but high enough so that the rotation does not precisely follow the field. As the dipole reorients to align itself with the field, the field is already changing, and a phase difference causes energy to be lost from the dipole in random collisions. Thus, giving rise to dielectric heating (Figure 11.1).

11.3.2 CONDUCTION MECHANISM

The conduction mechanism generates heat through resistance to an electric current. The oscillating electromagnetic field generates an oscillation of electrons or ions in a conductor resulting in an electric current. This current faces internal resistance, which heats the conductor.

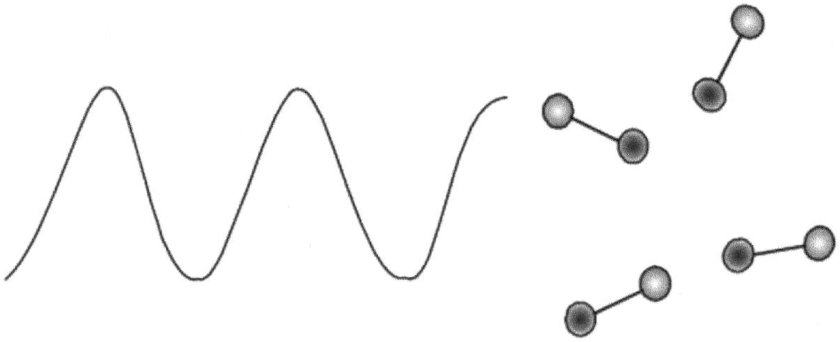

FIGURE 11.1 Microwave heating by dipolar polarization mechanism.

FIGURE 11.2 Conduction mechanism.

When the irradiated sample is an electrical conductor, the charge carriers (electrons, ions etc.) are moved through the material under the influence of the electric field, resulting in a polarization. These induced currents will cause heating in the sample due to electrical resistance.

11.3.3 MICROWAVE CHEMISTRY APPARATUS

Microwave synthesis has started with a kitchen microwave oven with good results. Nowadays, many types of advanced microwave ovens have been introduced in the market. These consist of a microwave source (Magnetron), a microwave cavity or an applicator (multimode cavity or single-mode cavity), mode stirrer, sensor probe (thermocouples or IR sensor) and software with digital display.

Two types of reactors are used in microwave-assisted organic synthesis. These are multimode and monomode reactors. The differentiating feature of a single-mode apparatus is its ability to create a new standing wave pattern, which is generated by the interference of fields that have the same amplitude but different oscillating directions. This interface generates an array of nodes where microwave intensity is zero and an array of antinodes where the magnitude of microwave energy is at its highest. Therefore, sample should be placed at the antinode at appropriate distance from the magnetron.

Multimode reactors (domestic microwave ovens) are the most common instruments used in organic synthesis since they are comparatively inexpensive and readily available. A lot of satisfying organic synthesis has been done with domestic microwave oven. Multimode reactors provide a field pattern with areas of high and low field strength, commonly referred to as "hot and cold spots." This nonuniformity of the field leads to the heating efficiency varying drastically between different positions of the sample. This drawback is overcome by using mode stirrer.

The mode stirrer is a periodically moving metal vane that continuously changes the instantaneous field pattern inside the cavity, and therefore, the field intensity is homogeneous everywhere throughout the cavity. Thus, samples can be placed anywhere inside the cavity, because the field is homogeneous throughout the cavity. In modern microwave reactors, preinstalled digital thermometers (sensors and probes) are used for temperature measurement. Moreover, some sophisticated ovens are equipped with computers also (Ranger et al., 1995; Barlow and Marder, 2003).

11.3.4 REACTION VESSELS AND REACTION MEDIUM

The reaction vessel must be transparent to the microwaves. These are preferably being made up of teflon, polystyrene, and glass (Strauss and Trainor, 1995). Metallic containers are not used as it gets heated soon due to preferential absorption and reflection of rays. For reactions in solvents, the solvent must have a dipole moment and a boiling point higher than the desired reaction temperature and a dielectric constant. Some of the solvents used commonly as microwave absorber are N,N-dimethyl formamide or DMF (b.p.154°C, $\varepsilon = 36.7$), formamide (b.p. 216 °C, $\varepsilon = 111$), methanol (b.p. 65°C, $\varepsilon = 32.7$), ethanol (b.p. 78°C, $\varepsilon = 24.6$), chlorobenzene (b.p. 132°C, $\varepsilon = 5.6$), 1,2-dichlorobenzene (b.p. 180°C, $\varepsilon = 10.5$), 1,2-dichloroethane (b.p. 83°C), ethylene glycol (b.p. 196°C, $\varepsilon = 37.7$), dioxane (b.p. 101°C, $\varepsilon = 2.25$) and diglyme (b.p. 162°C, $\varepsilon = 7.23$). The presence of salts in polar solvents can frequently enhance microwave coupling. Hydrocarbon solvents such as toluene ($\varepsilon = 2.4$), hexane ($\varepsilon = 1.9$), benzene ($\varepsilon = 2.3$), and xylene ($\varepsilon = 2.2$-2.57), because of less dipole moment, are unsuitable as they absorb microwave radiation poorly. For solid-state reactions, mineral oxides, such as zeolite, alumina, silica, montmorillonite K-10 clay are used as absorbents.

11.3.5 MICROWAVE EFFECT

The microwave effect applies to a range of observations in microwave chemistry. These may be classified into two categories: specific microwave and nonthermal microwave effects (Stuerga and Gaillard, 1996a, 1996b).

- **Specific microwave effects**

These are effects that cannot be easily done by conventional heating methods. Examples include: (i) selective heating of specific reaction components, (ii) rapid heating rates and temperature gradients, (iii) elimination of wall effects, and (iv) superheating of solvents

- **Nonthermal microwave effects**

Excitation with microwave radiation results in the molecules aligning their dipoles within the external field. Strong agitation, provided by the reorientation of molecules, in phase with the electrical field excitation, causes an intense internal heating. The question of whether a nonthermal process is operating can be answered simply by comparing the reaction rates between the cases where the reaction is carried out under irradiation versus under conventional heating. In fact, no nonthermal effect has been found in the majority of reactions (De la Hoz et al., 2005), and the acceleration is attributed to superheating alone. It is clear that nonthermal effects do play a role in some reactions.

11.3.6 COMPARISON BETWEEN MICROWAVE HEATING AND CONVENTIONAL HEATING

Microwave radiation provides rapid and homogeneous heating, which has certain advantages, such as reaction rate acceleration, milder reaction conditions, and higher chemical yields. In short, microwave-enhanced chemical reactions are safer, faster, cleaner, and more economical than conventional reactions. It helps in developing cleaner and greener synthetic routes (Chemat and Esveld, 2001; Lidstrom et al., 2001; Kuhnert, 2002; Nuchter et al., 2003;

Nuchter et al., 2004; Dudley and Stiegman, 2018, Prieceland Lopez-Sanchez, 2019; Yadav et al., 2020). The microwave heating can change kinetics of reaction in a different way as observable under conventional heating.

11.3.6.1 INCREASED RATE OF REACTION

Microwave heating enhances the rate of certain chemical reactions by 10–1000 times compared with conventional heating. This is due to its ability to increase the temperature of a reaction. For instance, synthesis of fluorescein, which usually takes about 10 h by conventional heating methods, can be completed in only 35 min by means of microwave heating. The rate acceleration caused by microwaves has been attributed to superheating of solvents (liquid-phase reactions) and high temperature on the surface of catalyst or other solid reactants. The water molecule is the target for microwave ovens in the home. Like any other molecule with a dipole, it absorbs microwave radiation. Microwave radiation is converted into heat with high efficiency, so that "superheating" becomes possible at ambient pressure. Enormous accelerations in reaction time can be achieved, if superheating is performed in closed vessels under high pressure; a reaction that takes several hours under conventional conditions can be completed over the course of minutes.

- **Efficient source of heating**

Microwave-assisted heating is a highly efficient process and results in a significant energy saving. This is because microwaves heat up just the sample and not the apparatus, and therefore, energy consumption is less.

- **Higher yields**

In certain chemical reactions, microwave radiation produces higher yields as compared with conventional heating methods. For example, microwave synthesis of fluorescein results in an increase in the yield of the product from 70 to 82%.

- **Uniform heating**

Microwave radiation provides uniform heating throughout a reaction mixture unlike conventional heating methods. It is because in conventional heating, the walls of the oil bath gets heated first and then the solvent. As a result of this, there is always a temperature difference between the walls and the solvent. In the case of microwave heating, only the solvent and the solute particles are excited, which results in uniform heating of the solvent.

- **Selective heating**

Selective heating is based on the principle that different materials absorb microwaves to different extent. Some materials are transparent whereas others absorb microwaves. Therefore, microwaves can be used to heat a combination of such materials, for example, the production of metal sulfide with conventional heating requires weeks because of the volatility of sulfur vapors while rapid heating of sulfur in a closed tube results in the generation of sulfur fumes, which can cause an explosion. However, in microwave heating, as sulfur is transparent to microwaves, only the metal gets heated. Therefore, the reaction can be carried out at a much faster rate with rapid heating, without the threat of an explosion.

- **Environmental-Friendly chemistry**

Reactions conducted through microwaves are cleaner and more environment friendly than conventional heating methods. Microwaves heat the compounds directly, and therefore, the use of solvents in the chemical reaction can be reduced or eliminated. An approach was developed to carry out a solvent-free chemical reaction on solid support, such as clay, alumina, and zeolite. The reactants adsorbed on solid support under microwave react at a faster rate than conventional heating. The use of microwaves has also reduced the extent of purification required for the end products of chemical reactions involving toxic reagents.

- **Greater reproducibility of chemical reactions**

Reactions under microwave irradiation show more reproducibility as compared with conventional heating because of uniform heating and better control of process parameters. The temperature of chemical reactions can also be easily monitored. This is of particular relevance in the lead optimization phase of the drug development process in pharmaceutical companies.

11.3.7 LIMITATIONS OF MICROWAVE CHEMISTRY

The limitations associated with microwave heating are its scalability, limited application, and the hazards involved in its use.

- **Lack of scalability**

The yield obtained by using domestic microwave apparatus is limited to a few grams. Although there have been developments in the recent past relating to the scalability of microwave equipment, still there is a gap that needs to be spelled to make this technology scalable.

- **Limited applicability**

Microwaves can be used for heating only those materials, which absorb them. Microwaves cannot heat materials, which are transparent to these radiations.

- **Safety hazards relating to the use of microwave heating apparatus**

Although manufacturers of microwave heating apparatus have directed their research to make microwaves a safe source of heating. Uncontrolled reaction conditions may result in undesirable results, for example, chemical reactions involving volatile reactants under superheated conditions may result in explosive conditions. Moreover, improper use of microwave heating for rate enhancement of chemical reactions involving radioisotopes may result in uncontrolled radioactive decay.

- **Health hazards**

Health hazards related to microwaves are caused by the penetration of microwaves. The microwaves operating at a low frequency range are only able to penetrate the human skin, while higher frequency range microwaves can reach body organs. Research has proved that prolonged exposure to microwaves may result in the complete degeneration of body tissues and cells. It has also been established that constant exposure of DNA to high frequency microwaves during a biochemical reaction may result in complete degeneration of the DNA strand.

11.4 CLASSIFICATION OF MICROWAVE REACTIONS

Broadly microwave-assisted organic synthesis can be classified into solvent-assisted and solvent-free synthesis

11.4.1 SOLVENT-ASSISTED SYNTHESIS

Solvent-assisted synthesis proceeds in the presence of solvent with good polarity, high boiling point, and sufficient chemical stability. N, N-Dimethylformamide

(DMF) is a good solvent for microwave-assisted organic synthesis as it has a high boiling point (154°C), high dielectric constant, and is miscible with water, making it relatively easy to be removed from the reaction mixture.

Erdelyi and Gogoll (2001) reported reaction of aryl halides with trimethylsilyl acetylene in DMF as solvent under microwave radiations. They reported 85–95% yield of product in 5–25 min.

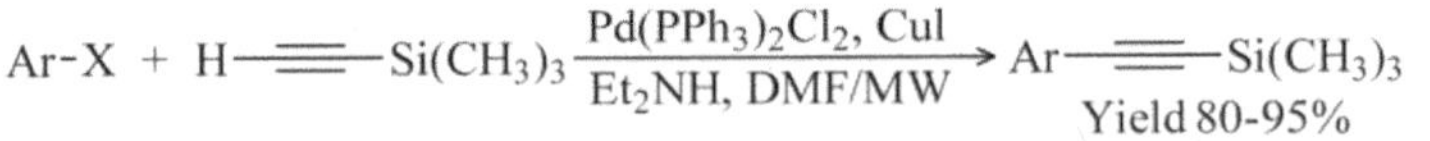

Hu et al. (1999) carried out the substitution of nitro group in the presence of tetrahydrofuran in good yields (83%).

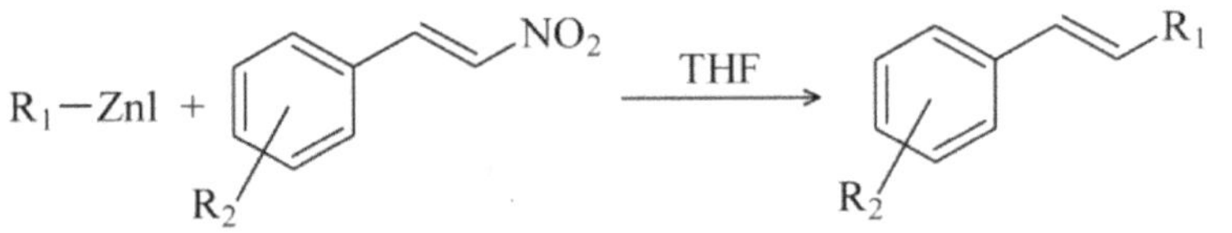

A rapid and efficient methods for the preparation of isoxazole, pyrazole, and pyrimidine derivatives of imidazolinone and quinazolinone under MWI has been reported by Vyas et al. (2008). Microwave-assisted synthesis and characterization of some new annulated pyrimidinone derivatives have been carried out by Sancheti et al. (2007).

11.4.2 SOLVENT-FREE SYNTHESIS

Solvent-free method offers safe and efficient reaction pathways, which are time and money saving and often enable the elimination of waste treatment. Solvent-free syntheses are divided into solid-support reaction and neat reactions.

- **Solid-support reactions**

Microwave-assisted solid-support reaction increases the rate of reactions and decreases the reaction time. It also gives high conversion with high selectivity, avoiding solvent in most of cases, and therefore, workup also becomes easier. The MW promoted deprotection, condensation, rapid one-pot synthesis, synthesis of heterocyclic compounds using recyclable mineral oxides supported reagents, such as $Fe(NO_3)_3$-clay (clayfen), NH_2OH-clay, $PhI(OAc)_2$-alumina, $NaIO_2$-silica, CrO_3-alumina, MnO_2-silica and $NaBH_2$-clay, and zeolite. It has proved as a base for the growth of a newer branch of green chemistry. Kann (2010) made the use of polymer-supported organometallic catalysts in organic synthesis. KF-alumina-mediated Bargellini reaction has been reported by Rohman and Myrboh (2010).

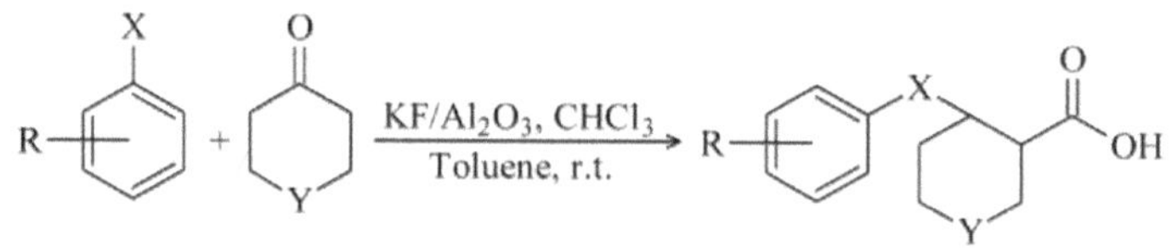

Lalitha and Sivakamasundari (2010) studied the synthesis of few vinyl quinolones on solid support. Self-condensation of hydroxybenzene derivatives under microwave using heterogenous has been reported by Gomez et al. (2010).

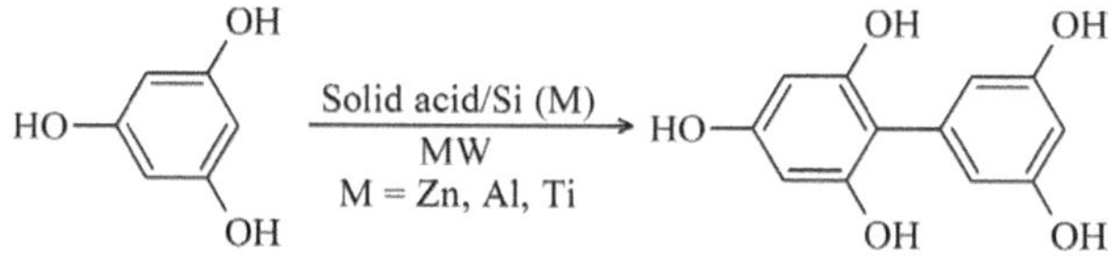

Yang et al. (2012) described microwave-assisted solid-phase synthesis, biological evaluation and molecular docking of angiotensin i-converting enzyme inhibitors. Bouasla et al. (2017) synthesized 7-hydroxy-4-methylcoumarin and 4-methylcoumarin using resorcinol, phenol, and ethyl acetoacetate in a solvent-free condition and heterogeneous solid acid catalysts, such as Amberlyst-15 under microwave irradiation.

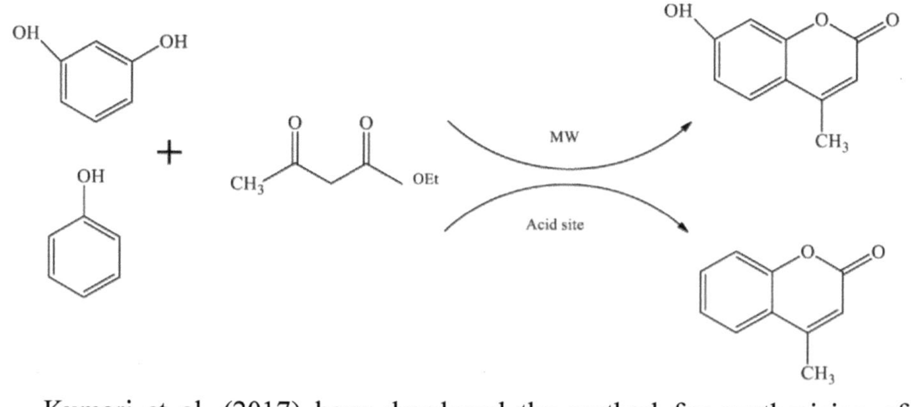

Kumari et al. (2017) have developed the method for synthesizing of bis-(*N,N*-dialkyl)/*O*-aryl *N,N*-dialkyl-2-(1-methyl/phenyl-2-oxopropylidene) phosphoro hydrazido oximes from their corresponding hydrazides by using magnesium sulfate as a solid support under microwave irradiation.

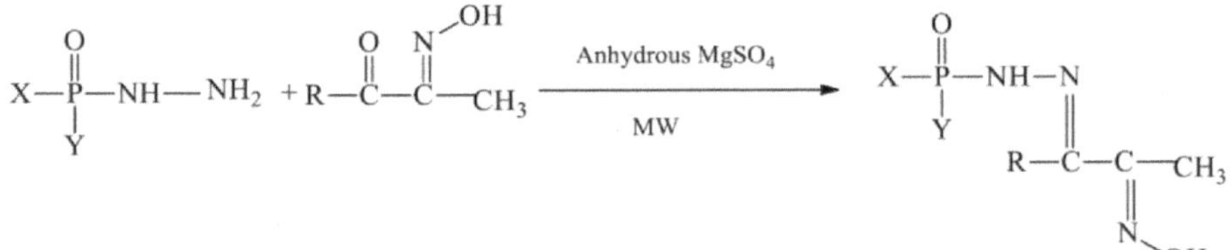

X = $N(C_2H_5)_2$, $N(C_3H_7)_2$
Y = $N(C_2H_5)_2$, $N(C_3H_7)_2$, N(CH3)(C4H9), OC_6H_5, (3-CH_3)-OC_6H_4, (4-CH_3)-OC_6H_4
R = CH3, C_6H_5

- **Neat reactions**

Such reactions are performed between neat reagents. Here, at least one of the reagents must be a polar liquid. These are liquid–liquid or liquid–solid systems; the latter implies that the solid is soluble in the liquid phase or at least the liquid counterpart is adsorbed on the solid, and hence, the reaction occurs at the interface. Ameta et al. (2011) reported solvent-free synthesis of pthalimide derivatives under microwave irradiation. A series of NH-pyrazoles were efficiently synthesized from the reaction of dimethylaminovinylketones and hydrazine hydrate in solid state by Longhi et al. (2010).

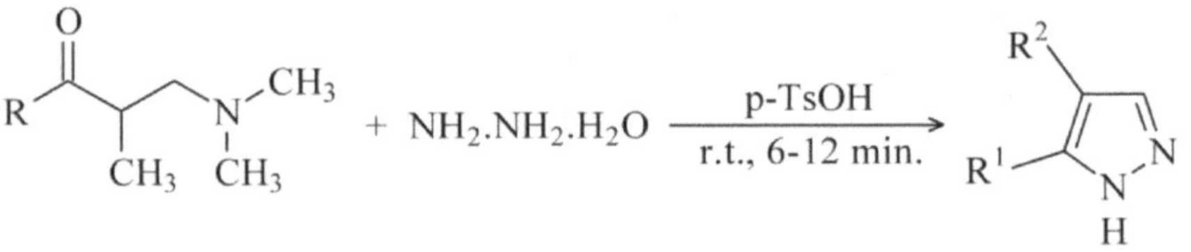

Ghorbani-Vaghei and Malaekehpour (2010) investigated efficient and solvent-free synthesis of 1-amidoalkyl-2-naphthols using N,N, N', N',-tetrabromobenzene-1,3-disulfonamide. LiBr has been used as a catalyst for solvent-free synthesis of fused thiazoloquinazoline derivatives (Ameta et al., 2010). Chang et al. (2010) carried out efficient solvent-free catalytic hydrogenation of solid alkenes and nitro-aromatics using Pd nanoparticles entrapped in aluminum oxy-hydroxide. $AlCl_3$ has been used as a catalyst for microwave-enhanced and solvent-free green protocol for the production of dihydropyrimidine-2-(1H)-ones (Kumar et al., 2010).

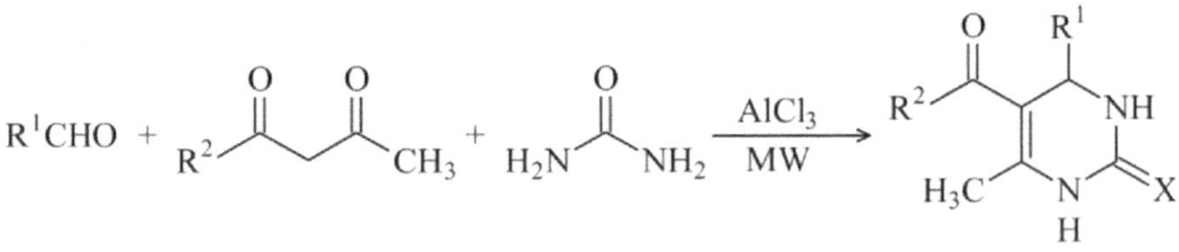

Microwave-assisted solvent-free synthesis of 1,3-diphenylpropenones has been reported (Kakati and Sarma, 2011). Solvent-free microwave-assisted synthesis of E-(1)-(6-benzoyl-3,5-dimethylfuro [3′,2′:4,5] benzo[b] furan-2-yl)-3-(aryl)-2-propen-1-ones and their antibacterial activity was studied by Ashok et al. (2011).

Yaragorla et al. (2014) reported an efficient solvent-free synthesis of amides by Ca(II) catalyzed Ritter reaction in the presence of microwave irradiation. This is a green protocol, which tolerates the diversity of substrate offering high yields of amides. It only utilized minimal loading of inexpensive and more abundant Ca(II) catalyst.

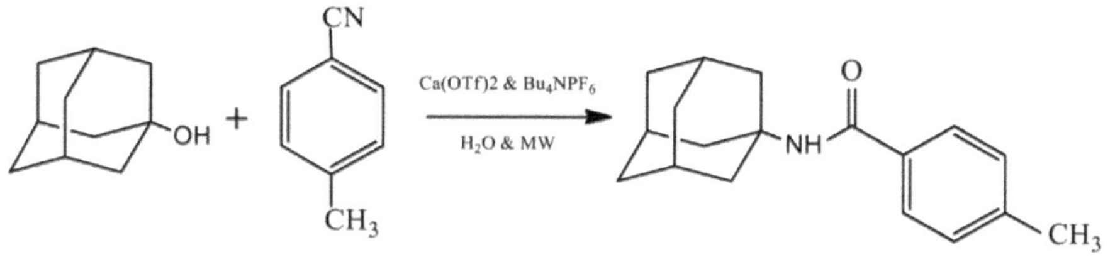

Basha et al. (2016) carried out microwave-assisted synthesis of phosphinates and diethyl/ethylphenyl {2-(benzo[d]thiazol-2-yl)phenylamino} phosphonates in neat conditions with high yields (80–93%) via Kabachnik-Fields reaction.

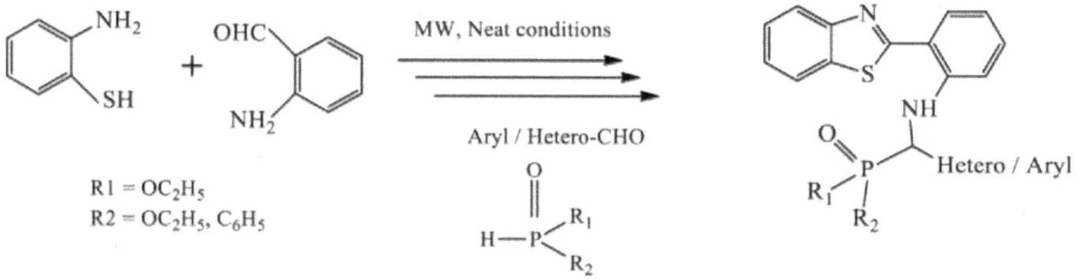

11.5 APPLICATIONS IN ORGANIC SYNTHESIS

Green chemistry involves design of chemical synthesis to prevent pollution and thereby solve the environmental problems. The microwave chemistry is the current approach in green chemistry, as it follows a number of principles of green chemistry. Here, some applications of microwave-assisted organic synthesis are reported.

11.5.1 OXIDATION REACTIONS

A remarkably fast microwave-assisted selective oxidation of benzylic alcohols with calcium hypochlorite under solvent-free conditions has been reported (Lukasiewicz et al., 2006). Alcohols are readily adsorbed on clayfen and rapidly oxidized under microwave irradiation into corresponding carbonyl compounds without using any solvent.

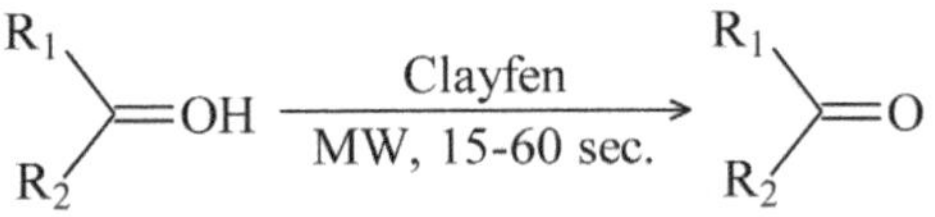

Primary and secondary alcohols are selectively oxidized under microwave irradiation to the corresponding aldehydes and ketones within 5–25 s using commercially available and magnetically retrievable Magtrieve™ (Mojtahedi et al., 2000).

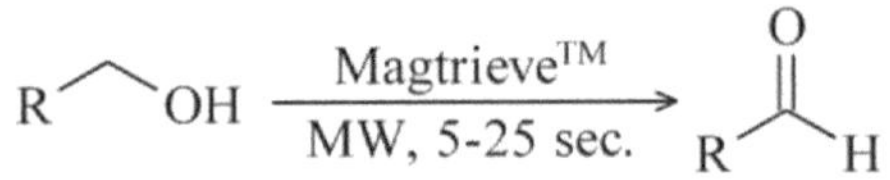

The oxidations of some simple secondary alcohols by hydrogen peroxide has been reported by hydrogen peroxide–urea adduct using microwaves as an energy source (Bogdal et al., 2003). Tetravalent chromium dioxide has been shown to be a very useful oxidant (as a magnetically retrievable oxidant) for microwave-assisted and conventional transformation of aromatic and alkyl aromatic molecules into the corresponding aryl ketones, quinones, or lactones (Lukasiewicz et al., 2004). Liu et al. (2010) studied size effect of silica-supported gold clusters in the microwave-assisted oxidation of benzyl alcohol with H_2O_2.

An efficient but mild methodology was developed by Mohammadi (2013) for the oxidation of organic compounds under microwave irradiation. They used cetyltrimethylammonium chlorochromate (CTMACC) as the oxidant. The reactions were controlled so as to stop at aldehyde stage, and no over-oxidation and formation of side products is there. This reaction has simple workup, easy procedure, in excellent yields and that too short reaction times.

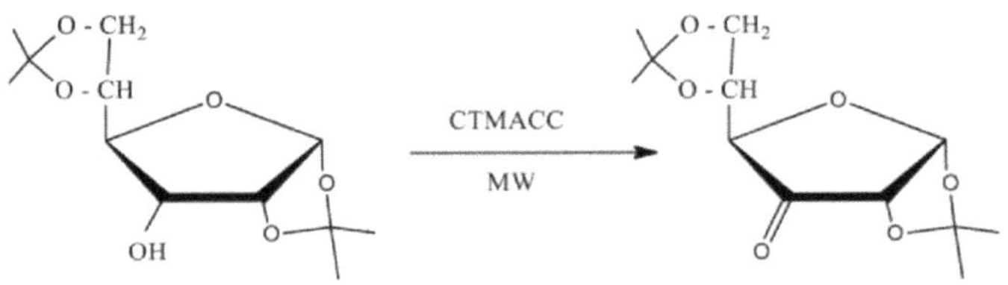

Graphene oxide (GO) was used as a catalyst by Bermudez et al. (2015) in the oxidizing alkenes and under microwave exposure. They observed that 25 wt.% loadings of catalyst and shorter reaction times of 1 h was necessary in the presence of a combination of GO and H_2O_2 under microwave irradiation for optimum oxidation capacity to oxidize any double bond to different alcohols and aldehydes.

The oxidation of 5-hydroxymethylfurfural to 2,5-furandicarboxylic acid (FDCA) was carried out by Zhao et al. (2020) under microwave (MW)-assisted heating using hydrogen peroxide as an oxidant. The use of commercial 5% Ru/C afforded an excellent yield of FDCA (88%) in 30 min only.

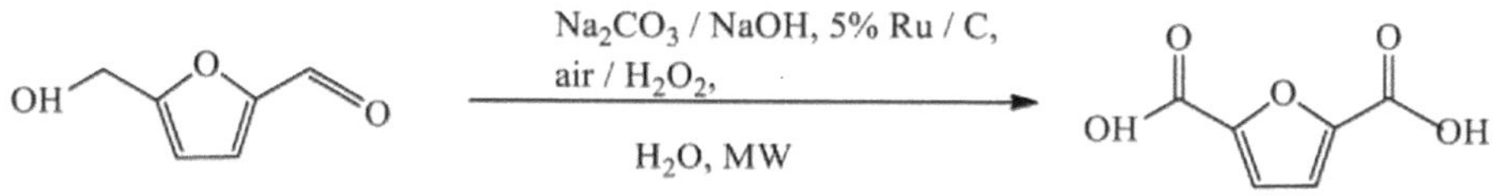

11.5.2 REDUCTION REACTIONS

Bose et al. (1993, 1999) were first to report the use of microwaves for transfer hydrogenation in organic synthesis. A series of β-lactam derivatives were hydrogenated using formates and Pd/C or Raney Ni as catalyst. Schmoger et al. (2011) carried out microwave-assisted reduction of organic compounds using catalytic and stoichiometric reactions. Ethylene glycol was used as the solvent and electron source for the microwave-assisted reduction reaction. An improved microwave-assisted one-pot tandem Staudinger/aza–Witting/reduction was reported by Chen et al. (2011) for the synthesis and biological activity of novel 5'–arylamino-nucleosides. Wada et al. (2000) carried out the reductive dehalogenation of chlorinated phenols to phenol, cyclohexanol and other chlorine-free compounds within 20 min using microwave irradiation. The carbonyl compounds were reduced in solid state using $NaBH_4$ (supported on alumina) by Varma and Saini (1997).

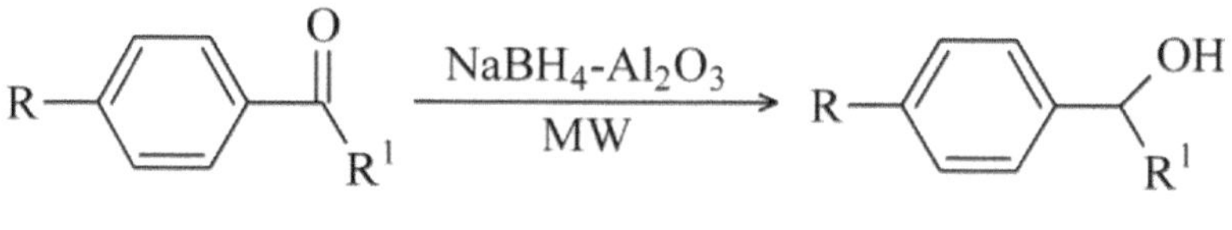

11.5.3 ALKYLATIONS

MW-assisted selective N-alkylation of 6-amino-2-thiouracil in DMF using different alkyl halides has been investigated by Loupy et al. (2001) whereas no reaction was observed under the same conditions in a thermoregulated oil bath.

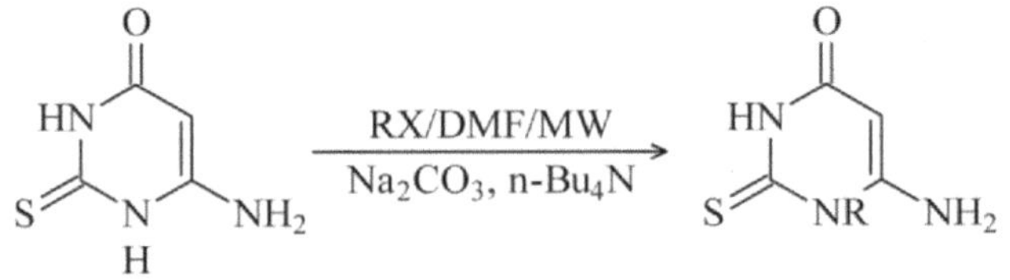

A series of diethers have been obtained by alkylation of dianhydrohexanes under MW conditions at 140°C in 90% yield (Chatti et al., 2001). Matondo et al. (2003) carried out O-alkylation of o-bromophenol under MW conditions using TBAB as catalyst. The reaction of diethyl ethoxycarbonylmethylphosphonate with a series of alkyl halides under microwave and solventless conditions has been studied at 120°C in the presence of Cs_2CO_3. In the absence of a phase-transfer catalyst, it afforded the corresponding monoalkylated products with yields of more than 70%. The thermal variant carried out in boiling acetonitrile was slow and led to incomplete conversions. In the MW method, the phase-transfer catalyst is substituted by MW irradiation and there is no need for a solvent (Grun et al., 2012).

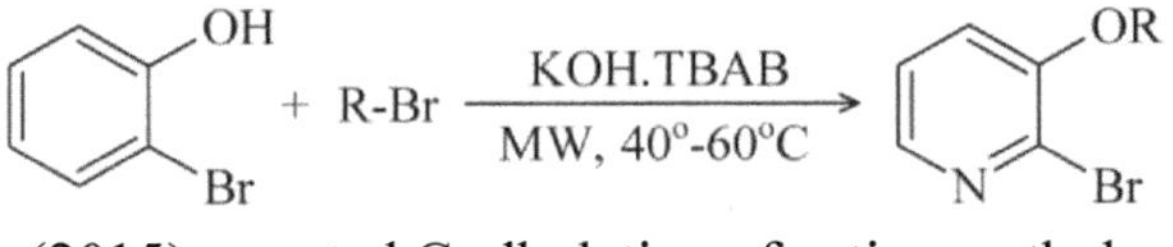

Grun et al. (2015) reported C-alkylation of active methylene compounds (solid–liquid phase) under microwave irradiation or phase-transfer catalysis, which contains P=O or C=O functions.

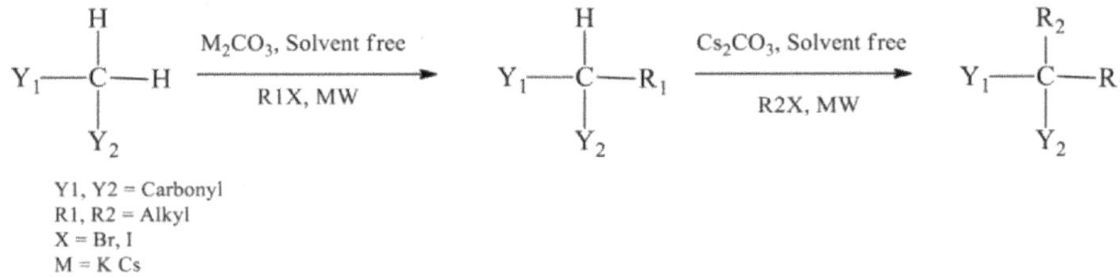

Jha and Jain (2016) observed that copper catalyzed regioselective cross-coupling of *N*-tosylhydrazones with azine *N*-oxides yields *ortho*-alkylated products under microwave irradiations.

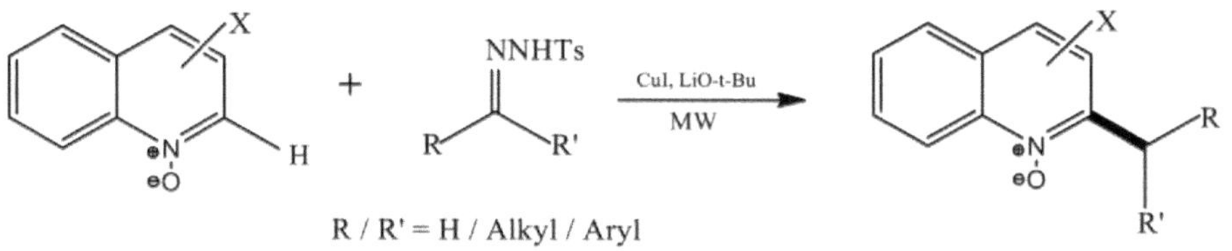

Gour et al. (2019) reported that N-alkylation of aromatic amines by using aromatic alcohols in the presence of SmI_2 as a catalyst under MW conditions afforded high selective monoalkylated amines.

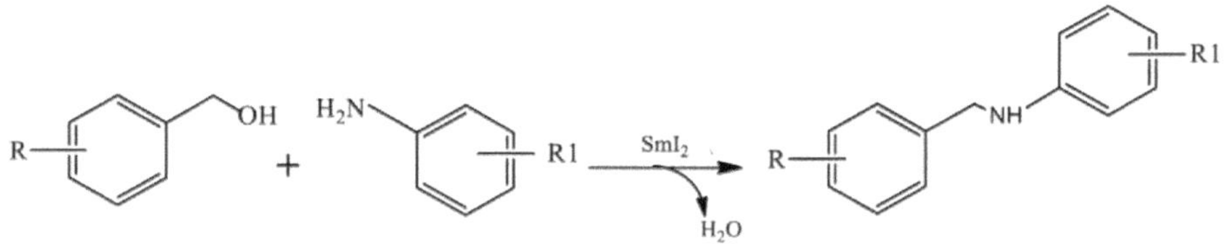

11.5.4 REARRANGEMENTS

A general and efficient procedure for the selective Meyer–Schuster isomerization of both, terminal and internal alkynols has been developed by Antinolo et al. (2012) using catalytic amounts of the readily accessible oxovanadium (V) complex $[V(O)Cl(OEt)_2]$. Reactions proceeded smoothly in toluene at 80°C under microwave irradiation to provide the corresponding α,β-unsaturated carbonyl compounds in excellent yields and short times without the assistance of any additive. Li et al. (2006) carried out Beckmann rearrangement in acetone under microwave irradiation using silica sulfate as an efficient and recyclable catalyst.

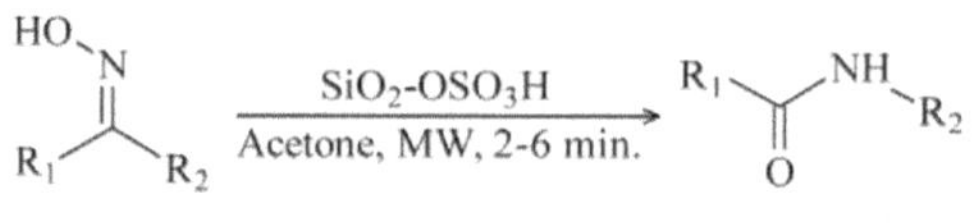

A microwave-assisted Domino rearrangement of propargyl vinyl ethers to multifunctionalized aromatic platforms has been described by Tejedor et al. (2011). An aza-Cope rearrangement of N-allylanilines using $BF_3 - OEt_2$

under microwave activation (in few min) has been studied by Gonzalez et al. (2008). Microwave-assisted Hoffmann rearrangement has been reported by Miranda et al. (2011).

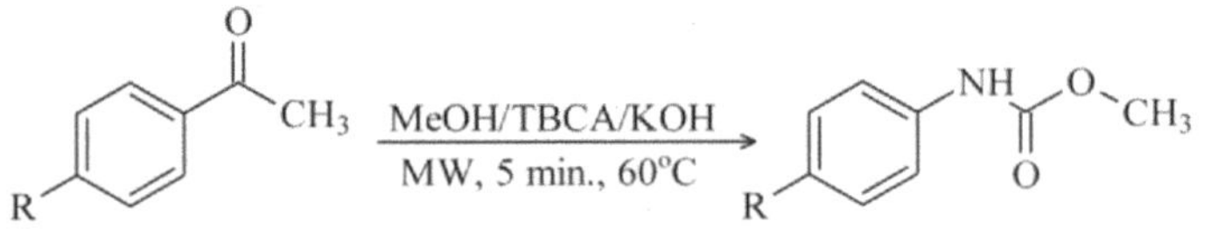

Egami et al. (2018) carried out Johnson–Claisen rearrangement of allyl alcohol and triethylorthoacetate in a microwave reactor. It was reported that this could be carried out without any solvent, and it required a catalytic amount of acetic acid. They obtained highest yield of the desired γ,δ-unsaturated ester and the productivity was about 89.5 g h^{-1} under the optimal conditions. According to it, 2.1 kg of the product could be obtained theoretically per day. Other allylic alcohols were also used to get the products in high yields.

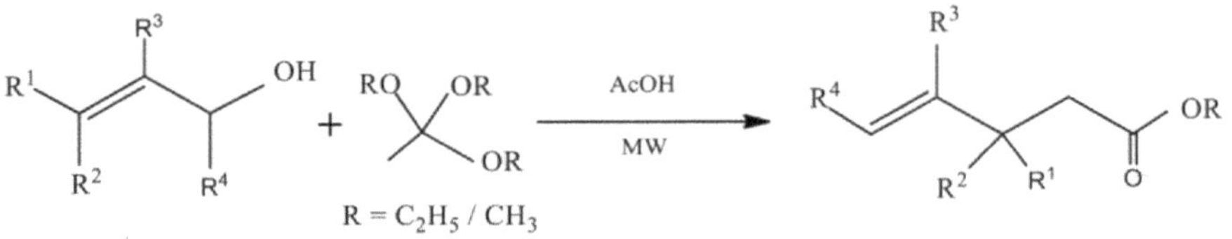

Tjeng et al. (2020) reported stereocontrolled microwave-assisted Domino [3,3]-sigmatropic reactions; a Winstein Overman reaction for the formation of two contiguous CN bonds.

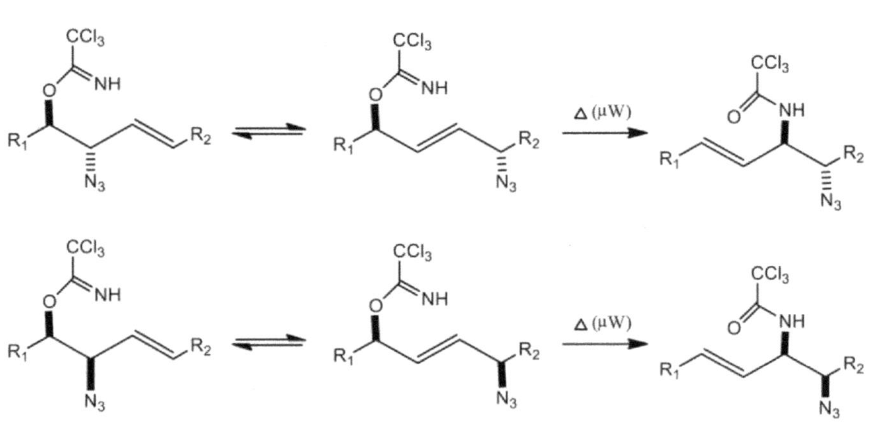

Hui et al. (2020) observed that catalysts phosphomolybdic acid (PMA) accelerated Claisen rearrangement reaction of some allyl aryl ethers in the temperature range (220–300°C) under microwave irradiation.

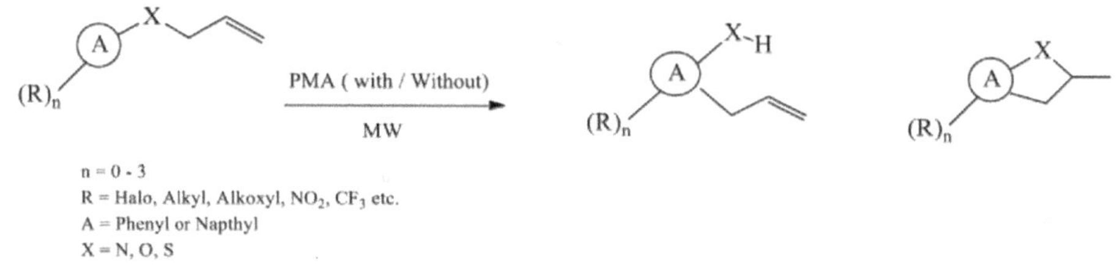

11.5.5 CYCLOADDITIONS

Microwave-assisted Diels-Alder cycloaddition reaction between 2-fluoro-3-methoxy-1,3 butadiene and ethylene was reported by Patrick et al. (2007).

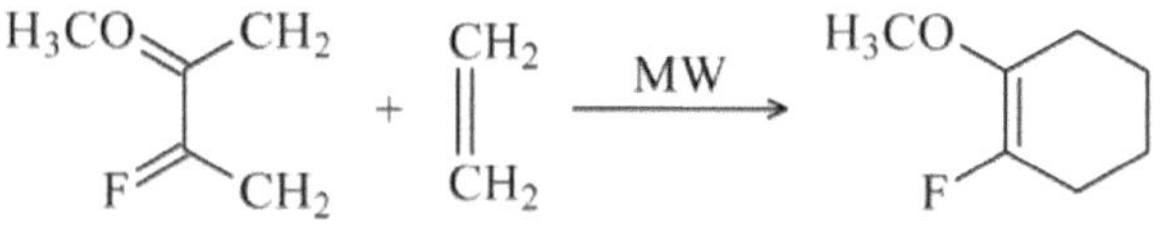

This reaction gives moderate yield under thermal conditions, but very good yield under microwave radiations. Similarly, intramolecular hetero Diels-Alder cycloaddition of acetylenic pyrimidines to bicyclic pyridine was reported by Shao (2005). The microwave-assisted functionalization of carbon nanohorns (CNHs) via [2 + 1] nitrenes cycloaddition, which provided well dispersible hybrid materials possessing aziridino-rings covalently grafted onto the graphitic network of CNHs (Karousis et al., 2011). Microwave-assisted stereoselective 1,3-dipolar cycloaddition of C,N-diarylnitrone and bis(arylmethylidene)acetone was reported by Paul et al. (2012).

A green synthesis of cyclic carbonates was reported by Tharun et al. (2013) using epoxides and CO_2. They used HCOOH/KI catalytic system in a microwave reactor. Various epoxide substrates were used in this microwave-irradiated cycloaddition. The influence of different operational parameters, such as microwave power, catalyst composition, CO_2 pressure, and reaction time was observed. It was observed that there was a synergistic effect of COOH/KI (catalyst) as compared with OH/KI system.

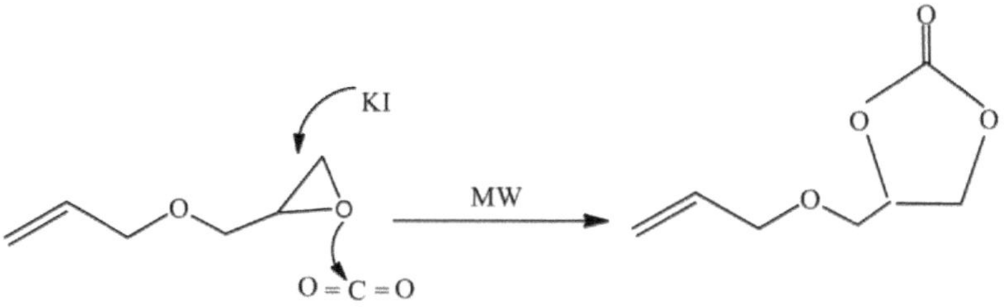

Taher et al. (2015) reported a solvent-free 1,3-dipolar cycloaddition reaction between azides and terminal alkynes to prepare 1,2,3-triazoles in the presence of polymer-supported Cu(I) as a catalyst under microwave irradiation.

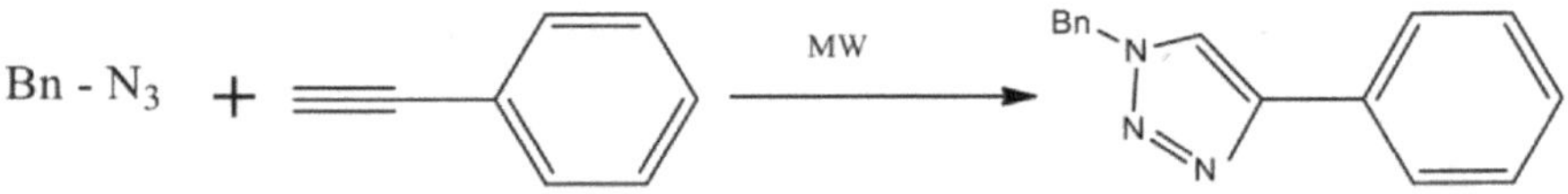

Nesaragi et al. (2021) synthesized quinolin-3-yl-methyl-1,2,3-triazolyl-1,2,4-triazol-3(4*H*)-ones by *click chemistry* as [3 + 2] cycloaddition of terminal alkynes with azides. The use of copper sulfate and THF/water promoted this [3 + 2] cycloaddition reactions with high yield of products.

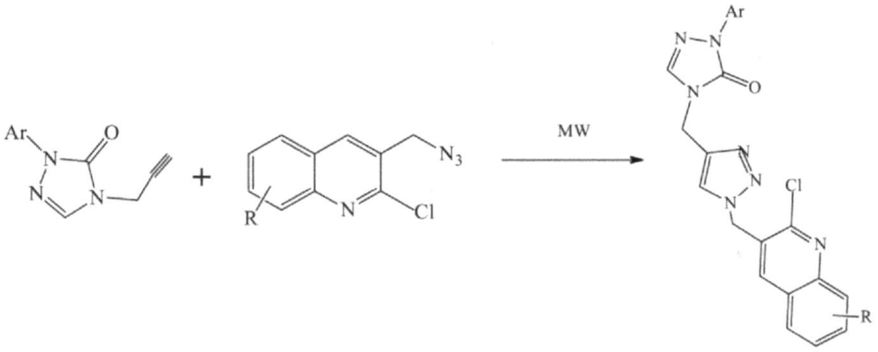

11.5.6 CONDENSATION

The Claisen condensation and its intramolecular variant, the Dieckmann condensation are classic reactions studied in organic chemistry because of their importance in organic synthesis and biochemical transformations. The growth in the use of microwave technology in both the synthetic and teaching laboratories warrants the modification of existing methodologies to incorporate this technology. Simple microwave-assisted procedures were developed for carrying out Claisen and Dieckmann condensation reactions that are suitable for organic chemistry. Although solvents can be used, the procedure is amenable to solvent-free conditions that promote green chemistry (Horta et al., 2011). Ighilahriz et al. (2008) synthesized 4(3H)–quinazolinones by cyclocondensation of anthranilic acid, aniline, and orthoester (or formic acid) in the presence of HPA as a catalyst under microwave irradiation.

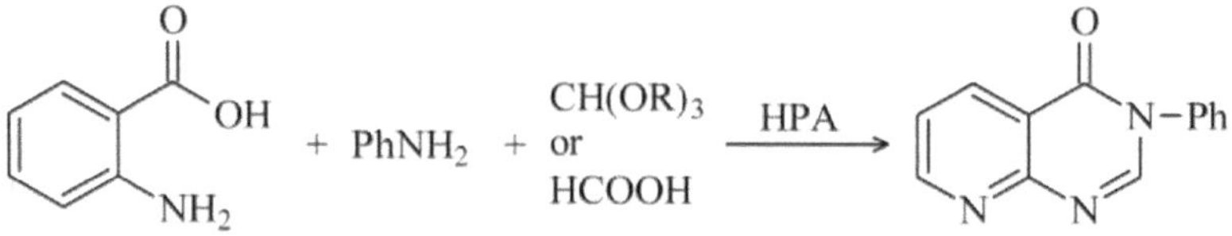

Agrawal and Joshipura (2005) studied Friedlander condensation under microwave irradiation. They also made a comparison over conventional method. In Friedlander condensation, a mixture of 2–aminobenzophenone and ethyl acetoacetate was irradiated for 5–7 min without catalyst and the products obtained were in good yields (80 – 91%).

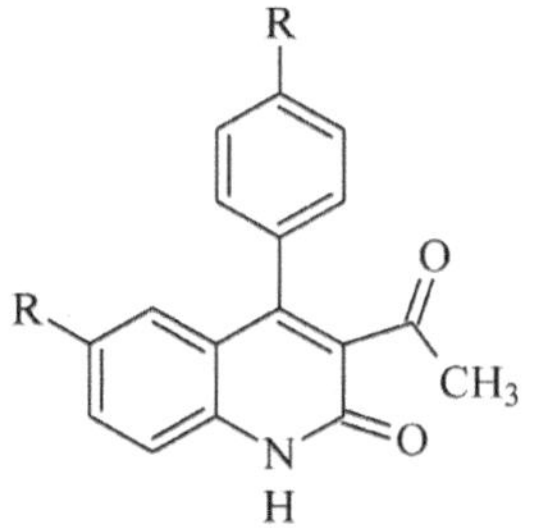

Thiourea is an inexpensive, efficient, and mild catalyst for the synthesis of Knoevenagel condensation of pyrozoles derivatives. In the presence of 10 mol% of thiourea, pyrazole aldehyde reacts with active methylene compound under microwave-assisted solvent-free conditions at 300 W for 2–5 min to give corresponding products in good yields (Li et al., 2011).

Crossed aldol condensation has been carried out by Handayani et al. (2017) in the presence of MW irradiation, which yielded dibenzylidene cyclohexanone derivatives. This reaction was performed by reacting benzaldehyde and cyclohexanone (in molar ratio of 2:1) in the presence of NaOH (catalyst) for 2 min. It was observed that the desired concentration (optimum) of NaOH was 5 mmole and the corresponding compounds dibenzylidene cyclohexanones were obtained with 93–100% yields.

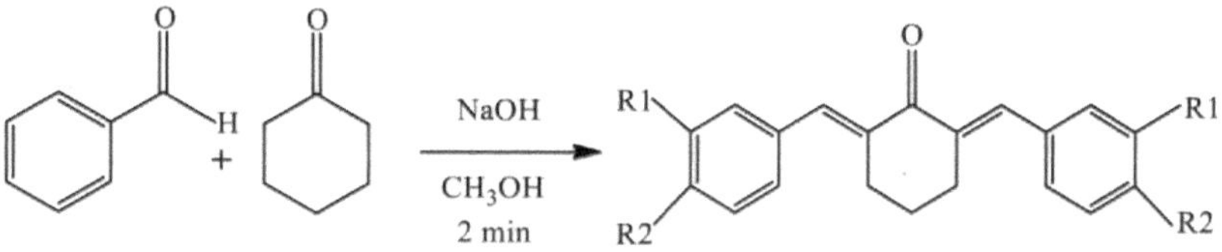

$R1 = OCH_3$, $R2 = H$
$R1 = R2 = H$
$R1 = R2 = OCH_3$

An amine-functionalized MOF was used by Taher et al. (2020), which is highly efficient and recyclable heterogeneous catalyst for Knoevenagel condensation of different aromatic ketones and aldehydes in ethanol. A high conversion of the reactants was achieved with 100% selectivity under microwave irradiation with significantly reduced reaction time than conventional heating. The MOF used was found to have same structural integrity after catalytic reactions is complete and it could be reused more number of times without any significant loss of activity.

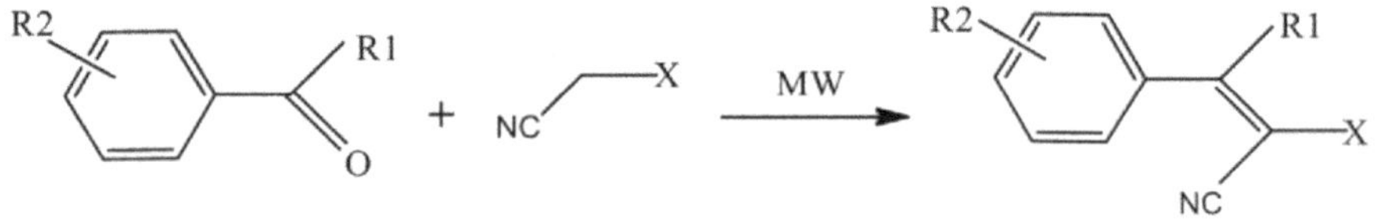

R1 = H, CH_3
X = CN, CO_2Et
R2 = CH_3, OCH_3, Cl, Br, HO, NO_2

Abirami et al. (2021) synthesized a new series of pyrano[2,3-b]quinolines from 3-formylquinolin-2(1H)- ones via Knoevenagel condensation. It was followed by cyclization reaction of ethyl cyanoacetate using DMSO as solvent with microwave exposure protocol. This protocol is simple, efficient, fast, clean, and eco-friendly method consuming less time in reaction with an improved yield and purity of the product obtained.

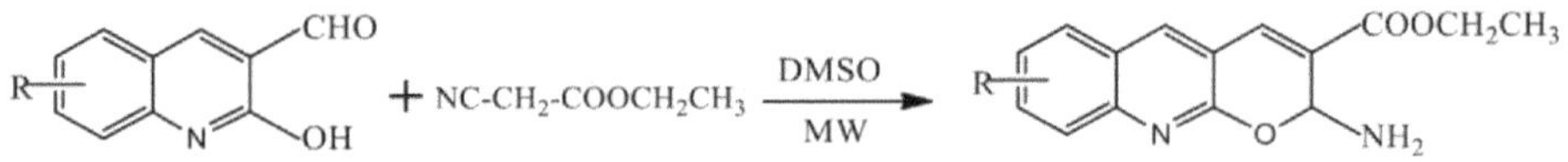

R= H, CH_3, OCH_3, Cl, Br

11.5.7 ESTERIFICATIONS

A new "on water" MW-assisted esterification protocol has been proposed by Oliverio et al. (2016). They used fatty chlorides and acetonide hydroxytyrosol as the starting materials. The formation of ester is a one-pot multistep reaction without using any catalyst. Such reaction occur at the interface of oil/water. Hence, it is strictly dependent on the lipophilicity of the fatty acid acyl chloride used. A potential scale-up of the proposed method has also been reported.

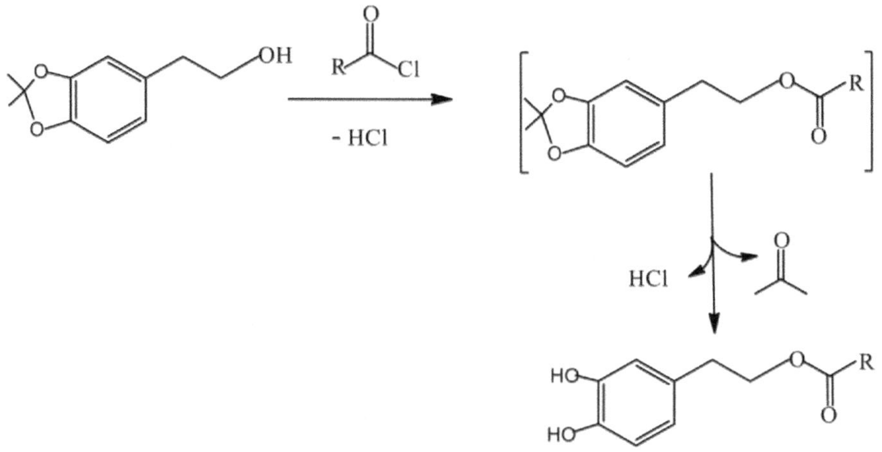

Werth et al. (2015) reported the reactive distillation for the homogeneously catalyzed trans- esterification of dimethyl carbonate with ethanol.

11.5.8 PROTECTION REACTIONS

Microwave in addition to its massive applications in rapid organic synthesis may also be used for simple and fast protection reactions. The selective protection of glutaraldehyde and its derivation (monoacetals and thioacetals) was carried out Flink et al. (2010) under microwave irradiation. They reported that reactions proceed efficiently and only traces of deprotected materials were left.

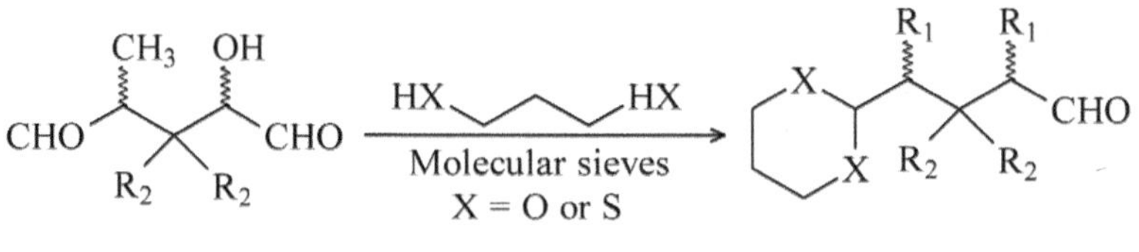

Liu et al. (2005) carried out graft polymerization of-caprolactone on chitosan under microwave irradiation via a protection-graft deprotection procedure with phthaloylchitosan using a catalyst (stannous octoate). After deprotection, the phthaloyl group was removed and amino group was regenerated. Hur et al. (2011) prepared microwave irradiation of unprotected as well as protected ferrocenoylamidoamino acids mediated by N-acylbenzotriazole. These amino acids undergo substitution reactions with 1-(ferrocenylcarbonyl)-1H-benzotriazole in the presence of microwave.

Walia et al. (2013) reported that N-substituted 2,5-dimethylpyrroles (primary amines) can be doubly protected. The reaction time taken for protection and deprotection under microwave can be decreased.

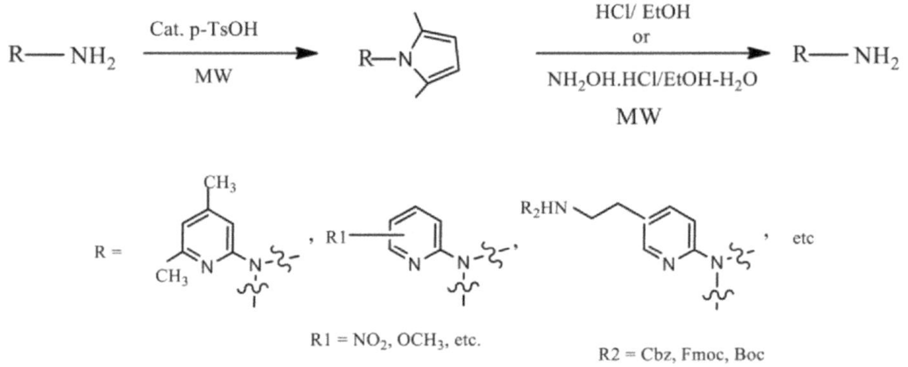

11.5.9 TRANSITION METAL-CATALYZED COUPLING REACTIONS

Organometallic catalysis in general and palladium catalyzed reactions in particular have emerged as the success stories in the growing heterogeneous field of microwave-assisted chemistry (Larhed et al., 2002; Olofsson et al., 2002). The first MW-promoted Suzuki coupling was published in 1996 (Larhed and Hallberg). Microwave has accelerated metal-catalyzed transformation of aryl and vinyl halides by carboxylations. Heck and Sonogashira reactions and cross-coupling have been reported by Nilsson et al. (2006).

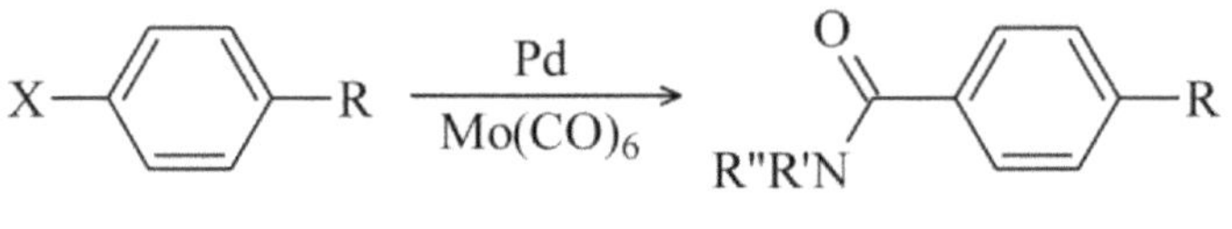

Perbenzylated pyranoid glycals couples with aryl bromides under microwave irradiation in the presence of 5% mol palladium (II) acetate in DMF to produce 2’, 3’-unsaturated C-aryl-d-glycopyranosides in a rapid and stereospecific manner (Lei et al., 2009).

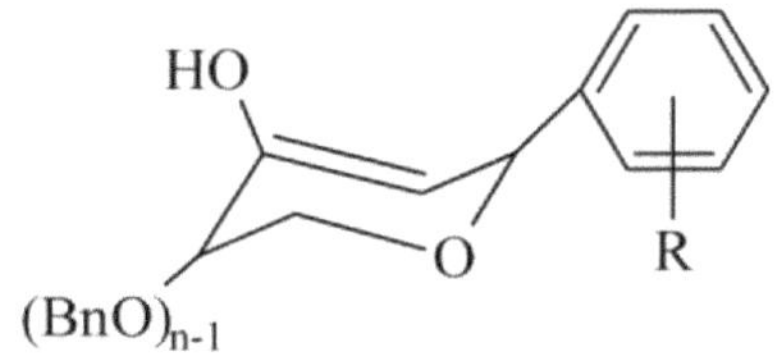

Leadbeater (2005) and Leadbeater and Marco (2003) reported Suzuki-Miyaura coupling in aqueous medium and isolated products in good yields. Fluoride-free cross-coupling reactions under ligand-free conditions with low Pd loadings in water using NaOH under microwave heating has been carried out by Alacid and Najera (2008).

Massaro et al. (2014) reported the use of modified halloysite nanotubes in the Suzuki reaction under microwave irradiation at 120 °C. They used K_2CO_3 as base. The palladium catalyst exhibited good activity in 0.1 mol%.

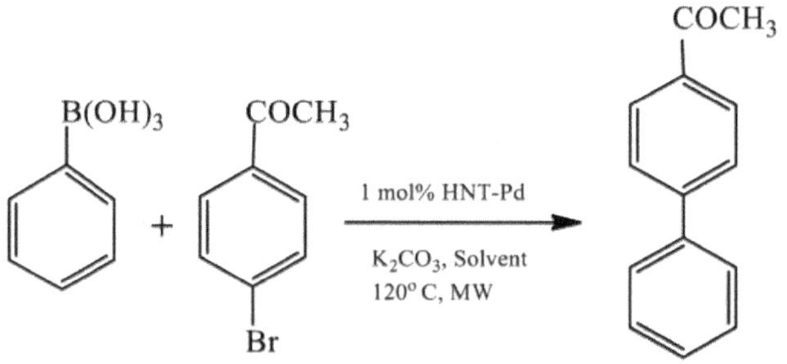

Jablonkai and Keglevich (2015) reported microwave-assisted PC coupling reactions with different halo benzoic acid and diarylphosphine oxides in water (solvent) without using any catalyst.

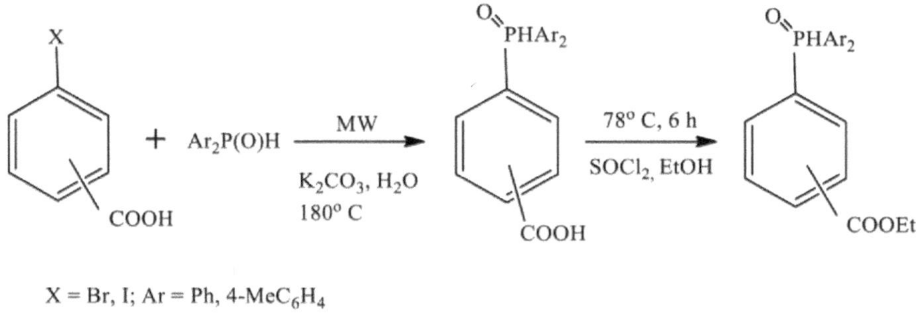

11.5.10 SYNTHESIS OF HETEROCYCLES

The efficient use of MW heating approach in several organic synthesis as an emerging green technique has been proposed by several workers (Molteni and Ellis, 2005; Polshettiwar and Varma, 2008). A rapid, efficient and environmental benign methodology for the preparation of 2,5-disubstituted indole analogues was developed (Biradar and Sasidhar, 2011). The synthesis of pyrroles by reaction of hexane–2,5–dione with primary amines has been carried out by Danks (1999) under microwave irradiation.

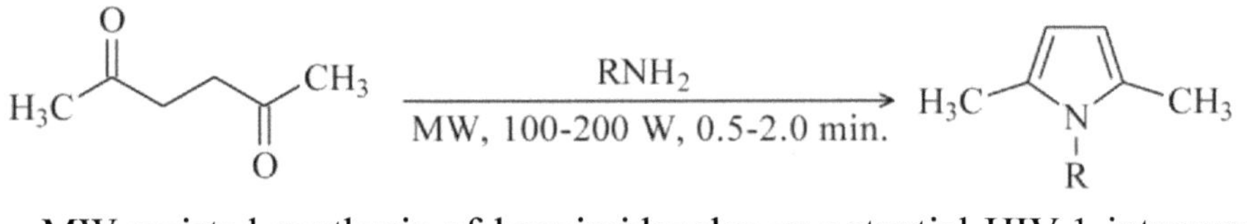

MW-assisted synthesis of benzimidazoles as potential HIV-1 integrase inhibitors by condensation of R-hydroxycinnamic acids with 1,2-phenylenediamine in water was performed by Ferro et al. (2004). A simple, high-yielding synthesis of 2,4,5–trisubstituted imidazoles from 1,2–diketones and aldehydes in the presence of ammonium acetale was reported by Wolkenberg et al. (2004) under microwave irradiation. Microwave-assisted three-component reaction has been used by Zhu et al. (2014) for the synthesis of benzodiazepines. It was reported that this reaction was promoted by acetic acid in water in the presence of microwaves. They used commonly available and low-cost starting materials for this purpose. It is a green synthesis involving water as the reaction solution, short reaction periods, concise one-pot conditions, easy purification, and less production of waste avoiding the use of any strong acids or metal promoters.

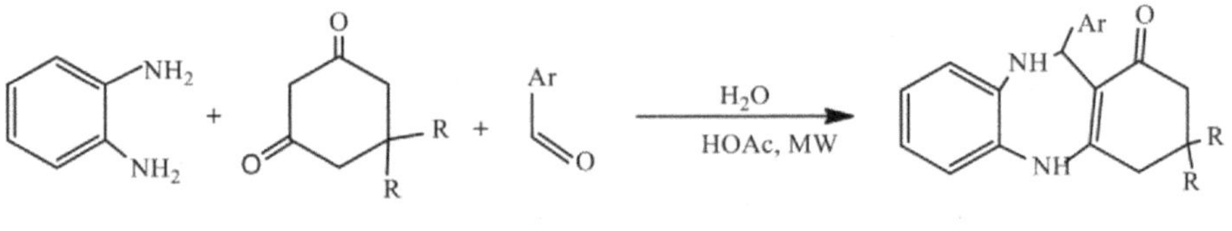

Cai et al. (2015) reported that microwave-assisted iminyl radical cyclizations can be terminated by trapping with TEMPO, and as a result, functionalized adducts were obtained. Alkynes were used as radical acceptors to provide a range of 2-acylpyrroles in good yields and there is no requirement of any toxic and hazardous reagents, which are commonly used in radical reactions. Here, this reaction is tin-free and no initiator is required.

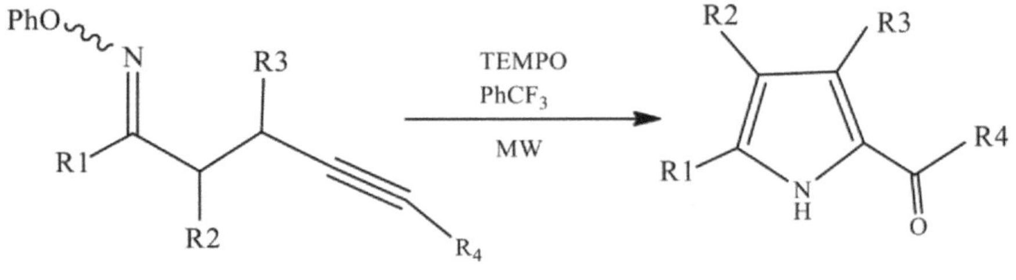

Kaur (2015) synthesized N-hetero cycles, such as three- and four-nitrogen-containing six-member under microwave irradiations.

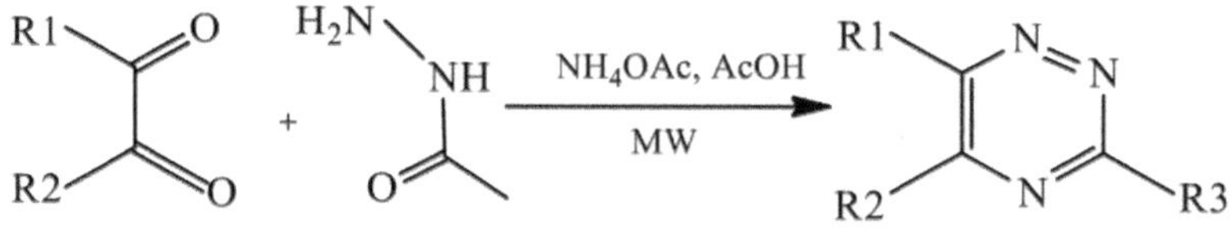

Kokel and Török (2017) reported an environmentally benign approach under microwave irradiation for solid-phase diazotization for the conversion of *o*-phenylenediamines to substituted benzotriazoles. They obtained excellent yields. The K-10 montmorillonite was used as a catalyst as well as medium, as it has strong microwave absorption capability. It was indicated that the catalyst was recyclable, and this reaction is highly efficient and no harmful waste was produced.

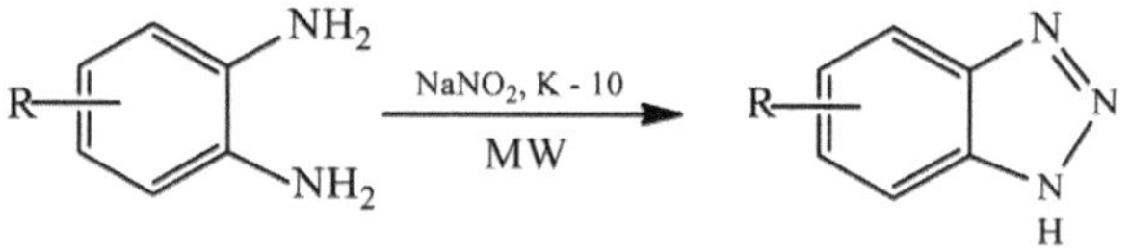

Kerru et al. (2021) reported a highly efficient green protocol for synthesis of a series of 1,2,4-triazole-tagged 1,4-dihydropyridine analogs. It was carried out as one-pot process using four-components 1*H*-1,2,4-triazol-3-amine, different aldehydes, diethyl acetyleneicarboxylate, and active methylene compounds under microwave irradiation in water under catalyst-free conditions. Excellent yields of product were obtained (94–97%) with high selectivity in a short reaction time (< 12 min) at room temperature. The advantages of this protocol are simple workup, no column chromatography, impressive yields, rapid reaction green solvent, and excellent functional group tolerance.

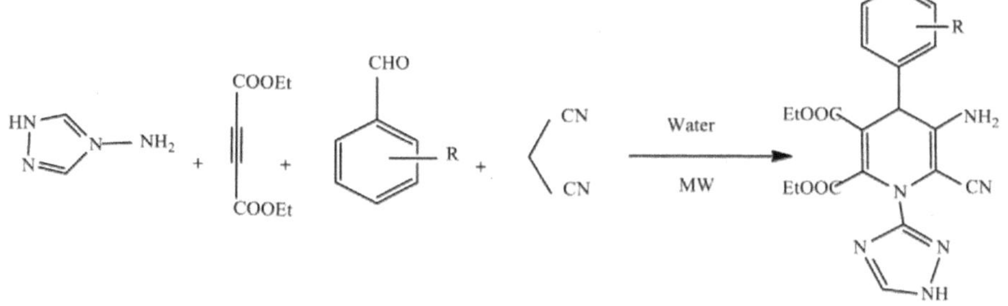

11.5.11 SYNTHESIS OF NANOCOMPOSITES

Wang and Lee (2005) carried out microwave-assisted synthesis of SnO_2-graphite nanocomposite for Li-ion battery applications. Polyacrylamide–metal nanocomposites were obtained by Zhu and Zhu (2006) via microwave heating. These nanocomposites contained metal nanoparticles dispersed homogeneously dispersed in the polymer matrix. They used metal salt and acrylamide monomer using ethylene glycol as a solvent. Mallikarjuna and Varma described (2007) shape-controlled bulk synthesis of noble nanocrystals and their catalytic properties under microwave irradiation. A single-step microwave-assisted route was used for synthesizing small gold nanoclusters, Au16NCs@BSA, which were used for detecting silver (I) ions as fluorescence-enhanced sensor with high sensitivity and selectivity (Yue, et al., 2012). UV curable inorganic/organic nanocomposites (hybrid) with moderate hardness, high refractive indices, excellent gas blocking performances, and good adhesive strength were rapidly synthesized by in situ microwave heating process (Lin et al., 2011). Chen and Wang (2010) reported microwave-assisted synthesis of a Co_3O_4 graphene sheet-on-sheet nanocomposite as an excellent material for anode in Li-ion batteries. In addition, a number of publications related to microwave-assisted synthesis of nanocomposites have appeared in recent past years (Ma et al., 2010; Zhang et al., 2010; Jia et al., 2011; Li et al., 2012; Motshekga et al., 2012).

Microwave irradiation has gained importance in the field of synthesis and treatment of nanomaterial, as it is a fast, cleaner, and cost-effective technique as compared with other methods (conventional and wet chemical). ZnO/graphene nanocomposites were prepared by Kim et al. (2017) and then treated with MW irradiation. They observed responses of MW-irradiated ZnO/graphene nanocomposites as sensors toward various gases/liquids including NO_2, CO, ethanol, acetone, and toluene. It was reported that the MW-irradiated sensor exhibited higher response particularly to NO_2 with higher selectivity and shorter response/recovery times than ZnO/graphene sensors and pristine ZnO without MW irradiation.

Pure magnetic nickel nanoparticles were synthesized by Zuliani et al. (2017) via a simple and fast microwave-assisted method. They used nickel chloride as precursor of Ni and mixture of ethylene glycol and ethanol as solvent and reducing agent, respectively. It was reported that highest performance (71% yields) could be achieved at 250°C in 5 min under microwave irradiation. These NPs exhibited are good catalysts for the hydrogenolysis of benzyl phenyl ether, in a microwave-assisted environmental-friendly reaction, which is eco-friendly in nature.

May-Masnou et al. (2018) fabricated small anatase titanium dioxide nanoparticles attached to larger anisotropic gold morphologies by a rapid process involving simple two-step microwave-assisted synthesis. These TiO_2/Au NPs were synthesized through a polyol approach using polyvinylpyrrolidone (PVP) as capping, reducing, and stabilizing agent. They removed PVP by calcination at mild temperatures just to optimize the contact between titania and gold and facilitate electron transfer. They used these NPs for photocatalytic generation of hydrogen from ethanol/water mixtures in gas phase at room temperature generating 5.3 mmol. g^{-1} cat. h^{-1}(7.4 mmol. g^{-1} TiO_2. h^{-1}) of hydrogen.

11.5.12 SYNTHESIS OF IONIC LIQUIDS

A microwave-assisted preparation of a series of ambient temperature ionic liquid, 1-alkyl-3-methylimidazolium halides via effective reaction of alkyl halide with 1-methylimidazole under solvent-free conditions has been described by Varma and Namboodiri (2001). Khadilkar and Rebeiro (2002) reported the synthesis of various alkyl pyridinium and 1-alkyl-3-methyl-imidazolium halide under microwave exposure in a closed vessel.

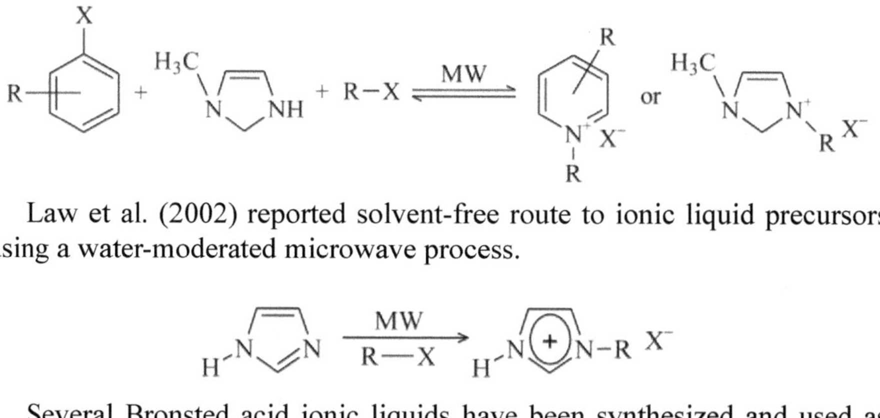

Law et al. (2002) reported solvent-free route to ionic liquid precursors using a water-moderated microwave process.

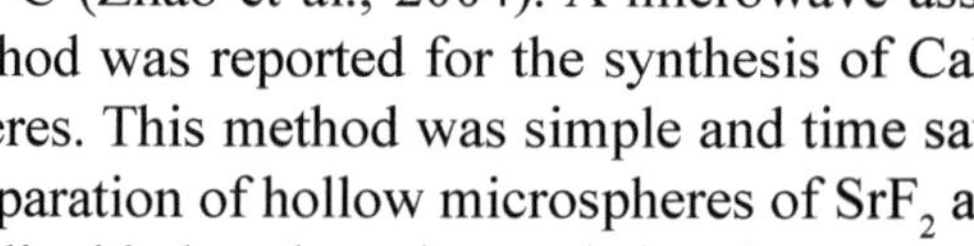

Several Bronsted acid ionic liquids have been synthesized and used as solvents and catalysts for three-component Mannich reactions of aldehydes and amines at 25°C (Zhao et al., 2004). A microwave-assisted ionic liquid solvothermal method was reported for the synthesis of CaF_2 double-shelled hollow microspheres. This method was simple and time saving and can also be used in the preparation of hollow microspheres of SrF_2 and MgF_2 (Xu and Zhu, 2012). Ionic liquid played a unique role in microwave-assisted synthesis of monodispersed magnetic nanoparticles has been studied (Hu et al., 2010).

Albalawi et al. (2018) synthesized various antibacterial imidazolium-based ionic liquids by both conventional as well as microwave irradiation method.

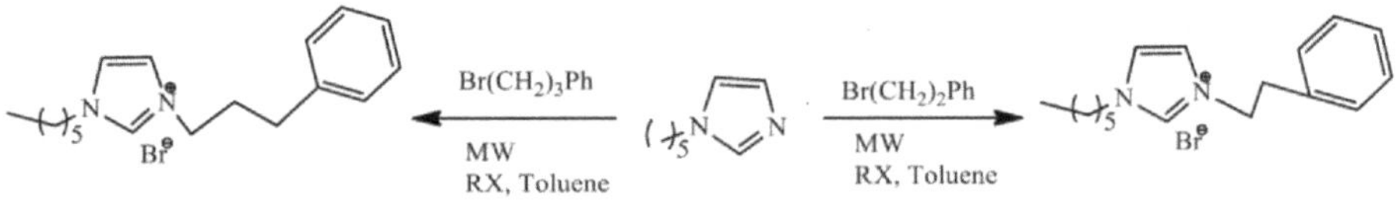

Aljuhani et al. (2019) prepared different substituted pyridinium-based ionic liquids under microwave irradiation. These ILs have good binding affinities with DNA and show anticancer activities.

11.5.13 MULTICOMPONENT REACTIONS

Belhani et al. (2018) reported a greener three-component syntheses of α-sulfamidophosphonates, using aldehyde, sulfonamides, and trimethyl-phosphite. The reaction was completed under microwave irradiation as well as catalyst and solvent-free conditions and generated lower amount of waste creating less pollution.

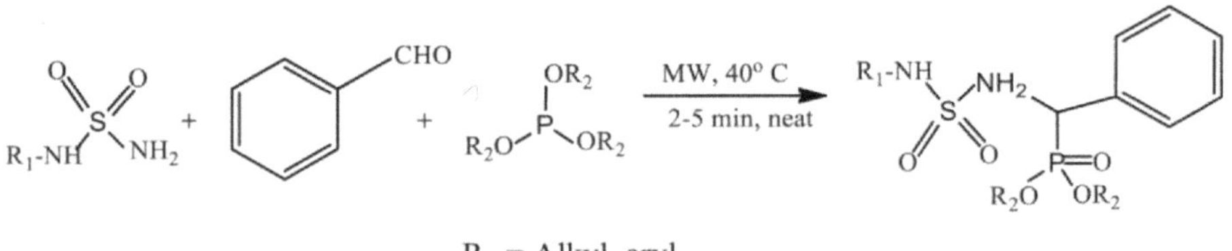

R_1 = Alkyl, aryl
R_2 = CH_3, C_2H_5

A three-component reaction of nitroalkanes, sodium azide and aldehydes for synthesizing N-unsubstituted-1,2,3-triazoles has been reported by Garg et al. (2020) under microwave irradiation. They used anthranilic acid (organocatalyst) and metal-free condition.

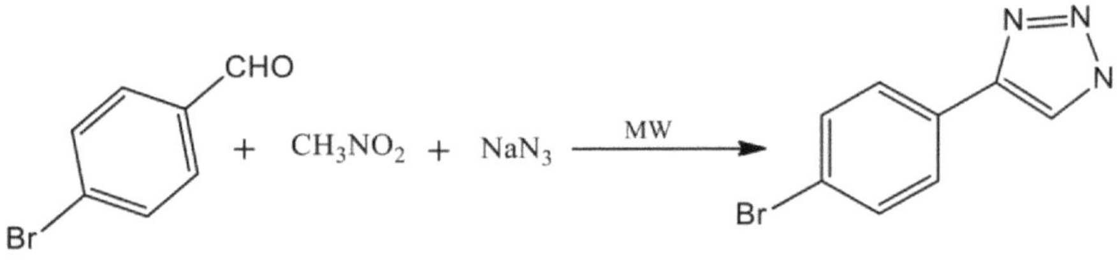

Abu-Melha et al. (2021) synthesized 1-thiazolyl-pyridazinedione derivatives via multicomponent synthesis under microwave irradiation as eco-friendly energy source and using naturally chitosan basic catalyst with high/efficient yields in shorter reaction time. All the synthesized compounds showed biological activities.

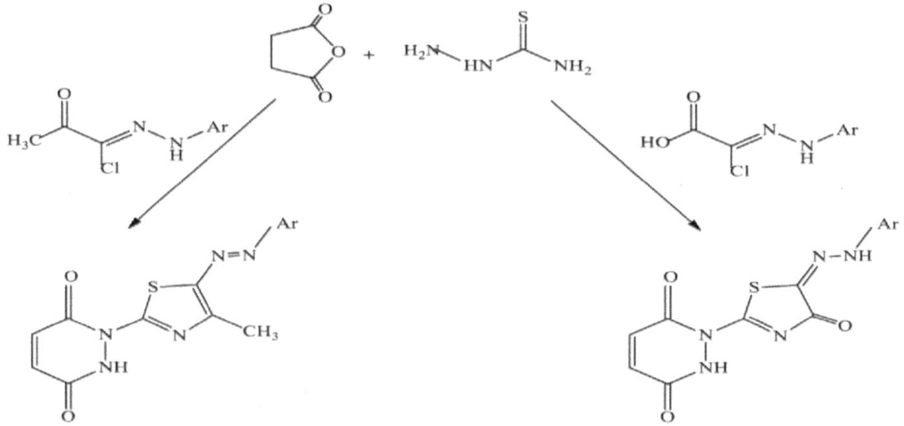

El-borai et al. (2012) synthesized substituted pyrazolo[3,4-b]pyridine from the reaction of 5-amino-1-phenyl-3-(pyridin-3-yl)-1H-pyrazole with 4-anisaldehyde and p-substituted -ketonitriles or with pyruvic acid and some aromatic aldehydes in acetic acid medium by microwave irradiation as well as conventional heating.

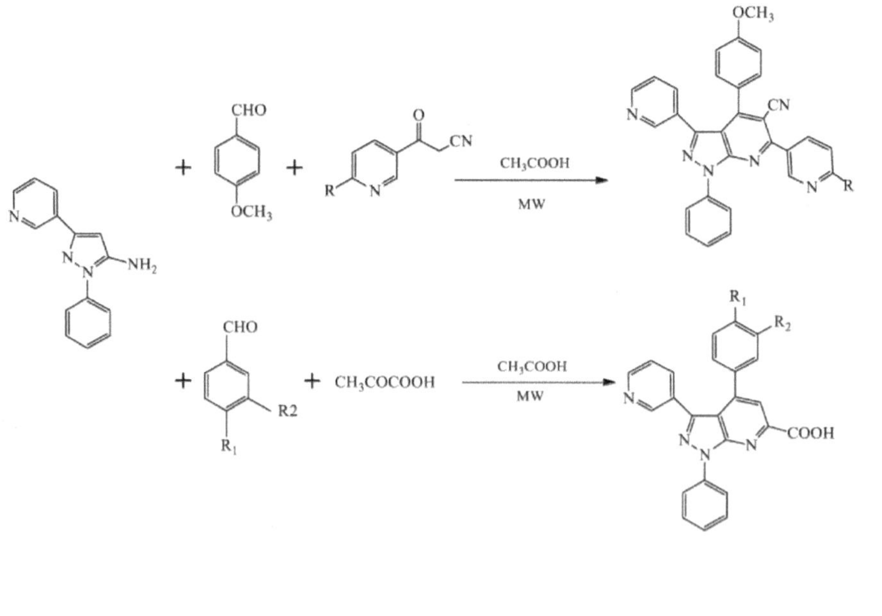

11.5.14 METAL ORGANIC FRAMEWORK (MOF)

Kurisingal et al. (2018) reported synthesis of a bimetallic MOF in aqueous, where Co and Ni were there as active metal centers. Benzene-1,4-dicarboxylic acid was used as linker under microwave irradiation, which afforded good yields. As-synthesized MOF was used for its catalytic efficacy in synthesizing cyclic carbonates from epoxides and CO_2. It was revealed that high rates of conversion of epoxides to cyclic carbonates on using NiCo–MOF with more than 99% selectivity under solvent-free conditions. This MOF was found to be useful for chemical fixation (catalytic) of carbon dioxide.

Kim et al. (2011) investigated cyanation of the Zr^{4+}-based UiO-66-Br with CuCN under microwave irradiation to produce UiO-66-CN. This is an example of post-synthetic modification PSM on an aryl halide MOF producing a cyano-functionalized metal–organic framework.

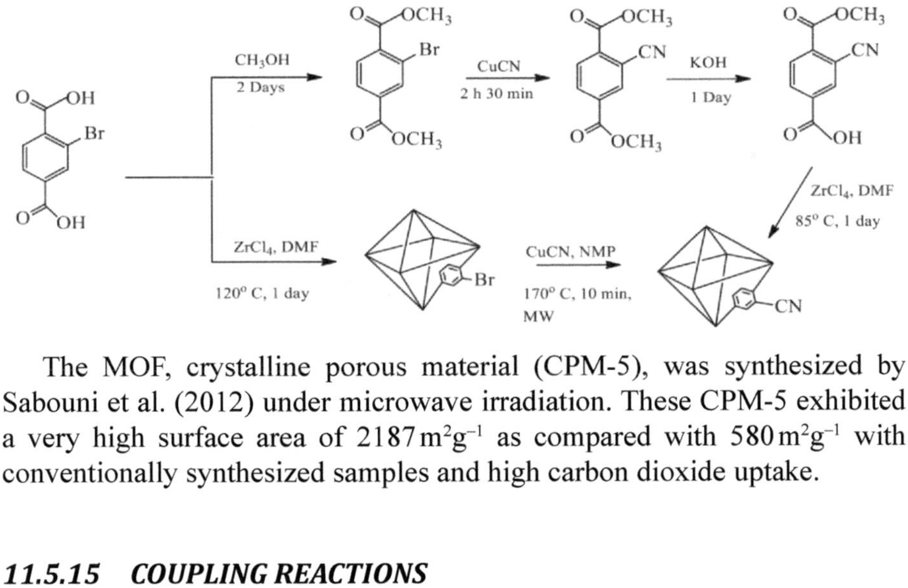

The MOF, crystalline porous material (CPM-5), was synthesized by Sabouni et al. (2012) under microwave irradiation. These CPM-5 exhibited a very high surface area of $2187\,m^2g^{-1}$ as compared with $580\,m^2g^{-1}$ with conventionally synthesized samples and high carbon dioxide uptake.

11.5.15 COUPLING REACTIONS

Dhara et al. (2019) synthesized glucose-stabilized palladium nanoparticles (PdNPs). They used these PdNPs as catalyst in Suzuki and Heck reactions to couple different boronic acids and styrene and obtained a variety of substituted alkenes and biaryl compounds.

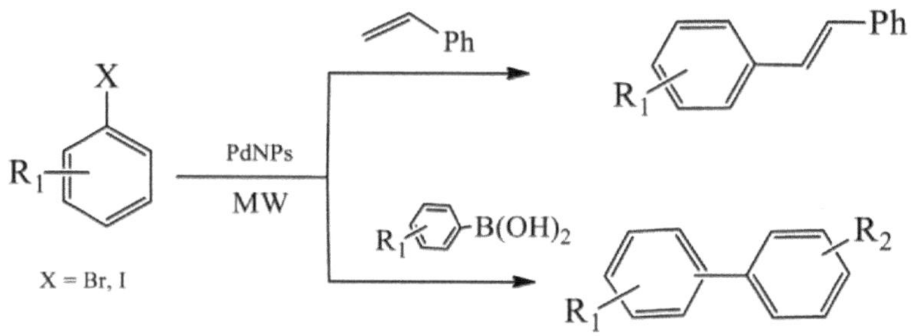

11.6 MISCELLANEOUS

Microwave heating process have been employed for regioselective, chemoselective synthesis, polymer synthesis, ceramic products, intercalation products, organometallic and coordination compounds, macromolecules, radiopharmaceuticals, etc. Faghihi et al. (2004) reported the synthesis of poly(amide-imide)s (optically active) containing N, N'-(pyromellitoyl)bisl-phenylalanine diacid chloride and 5,5-disubstituted hydantoin derivatives under microwave irradiation.

Various molecules like carboranes (Armstrong and Valliant, 2007), phenol–formaldehyde resole (Deetz et al., 2001), phenothiazine (Bajia et al., 2007), 4,5-disubstituted pyrazolopyrimidines (Gaina et al., 2007), and curcumin analogs (Wu et al., 2003) have been synthesized by microwave heating method with high efficiency. The synthesis of a variety of coordination compounds of transition metals under atmospheric conditions has been carried out under MW radiations (Nichols et al., 2006). A particularly useful application of microwave-assisted synthesis is the preparation of radiopharmaceuticals at elevated pressure with short half-lived isotopes, such as C11 and F18 (Haung et al., 1987; Thorell et al., 1992). In addition, microwave-assisted synthesis of macromolecules, such as epoxy resins (Bogdal and Gorczyk, 2003), polyesters (Chatti et al., 2003), Poly (aspartic acid) (Pielichowski et al., 2003), and poly-phosphazene (Burazyk et al., 2003) have also been reported.

Microwave-assisted chemical synthesis has advantages, such as reaction acceleration, yield improvement, enhanced physicochemical properties, and the evolvement of new material phases. Shi and Hwang (2003) demonstrated the significance of these advantages in industrial applications. Looking to the importance of microwave irradiation in synthetic chemistry, it can be concluded that time is not far off, when microwave heating technique will almost replace the conventional techniques, not only in laboratory but also in industries. They will truly become the Bunsen burner of the 21st century. But still, there are rooms to investigate this field and to develop greener routes to synthesize other useful compounds for the benefit of mankind.

Ultimately, it may be concluded that microwave-mediated synthesis is a green chemical technology because microwave not only accelerates chemical processes but also improves yield, selectivity, reduces pollution, and enable reactions to occur in solvent-free conditions. Microwave-assisted organic synthesis is a green chemical technique and reactions can be completed in less time, with increased yield and more purity. Many of these reactions can be carried out without solvents or on solid support. Time is not far off, when these microwave-assisted reactions will replace the majority of organic reactions, even on industrial scale.

11.7 RECENT DEVELOPMENT

Castro et al. (2022) converted fructose to hydroxymethylfurfural (HMF) with 74% yield. They isolated HNF with purity more than 97% using a biphasic system (ethyl acetate and water/NaCl). They used 1 mol% of *p*-sulfonic acid calix[4]arene ($CX4SO_3H$)) as organocatalyst, at 140°C for 10 min under microwave irradiation (75 W). It was revealed that aqueous phase (water/NaCl and $CX4SO_3H$) was recyclable for five reaction cycles without any significant decrease in yields. Apart from it, ethyl acetate may be recovered at the end of the process, and it can be reused.

Nardi et al. (2022) reported microwave-assisted synthesis of 1,2-disubstituted benzimidazoles by combining 1:1 molar ratio of benzaldehyde and N-phenyl-*o*-phenylenediamine. They used 1% mol of $Er(OTf)_3$ as efficient and environmental-friendly compound for preparing a variety of benzimidazoles under solvent-free conditions. It was reported that desired products were obtained in a short time and that too with very high selectivity.

George and Yadav (2022) synthesized 1,3-diphenyl-2-buten-1-one (dypnone), an intermediate used in the manufacture of a large range of compounds, through self-condensation of acetophenone using microwave irradiation under solvent-free conditions. They used different solid acid catalysts, out of which cesium (Cs)-substituted dodecatungstophosphoric acid [($Cs_{2.5}H_{0.5}PW_{12}O_{40}$) DTP] 20% (w/w) supported on K-10 clay (Cs-DTP/K-10) was found to be the best catalyst. It was reported that conversion of acetophenone and selectivity for trans-dypnone using Cs-DTP/K-10 (0.10 g/ cm^{-3})

were 56% and 92%, respectively at 413 K. It was also revealed that this catalyst can be recycled three times with only a marginal decrease in its activity, but no loss in the selectivity.

Abubakar et al. (2022) synthesized transition metal complex of Cu(II) with Schiff base N-salicylidene-4-chloroaniline using microwaves. It was reported that salicylaldehyde (0.01 mol) and 4-chloroaniline (0.01 mol) were taken in 50 mL beaker and mixed thoroughly. Then this reaction mixture was irradiated in a microwave oven (160 W) for 3 min. They used different solvents, such as dimethyl formamide (DMF), dimethyl sulfoxide (DMSO), ethanol, chloroform, and water. They also evaluated antibacterial activity of Schiff base and its metal(II) complex against *Candida albicans* and *Bacillus cereus* and found moderate activity as compared with Fluconazol and Ciprofloxacin, and this activity increases on increasing concentration.

Mohamad et al. (2022) utilized microwave-assisted methodology to synthesize acid hydrazides in a greener and single step with high yields. The conventional method requires two steps: Esterification and treatment with hydrazine hydrate. They carried out the synthesis of acid hydrazides of indomethacin, diclofenac, mefenamic acid, and ibuprofen using microwave radiation. It was revealed that they could get 86.7, 40.9, and 65.5% acid hydrazides of indomethacin, diclofenac, and mefenamic acid in 10, 3, and 3 min, respectively.

Badiger et al. (2022) synthesized tetrahydrobenzo[b]pyran derivatives via microwave-assisted condensation reaction of aromatic aldehyde, ethyl cyanoacetate, and dimedone in agro waste as a solvent. They used ethanol as a co-solvent and water extract of watermelon fruit peel ash as a green catalyst. This method is simple, low cost, fast, high yield, mild reaction conditions, and avoiding the use of toxic solvent and metals. Along with it, this can be recycled.

KEYWORDS

- **microwaves**
- **organic synthesis**
- **nanocomposites**
- **ionic liquids**
- **heterocycles**

REFERENCES

Abirami, M.; Selvi, S. T.; Nadaraj, V. Microwave Irradiation: Synthesis and Characterization of Substituted Pyranoquinolines. *Lett. Org. Chem.* **2021,** *18* (2), 156–159.

Abramovitch, R. A.; Shi Q.; Bogdal, D. Microwave-Assisted Alkylations of Activated of Activatived Methylene Groups. *Synth. Commun.* **1995,** *25*, 1–8.

Abubakar, T.; Haruna, A.; Isyaku, S.; Imam, M. Microwave-Assisted Synthesis, Characterization and Antimicrobial Activity of Copper (II) Complex with Salicylaldehyde and P-Chloroaniline Schiff Base. *Bayero J. Pure Appl. Sci.* **2022,** *13* (1), 117–122.

Abu-Melha, S.; Gomha, S. M.; Abouzied, A. S.; Edrees, M. M.; Dena, A. S. A.; Muhammad, Z. A. Microwave-Assisted One Pot Three-Component Synthesis of Novel Bioactive Thiazolyl-Pyridazinediones as Potential Antimicrobial Agents Against Antibiotic-Resistant Bacteria. *Molecules* **2021,** *26* (14). DOI: 10.3390/molecules26144260.

Agrawal Y. K.; Joshipura, H. M. Microwave Assisted Friedländer Condensation: A Comparative Study of Conventional Versus Microwave Mediated Solvent-Free Methodologies. *Indian J. Chem.*; 2005, *44B*, 1649–1652.

Alacid E.; Najera, C. Aqueous Sodium Hydroxide Promoted Cross-Coupling Reactions of Alkenyltrialkoxysilanes Under Ligand-Free Conditions. *J. Org. Chem.* **2008,** *73*, 2315–2322.

Albalawi, A. H.; El-Sayed, W. S.; Aljuhani, A.; Almutairi, S. M.; Rezki, N.; Aouad, M. R.; Messali, M. Microwave-Assisted Synthesis of Some Potential Bioactive Imidazolium-Based Room-Temperature Ionic Liquids. *Molecules* **2018,** *23* (7), 1727.

Aljohani, G.; Said, M. A.; Lentz, D.; Basar, N.; Albar, A.; Alraqa, S. Y. et al. Microwave-Assisted Synthesis of Mono- and Disubstituted 4-Hydroxyacetophenone Derivatives via Mannich Reaction: Synthesis, XRD and HS-Analysis. *Molecules* **2019,** *24* (3). DOI: 10.3390/molecules24030590

Aljuhani, A.; Aouad, M. R.; Rezki, N.; Aljaldy, O. A.; Al-Sodies, S. A.; Messali, M. et al. Novel Pyridinium Based Ionic Liquids with Amide Tethers: Microwave Assisted Synthesis, Molecular Docking and Anticancer Studies. *J. Mol. Liquids* **2019,** *285*, 790–802.

Al-obeidi, F.; Austin, R. E.; Okonoya J. F.; Bond, D. R. S. Microwave-Assisted Solid-Phase Synthesis (MASS): Parallel and Combinatorial Chemical Library Synthesis. *Mini-Rev. Med. Chem.* **2003,** *3*, 449–460.

Ameta, C.; Ameta, R.; Tiwari, U.; Punjabi P. B.; Ameta, S. C. Enviro-Economic Synthesis of Some Phthalimide Derivatives (VII) Using Microwave Irradiation. *J. Indian Chem. Soc.* **2011,** *88*, 827–833.

Ameta, C.; Ameta, R.; Tiwari, U.; Punjabi P. B.; Ameta, S. C. Microwave Induced Synthesis of Some Fused Thiazoloquinazoline Derivatives Under Solvent Free Conditions Using LiBr as a Catalyst. *Afinidad.* **2010,** *LXVII*, 61–68.

Ameta, C.; Sitha, D.; Ameta R.; Ameta, S. C. Synthesis and Antimicrobial Activity of Thiazole Derivatives Containing Triazole Moiety Using LiBr as a Catalyst. *Indo. J. Chem.* **2010,** *10*, 376–381.

Anastas, P. T.; Warner, J. C. *Green Chemistry: Theory and Practice*; Oxford University Press: Oxford, 2000.

Antinolo, A.; Carrillo-Hermosilla, F.; Cadierno, V.; García-Álvarez J.; Otero, A. MW-Assisted Meyer-Schuster Rearrangement of Propargylic Alcohols Catalyzed by the Oxovanadate Complex [V (O)Cl $(OEt)_2$]. *Chem. Cat. Chem.* **2012,** *4*, 123–128.

Apsunde, T. D.; Trudell, M. L. Solvent-Free, Base-Free Microwave-Mediated Iridium-Catalyzed N-Alkylation of Amides with Alcohols. *Synthesis.* **2014,** *46* (02), 230–234.

Arevela R. K.; Leadbeater, N. E. Rapid, Easy Cyanation of Aryl Bromides and Chlorides Using Nickel Salts in Conjunction with Microwave Promotion. *J. Org. Chem.* **2003,** *68*, 9122–9125.

Armstrong A. F.; Valliant, J. F. Microwave-Assisted Synthesis of Tricarbonyl Rhenacarboranes: Steric and Electronic Effects on the 1, 2→ 1, 7 Carborane Cage Isomerization. *Inorg. Chem.* **2007,** *46*, 2148–2158.

Artem'ev, A. V.; Korocheva, A. O.; Vashchenko, A. V.; Smirnov, V. I.; Malysheva, S. F. The Direct Phosphorylation of 2-,3-, and 4-Methylstyrenes and 2,4,6-Trimethylstyrene with Elemental Phosphorus VIA Trofimov–Gusarova Reaction. *Phosphorus, Sulfur, Silicon Relat. Elem.* **2015,** *190* (9), 1455–1463.

Ashok, D.; Sudershan K.; Khalilullah, M. Microwave Assisted Synthesis of 3- (2-Benzoyl-6-Hydroxy-3-Methylbenzo[b]Furan-5-yl)-5-(aryl)-4,5-Dihydro-1h-Pyrazole Carbothioamides and Their Antibacterial Activity. *Heterocycl. Lett.* **2011,** *1*, 311–317.

Babu, R.; Kim, S. H.; Kathalikkattil, A. C.; Kuruppathparambil, R. R.; Kim, D. W.; Cho, J. S.; Park, D. W. Aqueous Microwave-Assisted Synthesis of Non-Interpenetrated Metal-Organic Framework for Room Temperature Cycloaddition of CO_2 and Epoxides. *Appl. Catal. A: Gen.* **2017,** *544*, 126–136.

Badiger, K. B.; Sannegowda, L. K.; Kamanna, K. Microwave-Assisted One-Pot Synthesis of Tetrahydrobenzo [b] Pyrans in the Presence of WEWFPA and Their Electrochemical Studies. *Org. Commun.* **2022,** *15* (2), 148–166.

Bajia, S. C.; Swarnkar, P.; Kumar S.; Bajia, B. Microwave Assisted Synthesis of Phenol-Formaldehyde Resole. *J. Chem.* **2007,** *4*, 457–460.

Banik, B. K.; Manhas, M. S.; Kaluza, Z.; Barakat K. J.; Bose, A. K. Microwave-Induced Organic Reaction Enhancement Chemistry: Convenient Synthesis of Enantiopure Hydroxy-β-Lactams. *Tetrahedron Lett.* **1992,** *33*, 3603–3606.

Banik, B. K.; Manhas, M. S.; Newaz S. N.; Bose, A. K. Facile Preparation of Carbapenem Synthons via Microwave Induced Rapid Reaction. *Bioorg. Med. Chem. Lett.* **1993,** *3*, 2363–2368.

Barlow S.; Marder, S. R. Single-Mode Microwave Synthesis in Organic Materials Chemistry. *Adv. Funct. Mater.* **2003,** *1B*, 517–518.

Basha, S. T.; Sudhamani, H.; Rasheed, S.; Venkateswarlu, N.; Vijaya, T.; Raju, C. N. Microwave-Assisted Neat Synthesis of α-Aminophosphonate/Phosphinate Derivatives of 2- (2-Aminophenyl)Benzothiazole as Potent Antimicrobial and Antioxidant Agents. *Phosphorus, Sulfur, Silicon Relat. Elem.* **2016,** *191* (10), 1339–1343.

Belhani, B.; Bechlem, K.; Grib, I.; Cheloufi, H,; Bouasla, R.; Berredjem, M. A Green, One-Pot, Three-Component and Microwave Assisted Synthesis of α-Sulfamidophosphonates. *J. Mater. Environ. Sci.* **2018,** *9* (2), 613–618.

Bermudez, J. M.; Menendez, J. A.; Arenillas, A.; Martínez-Palou, R.; Romero, A. A.; Luque, R. Graphene Oxide-Catalysed Oxidation Reaction of Unsaturated Compounds Under Microwave Irradiation. *Catal. Commun.* **2015,** *72*, 133–137.

Biradar J. S.; Sasidhar, B. S. Solvent-Free, Microwave Assisted Knoevenagel Condensation of Novel 2,5-Disubstituted Indole Analogues and Their Biological Evaluation. *Eur. J. Med. Chem.* **2011,** *46*, 6112–6118.

Blanco-Vega, M. P.; Guzmán-Mar, J. L.; Villanueva-Rodríguez, M.; Maya-Treviño, L.; Garza-Tovar, L. L.; Hernández-Ramírez, A. et al. Photocatalytic Elimination of Bisphenol A Under Visible Light Using Ni-Doped TiO_2 Synthesized by Microwave Assisted Sol-Gel Method. *Mater. Sci. Semicond. Process.* **2017,** *71*, 275–282.

Bogdal D.; Gorczyk, S. Synthesis of Solid Epoxy Resins Under Microwave Irradiation. *Polish J. Chem. Tech.* **2003,** *5*, 88–89.

Bogdal, D.; Lukasiewicz, M.; Pielichowski, J.; Miciak A.; Bednarz, S. Microwave Assisted Oxidation of Alcohols Using Magtrieve™. *Tetrahedron.* **2003,** *59*, 649–653.

Borade, R. M.; Somvanshi, S. B.; Kale, S. B.; Pawar, S. B.; Jadhav, K. M. Spinel Zinc Ferrite Nanoparticles: An Active Nanocatalyst for Microwave Irradiated Solvent Free Synthesis of Chalcones. *Mater. Res. Exp.* **2020,** *7* (1). doi.org/10.1088/2053–1591/ab6c9c

Bose, A. K.; Banik, B. K.; Barakat K. J.; Manhas, M. S. Simplified Rapid Hydrogenation Under Microwave Irradiation: Selective Transformations of β-Lactams. *Synlett* **1993,** 575–576.

Bose, A. K.; Banik, B. K.; Wagle D. R.; Manhas, M. S. Microwave Assisted Rapid and Simplified Hydrogenation. *J. Org. Chem.* **1999,** *64*, 5746–5753.

Bose, A. K.; Manhas, M. S.; Banik B. K.; Robb, E. W. Microwave-Induced Organic Reaction Enhancement (MORE) Chemistry: Techniques for Rapid, Safe, and Inexpensive Synthesis. *Res. Chem. Intermed.* **1994,** *20*, 1–12.

Bose, A. K.; Manhas, M. S.; Ghosh, M.; Shah, M.; Raju, V. S.; Bari, S. S. et al. Microwave-Induced Organic Reaction Enhancement Chemistry. 2. Simplified Techniques. *J. Org. Chem.* **1991,** *56*, 6968–6970.

Bouasla, S.; Gahete, J. A.; Esquivel, D.; López, M. I.; Sanchidrián, C. J.; Salguero, F. J. R. Coumarin Derivatives Solvent-Free Synthesis Under Microwave Irradiation Over Heterogeneous Solid Catalysts. *Molecules* **2017,** *22* (12), 2072. doi.org/10.3390/molecules 22122072.

Božinović, N.; Šolaja, B. A.; Opsenica, I. M. Microwave-Assisted Synthesis of Azepines via Nucleophilic Aromatic Substitution, *J. Serb. Chem. Soc.* **2016,** *81* (11), 1225–1230.

Bradley, D. The Nuke's the Thing for Synthesis. *Mod. Drug Discov.* **2001,** *4*, 32–36.

Burazyk, A.; Bogdal, D.; Pielichowski, J. Synthesis of Poly (Aspartic Acid) Under Microwave Irradiation. *Polish J. Chem. Tech.* **2003,** *5*, 3–4.

Burton, G.; Cao, P.; Li G.; Rivero, R. Palladium-Catalyzed Intermolecular Coupling of Aryl Chlorides and Sulfonamides Under Microwave Irradiation. *Org. Lett.* **2003,** *5*, 4373–4376.

Cai, Y.; Jalan, A.; Kubosumi, A. R.; Castle, S. L. Microwave-Promoted Tin-Free Iminyl Radical Cyclization with TEMPO Trapping: A Practical Synthesis of 2-Acylpyrroles. *Org. Lett.* **2015,** *17* (3), 488–491.

Calinescu, I.; Calinescu, R.; Martin D. I.; Radoiv, M. T. Microwave-Enhanced Dechlorination of Chlorobenzene. *Res. Chem. Intermed.* **2003,** *29*, 71–81.

Castro, G. A. D.; Fernandes, S. A. Green Synthesis of 5-Hydroxymethylfurfural in a Biphasic System Assisted by Microwaves. *Catal. Lett.* **2022**. DOI: 10.1007/s10562–022–04043-x.

Chakraborty, B.; Sharma, P. K.; Kafley, S. Microwave Assisted Synthesis of Novel Spiro Isoxazolidine Derivatives. *Green Chem. Lett. Revs.* **2013,** *6* (2), 141–147.

Chang, F.; Kim, H.; Lee, B.; Park S.; Park, J. Highly Efficient Solvent–Free Catalytic Hydrogenation of Solid Alkenes and Nitro-Aromatics Using Pd Nanoparticles Entrapped in Aluminum Oxy-Hydroxide. *Tetrahedron Lett.* **2010,** *51*, 4250–4252.

Chatti, S.; Bartolussi, M.; Loupy, A.; Blais, J. C.; Bogdal, D.; Rogr, P. Synthesis of New Polyethers Derived from Isoiodide Under Phase-Transfer Catalysis: Reactivity and Selectivity Under Microwaves and Classical Heating. *J. Appl. Polym. Sci.* **2003,** *90*, 1255–1266.

Chatti, S.; Bortolussi M.; Loupy, A. Cation and Leaving Group Effects in Isosorbide Alkylation Under Microwave in Phase Transfer Catalysis. *Tetrahedron.* **2001,** *57*, 4365–4370.

Chemat F.; Esveld, E. Microwave Super-Heated Boiling of Organic Liquids: Origin, Effect and Application. *Chem. Eng. Tech.* **2001,** *7*, 735–744.

Chen S. Q.; Wang, Y. Microwave-Assisted Synthesis of a Co_3O_4 Graphene Sheet-on-Sheet Nanocomposite as a Superior Anode Material for Li-Ion Batteries. *J. Mater. Chem.* **2010,** *20*, 9735–9739.

Chen, H.; Zhao, J.; Li, Y.; Shen, F.; Li, X.; Yin, Q.; Qin, Z.; Yan, X.; Wang, Y.; Zhang P.; Zhang, J. Synthesis and Biological Activity of Novel 5'-Arylamino-Nucleosides by Microwave Assisted One-Pot Tandem Staudinger/Aza-Wittig/Reduction. *Bioorg. Med. Chem. Lett.* **2011,** *21*, 574–576.

Chen, J.; Ding, S.; Ji, Y.; Ding, J.; Yang, X.; Zou, M. et al. Microwave-Enhanced Hydrolysis of Poultry Feather to Produce Amino Acid. *Chem. Eng. Process.* **2015,** *87*, 104–109.

Chen, Z.; Dou, X.; Zhang, Y.; Yang, M.; Wei, D. Rapid Thermal-Acid Hydrolysis of Spiramycin by Silicotungstic Acid Under Microwave Irradiation. *Environ. Pollut.* **2019,** *249*, 36–44.

Cochrane, A. R.; Kerr, W. J.; Paterson; L. C.; Pearson; C. M.; Shaw, P. Advances in the Cobalt-Catalysed Pauson-Khand Reaction: Development of a Sulfide-Promoted, Microwave-Assisted Protocol. *Tetrahedron.* **2021,** *78*. DOI: 10.1016/j.tet.2020.131805

Danks, T. N. Microwave Assisted Synthesis of Pyrroles. *Tetrahedron Lett.* **1999,** *40*, 3957–3960.

Das E.; Basak A.; Regioselective Synthesis of Benzo-Fused Tetrahydroisoquinoline-Based Biaryls Through a Tandem One-Pot Halogenation of p-Benzynes from Enediynes and Suzuki-Miyaura Coupling. *J. Org. Chem.* **2020,** *85* (4), 2697–2703.

De la Hoz, A.; Diaz-Ortiz A.; Moreno, A. Microwaves in Organic Synthesis: Thermal and Non-Thermal Microwave Effects. *Chem. Soc. Rev.* **2005,** 164–178.

Deetz, M. J.; Malerich, J. P.; Beatty A. M.; Smith, B. D. One-Step Synthesis of 4 (*3H*)-quinazolinones. *Tetrahedron Lett.* **2001,** *42*, 1851–1854.

Dev, D.; Kalita, T.; Mondal, T.; Mandal, B.; Ethyl2-Cyano-2-(2-Nitrobenzenesulfonyloxyimino) Acetate (Ortho-NosylOXY)-Mediated Double Beckmann Rearrangement of Ketoximes Under Microwave Irradiation: A Mechanistic Perception. *Adv. Synth. Catal.* **2021,** *363* (5), 1427–1435.

Dhara, K.; Parasar, B.; Patil, A. J.; Dash, J. Microwave Assisted Cross-Coupling Reactions Using Palladium Nanoparticles in Aqueous Media. *Synth. Comm.* **2019,** *49* (6), 859–868.

Diaz-Ortiz, A.; Hoz, A. D.; Merrero, M. A.; Prieto, P.; Sanchez-Migallon, A.; Cassio, F. P.; Arriela, A.; Vivanco S.; Foces, C. Reproducibility and Scalability of Solvent-Free Microwave Assisted Reactions. *Mol. Divers.* **2003,** *7*, 165–169.

Díaz-Ortiz, Á.; Prieto, P.; De La Hoz, A. A Critical Overview on the Effect of Microwave Irradiation in Organic Synthesis. *Chem. Record.* **2019,** *19* (1), 85–97.

Dudley, G. B.; Stiegman, A. E. Changing Perspectives on the Strategic Use of Microwave Heating in Organic Synthesis, *The Chemical Record*, **2018,** *18* (3), 381–389.

Egami, H.; Tamaoki, S.; Abe, M.; Ohneda, N.; Yoshimura, T.; Okamoto, T.; Odajima, H.; Mase, N.; Takeda, K.; Hamashima, Y. Scalable Microwave-Assisted Johnson–Claisen Rearrangement with a Continuous Flow Microwave System. *Org. Process Res. Dev.* **2018,** *22* (8), 1029–1033.

El-Borai, M. A.; Rizk, H. F.; Abd-Aal, M. F.; El-Deeb, I. Y. Synthesis of pyrazolo [3, 4-b] Pyridines Under Microwave Irradiation in Multi-Component Reactions and Their Antitumor and Antimicrobial Activities–Part 1. *Eur. J. Med. Chem.* **2012,** *48*, 92–96.

Erdelyi M.; Gogoll, A. Rapid Homogeneous-Phase Sonogashira Coupling Reactions Using Controlled Microwave Heating. *J. Org. Chem.* **2001,** *66*, 4165–4169.

Faghihi, K.; Zamani, K.; Mirgamie A.; Mallakpour. S. Facile Synthesis of Novel Optically Active Poly (Amide-Imide)s Containing *N,N'*- (Pyromellitoyl)-Bis-l-Phenylalanine Diacid Chloride and 5,5-Disubstituted Hydantoin Derivatives Under Microwave Irradiation. *J. Appl. Polym. Sci.* **2004,** *91*, 516–524.

Fardood, S. T.; Ramazani, A.; Ayubi, M.; Moradnia, F.; Abdpour, S.; Forootan, R. Microwave Assisted Solvent-Free Synthesis of 1-Phenyl-1, 2-Dihydro-3H-Naphtho[1, 2-e][1, 3] Oxazin-3-One Catalyzed by $FeCl_3$. *Chem. Methodol.* **2019,** *3* (5), 519–525.

Ferro, S.; Rao, A.; Zappala, M.; Chimirri, A.; Barreca, M. L.; Witwouw, M. et al. Synthesis of New Potential HIV-1 Integrase Inhibitors. *Heterocycles* **2004,** *63*, 2727–2734.

Flink, H.; Putkonen, T.; Sipos A.; Jokela, R. Microwave Assisted Selective Protection of Glutaraldehyde and Its Symmetrical Derivatives as Monoacetals and Thioacetals. *Tetrahedron.* **2010,** *66*, 887–890.

Fuchibe K.; Imaoka H.; Ichikawa J. Synthesis of Pinpoint-Fluorinated Polycyclic Aromatic Hydrocarbons: Benzene Ring Extension Cycle Involving Microwave-Assisted SNAr Reaction. *Chem. Asian J.* **2017,** *12* (18), 2359–2363.

Gaina, L.; Cristea, C.; Moldovan, C.; Porumb, D.; Surducan, E.; Deleanu, C.; Mahamoud, A.; Barbe J.; Silberg, I. A. Microwave Assisted Synthesis of Phenothiazine and Quinoline Derivatives. *Int. J. Mol. Sci.* **2007,** *8*, 70–80.

Garg, A.; Sarma, D.; Ali, A. A. Microwave Assisted Metal-Free Approach to Access 1,2,3-Triazoles Through Multicomponent Synthesis. *Curr. Res. Green Sustain. Chem.* **2020,** *3*. DOI: 10.1016/j.crgsc.2020.100013

Gedye, R.; Smith, F.; Westway, K.; Baldisera, H. A. L.; Laberage L.; Rousell, J. The Use of Microwave Ovens for Rapid Organic Synthesis. *Tetrahedron Lett.* **1986,** *27*, 279–282.

George, G.; Yadav, G. D. Self-Condensation of Acetophenone to Dypnone: Synergism of Microwave and Solid Acid Catalysis. *Cat. Green Chem. Eng.* **2022,** *5* (1). 29–39.

Ghorbani-Vaghei R.; Malaekehpour, S. M. Efficient and Solvent-Free Synthesis of 1-Amidoalkyl-2-Naphthols Using *N,N,N',N'*-Tetrabromobenzene-1,3-Disulfonamide. *Central Eur. J. Chem.* **2010,** *8*, 1086–1089.

Giguere, R. J.; Bray, T. L.; Duncan S. M.; Majetich, G. Application of Commercial Microwave Ovens to Organic Synthesis. *Tetrahedron Lett.* **1986,** *27*, 4945–4948.

Gomez, M. V.; Moreno, A.; Vazquez, E.; De la Hoz, A.; Aranda A. I.; Diaz-Ortiz, A. Synergy Between Microwave Irradiation and Heterogeneous Catalysis in an Environmentally Friendly Self-Condensation of Hydroxybenzene Derivatives. *ARKIVOC*, **2010,** *iii*, 264–273.

Gonzalez, I.; Bellas, I.; Souto, A.; Rodriguez R.; Cruces, J. Microwave-Assisted Aza-Cope Rearrangement of *N*-Allylanilines. *Tetrahedron Lett.* **2008,** *49*, 2002–2004.

González-Liste, P. J.; Cadierno, V.; García-Garrido, S. E. Catalytic Rearrangement of Aldoximes to Primary Amides in Environmentally Friendly Media Under Thermal and Microwave Heating: Another Application of the Bis (Allyl)-Ruthenium (IV) Dimer [{RuCl (μ-Cl) (η3: η3-$C_{10}H_{16}$)}$_2$]. *ACS Sustain. Chem. Eng.* **2015,** *3* (11), 3004–3011.

Gospondinova, M.; Gredard, A.; Jeannin, M.; Chitanv, G. C.; Carpov, A.; Thiery V. et al. Efficient Microwave Phosphorylation of Microcrystalline Cellulose Under Heterogeneous Conditions. *Green Chem.* **2002,** *4*, 220–222.

Goswami, A.; Rathi, A. K.; Aparicio, C.; Tomanec, O.; Petr, M.; Pocklanova, R. In situ Generation of Pd–Pt Core–Shell Nanoparticles on Reduced Graphene Oxide (Pd@Pt/rGO) Using Microwaves: Applications in Dehalogenation Reactions and Reduction of Olefins. *ACS Appl. Mater. Interfaces.* **2017,** *9* (3), 2815–2824.

Gour, J.; Gatadi, S.; Malasala, S.; Yaddanpudi, M. V.; Nanduri, S. A Microwave-Assisted SmI_2-Catalyzed Direct *N*-Alkylation of Anilines with Alcohols. *J. Org. Chem.* 2019, *84* (11), 7488–7494.

Grau, B. W.; Tsogoeva, S. B. Iron-Catalyzed Carbonyl–Alkyne and Carbonyl–Olefin Metathesis Reactions. *Catalysts* **2020,** *10* (9). DOI: 10.3390/catal10091092.

Grun, A.; Balint, E.; Keglevich, G. Solid-Liquid Phase C-Alkylation of Active Methylene Containing Compounds Under Microwave Conditions. *Catalysts.* **2015,** *5* (2), 634–652.

Grun, A.; Blastik, Z.; Drahos, L.; Keglevich, G. Microwave-Assisted Alkylation of Diethyl Ethoxycarbonylmethylphosphonate Under Solvent Less Conditions. *Heteroatom Chem.* **2012,** *23*, 241–246.

Gupta, S.; Pathak, A. K.; Ameta, C.; Punjabi, P. B. Microwave-Induced Expeditious Synthesis of Biologically Active Substituted Imidazoles Using CuO-TiO_2-GO Nanocomposite as a Recyclable Catalyst. *Lett. Org. Chem.* **2021,** *18* (4), 318–333.

Han, P.; Zhou, W.; Chen, M.; Wang, Q. Microwave-Assisted Synthesis of Polymethoxychalcone Mannich Bases and Their Antiproliferative Activity. *Lett. Org. Chem.* **2019,** *16* (2), 117–121.

Han, S. Y.; Gong, Y. D. Highly Efficient Enantioselective Synthesis of 1,3-Disubstituted 2,5-Diketopiperazine Derivatives via Microwave Irradiation. *Int. J. Rapid Commun. Synth. Org. Chem.* **2019,** *49* (24), 3426–3434.

Handayani, S.; Budimarwanti, C.; Haryadi, W. Microwave-Assisted Organic Reactions: Eco-Friendly Synthesis of Dibenzylidene cyclohexanone Derivatives via Crossed Aldol Condensation. *Indo. J. Chem.* **2017,** *17* (2), 336–341.

Haung, D. R.; Moerlein, S. M.; Lang, L.; Welch, M. J. Application of Microwave Technology to the Synthesis of Short-lived Radiopharmaceuticals. *J. Chem. Soc. Chem. Commun.* **1987,** *23*, 1799–1801.

Hayes, B. L.; *Microwave Synthesis: Chemistry at the Speed of Light*; CEM Publishing: Matthews, NC, 2002.

Horikoshi, S.; Watanabe, T.; Kamata, M.; Suzuki, Y.; Serpone, N. Microwave-Assisted Organic Syntheses: Microwave Effect on Intramolecular Reactions—The Claisen Rearrangement of Allylphenyl Ether and 1-Allyloxy-4-Methoxybenzene. *RSC Adv.* **2015,** *5*, 90272–90280.

Horta, J. E. Simple Microwave-Assisted Claisen and Dieckmann Condensation Experiments for the Undergraduate Organic Chemistry Laboratory. *J. Chem. Educ.* **2011,** *88*, 1014–1015.

Hosseinpour, R.; Pineda, A.; Ojeda, M.; Garcia, A.; Romero, A. A.; Luque, R. Microwave-Assisted Oxidation of Benzyl Alcohols Using Supported Cobalt Based Nanomaterials Under Mild Reaction Conditions. *Green Proc. Synth.* **2014,** *3* (2), 133–139.

Hu, H. L.; Yu, J. H.; Yang, S. Y.; Wang J. X.; Yin, Y. A. Substitution Reaction of Nitro Group on α-Nitrostyrene by Organozinc Halides Under Microwave Irradiation. *Synth. Commun.* **1999,** *29*, 1157–1164.

Hu, H.; Yang, H.; Heng, P.; Cui, D.; Peng, Y.; Zhang, J.; Lu, F.; Lian J.; Shi, D. Unique Role of Ionic Liquid in Microwave-Assisted Synthesis of Monodisperse Magnetic Nanoparticles. *Chem. Commun.* **2010,** *46*, 3866–3868.

Hui, Z.; Jiang, S.; Qi, X.; Ye, X. Y.; Xie, T. Investigating the Microwave-Accelerated Claisen Rearrangement of Allyl Aryl Ethers: Scope of the Catalysts, Solvents, Temperatures, and Substrates. *Tetrahedron Lett.* **2020,** *61* (24). DOI: 10.1016/j.tetlet.2020.151995.

Hull, A. W. The Effects of a Uniform Magnetic Field on the Motion of Electrons Between Coaxial Cylinders. *Phys. Rev.* **1921,** *18* (1), 31–57.

Hull, A. W. The Magnetron. *J. Am. Inst. Electrical Eng.* **1921,** *40 (9)*, 715–723.

Hur, D.; Dal, S. F. E.; Varol, G. A.; Hur, E. N-Acylbenzotriazole Mediated Microwave Assisted Synthesis of Protected and Novel Unprotected Ferrocenoylamidoamino Acids. *J. Organomet. Chem.* **2011,** *696* (13), 2543–2548.

Ighilahriz, K.; Bouteameur, B.; Chami, F.; Rabia, C.; Hamdi M.; Hamdi, S. M. A Microwave-Assisted and Heteropolyacids-Catalysed Cyclocondensation Reaction for the Synthesis of 4 (3h)-Quinazolinones, *Molecules* **2008,** *13*, 779–789.

Inagaki, T.; Fukuhara, T.; Hara, S. Effective Fluorination Reaction with $Et_3N•3HF$ Under Microwave Irradiation. *Synthesis.* **2003,** *8,* 1157–1159.

Iwasaki, M.; Xu, J.; Tani, Y.; Fu, L.; Ikemoto, Y.; Ura, Y. et al. Copper-Catalyzed Egioselective Chloroamination of Alkenes with Chlorotrimethylsilane and N-Fluorobenzenesulfonimide Under Microwave-Assisted Conditions. *Chem. Lett.* **2019,** *48* (3), 281–284.

Kurisingal, J. F.; Babu, R.; Kim, S. H.; Li, Y. X.; Chang, J. S.; Cho, S. J. et al. Microwave-Induced Synthesis of a Bimetallic Charge-Transfer Metal Organic Framework: A Promising Host for the Chemical Fixation of CO_2. *Catal. Sci. Technol.* **2018,** *8,* 591–600.

Jablonkai, E.; Keglevich, G. Catalyst-Free P–C Coupling Reactions of Halobenzoic Acids and Secondary Phosphine Oxides Under Microwave Irradiation in Water. *Tetrahedron Lett.* **2015,** *56* (13), 1638–1640.

Jha, A. K.; Jain, N. The Microwave-Assisted *Ortho*-Alkylation of Azine *N*-Oxides with *N*-Tosylhydrazones Catalyzed by Copper (i) Iodide. *Chem. Commun.* **2016,** *52,* 1831–1834.

Jia, N.; Li, S. M.; Ma, M. G.; Sun R. C.; Zhu, L. Green Microwave-Assisted Synthesis of Cellulose/Calcium Silicate Nanocomposites in Ionic Liquids and Recycled Ionic Liquids. *Carbohydr. Res.* **2011,** *346,* 2970–2974.

Jin, Y.; Su, G.; Yu, J. Mannich Reaction as a Key Strategy for the Synthesis of Trifluoroethyl Derived Tertiary and Secondary Amine. *Bull. Korean Chem. Soc.* **2021,** *42* (6), 836–839.

Kakati D.; Sarma, J. C. Microwave Assisted Solvent Free Synthesis of 1,3-Diphenylpropenones. *Chem. Central J.* **2011,** *5,* 8–12.

Kann, N. Recent Applications of Polymer Supported Organometallic Catalysts in Organic Synthesis. *Molecules* **2010,** *15,* 6306–6331.

Karamthulla, S.; Pal, S.; Khan, M. N.; Choudhury, L. H. "On-Water" Synthesis of Novel Trisubstituted 1, 3-Thiazoles via Microwave-Assisted Catalyst-Free Domino Reactions. *RSC Adv.* **2014,** *4* (71), 37889–37899.

Karamthulla, S.; Pal, S.; Md. Khan, N.; Choudhury, L. H. "On-Water" Synthesis of Novel Trisubstituted 1,3-Thiazoles via Microwave-Assisted Catalyst-Free Domino Reactions. *RSC Adv.* **2014,** *4,* 37889–37899

Karousis, N.; Ichihashi, T.; Yudasaka, M.; Iijima S.; Tagmatarchis, N. Microwave-Assisted Functionalization of Carbon Nanohorns *via* [2+1] Nitrenes Cycloaddition. *Chem. Commun.* **2011,** *47,* 1604–1606.

Karu, R.; Gedu, S.; Microwave-Assisted Domino Heck Cyclization and Phosphorylation: Synthesis of Phosphorus Containing Heterocycles. *Eur. J. Org. Chem.* **2019,** *2019* (24), 3856–3866.

Kaur, N. Six-Membered Heterocycles with Three and Four N-Heteroatoms: Microwave-Assisted Synthesis. *Syn. Comm.* **2015,** *45* (2), 151–172.

Kerru, N.; Maddila, S.; Jonnalagadda, S. B. A Facile and Catalyst-Free Microwave-Promoted Multicomponent Reaction for the Synthesis of Functionalised 1,4-Dihydropyridines with Superb Selectivity and Yields. *Front. Chem.* **2021,** *9.* DOI: org/10.3389/fchem.2021.638832

Khadilkar B. M.; Rebeiro, G. L. Microwave-Assisted Synthesis of Room-Temperature Ionic Liquid Precursor in Closed Vessel. *Org. Proc. Res. Dev.* **2002,** *6,* 826–828.

Kiasat, A. R.; Kazemi F.; Rafati, M. Microwave Promoted Rapid Oxidation of Alcohols Using Cobalt Nitrate Hexahydrate Supported on Silica Gel Under Solvent Free Conditions. *Synth. Commun.* 2003, *33,* 601–605.

Kim, H. W.; Kwon, Y. J.; Mirzaei, A.; Kang, S. Y.; Choi, M. S.; Bang, J. H.; Kim, S. S. Synthesis of Zinc Oxide Semiconductors-Graphene Nanocomposites by Microwave Irradiation for Application to Gas Sensors. *Sens. Actuat. B: Chem.* **2017,** *249,* 590–601.

Kim, M.; Garibay, S. J.; Cohen, S. M. Microwave-Assisted Cyanation of an Aryl Bromide Directly on a Metal–Organic Framework. *Inorg. Chem.* **2011,** *50* (3), 729–731.

Kingston, H. M.; Haswell, S. J., Eds. *Microwave—Enhanced Chemistry. Fundamentals, Sample Preparation and Applications*; American Chemical Society: Washington, DC, 1997.

Kokel, A.; Török, B. Microwave-Assisted Solid Phase Diazotation: A Method for the Environmentally Benign Synthesis of Benzotriazoles. *Green Chem.* **2017,** *19*, 2515–2519.

Krömer, M.; Klečka, L.; Slavětínská, B.; Klepetářová, M. H. Chemoselective Synthesis of 4,5-Diarylpyrrolo [2,3-d] Pyrimidines (6,7-Diaryl-7-Deazapurines) by Consecutive Suzuki and Liebeskind–Srogl Cross-Couplings. *Eur. J. Org. Chem.* **2014,** *2014* (32), 7203–7210.

Kugelgen, S.; Piskun, I.; Griffin, J. H.; Eckdahl, C. T.; Jarenwattananon, N. N.; Fischer, F. R. Templated Synthesis of End-Functionalized Graphene Nanoribbons Through Living Ring-Opening Alkyne Metathesis Polymerization. *J. Am. Chem. Soc.* **2019,** *141* (28), 11050–11058.

Kuhnert, N. Microwave-Assisted Reactions in Organic Synthesis—Are There Any Nonthermal Microwave Effects? *Angew. Chem. Int. Ed.* **2002,** *41*, 1863–1866.

Kumar, A.; Kuang, Y.; Liang, Z.; Sun, X. Microwave Chemistry, Recent Advancements, and Eco-Friendly Microwave-Assisted Synthesis of Nanoarchitectures and Their Applications: A Review. *Mater. Today Nano.* **2020,** *11*. DOI: 10.1016/j.mtnano.2020.100076.

Kumar, D.; Suresh; Sandhu, J. S. Microwave Enhanced, Solvent Free Green Protocol for the Production of 3,4-Dihyropyrimidine-2- (1H)-Ones Using $AlCl_3.6H_2O$ as a Catalyst. *Indian J. Chem.* **2010,** *49B*, 360–363.

Kumar, G. J.; Ajithabai, M. D.; Santhosh, B.; Veena C. S.; Nair, M. S. Microwave Assisted Acetylation and Deacetylation Deacetylation Studies on the Triterpenes Isolated from *Dysoxylum malabaricum* and *Dysoxylum beddomeri*. *Indian J. Chem.* **2003,** *42B*, 429–431.

Kumari, N.; Vyas, N.; Gupta, A. K. Microwave-Assisted Synthesis of Bis- (*N,N*-dialkyl)/*O*-Aryl *N,N*-Dialkyl-2- (1-Methyl/Phenyl-2-Oxopropylidene) Phosphorohydrazido Oximes on a Solid Support under Solvent Free Conditions: Marine Fish Toxin Analogues. *Phosphorus, Sulfur, Silicon Relat. Elem.* **2017,** *192* (7), 831–834.

Lalitha P.; Sivakamasundari, S. Solid Supports in the Synthesis of Few Vinyl Quinolones. *J. Chem. Pharm. Res.* **2010,** *2*, 387–393.

Larhed M.; Hallberg, A. Microwave-Promoted Palladium-Catalyzed Coupling Reactions. *J. Org. Chem.* **1996,** *61*, 9582–9584.

Larhed, M.; Moberg C.; Hallberg, A. Microwave-Accelerated Homogeneous Catalysis in Organic Chemistry. *Acc. Chem. Res.* **2002,** *35*, 717–727.

Law, M. C.; Wong K. Y.; Chan, T. H. Solvent-Free Route to Ionic Liquid Precursors Using a Water-Moderated Microwave Process. *Green Chem.* **2002,** *4*, 328–330.

Leadbeater N. E.; Marco, M. Rapid and Amenable Suzuki Coupling Reaction in Water Using Microwave and Conventional Heating. *J. Org. Chem.* **2003,** *68*, 888–892.

Leadbeater, N. E. Fast, Easy, Clean Chemistry by Using Water as a Solvent and Microwave Heating: The Suzuki Coupling as an Illustration. *Chem. Commun.* **2005,** *745*, 2881–2902.

Lei, M.; Gao L.; Yang, J. S. Microwave-Assisted Palladium-Catalyzed Cross-Coupling Reactions Between Pyranoid Glycals and Aryl Bromides. Synthesis of 2′-Deoxy *C*-Aryl-β-Glycopyranosides. *Tetrahedron Lett.* **2009,** *5*, 5135–5138.

Lerestif, J. M.; Perocheav, J.; Tonnard, F.; Bazareav J. P; Hamelin, J. 1,3-Dipolar Cycloaddition of Imidate Ylides On Imino-Alcohols: Synthesis of New Imidazolones Using Solvent Free Conditions. *Tetrahedron.* **1995,** *51*, 6757–6774.

Li, J. P.; Qiu, J. K.; Li H. J.; Zhang, G. S. Thiourea: An Efficient and Inexpensive Catalyst for the Knoevenagel Condensation of Pyrazole Derivates. *J. Chinese Chem. Soc.* 2011, *58* (2), 268–271.

Li, L.; Guo, Z.; Du A.; Liu, H. Rapid Microwave-Assisted Synthesis of Mn_3O_4 Graphene Nanocomposite and Its Lithium Storage Properties. *J. Mater. Chem.* **2012,** *22*, 3600–3605.

Li, Z.; Ding, R.; Lu, Z.; Xiao S.; Ma, X. Silica Sulfate as a Recyclable and Efficient Catalyst for Beckmann Rearrangement Under Microwave Irradiation. *J. Mol. Catal A : Chem.* **2006,** *250* (1–2), 100–103.

Liang, W.; D'Alessandro, D. M. Microwave-Assisted Solvothermal Synthesis of Zirconium Oxide Based Metal–Organic Frameworks. *Chem. Commun.* **2013,** *49*, 3706–3708

Lidstrom, P.; Tiernery, J. P. *Microwave Assisted Organic Synthesis*; Blackwell Publishing: Oxford, 2005.

Lidstrom, P.; Tierney, J.; Wathey B.; Westman, J. Microwave Assisted Organic Synthesis—A Review. *Tetrahedron.* **2001,** *57*, 9225–9283.

Lin, J. S.; Chung, M. H.; Chen, C. M.; Juang F. S.; Liu, L. E. Microwave-Assisted Synthesis of Organic/Inorganic Hybrid Nanocomposites and Their Encapsulating Applications for Photoelectric Devices. *J. Phys. Org. Chem.* 2011, *24* (3), 193–202.

Liu W.; Li Y.; Wang R.; Jiang Y.; Kuang C. Microwave Assisted Direct Arylation of 1-Benzyl-1,2,3-Triazole. *Tetrahedron Lett.*. **2020,** *61* (2). DOI: 10.1016/j.tetlet.2019.151390.

Liu, L.; Li, Y.; Fong, Y.; Chen, L. Microwave-Assisted Graft Copolymerization of ε-Caprolactone Onto Chitosan via the Phthaloyl Protection Method. *Carbohydr. Polym.* **2005,** *60*, 351–356.

Liu, N.; Guo, Y.; Yang, X.; Lin, H.; Yang, L.; Shi, Z. et al. Microwave-Assisted Reactant-Protecting Strategy Toward Efficient MoS_2 Electrocatalysts in Hydrogen Evolution Reaction. *ACS Appl. Mater. Interfaces.* **2015,** *7* (42), 23741–23749.

Liu, R. Y.; Zhou, Y.; Yang, Y.; Buchwald, S. L. Enantioselective Allylation Using Allene, a Petroleum Cracking Byproduct. *J. Am. Chem. Soc.* **2019,** *141* (6), 2251–2256.

Liu, X.; Wan, Q.; Zhao, Z.; Liu, J.; Zhang, Z.; Deng, F. et al. Microwave-Assisted Diels-Alder Reaction for Rapid Synthesis of Luminescent Nanodiamond with AIE-Active Dyes And Their Biomedical Applications. *Mater. Chem. Phys.* **2017,** *197*, 256–265.

Liu, Y.; Tsunoyama, H.; Akita T.; Tsukuda, T. Size Effect of Silica-Supported Gold Clusters in the Microwave-Assisted Oxidation of Benzyl Alcohol with H_2O_2. *Chem. Lett.* **2010,** *39*, 159–161.

Longhi, K.; Moreira, D. N.; Marzari, M. R. B.; Floss, V. M.; Bonacorso, H. G.; Zanatta N. et al. An Efficient Solvent-Free Synthesis of *NH*-pyrazoles from β-Dimethylaminovinylketones and Hydrazine on Grinding. *Tetrahedron Lett.* **2010,** *51*, 3193–3196.

Loudy, C. M.; Chasvised, S.; Paybou, C.; Courreges, C.; Allouche, J.; Martinez, H.; Revealing Surface Functionalities via Microwave for the Para-Fluoro-Thiol Click Reaction. *Polymer* **2020,** *202*, 122675.

Loupy, A. *Microwaves in Organic Synthesis*, 2nd ed.; Wiley-VCH: Weinheim, 2006.

Loupy, A.; Cabrales, N.; Lam, A.; Suarez, M.; Perez R.; Rodriguez, H. Alkylation of Some Pyrimidine and Purine Derivatives Using Microwave-Assisted Methods. *Heterocycles.* **2001,** *55*, 291–301.

Lourenço, M. A. O.; Siegel, R.; Mafra, L.; Ferreira, P. Microwave Assisted N-Alkylation of Amine Functionalized Crystal-Like Mesoporous Phenylene-Silica. *Dalton Trans.* **2013,** *42* (16), 5631–5634.

Lukasiewicz, M.; Bogdal D.; Pielichowski, J. Microwave Assisted Oxidation of Some Aromatics by Hydrogen Peroxide at Supported Tungsten Catalyst. *Mol. Divers.* **2006,** *10*, 491–493.

Lukasiewicz, M.; Bogdal D.; Pielichowski, J. Microwave-Assisted Oxidation of Alcohols Using Urea Hydrogen Peroxide. *Intern. Conf. Syn. Org. Chem. ECSOC-8*, 2004.

Ma, M. G.; Zhu, J. F.; Jia, N.; Li, S. M.; Sun, R. C.; Cao S. W.; Chen, F. Rapid Microwave-Assisted Synthesis and Characterization of Cellulose-Hydroxyapatite Nanocomposites in N,N-Dimethylacetamide Solvent. *Carbohydr. Res.* **2010,** *345*, 1046–1050.

Maddocks, C. J.; Aathimanikandan, S. V.; Richardson, J.; Ruble, J. C. Quinolin-8-yl Formate: A New Option for Small-Scale Carbonylation Reactions in Microwave Reactors. *Synlett.* **2020,** *31* (16), 1608–1612.

Mallikarjuna N. N.; Varma, R. S. Microwave-Assisted Shape-Controlled Bulk Synthesis of Noble Nanocrystals and Their Catalytic Properties. *Crys. Growth.* **2007,** *7*, 686–690.

Marx, S.; Ndabab, B. Rapid Microwave-Assisted Liquid Phase Conversion of Bio-Ethanol to N-Butanol Over a Heterogeneous Catalyst. *Biofuels.* **2021,** *12* (7), 861–868.

Massaro, M.; Riela, S.; Lazzara, G.; Gruttadauria, M.; Milioto, S.; Noto, R. Green Conditions for the Suzuki Reaction Using Microwave Irradiation and a New HNT-Supported Ionic Liquid-Like Phase (HNT-SILLP) Catalyst. *Appl. Organometal. Chem.* **2014,** *28* (4), 234–238.

Matondo, H.; Baboulene M.; Rico-Lattes, I. Synthesis of Novel Amphiphilic Pyridinylboronic Acids. *Appl. Organometal. Chem.* **2003,** *17*, 239–243.

Mavoral, J. A.; Cativicla, C.; Garcia, J. I.; Pires, E.; Rovo A. J.; Figueras, F. Diels-Alder Reactions of α-Amino Acid Precursors by Heterogeneous Catalysis: Thermal vs. Microwave Activation. *Appl. Catal.***1995,** *131*, 159–166.

May-Masnou, A.; Soler, L.; Torras, M.; Salles, P.; Llorca, J.; Roig, A. Fast and Simple Microwave Synthesis of TiO_2/ Au Nanoparticles for Gas-Phase Photocatalytic Hydrogen Generation. *Front. Chem.* **2018,** *6.* DOI: 10.3389/fchem.2018.00110

Mccarroll, A. J.; Sandham, D. A.; Titumb, L. R.; Lewis, A. K. D.; Cloke, F. G. N.; Davies, B. P.; Desantand, A. P.; Hiller W.; Caddicks, S. Studies on High-Temperature Amination Reactions of Aromatic Chlorides Using Discrete Palladium-N-Heterocyclic Carbene (NHC) Complexes and In Situ Palladium/Imidazolium Salt Protocols. *Mol. Divers.* **2003,** *7*, 115–123.

Mehra M. K.; Tantak M. P.; Kumar I.; Kumar D. Microwave-Assisted Metal- and Ligand-Free O-Arylation of Quinolones Using Diaryliodonium Salts: An Easy and Rapid Synthesis of Aryloxyquinolines. *Synlett* **2016,** *27* (04), 604–610.

Miljanic, O. S.; Volhardt K. P. C.; Whitener, G. D. An Alkyne Metathesis-Based Route to Ortho-Dehydrobenz annulenes. *Synlett.* **2003,** *1*, 29–34.

Miranda, L. S. M.; La Silva, T. R.; Crespo, L. T.; Esteves, P. M.; De Matos, L. F.; Diederichs C. C. et al. TBCA Mediated Microwave-Assisted Hofmann Rearrangement. *Tetrahedron Lett.* **2011,** *52*, 1639–1640.

Moghaddam, F. M.; Sharifi, A. Microwave Promoted Acetalization of Aldehydes and Ketones. *Synth. Commun.* **1995,** *25*, 2457–2461.

Mogilaiah, K.; Kavitha S.; Babu, H. R. Microwave Assisted Addition-Elimination Reactions of 3-Phenylimino-2-Indolinones with 2-Hydrazino-3- (p-Chlorophenyl)-1,8-Naphthyridine: A Simple and Facile Synthesis of 3- (3-P-chlorophenyl-1,8-Naphthyridin-2-Ylhydrazono)-2-Indolinones. *Indian J. Chem.* **2003,** *42B*, 1750–1752.

Mohamad, S.; Shantier, S.; Garalnabi, E. Microwave-Assisted Single-Step Synthesis of Acid Hydrazides from Corresponding Acids Utilizing Newly Designed Apparatus. *J. Appl. Pharm. Res.* **2022,** *10* (1), 29–37.

Mohammadi, M. K. Microwave-Assisted Oxidation of Organic Compounds with Cetyltrimethylammonium Chlorochromate. *Open J. Synth. Theory Appl.* **2013,** *2*, 87–90.

Mojtahedi, M. M.; Saidi, R.; Bolourchian, M.; Shirazi, J. S. Microwave Assisted Selective Oxidation of Benzylic Alcohols with Calcium Hypochlorite Under Solvent-Free Conditions. *Monatshefte fur chemie*. **2000,** *132*, 655–658.

Molteni V.; Ellis, A. D. Recent Advances in Microwave-Assisted Synthesis of Heterocyclic Compounds. *Curr. Org. Syn.* **2005,** *2*, 333–375.

Monteil-Rivera, F.; Paquet, L. Solvent-Free Catalyst-Free Microwave-Assisted Acylation of Lignin. *Ind. Crops Prod.* **2015,** *65*, 446–453.

Monterde, C.; Navarro, R.; Iglesias, M.; Sanchez, F. Fluorine-Phenanthroimidazole Porous Organic Polymer: Efficient Microwave Synthesis and Photocatalytic Activity. *ACS Appl. Mater. Interfaces*. **2018,** *11* (3), 3459–3465.

Motorina, I. A.; Parly F.; Grierson, S. Selective O-Allylation of Amidoalcohols on Solid Support. *Synlett.* **1996,** *4*, 389–391.

Motshekga, S. C.; Pillai, S. K.; Ray, S. S.; Jalama K.; Krause,R. W. M. Recent Trends in the Microwave-Assisted Synthesis of Metal Oxide Nanoparticles Supported on Carbon Nanotubes and Their Applications. *J. Nanomater.* **2012,** *2012*. DOI: 10.1155/2012/691503.

Murata, Y.; Sugawara, Y.; Matsumura, M.; Kakusawa, N.; Yasuike, S. Microwave-Assisted Debromination of α-Bromoketones with Triarylstibanes in Water. *Chem. Pharm. Bull.* **2017,** *65* (11), 1081–1084.

Naeimi, H.; Raeisi, A.; Moradian, M. Microwave Assisted Chemistry: A Rapid and Regioselective Route for Direct Ortho-Acylation of Phenols and Naphthols by Methanesulfonic Acid as Catalyst. *Arab. J. Chem*. **2017,** *10* (2), S2723–S2728.

Nardi, M.; Bonacci, S.; Herrera Cano, N.; Oliverio, M.; Procopio, A. The Highly Efficient Synthesis of 1, 2-Disubstituted Benzimidazoles Using Microwave Irradiation. *Molecules* **2022,** *27* (5). DOI: 10.3390/molecules27051751.

Nesaragi, A. R.; Kamble, R. R.; Bayannavar, P. K.; Shaikh, S. K. J.; Hoolageri, S. R.; Kodasi, B. et al. Microwave Assisted Regioselective Synthesis of Quinoline Appended Triazoles as Potent Anti-Tubercular and Antifungal Agents via Copper (I) Catalyzed Cycloaddition. *Bioorg. Med. Chem. Lett.* **2021,** *41*. DOI: 10.1016/j.bmcl.2021.127984

Nichols, C. E.; Youssef, D.; Harris R. G.; Jha, A.; Microwave- Assisted Synthesis of Curcumin Analogs. *ARKIVOC*. **2006,** *13*, 64–72.

Nilsson, P.; Olofssen K.; Larhed, M. Microwave-Assisted and Metal-Catalyzed Coupling Reactions. *Chem. Mat. Sci.* **2006,** *266*, 103–144.

Novanna, M.; Kannadasan, S.; Shanmugam, P. Microwave-Assisted N-Allylation/Homoallylation-RCM Approach: Access to Pyrrole-, Pyridine-, or Azepine-Appended (Het)Aryl Aminoamides. *ACS Omega.* **2020,** *5* (15), 8515–8522.

Nuchter, M.; Muller, U.; Ondruschka, B.; Tied A.; Lautenschlager, W. Microwave-assisted chemical reactions. *Chem. Eng. Technol.* **2003,** *26*, 1208–1216.

Nuchter, M.; Ondruschka, B.; Bonrath W.; Gum, A. Microwave Assisted Synthesis—A Critical Technology Overview. *Green Chem*. **2004,** *6*, 128–141.

Oliverio, M.; Nardi, M.; Cariati, L.; Vitale, E.; Bonacci, S.; Procopio, A. "On Water" MW-Assisted Synthesis of Hydroxytyrosol Fatty Esters. *ACS Sustain. Chem. Eng.* **2016,** *4* (3), 661–665.

Olofsson, K.; Hallberg A.; Larhed M.; Loupy, A. *Microwaves in Organic Synthesis*; Wiley: Weineim, 2002.

Ortiz, A. D.; Prieto, P.; Hoz, A. A Critical Overview on the Effect of Microwave Irradiation in Organic Synthesis. *Chem. Rec.* **2018,** *19* (1), 85–97.

Pathak, A. K.; Gupta, S.; Punjabi, P. B.; Ameta, G.; Ameta, C. Synthesis of Some Substituted Pyrazolopyrimidine Derivatives: An Environmentally Benign Approach. *J. Heterocycl. Chem.* **2019,** *56* (7), 2056–2062.

Patrick, T. B.; Gorrell K.; Rogers, J. Microwave Assisted Diels-Alder Cycloaddition of 2-Fluoro-3-Methoxy-1,3-Butadiene. *J. Fluorine Chem.* **2007,** *128*, 710–713.

Paul, N.; Kaladevi S.; Muthusubramanian, S. Microwave-Assisted Stereoselective 1,3-Dipolar Cycloaddition of *C,N*-Diarylnitrone (*i.e., N*- (Arylmethylidene)Benzenamine *N*-Oxide) and Bis (Arylmethylidene)Acetone (=1,5-Diarylpenta-1,4-Dien-3-One): NMR and Crystal Analysis of Diastereoisomeric Bis (Isoxazolidines). *Helvetica Chimica Acta* **2012,** *95*, 173–184.

Peipei, H.; Wenhua, Z.; Mingxia, C.; Qiuan, W. Microwave-Assisted Synthesis of Polymethoxy Chalcone Mannich Bases and Their Antiproliferative Activity. *Lett. Org. Chem.* **2019,** *16* (2), 117–121.

Perreux L.; Loupy, A. Microwave Effects: Athermal ("Specific") Effects. *Tetrahedron.* **2001,** *57*, 9199–9223.

Pielichowski, J.; Dziki, E.; Polaczek, J. Synthesis of Poly (Aspartic Acid) Under Microwave Irradiation. *Polish J. Chem. Tech.* **2003,** *5*, 3–4.

Polshettiwar P. V.; Varma, R. S. Microwave-Assisted Organic Synthesis and Transformations Using Benign Reaction Media. *Acc. Chem. Res.* **2008,** *41*, 629–639.

Portero-Barahona, P.; Carvajal-Barriga. E. J.; Martín-Gil, J.; Martín-Ramos, P. Sugarcane Bagasse Hydrolysis Enhancement by Microwave-Assisted Sulfolane Pretreatment. *Energies* **2019,** *12* (9). DOI: 10.3390/en12091703.

Priecel, P.; Lopez-Sanchez, J. A. Advantages and Limitations of Microwave Reactors: From Chemical Synthesis to the Catalytic Valorization of Bio Based Chemicals. *ACS Sustainable Chem. Eng.* **2019,** *7* (1), 3–21.

Rajbongshi, K. K.; Ambala, S.; Govender, T.; Kruger, H. G.; Arvidsson, P. I.; Naicker, T. Microwave-Accelerated N-Acylation of Sulfoximines with Aldehydes Under Catalyst-Free Conditions. *Synthesis.* **2020,** *52* (08), 1279–1286.

Ranger, K. D.; Strauss, C. R.; Trainer R. W.; Thorn, J. S. A New Microwave Reactor for Batch Wise Organic Synthesis. *J. Org. Chem.* **1995,** *60*, 2456–2460.

Rohman M. R.; Myrboh, B. KF-Alumina Mediated Bargellini Reaction. *Tetrahedron Lett.* **2010,** *51*, 4772–4775.

Roy, I.; Gupta, M. N. Non-Thermal Effects of Microwaves on Protease-Catalyzed Esterification and Transesterification. *Tetrahedron.* **2003,** *59*, 5431–5436.

Rubiño, M. E. G.; Carretero, M. C. N.; Lazarte, D. C.; Ruiz, J. M. G.; Madrid, Y.; Campos, J. M. Stereospecific Alkylation of Substituted Adenines by the Mitsunobu Coupling Reaction Under Microwave-Assisted Conditions. *RSC Adv.* **2014,** *4*, 22425–22433.

Sabouni, R.; Kazemian, H.; Rohani, S. Microwave Synthesis of the CPM-5 Metal Organic Framework. *Chem. Eng. Technol.* **2012,** *35* (6), 1085–1092.

Sancheti, A.; Swarnkar, N.; Soni, M. D.; Vardia, J.; Punjabi P. B.; Ameta, S. C. Microwave Assisted Synthesis and Characterization. *J. Indian Chem. Soc.* **2007,** *84*, 1234–1238.

Santagada, V.; Fiorino, F.; Perissuti, E.; Severino, B.; Terracciano, S.; Cirino G.; Caliendo, G. The Effect of Microwave Irradiation on Mo (VI) Catalyzed Transformations of Reducing Saccharides. *Tetrahedron Lett.* **2003,** *5*, 2131.

Sawant D. N.; Bhanage B. C. M. Pd $(OAc)_2$/DPPF-Catalysed Microwave-Assisted Cyanide-Free Synthesis of Aryl Nitriles. *J. Chem. Sci.* **2014,** *126* (2), 319–324.

Sawant R.T.; Stevens M. Y.; Odell L. R. Microwave-Assisted Aza-Friedel–Crafts Arylation of N-Acyliminium Ions: Expedient Access to 4-Aryl 3,4-Dihydroquinazolinones. *ACS Omega* **2018,** *3* (10), 14258–14265.

Schmoger, C.; Stolle, A.; Bonrath W.; Ondruschka, B. Microwave-Assisted Organic Reduction Reactions. *Curr. Org. Chem.* **2011,** *15*, 151–167.

Schultze, B.; Schmidt, B. Functionalized Benzofurans Via Microwave-Promoted Tandem Claisen-Rearrangement/5-Endo-Dig Cyclization. *J. Heterocyclic Chem.* **2019,** *56* (9), 2619–2629.

Seubert, P.; Freund, M.; Rudolf, R. D.; Lin, Y.; Altevogt, L.; Bilitewski, U. et al. Buchwald–Hartwig Versus Microwave-Assisted Amination of Chloroquinolines: En Route to the Pyoverdin Chromophore. *Synlett.* **2020,** *31* (12), 1177–1181.

Shah, J. J.; Mohanraj, K. Comparison of Conventional and Microwave-Assisted Synthesis of Benzotriazole Derivatives. *Indian J. Pharm. Sci.* **2014,** *76* (1), 46–53.

Shao, B. Synthesis of Fused Bicyclic Pyridines with Microwave-Assisted Intramolecular Hetero-Diels–Alder Cycloaddition of Acetylenic Pyrimidines. *Tetrahedron Lett.* **2005,** *46*, 3423–3427.

Shi, S.; Hwang, J. Y. Microwave-Assisted Wet Chemical Synthesis: Advantages, Significance, and Steps to Industrialization. *J. Miner. Mater. Character. Eng.* **2003,** *2*, 101–110.

Shinde, S. D.; Yadav, G. D. Microwave Irradiated Immobilized Lipase Catalyzed Synthesis of Alkyl Benzoate Esters by Trans Esterification: Mechanism and Kinetic Modeling. *Ind. Eng. Chem. Res.* **2014,** *53* (21), 8706–8713.

Shrimali, K.; Sitha, D.; Vardia J.; Ameta, S. C. Microwave Induced Synthesis and Characterization of Some Thiazolidinone Derivatives Bearing Benzotriazole Moeity. *Afinidad* **2009,** *66*, 173–176.

Simin, S.; Chong, T.; Jiaxi, X. Microwave Thermal Effect on Diels-Alder Reaction of Furan and Maleimide. *Curr. Microw. Chem.* **2020,** *7* (1), 67–73.

Singh, A.; Singh, S.; Sewariya, S.; Singh, N.; Singh, P.; Kumar, A. et al. Stereospecific *N*-Acylation of Indoles and Corresponding Microwave Mediated Synthesis of Pyrazinoindoles Using Hexafluoroisopropanol. *Tetrahedron.* **2021,** *84*. DOI: 10.1016/j.tet.2021.132017

Sitha, D.; Ameta, R.; Punjabi P. B.; Ameta, S. C. Efficient synthesis of 4-Aryl/Alkyl-3, 4-Dihydro-2 (1H)-Pyrimidone Esters and 5-Acetyl-4-aryl/alkyl-6-methyl-3, 4-Dihydropyrimidin-2 (1)-Ones Using Montmorillonite Clay-Cu $(NO_3)_2$. 2.5H_2O System Under Microwave Irradiation. *Int. J. Chem. Sci.* **2010,** *8 (3)*, 1973–1982.

Srikrishna A.; Kumar, P. P. Naphthalenes via Microwave Irradiation Induced Rearrangement on Montmorillonite K-10. *Tetrahedron Lett.* **1995,** *36*, 6313–6316.

Stadler A.; Kappe, C. O. Microwave-Mediated Biginelli Reactions Revisited. On the Nature of Rate and Yield Enhancements. *J. Chem. Soc. Perkin Trans.* **2000,** *2*, 1363–1368.

Strauss C. R.; Trainor, R. W. Developments in Microwave-Assisted Organic Chemistry. *Aust. J. Chem.* **1995,** *48*, 1665–1692.

Strauss, C. R. Microwave-Assisted Reactions in Organic Synthesis—Are There Any Nonthermal Microwave Effects? Response to the Highlight by N. Kuhnert. *Angew. Chem. Int. Ed.* **2002,** *41*, 3589–3591.

Stuerga D.; Gaillard, P. Microwave Athermal Effects in Chemistry—A Myths Autopsy Part I: Historical Background and Fundamentals of Wave-Matter Interaction. *J. Microwave Power Electromagn. Energy.* **1996a,** *31*, 87–99.

Stuerga D.; Gaillard P. Microwave Athermal Effects in Chemistry—A Myths Autopsy, Part II: Orienting Effects and Thermodynamic Consequences of Electric field. *J. Microwave Power Electromagn. Energy.* **1996b,** *31*, 101–113.

Sun, S.; Teng, C.; Xu, J. Microwave Thermal Effect on Diels-Alder Reaction of Furan and Maleimide. *Curr. Microwave Chem.* **2020,** *7* (1), 67–73.

Taher, A.; Nandi, D.; Islam, R. U.; Choudhary, M.; Mallick, K. Microwave Assisted Azide–Alkyne Cycloaddition Reaction Using Polymer Supported Cu (I) as a Catalytic Species: A Solvent Less Approach. *RSC Adv.* **2015,** *5* (59), 47275–47283.

Taher, A.; Lumbiny, B. J.; Lee, I. M. A Facile Microwave-Assisted Knoevenagel Condensation of Various Aldehydes and Ketones Using Amine-Functionalized Metalorganic Frameworks. *Inorg. Chem. Communi.* **2020,** *119*, 108092.

Tankam, T.; Srisa, J.; Sukwattanasinitt, M.; Wacharasindhu, S. Microwave-Enhanced on-Water Amination of 2-Mercaptobenzoxazoles to Prepare 2-Aminobenzoxazoles. *J. Org. Chem.* **2018,** *83* (19), 11936–11943.

Tejedor, D.; Mendez-Abt, G.; Cotos, L.; Ramirez M. A.; García-Tellado, F. A Microwave-Assisted Domino Rearrangement of Propargyl Vinyl Ethers to Multifunctionalized Aromatic Platforms. *Chem. A European J.* **2011,** *17*, 3318–3321.

Tharun, J.; Mathai, G.; Kathalikkattil, A. C.; Roshan, R.; Kwak, J. Y.; Park, D. W. Microwave-Assisted Synthesis of Cyclic Carbonates by a Formic Acid/KI Catalytic System. *Green Chem.* **2013,** *15* (6), 1673–1677.

Thorell, J. O.; Stone-Elander S.; Elander, N. Use of a Microwave Cavity to Reduce Reaction Times in Radiolabelling with [^{11}C]- Cyanide. *J. Label. Compds. Radiopharm.* **1992,** *31 (3),* 207–217.

Tjeng, A. A.; Handore, K. L.; Batey, R. A. Stereocontrolled Microwave-Assisted Domino [3,3]-Sigmatropicreactions: A Winstein–Overman Rearrangement for the Formation of Differentiated Contiguous C–N Bonds. *Org. Lett.* 2020, *22* (8), 3050–3055.

Varma R. S.; Namboodiri, V. V. Solvent-Free Accelerated Organic Syntheses Using Microwaves. *Chem. Commun.* **2001,** *7*, 643–644.

Varma R. S.; Saini, R. K. Microwave-Assisted Reduction of Carbonyl Compounds in Solid State Using Sodium Borohydride Supported on Alumina. *Tetrahedron Lett.* **1997,** *38*, 4337–4338.

Vidal, T.; Pefit, A.; Loupy A.; Gedye, R. N. Re-examination of Microwave-Induced Synthesis of Phthalimides. *Tetrahedron.* **2000,** *56*, 5473–5478.

Viola, A.; Ferrazzano, L.; Martelli, G.; Ancona, S. L.; Gentilucci, A. Tolomelli, An Improved Microwave Assisted Protocol for Yonemitsu-Type Trimolecular Condensation. *Tetrahedron* **2014,** *70* (38), 6781–6788.

Vu, Z. T.; Liu L. J.; Zhuo, R. X. Microwave-Improved Polymerization of ϵ-Caprolactone Initiated by Carboxylic Acids. *Polym. Chem. Ed.* **2003,** *41*, 13–21.

Vyas, R.; Swarnkar, N.; Sancheti, A.; Vardia J.; Punjabi, P. B. Heterocyclization of Some Chalcones to Isoxazoles, Pyrazoles and Pyrimidine Nuclei Under Microwave Irradiation and Their Biological Profile. *J. Indian Chem. Soc.* **2008,** *85*, 1217–1226.

Wada, Y.; Yin, H. B.; Kitamura T.; Yanagida, S. Microwave-Assisted Hydrogenation of Chlorinated Phenols for Complete Dechlorination. *Chem. Lett.* **2000,** *29*, 632–633.

Wali, A.; Paillai S. M.; Satish, S. Heterogeneous Pd Catalysts and Microwave Irradiation in Heck Arylation. *Indian Petrochem. Corp. Ltd.* **1995,** 294.

Walia, A.; Kang, S.; Silverman, R. B. Microwave-Assisted Protection of Primary Amines as 2, 5-Dimethylpyrroles and Their Orthogonal Deprotection. *J. Org. Chem.* **2013,** *78* (21), 10931–10937.

Wang Y.; Lee, J. Y. Microwave-Assisted Synthesis of SnO_2–Graphite Nanocomposites for Li-Ion Battery Applications. *J. Power Sources*. **2005,** *144*, 220–225.

Watkins, K. J. Fighting the Clock. *Chem. Eng. News*. **2002,** *80*, 27–34.

Werth, K.; Lutze, P.; Kiss, A. A.; Stankiewicz, A. I.; Stefanidis, G. D.; Górak, A. A Systematic Investigation of Microwave-Assisted Reactive Distillation: Influence of Microwaves on Separation and Reaction. *Chem. Eng. Proc. Proc. Intensif.* **2015,** *93*, 87–97.

Wolkenberg, S. E.; Wisnoski, D. D.; Leister, W. H.; Wang, Y.; Zhao Z.; Lindsley, C. W. Efficient Synthesis of Imidazoles from Aldehydes and 1,2-Diketones Using Microwave Irradiation. *Org. Lett.* **2004,** *6*, 1453–1456.

Wu, S. W.; Liu, J. L.; Liu, F. Metal-Free Microwave-Assisted Decarboxylative Elimination for the Synthesis of Olefins. *Org. Lett.* **2016,** *18* (1). DOI: 10.1021/acs.orglett.5b03069.

Wu, T. Y. H.; Schultz P. G.; Ding, S. Solid-Phase Synthesis of 2,3,5-Trisubstituted Indoles. *Org. Lett.* **2003,** *5*, 3827–3830.

Xu J. S. and Zhu,Y. J. Microwave-Assisted Ionic Liquid Solvothermal Rapid Synthesis of Hollow Microspheres of Alkaline Earth Metal Fluorides (MF 2, M= Mg, Ca, Sr). *Cryst. Eng. Commun.* **2012,** *14* (8), 2630–2634.

Xue, F.; Song, X.; Lin, T. T.; Munkerup, K.; Albawardi, S. F.; Huang, K. W. Dimerization of Terminal Aryl Alkynes Catalyzed by Iron (II) Amine-Pyrazolyl Tripodal Complexes with *E*/*Z* Selectivity Controlled by *Tert*-Butoxide. *ACS Omega* **2018,** *3* (5), 5071–5077.

Yadav, A. R.; Shrinivasan, M. K.; Magdum, C. S. Comparative Study of Conventional and Microwave Assisted Synthesis of Some Organic Reactions. *Asian J. Pharm. Res.* **2020,** *10* (3), 217–220.

Yamazaki K.; Kondo, V. Palladium-Catalyzed Synthesis of Indole 3-Carboxylates on a Solid Polymer Support. *J. Comb. Chem.* **2002,** *4 (3)*, 191–192.

Yang, S.; Da-Wei, H.; Xiao-Hui, L.; Jian-En H.; Zhi-Long, X. Microwave-Assisted Solid-Phase Synthesis, Biological Evaluation and Molecular Docking of Angiotensin I-Converting Enzyme Inhibitors. *Chem. Res. Chinese Univ.* **2012,** *28*, 108–113.

Yaragorla, S.; Singh, G.; Saini, P. L.; Reddy, M. K. Microwave Assisted, Ca (II)-Catalyzed Ritter Reaction for the Green Synthesis of Amides. *Tetrahedron Lett.* **2014,** *55* (33), 4657–4660.

Yıldız, Y.; Esirden, I.; Erken, E.; Demir, E.; Kaya, M.; Sen, F. Microwave (MW)-Assisted Synthesis of 5-Substituted 1H-Tetrazoles via [3+2] Cycloaddition Catalyzed by MW-Pd/Co Nanoparticles Decorated on Multi-Walled Carbon Nanotubes. *Chem. Select.* **2016,** *1* (8), 1695–1701.

Yin, D.; Jin, Z.; Liu, M.; Gao, T.; Yuan, H.; Xiao, D. Microwave-Assisted Synthesis of the Cobalt-Iron Phosphates Nanosheets as an Efficient Electrocatalyst for Water Oxidation. *Electrochim. Acta.* **2018,** *260*, 420–429.

Yuan, S.; Feng, L.; He, A.; Liu, L.; Liu, B.; Chen, Y.; Li, X. Triazine-Functionalized Highly Ordered Hierarchically Porous Organic Polymer with High CO_2 Uptake Capacity and Catalytic Activity for Microwave-Assisted Knoevenagel Condensation Reaction. *Colloids Sur. A: Physicochem. Eng. Aspects.* **2020,** *607*, 125475. DOI: 10.1016/j.colsurfa.2020.125475

Yue, Y.; Liu, T. Y.; Li, H.-W.; Liu Z.; Wu, Y. Microwave-Assisted Synthesis of BSA-Protected Small Gold Nanoclusters and Their Fluorescence-Enhanced Sensing of Silver (I) Ions. *Nanoscale* **2012,** *4*, 2251–2254.

Zhang, W.; Chen, J.; Swiegers, G. F.; Ma Z. F.; Wallace, G. G. Microwave-Assisted Preparation of Carbon Nanotubes with Versatile Functionality. *Nanoscale.* **2010,** *2*, 282–286.

Zhao, D. R. Padron, K. S. Triantafyllidis, Y. Wang, R. Luque, C. Len, Microwave-Assisted Oxidation of Hydroxymethyl Furfural to Added-Value Compounds Over a Ruthenium-Based Catalyst. *ACS Sustain. Chem. Eng.* **2020,** *8* (8), 3091–3102.

Zhao, G.; Jiang, T.; Gao, H.; Han, B.; Huang J.; Sun, D. Mannich Reaction Using Acidic Ionic Liquids as Catalysts and Solvents. *Green Chem.* **2004,** *6*, 75–77.
Zhu J. F.; Zhu, Y. J. Microwave-Assisted One-Step Synthesis of Polyacrylamide–Metal (M = Ag, Pt, Cu) Nanocomposites in Ethylene Glycol. *J. Phys. Chem.* **2006,** *110B*, 8593–8597.
Zhu, X. T.; Liu, J. Y.; Jiang, Tu, S. J. Microwave-Assisted Aqueous Reactions: An Efficient Route to Benzodiazepines. *J. Hetero. Chem.* **2014,** *52* (1), 92–96.
Zuliani, A.; Balu, A. M.; Luque, R. Efficient and Environmentally Friendly Microwave-Assisted Synthesis of Catalytically Active Magnetic Metallic Ni Nanoparticles. *ACS Sustain. Chem. Eng.* **2017,** *5* (12), 11584–11587.

CHAPTER 12

Green Composites

PRIYANKA JHALORA[1], NARENDRA PAL SINGH CHAUHAN[2], YASMIN[3], and ROHIT AMETA[4]

[1]*Department of Chemistry, PAHER University, Udaipur, India*

[2]*Department of Chemistry, B. N. University, Udaipur, India*

[3]*Department of Chemistry, Geetanjasli Institute of Technical Studies, Udaipur, India*

[4]*R&D Section, Apollo Tyres, Chennai, India*

ABSTRACT

There is a growing demand for eco-friendly materials, which has led to development and design of biodegradable composite materials. Green chemistry concepts may be utilized to prepare such materials, so as to minimize the toxic effects of organic compounds, which pose potential risks to human health. Nature is already producing such composite materials like corn husk packing, leaf network, cotton kenaf, etc. Polymeric natural sources are also there such as polylactic acid, polyhydroxy alkanoates, starch, etc. Natural biocomposites have been prepared based on cellulose, jute, flax and hemp, etc. These biocomposites have a great scope of application in automobiles, aircrafts, ships, trains, packaging, computers, mobile phones, and so on. These are biodegradable also and therefore, fall in the category of green materials. This has been discussed in this chapter.

12.1 INTRODUCTION

The demand of environment-friendly materials is growing and will continue to grow by growing public concern about environmental pollution, which has

Green Chemistry, 2nd Edition: Fundamentals and Applications. Suresh C. Ameta & Rakshit Ameta (Eds.)

led to development and design of biodegradable composite materials. Green chemistry concepts may be utilized for environment-friendly materials and to minimize the toxic effects of organic compounds, which pose potential risks to human health. The serious efforts are needed for the development of green composites for the minimization of environmental impact of polymer composite production.

Assessment of the environmental impact arising from activities such as construction, packaging, or transport is an essential activity, when attempting to design sustainable approaches to our developmental needs. In this respect, life cycle assessment (LCA) has emerged as a widely accepted technique for evaluating the environmental aspects associated with a wide variety of products, processes, or activities from initial synthesis to final disposal for a sustainable environment. The regulatory assessment and monitoring procedures must be updated time to time depending upon the composition, intended usage conditions to promote clean processing, applications, biodegradation, recycling, and reprocessing.

Most of the commercial fibers and resins like plastics and polymers are derived from petroleum feedstock. The major problem associated with this is the high rate of depletion of petroleum resources. By one estimate, the current consumption rate of petroleum is about 100,000 times the rate at which the earth can generate it. Another problem is most of the composites and plastics derived from petroleum are nonbiodegradable under normal environmental conditions. Composites made by thermoset resins cannot be reused or recycled and most of these composites end up in the landfills at the end of their life. They last for several decades without degrading and make that land unusable. Incineration produces large amount of the toxic gases that require costly scrubbers; hence both land filling and incineration are expensive and environmentally undesirable.

Growing global environmental and social concern, the high rate of depletion of petroleum resources, and now environmental regulations have forced the search for green composites, compatible with the environment. Green composite combines plant fibers with natural resins to form natural composite materials. Natural fibers such as kenaf, flax, jute, hemp, sisal, banana fiber, cotton, kapok, abaca, and coir are emerging as low cost, light weight, and eco-friendly alternative to synthetic fibers such as carbon fibers, glass fibers, and kevlar fibers. The resin and fibers used in green composites are biodegradable, when they are dumped and decomposed by the action of microorganisms. They are converted into H_2O and CO_2, which are absorbed into the plant system. Moreover because of their moderate mechanical strength, they can be used to reinforce plastics and fabricate composites for various applications, for example, packaging, product casing, housing and automotive panels, furniture etc.

Composites are one of the most widely needed materials as these have certain advantages such as adaptability to varied situations and easy combination to fabricate with other materials to be used for different purposes exhibiting desirable properties. Green composites are environment friendly in nature. Green composites are developing as a specific class of biocomposites. Here, a bio-based polymer matrix is reinforced by natural fibers. The applications of green composites are increasing day by day in the fields such as automotive sector, transportation, sports goods, household applications, construction, etc.

12.2 DESIGNING FOR COMPOSITES

The best way to get the ideas of designing for composites is from natural processes. It has made functional products for millennia.

A leaf is a good example of natural processes. A real leaf is passing through different stages in its life, that is, changing form, changing process, response to adverse conditions, damage and attack, aerodynamics, thermal control, chemistry, physics, deployment, retraction, disposability, and recycling, even the role of environment for other life forms after it shades off from tree. Natural processes give ideas about what is valuable and what enhances the environment and life processes? Technology provides us with the tools to see these attributes of nature and to design composite material properties from nano to micro and micro to macro scale in terms of fabrication and assembly techniques.

The observable visual aspects of the physical formation and patterns that are characteristics of natural material engineering are network construction, for example, bird nest, leaf, and web. Interfacial issues like differential properties are arranged in 3D to accommodate complex characteristics, for example, tendon, spider web, foot, braiding, twisting, binding, orientation of fibers, layers of fiber orientation, combination of different fiber materials, fiber matrix relation, etc. and it helps us to conceptualize the principles of our design, that is, no waste, related to environment and local available resources. Let us discuss some examples to learn from natural material structure, in which the visible shapes of the object, surface detail, texture, and structural features display some of the shapes of essential functional attributes.

i) Leaf networks (Figure 12.1). Form fluid variation
ii) Crab shell architecture (Figure 12.2). For improving damage control
iii) Spider web (Figure 12.3). Three dimensional arrangement to accommodate complex characters.
iv) Corn husk leaf packaging (Figure 12.4). A nature food wrapper.

FIGURE 12.1 Leaf networks.

FIGURE 12.2 Crab shell structure.

FIGURE 12.3 Spider web.

FIGURE 12.4 Corn husk packing.

12.3 LIFE CYCLE ASSESSMENT

The life cycle assessment is a technique for evaluating environmental impact at all stages of a product's life from initial synthesis to final disposal such as recycling incineration and disposal for a sustainable environment. The sustainability parameter includes energy use, price, transportation, health hazards, renewable resources, waste prevention, biodegradability, recycling, etc. In recent years, life cycle assessment work is easily supported by using a number of commercial LCA software packages as well as authoritative public access datasets, for example, ISO standard 14040–14043 provides detailed guidelines for conducting LCA, but attention to the selection of appropriate data for the goal and scope of the study will always remain an important issue in LCA.

LCA methodology consists of four independent elements (ISO 14040 series), that is:

i) The definition of goal and scope.
ii) The life cycle inventory analysis.
iii) The life cycle impact assessment, and
iv) The life cycle interpretation of results.

12.4 NATURAL FIBER SOURCES

The natural composite materials are created by combining plant fibers with natural resins. Natural fibers are emerging as low cost, light weight, and recyclable materials. Easy availability, thermal insulation, CO_2 neutrality, acceptable strength, stiffness, and biodegradability are the added advantages that make biocomposites environmentally superior alternative to synthetic fibers.

- Fibers such as jute, cotton kenaf, banana fiber, pineapple leaf, sisal, henequen, kapek, flax, coir, etc. are being used as biofiller in thermoplastic and thermosetting polymers to develop composites through various techniques.
- Chicken feather, bone, scales of marine animals, silk, lamb wool, etc. are also used as biofiller in composites.
- Recently, agricultural waste fibers like wheat straw, grass fibers, soy stalk, corn stalk like miscanthus, etc. are also promoted as reinforcing filler in biocomposites.

Natural fibers are abundant, low cost, biocompatible, and biodegradable, although there are certain limitations such as poor fire resistance, high moisture absorption, subsequent swelling and degradation, nonhomogeneity of their mechanical properties, and poor interfacial interactions.

i) Kenaf fiber

In recent years, cellulose materials are also used as reinforcement fibers, not only for ecological and economical reasons, but also because of their high mechanical and thermal performance. Kenaf is well known as a cellulose source and it can be a good alternative to glass fibers because:

i) It fixes CO_2 effectively.
ii) Kenaf absorbs N_2 and P found in the soil and in waste water (Abe and Ozaki, 1998).
iii) It has low density and high specific mechanical properties and it is biodegradable. (Inagaki, 2002).

Han et al. (1999) concluded that kenaf has been used as an alternative raw material to wood in the pulp and paper industries to avoid the destruction of forests. Serizawa et al. (2006) reported that adding kenaf fiber to poly(lactic acid) (PLA) greatly increases its heat resistance and modulus and also its crystallization and therefore, the ease of molding this material is improved. They also reported that PLA/ kenaf fiber and PLA/ kenaf fiber/flexibilizers (which is a copolymer of lactic acid and aliphatic polyester) show good practical characteristics for housing materials of electronic products as compared to petroleum-based plastic used in housing such as glass-fiber reinforced acrylonitrile-butadiene-styrene (ABS) resin. Khristova et al. (1998) reported that the soda-AQ pulp blends from kenaf and sunflower stalks result in considerable improvement of the sunflower-pulp properties.

ii) Micro-fibrillated materials: Taniguchi and Okamura (1998) developed new type of micro fibrillated materials (MM) from natural fibers such as wood pulp fibers, cotton fibers, tunicin cellulose, chitosan, silk fibers, collagen, etc. by a super grinding method. They made films using MM from natural fibers and the films obtained had 3–100 μm thickness and were homogeneous, strong, and translucent. Their tensile strength was much superior to those of commercial print grade papers.

iii) Micro-fiber bundles: Yano and Nakahara (2004) used plant micro-fiber bundles with a nanometer unit web like network to obtain a molded product and found that the moldings have a combination of environmentally friendly and high strength properties.

iv) Silkworm silk: Silkworm silk is also a natural fiber. Shao and Vollrath (2002) reported that the mechanical properties of silkworm silks can approach those of spider dragline silk, when reeled under controlled conditions.

v) **Oil palm empty fruit bunch fiber:** A comparative study of polypropylene composites reinforced with oil palm empty fruit bunch fiber and oil palm derived cellulose was made by Khalid et al. (2008). The structure and mechanical properties of sheets prepared from cellulose were studied by Nishi et al. (1990).

vi) **Spider dragline silks:** Spider dragline silks are exceptionally strong and extensible (Kaplan et al., 1994). In fact, their toughness equals that of commercial polyaramid (aromatic nylon) filaments, which is used to make materials ranging from radial tires and bullet proof clothing to reinforced composites for aircraft panels. Heslot (1998) reported that the exceptional toughness of dragline silk is achieved under benign conditions in contrast to current techno polymers based on petrochemicals. The spider spins its totally recyclable fibers at ambient temperature, low pressures, and with water as a solvent. Tirrell et al. (1994) concluded that genetically engineered silk with specially tailored properties and spun using "green" processes could replace the ubiquitous plastics, which are often detrimental to the environment in both the production and disposal.

vii) **Wood fibers and paper fibers (Alternative natural fibers):** Wood fibers are cellulose fibers made by mechanical and chemical methods, whereas paper fibers are mechanically ground and chemically made organic material from trees and used in paper industry. In paper industry, both fibers are mixed to combine their properties.

Paper fibers contribute too many properties of composites like renewability, biodegradability, low price, found in abundance and strength. The wood fibers sources are eucalyptus, pine, spruce, larch, beech, white beech, oak, aspen, linden trees, etc.

Wood fiber consists mainly of cellulose, hemicellulose, and lignin.

i) Cellulose forms the frame of the cell wall and it is the main component of wood fiber, which is responsible for the properties of fibers and makes it possible to use them in paper- making industry.
ii) Hemicellulose and lignin forms the surrounding intercellular substance and hemicellulose affects the ability of fibers to form bonding between each other.
iii) Lignin bonds fibers and gives stiffness to wood.

Wood can be used in various forms as reinforcement for example in the form of wood sheets, wood flours, or fiber (paper). The mechanical properties of the wood fibers can be compared with those of synthetic fibers, for example, glass fiber. Cellulose fibers and plastic composites could offer higher specific strength than glass fiber at a low cost.

The mechanical properties of the composites can be improved by using wood fibers. The properties of the composites are affected by the wood fiber form, dimensions, treatment of fibers and fillers. The strength of the composites depends on the length of fiber, the longer is the fiber, the better will be the strength, whereas shorter fibers should disperse more easily and as a consequence, increase properties of the fiber more than adhesion (Bolton, 1994).

The most important factor to affect the properties of the composites is the adhesion between fibers and matrix. The lignin forms better adhesion between hydrophobic and nonpolar polymers such as polyolefins, if used in wood plastic composites. If adhesion is poor, wood fibers work as filler in the matrix and therefore, the tensile strength can further decrease. When the adhesion is good, the wood fiber works as reinforcement in the matrix. Adhesion can be improved by coupling agent e.g. silanes.

Low melting point materials such as polyolefins, thermosets, and polystyrenes are used as matrix in wood fiber composites. The composites made of polyolefins and wood fiber are likely to absorb less moisture than commercial hardboards, where wood flour can be used as a filler. The benefits of using polyolefins as a matrix material are that they are easily available and having low processing temperature so that wood fiber cannot degrade.

Sameni et al. (2002) used rubber matrix and biopolymer matrix for their studies. Amnuay et al. (2011) used cellulose fibers from recycled newspaper as reinforcement for thermoplastic starch in order to improve its mechanical, thermal, and water resistance properties. They prepared composites from corn starch plasticized by glycerol as matrix, which was reinforced with microcellulose fibers, obtained from used newspaper. They reported that thermoplastic starch reinforced with recycled newspaper cellulose fibers could be fruitfully used as commodity plastic being strong, low cost, abundant, and recyclable.

12.5 NATURAL POLYMER SOURCES

The challenge of green composites involves basically obtaining 'green' polymers that are used as a matrix for the production of the composites. Polymer is said to be green, when it possesses environmentally favorable properties such as renewability and degradability. Biodegradation implies degradation of a polymer in natural environment that includes changes in the chemical structure, loss of mechanical and structural properties, and changing into other compounds that are beneficial to the environment (Pandey et al., 2007, Jamshidian et al., 2010).

Polymers from natural sources, such as starch, lignin, cellulose acetate, polylactic acid (PLA), polyhydroxy alkanoates (PHA), polyhydroxy butyrate (PHB), and some other synthetic sources such as aliphatic and aromatic polyesters, polyvinyl alcohol, modified polyolefins, etc. are degradable and classified as biopolymers. The most important biopolymers from an overall market perspective are–

i) Polylactic acid (PLA)

Avella et al. (2009) briefly reviewed most suitable and commonly used biodegradable polymer matrices and natural fiber reinforcement in eco-composites and nanocomposites with a special focus on PLA-based material.

It is a class of crystalline biodegradable thermoplastic polymers with relatively high melting point and excellent mechanical properties. All PLA resins are manufactured using renewable agricultural resources such as corn and sugar beets. Corn has the advantage of providing the required high purity lactic acid than other sources. PLA is synthesized by condensation polymerization of D- or L-lactic acid or ring- opening polymerization of the lactide (Fang and Hanna, 1999; Garlotta, 2002).

PLA is commercially interesting because of its good strength properties, film transparency, biodegradability, biocompatibility, and availability from renewable resources. Under specific environmental conditions, pure PLA can degrade to CO_2, H_2O, and CH_4 over a period of several months to two years, a distinct advantage compared to other petroleum plastics that need much longer periods. The final properties of PLA strictly depend on its molecular weight and crystallinity. PLA has been studied as a biomaterial in medicine, but only recently, it has been used as a polymer matrix in composites.

In last few years, different natural fibers have been employed in order to modify the properties of PLA. The most studied natural fiber reinforcements for PLA were kenaf (Huda et al., 2009; Avella et al., 2009), flax (Bax and Mussig, 2008), hemp (Hu and Lim, 2007), bamboo (Tokoro et al., 2008), jute (Shikamoto et al., 2007), and wood fibers (Huda et al., 2006). Besides conventional natural fibers, reed fibers have been tested in appropriate PLA composites to improve the tensile modulus and strength (Huda et al., 2008a).

ii) Polyhydroxy alkanoates (PHA)

Poly-R-3-hydroxy butyrate (PHB) is the simplest family of PHA. PHAs are synthesized biochemically by microbial fermentation and represent natural polyesters. Bacteria are still the only source of these polyesters. *Alcaligenes eutrophus* has been used to produce the commercial product, a polyhydroxy butyrate covalerate PHB (commercial name Biopol®). It is biotechnologically prepared polyester that constitutes a carbon reserve in a wide variety of bacteria (Avella et al., 1996) and has attracted much attention as a biodegradable thermoplastic polyester (Doi, 1990). A large number of PHBV random copolymers can be produced from *A. eutrophus* depending upon the carbon substrate. Carbon sources include propionic acid, pentanoic acid, 4-hydroxybutyric acid, 1,4- butanediol etc. following the formation process, the dilute aqueous broth is extracted to obtain biopolymers that must be isolated and purified. PHB of very high purity can be produced by continuous fermentation in combination with advanced isolation procedures.

One of the main commercial developments in PHA technology has been the production of PHBV in the form of Biopol®. The first products from this polymer were shampoo bottles and cosmetics containers. Future applications for PHB-based polymers could be in disposable products such as fast food utensils, garbage bags, and diapers. Hocking and Marchessault (1994) reviewed the use of PHB or PHBV in a variety of products such as films, bottles, and containers.

iii) Starch

Starch is produced in plants and some microorganisms. It is a mixture of linear amylase and branched amylopectin. The amount of amylase and amylopectin and the size and shape of starch granules depends upon its source. Amylase is the minor component of starch ranging from 20 to 30%. The amylopectin is responsible for crystalline properties of starches. The relative proportions of amylopectin and amylase in starch are determined by genetic and environmental control during biosynthesis and as a result, wide variations are found among plant raw materials. Starch is one of the low-cost biodegradable materials available in the market today. It is a versatile biopolymer with immense potential for use in the non-food industries.

The primary source of starch is maize and other sources are rice, wheat, potato, etc. Starch is an interesting alternative to thermoplastic material in situation, where long- term durability is not required and rapid degradation is an advantage. Shogren (1998) has reviewed the properties and applications of starch. Starch can be made thermoplastic through destructurization process in specific extrusion conditions (Bastioli, 1998). In destructurization of starch, the semicrystalline starch granules are converted into a homogeneous amorphous polymer matrix. The mechanical properties of destructurized starch depend upon the degree of destructurization that is attained. As destructurization increases, tensile strength and elongation also increases but the elastic modulus is reduced. This means that the material becomes increasingly flexible.

The properties of thermoplastic starch can be improved by adding plasticizers (e.g., water, urea, ammonia, diethylene glycol, citric acid, etc.), lubricants (e.g., lipids, fatty acid, talc, silicon, etc.), and fillers (e.g., proteins, water-soluble polysaccarides, and water-soluble polymers). Thermoplastic starch products with different viscosity, water absorption properties, and water solubility have been prepared by altering the moisture content, the amylase/amylopectin ratio of the raw material, temperature and pressure in the extruder (Mercier and Feillet, 1975; Donovan, 1979). Thermoplastic starch is sensitive to humidity and is, therefore, unsuitable for most food packaging applications. The thermoplastic starch alone is mainly used in soluble compostable foams such as loose fillers, expanded trays, expanded layers, and shape molded parts as a replacement for polystyrene.

A rational method was developed by Damian et al. (2020) to harness a triglyceride-based by-product, which contains chicken fat traces. They used methacrylated linseed oil as photo-reactive monomer, which can catch grease molecules. It resulted in a polymeric network (PFrec), which was embedded in starch/poly(vinyl alcohol) (St/PVA)-based composites, either with or without glycerol with enhanced properties. It was reported that the flexibility was improved on association of polymeric network with glycerol. Hydrophobic PFrec was found to increase the water resistance almost 40%.

The porous thermoplastic starch/cellulose nanofiber composites were prepared by Nasri-Nasrabadi et al. (2014). The diameter of 70% nanofibers was found to be in range of 40–90 nm. It was also observed that they exhibited good mechanical properties for cartilage tissue engineering applications, when total porogen contents was 70 wt.%. It was also observed that water uptake ratio of as-prepared nanocomposites was increasing significantly on adding 10% cellulose nanofibers. It was partially destroyed after more than 20 weeks.

The effect of the addition of carboxymethylcellulose and methylcellulose was investigated by Kibar and Us (2013) on the mechanical, thermal, and water adsorption properties of corn starch-based films, which were plasticized with polyethylene glycol (PEG) or glycerol. It was observed that these films were found more resistant to break, and have higher TS values on increasing methylcellulose and carboxymethylcellulose proportion. It was also concluded that properties of by film formation by starch can be increased on introducing carboxymethylcellulose and methylcellulose to the polymer matrix.

With their advanced properties, starch-based materials have aroused interest in scientists (Table 12.1).

12.6 GREEN COMPOSITES

A composite is a material made of two different phases that are a matrix phase and a disperse phase. The disperse phase may consist of synthetic material (e.g., fibers) or natural materials (e.g., natural fibers). The matrix phase can be a synthetic or a natural polymer; the matrix can also be classified as thermoplastic or thermoset. Thermoplastics are recyclable and biodegradable, whereas thermosets are biodegradable only and are not recyclable. The combination of a plastic matrix and reinforcing fibers gives rise to composites having the best properties of each component, since plastics are soft, flexible, and light weight as compared to fibers, their combination provides a high strength-to-weight ratio for the resulting composite for various applications such as packaging, product casing, housing, automotive panels, furniture, etc.

12.6.1 NATURAL FIBER BIOCOMPOSITES

Natural composites or green composites are emerging as a viable alternative to glass fiber reinforced composites, especially in automotive and building product applications. It can also be effectively used as a material for structural, medical, and electronic applications.

Natural composites reinforced with different natural fibers are as follows:

i) Cellulose Fiber-Based Biocomposites

Cellulose materials are used in the polymer industry for a wide range of applications, including laminates, panel products, fillers, alloys, blends, composites, cellulose derivatives (Maldas and Kokta, 1993). The mechanical properties of cellulose-polymer composites can be improved by using graft copolymers of the matrix material and by the addition of a polar group (Felix and Gatenholm, 1991). Cellulose fibers can also reinforce thermoset polymers

TABLE 12.1 Starch-Based Green Materials with Different Properties.

Starch-based green composite	Properties	References
Starch nanoparticles-polydimethylsiloxane	Superhydrophobic property	Wang et al. (2021)
Starch magnetite nanoparticles- poly(vinyl alcohol)	Optical and photoelectric properties	Ashraf et al. (2021)
Starch -β-cyclodextrin composite	Extraction of polycyclic aromatic hydrocarbons from water samples	Yazdanpanah and Nojavan (2021)
High-amylose corn starch-dihydromyricetin	Molecular interaction–particle characteristics–gel properties	Geng et al. (2021)
Starch/polyvinyl alcohol/glyceryl	Mechanical property	Yang et al. (2021)
Magnetic dialdehyde starch -aspartame	Ion exchange property	Heydari (2021)
Starch/ pectin-ZrO_2 nanoparticles- microencapsulated Zataria multiflora essential oil	Antimicrobial properties	Sani et al. (2021)
Thermoplastic starch-poly(3-hydroxybutyrate)	Excellent thermal stability	Haeldermans et al. (2021)
Thermoplastic starch-beta-tricalcium phosphate	Biomedical properties	Taherimehr et al. (2021)
Starch-albumin/MgO	antibacterial, antioxidant properties	Hosseini et al. (2021)
Starch/chitosan film incorporated with lemongrass essential oil	Antimicrobial activity	Perdana et al. (2021)
Avocado seed starch-microfibrillated cellulose	Excellent tensile properties	Fu and Netravali (2021)
Arrowroot starch-g-poly (acrylic acid-acrylamide)/zeolite hydrogel	Ion release property	Sjaifullah et al. (2020)
Starch-TiO_2	Superabsorbent and excellent thermal properties	Shirsath et al. (2020)
Starch/ TiO_2- polyvinylidene difluoride nanocomposite	Antibacterial activity	Ansarizadeh et al. (2020)
Arracacha starch –Chitosan	Excellent tensile and elongation at break and barrier properties	Garcia et al. (2020)
Starch and poly(butylene adipate-co-terephthalate)	Good mechanical properties, water resistance and antimicrobial activity against pathogens	Spiridon et al. (2020)
Starch-poly(vinyl alcohol) glycerol-citric acid	Antibacterial activity	Das et al. (2020)
Thermoplastic acetylated starch –poly(lactic acid)	Good biodegradability	Yu et al. (2020)

TABLE 12.1 *(Continued)*

Starch-based green composite	Properties	References
Starch- Fe_3O_4@SiO_2- carbon	Catalytic activity	Cai et al. (2020)
Hydrolyzed starch-chitosan loaded with ciprofloxacin hydrochloride	Antibiotic properties	Shehabeldine and Hasanin (2019)
Corn distarch phosphate-nanocrystalline cellulose films	Antimicrobial properties	Sun et al. (2019)
Starch-PVA- zinc-oxide	antimicrobial properties	Jayakumar et al. (2019)
Pea starch-ZnO	Antimicrobial properties	Wu et al. (2019)
Porous starch-carbon black-natural rubber	Resistance and hysteretic property	Du et al. (2019)
Starch-poly(lactide)	Mechanical properties and biodegradability	Rogovina et al. (2018)
Starch-poly(3-hydroxybutyrate)-poly(lactic-co-glycolide)	Antimicrobial properties	Mlalila et al. (2018)
Starch-silver nanoparticles	Antimicrobial properties	Ortega et al. (2017)
Thermoplastic starch-based composites	Thermal and thermo-mechanical properties	Espigulé et al. (2013)
Starch-g-poly(acrylic acid)-organo-zeolite	Swelling Properties	Zhang et al. (2016)
Potato starch	Antibacterial properties	Wang et al. (2016)
Carboxymethyl starch/montmorillonite	Pollution reduction property	Wilpiszewska et al. (2016)
Starch-PVA nanocomposites	Excellent mechanical properties	Guimarães Jr et al. (2015)
Starch-graphene in PVA	Thermomechanical properties	Jose et al. (2015)
Dialdehyde starch-reduced graphene oxide-polyaniline composite	Good electrochemical properties	Wu et al. (2014)
Starch-poly(caprolactone)	mechanical properties	Liao et al. (2014)
Thermoplastic starch -poly(butylene succinate)	Mechanical, thermal and barrier properties	Boonprasith et al. (2013)
Plasticized starch-lignosulfonated layered double hydroxide	Mechanical properties and oxygen barrier properties	Privas et al. (2013)
Corn starch –poly(propylene)	Good mechanical properties	Qiu et al. (2013)
Corn starch -poly(acrylic acid) Excellent	Mechanical properties	Dang et al. (2017)

like polyester, epoxy, amino, and phenolic resins (Flodin and Zadorecki, 1986). Short cellulosic fiber-reinforced elastomer composites have gained practical and economic interest in the rubber industry (Setua, 1986).

The cellulose-based composites are not fully biodegradable because of nonbiodegradable synthetic matrix components. The processing and properties of biodegradable composites of bacteria-produced polyesters (Biopol®) reinforced with wood cellulose have been reported by Gatenholm et al. (1992) although cellulose fibers improved the strength and stiffness of the (PHB), but the composites were very brittle. The effect on the tensile modulus by the incorporation of cellulose fibers into three different thermoplastics like polypropylene, polystyrene, and Poly-R-3-hydroxy butyrate (PHB) has also been investigated, which revealed that the tensile modulus increased for each composite with increasing fiber content. The stiffing effect of cellulose fiber in PHB was in the same order as in polystyrene. Curvelo et al. (2001) proposed that the properties of thermoplastic starch composites can be improved using reinforcing fibers and fillers particularly cellulose fibers.

Thermoplastic starch/cellulose fiber composites have been prepared by using fibers from different sources, such as flax and ramie fibers (Wollerdarfer and Bader, 1998), potato pulp fibers (Dufresne and Vignon, 1998), bleached leaf wood fibers (Averous et al., 2001), and wood pulp fibers (Carvalho et al., 2002). Most of these authors have shown an improvement of the mechanical properties of the composites that was attributed to the chemical compatibility between the two polysaccharides, that is, starch and vegetal fibers (Wallerdorfer and Bader, 1998; Curvelo et al., 2001; Averous and Boquillon, 2004). As a result, water resistance of the composites substantially increased (Dufresne et al., 2000; Ma et al., 2005) as a direct consequence of the addition of the less hydrophilic fibrous filler (Averous and Boquillon, 2004; Ma et al., 2005).

Courgneau et al. (2013) prepared low odor-emissive polylactide/cellulose fiber biocomposites, which can be used for car interior. They also investigated impact of the different stages of processing (compounding, drying cycles, injection molding) on the extent of degradation of polylactide as well as biocomposite properties.

Biocomposite films based on alginate and cellulose were prepared by Sirviö et al. (2014). They used unmodified birch pulp, nanofibrillated cellulose (NFC), microfibrillated cellulose (MFC), and birch pulp. The tensile strength of the film was increased from 70.02 to 97.97 MPa on addition of 15% of NFC. As-prepared biocomposite films have better grease barrier properties and also decrease water vapor permeability (WVP).

Heidarian et al. (2017) prepared a bio-based composite from cross-linked cellulose nanofibril (CNF)/starch/polyvinyl alcohol, which can be used for film packaging. The CNF was initially isolated from aspen wood sawdust (AWS) as reinforcing phase, with chemo-mechanical treatments. They also investigated morphological, mechanical, biodegradability, and barrier properties of these biocomposites.

A biocomposite material was obtained by Revin et al. (2021). It was reported that thickness of fiber of bacterial cellulose was in the range of 6090 nm which was formed by the *K. sucrofermentans* H-110 strain. This composite material can be used for removal of fluoride ions from water due to high sorption ability. A maximum adsorption capacity was determined to be 80.1 mg g^{-1}.

The recent progresses on cellulose-based composites are tabulated in Table 12.2.

ii) Jute-Based Biocomposites

There are many reports about the use of jute as reinforcing fibers, a thermoplastics (Karmaker and Hinrichseh, 1991; Karmaker and Youngquist, 1996; Rana et al., 1998), and thermoset (Bledzki and Gassan, 1999; Sahoo et al., 1999). Jute fibers are having high tensile modulus and low elongation at break. The specific modulus of jute is superior to glass fibers and on a modulus per cost basis, jute is far superior. The specific strength per unit cost of jute is almost equal to that of glass fiber.

Jute reinforced thermoplastic, thermosets, and rubber-based composites have been reviewed by Mohanty and Misra (1995). It is essential to pretreat jute so that the moisture absorption would be reduced and wettability of the matrix polymer would be improved. Mitra et al. (1998) have reported jute reinforced composites, their limitations, and some solutions through chemical modifications of fibers. The effect of different additives on performance of biodegradable jute fabric-Biopol® composites has been reported by Khan et al. (1999). To study effects of additives, the jute fabrics were soaked with several additives solution of different concentrations, dicumyl peroxide was used as the initiator during the treatments. Mohanty et al. (2000) studied the surface modifications of jute, involving dewaxing, alkali treatment, cyanoethylation and grafting are made with the aim of improving the hydrophobicity of the fiber to obtain good fiber-matrix adhesion in the resulting composites.

The superior strength of alkali-treated jute may be attributed to the fact that alkali treatment improves the adhesive characteristics of jute surface

TABLE 12.2 Cellulose-Based Composites and Their Properties.

Cellulose-based composite	Properties	References
Kenaf cellulosic fiber biocomposites	Excellent mechanical performance	Asyraf et al. (2021)
Microcrystalline cellulose-polylactic acid-polypropylene biocomposites	Morphological, mechanical, thermal and rheological properties	Bhasney et al. (2020)
Polylactide-cellulose fibers	Tensile strength, impact strength and hardness	Fabijanski (2020)
Biocomposites based on oriented holocellulose	Tensile strength, optical transmittance, mechanical properties	Yang et al. (2019)
Cellulose fibers- hydroxyethyl cellulose	Mechanical and dynamic thermomechanical properties	Sirviö et al. (2018)
Cellulose Microfiber-Hibiscus Sabdariffa (HS) fibers/polyester biocomposites	Tensile retention properties	Aseer and Sankara-narayanasamy (2017)
Poly(vinyl alcohol)-starch-cellulose	Mechanical, morphological, biodegradability, and barrier properties	Heidarian et al. (2017)
Polylactide-cellulose nanofiber biocomposites	Compatibility and rheological properties	Safdari et al. (2017)
Cellulose fiber-reinforced PLA composites	Tensile and flexural and bio-compatibility properties	Graupner et al. (2017)
New bacterial cellulose/chitosan nanocomposite	Physicochemical and antibacterial properties	Savitskaya et al. (2017)
Cellulose nanofibers-unsaturated polyester matrix	Good mechanical properties	Ansari et al. (2015)
All-cellulose composites	Excellent Flexural strength, puncture impact strength and impact strength	Huber et al. (2013)

removing natural and artificial impurities, thereby producing a rough surface topography (Bisanda et al., 1991). In addition, alkali treatment leads to fiber fibrillation, that is, breaking down of fabrics fiber bundle into smaller fibers. This increases the effective surface area available for the contact with the matrix polymer.

Ma et al. (2018) fabricated four types of specimens of composites (i.e., untreated jute fabric/epoxy, untreated basalt fiber/epoxy, alkali-treated jute, and silane-treated jute fabric/epoxy). Then these were kept for different ageing conditions for almost 3 months. Weight gains of these three plant-based natural jute/epoxy specimens ranged from 5.0 to 8.5% as compared to mineral-based natural basalt/epoxy composite, which was lower; from 1.1 to 2.2%). The water absorption was reduced and tensile strength was enhanced of the resulting jute fabric/epoxy composites on alkali during silane treatments of jute fiber. Ejaz et al. (2020) used flax and jute natural fibers individually and also as hybrid reinforcement into polylactic acid (PLA) matrix. It was reported that composites may be suitable as biodegradable products in packaging and automobile industries.

It was observed by Sanvezzo and Branciforti (2021) that the presence of industrial waste can improve the elastic modulus of the PP matrix and polypropylene. It also decreased deformation at break from 435 to 5%. An antibacterial/antioxidant film was prepared by Hossain et al. (2020) based on modified starch and albumin, where magnesium oxide nanoparticles (05 w/w%) were used. It was observed that thickness and antioxidant activity of films increased on addition of magnesium oxide nanoparticles, while moisture content, water solubility, and water vapor permeability decreased. Table 12.3 summarizes recent progress of jute-based green composites and their properties.

iii) Flax and Hemp-Based Biocomposites

Flax and hemp are also used as reinforcement in composites. It was suggested that flax and sisal-based composites are used for making vehicle interior parts (Haager, 1995). The reinforcement of polyisocyanate-bonded particle boards with flax fibers led to products comparable to those of carbon and glass fiber reinforced particle boards (Barbu and Fritz, 1996). Biocomposites containing natural fibers and biodegradable matrices are patented for applications as building materials (Herrman et al., 1994). These materials contain natural fibers, for example, flax, hemp, ramie, sisal or jute, and biodegradable matrix such as cellulose diacetate, or a starch derivative.

TABLE 12.3 Recent Progress of Jute-Based Composites with Their Properties.

Jute composite	Properties	References
Jute fiber composites	Excellent compatibility	Hasan et al. (2021)
Jute fiber-graphene-composite	Biodegradability, recyclability, tensile, and interfacial properties	Karim et al. (2021)
Jute-carbon fabric reinforced epoxy hybrid composite	Tensile property	Khalid et al. (2020)
Jute-epoxy-glass composite	Water absorption property	Borah and Samanta (2020)
Jute-PP-LLDPE	Physico-mechanical properties	Rahaman et al. (2019)
Jute fiber-glass hybrid	Tensile and flexural mechanical strength	Arasu et al. (2019)
Jute fiber-banana-reinforced epoxy-based hybrid	Thermal properties	Devireddy and Biswas (2018)
Jute-cellulose-nanofibrils-hydroxypropylmethylcellulose nanocomposite	Excellent tensile properties	Orasugh et al. (2018)
Jute-hemp-flax-epoxy hybrid composite	Highest tensile strength and maximum flexural strength	Chaudhary et al. (2018)
Jute fiber-poly(lactic acid) composites	Mechanical and flammability properties	Yu et al. (2017)
Jute-epoxy laminar composites	Strength and fracture toughness properties	Pinto et al. (2016)

12.6.2 NATURAL POLYMER BIOCOMPOSITES

i) Thermoplastic and thermoset biocomposites

Thermoplastic composites are composites that make use of a thermoplastic polymer as a matrix. The properties of these polymers are toughness, chemically inert, and recyclability. The advantage of thermoplastic is that they can be rapidly heated and cooled without any damaging effects on their microstructure. In thermoplastics, most of the work reported deals with polymers such as polyethylene, polypropylene, polystyrene, and polyvinyl chloride. The natural fibers used to reinforce thermoplastics include wood, cotton, flax, hemp, jute, sisal, banana, pineapple, and sugarcane fibers (Yao et al., 2008). Thermoset polymers are also used as a matrix material for most of the structural composite materials. The single biggest advantage of thermoset polymers is that they have a very low viscosity and therefore they can be introduced into fibers at low pressures. These composite materials are chemically cured to a highly cross-linked 3D network structure and are highly solvent resistant, tough, and creep resistant. The common thermosetting materials are epoxy resins and unsaturated polyesters, phenolic resins, amino resins, and polyurethane.

a) Thermoplastic starch-based composites

Carvalho et al. (2001) first reported the use of thermoplastic starch for the production of composites by melt intercalation in a twin screw extruder. The composites were prepared with regular corn starch plasticized with glycerin and reinforced with hydrated kaolin. Biotech of Germany has conducted R and D along the lines of starch-based thermoplastic materials. The company's three product lines are Bioplast granules, Bioflex film, and Biopur foamed starch.

Novamont of Italy produces four classes of biodegradable materials Z, Y, V, and A under the Mater-Bi trademark. All four classes of materials containing starch and differing in synthetic grades have been developed to meet the requirements of specific applications. The physicol–mechanical properties of Mater-Bi are similar to those of conventional plastics like plates, cutlery, cup lids, etc. packaging like wrapping films, film for dry food packaging, board lamination, etc.; stationary like pens, cartridges, pencil, etc., personal care and hygiene and others like toys, bags, etc. Different starch plastics with different trade names are now available in the market. In comparison to thermoplastic biodegradable products based on starch still had many disadvantages such as hydrophilic character of starch polymers.

Despite many advantages, thermoplastic starch-based materials are still at an early stage of development and the markets for such products are expected to grow in the future as the properties are more improved, prices still declined, and an infrastructure for composting becomes more established.

b) Polylactic acid (PLA)-based composites

PLA-based materials are a new class of materials of interest developed in recent years due to the continuously increasing environmental awareness throughout the world. They can be considered the 'green' evolution of the more traditional eco-composites, essentially consisting of synthetic polymers-based composites reinforced with natural fibers or other micro or nanofiller.

In the past years, different natural fibers have been employed to modify the properties of PLA. Up to now, the most studied natural fiber reinforcements for PLA were kenaf (Huda et al., 2008b; Avella et al., 2008), flax (Bax and Mussig, 2008), hemp (Hu and Lim, 2007), bamboo (Tokoro et al., 2008), jute (Shikamoto et al., 2007), and wood fibers (Huda et al., 2006). Huda et al. (2008a, 2008b) worked on kenaf fiber reinforced PLA laminated composites prepared by compression molding using the film-stacking method. It was found by them that standard PLA resins are suitable for the manufacture of kenaf fiber reinforced laminated biocomposites with useful engineering properties. The mechanical properties of PLA composites reinforced with cordenka rayon fibers and flax fibers (examples completely biodegradable composites) were tested and compared. The promising impact properties of the presented PLA/cordenka composites show their potential as an alternative to traditional composites (Bax and Mussig, 2008).

The effects of the alkali-treated natural fibers on the mechanical properties of PLA/hemp fibers were studied by Hu and Lim (2007). The results show that the composite with 40% volume fraction of alkali-treated fiber possessed the best mechanical properties. Bamboo fiber reinforced PLA composites were prepared to improve the impact strength and heat resistance of PLA (Tokoro et al., 2008). The suitability of wood fibers as natural reinforcement in PLA-based composites has been also demonstrated in comparison with wood fiber reinforced polypropylene composites (Huda et al., 2006). Silkworm silk fibers are recognized as reinforcements in PLA natural fiber composites for tissue engineering application due to the fact that the silk fiber surface bonds well with the polymer matrix. These fibers can be good material, as reinforcements for the development of polymeric scaffolds for tissue engineering applications (Cheung et al., 2008). The nanocomposites based on PLA are of special interest for medical purposes. Special attention has been paid to novel nanomaterials capable of facilitating the biorecognition of anticancer drugs. These novel nanocomposites imply some potential valuable application as a kind of drug carriers in view of the respective good biocompatibility of PLA and large surface area of the nanoparticles (Song et al., 2008).

c) Cellulose-Based Biocomposites

Cellulose from agricultural products has been identified as a source of biopolymer that can replace synthetic polymer. Cellulose acetate is considered potentially useful polymers in biodegradable applications (Rivard et al., 1992; Buchanan et al., 1993). Green nanocomposites have been successfully produced from cellulose acetate, triethyl citrate plasticizer, and originally modified clay via melt compounding (Misra et al., 2004).

A cellulose nanocomposite material has been investigated as a flexible humidity and temperature sensor (Mahadeva et al., 2011). Cellulose was obtained from cotton pulp via acid hydrolysis using a solution of lithium chloride and N,N-dimethylacetamide. An active antimicrobial packaging material has been developed using methyl cellulose as the base material with montomorillionile as reinforcement (Tune and Duman, 2011). Researchers have evaluated the use of cellulose-based nanocomposite with hydroxylapatite for medical applications (Zimmermann et al., 2011; Zadegan et al., 2011).

d) Miscellaneous Biocomposites

Soy-based polyurethane can be used as a matrix for the production of bio-based nanocomposites (Tate et al., 2010). Relatively water-resistant biodegradable soy protein composite resulted through blending of special bioabsorbable polyphosphate fillers, biodegradable soy protein isolate, plasticizer, and adhesion promoter in a high shear mixer followed by compression molding (Otaigbe, 1998). The degradable composite films composed of soy protein isolate and fatty acids as well as soy protein isolate and propylene glycolalginate have been prepared by Rhim et al. (1999). Luo and Netravali (1999) have reported the mechanical and thermal properties of bio/green composites obtained from pineapple leaf fibers (28% fiber content) and Biopol®, that is, PHBV resin.

The blending of two or more polymers to achieve a polymer that is biodegradable has drawn such research interest. These polymers have been tested for their degradability and mechanical properties and thus, recommended for use as degradable polymers for composite applications. Such polymer blend starch/PLA blends, polybutylenes succinate/cellulose acetate blends, starch/modified polyester blends, polycarprolactone/polyvinyl alcohol blends, and thermoplastic starch/polyesteramide blends have been reported (Ke and Sun, 2001; Uesaka et al., 2000; Kesel et al., 1997; Averous et al., 2000; Willett and Shogren, 2002; Martin and Averous, 2001). The use of binary and ternary blends of PLA, polycaprolactone, and thermoplastic starch (TPS) as composites has been reported by Sarazin et al. (2008).

12.7 PROPERTIES OF BIOCOMPOSITES

The properties of the composite are determined by:

i) Properties of fiber,
ii) Properties of resin,
iii) Geometry and orientation of the fibers in the composite, and
iv) Surface interaction of fiber and resin.

i) Properties of fiber

In fiber-reinforced composites, the strength of the composite is determined by the strength of the fiber and by the ability of the matrix to transmit stress to the fiber plant. Fibers consist of microfibrils, which are interconnected via lignin and hemicellulose fragments. The more parallel, microfibrils are arranged to the fiber axis, the higher will be the fiber strength. The microfibril angle is one of the major factors in determining the mechanical properties of the fiber. In the spiral structure, the microfibril angle is the angle between the cellulose microfibrils and the longitudinal cell axis. The tensile strength and the Young's modulus decreased with the increase in the microfibril angle. Different fibers used in composites have different ways. The mechanical properties of natural fibers such as jute, hemp, flax, and sisal are very good and may compete with glass fiber in specific strength and modulus. Natural fibers show higher elongation to break than glass or carbon fibers, which may enhance composite performance. Thermal conductivity of natural fibers is low and as a consequence, they make a good thermal barrier. The stiffest and strongest composites are based on unidirectional aligned natural fiber bundles impregnated with synthetic or natural resins.

To maximize the composite strength and stiffness and also to transfer stress effectively between natural fibers and resin matrix, the following factors should be considered:

a) Fibers should contain high cellulose content.
b) The winding angle of cellulose in plant cell with respect to fiber axis should be small.
c) Fibers should be disposed in a direction parallel to an applied uniaxial stress but should be cross laminated, if the stresses are bidirectional.
d) The plant fiber should be surface treated to a good interfacial bond between fiber and matrix.

The resulting properties of the composites are influenced by the manufacturing stages of plant fiber composites because the exact nature of the interaction between plant fiber and polymer matrix is complex.

ii) Properties of resin

a) Resin should be a thermosetting polymer to avoid plastic deformation at low stresses.
b) Most natural resins are hydrophilic and are susceptible to changes in strength, chemistry, or dimensions with water, so a composite will have an application period during which no such change occurs, followed by onset of degradation after the required lifetime.
c) Composites with the same polymer in both fiber and matrix offer strong interfacial adhesion due to transcrytallinity in the matrix near fibers and also due to melting or dissolution of some of the fiber in the molten matrix near the fiber surface and as a consequence, the interfacial strength is increased.
d) Resin should have good mechanical, adhesive, and toughness properties.
e) Resin should have good resistance to environmental degradation.
f) Thermally stable, nonyielding composites require thermosetting rather than thermoplastic matrix.

iii) Geometry and orientation of the fibers in the composite

a) Orientation of the fibers is of advantage in providing maximum properties in the direction of orientation. Woven fibers are expected to provide the best strength properties, but they are expensive than nonwoven.

   Jordan et al. (2003) stated that a simple weave will have maximum mechanical properties in the two perpendicular directions of the fiber while in the diagonal and other directions, these properties will decrease. The weaving of the fibers provides an interlocking that increases the strength better than can be achieved by fiber matrix adhesion.

b) Fiber diameter is an important factor. More expensive smaller diameter fibers provide higher fiber surface areas, spreading the fiber/matrix interfacial loads.
c) The geometry of the fibers in a composite is also important since fibers have their highest mechanical properties along their length, rather than across their widths. This leads to the highly anisotropic properties of composites. This means that it is very important, when considering the use of composites to understand at the design stage; both the magnitude and the direction of the applied loads.

iv) Surface interaction of fiber and resin

The properties of composites mainly depend on the interface. The ultimate mechanical properties of fiber-reinforced polymeric composites depend not only on the properties of the fibers and the matrix but also on the degree of interfacial adhesion between the fiber and the polymer matrix (Hong et al., 2008).

The following factors affect strength, toughness, and stiffness of the natural fiber composites–

a) The surface energies of the fiber and matrix phases
b) The nature of fiber chemical pretreatment, for example, alkalization, acetylation, and silane treatment.
c) The addition of adhesion promoters to the resin, for example, maleic anhydride modified polypropylene.
d) The degree of shrinkage of the matrix onto the fibers during manufacture of composites.

12.8 APPLICATIONS OF GREEN COMPOSITES

Eco-friendly biocomposites from natural fibers and bioplastic are novel engineering materials of the twenty-first century and would be of great importance to the material world, not only as a solution to growing environmental threat but also as a solution to the uncertainty of petroleum supply. The use of materials from renewing resources is attaining increased importance, and the world's leading industries and manufacturers are therefore interested in composites derived from natural fibers and polymers.

Green composites have several applications ranging from leisure goods to construction. Some of the areas in which the green composites find applications are as follows.

i) Automobiles

Composites made from natural fibers are attractive because these are light in weight and have good mechanical properties comparable to widely used glass fiber reinforced plastics. In Europe, a large and still expanding market for natural fiber reinforced composites plastics is in automotive applications. It was estimated that more than 20,000 tones of natural fiber, mainly flax and hemp, were used in automotive components (Karus et al., 2000). Green composite materials were first used extensively in production and its construction was the Eastern European Trabant. The body of this car was composed of panels of a natural fiber reinforced plastic composite called Doroplast, which were screwed to the galvanized steel frame of the substructure.

Natural fibers have been used for thermal and acoustic insulation, that is, low-performance applications in interior automotive situations. The introduction of natural fiber composites for door panels in the Mercedes-Benz E-Class provided a step toward higher performance applications. In this particular application, a flax/sisal mat was used as reinforcement in an epoxy matrix. A weight reduction of approximately 20% was claimed over the existing wood fiber material (Schuh, 1999). Wood plastic composites were also used for construction applications and natural fiber-based synthetic polymers for automotive applications. Another example of application of natural composites in automotives is a resin made out of soy bean oil on reinforcement with glass fiber to be used in parts of newest tractors produced by John Deere.

ii) Aircrafts, ships, and trains

The green composites are used in aircrafts and ships because of their light weights and also biodegradable nature. It is known that the fuel consumption will come down certainly, if the weight of the vehicle is reduced. These types of green composites are also used in trains for the above reason. One of the first, true synthetic composites, potentially capable of being used in structural applications was 'Gordon-Aerolite' (McMullen, 1984). This was a composite consisting of unidirectionally aligned unbleached flax thread impregnated with phenolic resin. Since biocomposites are organic materials, they are combustible. So, one of the most important requirements for biocomposites is its use for paneling in railways or aircraft as it has a certain degree of flame resistance. The new aspects of eco-friendly polymer flame retardant systems have been reported (Zaikov and Lomakin, 1997).

Balakrishnan et al. (2016) reported that fiber-reinforced polymer composite materials are potential materials for construction of aircrafts and spacecrafts. These days, fiber polymer composites alongside aluminum alloys are commonly used materials in aircraft structures. The main advantages of using these composites are light weight, increased specific stiffness and strength, eco-friendliness, extended fatigue life, nonhazardous nature, and almost free from corrosion problems.

The use of lightweight materials has become more common these days as transport vehicle manufacturers are looking for reducing weight of vehicle to improve its performance, reduced oil and fuel consumption, and also emissions or use of fuel cells. Hence, a newer generation of polymer composites is quite commonly used in the transport industry. Fan and Njuguna (2016) reviewed composite materials with light weight and their possible use in transport industry, mainly elastomers, thermosets, thermoplastic, and core materials.

iii) Packaging

Green composites have barrier properties, chemical resistance, and surface appearance and these properties make it an excellent material for packaging applications such as in beer and carbonated drinks bottles and paper board for fruit and dairy. The pallets, which have traditionally been made from low-grade timber, are now being manufactured from wood fiber-plastic composites to fulfill the demands of hygiene, safety, and longevity. Crates and boxes are other examples of applications for wood fiber-plastic composite in packaging.

As awareness is regularly increasing about increased pollution, growing demand for biodegradable materials, need for material with CO_2 neutrality, lower emissions of greenhouse gases, regularly updated new environmental laws and regulations and these have forced manufacturers to search for novel environmental friendly materials. Soy protein isolate (SPI) is a protein, which has good biocompatibility, reproducible resource, biodegradability, and processability and as a result, it has a potential for use in the food industry, bioscience, agriculture, and biotechnology. Koshy et al. (2015) reviewed soy protein isolate as an organic and inorganic fillers in the macro- micro-, and nano-scale and discussed some applications of these materials in the field of food preservation and packaging technology.

Petroleum-based thermoplastics are used in packaging, but their use has resulted in alarming pollutant emissions. Thus, efforts were made to search for environmentally friendly alternative packaging materials, which can be recycled as well as these are biodegradable. Natural fibers have excellent mechanical properties and hence, these are used to strengthen biopolymers to produce some biodegradable composites. Recyclability impact resistance, deflection temperature, and biodegradability are some important functional properties required for packaging materials.

Fazita et al. (2016) discussed several such functional properties for packaging applications. The forestry and agricultural market have seen an outstanding growth because of benefits of green composites as these are cost effective, excellent mechanical properties, and environment friendly in nature. Hermawan et al. (2019) reported the characterization of microfiber handsheet-molded thermoplastic green composites, which were developed by a combination of oil palm empty fruit bunch (OPEFB)-based microfiber pulp as filler. The polyester polypropylene was used as a matrix. Then they mixed refined alkaline extracted OPEFB pulp fiber at different layered compositions of the composite of grafted polypropylene, which showed significant improvement in physical properties and mechanical properties of the handsheet-molded composite. This composite exhibited an improved thermal stability as compared to neat composite.

Bugatti et al. (2019) reported recovery and upgrading tomato processing residues. They prepared some new green composites, which are based on natural halloysite nanotubes (HNTs) and tomato peels (TPs), which were loaded with carvacrol (an antibacterial agent found in nature). No chemical modifications of HNTs are required for loading of carvacrol into HNTs. As-obtained composites released antimicrobial agent for a longer time, which suggest that they can be considered promising materials for packaging of food.

Kamble and Behera (2021) reported some composites (thermoset) stiffened with polyester and cotton fibers, which were recovered from textile waste. They used carded web of cotton, polyester fibers, epoxy resin, and as-developed composites by the compression molding technique. As-prepared epoxy/polyester composites exhibited average tensile and impact strength relatively higher than cotton/epoxy composites. It was revealed that bearing strength in a pinned joint for polyester/epoxy composites was higher than cotton/epoxy composites but low flexural strength higher than epoxy/ polyester composites.

iv) Construction and building products

Green composites may find use in applications, where timber, synthetic polymers, or synthetic composites are currently used. In North America, wood fiber-plastic composites frequently occupy the position of timber products specially decking. Wood fiber-plastic composites have several advantages over wood. These include easy processing and treatment, no splintering, good appearance, improved resistance to biodegradation, no termite attack, low maintenance, etc. Good dimensional stability is a further attractive feature of wood fiber-plastic composites. Other building products such as railing, fencing, window, floor profiles, siding, and shingles are manufactured from this material.

Hoiby and Netravali (2015) discussed a virtual model for the construction of a green cabin with two types of composites: (i) Use of soy protein-based resin with natural plant-based fibers with mechanical properties almost comparable to wood and wood products, and (ii) use of liquid crystalline cellulose fibers with soy protein-based resin having properties comparable to high strength steel. Moderate strength green composites may be used for molded walls while load-bearing framework of cabin can have advanced green composites.

Hassan et al. (2020) investigated mechanical, acoustic, and thermal properties of natural fiber waste strengthened green epoxy composites. They used three different types of fiber wastes such as coconut, cotton, and sugarcane with epoxy resin. The sound absorption coefficient was found to increase on increasing the fiber content. The coconut fiber-based composites exhibited a more sound absorption coefficient than others. Cotton fiber-based composite also exhibited higher impact and flexural strength in comparison to other samples. A higher coefficient of thermal expansion was there in case of cotton fiber-based composite as compared to other composites. These composites (natural fiber-based) can be used in household furniture and building interiors.

Concrete and steel have high strength and stress and hence, these are used as traditional building materials. These are commonly used in structures of bridges, civil construction, etc. Huyen (2021) fabricated green composites, which are based on the high performance of epoxy combined with bamboo to have a high flexural strength. Some cellulosic impurities in bamboo were removed by treatment with: (i) Mixture of NaOH and Na_2SO_3 and (ii) H_2O_2 solution. Structure and composition of bamboo were changed after chemical treatment. An outstanding growth in tensile and flexural strength of about 300–500% was observed as compared to the original bamboo raw material and it proved that green composite materials and better. Therefore, there is a potential of using chemically treated bamboo-reinforced composites in replacing steel in green building materials slowly.

v) Mobile phones and computers

Green composites are used for mobile phones body. Kenaf and PLA composites are used in mobile phones part in Japan to reduce the amount of CO_2 emissions during fabrications. NTT Docomo is one of the models of mobile phones in Japan, where green composites are used for such purpose. Components such as the covers for mobile phones, casing for computers, and monitors could all be produced from biodegradable composite materials.

vi) Miscellaneous applications

Green composites are used for indoor structural applications in housing. The composite used for the interior decoration is banana fiber and its composites. The walls and flooring can be covered with the boards, which will be attractive and will decrease the cost of construction. Fiber-reinforced composites have unique properties in biomedical applications, such as being transparent to X-rays. Titanium implants vs. fiber-reinforced composite implants were also studied. The stress range of the fiber-reinforced composite implant was close to the stress level for optimal bone growth and the stress at the bone around the fiber reinforced implant was more even than that of the Ti implant.

The Nafion proton exchange membranes are used in vanadium redox flow batteries (VRBs), but they have poor ion selectivity and significantly high cost. Ye et al. (2019) used a green biopolymer (lignin) as an additive in the pristine sulfonated poly(ether ether ketone) (SPEEK) membrane instead of Nafion membrane. It increases significantly the proton conductivity as well as ion selectivity of the membrane. The VRB single cell based on lignin/SPEEK membrane exhibited 95.95 and 71.47% capacity retention even 100 and 300 cycles under 120 mA cm^{-2}, respectively. The low cost, excellent stability, and remarkable performance of this composite membrane make it a promising candidate in VRB applications as the next generation on a larger scale.

Qiao et al. (2021) introduced antimicrobial biomolecules (nisin) into rationally designed covalent organic frameworks (COFs), which resulted in a new type of "smart formulation. It was reported that it could inhibit microbial contamination apart from ensuring orderly progression of the fermentation process. As-prepared encapsulated biomolecules were found to retain their activity with improved stability and pH-responsive releasing process (Almost 100% bacteriostatic efficiency at pH 3). It was revealed that nisin@COF composites will not affect the fermentation strains and played a major role in sustainable biomanufacturing.

Nanocellulose also has a potential for use as sustainable and eco-friendly nanomaterial for green and renewable electronics. Lasrado et al. (2020) reviewed recent advances in integrating nanocellulose with some active materials to form a flexible film/aerogel/3D structures. As-prepared nanocellulose-based composites found varied applications in energy conversion devices (solar cells, lithium ion batteries, and piezoelectric materials). It was revealed that the piezoelectric performance of piezoelectric materials as well as power conversion efficiency of solar cells can be improved by reinforcing matrix with nanocellulose. The nanocellulose-based composites may also be used in binders, separators, supercapacitors, and CO2 separators.

12.9 REUSE, RECYCLING, AND DEGRADATION OF COMPOSITES

Recycling and reuse of materials is by no means a new concept, since over the last few decades and earlier, waste newspaper, papers, cardboards, and glass were recycled and reused. However, polymer-based products gained popularity toward the end of seventies due to significant increase in costs of materials because of the unexpected rise in crude oil price. In recent years, there had been an increase in the use of composite products, particularly in the construction and automobile industries. These consume nearly half of all composites manufactured and therefore, the issue of composite recycling and use is becoming very important. In case of advanced composite materials, with an increasing number of additives and reinforcement materials, the only solution to reducing waste is to crush the compound and to recycle. Recycle means to recover a product at the end of its useful life, break it down into its constituent components, and reincorporate it into new product that has an inherent value equal to the original product.

Recycling of plastics consists of four phases of activity –

(i) Collection, (ii) Separation, (iii) Processing, and (iv) Marketing

In recycling thermoplastics, a mixed waste stream occurs due to problem in collection and sorting, associated with visual similarity, and similar physical properties of commonly used polymers. Many polymers are not compatible and yield low properties, when blended and molded directly. To overcome these problems, chemical compatibilizers are required.

Processing methods of recycled composites were reported to have a large effect on the quality of final product. The recycling of thermoplastic and their composites are much easier compared to the recycling of thermosets, so the widespread use of thermoplastic composites in different industries over the last decades is expected to have a more favorable environmental impact.

Recycled thermoplastic composites show a degradation of their mechanical performance and the extent, which depends on the recycling process and on the service conditions history. Furthermore, the possibility of recycling offers sound economical benefits because of the high price of the virgin material (Papanicolaou et al., 2008). The recycling processes do not cause very significant changes in flexural strength and thermal stability of the composites, particularly polypropylene-based composites reinforced with kenaf fibers are less sensitive to reprocessing cycles with respect to rice hulls-reinforced polypropylene-based composites.

The properties of these composites remain unchanged after recycling processes and the recycled composites are suitable as construction materials for indoor applications (Srebrenkoska et al., 2008). Bourmaud and Baley (2007) have investigated the mechanical properties of sisal/polypropylene and hemp/polypropylene as a function of recycling and they found that both tensile modulus and strength were quite stable even after up to seven injection cycles, but the initial value was quite low due to the relatively poor mechanical properties of polypropylene.

Duigou et al. (2008) observed the recyclability of flax/ PLLA biocomposites elaborated with the injection molding process, and compared their behavior with that of polypropylene composites. It was found that repeated injection cycles had shown to influence many parameters such as reinforcement geometry, mechanical properties, molecular weight of PLLA, thermal behavior, and rheological behavior.

Although biocomposites became more brittle on recycling, they retain a large part of their properties, at least until the third injection cycle. In an industrial situation, 100% of the recycled biocomposite is not used but the recycled material is always mixed with virgin material. Thermoset matrix recycling is unfeasible because of the thoroughly cross-linked nature and the inability to be remolded. Nevertheless, technologies are now being developed that can reprocess scrap composites to recover some of the value in the material and then prevent disposing the solid in a landfill, which has been the only option earlier.

The four main processes in recycling of thermosetting composites are: (i) grinding, (ii) chemical degradation and fibers recovery, (iii) pyrolysis, and (iv) incineration.

Recycling of thermosetting composites by grinding enables reuse of glass fibers, $CaCO_3$, and polymeric matrix without separation of the components. The composite is shredded, granulated to small fiber, and used as filler in a new process of manufacture. These products have the same or even better mechanical properties as the virgin composite materials (Petterson and Nilsson, 1994; Inoh et al., 1994). In addition, grinded recyclate has a lower density and contributes to reduction in weight, when compared to conventional fibers.

i) **Grinding:** In this process, the composite material is shredded to a convenient size using shredders designed for high torque and low speed. Hammer mills are used to reduce the size of recyclate further.
ii) **Chemical degradation of composites:** It involves partial or selective degradation of polyester/styrene polymer network in the presence of water, ethanol, KOH, and various amides (Winter et al., 1995). This process is inferior, when compared with the quality of grinding recyclate. A neutralization step is also required, which generates large quantities of waste water and adds to the cost. Fluidized bed thermal processing technique has been developed to recover energy and fibers in a form suitable for recycling into high-value products. These techniques are suitable for contaminated and mixed scrap material from end of life applications, especially in the automotive industry.
iii) **Pyrolysis:** It is in a simple and well-controlled process that recovers a good part of glass fibers and $CaCO_3$, which can then be reused as filler and reinforcement. The process separates organics from inorganics, due to a significant difference between the temperatures of their thermal decompositions. Low-temperature pyrolysis ($< 200°C$) is applicable to thermoplastic composites and yields excellent recovery of glass fibers and inorganic fillers. High-temperature pyrolysis ($\sim 750°C$) leads to a considerable reduction in the strength of glass fibers, which prevents their reuse as a high-quality reinforcement.
iv) **Incineration:** Incineration of plastics and composites involves annihilation of the material, with inevitable air and land pollution, due to poor combustion and to gaseous/liquid/solid products of the process.

The advantages of an increased content of plastic waste for incineration are –

a) Higher temperature of incineration due to volatiles, which reduce leachability of the fly ash and
b) Shorter combustion zones and more intensely burning fire.

The disadvantages from increased plastic waste are –

a) Increased formation of NO_x with an increase in incineration temperatures, which automatically contradicts the above-listed advantages,
b) Increased CO emissions, and
c) Other excessive emissions such as Cl_2, dioxins, and furans.

12.10 BIODEGRADATION OF COMPOSITES

Biodegradable polymers have offered a possible solution to waste-disposal problems associated with traditional petroleum-derived plastics. Biodegradation is defined as a process, which takes place through the action of enzymes and/or chemical decomposition associated with living organisms like bacteria, fungi, etc. and their secretion products (Albertsson and Karlsson, 1994). Also abiotic reactions like photodegradation, oxidation, and hydrolysis may alter the polymer before, during, or instead of biodegradation because of environmental factors.

Almost all biosynthetic polymers are biodegradable within a reasonable time scale. There are four biodegradation environments for polymers and plastic products. These are (i) Soil, (ii) Aquatic, (iii) Landfill, and (iv) Compost. Each environment contains different microorganisms and has different conditions for degradation (Wypych, 2003). In soil, fungi are mostly responsible for the degradation of organic polymers. The aquatic environment contains two types of bacteria, on the surface and in the sediment. The bacterial concentration in water decreases with increasing depth. Microorganisms biodegrade organic materials by the use of their enzymes; however, microorganisms do not synthesize polymer-specific enzymes capable of degrading and consuming synthetic polymers of recent origin. Enzyme activity is inhibited by the hydrophobic nature of the plastic and high molecular weight. The amount of nonbiodegradable plastic waste can be greatly reduced by proper development of biodegradable polymers and composites for short-term products.

Biodegradation can take place by two mechanisms –

(i) Hydro-biodegradation and (ii) Oxo-biodegradation

Hydro-biodegradation is much more important in case of hydrophilic natural polymers such as cellulose, starch, and polyesters, whereas the oxo-biodegradation predominates in the case of other natural polymers such as rubber and lignin. Synthetic flexible polymers do not hydrolyze under normal environmental conditions but biodegrade readily in the presence of a variety of thermophilic microorganisms in the surface layers of the polymer after transition metal catalyzed thermal peroxidation. The surface of the polymer after biological attack is physically weak and readily disintegrates under mild pressure (Bonhomme et al., 2003).

Polymers containing mainly covalent bonds in its main chain show little or no susceptibility to enzyme-catalyzed degradation reactions, especially those with higher molecular weights. To overcome this problem, "weak-links" are inserted in the backbones of such polymers by the insertion of functional groups in the main chain, especially ester groups, which can be cleaved by chemical hydrolysis, and insertion of functional groups in, or on, the main chain that can undergo photochemical chain cleavage reactions, particularly of carbonyl groups. These can then be utilized and consumed by microorganisms through biodegradation processes.

Liu et al. (2017) developed an effective method for mild chemical recycling of CFRP with a Tg of ~210°C using $ZnCl_2$/ethanol catalyst system. The decomposed matrix polymer (DMP) was found to be in an oligomer form and it contained multifunctional reactive groups. When it was used as a reactive ingredient and added up to 15 wt.% in the fabrication of new epoxy materials, then it was observed that resulting cross-linked polymers still retained the high strength and modulus as compared to the neat polymer without DMP.

A composite consisting of isocyanate derived from palm oil and polyurethane and polyurethane/nanosilica based on trans-esterified castor oil was investigated by Das et al. (2017) for a period of 3 months under composting condition. They isolated and identified fungal and bacterial colonies from the surface of the degraded sample. The biodegradation was also confirmed through the appearance of cracks, corrosive structures, and surface pitting. Degradation of polyurethane/nanosilica composite was rapid, may be due to its lower crystallinity, surface hydrophobicity, and surface hydroxyl groups that helps in the attack of microorganisms.

A carbon fiber reinforced polymer (CFRP) is a structural composite, which is very widely used in automobiles, sporting goods, aerospace industry, etc. due to its corrosion resistances, high strength-to-weight ratio, and excellent fatigue. But it is difficult to recycle it because of permanent cross-linked thermosetting matrices. It is known that carbon fibers (CFs) can be reclaimed easily retaining their original high value and performance. It was revealed that degraded thermosetting matrices may be reused again.

Wang et al. (2019) discussed synthetic routes and recycling mechanisms of degradable thermosets, along with recovery and reuse of CFs and resins.

Date palm waste is used as a filler in a linear-low density polyethylene (LLDPE) matrix, which was recycled to obtain green composites by Alshabanat (2019). Two types of these LLDPE, based on basic additives (UV stabilizer and the slip and antiblock), were used. It was observed that the biodegradation of composites was accelerated in the presence of the bio-filler.

Green composites were prepared by Yorseng et al. (2019) using poly(3-hydroxybutyrate-co-3-hydroxyvalerate) (PHBV) and spent coffee bean powder (SCBP) as matrix and a filler, respectively. It was observed that

SCBP filler exhibited good thermal properties than matrix and composites. It was revealed that tensile properties of composites were comparable with other polymers such as LDPE, HDPE, etc. and recycling of films did not affect mechanical and thermal properties significantly. It was claimed that replacing the conventional synthetic nonbiodegradable polymers can be replaced by these green composites for packaging applications.

Muniyasamy et al. (2019) studied biodegradation and thermal-chemical behavior of green composites, which are based on flax fibers (untreated and treated) treated with alginic acid and polyhydroxybutyrate-co-valerate (PHBV) under conditions of composting. The biodegradability of PHBV and PHBV composites was evaluated by observing production of carbon dioxide from polymeric carbon in controlled aerobic composting conditions. It was reported that alginic acid-treated flax/PHBV composites exhibited increased rate of degradation as compared to untreated flax/PHBV composite and neat PHBV. It was revealed that biodegradation of PHBV composites were readily attacked by fungus while PHBV degradation was there by bacteria.

Poly(lactic acid) (PLA) is a biodegradable, brittle, and high-cost polymer. It can be used green packaging and applied over structural components. Macedo et al. (2019) reinforced PLA with natural cotton (10 wt.%) and thermoplastic starch (TPS; 3 wt.%) and obtained a biodegradable and lower cost composite. TPS was incorporated in the following three ways: (i) Blending, (ii) Coating, and (iii) Blending and coating. The adaptability of TPS in the improvement of matrix strengthens adhesion and an increase in the tensile behavior was observed without compromising in biodegradation. Such a formulation had thermal stability, higher glass-transition temperature, storage modulus, ductility, and wettability.

Scaffaro et al. (2019) reviewed degradation behavior of some biodegradable polymers and related composites. It was reported that polylactic acid (PLA), polybutylene adipate-co-terephthalate (PBAT), and polycaprolactone (PCL) are the commonly used biodegradable polymers, but these are susceptible to hydrolytic degradation in course of processing. Their environmental degradation by enzymes takes place within weeks, but it may take months or even to years in water.

Sun et al. (2019) prepared biodegradable copolymer-based composites, which were produced from straw fiber (SF) and hydrolyzed soybean protein isolate (HSPI)/urea (U)/formaldehyde (F) (HSPI/U/F) copolymer-based adhesive. It can find application for biocomposite flowerpots (BFP). As-prepared BFP showed good biodegradation, and the introduction of HSPI was found to increase the degradation rate of BFP, which was about 50% in 24 months. It was revealed that CO2 release accumulation of BFP could reach 24 g after a month of controlled composting. It was observed that number of fungi and bacteria on the surface was higher as compared to actinomyces, which indicated that the bacteria and fungi were selectively accumulated on the BFP composites surface to enhance its degradation. Nitrate ion could also be formed by nitrobacteria promoted by plant growth.

The biodegradation (aerobic) of modified poly(lactic acid) (PLA)-based biocomposites were evaluated by Kalita et al. (2019). under composting conditions. Different samples (Neat PLA (NPLA), PLA/Gum Arabic 5% (modified) (PLA/Gum), PLA/CNC SO4 (1%) (PLA/CNC), and PLA/chitosan 5% (modified) (PLA/Chitosan)) were used for biodegradation under controlled composting condition without any inoculums added to the compost. It utilized thermophilic microflora, which was already present in that compost. It was confirmed that modified PLA films are biodegradable.

The bamboo/kenaf fiber reinforced epoxy hybrid composites were prepared by Chee et al. (2019) with the hand lay-up technique, with loading of 40% fibers for all the composites. Three hybrid composites were prepared in mixing different ratios of bamboo fibers (B) and kenaf fibers (K) as: (i) 70:30, (ii) 50:50, and (iii) 30:70. The effects of soil burial on as-developed composites were evaluated for different time periods, that is, 3, 6, and 12 months. Thermal stability, oxidation stability, and dynamic mechanical properties were evaluated. It was indicated that biodegradability of these hybrid composites at higher kenaf loading was increased on increasing soil burial period. It was also revealed that the hybrid composite formulation (B:K:50:50) presented a balance between environmental effects on resistance, along with maintaining the biodegradability characteristics.

Xu et al. (2019) fabricated a homogeneous C/PLA composite film by dissolving PLA in N,N-dimethylformamide (DMF). This was then followed by the addition of cellulose and 1-butyl-3-methylimidazolium acetate, [bmim]Ac. It was revealed that this C/PLA composite film exhibited significantly improved biodegradability, biocompatibility, and tensile strength. The demand of carbon fiber-reinforced polymers is expected to reach 194 ktons by 2022 as these are able to conjugate lightweight and superior mechanical resistance. As a consequence, they found widespread applications in varied range of products from automotive, wind turbines, aerospace, and sporting goods.

Giorgini et al. (2020) reviewed some recent advances of recycling cumulative composite wastes, but there are still many issues and problems.

Pantaloni et al. (2020) developed fully green composites by reinforcing three bio-based and biodegradable matrices. These are poly(hydroxyalkanoate)

(PHA), poly(butylene-succinate) (PBS), and poly(lactide) (PLA) with nonwoven flax fiber preforms. These composites were then buried in garden compost and it was revealed that surface degradation dominated in the case of PHA and PBS biocomposites, while rapid of degradation fiber–matrix interface was observed in case of PLA biocomposites. It was interesting to note that all the composites retained tensile strengths to half of their initial values even after six months.

The CFRP composites can be categorized into two classes: (i) Thermoset composites and (ii) Thermoplastic composites. The recycling is different in case of industrial thermoset composites and it is still an existing challenge, may be due to a highly cross-linked thermosetting polymer matrix. A facile strategy was developed by Zhao et al. (2020) for recycling amine-cured CFRP composites, where the degradation ratio of CFRP composites could reach 98.82% after processing for 90 min at 160°C and it is accompanied by a small reduction in the tensile strength (6.5%).

Thermoplastic cassava starch (TPCS) has a great potential as a substitute for nonbiodegradable petroleum-based polymer. It is abundant, recyclable, sustainable, and biodegradable in nature. Jumaidin et al. (2020) incorporated cogon grass fiber (CGF) into TPCS. It was reported that the incorporation of CGF increased flexural and tensile properties of the TPCS composites, but impact strength and elongation were found to be reduced. It was concluded that CGF/TPCS biopolymer composites are promising materials as eco-friendly material that is biodegradable as well as renewable.

Zabihi et al. (2020) introduced microwave-assisted chemical oxidation as a low-cost, sustainable, and effective method, which can be used recycling of the glass fiber (GF) using GF reinforced epoxy polymer (GFRP) waste. It was observed that a mixture of hydrogen peroxide (green oxidizer) and tartaric acid (TA) could be used to decompose the epoxy matrix of a waste GFRP with high yield (90%). As-recycled GFs could retain about ~92.7% tensile strength, ~99.0% Young's modulus, and ~96.2% strain-to-failure retentions. Lo et al. (2018) reported a different approach for the degradation of blended epoxy resin/benzoxazine under mild oxidizing conditions. The thermosetting resin depolymerized catalytically through abstraction of hydride using a ruthenium catalyst.

Every material has its own characteristics property and therefore, composites are prepared, which can have different properties than its components or many have hybrid property of its components. Green composite is the requirement of the day and the nature is doing this job from time immemorial.

12.11 RECENT DEVELOPMENTS

Dolza et al. (2022) developed green composites from bio-based polymeric matrix (Bio PBSA) and introduced 30 wt.% short hemp fibers as a natural reinforcement. They added different additives derived from bio-based itaconic acid to Bio PBSA/HEMP composite, such as a dibutyl itaconate (DBI), a (PBSA-*g*-IA) copolymer of PBSA grafted with itaconic acid. A different copolymer of PBSA grafted with maleic anhydride (PBSA-*g*-MA) was also used. It was reported that incorporation of hemp fibers increases the stiffness of the base polymer and tensile modulus was also increased from 281 to 3482 MPa with 30% fiber.

Scaffaro et al. (2022) produced green composite by the addition of *Hedysarum coronarium* (HC) flour to a starch-based biodegradable polymer (Mater-Bi®, MB). This flour was prepared by grinding stems, leaves, and flowers together and sieving in a fraction from 75 to 300 μm. Four different formulations were obtained by compression molding (CM) and fused deposition modeling (FDM) with the addition of 5, 10, 15, and 20% of HC to MB. It was reported that 5 and 10% HC-filled composites were found to be easily printable. Tensile strengths and Young's modulus of composites increased from 18.6 to 33.4 MPa, when 20% HC was added.

Yorseng et al. (2022) fabricated synthetic epoxy and bioepoxy hybrid composites using: Bamboo (BB)/basalt: (BS)/carbon (CB). They used laminate stacking BSBBBS (basalt + bamboo + basalt) and CBBBCB (carbon + bamboo + carbon) hybrid composites. It was reported that hybrid composites excellently retained their mechanical properties even after the accelerated weathering test. It was also revealed that the bioepoxy/bamboo/basalt/carbon hybrid composites can be used as substitutes with synthetic epoxy composites in automobile industries and structural applications in all weather conditions.

A series of biocomposites were developed by Nandi and Das (2022) using nettle (*Girardinia diversifolia*) woven fabric and poly(lactic acid) fibers. It was reported that biocomposite prepared with equal mass fractions of nettle and poly(lactic) acid exhibited the highest mechanical properties: (i) Young's modulus (4.37 GPa), (ii) tensile strength (39.87 MPa), (iii) impact strength (35.95 kJ m^{-2}), and (iv) flexural strength (50.96 MPa). This biocomposite was light-weight and displayed an excellent thermal stability with almost no weight loss up to 309°C. It was revealed that very good biodegradability is there with 10.57% weight loss and loss of 37.42% strength even after 20 days of burial in soil.

Abukhadra et al. (2022) synthesized green nanocomposite of ZnO supported into polyaniline/bentonite hybrid structure (Zn@/PA/BE). Its band gap was found to be 1.86 eV. This catalyst was used for oxidizing levofloxacin (LV) in water under visible light. A significant improvement in photocatalytic activity of Zn@/PA/BE was observed. It was reported that there was a significant decrease in the TOC content in 45 min, which indicated complete degradation and mineralization of LV.

Zinc ferrite was synthesized by Welter et al. (2022) and supported on different masses of chitin biochar (ZnFO1/B1, ZnFO1/B3, and ZnFO3/B1). It was reported that 20, 85, 70, and 100% rhodamine B degradation was obtained on using $ZnFe_2O_4$, ZnFO1/B3, ZnFO1/B1, and ZnFO3/B1, respectively, in 1 h under visible light. It was reported that the ZnFO1/B1 composite was able to fully degrade rhodamine B dye and the catalyst also remained efficient even after eight cycles of photo-Fenton reactions.

Mousa et al. (2022) used date palm rachis (DPR) waste as a filler (30, 40, and 50 wt.%) to obtain a biodegradable composite with polylactic acid (PLA). DPR–PLA composites were prepared using a melt-mixing extruder at 180°C with variation in DPR composition, mixing time, plasticizer type, and its composition. They considered 30 wt.% DPR–PLA composite as the optimal composite having a lowest melt flow index (16 g 10 min^{-1}). It was also reported that the addition of 10 wt.% of triethyl citrate (TEC) elongation at break changes from 2.10 to 4.20% as compared to polybutylene adipate terephthalate (PBAT); however, the addition of 10 wt.% of PBAT resulted in a lower tensile strength (21.80 MPa) as compared to composite with 10 wt.% of TEC (33.20 MPa).

KEYWORDS

- **green composite**
- **biocomposite**
- **cellulose**
- **starch**
- **polylactic acid**

REFERENCES

Abe, K.; Ozaki, Y. Comparison of Useful Terrestrial and Aquatic Plant Species for Removal of Nitrogen and Phosphorus from Domestic Wastewater. *Soil Sci. Plant Nutr.* **1998,** *44*, 599–607.

Abukhadra, M. R.; Saad, I.; Khim, J. S.; Ajarem, J. S.; Allam, A. A. Enhanced Oxidation of Antibiotic Residuals (Levofloxacin) Using a Green Composite of ZnO@ Polyaniline/Bentonite (Zn@ PA/BE) as Multifunctional Photocatalyst Under Visible Light. *Int. J. Environ. Anal. Chem.* **2022**. DOI: 10.1080/03067319.2022.2032010.

Albertsson, A. C.; Karlsson, S. *Chemistry and Biochemistry of Polymer Biodegradation. Chemistry and Technology of Biodegradable Polymers*; Blackie: Glasgow, 1994; pp 7–17.

Alshabanat, M. Morphological, Thermal, and Biodegradation Properties of LLDPE/Treated Date Palm Waste Composite Buried in a Soil Environment. *J. Saudi Chem. Soc.* **2019,** *23*, 355–364.

Amnuay, W.; Katavut, P.; Supranee, K.; Pichan, S.; Claudio M. *Songklanakarin J. Sci. Technol.* **2011,** *33* (4), 461–167.

Ansari, F.; Skrifvars, M.; Berglund, L. Nanostructured Biocomposites Based on Unsaturated Polyester Resin and a Cellulose Nanofiber Network. *Compos. Sci. Technol.* **2015,** *117*, 298–306.

Ansarizadeh, M.; Haddadi, S. A.; Amini, M.; Hasany, M.; Ramazani S. A. A. Sustained Release of CIP from TiO2-PVDF/Starch Nanocomposite Mats with Potential Application in Wound Dressing. *J. Appl. Polymer Sci.* **2020,** *137* (30). DOI: 10.1002/app.48916.

Arasu, P. M.; Karthikayan, A.; Venkatachalam, R. Mechanical and Thermal Behavior of Hybrid Glass/Jute Fiber Reinforced Composites with Epoxy/Polyester Resin. *Polimery* **2019,** *64* (7–8), 504–508.

Aseer, J. R.; Sankaranarayanasamy, K. Effect of Fiber Content on Tensile Retention Properties of Cellulose Microfiber Reinforced Polymer Composites for Automobile Application. *IOP Conf. Ser.: Mater. Sci. Eng.* **2017,** *272*. DOI: 10.1088/1757–899X/272/1/012020.

Ashraf, I. M.; El-Zahhar, A. A.; Idris, A. M.; Ahmed, F. B. Sensitive, Stable, and Biocompatible Photodetector Based on Poly (Vinyl Alcohol)-Starch/Magnetite Nanocomposite. *Optik* **2021,** *242*. DOI: 10.1016/j.ijleo.2021.167247.

Asyraf, M. R. M.; Rafidah, M.; Azrina, A.; Razman, M. R. Dynamic Mechanical Behaviour of Kenaf Cellulosic Fibre Biocomposites: A Comprehensive Review on Chemical Treatments. *Cellulose* **2021,** *28*, 2675–2695.

Avella, M.; Bogoeva-Gaceva, G.; Buzarvska, A.; Errico, M. E.; Gentile, G.; Grozdanov, A. J. Poly (Lactic Acid)-Based Biocomposites Reinforced with Kenaf Fibers. *Appl. Polym. Sci.* **2008,** *108*, 3542–3551.

Avella, M.; Buzarovska, A.; Errico, M. E.; Gentile, G.; Grozdanov, A. Eco-Challenges of Bio-Based Polymer Composites. *Materials* **2009,** *2* (3), 911–925.

Avella, M.; Immirzi, B.; Malinconico,M.; Martuscelli, E.; Volpe, M. G. Reactive Blending Methodologies for Biopol. *Polym. Int.* **1996,** *39*, 191–204.

Averous, L.; Boquillon, N. Biocomposites Based on Plasticized Starch: Thermal and Mechanical Behaviours. *Carbohyd. Polym.* **2004,** *56*, 111–122.

Averous, L.; Fauconnier, N.; Moro, L. Blends of Thermoplastic Starch and Polyesteramide: Processing and Properties. *J. Appl. Polym. Sci.* **2000,** *76*, 1117–1128.

Averous, L.; Frigant, C.; Moro, L. *Polymers* **2001,** *42* (15), 6565–6572.

Balakrishnan, P.; John, M. J.; Pothen, L.; Sreekala, M. S. Thomas, S.; Natural Fibre and Polymer Matrix Composites and Their Applications in Aerospace Engineering. In *Advanced Composite Materials for Aerospace Engineering*; Woodhead Publishing: Sawston, 2016; 365–383.

Barbu, M.; Fritz, T. Flex Fibre-Reinforced Particleboard: In Symp.: Pakroky Vyrobe Pouziti Lepidiel Drevopriem **1995,** 12th; Chem. Abstr. **1996,** *124*, 292690s.

Bastioli, C. Properties and Applications of Mater-Bi Starch-Based Materials. *Polym. Degrad. Stab.* **1998,** *59*, 263–272.
Bax, B.; Mussig, J. Impact and Tensile Properties of PLA/Cordenka and PLA/flax Composites. *Compos. Sci. Technol.* **2008,** *68*, 1601–1607.
Behera, A. K.; Avancha, S.; Basak, R. K.; Sen, R.; Adhikari, B. Fabrication and Characterizations of Biodegradable Jute Reinforced Soy Based Green Composites. *Carbohydr. Polym.* **2012,** *88* (1), 329–335.
Bhasney, S. M.; Kumar, A.; Katiyar, V. Microcrystalline Cellulose, Polylactic Acid and Polypropylene Biocomposites and Its Morphological, Mechanical, Thermal and Rheological Properties. *Compos. B. Eng.* **2020,** *184*. DOI: 10.1016/j.compositesb.2019.107717.
Bisanda, E. T. N.; Ansell, M. P. The Effect of Silane Treatment on the Mechanical and Physical Properties of Sisal-Epoxy Composites. *Compos. Sci. Technol.* **1991,** *41*, 165–178.
Bledzki, A. K.; Gassan, J. Composites Reinforced with Cellulose Based Fibres. *Prog. Polym. Sci.* **1999,** *24* (2), 221–274.
Bolton, A. J. Natural fibers for Plastic Reinforcement. *Mater. Technol.* **1994,** *9*, 12–20.
Bonhomme, S.; Cuer, A.; Delort, A. M.; Lemaire, J.; Sancelme, M.; Scott, G. Environmental Biodegradation of Polyethylene. *Polym. Degrad. Stab.* **2003,** *81* (3), 441–52.
Boonprasith, P.; Wootthikanokkhan, J.; Nimitsiriwat, N. Mechanical, Thermal, and Barrier Properties of Nanocomposites Based on Poly (Butylene Succinate)/Thermoplastic Starch Blends Containing Different Types of Clay. *J. Appl. Polym. Sci.* **2013,** *130* (2), 1114–1123.
Borah, P.; Samanta, S. Wear Behaviour of Glass/Jute Hybrid Epoxy Composites with Addition of Fillers. Adv. Mater. Process. Technol. **2020.** DOI: 10.1080/2374068X.2020.1793270.
Bourmaud, A.; Baley, C. Investigations on the Recycling of Hemp and Sisal Fibre Reinforced Polypropylene Composites. *Polym. Degr. Stab.* **2007,** *92*, 1034–1045.
Buchanan, C. M.; Gardner, R. M.; Komarek, R. J. J. Aerobic Biodegradation of Cellulose Acetate. *Appl. Polym. Sci.* **1993,** *47*. DOI: 10.1002/app.1993.070471001.
Bugatti, V.; Brachi, P.; Viscusi, G.; Gorrasi, G. Valorization of Tomato Processing Residues Through the Production of Active Bio-Composites for Packaging Applications. *Front. Mater.* **2019,** *6*. DOI: 10.3389/fmats.2019.00034.
Cai, X.; Zhang, Y.; Hu, H.; Huang, Z.; Yin, Y.; Liang, X.; et al., Valorization of Manganese Residue to Prepare a Highly Stable and Active Fe_3O_4@ SiO_2/Starch-Derived Carbon Composite for Catalytic Degradation of Dye Waste Water. *J. Clean. Prod.* **2020,** *258*. DOI: 10.1016/j.jclepro.2020.120741.
Carvalho, A. J. F. D.; Curvelo, A.A. S.; Agnelli, J. A. M. Wood Pulp Reinforced Thermoplastic Starch Composites. *Int. J. Polym. Mater.* **2002,** *51*, 647–660.
Carvalho, A.J.F.; Curvelo, A. A. S.; Agnelli, J. A. M. A First Insight on Composites of Thermoplastic Starch and Kaolin. A. *Carbohyd. Polym.* **2001,** *45*, 189–194.
Chaudhary, V.; Bajpai, P. K.; Maheshwari, S. Studies on Mechanical and Morphological Characterization of Developed Jute/Hemp/Flax Reinforced Hybrid Composites for Structural Applications. *J. Natural Fibers* **2018,** *15* (1), 80–97.
Chee, S. S.; Jawaid, M.; Sultan, M. T. H.; Alothman, O. Y.; Abdullah, L. C. Accelerated Weathering and Soil Burial Effects on Colour, Biodegradability and Thermal Properties of Bamboo/Kenaf/Epoxy Hybrid Composites. *Polym. Test.* **2019,** *79*. DOI: 10.1016/j.polymertesting.2019.106054.
Cheung, H. Y.; Lau, K.T.; Tao, X. M.; Hui, D. A Potential Material for Tissue Engineering: Silkworm Silk/PLA Biocomposite. *A. Compos. B. Eng.* **2008,** *39*, 1026–1033.

Corbiere-Nicollier, T.; Gfeller Laban, B.; Lundquist, L.; Leterrier, Y; Manson, J. A. E.; Jolliet, O. *Resour. Conserv. Recycl.* **2001,** *33* (4), 267–287.

Courgneau, C.; Rusu, D.; Henneuse, C.; Ducruet, V.; Lacrampe, M. F.; Krawczak, P. Characterisation of Low-Odour Emissive Polylactide/Cellulose Fibre Biocomposites for Car Interior. *Express Polym. Lett.* **2013,** *7* (9), 787–804.

Curvelo, A. A. S.; Carvalho, A. J. F. Agnelli, J. A. M. Thermoplastic Starch–Cellulosic Fibers Composites: Preliminary Results. *Carbohyd. Polym.* **2001,** *45* (2), 183–8.

Damian, C. M.; Ott, C.; Stan, R.; Hanganu, A.; Trusca, R.; Balanuca, B. Harnessing a Byproduct from Wastewater Treatment to Obtain Improved Starch/Poly (Vinyl Alcohol) Composites. *Carbohyd. Polym.* **2020,** *238*. DOI: 10.1016/j.carbpol.2019.115777.

Dang, X.; Chen, H.; Shan, Z. Preparation and Characterization of Poly (Acrylic Acid)—Corn Starch Blend for Use as Chemical Sand-Fixing Materials. *Mater. Res. Express* **2017,** *4* (7). DOI: 10.1088/2053–1591/aa760f.

Das, A.; Bhattacharyya, S.; Uppaluri, R.; Das, C. Optimality of Poly-Vinyl Alcohol/Starch/Glycerol/Citric Acid in Wound Dressing Applicable Composite Films. *Int. J. Biol. Macromol.* **2020,** *155*, 260–272.

Das, S.; Pandey, P.; Mohanty, S.; Nayak, S. K.; Evaluation of Biodegradability of Green Polyurethane/Nanosilica Composite Synthesized from Transesterified Castor Oil and Palm Oil Based Isocyanate. *Int. Biodeterior. Biodegr.* **2017,** *117*, 278–288.

Devireddy, S. B. R.; Biswas, S. Thermo-Physical Properties of Short Banana–Jute Fiber-Reinforced Epoxy-Based Hybrid Composites. *Proc. Inst. Mech. Eng. L. J. Mater.* **2018,** *232* (11), 939–951.

Dicker, M. P.; Duckworth, P. F.; Baker, A. B.; Francois, G.; Hazzard, M. K.; Weaver, P. M. Green Composites: A Review of Material Attributes and Complementary Applications. *Compos. A: Appl. Sci. Manuf.* **2014,** *56*, 280–289.

Doi, Y. *Microbial Polyesters*; VCH Publishers: New York, 1990.

Dolza, C.; Gonga, E.; Fages, E.; Tejada-Oliveros, R.; Balart, R.; Quiles-Carrillo, L. Green Composites from Partially Bio-Based Poly (Butylene Succinate-Co-Adipate)-PBSA and Short Hemp Fibers with Itaconic Acid-Derived Compatibilizers and Plasticizers. *Polymers* **2022,** *14* (10). DOI: 10.3390/polym14101968.

Donovan, J. W. Phase Transitions of the Starch–Water System. *Biopolymers* **1979,** *18*, 263–275.

Du, X.; Zhang, Y.; Pan, X.; Meng, F.; You, J.; Wang, Z. Preparation and Properties of Modified Porous Starch/Carbon Black/Natural Rubber Composites. *Comp. B. Eng.* **2019,** *156*. DOI: 10.1016/j.compositesb.2018.08.033.

Dufresne, A.; Dupeyre, D.; Vignon, M. R. Cellulose Microfibrils from Potato Tuber Cells: Processing and Characterization of Starch–Cellulose Microfibril Composites. *J. Appl. Polym. Sci.* **2000,** *76*, 2080–2092.

Dufresne, A.; Vignon, M. R. Improvement of Starch Film Performances Using Cellulose Microfibrils. *Macromolecules* **1998,** *31* (8), 2693–2696.

Duigou, A. L.; Pillin, I.; Bourmaud, A.; Davies, P.; Baley, C. Manufacturing, Characterization and Modelling of Biodegradable Composite Materials. **2008,** *39*, 1471–1478.

Ejaz, M.; Azad, M. M.; Shah, A. U. R.; Afaq, S. K.; Song, J. I. Mechanical and Biodegradable Properties of Jute/Flax Reinforced PLA Composites. *Fibers Polym.* **2020,** *21* (11), 2635–2641.

Espigulé, E.; Puigvert, X.; Vilaseca, F.; Mendez, J. A.; Mutjé, P.; Girones, J. Thermoplastic Starch-Based Composites Reinforced with Rape Fibers: Water Uptake and Thermomechanical Properties. *BioResources* **2013,** *8* (2), 2620–2630.

Fabijanski, M. Polymer Biocomposites Based on Polylactide and Cellulose Fibers. *Przem. Chem.* **2020,** *99* (6), 923–926.

Fan, J.; Njuguna, J. An Introduction to Lightweight Composite Materials and Their Use in Transport Structures. In *Lightweight Composite Structures in Transport*; Woodhead Publishing: Sawton, 2016; pp 3–34.

Fang, Q.; Hanna, M. A. Rheological Properties of Amorphous and Semicrystalline Polylactic Acid Polymers. *Ind. Crop Prod.* **1999,** *10*, 47–53.

Fazita, M. R. N.; Jayaraman, K.; Bhattacharyya, D.; Haafiz, M. K. M.; Saurabh, C. K.; Hussin, M. H.; HPS, A. K. Green Composites Made of Bamboo Fabric and Poly (Lactic) Acid for Packaging Applications—A Review. *Materials* **2016,** *9* (6). DOI: 10.3390/ma9060435.

Felix, J. M.; Gatenholm, P. The Nature of Adhesion in Composites of Modified Cellulose Fibers and Polypropylene. *J. Appl. Polym. Sci.* **1991,** *42* (3), 609–620.

Flodin, P.; Zadorecki, P. *Composite Systems from Natural and Synthetic Polymers*; Elsevier Amsterdam, 1986.

Fu, D.; Netravali, A. N. 'Green' Composites Based on Liquid Crystalline Cellulose Fibers and Avocado Seed Starch. *J. Mater. Sci.* **2021,** *56* (10), 6204–6216.

Garcia, O. R.; Pinzón, M. I.; Villa, C. C. Analysis and Modeling of Mechanical and Barrier Properties of Arracacha Starch-Chitosan Composite Biodegradable Films. *J. Polym. Environ.* **2020,** *28*, 2253–2262.

Garlotta, D. A. A Literature Review of Poly (Lactic Acid). *J. Pol. Env.* **2002,** *9*, 63–84.

Gatenholm, P.; Kubat, J.; Mathiasson, A. Biodegradable Natural Composites. I. Processing and Properties. *J. Appl. Polym. Sci.* **1992,** *45*, 1667–1677.

Geng, S.; Liu, X.; Ma, H.; Liu, B.; Liang, G. Multi-Scale Stabilization Mechanism of Pickering Emulsion Gels Based on Dihydromyricetin/High-Amylose Corn Starch Composite Particles. *Food Chem.* **2021,** *355*. DOI: 10.1016/j.foodchem.2021.129660.

Giorgini, L.; Benelli, T.; Brancolini, G.; Mazzocchetti, L. Recycling of Carbon Fiber Reinforced Composite Waste to Close Their Life Cycle in a Cradle-to-Cradle Approach. *Curr. Opin. Green Sustain. Chem.* **2020,** *26*. DOI: 10.1016/j.cogsc.2020.100368.

Graupner, N.; Labonte, D.; Müssig, J. Rhubarb Petioles Inspire Biodegradable Cellulose Fibre-Reinforced PLA Composites with Increased Impact Strength. *Compos. A Appl. Sci. Manuf.* **2017,** *98*, 218–226.

Guimarães Jr.; M.; Botaro, V. R.; Novack, K. M.; Teixeira, F. G.; Tonoli, G. H. D. Starch/PVA-Based Nanocomposites Reinforced with Bamboo Nanofibrils. *Ind. Crops Prod.* **2015,** *70*, 72–83.

Haager, S. Mit der Natur im Bund: Pflanzenfasern erobern das Automobil. *Energie* **1995,** *47* (5), 50–53.

Haeldermans, T.; Samyn, P.; Cardinaels, R.; Vandamme, D.; Vanreppelen, K.; Cuypers, A.; Schreurs, S. Bio-Based Poly (3-Hydroxybutyrate)/Thermoplastic Starch Composites as a Host Matrix for Biochar Fillers. *J. Polym. Environ.* **2021,** *29* (8), 2478–2491.

Han, J. S.; Miyashita, E. S.; Spielvogel, S. J. In *Kenaf Properties, Processing and Products*; Sellers, T., Reichert, N. A., Eds.; Mississippi State University, 1999; pp 267–83.

Hasan, M.; Saifullah, A.; Dhakal, H. N.; Khandaker, S.; Sarker, F. Improved Mechanical Performances of Unidirectional Jute Fibre Composites Developed with New Fibre Architectures. *RSC Adv.* **2021,** *11* (37), 23010–23022.

Hassan, T.; Jamshaid, H.; Mishra, R.; Khan, M. Q.; Petru, M.; Novak, J. et al., Acoustic, Mechanical and Thermal Properties of Green Composites Reinforced with Natural Fibers Waste. *Polymers* **2020,** *12* (3). DOI: 10.3390/polym12030654.

Heidarian, P.; Behzad, T.; Sadeghi, M. Investigation of Cross-Linked PVA/Starch Biocomposites Reinforced by Cellulose Nanofibrils Isolated from Aspen Wood Sawdust. *Cellulose* **2017,** *24* (8), 3323–3339.

Hermawan, D.; Hazwan, C. M.; Owolabi, F. A. T.; Gopakumar, D.A.; Hasan, M.; Rizal, S.; et al., Oil Palm Microfiber-Reinforced Handsheet-Molded Thermoplastic Green Composites for Sustainable Packaging Applications. *Prog. Rubber, Plastics Recycl. Technol.* **2019,** *35* (4), 173–187.

Herrmann, A.S.; Hanselka, H.; Niederstadt, G. *Eur. Pat. Appl. EP 687, 711, 20* Dec. 1995, DE Appl. 4,420, *817* 16 June 1994.

Heslot, H. Artificial Fibrous Proteins: A Review. *Biochimie* **1998,** *80*, 19–31.

Heydari, S. Development of Aspartame Functionalized Magnetic Polymer Nanocomposites for the Zinc (II) Removal from Aqueous Solution. *Iran. J. Sci. Technol. Trans. A. Sci.* **2021,** *45*, 1311–1318.

Hinrichsen, G.; Karmaker, A. C.; Wuttke, B. *Proc. BMFT Seminar*; Bonn-Bad Godesberg, FRG, 1991; p 285.

Hocking, P. J.; Marchessault, R. H. Biopolyesters. In *Chemistry and Technology of Biodegradable Polymers*, Griffin, G. J. L., Ed.; Blackie: Glasgow, 1994, pp 48–96.

Hoiby, J. C.; Netravali, A. N. Can We Build with Plants? Cabin Construction Using Green Composites. *J. Renew. Mater.* **2015,** *3* (3), 244–258.

Hong, C. K.; Hwang, I.; Kim, N.; Park, D. H.; Hwang, B. S.; Nah, C. Mechanical Properties of Silanized Jute–Polypropylene Composites. *J. Ind. Eng. Chem.* **2008,** *14*, 71–6.

Hossain, M. S.; Razzak, M.; Uddin, M. B.; Chowdhury, A. S.; Khan, R. A. Physico-Mechanical Properties of Jute Fiber-Reinforced LDPE-Based Composite: Effect of Disaccharide (Sucrose) and Gamma Radiation. *Radiat. Eff. Defects Solids* **2020,** *175* (5–6), 516–528.

Hosseini, S. N.; Pirsa, S.; Farzi, J. Biodegradable Nano Composite Film Based on Modified Starch-Albumin/MgO; Antibacterial, Antioxidant and Structural Properties. *Polym. Test.* **2021,** *97*. DOI: 10.1016/j.polymertesting.2021.107182.

Hu,R.; Lim, J. K. Fabrication and Mechanical Properties of Completely Biodegradable Hemp Fiber Reinforced Polylactic Acid Composites. *J. Compos. Mater.* **2007,** *41*, 1655–1669.

Huber, T.; Bickerton, S.; Müssig, J.; Pang, S.; Staiger, M. P. Flexural and Impact Properties of All-Cellulose Composite Laminates. *Compos. Sci. Technol.* **2013,** *88*, 92–98.

Huda, M. S.; Drzal, L. T.; Mohanty, A. K.; Misra, M. Effect of Fiber Surface-Treatments on the Properties of Laminated Biocomposites from Poly (Lactic Acid) (PLA) and Kenaf Fibers. *Compos. Sci. Technol.* **2008a,** *68*, 424–432.

Huda, M. S.; Drzal, L. T.; Mohanty, A. K.; Misra, M. Effect of Chemical Modifications of the Pineapple Leaf Fiber Surfaces on the Interfacial and Mechanical Properties of Laminated Biocomposites. *Compos. Interfaces* **2008b,** *15*, 169–191.

Huda, M. S.; Drzal, L. T.; Mohanty, A. K.; Misra, M. Wood-Fiber-Reinforced Poly (Lactic Acid) Composites: Evaluation of the Physicomechanical and Morphological Properties. *J. Appl. Polym. Sci.* **2006,** *102*, 4856–4869.

Huyen, N. T. B. Microstructural Characteristics and Mechanical Properties of the Green Composites Reinforced Bamboo Optimized the Treated Parameters. *J. Polym. Compos.* **2021,** *9* (1), 28–38.

Inagaki, H. Kenaf: Plant of Environmental Conservation and Fiber Materials. *Kobunshi* **2002,** *51*, 597–602.

Inoh, T.; Yokoi, T; Sekiyama, K. I.; Kawamura, N. and Mishima, Y. SMC Recycling Technology. *J. Thermoplast. Compos. Mater.* **1994,** *7* (1), 42–55.

Jamshidian, M.; Tehrany, E. A.; Imran, M.; Jacquot, M.; Desobry, S. Poly-Lactic Acid: Production, Applications, Nanocomposites, and Release Studies. *Compr. Rev. Food Sci. Food Saf.* **2010,** *9*, 552–571.

Jayakumar, A.; Heera, K. V.; Sumi, T. S.; Joseph, M.; Mathew, S.; Praveen, G. Starch-PVA Composite Films with Zinc-Oxide Nanoparticles and Phytochemicals as Intelligent pH Sensing Wraps for Food Packaging Application. *Int. J. Biol. Macromol.* **2019,** *136*, 395–403.

Jensen, A.; Hoffmann, L.; Mollar, B.;Schmidt, A.; Christiansen, K.;Elkington, J.; Van Dijk, F. Life Cycle Assessment, a Guide to Approaches, Experiences, and Information Sources; Environmental Issues Series No. 6; European Environment Agency: Copenhagen, 1997.

Jordan, N. D.; Bassett, D. C.; Olley, R. H.; Hine, P. J.; Ward, I. M. The Hot Compaction Behaviour of Woven Oriented Polypropylene Fibres and Tapes. II. Morphology of Cloths Before and After Compaction. *Polymer*, **2003,** *44*, 1133–43.

Jose, J.; Al-Harthi, M. A.; AlMa'adeed, M. A. A.; Bhadra Dakua, J.; De, S. K. Effect of Graphene Loading on Thermomechanical Properties of Poly (Vinyl Alcohol)/Starch Blend. *J. Appl. Polym. Sci.* **2015,** *132* (16). DOI: 10.1002/app.41827.

Jumaidin, R.; Khiruddin, M. A. A.; Saidi, Z. A. S.; Salit, M. S.; Ilyas, R. A. Effect of Cogon Grass Fibre on the Thermal, Mechanical and Biodegradation Properties of Thermoplastic Cassava Starch Biocomposite. *Int. J. Biol. Macromol.* **2020,** *146*, 746–755.

Kalita, N. K.; Nagar, M. K.; Mudenur, C.; Kalamdhad, A.; Katiyar, V.; Biodegradation of Modified Poly (Lactic Acid) Based Biocomposite Films Under Thermophilic Composting Conditions. *Polym. Test.* **2019,** *76*, 522–536.

Kamble, Z.; Behera, B. K. Fabrication and Performance Evaluation of Waste Cotton and Polyester Fiber-Reinforced Green Composites for Building and Construction Applications. *Polym. Compos.* **2021,** *42*, 3025–3037.

Kandpal, B. C.; Chaurasia, R.; Khurana, V.; Recent Advances in Green Composites—A Review. *Int. J. Technol. Res. Eng.* **2015,** *2* (7), 742–747.

Kaplan, D. L.; Adams, W. W.; Vincy, C.; Farmer, B. L. *Silk Polymers: Materials Science and Biotechnology*; ACS Books: Washington, DC, 1994.

Karim, N.; Sarker, F.; Afroj, S.; Zhang, M.; Potluri, P.; Novoselov, K. S. Sustainable and Multi-functional Composites of Graphene-Based Natural Jute Fibers. *Adv. Sustain. Syst.* **2021,** *5* (3). DOI: 10.1002/adsu.202000228.

Karmaker, A. C.; Hinrichsen, G. Processing and Characterization of Jute Fiber Reinforced Thermoplastic Polymers. *Polym. Plast. Technol. Eng.* **1991,** *30*. DOI: 10.1080/0360255910 8019223.

Karmaker, A. C.; Youngquist, J. Injection Molding of Polypropylene Reinforced with Short Jute Fibers. *J. Appl. Polym. Sci.* **1996,** *62*, 1147–1151.

Karus, M.; Kaup, M.; Lohmeyer, D. Study on Markets and Prices for Natural Fibres (Germany and EU). Nova Institute GmbH, 2000. http://www.nova-institut.de (Accessed on 26 January 2004).

Kesel, C. D.; Wauven, C. V.; David, C. Biodegradation of Polycaprolactone and Its Blends with Poly (Vinylalcohol) by Micro-Organisms from a Compost of House-Hold Refuse. *Polym. Degrada. Stab.* **1997,** *55*, 107–113.

Khalid, M. Y.; Nasir, M. A.; Ali, A.; Al Rashid, A.; Khan, M. R. Experimental and Numerical Characterization of Tensile Property of Jute/Carbon Fabric Reinforced Epoxy Hybrid Composites. *SN Appl. Sci.* **2020,** *2* (4). DOI: 10.1007/s42452–020–2403–2.

Khalid, M.; Ratnam, C. T.; Chuah, T. G.; Salmiaton, A.; Thomas, S. Y. C. Comparative Study of Polypropylene Composites Reinforced with Oil Palm Empty Fruit Bunch Fiber and Oil Palm Derived Cellulose. *Mater. Design* **2008,** *29*, 173–178.

Khan, M. A.; Ali, K. M. I.; Hinrichsen, G.; Kopp, C.; Kropke, S. Study of Tensile Properties of Biopol-Jute Composites. *Polym. Plast. Technol. Eng.* **1999,** *38* (1), 99.

Khristova, P.; Bentcheva, S.; Karar, I. Soda-AQ Pulp Blends from Kenaf and Sunflower Stalks. *Bioresour. Technol.* **1998,** *66*, 99–103.

Kibar, E. A. A.; Us, F. Thermal, Mechanical and Water Adsorption Properties of Corn Starch–Carboxymethylcellulose/Methylcellulose Biodegradable Films. *J. Food Eng.* **2013,** *114* (1), 123–131.

Koronis, G.; Silva, A.; Fontul, M. Green Composites: A Review of Adequate Materials for Automotive Applications. *Compos. B. Eng.* **2013,** *44* (1), 120–127.

Koshy, R. R.; Mary, S. K.; Thomas, S.; Pothan, L. A. Environment Friendly Green Composites Based on Soy Protein Isolate–A Review. *Food Hydrocoll.* **2015,** *50*, 74–192.

Lam, T. B. T.; Liyama, K. Structural Details of Kenaf Cell Walls and Fixation of Carbon Dioxide. In Abstract of the 2000 International Kenaf Symposium, 2000; p 14.

Lasrado, D.; Ahankari, S.; Kar, K. Nanocellulose-Based Polymer Composites for Energy Applications—A Review. *J. Appl. Polym. Sci.* **2020,** *137* (27). DOI: 10.1002/app.48959.

Liao, J.; Luo, Z.; Zhang, Y.; Zhang, X.; Cheng, J.; Wu, Q. Effects of a Novel Compatible Interface Structure on the Properties of Starch–PCL Composites. *New J. Chem.* **2014,** *38* (6), 2522–2529.

Liu, T.; Zhang, M.; Guo, X.; Liu, C.; Liu, T.; Xin, J.; Zhang, J. Mild Chemical Recycling of Aerospace Fiber/Epoxy Composite Wastes and Utilization of the Decomposed Resin. *Polym. Degrad. Stab.* **2017,** *139*, 20–27.

Lo, J. N.; Nutt, S. R.; Williams, T. J.; Recycling Benzoxazine–Epoxy Composites via Catalytic Oxidation. *ACS Sust. Chem. Eng.* **2018,** *6* (6), 7227–7231.

Luo, S.; Netravali, A. N. Mechanical and Thermal Properties of Environment-Friendly "Green" Composites Made from Pineapple Leaf Fibers and Poly (Hydroxybutyrate-Co-Valerate) Resin. *Polymer. Compos.* **1999,** *20* (3), 367–378.

Ma, G.; Yan, L.; Shen, W.; Zhu, D.; Huang, L.; Kasal, B. Effects of Water, Alkali Solution and Temperature Ageing on Water Absorption, Morphology and Mechanical Properties of Natural FRP Composites: Plant-Based Jute vs. Mineral-Based Basalt. *Compos. B. Eng.* **2018,** *153*, 398–412.

Macedo, J. R. N.; dos Santos, D. J.; d. S. Rosa, D.; Poly (Lactic Acid)–Thermoplastic Starch–Cotton Composites: Starch-Compatibilizing Effects and Composite Biodegradability. *J. Appl. Polym. Sci.* **2019,** *136* (21). DOI: 10.1002/app.47490.

Mahadeva, S. K.; Yun, S.; Kim, J. Flexible Humidity and Temperature Sensor Based on Cellulose–Polypyrrole Nanocomposite. *Sens. Actuat. Phys.* **2011,** *165*, 194–199.

Maldas, D.; Kokta, B. V. *TRIP* **1993,** *1* (6), 174.

Martin, O.; Averous, L. Poly (Lactic Acid): Plasticization and Properties of Biodegradable Multiphase Systems. *Polymer* **2001,** *42*, 6209–6219.

McMullen, P. Fibre/Resin Composites for Aircraft Primary Structures: A Short History. *Composites* **1984,** *15* (3), 222–230.

Mercier, C. and Feillet, P. Modification of Carbohydrate Components by Extrusion-Cooking of Cereal Products. *Cereal Chem.* **1975,** *52*, 283–297.

Misra, M.; Park, H.; Mohanty, A.K.; Drzal, L. T. Injection Molded Green Nanocomposite Materials from Renewable Resources. Presented at the Global Plastics Environmental Conference, Detroit, MI, USA, 2004; pp 18–19.

Mitra, B. C.; Basak, R. K.; Sarkar, M. Studies on Jute-Reinforced Composites, Its Limitations, and Some Solutions Through Chemical Modifications of Fibers. *J. Appl. Polym. Sci.* **1998,** *67*, 1093–1100.

Mlalila, N.; Hilonga, A.; Swai, H.; Devlieghere, F.; Ragaert, P. Antimicrobial Packaging Based on Starch, Poly (3-Hydroxybutyrate) and Poly (Lactic-Co-Glycolide) Materials and Application Challenges. *Trends Food Sci. Technol.* **2018,** *74*. DOI: 10.1016/j.tifs.2018.01.015.

Mohanty, A. K.; Khan, M. A.; Hinrichsen, G. Surface Modification of Jute and Its Influence on Performance of Biodegradable Jute-Fabric/Biopol Composites. *Compos. Sci. Technol.* **2000,** *60*, 1115–1124.

Mohanty, A. K.; Misra, M. Studies on Jute Composites—A Literature Review. *Polym. Plast. Technol. Eng.* **1995,** *34*, 729–792.

Mousa, N.; Galiwango, E.; Haris, S.; Al-Marzouqi, A. H.; Abu-Jdayil, B.; Caires, Y. L. A New Green Composite Based on Plasticized Polylactic Acid Mixed with Date Palm Waste for Single-Use Plastics Applications. *Polymers* **2022,** *14* (3). DOI: 10.3390/polym 14030574.

Muniyasamy, S.; Ofosu, O.; Thulasinathan, B.; Rajan, A. S. T.; Ramu, S. M.; Soorangkattan, S. et al., Thermal-Chemical and Biodegradation Behaviour of Alginic Acid Treated Flax Fibres/Poly (Hydroxybutyrate-Co-Valerate) PHBV Green Composites in Compost Medium. *Biocatal. Agric. Biotechnol.* **2019,** *22*. DOI: 10.1016/j.bcab.2019.101394.

Nandi, P.; Das, D. Mechanical, Thermo-Mechanical and Biodegradation Behaviors of Green-Composites Prepared from Woven Structural Nettle (Girardinia diversifolia) Reinforcement and Poly (Lactic Acid) Fibers. *Ind. Crops Prod.* **2022,** 175. DOI: 10.1016/j. indcrop.2021.114247.

Nasri-Nasrabadi, B.; Mehrasa, M.; Rafienia, M.; Bonakdar, S.; Behzad, T.; Gavanji, S. Porous Starch/Cellulose Nanofibers Composite Prepared by Salt Leaching Technique for Tissue Engineering. *Carbohyd. Polym.* **2014,** *108*, 232–238.

Netcomposites: Newsroom: Eastern European Icon Set for Africa. http://www.netcomposites.com/news.asp/1680 (Accessed on 26 January 2004).

Nishi, Y.; Uryu, M.; Yamanaka, S.; Watanabe, K.; Kitamura, N.; Iguchi, M. et al., The Structure and Mechanical Properties of Sheets Prepared from Bacterial Cellulose. *J. Mater. Sci.* **1990,** *25*, 2997–3001.

Orasugh, J. T.; Saha, N. R.; Rana, D.; Sarkar, G.; Mollick, M. M. R.; Chattoapadhyay, A. et al., Jute Cellulose Nano-Fibrils/Hydroxy Propylmethyl Cellulose Nanocomposite: A Novel Material with Potential for Application in Packaging and Transdermal Drug Delivery System. *Ind. Crop Prod.* **2018,** *112*, 633–643.

Ortega, F.; Giannuzzi, L.; Arce, V. B.; García, M. A. Active Composite Starch Films Containing Green Synthetized Silver Nanoparticles. *Food Hydrocoll.* **2017,** *70*, 152–162.

Otaigbe, J. U. Controlling the Water Absorbency of Agricultural Biopolymers. *Plast. Eng.* **1998,** *54* (4), 37–39.

Pandey, J. K.; Chu, W. S.; Lee, C. S.; Ahn, S. H. Presented at the International Symposium on Polymers and the Environment: Emerging Technology and Science, Bio Environmental Polymer Society (BEPS), Vancouver, WA, USA, 17–20 October 2007.

Pantaloni, D.; Shah, D.; Baley, C. Bourmaud, A.; Monitoring of Mechanical Performances of Flax Non-Woven Biocomposites During a Home Compost Degradation. *Polym. Degrad. Stab.* **2020,** *177*. DOI: 10.1016/j.polymdegradstab.2020.109166.

Papanicolaou, G. C.; Karagiannis, D.; Bofilios, D. A.; Van Lochem, J. H.; Henriksen, C.; Lund, H. H. Impact Strength of Recycled Thermoplastic Composites Subjected to Corrosive Environment. *Polym. Compos.* **2008,** *29*, 1026–1035.

Perdana, M. I.; Ruamcharoen, J.; Panphon, S.; Leelakriangsak, M. Antimicrobial Activity and Physical Properties of Starch/Chitosan Film Incorporated with Lemongrass Essential Oil and Its Application. *LWT-Food Sci. Technol.* **2021,** *141*. DOI: 10.1016/j.lwt.2021.110934.

Petterson J.; Nilsson P. J. Recycling of SMC and BMC in Standard Process Equipment. *J. Thermoplast. Compos. Mater.* **1994,** *7* (1), 56–63.

Pinto, M.; Chalivendra, V. B.; Kim, Y. K.; Lewis, A. F. Improving the Strength and Service Life of Jute/Epoxy Laminar Composites for Structural Applications. *Compos. Struct.* **2016,** *156*, 333–337.

Privas, E.; Leroux, F.; Navard, P. Preparation and Properties of Blends Composed of Lignosulfonated Layered Double Hydroxide/Plasticized Starch and Thermoplastics. *Carbohyd. Polym.* **2013,** *96* (1), 91–100.

Qiao, S.; Duan, W.; Yu, J.; Zheng, Y.; Yan, D.; Jin, F. et al., Fabrication of Biomolecule–Covalent-Organic-Framework Composites as Responsive Platforms for Smart Regulation of Fermentation Application. *ACS Appl. Mater. Interfaces* **2021,** *13*, 32058–32066.

Qiu, J. H.; Song, P. G.; Fu, S. Y.; Ge, X. X.; Huang, B.; Feng, J. B. Compatibilization of Polypropylene/Corn Starch Plasticized with Diethanol Amine. *Adv. Mater. Res.* **2013,** *610*, 475–479.

Rahaman, M. N.; Hossain, M. S.; Razzak, M.; Uddin, M. B.; Chowdhury, A. S.; Khan, R. A. Effect of Dye and Temperature on the Physico-Mechanical Properties of Jute/PP and Jute/LLDPE Based Composites. *Heliyon* **2019,** *5* (6). DOI: 10.1016/j.heliyon.2019.e01753.

Rana, A. K.; Mandal, A.; Mitra, B. C.; Jacobson, R.; Rowell, R.; Banerjee, A. N. Short Jute Fiber-Reinforced Polypropylene Composites: Effect of Compatibilizer. *J. Appl. Polym. Sci.* **1998,** *69*, 329–338.

Revin, V. V.; Dolganov, A. V.; Liyaskina, E. V.; Nazarova, N. B.; Balandina, A. V.; Devyataeva, A. A. et al., Characterizing Bacterial Cellulose Produced by Komagataeibacter Sucrofermentans h-110 on Molasses Medium and Obtaining a Biocomposite Based on It for the Adsorption of Fluoride. *Polymers* **2021,** *13* (9). DOI: 10.3390/polym13091422.

Rhim, J. W.; Wu,Y.; Weller, C. L.; Schnepf, M. Physical Characteristics of Emulsified Soy Protein-Fatty Acid Composite Films. *Sci. Aliments* **1999,** *19* (1), 57–71.

Rivard, C. J.; Adney, W. S.; Himmel, M. E.; Mitchell, D. J.; Vinzant, T. B.; Grohmann, K.; et al., Effects of Natural Polymer Acetylation on the Anaerobic Bioconversion to Methane and Carbon Dioxide. *Appl. Biochem. Biorech.* **1992,** *34/35*, 725–736.

Rogovina, S. Z.; Aleksanyan, K. V.; Loginova, A. A.; Ivanushkina, N. E.; Vladimirov, L. V.; Prut, E. V. Influence of PEG on Mechanical Properties and Biodegradability of Composites Based on PLA and Starch. *Starch-Stärke* **2018,** *70* (7–8). DOI: 10.1002/star.201700268.

Safdari, F.; Carreau, P. J.; Heuzey, M. C.; Kamal, M. R. Effects of Poly (Ethylene Glycol) on the Morphology and Properties of Biocomposites Based on Polylactide and Cellulose Nanofibers. *Cellulose* **2017,** *24* (7), 2877–2893.

Sahoo, S.; Pati, D.; Misra, M.; Tripathy, S. S.; Nayak, S. K.; Mohanty, A. K. International Symposium: Polymers 1999; Society of Polymer Science, India, January 1999; pp 542–545.

Sameni, J. K.; Ahmad, S. H.; Zakaria, S. Plastics, Rubber and Composites (UK), **2002,** *31/4*, 162–166.

Sani, I. K.; Geshlaghi, S. P.; Pirsa, S.; Asdagh, A. Composite Film Based on Potato Starch/Apple Peel Pectin/ZrO_2 Nanoparticles/Microencapsulated Zataria Multiflora Essential Oil; Investigation of Physicochemical Properties and Use in Quail Meat Packaging. *Food Hydrocoll.* **2021,** *117*. DOI: 10.1016/j.foodhyd.2021.106719.

Sanvezzo, P. B.; Branciforti, M. C. Recycling of Industrial Waste Based on Jute Fiber-Polypropylene: Manufacture of Sustainable Fiber-Reinforced Polymer Composites and Their Characterization Before and After Accelerated Aging. *Ind. Crop Prod.* **2021,** *168*. DOI: 10.1016/j.indcrop.2021.113568.

Sarazin, P.; Li, G.; Orts, W. J.; Favis, B. D. Binary and Ternary Blends of Polylactide, Polycaprolactone and Thermoplastic Starch. *Polymer* **2008,** *49*, 599–609.

Savitskaya, I. S.; Kistaubayeva, A. S.; Digel, I. E.; Shokatayeva, D. H. Physicochemical and Antibacterial Properties of Composite Films Based on Bacterial Cellulose and Chitosan for Wound Dressing Materials. *Eurasian Chem.-Technol. J.* **2017,** *19* (3), 255–264.

Scaffaro, R.; Citarrella, M. C.; Gulino, E. F.; Morreale, M. Hedysarum Coronarium-Based Green Composites Prepared by Compression Molding and Fused Deposition Modeling. *Materials* **2022,** *15* (2). DOI: 10.3390/ma15020465.

Scaffaro, R.; Maio, A.; Sutera, F.; Gulino, E. F. Morreale, M.; Degradation and Recycling of Films Based on Biodegradable Polymers: A Short Review. *Polymers*, **2019,** *11* (4). DOI: 10.3390/polym11040651.

Schuh, T. G. Natural Fibres Performance Forum, Copenhagen 27–28 May 1999.

Serizawa, S.; Kazuhiko, I.; Masatoshi, I. Kenaf-Fiber-Reinforced Poly (Lactic Acid) Used for Electronic Products. *J. Appl. Polym. Sci.* **2006,** *100*, 618–624.

Setua D.K. (1986) Short Fibre-Rubber Composites. In *Renewable-Resource Materials*; Carraher, C. E.; Sperling, L. H., Eds.; Springer: Boston, MA. DOI: 10.1007/978–1-4613–2205–4_22

Shao, Z. Z.; Vollrath, F. Surprising Strength of Silkworm Silk. *Nature* **2002,** *418*. DOI: 10.1038/418741a.

Shehabeldine, A.; Hasanin, M. Green synthesis of Hydrolyzed Starch–Chitosan Nano-Composite as Drug Delivery System to Gram Negative Bacteria. *Environ. Nanotechnol. Monit. Manag.* **2019,** *12*. DOI: 10.1016/j.enmm.2019.100252.

Shikamoto, N.; Ohtani, A.; Leong, Y. W.; Nakai, A. In 22nd Technical Conference of the American Society for Composites 2007: Composites: Enabling a New Era in Civil Aviation; Curran Associates, Inc; Red Hook, NY, USA, 2007; pp 151:1–151:10.

Shirsath, N. B.; Roy, C. B.; Meshram, J. S. TiO_2/Starch-Based Biodegradable Copolymer Nanocomposite as a Plant Growth Regulator. *Green Mater.* **2020,** *9*, 69–77.

Shogren, R. L. In Kaplan, D. L. *Biopolymers from Renewable Resources, Macromolecular Systems-Materials Approach*; Springer-Verlag: Berlin, 1998; pp 30–46.

Sirviö, J. A.; Kolehmainen, A.; Liimatainen, H.; Niinimäki, J.; Hormi, O. E. Biocomposite Cellulose-Alginate Films: Promising Packaging Materials. *Food Chem.* **2014,** *151*, 343–351.

Sirviö, J. A.; Visanko, M.; Ukkola, J.; Liimatainen, H. Effect of Plasticizers on the Mechanical and Thermomechanical Properties of Cellulose-Based Biocomposite Films. *Ind. Crop. Prod.* **2018,** *122*, 513–521.

Sjaifullah, A.; Faidza, L. Z.; Mitoma, Y. Arrowroot Starch-G-Poly (Acrylic Acid-Acrylamide)/Zeolite Hydrogel Composite as Matrix for CRF of Nitrogen, Phosphorous and Kalium. *AIP Conf. Proc.* **2020,** 2278. DOI: 10.1063/5.0015312

Song, M.; Pan, C.; Chen, C.; Li, J.; Wang, X.; Gu, Z. The Application of New Nanocomposites: Enhancement Effect of Polylactide Nanofibers/Nano-TiO2 Blends on Biorecognition of Anticancer Drug Daunorubicin. *Appl. Surf. Sci.* **2008,** *255*, 610–612.

Spiridon, I.; Anghel, N. C.; Darie-Nita, R. N.; Iwańczuk, A.; Ursu, R. G.; Spiridon, I. A. New Composites Based on Starch/Ecoflex®/Biomass Wastes: Mechanical, Thermal, Morphological and Antimicrobial Properties. *Int. J. Biol. Macromol.* **2020,** *156*, 1435–1444.

Srebrenkoska, V.; Gaceva, G. B.; Avella, M.; Errico, M. E.; Gentile, G. Recycling of Polypropylene-Based Eco-Composites. *Polym. Int.* **2008,** *57*, 1252–1257.

Sun, E.; Liao, G.; Zhang, Q.; Qu, P.; Wu, G.; Huang, H. Biodegradable Copolymer-Based Composites Made from Straw Fiber for Biocomposite Flowerpots Application. *Compos. B. Eng.* **2019,** *165*, 193–198.

Sun, H.; Shao, X.; Zhang, M.; Wang, Z.; Dong, J.; Yu, D. Mechanical, Barrier and Antimicrobial Properties of Corn Distarch Phosphate/Nanocrystalline Cellulose Films Incorporated with Nisin and ε-polylysine. *Int. J. Biol. Macromol.* **2019,** *136*, 839–846.

Taherimehr, M.; Bagheri, R.; Taherimehr, M. In-Vitro Evaluation of Thermoplastic Starch/Beta-Tricalcium Phosphate Nano-Biocomposite in Bone Tissue Engineering. *Ceram. Int.* **2021,** *47* (11), 15458–15463.

Taniguchi, T.; Okamura, K. New Films Produced from Microfibrillated Natural Fibres. *Polym. Int.* **1998,** *47*, 291–294.

Tate, J. S.; Akinola, A. T.; Kabakov, D. Bio-Based Nanocomposites: An Alternative to Traditional Composites. *J. Technol. Stud.* **2010,** *1*, 25–32.

Tirrell, J. G.; Fournier, M. J.; Mason, T. L.; Tirrell, D. A. Biomolecular Materials. *Chem. Eng. News* **1994,** *72*, 40–51.

Tokoro, R.; Vu, D. M.; Okubo, K.; Tanaka, T.; Fujii, T.; Fujiura, T. How to Improve Mechanical Properties of Polylactic Acid with Bamboo Fibers. *J. Mater. Sci.* **2008,** *43*, 775–787.

Tune, S.; Duman, O. Preparation of Active Antimicrobial Methyl Cellulose/Carvacrol/Montmorillonite Nanocomposite Films and Investigation of Carvacrol Release. *LWT-Food Sci. Technol.* **2011,** *44*, 465–472.

Uesaka, T.; Nakane, K.; Maeda, S.; Ogihara, N. Structure and Physical Properties of Poly (Butylene Succinate)/Cellulose Acetate Blends. *Polymer* **2000,** *41*, 8449–8454.

Wang, B.; Ma, S.; Yan, S.; Zhu, J.; Readily Recyclable Carbon Fiber Reinforced Composites Based on Degradable Thermosets: A Review. *Green Chem.* **2019,** *21* (21), 5781–5796.

Wang, F.; Chang, R.; Ma, R.; Tian, Y. Eco-Friendly and Superhydrophobic Nano-Starch Based Coatings for Self-Cleaning Application and Oil-Water Separation. *Carbohyd. Polym.* **2021,** *271*. DOI: 10.1016/j.carbpol.2021.118410.

Wang, L.; Ma, S.; Jia, C.; Patel, B.; Campanella, O.; You, L.; Yang, S. The Effects of Calcium Propionate and Cinnamaldehyde on the Mechanical, Physical and Antimicrobial Properties of Composite Films Based on Potato Starch. *J. Biobased Mater. Bioenergy* **2016,** *10* (3), 176–183.

Washington State University Engineered Wood Composites for Waterfront Facilities. http://composites.wsu.edu/navy/ (accessed on 27 January 2004).

Welter, N.; Leichtweis, J.; Silvestri, S.; Sánchez, P. I. Z.; Mejía, A. C. C.; Carissimi, E. Preparation of a New Green Composite Based on Chitin Biochar and $ZnFe_2O_4$ for Photo-Fenton Degradation of Rhodamine B. *J. Alloys Compd.* **2022,** 901. DOI: 10.1016/j.jallcom.2022. 163758.

Willett, J. L.; Shogren, R. L. Processing and Properties of Extruded Starch/Polymer Foams. *Polymer* **2002,** *43*, 5935–5947.

Wilpiszewska, K.; Spychaj, T.; Paździoch, W. Carboxymethyl Starch/Montmorillonite Composite Microparticles: Properties and Controlled Release of Isoproturon. *Carbohyd. Polym.* **2016,** *136*, 101–106.

Winter, H.; Mostert, H. A. M.; Smeets, P.J. H. M. Paas, G. Recycling of Sheet-Molding Compounds by Chemical Routes. *J. Appl. Polym. Sci.* **1995,** *57*, 1409–1417.

Wollerdorfer, S. F.; Bader, H. Influence of Natural Fibres on the Mechanical Properties of Biodegradable Polymers. *Ind. Crop. Prod.* **1998,** *8* (2), 105–12.

Wu, W.; Li, Y.; Yang, L.; Ma, Y.; Pan, D.; Li, Y. A Facile One-Pot Preparation of Dialdehyde Starch Reduced Graphene Oxide/Polyaniline Composite for Supercapacitors. *Electrochim. Acta* **2014,** *139*, 117–126.

Wu, Y.; Gu, Y.; Tong, L.; Chen, R.; Xie, N. Electrochemical Synthesis of ZnO Nanoparticles and Preparation of Pea Starch/ZnO Composite for Active Food Packaging Application. *Int. J. Electrochem. Sci.* **2019,** *14*, 10745–10753.

Wypych, G. Biodegradation. In *Handbook of Material Weathering*, 3rd ed.; ChemTech Publishing, 2003; Chapter 19, pp 523–33.

Xu, A.; Wang, Y.; Gao, J.; Wang, J.; Facile Fabrication of a Homogeneous Cellulose/Polylactic Acid Composite Film with Improved Biocompatibility, Biodegradability and Mechanical Properties. *Green Chem.* **2019,** *21* (16), 4449–4456.

Yang, G.; Xia, Y.; Lin, Z.; Zhang, K.; Fatehi, P.; Chen, J. Physicochemical Impact of Cellulose Nanocrystal on Oxidation of Starch and Starch Based Composite Films. *Int. J. Biol. Macromol.* **2021,** *148*, 42–49.

Yang, X.; Berthold, F.; Berglund, L. A. High-Density Molded Cellulose Fibers and Transparent Biocomposites Based on Oriented Holocellulose. *ACS Appl. Mater. Interfaces* **2019,** *11* (10), 10310–10319.

Yano, H.; Nakahara, S. Bio-Composites Produced from Plant Microfiber Bundles with a Nanometer Unit Web-Like Network. *J. Mater. Sci.* **2004,** *39*, 1635–1638.

Yao, F.; Wu, Q.; Lei, Y.; Xu, Y. Rice Straw Fiber-Reinforced High-Density Polyethylene Composite: Effect of Fiber Type and Loading. *Ind. Crops. Prod.* **2008,** *28*, 63–72.

Yazdanpanah, M.; Nojavan, S. Cyclodextrin-Starch Hard Gel as an Efficient Green Sorbent for Dispersive Micro Solid-Phase Extraction of Eight Polycyclic Aromatic Hydrocarbons from Environmental Water Samples. *Microchem. J.* **2021,** *168*. DOI: 10.1016/j.microc.2021.106509.

Ye, J.; Cheng, Y.; Sun, L.; Ding, M.; Wu, C.; Yuan, D.; et al., A Green SPEEK/Lignin Composite Membrane with High Ion Selectivity for Vanadium Redox Flow Battery. *J. Membr. Sci.* **2019,** *572*, 110–118.

Yorseng, K.; Rajini, N.; Siengchin, S.; Ayrilmis, N.; Rajulu, V. Mechanical and Thermal Properties of Spent Coffee Bean Filler/Poly (3-Hydroxybutyrate-Co-3-Hydroxyvalerate) Biocomposites: Effect of Recycling. *Proc. Safety Environm. Prot.* **2019,** *124*, 187–195.

Yorseng, K.; Rangappa, S. M.; Parameswaranpillai, J.; Siengchin, S. Towards Green Composites: Bioepoxy Composites Reinforced with Bamboo/Basalt/Carbon Fabrics. *J. Clean. Prod.* **2022,** *363*. DOI: 10.1016/j.jclepro.2022.132314.

Yu, M.; Zheng, Y.; Tian, J. Study on the Biodegradability of Modified Starch/Polylactic Acid (PLA) Composite Materials. *RSC Adv.* **2020,** *10* (44), 26298–26307.

Yu, T.; Ding, D.; Sheng, C.; Tuerhongjiang, T.; Li, Y. Enhanced Mechanical Properties and Flame Retardancy of Short Jute Fiber/Poly (Lactic Acid) Composites with Phosphorus-Based Compound. *Sci. China Technol. Sci.* **2017,** *60* (11), 1716–1723.

Zabihi, O.; Ahmadi, M.; Liu, C.; Mahmoodi, R.; Li, Q.; Ferdowsi, M. R. G.; Naebe, M. A. Sustainable Approach to the Low-Cost Recycling of Waste Glass Fibres Composites Towards Circular Economy. *Sustainability* **2020,** *12* (2). DOI: 10.3390/su12020641.

Zadegan, S.; Hosainalipour, M.; Rezaie, H. R.; Ghassai, H.; Shokrgozar, M. A. Synthesis and Biocompatibility Evaluation of Cellulose/Hydroxyapatite Nanocomposite Scaffold in 1-N-Allyl-3-Methylimidazolium Chloride. *Mater. Sci. Eng.* **2011,** *31*, 954–961.

Zaikov, G. E.; Lomakin, S. M. New Aspects of Ecologically Friendly Polymer Flame Retardant Systems. *Polym. Plast. Technol. Eng.* **1997,** *36* (4), 647–668.

Zhang, Y.; Zhao, L.; Chen, Y. Swelling Properties and Environmental Responsiveness of Superabsorbent Composite Based on Starch-G-Poly Acrylic Acid/Organo-Zeolite. *J. Macromol. Sci. B.* **2016,** *55* (7), 662–679.

Zhao, Q.; Jiang, J.; Li, C.; Li, Y. Efficient Recycling of Carbon Fibers From Amine-Cured CFRP Composites Under Facile Condition. *Polym. Degrad. Stab.* **2020,** *179*. DOI: 10.1016/j.polymdegradstab.2020.109268.

Zimmermann, K. A.; LeBlanc, J. M.; Sheets, K. T.; Fox, R. W.; Gatenholm, P. Biomimetic Design of a Bacterial Cellulose/Hydroxyapatite Nanocomposite for Bone Healing Applications. *Mater. Sci. Eng.* **2011,** *31*, 43–49.

CHAPTER 13

Green Manufacturing Processes

SHAMBHU LAL AGARWAL[1], JITENDRA VARDIA[2], DIPTI SONI[3], and RAKSHIT AMETA[3]

[1]*Department of Chemistry, PAHER University, Udaipur, India*

[2]*J. D. M. Scientific, Research Organisation, Vadodara, India*

[3]*Department of Chemistry, J. R. N. Rajasthan Vidhyapeeth, (Deemed to be University), Udaipur, India*

ABSTRACT

Green manufacturing processes involve production utilizing inputs with relatively low environmental impacts, which are highly efficient and generate little or no waste or take care of pollution. Thus, green manufacturing involves source reduction (minimization or prevention of waste or pollution), recycling, and green product design. Various green processes used in polymer and pharmaceutical industries have been presented with some stress on biocatalysis, which is a need of the day.

Green manufacturing involves production processes, which use inputs with relatively low environmental impacts, which are highly efficient and generate little or no waste or take care of pollution. Green manufacturing involves source reduction (also known as waste or pollution minimization or prevention), recycling, and green product design.

Source reduction is broadly defined to include any actions reducing the waste initially generated. Recycling includes using or reusing waste as ingredients in any process or as an effective substitute for a commercial product, or returning the waste to the original process, which generated it as a substitute for raw material feedstock. Green product design involves creating products, whose design, composition, and uses minimize the environmental impact throughout their life cycle. Green manufacturing goals are to conserve

Green Chemistry, 2nd Edition: Fundamentals and Applications. Suresh C. Ameta & Rakshit Ameta (Eds.)

natural resources for future generations. The benefit of green manufacturing is to create a great reputation to the public, save cost, and promote research and design.

The options for green manufacturing can be divided into four major areas.

13.1 PRODUCTION PROCESS CHANGES

Major process changes fall into the following categories:

i) **Changing dependence on human intervention:** Production dependence on active human intervention has a significant failure rate. This may lead to various problems ranging from off specification products to major accidents. A strategy that can reduce the dependence of production process on active human intervention is having machines to take over parts of what humans use to do? Automated process control, robots for welding purpose, and numerically controlled cutting tools all may reduce waste.

ii) **Use of continuous process instead of a batch process:** Continuous process causes less environmental impact than batch process. This is due to the reduction of residuals in the production machinery and thus to reduce need for cleaning and better opportunity for the process control allowing the improved resource, energy efficiency and decreasing off-specification products. There are opportunities for environmentally improved technology in batch processes. For chemical batch process for instance, the main waste prevention methods are –
   - Eliminate or minimize unwanted by-products possibly by changing reactants, processes, or equipment
   - Recycle the solvent used in the reactions and extractions
   - Recycle excess reactants

iii) **Changing the nature of steps in the production process:** Physical, chemical, or biological process can affect its environmental impact. Such changes may involve switching from one chemical process to another or from a chemical to a physical or biological process or vice versa. In general, using a selective production route such as through inorganic catalyst and enzymes, which will be environmentally beneficial by reducing inputs and their associated waste, for example, the banning of chlorofluorocarbon led to other ways of producing flexible polyurethane forms. Another example of an environmentally beneficial change in the physical nature of a process is using electrodynamics in spraying. A major problem of spraying process is that a significant amount of material misses its target. In such cases, waste may be generally reduced by giving the target and the spread material opposite electrical charges.

iv) **Eliminating steps in the production processes:** In this process, waste is prevented because each step typically creates waste. In a chemical industry, there is a trend to eliminate the neutralization steps, which generates the waste salts as by-products. This is mainly achieved by using a more selective type of synthesis.

v) **Changing the cleaning processes:** Cleaning is the source of considerable environmental impacts from the production processes. These impacts can partially be reduced by changing inputs in the cleaning process, for example, using water-based cleaners rather than a solvent. In other processes, reduced cleanliness is achieved by minimizing carry over from one process step to next. The switch from batch to continuous process will also usually reduce the need for cleaning.

13.2 CHANGES OF INPUTS IN THE PRODUCTION PROCESS

Changes in inputs are important tools in the green manufacturing. Both major and minor product ingredients, which contribute to the production without being incorporated in the end product, may be worth changing, for example, where changing a minor input in production may substantially reduce its environmental impact, for example, in car and airplanes. The introduction of powder- based and high solid paints substantially reduces the emission of volatile organic compounds. Substituting water-based coating by solvent-based coating may have a less environmental impact.

13.3 INTERNAL REUSE OF WASTE

The potential for internal reuse is often substantial with many possibilities for the reuse of water, energy, some chemicals, and metals. Washing, heating, and cooling in a counter current process will facilitate the internal reuse of energy and water in a closed-loop process. Water recycling, which replaces single-pass system, is usually economically attractive with both water and chemicals potentially being recycled. In some production processes, there may be possibilities for cascade-type reuse, in which water is used in one-process step; where quality requirement is less stringent.

13.4 BETTER HOUSEKEEPING

Good housekeeping refers to generally simple, routinized, and nonresource intensive measures that keep a facility in good working and environmental order. It includes segregating wastes, minimizing chemical and waste inventories, installing overflow alarms and automatic shutoff valves, eliminating leaks and drips, and putting collecting devices at places, where spills may occur, frequent inspection aimed at identifying environmental concerns and potential malfunctioning of the production process, instituting better controls on operating conditions (flow rate, temperature, pressure, etc.), regular fine tuning of machinery, and optimizing maintenance schedules. These types of actions often offer quick, easy, and inexpensive ways to reduce chemicals and wastes.

The green manufacturing processes have a wide scope in all industries, but in recent years, some of the main industries, which got special attention, are pharmaceutical, polymer, petroleum, fine chemicals industries, etc.

13.5 PHARMACEUTICAL INDUSTRY

The pharmaceutical industry is well known for its intensive use of many petrochemicals as starting materials, synthetic route with conventional techniques, high-energy requirements for industrial processes, high use of organic solvents for separation and purification of high-volume waste. Pharmaceutical industry produces more waste per kg of product than other chemical industries (petrochemical, bulk and fine chemical, polymer, etc). The pharmaceutical industry uses six to eight step organic synthetic routes and generates 25–100 kg of waste for every kg of product (Fortunak et al. 2007; Fortunak, 2009). The big pharmaceutical company in its manufacturing uses large amount of solvents and its known water liquid waste contents is about 85–90% organic solvents (Bruggink et al., 2003; Slater and Savefski 2007). In the last decade, pharmaceutical manufacturers embraced green chemistry ideas to promote their environmental credentials and increase the efficiency of their manufacturing processes. Slater et al. (2012) designed a green engineering program for the purification and recovery of isopropanol (IPA) in waste streams of celecoxib process.

Pharmaceutical processes use large amount of organic solvents and liquid waste is 85% nonaqueous in nature. The reduction in the use of organic solvent is an important issue in most of the pharmaceutical industries. Some new organic synthetic routes with minimum of zero solvent were in the research state (Sheldon, 2005). The solvents, which are more acceptable for organic synthetic processes, having low toxicity are normally acetone, ethanol, 2-propanol, ethyl acetate, isopropyl acetate, methylethylketone, 1-butanol, etc. Solvents that are used for their ability to dissolve other chemicals despite their toxicity are cyclehexane, n-hepane, toluene, methylhexane, acetonitrile, THF, DMSO, acetic acid, ethylene glycol, etc. The pharmaceutical industry has initiated many studies on the replacement of such a toxic solvent with solvents that are benign to human health and the environment (Jiménez-González et al., 2001, 2002).

Enzymes can accelerate a reaction, lower the use of energy, use alternative starting materials, reduce the use of solvents and the production of waste. Enzymes are biomaterials that can biodegrade under environmental conditions. The enzymes are used for biocatalysis of the basic steps in the synthesis, reducing the use of solvents by 90% and starting material by 50%. The company will reduce its industrial waste by 200.000 metric tons, compared to the old method.

Doramectin was synthesized as the antiparasitic drug. Changing into biocatalysis synthesis, the efficiency of the reaction was increased by 40% and the by-products are reduced. Biocatalysis improved the synthetic routes for the industrial production operation of these drugs; Oselravimit and Pelitrexol (Harrington et al., 2004; Hu et al., 2006). Biocatalysis has been primarily used in the production of chiral chemical intermediates required for the production of medicines. The "green" biocatalytic synthesis of the active substances like atrovastatin has been reported (Ma et al., 2010). An improved efficient method was developed to make ibuprofen using only three steps instead of the six steps used in earlier method. In this case, all starting materials are converted into products, reclaimed as by-product or completely recycled in the process. Thus, the generation of wastes has been practically eliminated (Cann and Connelly, 2000). Desai (2011) highlighted three generations of process research and development of industrial manufacture of sitagliptin phosphate, which is a leading drug for type 2 diabetes.

Bennett et al. (2019) discussed green manufacture of value-added chemicals with flow chemistry. It was reported that continuous manufacturing techniques are regularly increasingly as they reduce the energy as well as amount of material utilized in a process taking care of control enhanced process safety along with real-time analysis. Such time-efficient and material-efficient flow-screening platforms can be further used in future generation of process development.

13.6 POLYMER INDUSTRIES

Plastics based on petroleum are creating severe environmental hazards because of its less or nondegradability nature, which remained in the environment for a longer period and can cause severe damage to the ecosystem as a whole.

In polymer industry, useful biopolymers are developed from naturally occurring, renewable resources, as an alternative to the petrochemical-based sources, which is in common use presently. Utilizing lignocellulosic and other biomasses, especially residues from food and agricultural processes, is one of the ways to develop new polymer products from industrial wastes or recyclate streams. Hilonga et al. (2012) reported a two-step rapid route of synthesizing inexpensive mesoporous silica, which was used in the manufacturing of green tire. Zhuo et al. (2011) developed a versatile CNT synthesis process, where they have used waste solid polymers like polyethylene, polypropylene, polystyrene etc. Bertaud et al. (2012) have used natural phenolic polymers of tannins and lignin as substitutes of petro-based chemicals, which were used in wood panels to reduce formaldehyde emissions and to develop green adhesives. Chang and Abu-Zahra (2011) developed a robust and cost-effective method to maximize the use of recycled poly (vinyl chloride) (PVC) back into its virgin compounds. In this method, 70% of recycled/regrind PVC was successfully implemented in the extrusion process for manufacturing high-quality foam PVC profiles, which may be used in building industry.

The possibilities for producing eco-friendly bioplastics from leaf extracts of *Manihot esculenta* have been explored by Sakthivel et al. (2020) along with some other degradable blending materials. They effectively blended plant extract with 0.75, 1.125, and 0.565 g of glucose, gelatine, and agar, respectively, along with 1.8 mL of glycerol to produce a plastic-like layer on aluminum foil. This pealed film layer was a bioplastic layer and the isolate was investigated for its water-adsorbing potential. It was observed that the absence of water-adsorbing trait was there. The biodegradability of this bioplastic was investigated by the soil burial test for 15 days and it was found that it was effectively degraded (80 to 90%) as compared with petroleum-based plastics.

13.6.1 BIOPLASTICS

Some green processes have been developed for the synthesis of polyhydroxy-alkanoates (PHAs), a kind of non-petrochemical bioplastics. Bioplastics synthesized by living organisms are generally biodegradable. PHAs continue to attract increasing industrial interest as renewable, biodegradable, biocompatible, and extremely versatile thermoplastics (Steinbüchel and Lütke-Eversloh, 2003; Suriyamongkol et al., 2007). PHAs are the only water-proof thermoplastic materials available that are fully biodegraded both in aerobic and in anaerobic environments. Two classes of PHAs are distinguished according to their monomer composition: short-chain length (SCL) PHAs and medium-chain length (MCL) PHAs. SCL-PHAs are polymers of 3-hydroxyacid monomers with a chain length of three to five carbon atoms, such as poly(3-hydroxybutyrate) (PHB, the most common PHA), whereas MCL-PHAs contain 3-hydroxyacid monomers with six to sixteen carbon atoms. All of them are optically active *R*-(—) compounds. This versatility is partly due to the wide substrate range of the PHA-synthesizing enzymes, and gives PHAs an extended spectrum of associated properties, which is a clear advantage *vis-à-vis* to other bioplastics. Around 200 different monomer constituents were found in the polymers analyzed so far.

The versatile copolymer P(HB-*co*-HV) was initially manufactured as shampoo bottles and other cosmetic containers (Hocking and Marchessault, 1994). Later on, pens, cups, and packaging elements (e.g., films) made with PHAs also appeared in the market. PHAs are biocompatible and for this reason, they have also attracted attention as raw material to be used in medical devices. Being composed by *R*-(—) monomers, PHAs are source of chiral compounds with a high demand from the pharmaceutical industries also (Chen and, Wu, 2005). However, the manufacture of PHAs is carried out at small facilities and, as a consequence, it lacks the economic benefit of a large-scale production (Chanprateep, 2010).

13.7 OTHER INDUSTRIES

Tu et al. (2012) developed a green method for manufacturing $CuFe_2O_4$ from industrial Cu sludge. It was successfully applied for combustion of volatile organic compounds (VOCs) derived from isopropyl alcohol. Zhang et al. (2013) converted formic acid into glycerin with a yield of 31.0%. Their work may be explored for the production of formic acid from renewable biomass. Triammonium citrate was manufactured by using neutralizating of citric acid solution with liquid ammonia directly by Yang et al. (2012). Their method has advantages of 100% raw materials utilization ratio and totally zero emission during mass production over conventional route. Liu et al. (2011) demonstrated the versatile use of CO_2 in organic synthesis as the alternative carbonyl source of phosgene for the manufacturing of some compounds such as cyclic carbonates, oxa-zolidinones, ureas, isocyanates, and polymers. A green chemical approach was used to synthesize silica xerogels from sodium silicate through a cost-effective way by Pirzada et al. (2012). It was synthesized using the waste material (hexafluorosilicic acid, H_2SiF_6) of phosphate fertilizer industry and sodium silicate ($Na_2O \cdot SiO_2$). Cationic surfactant hexadecyl trimethyl ammonium bromide (CTAB) was used as the structure template.

Shanmuganathan et al. (2011) observed that the photocurable mixture of a multifunctional acrylate, a tetrafunctional thiol, and a photoinitiator can be processed into continuous fibers by *in situ* photopolymerization during electrospinning under ambient conditions. The prepared fibers were mechanically robust and have excellent chemical and thermal stability. A novel eco-sustainable catalytic pathway was developed by Falletta et al. (2011) for the synthesis of 3-hydroxypropionic acid from allyl alcohol. This method highlights the good potential of gold-based and bimetallic catalysts in the aerobic oxidation of allyl alcohol. Lovell et al. (2010) investigated green lubricant combinations, which were prepared by homogeneously mixing of nano-, submicrometer-, and micrometer-boric acid powder additives with canola oil in a vortex generator.

Pallavkar et al. (2010) studied the use of microwave energy to accomplish high-temperature destruction of p-xylene in a packed bed reactor, which was performed using a SiC (silicon carbide) foam, while Cambronero et al. (2009) manufactured Al-Mg-Si alloy foam using calcium carbonate as a foaming agent. The prepared foam showed a low degree of aluminum draining, no wall cell cracks and a good fine cell size distribution.

Menzler et al. (2010) manufactured anode-supported solid oxide fuel cells (SOFC) by different wet chemical powder processes and subsequent sintering at high temperatures. The cell was characterized for its slurry viscosity, green tape thickness, relative density, substrate strength, electrical conductivity, and shrinkage. Li et al. (2009) studied an environmentally benign and cost-effective production method of nanoscale zero-valent iron (nZVI) using a precision milling system.

Various solid acids like zeolites, ion-exchange resins, and mixed metal oxides are known catalysts in the esterification of dodecanoic acid with 2-ethylhexanol, 1-propanol, and methanol out of which sulfated zirconia is considered most promising material (Kiss et al., 2006). Chromium compound is an important chemical for many industries, which normally shows quite low utilization efficiency of resources and energy. Discharge of chromium-containing toxic solid wastes and gas results in serious pollution problems. A green manufacturing process of chromium compounds has been developed with the design objective of eliminating pollution at the source. this green process achieves higher resource utilization efficiency and zero emissions of chromium-containing waste residue by alteration of process chemistry, change of reactor, operation, regeneration and recycle of reaction media, and comprehensive use of resources (Zhang et al., 2005).

The vat dye manufacturing process is responsible for a large amount of contamination of both soil and groundwater. Recently, dyeing methods have been improved, with better engineering, better control, and the ability to hot wash, which has enabled dyers to process textiles in the right manner considering the environmental impacts. Solid acids, especially, which are based on micelle-templated silicas and other mesoporous high surface area support materials play an important role in the greening of fine chemicals manufacturing processes. The bamboo is considered a renewable and bio-degradable resource in textiles, as it is the growing demand for more comfortable, healthier, and environment-friendly products in textile industry (Rekha and Sudam, 2009).

Recently, novel biocatalytic pharmaceutical processes have been developed to replace chemical routes, which contain poorer process efficiency and higher manufacturing costs (Chen and Wu, 2007; Tao and Xu, 2009). Green chemistry challenges the innovators to take care to design and utilize matter and energy in a way that increases performance and value, while protecting human health and the environment (Manley et al., 2008). In present scenario, the key issues of switching to renewable resources, avoiding hazardous and polluting processes for manufacturing and using safe and environmentally compatible products, one needs to develop sustainable and green chemical product supply chains. Organic chemicals and materials need to operate under agreed and strict criteria and need to start with widely available, totally renewable, and low-cost carbon (the only source is biomass) and the conversion of biomass into useful products will be carried out in biorefineries (Clark, 2007).

Glycerol carbonate is a derivative of glycerol, which is commonly used in industrial applications. It was reported by Ochoa-Gomez et al. (2012) that transesterification of dimethyl carbonate or ethylene carbonate with glycerol using uncalcined CaO as a catalyst is the most suitable industrial process.

Waste valorization practices have attracted attention of industrialists in recent years with the aim of managing waste in a most sustainable way. Food waste is a large residue, which can be used to derive a variety of valuable chemicals (Luque and Clark, 2013).

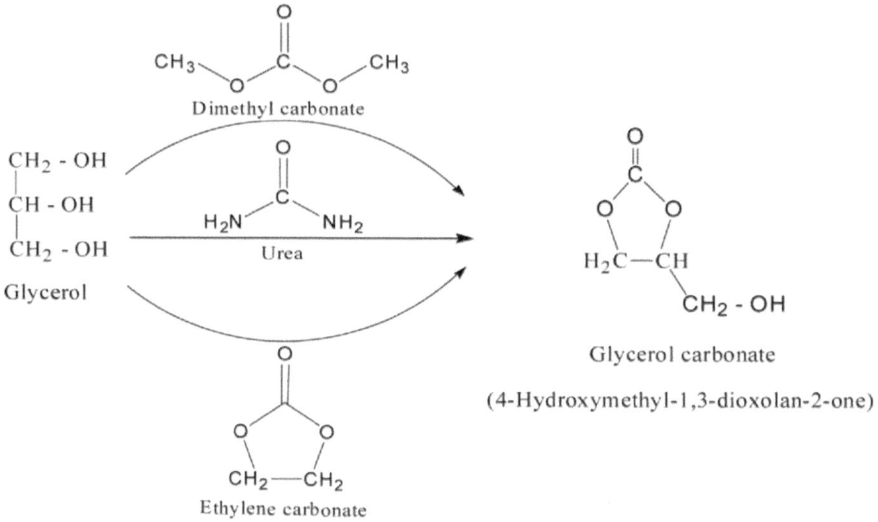

Valorization is a good strategy to convert waste biomass into useful chemicals. A redox economic approach is to convert carbohydrates into oxygenates either by chemocatalytic processes or by fermentation. Sheldon (2014) described two approaches: (i) Carbohydrate's fermentation to lower alcohols, diols, carboxylic acids, and hydrocarbons and (ii) hydrolysis of hexoses (acid catalyzed) to hydroxymethyl furfural and their further conversion to levulinic acid, γ-valerolactone, and furan dicarboxylic acid. Three possible routes are there for the production of polyethylene terephthalate, which is a bio-based equivalent of the large volume polymer. Valorization of waste protein can also provide amino acids such as L-glutamic acid and L-lysine.

Drummond et al. (2014) presented a possible industrial scale design for bioinspired green production of precipitated silica. This green process utilizes an additive, which allowed rapid formation of silica in less than 5 min at room temperature and that too at neutral pH. The green process has significant advantages such as production capacity without any heat requirements, which therefore reduced both emission of carbon dioxide and running costs.

Wang et al. (2015) studied reducing disposal issues of harmful chemicals using water-based ultra-polishing experiment with alumina abrasives. It was reported that the material removal rate (MRR) was quite sensitive to pH, slurry flow rate, and concentration of oxidant. The highest MRR was obtained as: Slurry flow rate (71.86 mL min^{-1}), oxidizer concentration (0.44%), and pH (7), under optimum conditions.

Improved antibacterial knitted fabrics were reported by Alay et al. (2016). Lyocell polylactic acid (PLA) and chitosan fibers were mixed to produce single-jersey knitted fabric composed of lyocell (15%), PLA (80%), and chitosan (5%). Biodegradation of this knitted fabric was tested by the soil burial method for a period of 9, 3, and 6 months.

It is very important to turn concept or experiments on laboratory scales into manufacture to industrial scales. Efforts are being made to use green chemical routes to develop green manufacturing processes, which are most desired but less taken care of.

13.8 RECENT DEVELOPMENTS

Biocatalysis has significantly contributed in manufacturing processes of pharmaceutical, flavor, food, vitamin, polymer, fragrance, agrochemical, specialty, and fine chemical industries. Alcantara et al. (2022) reported that the major benefits of using biocatalysis are:

- Lower costs
- More sustainable
- Lower E-factor/less waste
- Higher selectivity
- Lower CO_2 emission
- Better step economy

Galant et al. (2022) opined that scale-up of mechanochemical methods could play a major and transformative role in pharmaceutical manufacturing processes by avoiding the use of solvent. They have also evaluated some sustainability and green chemistry metrics in the production of nitrofurantoin, which is an active pharmaceutical ingredient (API), via conventional solvent-batch synthesis and mechanochemical continuous twin-screw extrusion (TSE) method. It was observed that there was a significant reduction in all metrics in this TSE such as climate change, energy, and human ecological health, reduced cost, avoiding solvents, and excess reactant consumption with high selectivity of product.

Omran and Baek (2022) trend in the valorization of zero-cost and readily available biodegradable agro-industrial biowaste to prepare green bio-nanosorbents and bio-nanocatalysts for wastewater treatment. They have also highlighted environmental, health impacts, and economic potential of valorizing agro-industrial biowaste to green nanomaterials.

Japanese green tea (Sencha) is normally manufactured by these six processing steps:

- Primary rolling,
- Steaming,
- Rolling,
- Secondary rolling,
- Final rolling, and
- Final drying

Qin et al. (2022) evaluated total antioxidant activity and total polyphenol content (TPC). It was found that folds or wrinkle-like microstructures present on the tea leaf surface were increased during processing, and particularly during the primary rolling (PR) step and leaf shape was also collapsed. It was indicated that the processing conditions in each step, and in particular, the steps for steaming, primary rolling, and rolling affected quality of the final tea leaf products along with functional properties of tea leaves and infusions.

Sathish et al. (2022) applied GlaxoSmithKline's solvents ranking model (GSK's SRM) to determine environmental efficiency of cyclic carbonate-based alkali neutralization process in leather making. These cyclic carbonates have attracted the attention of green chemist due to low toxicity, availability, and biodegradability. They used propylene carbonate (PC) as a deliming agent (an alkali neutralizing agent) in leather making, which avoided the generation of toxic ammonia gas. It was reported that 2.5 and 1.5% (w/w) of propylene carbonate is required to completely neutralize the calcium hydroxide (alkali) present in the cow hide, and goatskin, respectively. It was also observed that the rate of neutralization was faster as compared to ammonium salt-based process. It was claimed that a reduction of 96, 92, and 76.3% was there in chloride content, total Kjeldahl nitrogen (TKN), and reduction in total dissolved solids (TDS) in wastewater using as-developed process respectively. The environmental efficiency index (EEI) value of this neutralization (27.44) was about 1.5 times higher as compared to conventional ammonium salt-based system (17.93).

Modification of a polymer surface is a major step in manufacturing different parts for applications in catalysis, biotechnology, sensing, surgical, solar cells, etc. Robustness of the material is sometimes required as these have to face harsh chemical treatment, temperature, and high-energy consumption. Zimmermann et al. (2022) presented an approach for chemical functionalization of surfaces of polymers, which does not require harmful or toxic chemicals and it is energy efficient also. Embedding polyethylenimine-coated calcium carbonate particles into polymer melts is advantageous, which will allow these soluble particles for further modification of the polyethylenimine. This type of modification resulted in surfaces, which feature undercuts and have hierarchically structured making them robust against any wear and tear. Sidewise, it increases specific area available for functionalization. They demonstrated such fictionalizations with metal layers, dyes, and metal nanoparticles.

KEYWORDS

- **biocatalytic**
- **bioplastics**
- **green manufacturing**
- **polymers**
- **pharmaceutical drugs**

REFERENCES

Alay, E.; Duran, K.; Korlu, A. A Sample Work on Green Manufacturing in Textile Industry. *Sustain. Chem. Pharm.* **2016,** *3*, 39–46.

Alcantara, A. R.; Domínguez de María, P.; Littlechild, J. A.; Schürmann, M.; Sheldon, R. A.; Wohlgemuth, R. Biocatalysis as Key to Sustainable Industrial Chemistry. *ChemSusChem* **2022,** *15* (9). DOI: 10.1002/cssc.202102709.

Bennett, J. A.; Campbell, Z. S.; Abolhasani, M. Role of Continuous Flow Processes in Green Manufacturing of Pharmaceuticals and Specialty Chemicals. *Curr. Opin. Chem. Eng.* **2019,** *26*, 09–19.

Bertaud, F.; Tapin-Lingua, S.; Pizzi, A.; Navarrete, P.; Petit-Conil, M. Development of Green Adhesives for Fibreboard Manufacturing, Using Tannins and Lignin from Pulp Mill Residues. *Cellulose Chem. Technol.* **2012,** *46* (7–8), 449–455.

Bruggink, A.; Straathof, A. J. J.; van der Wielen, L. A. M. *A 'Fine' Chemical Industry for Life Science Products: Green Solutions to Chemical Challenges in Process Integration in Biochemical Engineering*; von Stockar, U., van der Wielen, L., Eds.; Springer: New York, 2003, pp 69–113.

Cambronero, L. E. G.; Ruiz-Roman, J. M.; Corpas, F. A.; Prieto, J. R. Manufacturing of Al–Mg–Si Alloy Foam Using Calcium Carbonate as Foaming Agent. *J. Mater. Process. Technol.* **2009,** *209* (4), 1803–1809.

Cann, M. C.; Connelly, M. E. *Real-World Cases in Green Chemistry*; American Chemical Society: Washington, DC, 2000.

Chang, H.; Abu-Zahra, N. One Step Forward to a Sustainable Green Solution for Extruded Foam PVC Building Products. *J. Vinyl Addit. Technol.* **2011,** *17* (1), 17–20.

Chanprateep, S. Current Trends in Biodegradable Polyhydroxyalkanoates. *J. Biosci. Bioeng.* **2010,** *110* (6), 621–632.
Chen Y.; Wu X. Biocatalysis in Drug Discovery and Development. *Prog. Chem.* **2007,** *19* (12), 1947–1954.
Chen, G. Q.; Wu, Q. Microbial Production and Applications of Chiral Hydroxyalkanoates. *Appl. Microbiol. Biotechnol.* **2005,** *67* (5), 592–599.
Clark, J. H. Green Chemistry for the Second Generation Biorefinery—Sustainable Chemical Manufacturing Based on Biomass. *J. Chem. Technol. Biotechnol.* **2007,** *82* (7), 603–609.
Desai, A. A. Sitagliptin Manufacture: A Compelling Tale of Green Chemistry, Process Intensification, and Industrial Asymmetric Catalysis. *Angew. Chem. Int. Ed.* **2011,** *50* (9), 1974–1976.
Drummond, C.; McCann, R.; Patwardhan, S. V. A Feasibility Study of the Biologically Inspired Green Manufacturing of Precipitated Silica. *Chem. Eng. J.* **2014,** *244*, 483–492.
Falletta, E.; Della Pina, C.; Rossi, M.; He, Q.; Kiely, C. J.; Hutchings, G. J. Enhanced Performance of the Catalytic conversion of Allyl Alcohol to 3-Hydroxypropionic Acid Using Bimetallic Gold Catalysts. *Faraday Disc.* **2011,** *152*, 367–379.
Fortunak, J. M. Current and Future Impact of Green Chemistry on the Pharmaceutical Industry. *Future. Med. Chem.* **2009,** 1, 571–575.
Fortunak, J. M.; Confalone P. N.; Grosso J. A. Strength and Honor Through the Pharmaceutical Industry's Embrace of Green Chemistry? *Curr. Opin. Drug Discov. Develop.* **2007,** *10*, 651–653.
Galant, O.; Cerfeda, G.; McCalmont, A. S.; James, S. L.; Porcheddu, A.; Delogu, F. et al. Mechanochemistry Can Reduce Life Cycle Environmental Impacts of Manufacturing Active Pharmaceutical Ingredients. *ACS Sustain. Chem. Eng.* **2022,** *10* (4), 1430–1439.
Harrington, P. J.; Brown, J. D.; Foderaro, T.; Hughes, R. C. Research and Development of a Second-Generation Process for Oseltamivir Phosphate, Prodrug for a Neuraminidase Inhibitor. *Org. Process Res. Dev.* **2004,** *8* (1), 86–91.
Hilonga, A.; Kim, J. K.; Sarawade, P. B.; Quang, D. V.; Shao, G. N.; Elineema, G.; et al. Synthesis of Mesoporous Silica with Superior Properties Suitable for Green Tire. *J. Ind. Eng. Chem.* **2012,** *18* (5), 1841–1844.
Hocking, P. J.; Marchessault, R. H. Biopolyesters. In *Chemistry and Technology of Biodegradable Polymers*; Griffin, G. J. L., Ed.; Blackie Academic & Professional: Glasgow, 1994.
Hu, S.; Kelly, S.; Lee, S.; Tao, J.; Flahive, E. Efficient Chemoenzymatic Synthesis of Pelitrexol via Enzymic Differentiation of a Remote Stereocenter. *Org. Lett.* **2006,** *8* (8), 1653–1655.
Jiménez-González, C.; Constable, D. J.; Curzons, A. D.; Cunningham, V. L. Developing GSK's Green Technology Guidance: Methodology for Case-Scenario Comparison of Technologies. *Clean Technol. Environ. Policy* **2002,** *4* (1), 44–53.
Jiménez-González, C.; Curzons, A. D.; Constable, D. J.; Overcash, M. R.; Cunningham, V. L. How Do You Select the "Greenest" Technology? Development of Guidance for the Pharmaceutical Industry. *Clean Technol. Environ. Policy* **2001,** *3* (1), 35–41.
Kiss, A. A.; Dimian, A. C.; Rothenberg, G. Solid Acid Catalysts for Biodiesel Production-Towards Sustainable Energy. *Adv. Synth. Catal.* **2006,** *348* (1–2), 75–81.
Li, S.; Yan, W.; Zhang, W. X. Solvent-Free Production of Nanoscale Zero-Valent Iron (nZVI) with Precision Milling. *Green Chem.* **2009,** *11* (10), 1618–1626.
Liu, A. H.; Li, Y. N.; He, L. N. Organic Synthesis Using Carbon Dioxide as Phosgene-Free Carbonyl Reagent. *Pure Appl. Chem.* **2011,** *84* (3), 581–602.
Lovell, M. R.; Kabir, M. A.; Menezes, P. L.; Higgs III, C. F. Influence of Boric Acid Additive Size on Green Lubricant Performance. *Phil. Trans. Royal Soc. A: Math. Phys. Eng. Sci.* **2010,** *368* (1929), 4851–4868.

Luque, R.; Clark, J. H. Valorisation of Food Residues: Waste to Wealth Using Green Chemical Technologies. *Sustain. Chem. Process.* **2013,** *1* (1). DOI: 10.1186/2043–7129–1-10.

Ma, S. K.; Gruber, J.; Davis, C.; Newman, L.; Gray, D.; Wang, A. et al. A Green-by-Design Biocatalytic Process for Atorvastatin Intermediate. *Green Chem.* **2010,** *12* (1), 81–86.

Manley, J. B.; Anastas, P. T.; Cue Jr, B. W. Frontiers in Green Chemistry: Meeting the Grand Challenges for Sustainability in R&D and Manufacturing. *J. Clean. Prod.* **2008,** *16* (6), 743–750.

Menzler, N. H.; Schafbauer, W.; Buchkremer, H. P. Influence of Processing Parameters on the Manufacturing of Anode-Supported Solid Oxide Fuel Cells by Different Wet Chemical Routes. *Mater. Sci. Forum.* **2010,** *638*, 1098–1105.

Ochoa-Gómez, J. R.; Gómez-Jiménez-Aberasturi, O.; Ramirez-Lopez, C.; Belsué, M. A Brief Review on Industrial Alternatives for the Manufacturing of Glycerol Carbonate, a Green Chemical. *Org. Proc. Res. Develop.* **2012,** *16* (3), 389–399.

Omran, B. A.; Baek, K. H. Valorization of Agro-Industrial Biowaste to Green Nanomaterials for Wastewater Treatment: Approaching Green Chemistry and Circular Economy Principles. *J. Environ. Manage.* **2022,** *311*. DOI: 10.1016/j.jenvman.2022.114806.

Pallavkar, S.; Kim, T. H.; Lin, J.; Hopper, J.; Ho, T.; Jo, H. J.; Lee, J. H. Microwave-Assisted Noncatalytic Destruction of Volatile Organic Compounds Using Ceramic-Based Microwave Absorbing Media. *Ind. Eng. Chem. Res.* **2010,** *49* (18), 8461–8469.

Pirzada, T.; Demirdoogen, R. E.; Shah, S. S. A Green Chemistry Approach to Synthesize Ctab Templated Silica Xerogels from Sodium Silicate. *J. Chem. Soc. Pak.* **2012,** *34*, 177–183.

Qin, W.; Yamada, R.; Araki, T.; Ogawa, Y. Changes in Morphological and Functional Characteristics of Tea Leaves During Japanese Green Tea (Sencha) Manufacturing Process. *Food Bioproc. Tech.* **2022,** *15* (1), 82–91.

Rekha, R.; Sudam, A. Bamboo: Green and Breathable Natural Fiber. *Man-Made Text. India* **2009,** *52*, 397–401.

Sakthivel, G.; Natarajan, D.; Kandasamy, G.; Kandasamy, S.; Vijayan, S.; Umadevi, G.; et al. Development of Ecofriendly Bio-Plastics from the Leaf Extract of Manihot Esculenta. *Chem. Sci. Rev. Lett.* **2020,** *9* (36), 978–985.

Sathish, M.; Thaikaivelan, P.; Rao, J. R. Application of GSK's Model in Leather Making: Quantification of the Environmental Efficiency of a Green Solvent Based Deliming Process. *ACS Sustain. Chem. Eng.* **2022,** *10* (15), 4943–4953.

Shanmuganathan, K.; Sankhagowit, R. K.; Iyer, P.; Ellison, C. J. Thiol–Ene Chemistry: A Greener Approach to Making Chemically and Thermally Stable Fibers. *Chem. Mater.* **2011,** *23* (21), 4726–4732.

Sheldon, R. A. Green and Sustainable Manufacture of Chemicals from Biomass: State of the Art. *Green Chem.* **2014,** *16* (3), 950–963.

Sheldon, R. A. Green Solvents for Sustainable Organic Synthesis: State of the Art. *Green Chem.* **2005,** *7* (5), 267–278.

Slater C. S.; Savefski M. A Method to Characterize the Greenness of Solvents Used in the Pharmaceutical Manufacturing. *Environ. Sci. Heath A, Environ. Sci. Eng. Toxic Hazard. Subst. Control* **2007,** *2*, 1595–1605.

Slater, C. S.; Savelski, M.; Hounsell, G.; Pilipauskas, D.; Urbanski, F. Green Design Alternatives for Isopropanol Recovery in the Celecoxib Process. *Clean Technol. Environ. Policy* **2012,** *14* (4), 687–698.

Steinbüchel, A.; Lütke-Eversloh, T. Metabolic Engineering and Pathway Construction for Biotechnological Production of Relevant Polyhydroxyalkanoates in Microorganisms. *Biochem. Eng. J.* **2003,** *16* (2), 81–96.

Suriyamongkol, P.; Weselake, R.; Narine, S.; Moloney, M.; Shah, S. Biotechnological Approaches for the Production of Polyhydroxyalkanoates in Microorganisms and Plants—A Review. *Biotechnol. Adv.* **2007,** *25* (2), 148–175.

Tao, J.; Xu, J. H. Biocatalysis in Development of Green Pharmaceutical Processes. *Curr. Opin. Chem. Biol.* **2009,** *13* (1), 43–50.

Tu, Y. J.; Chang, C. K.; You, C. F. Combustion of Isopropyl Alcohol Using a Green Manufactured $CuFe_2O_4$. *J. Hazard. Mater.* **2012,** *229*, 258–264.

Wang, Y. G.; Chen, Y.; Zhao, Y. W. Chemical Mechanical Planarization of Silicon Wafers at Natural pH for Green Manufacturing. *Int. J. Precis. Eng. Manuf.* **2015,** *16* (9), 2049–2054.

Yang, C. M.; Zhao, W. L.; Wang, H. B.; Zhang, X. F. A Green Route to Mass Production of Anhydrous Triammonium Citrate. *Adv. Mater. Res.* **2012,** *518*, 3908–3911.

Zhang, Y. L.; Zhang, M.; Shen, Z.; Zhou, J. F.; Zhou, X. F. Formation of Formic Acid from Glycerine Using a Hydrothermal Reaction. *J. Chem. Technol. Biotechnol.* **2013,** *88* (5), 829–833.

Zhang, Y.; Li, Z. H.; Qi, T.; Zheng, S. L.; Li, H. Q.; Xu, H. B. Green Manufacturing Process of Chromium Compounds. *Environ. Prog.* **2005,** *24* (1), 44–50.

Zhenming, C.; Jinhua, L.; Tao, J. Biocatalysis for Green Chemistry and Drug Development. *Prog. Chem.* **2007,** *19* (012), 1919–1927.

Zhuo, C.; Hall, B.; Levendis, Y.; Richter, H. A Novel Technology for Green (er) Manufacturing of CNTs via Recycling of Waste Plastics. *MRS Online Proc. Lib. (OPL)* **2011,** 1317. DOI: 10.1557/opl.2011.139.

Zimmermann, P.; Schlenstedt, K.; Schwarz, S.; Vehlow, D.; Blanke, M.; Fery, A.; Nagel, J. Green Approach for Manufacturing of Polymer Surface Structures with Microcavities Having Robust Chemically Functionalized Inner Surfaces. *ACS Appl. Poly. Mater.* **2022,** *4* (7), 5189–5198.

CHAPTER 14

Future Trends

SURESH C. AMETA

Department of Chemistry, PAHER University, Udaipur, India

ABSTRACT

Most of the green starting materials are either biomass or biomass-derived materials, but it is not possible to fulfill all the demands of the society such as food, polymers, textiles, healthcare products, cosmetics, energy sources, detergents, pesticides, paints, etc. These cannot be procured by biomass or biomass derivatives only. Therefore, it has become necessary to find out newer greener alternatives for starting materials to obtain these materials. This may be the future prospects of green chemistry. Green solvents, green catalysts, green reaction conditions, benign starting materials, etc. are required to prepare these products. Biodegradability is another concern of the final product.

The environment is a surrounding for all of us and it comprises a variety of physical and chemical components. One interacts with this environment and it is also a part of it. It is well established that science, particularly, chemical sciences is developing at a very fast pace. As a result, a huge amount of chemicals are used in manufacturing processes and they are increasing in the environment. These chemical components are increasing day by day in the environment, many of which are undegradable. This causes environmental pollution. Here, it may be concluded that the addition of these undegradable or recalcitrant substances or molecules is causing disorder, harm, discomfort, or instability to our ecosystem. It creates nothing but pollution.

Most of the green starting materials either come from biomass or biomass-derived materials, but it is not possible to fulfill all the requirements of useful chemicals in society, such as food materials, polymers, textiles, healthcare

Green Chemistry, 2nd Edition: Fundamentals and Applications. Suresh C. Ameta & Rakshit Ameta (Eds.)

products, cosmetics, energy sources, detergents, pesticides, paints. These may not be prepared by only biomass or biomass derivatives. It is therefore necessary to find out newer greener alternatives for starting materials and this may be the future prospects in this direction.

All the products used may not be biodegradable or may be recycled, and thus may create a problem with their disposal. Wood, iron, textile, etc. are being rapidly replaced by some or the other kind of polymer, and therefore, this is called a polymer era. This may be appreciated on the one hand, but generates a large amount of waste also on the other. Therefore, there is an urgent need to search for biodegradable polymers. The same is true for detergents, which are almost replaced by washing soap. It is predicted that more and more products will come in the future, which are either biodegradable or degraded into less harmful products.

Most of the chemical reactions are slow enough and require a catalyst for the enhancement of their reaction rate. Catalysts available at present are either metal or metal-based, which are toxic in nature. At this stage, enzymes are more preferred to drive a reaction, but these are quite specific or have their own limitations. So searching for newer green catalysts is a demand of the day, which are specific in catalyzing reactions and are less toxic as compared with the existing conventional catalysts. The next few decades will witness some more green catalysts.

Volatile organic solvents create a lot of pollution by their vapors. Thus, there is an urgent requirement of some high-boiling solvents, which will dissolve the majority of organic and inorganic reactants. This requirement is fulfilled by ionic liquids. One of the beauties of ionic liquid is that it can be designed as per our requirement. These are truly designated as designer solvents. We will witness some more interesting designed ionic liquids in the future.

Some chemical reactions can now be carried out in supercritical carbon dioxide ($scCO_2$) and supercritical water (scH_2O); however, the conditions are quite cumbersome. It is hoped that these supercritical fluids will find a proper place in the field of organic synthesis in years to come.

Water is a universal solvent as well as a green solvent also. Many other green solvents, such as cyclopentyl methyl ether (CPME), 2-methyltetrahydrofuran, polyethylene glycol, 1,3-dioxolane have found their place as green solvents, but the search is still incomplete. Efforts are being made to carry out many more reactions in water and other green solvents. Sidewise newer green solvents may be searched out, which fulfills the demand for any chemical reactions.

Photocatalysis is an emerging technology for wastewater treatment. It is an advanced oxidation process, which completely mineralizes most of the organic pollutants. It provides an electron–hole pair, which can be used for both oxidation and reduction. Newer photocatalysts can be synthesized with different energy levels of conduction and valence bands or the available semiconductor can be modified by sensitization, metallization, or doping to have better photocatalytic activities.

Photo-Fenton reagent is an eco-friendly reagent as ferrous and ferric ions are recycled. This reagent is more active than the Fenton reaction. It also degrades a number of pollutants. There is a lot more scope in developing photo-Fenton-like reagents where iron is replaced by some other metal ion with variable oxidation states.

Ultrasound has found many applications, but it has been used to carry out chemical reactions of synthetic importance only a few decades back. It creates a very high temperature and pressure to drive chemical reactions. Its combination with some other advanced oxidation processes may add a new chapter in the field of combating against ever-increasing water pollution. We can also see some more sonochemical reactions of importance in organic synthesis in the future.

Microwave-assisted organic synthesis (MAOS) has opened an avenue in the field of organic synthesis. This type of reaction is less time-consuming, provides more yield with greater purity, and needs less solvent, or it is almost solvent-free. The majority of reactions have been carried out under microwave irradiation, and the list is becoming quite long. MAOS has been successfully used in laboratory conditions, and in some cases, even on an industrial scale and time is not far off, when MAOS may substitute the majority of organic synthesis.

Nature has provided us with biocomposites and there are good examples of green composites like the networking of pipelines of water supply in plants, spider web, etc. Green composites have combined the properties of two materials of different properties, such as hardness, softness, flexibility, etc. but the most important is that these composites should be biodegradable. Some more new green composites will find their place in the coming decades.

Green chemical processes have many established examples on a laboratory scale, but only limited examples are available on an industrial scale. It is utmost necessary to bring these laboratory exercises and prove their worth on industrial level, the so-called green manufacturing processes. While designing a green manufacturing process, one has to take care of the environment along with its economic viability to make this process acceptable by society.

Marine biofouling is an ever-exciting problem that is because of different human marine activities and marine industries. Biofouling affects the environment, economy, ecology, and safety adversely. Initially, metal ions were utilized to prevent biological contamination, but their use also creates environmental pollution and is detrimental to the ecosystem. Therefore, efficient and environmentally friendly coatings are the need of the day, so marine devices can be prevented from biofouling.

Index

A

Acid black 1 (AB 1), 287
Acid red 17 (AR17), 386
Acrylonitrile- butadiene-styrene (ABS), 459
Active pharmaceutical ingredient (API), 517
Advanced oxidation processes (AOPs), 279, 323
Aircrafts, 479–480
Aldol condensation, 225
Alkylations, 202, 253, 417–418
2-Amino-2 chromones, 238–239
Apparatus, 405–406
Aspen wood sawdust (AWS), 469
Automobile shredder residue (ASR), 339

B

Band gap, 281–282
Better housekeeping, 510
Biocatalysis, 517–519
Biocomposite flowerpots (BFP), 489
Biodegradation, 487–491

C

Carbon dioxide (CO_2), 195–208, 205–208
  alkylation, 202
  coupling, 201–202
  Diel's-Alder reaction, 197
  esterification, 202–203
  extraction, 203–204
  Freidel–Crafts reaction, 197
  Heck reactions, 201
  hydrogenation, 200–201
  oxidation, 198–200
  photochemical, 202
  solubility, 204–205
  supercritical polymerization, 198
  synthesis, 205
Carbon fiber reinforced polymer (CFRP), 488
Carbon fibers (CFs), 488
Carbon nanotube (CNT), 297
Carbon quantum dots (CQDs), 288
Carboxylic acids, 330
Catalyst design, 245–246
Cationic surfactant hexadecyl trimethyl ammonium bromide (CTAB), 514
Cetylpyridinium chloride (CPC), 293
Cetyltrimethylammonium chlorochromate (CTMACC), 415
Claisen rearrangement, 224–225
Cogon grass fiber (CGF), 491
Composites, 465
  fiber biocomposites, 465–469
  flax, 471–472
  jute, 469–471
Compression molding (CM), 492
Cyclodextrins (CD), 291
Cyclopentyl methyl ether (CPME), 248–255, 254–255, 524
  alkylating agent, 249
  alkylation, 253
  enolate chemistry, 251
  Friedel-crafts, 249
  Grignard-type, 249–250
  oxidation, 252
  polymerization, 253
  presence of acid, 253
  reduction, 252–253
  solar cells, 255
  transformations, 251–252
  transition metal catalysts, 250

D

Date palm rachis (DPR), 493
Decarboxylation, 238
Decomposed matrix polymer (DMP), 488
Dibutyl itaconate (DBI), 492
Dimethyl acetylenedicarboxylate (DMAD), 225
Dimethyl sulfoxide (DMSO), 436
Dipolar polarization, 404
Dye-sensitized solar cells (DSSCs), 389

E

Electron hole pair
  fate, 283–284
Enolate chemistry, 251
Environmental efficiency index (EEI), 518
Environmental-friendly chemistry, 409

F

Fenton's reagent, 326–327
2,5-Furandicarboxylic acid (FDCA), 416
Fused deposition modeling (FDM), 492
Future trends, 523
  cyclopentyl methyl ether (CPME), 524
  microwave-assisted organic synthesis (MAOS), 525

G

Gamma-valerolactone (GVL), 262
Glass fiber (GF), 491
Glycerol, 247–248
  biocatalysis, 245
  catalyst design, 245–246
  condensation, 247
  coupling, 247
  cycloaddition, 247
  enhancing reaction, 244–245
  materials chemistry, 246
  nanoparticles, 246–247
  organic synthesis, 243–244
  separation, 246
Graphene oxide (GO), 288, 416
Green composites, 453
  acrylonitrile- butadiene-styrene (ABS), 459
  aircrafts, 479–480
  applications, 478–484
  aspen wood sawdust (AWS), 469
  automobiles, 479
  biocomposite flowerpots (BFP), 489
  biodegradation, 487–491
  building products, 481–482
  carbon fiber reinforced polymer (CFRP), 488
  carbon fibers (CFs), 488
  cellulose-based, 470
  chemical degradation, 486
  cogon grass fiber (CGF), 491
  composites, 465
    fiber biocomposites, 465–469
    flax, 471–472
    jute, 469–471
  compression molding (CM), 492
  computers, 482–483
  construction, 481–482
  date palm rachis (DPR), 493
  decomposed matrix polymer (DMP), 488
  degradation, 484–486
  designing, 455–457
  dibutyl itaconate (DBI), 492
  fiber, 476–477
  fused deposition modeling (FDM), 492
  geometry and orientation, 477–478
  glass fiber (GF), 491
  grinding, 486
  halloysite nanotubes (HNTs), 481
  hedysarum coronarium (HC), 492
  hydrolyzed soybean protein isolate (HSPI), 489
  incineration, 486
  kenaf fiber, 459
  life cycle assessment (LCA), 454, 457–458
  linear-low density polyethylene (LLDPE), 488
  micro-fiber bundles, 459
  microfibrillated cellulose (MFC), 468
  micro-fibrillated materials, 459
  miscellaneous, 483–484
  mobile phones, 482–483
  nanofibrillated cellulose (NFC), 468
  natural fiber sources, 458–461
  N-dimethylformamide (DMF), 490
  oil palm empty fruit bunch (OPEFB), 481
    fiber, 460
  packaging, 480–481
  poly(3- hydroxybutyrate-co-3-hydroxyvalerate) (PHBV), 488
  poly(butylene-succinate) (PBS), 491
  poly(lactic acid) (PLA), 459, 489
  polybutylene adipate terephthalate (PBAT), 489, 493
  polycaprolactone (PCL), 489
  polyethylene glycol (PEG), 464
  polyhydroxy alkanoates (PHA), 462, 463, 490
  polyhydroxy butyrate (PHB), 462

polyhydroxybutyrate-co-valerate (PHBV), 489
polylactic acid (PLA), 462
polymer, 461–465
poly-R-3-hydroxy butyrate (PHB), 463
properties, 476
pyrolysis, 486
recent developments, 492–493
recent progress, 472
recycling, 484–486
resin, 477
reuse, 484–486
ships, 479–480
silkworm silk, 459–460
spent coffee bean powder (SCBP), 488
spider dragline silks, 460
starch, 463–465
starch-based, 466–467
straw fiber (SF), 489
surface interaction, 478
thermoplastic, 473–476
thermoplastic cassava starch (TPCS), 491
thermoset, 473–476
tomato peels (TPs), 481
trains, 479–480
triethyl citrate (TEC), 493
water vapor permeability (WVP), 468
wood fibers and paper fibers (alternative natural fibers), 460
Green manufacturing processes, 507
active pharmaceutical ingredient (API), 517
batch process
use of, 508
better housekeeping, 510
biocatalysis, 517–519
cationic surfactant hexadecyl trimethyl ammonium bromide (CTAB), 514
changes of inputs, 509
cleaning, 509
eliminating steps, 509
environmental efficiency index (EEI), 518
green method, 513–517
human intervention
changing dependence, 508
internal reuse of waste, 509
isopropanol (IPA), 510
medium-chain length (MCL), 513
nature of steps, 508–509
pharmaceutical industry, 510–511
polymer industries, 512
bioplastics, 512–513
primary rolling (PR), 518
production process changes, 508–509
propylene carbonate (PC), 518
short-chain length (SCL), 513
solid oxide fuel cells (SOFC), 514
total dissolved solids (TDS), 518
total Kjeldahl nitrogen (TKN), 518
total polyphenol content (TPC), 518
twin-screw extrusion (TSE), 517
volatile organic compounds (VOCs), 513
Green method, 513–517
Green solvents
cyclopentyl methyl ether (CPME), 254–255
alkylating agent, 249
alkylation, 253
enolate chemistry, 251
Friedel-crafts, 249
Grignard-type, 249–250
oxidation, 252
polymerization, 253
presence of acid, 253
reduction, 252–253
solar cells, 255
transformations, 251–252
transition metal catalysts, 250
glycerol, 247–248
biocatalysis, 245
catalyst design, 245–246
condensation, 247
coupling, 247
cycloaddition, 247
enhancing reaction, 244–245
materials chemistry, 246
nanoparticles, 246–247
organic synthesis, 243–244
separation, 246
major solutions with
cyclopentyl methyl ether (CPME), 248–255
ethyl lactate, 258–259
gamma-valerolactone (GVL), 262
glycerol, 242–248
limonene, 262
2-methyltetrahydrofuran (2-METHF), 255–258

p-cymene, 261–262
perfluorinated (fluorous), 259–261
polyethylene glycol (PEG), 234–242
water, 222–234
2-methyltetrahydrofuran (2-METHF)
coupling reactions, 257–258
reduction, 258
polyethylene glycol (PEG), 242
addition, 239
2-amino-2 chromones, 238–239
decarboxylation, 238
Diels-Alder, 237
Friedel-craft, 241
Heck coupling, 237
heterocycles, 240–241
oxidation, 235–236
oxidative cyclization, 239
substitution, 236
recent developments, 263–264
water, 232–234
aldol condensation, 225
Barbier-Grignard, 229
benzoin, 225
carbanion equivalents, 229
Claisen rearrangement, 224–225
Claisen-Schmidt, 225–226
dehalogenation, 230
Diels-Alder reaction, 223–224
dimethyl acetylenedicarboxylate (DMAD), 225
Heck reaction, 226
heterocycle, synthesis, 231–232
hydroxylation, 232
Knoevenagel, 226
multicomponent, 230
oxidation, 226–228
phosphorylation, 231
photochemical, 228–229
quadricyclane with, 225
radicals, 229
reduction, 228
sulphonation, 230–231

H

Halloysite nanotubes (HNTs), 481
Haloacetonitriles (HANs), 298
Heck coupling, 237
Hedysarum coronarium (HC), 492
High yields, 242
Highest occupied molecular orbital (HOMO), 281
Human intervention
changing dependence, 508
Hydrodesulfurization (HDS), 378
Hydrolyzed soybean protein isolate (HSPI), 489
Hydroxymethylfurfural (HMF), 435

I

Isopropanol (IPA), 297, 510

K

Kenaf fiber, 459

L

Life cycle assessment (LCA), 454, 457–458
Limiting oxygen index (LOI), 376
Linear-low density polyethylene (LLDPE), 488
Liquid-liquid systems, 376
Lithium diisopropylamide (LDA), 382
Lowest unoccupied molecular orbital (LUMO), 281

M

Medium-chain length (MCL), 513
Meerwein-Ponndorf-Verley (MPV), 299
2-Mercaptobenzothiazole (MBT), 342
Metal organic framework nanocubes (MOF-5-NCs), 386
Metal organic framework (MOF), 433
Metal organic polymer (MOP), 387
2-Methyltetrahydrofuran (2-METHF), 255–258
coupling reactions, 257–258
reduction, 258
Metoprolol (MET), 339
Metronidazole (MTZ), 339
Micro-fiber bundles, 459
Microfibrillated cellulose (MFC), 468
Micro-fibrillated materials, 459
Microwave heating, 242
Microwave-assisted organic synthesis (MAOS), 399, 400, 525
cetyltrimethylammonium chlorochromate (CTMACC), 415

classification, 410
solvent-assisted synthesis, 410–411
solvent-free, 411–414
dimethyl sulfoxide (DMSO), 436
energy source, 401
environmental-friendly chemistry, 409
2,5-furandicarboxylic acid (FDCA), 416
graphene oxide (GO), 416
greater reproducibility, 409
health, 410
heating
efficient source of, 408
higher yields, 408
hydroxymethylfurfural (HMF), 435
lack of scalability, 410
limited applicability, 410
miscellaneous, 434–435
N-dimethylformamide (DMF), 410
neat reactions, 413–414
nonthermal, 407
organic synthesis, 414
alkylations, 417–418
condensation, 421–423
coupling, 433–434
cycloadditions, 420–421
esterifications, 423–424
heterocycles, 426–428
ionic liquids, 430–431
metal organic framework (MOF), 433
multicomponent, 431–432
nanocomposites, 429–430
organometallic catalysis, 425–426
oxidation reactions, 415–416
protection, 424–425
rearrangements, 418–420
reduction, 416
phosphomolybdic acid (PMA), 419
polyvinylpyrrolidone (PVP), 430
principle, 403–404
apparatus, 405–406
comparison, 407–409
conduction mechanism, 404–405
dipolar polarization, 404
effect, 407
limitations, 409–410
medium, 406–407
reaction vessel, 406–407
recent development, 435–436
safety hazards relating, 410
selective heating, 409
solid-support, 411–413
specific, 407
synthetic chemistry
tool for, 402–403
uniform heating, 408–409

N

Nanofibrillated cellulose (NFC), 468
Nanoparticles (NPs), 386
Nanorod (NR), 298
Natural clay powder (NCP), 342
Natural fiber sources, 458–461
N-dimethylformamide (DMF), 410, 490
Near critical water (NCW)
carboxylic acids
decarboxylation of, 193
Claisen rearrangement, 192
Fischer indole synthesis, 193
hydrolysis of amide, 193
pinacol-pinacolone, 192–193
Neat reactions, 413–414
5-Nitro-1,2,4-triazol-3-one (NTO), 300

O

Ofloxacin (OFX), 339
Oil palm empty fruit bunch (OPEFB), 481
fiber, 460
Organic synthesis, 414
alkylations, 417–418
condensation, 421–423
coupling, 433–434
cycloadditions, 420–421
esterifications, 423–424
heterocycles, 426–428
ionic liquids, 430–431
metal organic framework (MOF), 433
multicomponent, 431–432
nanocomposites, 429–430
organometallic catalysis, 425–426
oxidation reactions, 415–416
protection, 424–425
rearrangements, 418–420
reduction, 416
Organometallic catalysis, 425–426
Organometallic processes, 379–383

Ornidazole (ORZ), 339
Oxidative cyclization, 239

P

P-cymene, 261–262
Perfluorinated (fluorous), 259–261
Phenanthrene (PHE), 293
Phosphomolybdic acid (PMA), 419
Photocatalysis, 279
  acid black 1 (AB 1), 287
  advanced oxidation processes (AOPs), 279
  carbon nanotube (CNT), 297
  carbon quantum dots (CQDs), 288
  cetylpyridinium chloride (CPC), 293
  cyclodextrins (CD), 291
  different, 282
  graphene oxide (GO), 288
  haloacetonitriles (HANs), 298
  heterogeneous, 281
  highest occupied molecular orbital (HOMO), 281
  homogeneous, 280–281
  isopropanol (IPA), 297
  lowest unoccupied molecular orbital (LUMO), 281
  Meerwein-Ponndorf-Verley (MPV), 299
  nanoparticle (NP), 286
  nanorod (NR), 298
  5-nitro-1,2,4-triazol-3-one (NTO), 300
  phenanthrene (PHE), 293
  p-nitrophenol (PNP), 301
  polystyrene (PS), 286
  quantum dots (QDs), 288
  reactive black-5 (RB-5), 286
  recent developments, 307–308
  reduced graphene oxide (RGO), 299
  removal efficiency, 297
  rhodamine-B (RB), 285
  sequential ionic layer adsorption and reaction (SILAR), 288
  sodium dodecyl sulfate (SDS), 293
  substrate, 280
    band gap, 281–282
    electron hole pair, fate, 283–284
    mechanism, 282–283
    photocatalytic reactions, 280–281
    semiconductors, 281
  titania, 306–307
  titania nanotube (TNT), 297
  trihalomethanes (THMs), 298
  wastewater treatment processes, 284
    alcohols, 295–296
    carbonyl, 298–299
    dyes, 284–290
    halo compounds, 298
    hydrocarbons, 294–295
    hydrogen production, 302–303
    nitrogen-containing, 299–302
    pesticides, 290–292
    phenols, 296–298
    reduction of carbon dioxide, 303–306
    surfactants, 292–294
Photo-Fenton reactions, 323
  2, 4-dimethylphenol (DMP), 342
  advanced oxidation processes (AOPs), 323
  automobile shredder residue (ASR), 339
  carboxylic acids, 330
  drugs, 338–339
  dyes, 335–338
  Fenton's reagent, 326–327
  halo compounds, 332
  hydrocarbons, 329–330
  irradiation, 327–329
  2-mercaptobenzothiazole (MBT), 342
  metoprolol (MET), 339
  metronidazole (MTZ), 339
  natural clay powder (NCP), 342
  nitro, 332–333
  ofloxacin (OFX), 339
  ornidazole (ORZ), 339
  oxidative degradation, 339–343
  pesticides, 333–334
  phenols, 331–332
  poly(vinyl alcohol) (PVA), 342
  recent developments, 343–344
  tetracycline (TC), 339
P-nitrophenol (PNP), 301
Polarity, 242
Poly(3- hydroxybutyrate-co-3-hydroxyvalerate) (PHBV), 488
Poly(butylene-succinate) (PBS), 491
Poly(hydroxyalkanoate) (PHA), 490
Poly(lactic acid) (PLA), 459, 489
Poly(vinyl alcohol) (PVA), 342
Polybutylene adipate terephthalate (PBAT), 489, 493
Polycaprolactone (PCL), 489

Polyethylene glycol (PEG), 234–242, 242, 464
addition, 239
2-amino-2 chromones, 238–239
decarboxylation, 238
Diels-Alder, 237
Friedel-craft, 241
Heck coupling, 237
heterocycles, 240–241
oxidation, 235–236
oxidative cyclization, 239
substitution, 236
Polyhydroxy alkanoates (PHA), 462, 463
Polyhydroxy butyrate (PHB), 462
Polyhydroxybutyrate-co-valerate (PHBV), 489
Polylactic acid (PLA), 462
Polymer, 373
radical polymerization, 373–374
scission, 374–376
Polymer industries, 512
bioplastics, 512–513
Poly-R-3-hydroxy butyrate (PHB), 463
Polystyrene (PS), 286
Polyvinylpyrrolidone (PVP), 430
Primary rolling (PR), 518
Propylene carbonate (PC), 518

Q

Quantum dots (QDs), 288

R

Reaction vessel, 406–407
Reactive black-5 (RB-5), 286
Recycling, 484–486
Reduced graphene oxide (RGO), 299
Resin, 477
Rhodamine-B (RB), 285

S

Safety hazards relating, 410
Selective heating, 409
Sequential ionic layer adsorption and reaction (SILAR), 288
Short-chain length (SCL), 513
Silkworm silk, 459–460
Sodium dodecyl sulfate (SDS), 293
Solar cells, 389
Solid oxide fuel cells (SOFC), 514
Solvent-assisted synthesis, 410–411
Solvent-free, 411–414
Sonochemistry, 359
acid red 17 (AR17), 386
biological, 368–372
cavitation
phenomenon of, 361–362
classification, 360
dye-sensitized solar cells (DSSCs), 389
external pressure
effect of, 362
extraction, 385–386
factors affecting, 362
gas, presence, 362
hydrodesulfurization (HDS), 378
hydrogen production, 384–385
limiting oxygen index (LOI), 376
liquid-liquid systems, 376
liquid-solid, 376–379
lithium diisopropylamide (LDA), 382
metal organic framework nanocubes (MOF-5-NCs), 386
metal organic polymer (MOP), 387
nanoparticles (NPs), 386
organic synthesis, 363–368, 388–389
organometallic processes, 379–383
polymer, 373
radical polymerization, 373–374
scission, 374–376
principles, 360–361
probe system, 363
recent developments, 389–390
scale-up consideration, 384
solar cells, 389
sources, 363
submersible transducer, 363
temperature, 362
tube, 363
ultrasonic cleaning bath, 363
ultrasound frequency, 362
ultrasound-assisted extraction (UAE), 385
water treatment, 386–388
whistle reactor, 363
Spent coffee bean powder (SCBP), 488
Spider dragline silks, 460
Starch, 463–465
based, 466–467
Straw fiber (SF), 489
Submersible transducer, 363

Supercritical solvents
carbon dioxide (CO_2), 195–208
alkylation, 202
coupling, 201–202
Diel's-Alder reaction, 197
esterification, 202–203
extraction, 203–204
Freidel–Crafts reaction, 197
Heck reactions, 201
hydrogenation, 200–201
miscellaneous, 205–208
oxidation, 198–200
photochemical, 202
solubility, 204–205
supercritical polymerization, 198
synthesis, 205
near critical water (NCW)
carboxylic acids, decarboxylation, 193
Claisen rearrangement, 192
Fischer indole synthesis, 193
hydrolysis of amide, 193
pinacol-pinacolone, 192–193
recent developments, 209–211
Surface interaction, 478

T

Temperature, 362
Tetracycline (TC), 339
Thermoplastic, 473–476
Thermoplastic cassava starch (TPCS), 491
Thermoset, 473–476
Titania, 306–307
Titania nanotube (TNT), 297
Tomato peels (TPs), 481
Total dissolved solids (TDS), 518
Total Kjeldahl nitrogen (TKN), 518
Total polyphenol content (TPC), 518
Triethyl citrate (TEC), 493
Trihalomethanes (THMs), 298
Tube, 363
Twin-screw extrusion (TSE), 517

U

Ultrasonic cleaning bath, 363
Ultrasound frequency, 362
Ultrasound-assisted extraction (UAE), 385
Uniform heating, 408–409

V

Volatile organic compounds (VOCs), 513
Volatility and boiling point, 242

W

Wastewater treatment processes, 284
alcohols, 295–296
carbonyl, 298–299
dyes, 284–290
halo compounds, 298
hydrocarbons, 294–295
hydrogen production, 302–303
nitrogen-containing, 299–302
pesticides, 290–292
phenols, 296–298
reduction of carbon dioxide, 303–306
surfactants, 292–294
Water, 232–234
aldol condensation, 225
Barbier-Grignard, 229
benzoin, 225
carbanion equivalents, 229
Claisen rearrangement, 224–225
Claisen-Schmidt, 225–226
dehalogenation, 230
Diels-Alder reaction, 223–224
dimethyl acetylenedicarboxylate (DMAD), 225
Heck reaction, 226
heterocycle
synthesis of, 231–232
hydroxylation, 232
Knoevenagel, 226
multicomponent, 230
oxidation, 226–228
phosphorylation, 231
photochemical, 228–229
quadricyclane with, 225
radicals, 229
reduction, 228
sulphonation, 230–231
Water treatment, 386–388
Water vapor permeability (WVP), 468
Whistle reactor, 363
Wood fibers and paper fibers (alternative natural fibers), 460